Nonlinear Optics
and Quantum Electronics

Nonlinear Optics and Quantum Electronics

MAX SCHUBERT
BERND WILHELMI
Friedrich-Schiller-Universität
Jena

A Wiley-Interscience Publication

John Wiley & Sons

New York / Chichester / Brisbane / Toronto / Singapore

Library of Congress Cataloging-in-Publication Data:

Schubert, Max.
Nonlinear optics and quantum electronics.

(Wiley series in pure and applied optics,
ISSN 0277-2493)

"A Wiley-Interscience publication."
Bibliography: p.
Includes index.
1. Nonlinear optics. 2. Quantum electronics.
I. Wilhelmi, Bernd. II. Title. III. Series.

QC446.2.S39 1986 535′.2 85-20310

ISBN 0-471-08807-2

Printed in the United States of America

10 9 8 7 6 5 4 3 2 1

Preface

Since the invention of the laser, nonlinear processes have attained increasing importance in optics. As a result, quantum electronics and nonlinear optics have evolved as important and independent fields of science in which much progress has been made in recent years. This is especially true of the advances made in elucidating basic effects where nonlinearities are of decisive importance. Moreover, on this basis physicists have succeeded in designing new devices, in particular radiation sources and optical measuring apparatus. Fundamental effects as well as devices and techniques have found their applications in a variety of fields in science, medicine, and technology.

In nonlinear optics the main object to be studied is the interaction between radiation and matter; depending on the particular physical situation, there arise phenomena such as multiphoton absorption and multiphoton emission, the generation of harmonics, parametric effects, stimulated scattering, the generation of highly coherent and highly nonstationary radiation, the occurrence of field-dependent optical material parameters, optical bistability, phase conjugation, and saturation phenomena in one- and multiphoton processes. The investigation of these phenomena has led to their successful application in fields such as laser technology, laser spectroscopy, photophysics and photochemistry, and material processing.

This book is based on a course of lectures for advanced students. Part I is devoted to an introductory treatment of general concepts and methods to be used for describing nonlinear processes. Part II is concerned with the application of these concepts and methods to significant effects and processes, covering also the particular experimental arrangements, measuring methods, and empirical data connected with them.

In describing the methods and processes we shall use as far as possible the classical or semiclassical theory. However, in the treatment of important problems concerning fluctuations, coherence, and quantum processes of the field, the more complex fully quantum-theoretical description had to be employed. The authors have tried to render the book as self-sufficient as possible; in particular, Appendix A at the end is intended to provide a survey of the quantization of matter and of the electromagnetic field, with applicable formulas. Readers who are not familiar with this subject may read Appendix A as a whole and/or may use the particular references to Appendix A in the various chapters.

SI units are used throughout the book. To make the content easily accessible to the reader, general notation and important symbols are listed prior to Part I. Here also the marks characterizing vectors, tensors, operators, and Fourier transforms can be found, which allow us to denote one and the same fundamental physical quantity by the same letter symbol (e.g. the field-strength vector $E_\bullet$, the Hilbert-space operator $\hat{E}_\bullet$ of the field strength, the Fourier transform $\underset{\smile}{E}_\bullet$ of the field strength). This convention is of special advantage in nonlinear optics. Throughout the book the most important equations are marked with a rule on the left. Single-digit reference numbers refer to general references in the list at the end of the book; double-digit ones refer to the reference numbers at the end of the corresponding chapter.

The authors wish to thank, first of all, Mrs. E. Wagner, English lector, for her very effective cooperation with the authors in translating the manuscript.

The authors are grateful to John Wiley & Sons and in particular to Beatrice Shube, Editor, for her help in publishing this volume.

We gratefully acknowledge the permissions granted by publishers (The American Physical Society; North-Holland Physics Publishing; Optical Society of America; American Institute of Physics; Springer-Verlag Heidelberg; Chapman and Hall Ltd., London; Redakzia Kvantovoj Elektroniki, Moscow; Verlag Chemie G.m.b.H., Weinheim; The Institute of Physics, London; Redaktion d. Wissenschaftl. Zeitschr. der Univ. Jena; Institute of Electrical and Electronic Engineers, New York; VEB Deutscher Verlag d. Wissenschaften, Berlin; Pergamon Press Ltd., Oxford; Akademie-Verlag Berlin) and by authors to use figures from cited references for reproduction in the book.

The authors are greatly indebted to Professor G. Weber, Professor E. Wolf, Dr. R. Gase, Dr. H. Ponath, Dr. K. Süsse, Dr. W. Vogel, Dr. H. Wabnitz, Dr. D. Welsch, and Dipl. phys. M. Kaschke for valuable discussions, and to Mrs. U. Kaschlik, Mrs. E. Rauschelbach, and Mrs. K. Triebel for technical help in preparing the manuscript.

Finally we would like to thank our wives, Hannelore Schubert and Edeltraud Wilhelmi, for their help in preparing the manuscript and for steady encouragement.

MAX SCHUBERT
BERND WILHELMI

Jena, German Democratic Republic
April 1986

Contents

Notation and Symbols

Notation, abbreviations, and symbols used in this book are explained in detail within the text. Here we give a short survey of frequently used conventions.

GENERAL NOTATION AND ABBREVIATIONS

$X_{\bullet \cdots \bullet}$ Tensor: $X_\bullet$ vector, $X_{\bullet\bullet}$ 2nd-rank tensor

Tensor relations:

(a) $Y_\bullet = X_{\bullet\bullet\bullet} A_\bullet B_\bullet C_\bullet$ means

$$Y_i = \sum_{j,k,l} X_{jkl} A_j B_k C_l$$

(b) $X_\bullet Y_\bullet$ scalar product

(c) $X_\bullet \times Y_\bullet$ vector product

(d) $\nabla_\bullet X_\bullet$ divergence of the vector $X_\bullet$

(e) $\nabla_\bullet \times X_\bullet$ curl of the vector $X_\bullet$

(f) $\nabla_\bullet X$ gradient of the scalar X

$\mathrm{FT}\{X(t)\} \equiv \underset{\smile}{X}(\omega)$ Fourier transform of $X(t)$:

$$X(t) = \frac{1}{2\pi} \int_{-\infty}^{+\infty} d\omega \, \underset{\smile}{X}(\omega) e^{i\omega t}$$

$$\underset{\smile}{X}(\omega) = \int_{-\infty}^{+\infty} dt \, X(t) e^{-i\omega t}$$

$$X^{(+)}(t) = \frac{1}{2\pi} \int_{0}^{\infty} d\omega \, \underset{\smile}{X}(\omega) e^{i\omega t}, \quad X^{(-)} = \frac{1}{2\pi} \int_{-\infty}^{0} d\omega \, \underset{\smile}{X}(\omega) e^{i\omega t}$$

c.c. Complex conjugate part:

$$X + \text{c.c.} \equiv X + X^*$$

m.f. Part with "minus frequencies":

$$\mathrm{FT}\{X(t)\} + \text{c.c.} \equiv \underset{\smile}{X}(\omega) + \text{m.f.}$$

$\overset{\frown}{X}$	Amplitude of a monochromatic vibration:

$$X(t) = \tfrac{1}{2}\overset{\frown}{X}e^{-i\omega t} + \text{c.c.}$$

$\overset{\frown}{X}$(t)	Slowly varying (temporal) amplitude of a quasimonochromatic vibration
$\overline{X}$	Amplitude of a monochromatic wave:

$$X(r_\bullet, t) = \tfrac{1}{2}\overline{X}e^{-i(\omega t - k_\bullet r_\bullet)} + \text{c.c.}$$

$\overline{X}(r_\bullet, t)$	Slowly varying amplitude of a quasimonochromatic wave (pulse envelope):

$$X(r_\bullet, t) = \tfrac{1}{2}\overline{X}(r_\bullet, t)e^{-i(\omega t - k_\bullet r_\bullet)} + \text{c.c.}$$

$\hat{X}$	Operator of the Hilbert space
$\lvert\psi\rangle, \langle\psi\rvert$	State vectors in the Hilbert space: $\lvert\psi\rangle$ ket, $\langle\psi\rvert$ bra
h.c.	Hermitian conjugate:

$$\hat{X} + \text{h.c.} \equiv \hat{X} + \hat{X}^\dagger$$

$\text{Tr}\{\hat{X}\}$	Trace of the operator $\hat{X}$
$\hat{X}, \hat{X}_H, \hat{X}_I$	Operators in the Schrödinger picture (without index), Heisenberg picture (index H) and interaction picture (index I)
$[\hat{X}, \hat{Y}]$	Commutator:

$$[\hat{X}, \hat{Y}] \equiv \hat{X}\hat{Y} - \hat{Y}\hat{X}$$

$[\hat{X}, \hat{Y}]_+$	Anticommutator:

$$[\hat{X}, \hat{Y}]_+ \equiv \hat{X}\hat{Y} + \hat{Y}\hat{X}$$

$[X, Y]_{\text{PB}}$	Poisson bracket
$\langle\hat{X}\rangle$	Expectation value of the operator $\hat{X}$:

$$\langle\hat{X}\rangle = \text{Tr}\{\hat{\rho}\hat{X}\}$$

$\overline{X}^t$	Time average of X
$\overline{X}$	Ensemble average of X
$\overline{X}^0$	Orientation average of X
$\overline{X}^e$	Equilibrium value of X
$\equiv$	$X \equiv Y$, X is identically equal to Y

$\approx$	$X \approx Y$, X is approximately equal to Y
$\sim$	$X \sim Y$, X and Y are of equal order of magnitude
$\propto$	$X \propto Y$, X is proportional to Y
$\int dx$	$= \int_{-\infty}^{+\infty} dx$, where x is a real number
$\int d^2\alpha$	$= \int d(\mathrm{Re}\,\alpha) \int d(\mathrm{Im}\,\alpha)$, where α is a complex number
$\int d^3 r_{\bullet}$	$\equiv \int dx \int dy \int dz$
$\sum_n$	sum over all allowed values n

SYMBOLS USED OFTEN OR THROUGHOUT

Latin Symbols

a, $\hat{a}$, $\hat{a}^{\dagger}$	Complex normal amplitude, boson annihilation operator, boson creation operator
$A_{\bullet}$, $\hat{A}_{\bullet}$	Vector potential
$\hat{b}$, $\hat{b}^{\dagger}$	Annihilation and creation operators of fermions
$B_{\bullet}$, $\hat{B}_{\bullet}$	Magnetic induction
c	Light velocity in vacuo
$d_{\bullet}$, $\hat{d}_{\bullet}$	Electric dipole moment
$D_{\bullet}$, $\hat{D}_{\bullet}$	Electric displacement
$E_{\bullet}$, $\hat{E}_{\bullet}$	Electric field strength
$\mathscr{E}$	Energy
$e_{\bullet}$	Unit vector in polarization direction
$g(\omega)$	Line shape
h, $\hbar$	Planck's constant; $\hbar = h/2\pi$
H, $\hat{H}$	Hamiltonian function, Hamiltonian operator
$H_{\bullet}$, $\hat{H}_{\bullet}$	Magnetic field strength
i	Imaginary unit
I	Photon flux per unit area
$\hat{I}$	Identity operator
J	Radiation power per unit area, $J = \hbar\omega I$
$k_{\bullet}$	Wave vector, magnitude $k = 2\pi/\lambda$
$\mathscr{k}_B$	Boltzmann's constant
$\mathscr{k}_a$	Absorption coefficient
$\hat{N}$	Number operator
$P_{\bullet}$, $\hat{P}_{\bullet}$	Polarization

$r_\bullet$	Space vector
$S_\bullet$	Poynting vector
t	Time
T	Time, lifetime
$\mathscr{T}$	Temperature
$\hat{U}$	Operator of unitary transformation
V	Volume
x, y, z	Space coordinates

Greek Symbols

α_{ij}	Polarizability tensor
$\|\alpha\rangle, \|\beta\rangle$	Glauber states
γ	Number density (number of atomic systems per unit volume)
γ_I	Density of the occupation-number inversion
$\varepsilon, \varepsilon_0$	Dielectric permittivity, dielectric permittivity of vacuum
$\varkappa$	Susceptibility
λ	Wavelength
μ, μ_0	Permeability, permeability of vacuum
ν	Frequency
$\hat{\rho}$	Density operator
τ	Time, phase decay time
χ	Susceptibility at discrete frequencies
ω	(Angular) frequency
$\Delta\omega$	Line width

Nonlinear Optics
and Quantum Electronics

Part I

General Concepts and Methods of Nonlinear Optics

The explanation of nonlinear optical effects and their applications, as it will be given in Part II of this book, requires the knowledge of general concepts and methods concerning the description of the electromagnetic radiation, of its typical properties, and of its interaction with matter. Chapter 1 is devoted to the classical description of electromagnetic fields and their constitutive equations, including their nonlinear parts. As a rule the classical theory allows a simplified treatment and an easy interpretation of the results obtained. However, for an adequate description of the processes the quantum state of the radiation field is needed, which involves information about the elementary particles of the field, the photons. Therefore in Chapter 2 we will discuss the quantized isolated radiation field, and in Chapter 3 the interaction of the photons with the particles (and quasiparticles) of the atomic system as well as the interaction with atomic systems acted on by dissipative systems. The semiclassical description of nonlinear optics, where the interaction between classical electromagnetic fields and quantized atomic systems is considered, is presented in Chapter 4; the relationship between the polarization and the field strength contains the linear and nonlinear susceptibilities that are given as a function of the atomic (quantum) parameters. The statistical and coherence properties of the radiation fields and their typical measurement methods are the subject of Chapter 5. Nonstationary processes occurring in nonlinear optics in the case of nonresonant interaction are treated in Chapter 6.

1

Electromagnetic Fields. Classical Description

For more than 50 years it has been known that only quantum theory is capable of covering all aspects of the wave as well as the particle properties of electromagnetic radiation in an adequate manner. This has been confirmed by conclusive experiments and their interpretation. Why do we nevertheless start with a classical description? On the one hand there are procedural grounds: often, because of its simplicity, the classical description may serve as a model and guide in a quantum physical treatment. On the other hand, many experiments can be described with sufficient accuracy in the framework of classical physics. This leads to a simplification in treatment, and in addition it permits an easier interpretation of the results obtained. The merit of this approximate description, however, can only be assessed within the scope of a quantum-physical treatment. This is an aspect we will return to in the following chapters. As a rule the classical description facilitates work so significantly that in optics and quantum electronics two kinds of mixed descriptions are applied in the transition area between the classical and quantum theoretical description: on the one hand a classical treatment of the radiation field is carried out with a quantization of the atomic systems only; on the other hand the interaction between quantized radiation fields and classically described materials is investigated. The decision whether to employ the first or the second method of description is determined by the nature of the experiment to be described. This question will be taken into account in the applications to be presented later on, and in any case there will be chosen a reasonably simple method which will cover the most important features.

Our classical description will also include the discussion of some problems of linear optics that are needed for further treatment, for example the propagation of waves in resonators or in crystals.

1.1. ELECTROMAGNETIC FIELDS IN VACUO

After presenting the basic equations, we discuss the expansion of the electromagnetic field in modes.

1.1.1. Maxwell's Equations

In the absence of charged particles and conduction currents the electromagnetic field in vacuo can be described with Maxwell's equations

$$\begin{aligned} &\nabla_\bullet \times E_\bullet = -\frac{\partial}{\partial t} B_\bullet, \quad \text{(a)} && \nabla_\bullet \times H_\bullet = \frac{\partial}{\partial t} D_\bullet, \quad \text{(b)} \\ &\nabla_\bullet B_\bullet = 0, \quad \text{(c)} && \nabla_\bullet D_\bullet = 0, \quad \text{(d)} \\ &B_\bullet = \mu_0 H_\bullet, \quad \text{(e)} && D_\bullet = \varepsilon_0 E_\bullet, \quad \text{(f)} \end{aligned} \tag{1.1}$$

which connect the electric field strength $E_\bullet$, the electrical displacement $D_\bullet$, the magnetic induction $B_\bullet$ and the magnetic field strength $H_\bullet$. From Maxwell's equations the *wave equations**

$$\nabla_\bullet^2 E_\bullet = \frac{1}{c^2}\frac{\partial^2}{\partial t^2} E_\bullet \quad \text{and} \quad \nabla_\bullet^2 H_\bullet = \frac{1}{c^2}\frac{\partial^2}{\partial t^2} H_\bullet \tag{1.2}$$

can be derived where $\varepsilon_0 \mu_0 = 1/c^2$. For the *total energy* H of the electromagnetic field we have

$$H = \frac{1}{2}\int dV \left\{ \varepsilon_0 E_\bullet^2 + \frac{1}{\mu_0} B_\bullet^2 \right\}, \tag{1.3}$$

and for the *total momentum*

$$G_\bullet = \frac{1}{c^2}\int dV E_\bullet \times H_\bullet, \tag{1.4}$$

where the integration extends over the space filled by the field.

1.1.2. Expansion of the Radiation Field in Modes

The radiation field can be characterized by giving the time behavior of one of the field quantities at every point of the given space (i.e. in the $r_\bullet$-continuum). This requires an *uncountable* number of values. A considerable simplification of the description can be achieved by performing an expansion in appropriate sets of spatial functions in conformity with the physical problem. Then the characterization will be possible by means of a *countable* number of values. Two such ways of representing electromagnetic fields will be discussed. In doing so, certain boundary conditions in finite space have to be satisfied, which, however, do not affect the general validity of the method, because in the end the boundary can be shifted towards infinity, where the field-strength

*Here and throughout, the most important equations are designated by a vertical rule on the left.

functions are expected to vanish in any case. This is followed, in Section 1.1.3, by a discussion of the properties of electromagnetic fields within finitely extended structures, for example within metallic or dielectric resonators.

1.1.2.1. Modes of a Cavity. Let the walls S of a cavity have an infinite conductivity. In this case there hold the boundary conditions $E_{\parallel} = 0$ and $H_{\perp} = 0$ for the tangential component of the electric field and the normal component of the magnetic field, respectively (see Fig. 1.1). The general field functions $E_{\bullet}(r_{\bullet}, t)$ and $H_{\bullet}(r_{\bullet}, t)$ are expanded in the eigenfunctions of the cavity, the so-called (cavity) modes $E_{\mu\bullet}(r_{\bullet})$, which result as solutions of the time-independent wave equation. By substituting

$$E_{\bullet}(r_{\bullet}, t) = -\frac{1}{\sqrt{\varepsilon_0}} \sum_{\mu} p_{\mu}(t) E_{\mu\bullet}(r_{\bullet}), \tag{1.5a}$$

$$H_{\bullet}(r_{\bullet}, t) = \frac{1}{\sqrt{\varepsilon_0 \mu_0}} \sum_{\mu} q_{\mu}(t) \nabla_{\bullet} \times E_{\mu\bullet}(r_{\bullet}) \tag{1.5b}$$

as well as the corresponding expressions for $D_{\bullet}(r_{\bullet}, t)$ and $B_{\bullet}(r_{\bullet}, t)$ into Maxwell's equations, we obtain

$$p_{\mu}(t) = \dot{q}_{\mu}(t) \quad \text{and} \quad \dot{p}_{\mu}(t) = -\omega_{\mu}^2 q_{\mu}(t) \tag{1.6}$$

and

$$\ddot{q}_{\mu}(t) + \omega_{\mu}^2 q_{\mu}(t) = 0 \tag{1.7}$$

for the time-dependent expansion coefficients, and the *time-independent wave equation*

$$\nabla_{\bullet}^2 E_{\mu\bullet}(r_{\bullet}) + \frac{\omega_{\mu}^2}{c^2} E_{\mu\bullet}(r_{\bullet}) = 0, \tag{1.8}$$

where ω_{μ}^2 is a separation constant whose square root has the dimensions of

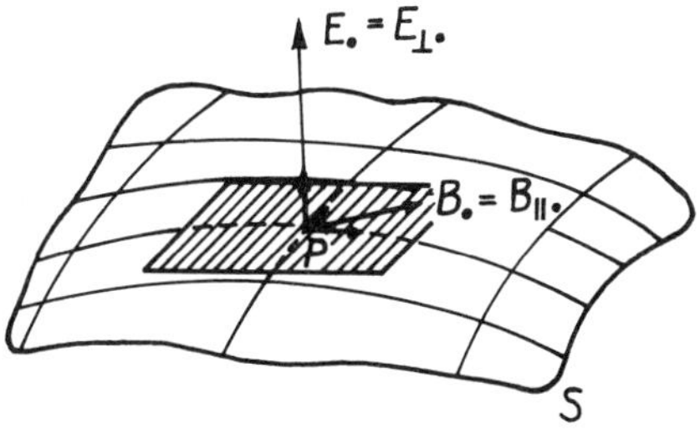

Figure 1.1. Part of the surface S of a cavity. The tangential plane touching the point P of S is shown. The vector of magnetic induction $B_{\bullet}$ lies in the tangential plane; the vector of the electric field $E_{\bullet}$ has the direction of the normal vector to S at P.

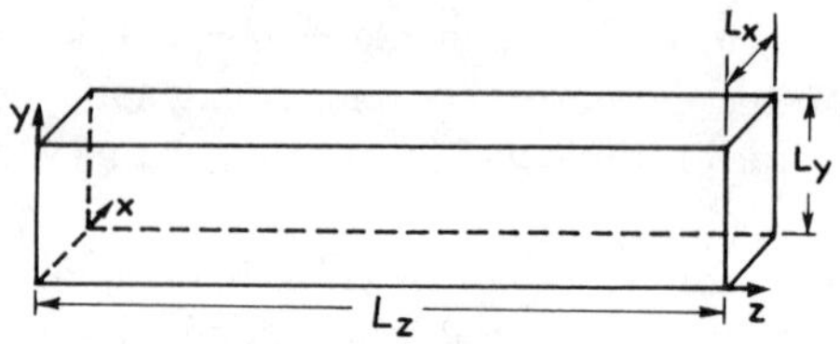

Fig. 1.2. Cavity with rectangular faces.

(angular) frequency. The cavity modes satisfy the condition of orthonormality,

$$\int dV E_{\mu\bullet}(r_\bullet) E_{\nu\bullet}(r_\bullet) = \delta_{\mu\nu}. \tag{1.9}$$

The values of ω_μ are given as eigenvalues of the partial differential equation (1.8) in connection with the boundary conditions. Thus the time-dependent expansion coefficients satisfy the equation of motion of a harmonic oscillator, with the solution

$$q_\mu = \hat{q}_\mu \cos(\omega_\mu t + \varphi_\mu). \tag{1.10}$$

The solution of the time-independent wave equation (1.8) turns out to be quite simple if a rectangular cavity (see Fig. 1.2) is used. Then the following modes result:

$$E_{\mu x}(x, y, z) = \sqrt{\frac{8}{V}}\, e_{l\sigma x}\cos(k_{lx}x)\sin(k_{ly}y)\sin(k_{lz}z),$$

$$E_{\mu y}(x, y, z) = \sqrt{\frac{8}{V}}\, e_{l\sigma y}\sin(k_{lx}x)\cos(k_{ly}y)\sin(k_{lz}z),$$

$$E_{\mu z}(x, y, z) = \sqrt{\frac{8}{V}}\, e_{l\sigma z}\sin(k_{lx}x)\sin(k_{ly}y)\cos(k_{lz}z), \tag{1.11}$$

where we have the conditions

$$k_{li} = \frac{l_i \pi}{L_i} \qquad (l_i = 0, 1, 2, \ldots,\ i = x, y, z), \tag{1.12}$$

$$k_{lx}^2 + k_{ly}^2 + k_{lz}^2 = \frac{\omega_l^2}{c^2} \quad \text{and} \quad V = L_x L_y L_z \tag{1.13}$$

for the parameters k_{lx}, k_{ly}, k_{lz} and the orthogonality relation

$$e_{l\sigma x}k_{lx} + e_{l\sigma y}k_{ly} + e_{l\sigma z}k_{lz} = 0 \tag{1.14}$$

between the wave vector $k_{l\bullet}$ and the *mode polarization vector* $e_{l\sigma\bullet}$. Each number triple l_1, l_2, l_3 is associated with two independent modes characterized by $\sigma = 1$ and 2 respectively, which differ in the direction of their polarization. Thus, the index μ is used as a symbol for (l_1, l_2, l_3, σ).

In optical problems the cavity dimensions are mostly very large compared with the wavelength. In our theoretical considerations presented at the beginning, the boundary had to be shifted even towards infinity. Under these conditions the number of modes in a definite wave-number or frequency interval and the corresponding mode densities will be determined. For this purpose a mode of a given direction of polarization will be represented by a point in (l_1, l_2, l_3) space (see Fig. 1.3). Then a given volume element of the l space (at $dl_i \gg 1$) contains

$$dZ = 2\,dl_x\,dl_y\,dl_z$$

modes. By use of (1.12) we obtain

$$dZ = 2\frac{V}{\pi^3}\,dk_x\,dk_y\,dk_z,$$

and on passing over to polar coordinates ($dk_x\,dk_y\,dk_z = k^2\,dk\,d\Omega$, with $d\Omega$ the solid-angle element around the wave vector),

$$dZ = 2\frac{V}{\pi^3}k^2\,dk\,d\Omega. \tag{1.15}$$

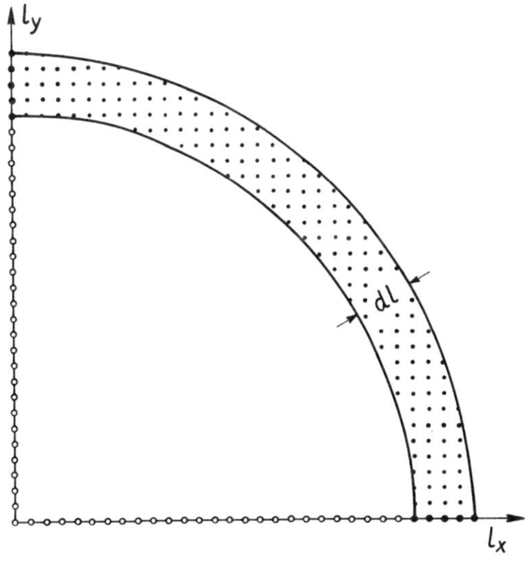

Fig. 1.3. Standing-wave modes in the l-space (in a plane l_z = const).

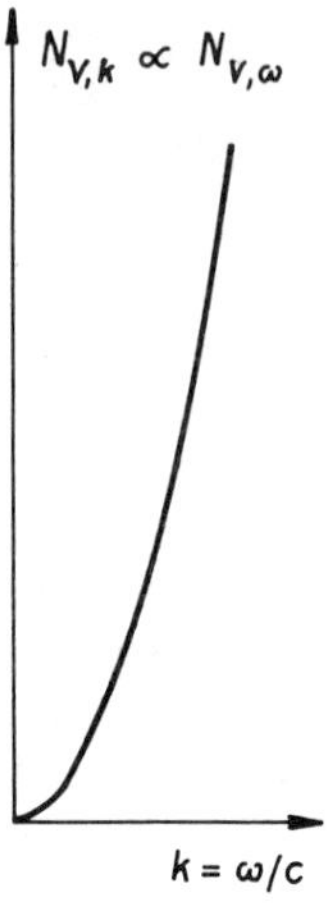

Figure 1.4. Number of modes $N_{V,k}$, per unit volume and unit wave number versus wave number k.

This equation yields the number of modes with a wave vector whose length is between k and $k + dk$ and whose direction is within the solid-angle element $d\Omega$. From this the *mode density* $Z_{V,k,\Omega}$ (that is, the number of modes per unit volume, wave number, and solid angle) results. It is

$$Z_{V,k,\Omega}\,dk\,d\Omega = 2\frac{1}{\pi^3}k^2\,dk\,d\Omega. \tag{1.16}$$

Since the mode density $Z_{V,k,\Omega}$ does not depend on Ω,

$$Z_{V,k}\,dk = \int d\Omega\, Z_{V,k,\Omega} = \frac{1}{\pi^2}k^2\,dk = \frac{1}{\pi^2 c^3}\omega^2\,d\omega \tag{1.17}$$

can be found by integrating over the solid angle (within the octant with $l_1, l_2, l_3, \geq 0$). It can be seen from (1.16) and (1.17) that the mode density increases with the square of the wave number k and the angular frequency $\omega = kc$ (see Fig. 1.4).

1.1.2.2. Modes of Propagating Waves. Let us describe the electromagnetic field by superposition of propagating *plane transverse waves*. To manage again with a countable number of expansion functions, a periodicity condition has to be imposed. Suppose the electromagnetic field to be given in a basic region consisting of a cube with edges of length L. By means of displacements along the three coordinate axes, a translational lattice filling all space is built up from the basic cube. Let the center of the nth cube have the coordinates $(n_x L + L/2, n_y L + L/2, n_z L + L/2)$, n denoting a triple of values (n_x, n_y, n_z) of arbitrary integers. Now the following periodicity condition of the field is imposed: let the electromagnetic field, in particular $E_{\bullet}$, have equal values at equivalent points of the various cubes. The periodic continuation of the field in the other cubes is physically irrelevant. In the following calculations the edge length L will have no upper limit. Hence the cube can be chosen to be so big as to let the basic region cover all arrangements and experimental situations. The periodicity condition is

$$E_\bullet(r_\bullet, t) = E_\bullet(r_\bullet + R_{n\bullet}, t), \tag{1.18}$$

where the vector $R_{n\bullet}$ has the components $(n_x L, n_y L, n_z L)$ and (n_x, n_y, n_z) is an arbitrary triple of integers. Because of this periodic continuation of the function from a finite region, the field can be easily expanded in a Fourier series. As is well known, plane waves with a definite angular frequency ω and a definite wave vector $k_\bullet$ can be considered as basic solutions to the wave equation (1.2). They have the structure $E_\bullet(r_\bullet, t) = E_\bullet(k_\bullet r_\bullet - \omega t)$. Then (1.2) leads to the dispersion relation

$$\omega = ck. \tag{1.19}$$

By superposition we obtain as a general ansatz for the electric field

$$E_\bullet(r_\bullet, t) = \sum_\mu \tfrac{1}{2} e_{\mu\bullet} \bar{E}_\mu e^{-i(\omega_\mu t - k_{\mu\bullet} r_\bullet)} + \text{c.c.}. \tag{1.20}$$

From the periodicity condition we have for the wave vector

$$k_{\mu\bullet} \equiv k_{l\bullet} = \left(\frac{2\pi}{L} l_x, \frac{2\pi}{L} l_y, \frac{2\pi}{L} l_z \right) \tag{1.21}$$

where l_x, l_y, and l_z are arbitrary integers (in contrast to the case of standing waves). The requirement that the set of transverse fields by means of which $E_\bullet(r_\bullet, t)$ is to be represented be complete necessitates the assumption of two orthogonal transverse fields for each k_l-vector, that is, two(plane-wave) modes. The two unit polarization vectors are perpendicular to $k_{l\bullet}$. This can be seen from substituting the field (1.20) into the Maxwell equation (1.1d). The polarization unit vectors $e_{\mu\bullet}$ can be chosen to be orthogonal to each other in order to guarantee the orthogonality of both modes, which will be denoted by $\sigma = 1, 2$. Hence, the mode index $\mu = (l, \sigma)$ characterizes the components of the wave vector by $l = (l_x, l_y, l_z)$ and the polarization direction of the mode by σ. The polarization unit vectors and the wave vector obey the orthogonality relations

$$e_{l\sigma\bullet} e_{l\sigma'\bullet} = \delta_{\sigma\sigma'} \quad \text{and} \quad e_{l\sigma\bullet} k_{l\bullet} = 0. \tag{1.22}$$

$\bar{E}_\mu$ is an amplitude factor (in general a complex one), which characterizes the intensity and phase of the mode excitation. In view of the quantization to be carried out later, it is convenient to represent the amplitude $\overline{E}_\mu$ in the form of

$$\overline{E}_\mu = 2i\left(\frac{\hbar\omega_\mu}{2\varepsilon_0 V}\right)^{1/2}\hat{a}_\mu, \tag{1.23}$$

where V is the volume of the basic region. Hence the contribution of the μth mode to the field energy is

$$H_\mu = \hbar\omega_\mu |\hat{a}_\mu|^2. \tag{1.24}$$

Considering the superposition of the fields of the two modes for one wave vector, we recognize that the resultant field is in general elliptically polarized and only in special cases linearly or circularly polarized. In addition to being represented by two modes of linearly polarized waves, an arbitrarily polarized wave of wave vector $k_{l\bullet}$ may be represented by two modes of circularly polarized waves. This representation, which is useful for certain applications, exhibits in the ansatz (1.20) the complex vectors

$$g_{l1\bullet} = \frac{1}{\sqrt{2}}(e_{l1\bullet} + ie_{l2\bullet}),$$

$$g_{l2\bullet} = \frac{1}{\sqrt{2}}(e_{l1\bullet} - ie_{l2\bullet}) \tag{1.25}$$

instead of the (real) unit vectors $e_{\mu\bullet} \equiv e_{l\sigma\bullet}$ ($\sigma = 1, 2$). These modes, too, are orthonormalized.

The *density of propagating modes* can be determined in the same way as it was in the case of standing waves. For the mode number in the wave-number interval dk and in the solid-angle element $d\Omega$ we obtain

$$dZ = 2\left(\frac{L}{2\pi}\right)^3 k^2\,dk\,d\Omega = VZ_{V,k,\Omega}\,dk\,d\Omega. \tag{1.26}$$

An integration over the entire solid angle leads to

$$Z_{V,k}\,dk = \frac{1}{\pi^2}k^2\,dk = \frac{1}{\pi^2 c^3}\omega^2\,d\omega, \tag{1.27}$$

which is in agreement with the result obtained in (1.17) for standing waves.

1.1.3. Radiation Fields in Real Resonators

In Section 1.1.2 we treated radiation fields within an empty cavity completely surrounded by walls of infinite conductivity and hence of reflectivity $R = 1$. This means that losses within the cavity and any coupling out of energy have been neglected. The electromagnetic field does not decay in such a cavity, and the time-dependent factor of the field of one resonator mode obeys the equation of a harmonic oscillator. In real resonators losses occur, caused by a reflectivity $R < 1$ and by incompletely closed walls of the cavity. Any electromagnetic field excited in such a cavity will decay within a finite time after excitation. Then, in general, the radiation field varies in dependence on space coordinates and time in a rather complicated way. We shall discuss the properties of electromagnetic fields in real resonators only in a rough manner, where the time variation of the whole field energy of one mode is approximated by an exponential decrease. More detailed treatments of the radiation fields in real resonators, including the description of the precise structure of spatial modes and of the diffraction of light at finite apertures, are given for instance in Refs. 5 and 1.1.*

1.1.3.1. Quality of the Resonator. A rough description of these damping processes of the field in mode μ can be given by the phenomenological introduction of a *loss* term $(2/T_A)\dot{q}_\mu(t)$ into (1.7). Then we have

$$\ddot{q}_\mu(t) + \frac{2}{T_A}\dot{q}_\mu(t) + \omega_\mu^2 q_\mu(t) = 0, \qquad (1.28)$$

where T_A is the *relaxation time of the field amplitude*. The damping term is assumed to be small so that $1/T_A$ is very small compared to the distance between adjacent modes. If we separate out in $q_\mu(t)$ a rapidly varying factor in the form

$$q_\mu(t) = \tfrac{1}{2}\bar{E}_\mu(t)e^{-i\omega t} + \text{c.c.}, \qquad (1.29)$$

where $\omega \approx \omega_\mu$, the differential equation

$$\dot{\bar{E}}_\mu + \frac{1}{T_A}\bar{E}_\mu - i(\omega - \omega_\mu)\bar{E}_\mu = 0 \qquad (1.30)$$

for the slowly varying amplitude factor $\bar{E}_\mu$ is obtained. This description of the one-mode field corresponds to the consideration of a simple equivalent electric circuit for the resonator mode: an electric circuit consisting of lumped elements (see Fig. 1.5). If we neglect the Ohmic resistance, the time dependence of the current or the voltage (or the electric field in the capacitor) will be given by the differential equation of the harmonic oscillator, which also describes the radiation field in the loss-free resonator. The resonator losses are equivalent to an Ohmic resistance. The field amplitude is seen to be exponentially damped

*Single-digit reference numbers refer to the references in the list at the end of the book; double-digit reference numbers refer to the references at the end of the chapter.

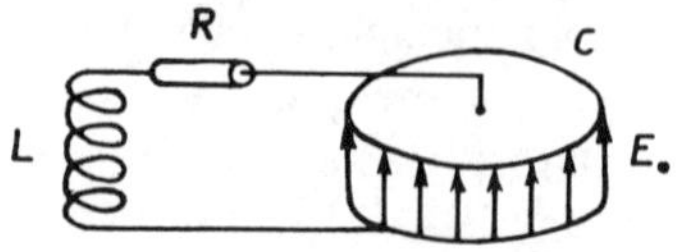

Fig. 1.5. Resonant electric circuit with inductance L, capacitance C, Ohmic resistance R. After excitation of an electric field of amplitude $\bar{E}$ within the circuit at $t = 0$, the field at time t is given by $E(t) = \frac{1}{2}\bar{E}e^{-t/2T_{\mathscr{E}}}e^{-i\omega_0 t}$ + c.c., where $T_{\mathscr{E}} = L/R$ and $\omega_0 = \sqrt{1/CL - (R/2L)^2}$. The Fourier transform of $E(t)$ is given by $\underset{\smile}{E}(\omega) \approx \frac{1}{2}\bar{E}^*/[-i(\omega_0 - \omega) + 1/(2T_{\mathscr{E}})]$, and the spectral density by $|\underset{\smile}{E}(\omega)|^2 \approx \frac{1}{4}\bar{E}^2/[(\omega - \omega_0)^2 + (\Delta\omega/2)^2]$, where $\Delta\omega = 1/T_{\mathscr{E}}$.

with a time constant T_A, and hence the field energy decreases according to

$$\mathscr{E}(t) = \mathscr{E}(0)e^{-t/T_{\mathscr{E}}}, \tag{1.31}$$

where $T_{\mathscr{E}} = T_A/2$. The losses are often characterized by the *quality Q of the cavity*, which is connected with $T_{\mathscr{E}}$ by

$$Q = \omega_0 T_{\mathscr{E}} \tag{1.32}$$

(ω_0 is the resonant frequency of the circuit or of the resonator mode under consideration).

In the ω-domain the mode is not characterized any longer by an infinitely sharp frequency but by a spectral distribution. In our model [represented by (1.28) or by the circuit of Fig. 1.5] this distribution is a Lorentzian line of half-width

$$\Delta\omega = \frac{1}{T_{\mathscr{E}}} = \frac{2}{T_A}. \tag{1.33}$$

1.1.3.2. Plane Fabry–Perot Resonator. To achieve an approximate description of a *real* laser resonator of the plane Fabry–Perot type, we start with a modification of the *ideal cavity* shown in Fig. 1.2:

1. The two ideally reflecting resonator faces in the xy plane (or at least one of them) are replaced by mirrors of transmission $\mathscr{T} > 0$ and reflectivity $R < 1$. This means that radiation leaves the resonator through these mirrors.
2. The four side faces are removed, and the radiation emerges from the resonator through these open windows.

We consider a plane Fabry–Perot resonator with infinitely extended mirrors (see Fig. 1.6) and calculate the stationary behavior and the decrease in radiation energy after suddenly switching off the light source. A plane wave of

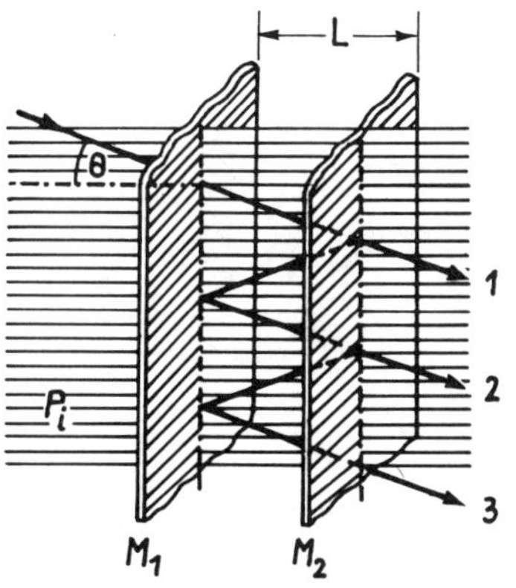

Figure 1.6. Plane Fabry–Perot interferometer. L is the distance between the mirrors, θ is the angle of incidence, and P_i is the plane of incidence.

frequency ω and amplitude $\bar{E}_0$ enters the Fabry–Perot resonator of length L. The wave vector forms an angle θ with the surface normal to the mirrors, that is, to the z axis. The amplitude $\bar{E}$ of the field behind the Fabry–Perot resonator is given by superposition of all the partial waves resulting from multiple reflection between the mirrors:

$$\frac{\bar{E}}{\bar{E}_0} = \mathscr{T} e^{i\delta'}(1 + Re^{i\delta} + R^2 e^{i2\delta} + \cdots + R^n e^{in\delta} + \cdots),$$

or

$$\frac{\bar{E}}{\bar{E}_0} = \mathscr{T} e^{i\delta'} \frac{1}{1 - Re^{i\delta}}, \tag{1.34}$$

where

$$\delta = \frac{2\pi}{\lambda} 2L\cos\theta \quad \text{and} \quad \delta' = \frac{2\pi}{\lambda} L \frac{1}{\cos\theta}.$$

[$\mathscr{T} = \sqrt{\mathscr{T}_1 \mathscr{T}_2}$, $R = \sqrt{R_1 R_2}$; $\mathscr{T}_1$, $\mathscr{T}_2$ and R_1, R_2 are the (energy) transmission and (energy) reflection coefficients of mirrors 1 and 2, respectively; if the resonator is filled with matter of refractive index n, then L has to be replaced by $\tilde{L} = nL$. The symbol used for the transmission should not be mixed up with the symbol for the temperature.] Now from (1.34) the transmission $\mathscr{T}_{FP}$ of the Fabry–Perot resonator is calculated to be:

$$\mathscr{T}_{FP} = \left|\frac{\bar{E}}{\bar{E}_0}\right|^2 = \left(\frac{\mathscr{T}}{1 - R}\right)^2 \left[1 + \frac{4R}{(1 - R)^2}\sin^2\left(\frac{\delta}{2}\right)\right]^{-1}. \tag{1.35}$$

Figure 1.7 shows this transmission plotted against the phase difference δ, which is proportional to the radiation frequency ω for constant cavity length

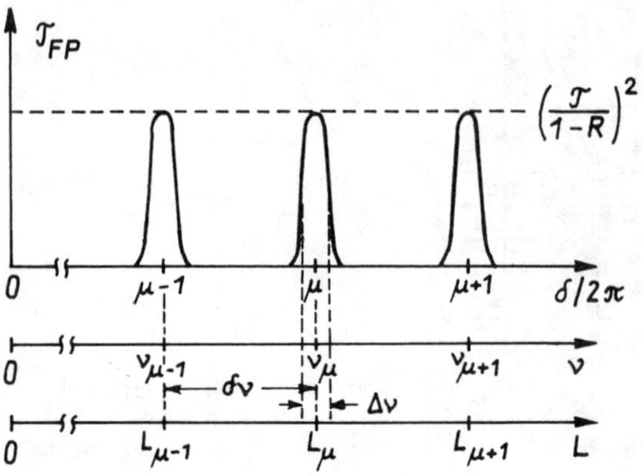

Fig. 1.7. Transmission of the Fabry–Perot interferometer.

and to the cavity length for constant frequency. Maxima of the transmission $\mathcal{T}_{FP}$ occur at $\delta = \mu \cdot 2\pi$ (μ being a natural number). On the frequency scale these maxima occur at

$$\nu_\mu = \frac{\omega_\mu}{2\pi} = \mu \frac{c}{2L\cos\theta}, \tag{1.36}$$

and the half-width of the transmission regions is given by

$$\Delta\nu = \frac{c}{2L\cos\theta}\,\frac{1-R}{\pi\sqrt{R}}. \tag{1.37}$$

The frequency difference between adjacent maxima,

$$\delta\nu = \frac{c}{2L\cos\theta}, \tag{1.38}$$

is called *dispersion region*; the ratio of $\delta\nu$ to $\Delta\nu$,

$$\frac{\delta\nu}{\Delta\nu} = \frac{\pi\sqrt{R}}{1-R}, \tag{1.39}$$

is the *finesse* $\mathcal{F}$ of the interferometer, and can be interpreted as an effective number of interfering waves. The *spectral resolution* of the interferometer is given by

$$\mathcal{A} = \frac{\nu}{\Delta\nu} = \mu\mathcal{F}. \tag{1.40}$$

The relations (1.37)–(1.39) have been derived by assuming the mirrors to be perfectly flat. Any deviation from the perfect flatness of the surfaces increases the half-width of the transmission peaks and therefore decreases the finesse and the spectral resolution. As an approximation the *total finesse* $\mathcal{F}_{\text{tot}}$ is calculated to be

$$\mathcal{F}_{\text{tot}}^{-1} = \mathcal{F}_{\text{refl}}^{-1} + \mathcal{F}_{\text{dev}}^{-1}, \tag{1.41}$$

where $\mathcal{F}_{\text{refl}}$ is the finesse caused by reflection losses and $\mathcal{F}_{\text{dev}}$ is the finesse caused by deviations from perfect flatness of the mirrors, which may result from any kind of roughness or curvature. The contribution $\mathcal{F}_{\text{dev}}$ depends on the shape and size of these deviations (see, e.g., Ref. 8). For the particular case of slight spherical curvature of the plates such that the optical length of the interferometer changes by one mth of the wavelength between the center and the edge of the mirrors, this contribution is given by

$$\mathcal{F}_{\text{dev}} = \frac{m}{2}. \tag{1.42}$$

$\mathcal{F}_{\text{dev}}$ of high-quality Fabry–Perot resonators can be as high as several hundred. If, then, the total finesse is to exceed 100, a reflectivity of $R > 0.99$ has to be chosen.

The half-width of the transmission peak may be further increased by the influence of the diffraction of light beams of finite diameter [see, e.g., Refs. 5 and 1.1].

Let us now deal with the decrease of the electromagnetic field after switching off a cw light source of frequency ω immediately in front of the interferometer at the time $-t_T$, where t_T is the transit time of the unreflected beam through the interferometer. At $t > 0$ the amplitude $\overline{E}$ of the electric field strength at the exit of the Fabry–Perot interferometer is no longer given by (1.34), because the first partial wave or several partial waves in this series do not contribute to the result. At the time $t_j = j(2L\cos\theta)/c$ $(j = 0, 1, 2, \ldots)$ the field strength $\overline{E}$ is given by

$$\frac{\overline{E}}{\overline{E}_0} = \mathcal{T}e^{+i\delta'} \sum_{p=j}^{\infty} (Re^{i\delta})^p = \mathcal{T}e^{i\delta'}(Re^{i\delta})^j \frac{1}{1 - Re^{i\delta}}, \tag{1.43}$$

and the square of its modulus by

$$\left|\frac{\overline{E}}{\overline{E}_0}\right|^2 = R^{2j}\left(\frac{\mathcal{T}}{1-R}\right)^2 \left[1 + \frac{4R}{(1-R)^2}\sin^2\left(\frac{\delta}{2}\right)\right]^{-1}. \tag{1.44}$$

This means that I, the radiation power per unit area behind the interferometer, decreases with time, as well as the energy $\mathcal{E}$ stored in the interferometer. This decrease may be approximated by

$$I(t) = I(0)e^{-t/T_{\mathscr{E}}} \quad \text{and} \quad \mathscr{E}(t) = \mathscr{E}(0)e^{-t/T_{\mathscr{E}}}. \tag{1.45}$$

Here

$$T_{\mathscr{E}} = -\frac{2L\cos\theta}{c}\frac{1}{2\ln R}$$

is again the energy relaxation time of the system. For a long resonator of high reflectivity, say $2L\cos\theta = 3$ m and $R = 0.99$ (commonly used, for example, in connection with long He–Ne lasers) a value of $T_{\mathscr{E}} \approx 500$ ns is obtained; whereas a very short resonator with low-reflection mirrors, say $2L\cos\theta = 3$ mm and $R = 0.5$ (used, for example, in connection with semiconductor lasers) has a value of $T_{\mathscr{E}} \approx 7$ ps.

The exponential decay of the radiation energy derived for the plane Fabry–Perot interferometer is a special case of the phenomenologically introduced relation (1.28). It can be easily seen that the slowly varying amplitude factor $\bar{E}(t)$ of (1.43) in reality obeys (1.30), and that the factor $q(t) = \frac{1}{2}\bar{E}(t)e^{-i\omega t} + \text{c.c.}$ with $\omega \approx \omega_\mu$ satisfies the relation (1.28). This means that the results obtained for the Fabry–Perot resonator with reflection losses agree with those gained by introducing a *phenomenological* loss term into the equations of the loss-free cavity. This is another justification for applying (1.28) and (1.30) to lossy cavities.

1.2. ELECTROMAGNETIC FIELDS WITH SOURCES

The aim of this section is the calculation of the dependence of the macroscopic polarization, magnetization, and current within the sample on the electromagnetic field of external sources. As discussed in the introduction, the mathematical structure of these relations is decisive with respect to whether phenomena can be described in the approximation of linear optics. We start from Maxwell's equations, taking into account the influence of the *microscopic distribution* of charges and currents. From this set of equations we then derive macroscopic relationships by appropriate averaging procedures.

1.2.1. Microscopic Relations

Any material consists of charged particles, the electrons and nuclei, which determine its properties by their motions and interactions. These interactions are caused by electromagnetic fields resulting from the charges themselves. In the empty space between these charge carriers the *local fields* $E'_\bullet$, $D'_\bullet$, $H'_\bullet$, $B'_\bullet$, which are composed of *external fields* acting on the material from outside and *internal fields* originating from the charges, are described by Maxwell's equations in the following way [see e.g. Refs. 4, 8]:

$$\nabla_\bullet \times H'_\bullet = \frac{\partial}{\partial t} D'_\bullet + j'_\bullet, \tag{1.46a}$$

$$\nabla_\bullet \times E'_\bullet = -\frac{\partial}{\partial t} B'_\bullet, \tag{1.46b}$$

$$\nabla_\bullet D'_\bullet = \rho', \tag{1.46c}$$

$$\nabla_\bullet B'_\bullet = 0, \tag{1.46d}$$

$$D'_\bullet = \varepsilon_0 E'_\bullet, \tag{1.46e}$$

$$B'_\bullet = \mu_0 H'_\bullet, \tag{1.46f}$$

where

$$\rho' = \rho'(r_\bullet, t) = \sum_n q_n \delta(r_{n\bullet} - r_\bullet) \tag{1.46g}$$

is the charge-density distribution of pointlike charges of charge q_n at the positions $r_{n\bullet}$,

$$j'_\bullet = j'_\bullet(r_\bullet, t) = \sum_n q_n \dot{r}_{n\bullet} \delta(r_{n\bullet} - r_\bullet) \tag{1.46h}$$

is the current density of these charges, and $\dot{r}_{n\bullet}$ is the velocity of the nth particle. In this way the local fields at $r_\bullet$, t are dependent on the positions and velocities of all charged particles, the positions and velocities of which on the other hand depend on the local electromagnetic fields via the force terms $q_n E'_\bullet(r_{n\bullet}, t)$ and $q_n\{\dot{r}_{n\bullet} \times B'_\bullet(r_{n\bullet}, t)\}$ in the equations of motion. The exact solution of the microscopic Maxwell equations (1.46) in conjunction with the (classical or quantum-mechanical) equations describing the motion of all the charged particles is impossible because of the very large number of degrees of freedom of the system and the limited knowledge of all the microscopic initial conditions at a certain time. This is the main reason for using macroscopic theories.

1.2.2. Macroscopic Relations

To deal with optical problems and their measurable results, we must give an adequate approximate description of the interaction between the electromagnetic fields and the microscopic particles. With this aim we subject the microscopic relationships between fields and sources to appropriate averaging procedures. The averaging with respect to the statistical ensemble of systems as well as with respect to the spatial (and temporal) structure, which varies rapidly on a microscopic scale, simplifies the problem under investigation and helps to make its solution possible. Both kinds of averaging procedure are closely linked with the real measuring process, either where relatively large samples are investigated, which contain a huge number of particles, or where, dealing with single atomic systems, a large number of measurements have to be carried out. In any case, we exploit the fact that in optical experiments the fields acting on the sample from outside as well as the fields emitted by the sample possess vacuum wavelengths large compared with the smallest distances a ($a \sim 10^{-10}$ m), where the microscopic fields undergo considerable changes. This means that in general, regions of extent large compared with a may be chosen where the external field can be considered as being approximately homogeneous. Moreover, we take into account that most of the charge carriers make up fairly stable structural units such as atoms, ions, molecules, and unit cells in solids. If the effect of external fields is not too strong (field strength $E \lesssim 10^9$ V/m), these structural units will be preserved during the interaction. (The destructive effect of high-power lasers requires a completely different description.) Therefore the material will be described by the parameters of these atomic systems (and, if required, by properties of the remaining free charge carriers). For this purpose the center-of-mass coordinates $r_{N\bullet}$ of the individual atomic systems (or of other small structural units) are introduced, and the coordinates of the charge carriers are represented by

$$r_{n\bullet} = r_{N\bullet} + r_{Nn\bullet}, \tag{1.47}$$

where $r_{Nn\bullet}$ is the vector from the Nth center to the nth charge carrier belonging to the system. Single charge carriers (e.g., quasifree electrons) that do not belong to such a subsystem are treated as independent systems. The expression (1.47) is used in the Maxwell equations (1.46e) and (146f), and the following formal expansion of the delta function is employed [cf. Ref. 4]:

$$\delta(r_{N\bullet} + r_{Nn\bullet} - r_\bullet) = \left[1 - r_{Nn\bullet}\nabla_\bullet + \tfrac{1}{2}(r_{Nn\bullet}\nabla_\bullet)^2 - \cdots\right]\delta(r_{N\bullet} - r_\bullet). \tag{1.48}$$

(This equation makes use of $\nabla_{Nn\bullet} = -\nabla_\bullet$, the differential operators relating to the vectors $r_{Nn\bullet}$ and $r_\bullet$, respectively.) The reason for this formal expansion becomes clear from applying it in an integral relation such as

$$\begin{aligned}
&\int dV' f(r'_\bullet)\delta(r_{N\bullet} + r_{Nn\bullet} - r'_\bullet)\\
&= \int dV' f(r'_\bullet)\delta(r_{N\bullet} - r'_\bullet)\\
&\quad - r_{Nn\bullet}\int dV' f(r'_\bullet)\nabla_\bullet\delta(r_{N\bullet} - r'_\bullet) + \cdots\\
&= f(r_{N\bullet}) + r_{Nn\bullet}\left[\nabla_\bullet f(r_\bullet)\right]_{r_\bullet = r_{N\bullet}} + \cdots.
\end{aligned} \tag{1.49}$$

In doing so it becomes obvious that the function $f(r_\bullet)$ is being expanded about the center of the atomic system. The convergence of this expansion depends on the ratio of the spatial frequencies in the function $f(r_\bullet)$ to $1/r_{Nn}$ (for example, if $f(r_\bullet)$ has the spatial structure of an optical wave $\hat{f}\cos k_\bullet r_\bullet$ with $k = 2\pi/\lambda$ then (1.49) corresponds to an expansion in terms of the parameter r_{Nn}/λ, which is assumed to be much less than 1). Using the representation of the delta function in (1.48), we obtain

$$\rho'(r_\bullet) = \rho^{(q)\prime}(r_\bullet) - \nabla_\bullet P'_\bullet(r_\bullet), \tag{1.50a}$$

where

$$\rho^{(q)\prime}(r_\bullet) = \sum_N \rho_N(r_\bullet), \tag{1.50b}$$

$$\rho_N(r_\bullet) = \sum_n q_n \delta(r_{N\bullet} - r_\bullet), \tag{1.50c}$$

$$P'_\bullet(r_\bullet) = \sum_N P_{N\bullet}(r_\bullet), \tag{1.50d}$$

$$P_{N\bullet}(r_\bullet) = (d_{N\bullet} - Q_{N\bullet\bullet}\nabla_\bullet + \cdots)\delta(r_{N\bullet} - r_\bullet), \tag{1.50e}$$

$$d_{N\bullet} = \sum_n q_n r_{Nn\bullet}, \tag{1.50f}$$

and

$$(Q_{Nij}) = \frac{1}{2}\sum_n q_n \begin{pmatrix} xx & xy & xz \\ yx & yy & yz \\ zx & zy & zz \end{pmatrix}_{Nn}. \tag{1.50g}$$

Obviously the tensor Q_{Nij} is formed by dyadic multiplication of the vectors $r_{Nn\bullet}$. The summation in (1.50c) and (1.50f) is performed over all charges n belonging to the Nth unit. Equation (1.50e) gives the contribution $P_{N\bullet}$ of the Nth atomic system to the polarization of the medium, $P'_\bullet$. As a rule $P_{N\bullet}$ is mainly determined by the *dipole moment* $d_{N\bullet}$; small additional contributions arise from the *quadrupole moment* $Q_{N\bullet\bullet}$ and *higher multipole moments* not explicitly given here. [It should be mentioned that the polarization may also be defined as the sum over the dipole contributions only (e.g. Ref. 7). Then higher multipole terms, which are not included in $P'_\bullet$, have to be added in (1.50a) and below in (1.51a).] Note the vanishing contribution of uncharged atomic systems to the charge density $\rho^{(q)\prime}(r_\bullet)$ and that of individual free charges to the polarization $P'_\bullet(r_\bullet)$.

In an analogous way the current density $j'_\bullet(r_\bullet)$ is obtained as

$$j'_\bullet(r_\bullet) = j^{(q)\prime}_\bullet(r_\bullet) + \frac{\partial}{\partial t}P'_\bullet(r_\bullet) + \frac{1}{\mu_0}\nabla_\bullet \times M'_\bullet(r_\bullet), \tag{1.51a}$$

where

$$j_\bullet^{(q)'}(r_\bullet) = \sum_N j_{N\bullet}(r_\bullet), \tag{1.51b}$$

$$j_{N\bullet}(r_\bullet) = \sum_n q_n \dot{r}_{N\bullet}\delta(r_{N\bullet} - r_\bullet), \tag{1.51c}$$

$$M_\bullet'(r_\bullet) = \sum_N M_{N\bullet}(r_\bullet), \tag{1.51d}$$

$$M_{N\bullet}(r_\bullet) = \mu_{N\bullet}^{(q)}\delta(r_{N\bullet} - r_\bullet) - \mu_0(\dot{r}_{N\bullet} \times P_{N\bullet}), \tag{1.51e}$$

and

$$\mu_{N\bullet}^{(q)} = \tfrac{1}{2}\mu_0 \sum_n q_n(r_{Nn\bullet} \times \dot{r}_{Nn\bullet}). \tag{1.51f}$$

The first term in (1.51a) is the current density caused by the motion of charged structural units or individual charged particles without taking into account their internal structure, and hence, it vanishes for uncharged atoms. The second and third terms are connected with the time derivative of the polarization $P_\bullet'(r_\bullet)$ and the magnetization $M_\bullet'(r_\bullet)$, which are built up from the contributions $P_{N\bullet}(r_\bullet)$ and $M_{N\bullet}(r_\bullet)$ of the atomic systems. The *magnetic moment* $\mu_{N\bullet}$ is given by the *orbital magnetic moment* $\mu_{N\bullet}^{(q)}$ of the Nth atom and a contribution from the motion of the whole atom with polarization $P_{N\bullet}$. The ratio of these two terms is of the order $\dot{r}_{Nn}/\dot{r}_N$ (the quotient of the velocity of the charges within the atom and the velocity of the atom), which is large in most cases of interest. (It is also possible to take into account in these relations the magnetic moment of the spins of the charged particles (see e.g. Ref. 4). But since this is a typical quantum-physical contribution, it will be discussed in Chapter 3.)

Substituting the expressions for the charge density of (1.50) and the current density of (1.51) into (1.46), Maxwell's equations for the *local fields* $E_\bullet'$ and $B_\bullet'$ are obtained in the following form:

$$\frac{1}{\mu_0}\nabla_\bullet \times B_\bullet' = \varepsilon_0 \frac{\partial}{\partial t} E_\bullet' + j_\bullet^{(q)'}(r_\bullet) + \frac{\partial}{\partial t} P_\bullet'(r_\bullet) + \frac{1}{\mu_0}\nabla_\bullet \times M_\bullet'(r_\bullet), \tag{1.52a}$$

$$\nabla_\bullet \times E_\bullet' = -\frac{\partial}{\partial t} B_\bullet', \tag{1.52b}$$

$$\varepsilon_0 \nabla_\bullet E_\bullet' = \rho^{(q)'}(r_\bullet) - \nabla_\bullet P_\bullet'(r_\bullet), \tag{1.52c}$$

$$\nabla_\bullet B_\bullet' = 0. \tag{1.52d}$$

This means that the local fields are calculated for the given microscopic source terms $\rho^{(q)\prime}$, $j_\bullet^{(q)\prime}$, $P'_\bullet$ and $M'_\bullet$, which are connected with the properties of atomic systems and free charges. In all cases of interest these source terms are not independent of the microscopic fields but are influenced by the fields at $r_\bullet, t$ as well as at the points $r'_\bullet$ in the vicinity of $r_\bullet$ and at earlier times t' (from $t' = -\infty$ until $t' = t - |r_\bullet - r'_\bullet|/c$). These dependences, like $P'_\bullet[E'_\bullet(r_\bullet, t)]$—the so-called constitutive relations or material equations—have to be derived from the (classical or quantum-mechanical) equations of motion for charged particles.

Before discussing the material equations more thoroughly let us *average* the microscopic fields and source terms. To do this we imagine a large number of microscopic realizations of the material investigated to be under the influence of external fields. In every realization the macroscopic parameters—for example, the fields of external sources, the temperature, pressure and number of particles—are assumed to be constant. The microscopic fields and source terms, however, are subjected to *stochastic fluctuations* among the members of the ensemble. From a large number of measurements performed at various realizations, statistical mean values $\langle X'(r_\bullet, t)\rangle$ of the fluctuating quantities $X'(r_\bullet, t)$ can be obtained, which may be denoted by

$$X(r_\bullet, t) = \langle X'(r_\bullet, t)\rangle. \tag{1.53}$$

The corresponding mean values, which in this sense can be understood as being measurable quantities, may be calculated within the framework of statistical physics. Thus it is obvious that this way of averaging does not require further assumptions to be made on the space–time structure of the microscopic fields or of the mean values of their ensemble averages. Applying the averaging to (1.52), we obtain Maxwell's equations for the statistically averaged fields and the source terms, where a formal analogy to (1.52) is obvious:

$$\frac{1}{\mu_0}\nabla_\bullet \times B_\bullet = \varepsilon_0 \frac{\partial}{\partial t} E_\bullet + j_\bullet^{(q)} + \frac{\partial}{\partial t} P_\bullet + \frac{1}{\mu_0}\nabla_\bullet \times M_\bullet, \tag{1.54a}$$

$$\nabla_\bullet \times E_\bullet = -\frac{\partial}{\partial t} B_\bullet, \tag{1.54b}$$

$$\varepsilon_0 \nabla_\bullet E_\bullet = \rho^{(q)} - \nabla_\bullet P_\bullet, \tag{1.54c}$$

$$\nabla_\bullet B_\bullet = 0. \tag{1.54d}$$

With the dielectric displacement

$$D_\bullet = \varepsilon_0 E_\bullet + P_\bullet \tag{1.55a}$$

and the magnetic field

$$H_\bullet = \frac{1}{\mu_0}(B_\bullet - M_\bullet), \tag{1.55b}$$

these equations attain the form

$$\nabla_\bullet \times H_\bullet = \frac{\partial}{\partial t} D_\bullet + j_\bullet^{(q)}, \tag{1.56a}$$

$$\nabla_\bullet \times E_\bullet = -\frac{\partial}{\partial t} B_\bullet, \tag{1.56b}$$

$$\nabla_\bullet D_\bullet = \rho^{(q)}, \tag{1.56c}$$

$$\nabla_\bullet B_\bullet = 0. \tag{1.56d}$$

These Maxwell equations have been obtained by statistical averaging from the microscopic relations (1.46). This procedure is advantageous in that it connects the quantities in the Maxwell equations with the statistical expectation values of atomic quantities according to (1.53) in conjunction with (1.50) and (1.51).

In any real measuring process a spatial and temporal averaging is carried out. In this connection it is important to note that (1.56) can be interpreted as the relationship between the corresponding measurable quantities of macroscopic electrodynamics (see Ref. 8). Let us discuss this topic for materials with small distances l (relative to the wavelength λ of an incident wave) between the atomic systems under investigation. In such cases it is possible to represent in a good approximation the number of atomic systems of type A in volume elements ΔV with linear dimensions L ($l \ll L \ll \lambda$) as a product of a number density $\gamma_A(r_\bullet, t)$ and ΔV, where $r_\bullet$ represents a certain point in the volume element. Furthermore, let us assume that the mutual interaction of the atomic systems arises from the electromagnetic fields only. Then the sums over the atomic systems in (1.50) and (1.51) can be transformed by approximation into integrals which can be easily evaluated by utilizing the properties of the delta function. Note that, when doing so, a spatial averaging is carried out in which terms with rapid spatial variation give vanishing contributions. If there is only one type of atomic system, the source terms of the macroscopic Maxwell equations can be written as

$$\rho^{(q)} = \left\langle \sum_N \rho_N^{(q)} \right\rangle = \gamma_A q_A, \tag{1.57a}$$

$$P_\bullet = \left\langle \sum_N P_{N\bullet} \right\rangle = \gamma_A \langle d_{A\bullet} \rangle - \langle Q_{A\bullet\bullet} \rangle \nabla_\bullet \gamma_A, \tag{1.57b}$$

$$j_\bullet^{(q)} = \left\langle \sum_N j_{N\bullet}^{(q)} \right\rangle = q_A \gamma_A v_{A\bullet}, \tag{1.57c}$$

$$M_\bullet = \left\langle \sum_N M_{N\bullet} \right\rangle = \langle \mu_{A\bullet}^{(q)} \rangle \gamma_A - \mu_0 \langle v_{A\bullet} \times P_{A\bullet} \rangle \gamma_A, \tag{1.57d}$$

where q_A and $v_{A\bullet}$ are the charge and the average velocity of the atomic systems, and $\langle d_{A\bullet}\rangle$, $\langle Q_{A\bullet\bullet}\rangle$, $\langle \mu^{(q)}_{A\bullet}\rangle$ are statistical mean values of the dipole moment, quadrupole moment, and magnetic dipole moment of these atomic systems, respectively. On using these continuously varying source terms, the Maxwell equations (1.54) and (1.56) no longer contain discontinuities. If atomic systems (including free charges) of various types are considered, it is necessary to introduce an appropriate density for each type, and instead of (1.57) there will be obtained relations for charge density, polarization, current density, and magnetization in which additional summations over these types (i.e. over A) are carried out.

1.2.3. Effective Fields and Field Corrections

As mentioned above, a consistent description of the interaction between electromagnetic fields and atomic systems requires not only the field equations —say in the form (1.54) or (1.56) in connection with (1.55) and (1.57)—but also the constitutive equations, which allow the calculation of the polarization $P_\bullet$, the magnetization $M_\bullet$, and the charge density and current density of free charge carriers, $\rho^{(q)}$ and $j^{(q)}_\bullet$, as functions of the field. These constitutive relations can be obtained from the microscopic equations of motion of the atomic systems with the electromagnetic fields acting on them. In Chapter 3 this approach will be discussed in greater detail in the framework of the quantum-theoretical description.

In this section we consider only an approximate method of calculating *local-field corrections*. In principle, for calculating the constitutive relations that result from the microscopic equations of motion, the microscopic fields are required. As mentioned in Section 1.2.2, an exact calculation of them is not possible in systems with many degrees of freedom. But it is possible to calculate the local electric and magnetic fields approximately by use of the *macroscopic fields* as well as of the *macroscopic polarization* and *magnetization*, which are given for a dense medium in (1.57). In what follows we consider the structure of the constitutive relations for a nonmagnetic material ($M_\bullet = 0$) without free charges ($\rho^{(q)} = 0$, $j^{(q)} = 0$). Thus only the relation $P_\bullet[E_\bullet]$ remains to be considered.

The *local field* $E_{\text{loc}\bullet}$ at the coordinate $r_\bullet$ of an atomic system is calculated in the following way. The atom is regarded as being surrounded by a sphere with radius R (R should be large in comparison with distances between the atomic systems, but small compared with the macroscopic dimensions of the sample). The local field equals the sum of the field E_{empty}, which would be present in the empty sphere with unchanged polarization $P_\bullet$ in the surrounding material, and the fields $E_{i\bullet}$ of the single atomic systems ($i = 1, \ldots, M$) inside the sphere. This gives

$$E_{\text{loc}\bullet} = E_{\text{empty}\bullet} + \sum_i^M E_{i\bullet}. \tag{1.58}$$

$E_{\text{empty}\bullet}$ can be calculated on the basis of macroscopic electrodynamics because the distances between the sources and $r_\bullet$ are large compared with atomic distances. There results the relation

$$E_{\text{empty}\bullet} = E_\bullet + \frac{1}{3\varepsilon_0} P_\bullet. \tag{1.59}$$

In many cases the fields of the atomic systems inside the sphere nearly cancel out. Full compensation will be achieved, for example, if the atomic systems occupy the vertices of a cubic lattice and if they can be represented by their dipole moments with the same orientation. In amorphous media the contribution of $\Sigma_i E_{i\bullet}$ can be shown to be negligibly small under similar conditions (see 1.2, 1.3). Hence we can suppose the relation

$$E_{\text{loc}\bullet} = E_\bullet + \frac{1}{3\varepsilon_0} P_\bullet \tag{1.60}$$

to be valid. It should be noted that in deriving this relation no assumption has been made on the dependence of $P_\bullet$ on the field. Thus (1.60) can be used in linear and in nonlinear optics.

For isotropic, linear optical media with negligible absorption and dispersion in the frequency range under consideration (i.e. far away from resonance frequencies) the Lorentz–Lorenz equation can be obtained from (1.60) in the following way. The *nonresonant* linear optical polarization $P_\bullet^{\text{LNR}}$ is given by

$$P_\bullet^{\text{LNR}} = \varepsilon_0 \{\gamma \chi^{\text{LNR}}\} E_{\text{loc}\bullet}, \tag{1.61}$$

where γ is the number of atomic systems per unit volume, and χ^{LNR} and $\{\gamma \chi^{\text{LNR}}\}$ represent the nonresonant susceptibility of one atom and of the ensemble, respectively. Using (1.60) and (1.61), we obtain

$$P_\bullet^{\text{LNR}} = \varepsilon_0 (\varepsilon^{\text{NR}} - 1) E_\bullet, \tag{1.62a}$$

where

$$\varepsilon^{\text{NR}} = \frac{1 + \frac{2}{3}\gamma\chi^{\text{LNR}}}{1 - \frac{1}{3}\gamma\chi^{\text{LNR}}}. \tag{1.62b}$$

The local field in (1.60) can now be expressed in the form

$$E_{\text{loc}\bullet} = \frac{\varepsilon^{\text{NR}} + 2}{3} E_\bullet. \tag{1.63}$$

Let us now consider a medium, the polarization of which is composed of nonresonant and resonant linear contributions $P_\bullet^{\text{LNR}}$ and $P_\bullet^{\text{LR}}$ and of a nonlinear term $P_\bullet^{\text{NL}}$:

$$P_\bullet = P_\bullet^{\text{LNR}} + P_\bullet^{\text{LR}} + P_\bullet^{\text{NL}}. \tag{1.64}$$

(The separation of a resonant part in the polarization is of course only possible with respect to a certain frequency region determined by the frequency content of the external fields and the resonance frequencies of the atomic systems. Such a separation will be performed explicitly in Chapter 4.) The substitution of (1.60) together with (1.64) into (1.61) results in

$$P_{\bullet}^{\mathrm{LNR}} = \varepsilon_0(\varepsilon^{\mathrm{NR}} - 1)E_{\bullet} + \tfrac{1}{3}(\varepsilon^{\mathrm{NR}} - 1)(P_{\bullet}^{\mathrm{LR}} + P_{\bullet}^{\mathrm{NL}}) \tag{1.65}$$

instead of (1.62a). This leads to the dielectric displacement $D_{\bullet} = \varepsilon_0 E_{\bullet} + P_{\bullet}$ in the form

$$D_{\bullet} = \varepsilon_0\varepsilon^{\mathrm{NR}}E_{\bullet} + \frac{\varepsilon^{\mathrm{NR}} + 2}{3}(P_{\bullet}^{\mathrm{LR}} + P_{\bullet}^{\mathrm{NL}}). \tag{1.66}$$

This means that the action of the *nonlinear* optical polarization and of the *resonant* part of the linear optical polarization is enhanced by a factor $(\varepsilon^{\mathrm{NR}} + 2)/3$. These enhanced contributions to the resonant linear and to the nonlinear polarization, which have to be substituted into Maxwell's equations, may be written for short as

$$P_{\mathrm{eff}\bullet}^{\mathrm{LR}} = \frac{\varepsilon^{\mathrm{NR}} + 2}{3}P_{\bullet}^{\mathrm{LR}}, \tag{1.67a}$$

$$P_{\mathrm{eff}\bullet}^{\mathrm{NL}} = \frac{\varepsilon^{\mathrm{NR}} + 2}{3}P_{\bullet}^{\mathrm{NL}}. \tag{1.67b}$$

Assuming $P_{\bullet}^{\mathrm{LR}}$ and $P_{\bullet}^{\mathrm{NL}}$ to be small, it is justified to use the dependence (1.63) between $E_{\mathrm{loc}\bullet}$ and $E_{\bullet}$ in the local constitutive relations $P_{\bullet}^{\mathrm{LR}}[E_{\mathrm{loc}\bullet}]$ and $P_{\bullet}^{\mathrm{NL}}[E_{\mathrm{loc}\bullet}]$. The nth-order polarization at frequency ω can be expressed by

$$\underset{\smile}{P}_{\bullet}^{(n)}(\omega) = \underset{\smile}{\varkappa}_{\bullet\bullet\cdots\bullet}^{(n)}(\omega; \omega_1, \ldots, \omega_n)\underset{\smile}{E}_{\mathrm{loc}\bullet}(\omega_1)\cdots\underset{\smile}{E}_{\mathrm{loc}\bullet}(\omega_n) \tag{1.68a}$$

or by

$$\underset{\smile}{P}_{\bullet}^{(n)}(\omega) = \left(\frac{\varepsilon^{\mathrm{NR}} + 2}{3}\right)^n \underset{\smile}{\varkappa}_{\bullet\bullet\cdots\bullet}^{(n)}(\omega; \omega_1, \ldots, \omega_n)\underset{\smile}{E}_{\bullet}(\omega_1)\cdots\underset{\smile}{E}_{\bullet}(\omega_n) \tag{1.68b}$$

where $\underset{\smile}{\varkappa}_{\bullet\bullet\cdots\bullet}^{(n)}(\omega; \omega_1, \omega_2, \ldots, \omega_n)$ is the nonlinear susceptibility of nth order, which connects the nth-order polarization at frequency $\omega = \omega_1 + \omega_2 + \cdots + \omega_n$ with the values of the electric field at the frequencies $\omega_1, \omega_2, \ldots, \omega_n$. The corresponding effectively acting polarization $\underset{\smile}{P}_{\mathrm{eff}\bullet}^{(n)}(\omega)$ can be written as

$$\underset{\smile}{P}_{\mathrm{eff}\bullet}^{(n)}(\omega) = \left(\frac{\varepsilon^{\mathrm{NR}} + 2}{3}\right)^{n+1} \underset{\smile}{\varkappa}_{\bullet\bullet\cdots\bullet}^{(n)}(\omega; \omega_1 \ldots, \omega_n)\underset{\smile}{E}_{\bullet}(\omega_1)\cdots\underset{\smile}{E}_{\bullet}(\omega_n). \tag{1.68c}$$

(For $n = 1$ this equation gives the relation for $P_{\bullet}^{\mathrm{LR}}$.)

Taking into consideration the dispersion of the dielectric permittivity, (1.68b) has to be replaced by

$$\underset{\smile}{P}^{(n)}_{\mathrm{eff}\bullet}(\omega) = C_{\mathrm{field}}\underset{\smile}{\varkappa}^{(n)}_{\bullet\bullet\cdots\bullet}(\omega;\omega_1,\ldots,\omega_n)\underset{\smile}{E}_{\bullet}(\omega_1)\cdots\underset{\smile}{E}_{\bullet}(\omega_n), \tag{1.69}$$

where

$$C_{\mathrm{field}} = \left(\frac{\varepsilon^{\mathrm{NR}}(\omega)+2}{3}\right)\left(\frac{\varepsilon^{\mathrm{NR}}(\omega_1)+2}{3}\right)\cdots\left(\frac{\varepsilon^{\mathrm{NR}}(\omega_n)+2}{3}\right)$$

is a correction factor by means of which the deviation between local and macroscopic fields is described.

As defined above, $E_\bullet$ is the *macroscopic field strength* within the material. It is often useful to calculate the polarization as a function of the field strength $E^{(0)}_\bullet$ of the external sources in vacuo, because this parameter is primarily given by the experimental conditions. If a plane wave is coupled into the material without reflection and absorption losses (which can be achieved by employing suitable antireflection coatings), then

$$(E^{(0)})^2 = \sqrt{\varepsilon^{\mathrm{NR}}}\,E^2 \tag{1.70}$$

because of the continuity of the Poynting vector at the surface. Thus instead of (1.69) we have to use

$$\underset{\smile}{P}^{(n)}_{\mathrm{eff}\bullet}(\omega) = C^{(0)}_{\mathrm{field}}\underset{\smile}{\varkappa}^{(n)}_{\bullet\bullet\cdots\bullet}(\omega;\omega_1,\ldots,\omega_n)\underset{\smile}{E}^{(0)}_{\bullet}(\omega_1)\cdots\underset{\smile}{E}^{(0)}_{\bullet}(\omega_n) \tag{1.71}$$

where

$$C^{(0)}_{\mathrm{field}} = \frac{\varepsilon^{\mathrm{NR}}(\omega)+2}{3}\prod_{i=1}^{n}\frac{1}{\sqrt[4]{\varepsilon^{\mathrm{NR}}(\omega_i)}}\left(\frac{\varepsilon^{\mathrm{NR}}(\omega_i)+2}{3}\right). \tag{1.72}$$

The equations (1.69) and (1.71) hold for isotropic optical materials. The corresponding relations for anisotropic media are given in Refs. 7, 1.4, and 1.5.

Finally, it should be mentioned once more that the Lorentz theory of *field correction*, which we have employed, simplifies the real interaction process. Objections to the application of the Lorentz theory are discussed for example in Ref. 1.6. In principle, a statistical theory of the ensemble of atomic systems has to be applied. Therefore the field corrections calculated in this subsection can only be considered as rough estimates, and in most cases the parameters involved have to be determined from experiments.

1.3. RELATIONSHIP BETWEEN POLARIZATION AND FIELD STRENGTH

In this section we start with the basic relationship between polarization and field strength in the time domain by introducing susceptibility functions dependent on time and space coordinates. Then we transform these relations as well as the corresponding susceptibilities to the frequency domain. Some general properties of linear and nonlinear optical susceptibilities will then be investigated.

1.3.1. Basic Equations

As described in Section 1.2, an electric field acting from outside on the sample under investigation produces dipole moments and multipole moments in the atomic system by changing the equilibrium distribution of the electrons and nuclei. These induced moments produce the polarization of the sample, which is the microscopic quantity to be measured. Thus we may say that the external electric field causes the polarization of the sample. Hence there exists a functional dependence between the field strength as the "cause" and the polarization as the "effect," which in general is a nonlinear one. Note that the polarization at time t and space point $r_\bullet$ can be affected not only by the value of the field at t and $r_\bullet$ but also by values at other times and space points. Hence the relation $P_\bullet[E_\bullet]$ exhibits a noninstantaneous and nonlocal structure, and its functional dependence is characterized by integrals over time and space coordinates. Whereas from the purely mathematical point of view the polarization at t and $r_\bullet$ may depend on the field strength at all times and space points, in fact causality imposes restrictions on such relations between the "cause" and its "effect" which will be discussed later. Under rather general conditions the nonlinear, noninstantaneous, and nonlocal dependence between the polarization and the electric field can be expressed by a Volterra expansion in the form

$$
\begin{aligned}
P_\bullet(r_\bullet, t) = P_\bullet^{(0)}(r_\bullet, t) + \sum_{n=1}^{\infty} \varepsilon_0 \int dV_1 \int dt_1 \cdots \int dV_n \int dt_n \\
\times \varkappa^{(n)}_{\bullet\bullet\cdots\bullet}(r_\bullet, t; r_{1\bullet}, t_1; \ldots, r_{n\bullet}, t_n) \\
\times E_\bullet(r_{1\bullet}, t_1) \cdots E_\bullet(r_{n\bullet}, t_n).
\end{aligned}
\tag{1.73}
$$

This representation is of use only if the expansion converges sufficiently fast. Note that this convergence depends on the field strength $E_\bullet$ and the atomic parameters. (Making use already of the knowledge of the microphysical structure of the medium and assuming the electromagnetic field to be far off resonance with atomic transition frequencies, we may say that the convergence depends on the ratio of the field strength E of the radiation to the atomic field strength, which is of the order of 10^{10} V/m. For instance, the electric field of the nucleus of a hydrogen atom acting on the electron at a distance equal to the Bohr radius $a_0 = 0.5 \times 10^{-10}$ m attains a value of about 5×10^{11} V/m.) We shall limit ourselves to moderately strong radiation fields and therefore to the first few terms of (1.73).

The term $P^{(0)}_{\bullet}$ may already exist even without external fields influencing the material. At zero frequencies $P^{(0)}_{\bullet}$ may arise from a static spatial arrangement of dipole moments. At optical frequencies a field-independent part of the polarization can only result from *fluctuations* of the atomic systems. Therefore its statistical average, which has to be used here, will vanish. (The influence of fluctuations on radiation processes will be discussed later in connection with the quantum-theoretical description.)

1.3.1.1. Linear Polarization. The first term in the sum in (1.73),

$$P^{(1)}_{\bullet}(r_\bullet, t) = \varepsilon_0 \int dV_1 \int dt_1\, \varkappa^{(1)}_{\bullet\bullet}(r_\bullet, t; r_{1\bullet}, t_1) E_\bullet(r_{1\bullet}, t_1) \qquad (1.74)$$

represents the *linear polarization*, where $\varkappa^{(1)}_{\bullet\bullet}$ is a *linear susceptibility*, which we also call the *linear response function*. In the language of system theory the relation between the *input signal* $E_\bullet(r_{1\bullet}, t_1)$ and the *output signal* $P^{(1)}_{\bullet}(r_\bullet, t)$ is described by the *pulse response* $\varkappa^{(1)}_{\bullet\bullet}(r_\bullet, t; r_{1\bullet}, t_1)$. In general, this relation exhibits nonlocal behavior as well as memory. This means that the polarization $P^{(1)}_{\bullet}(r_\bullet, t)$ depends not only on $E_\bullet(r_\bullet, t)$ but also on the values of $E_\bullet$ in the vicinity of $r_\bullet$ and at times $t_1 \leq t$. The time interval $\tau = t - t_1$ and the spatial distance $R = |r_\bullet - r_{1\bullet}|$, in which the influence of $E_\bullet(r_{1\bullet}, t_1)$ on $P^{(1)}_{\bullet}(r_\bullet, t)$ according to (1.74) has to be considered, are determined by the interaction processes between the atomic systems. Thus the *memory time* or *response time* is given by the *relaxation times* of the medium, during which a microscopic excitation is dying out (cf. Chapter 4), and the diameter of the region of nonlocality is given by the distance over which a microscopic excitation can be transferred during the relaxation time.

Often the diameter of the region of nonlocality is small compared with the wavelength. Then (1.74)—as well as (1.73)—reduces to local relations between $P_\bullet(r_\bullet, t)$ and $E_\bullet(r_\bullet, t_1)$, and we may omit the space coordinate in the following relations. Besides, the pulse response of the medium depends only on the time difference $t - t_1$, and therefore we may write

$$P^{(L)}_{\bullet}(t) = \varepsilon_0 \int_{-\infty}^{+\infty} dt_1\, \varkappa^{(1)}_{\bullet\bullet}(t - t_1) E_\bullet(t_1). \qquad (1.75)$$

The response function $\varkappa^{(1)}_{\bullet\bullet}(t)$ is a second-rank tensor characterizing the polarizability of the material. Taking into account the causality of the relation between polarization and field strength, due to which the polarization at time t can only be caused by the electric field at times $t' \leq t$, we conclude that the susceptibility function $\varkappa^{(1)}_{\bullet\bullet}(t - t_1)$ vanishes for $t_1 > t$. Thus, with the substitution $\tau_1 = t - t_1$ we obtain

$$P^{(L)}_{\bullet}(t) = \varepsilon_0 \int_0^{\infty} d\tau_1\, \varkappa^{(1)}_{\bullet\bullet}(\tau_1) E_{\bullet}(t - \tau_1). \tag{1.76}$$

The memory effect becomes obvious if we consider the action of a δ-like field strength pulse $E_{\bullet}(t) = \check{E}_{\bullet}\delta(t - t_0)$; it causes a polarization

$$P_{\bullet}(t) = \varepsilon_0 \varkappa^{(1)}_{\bullet\bullet}(t - t_0)\check{E}_{\bullet}, \tag{1.77}$$

which is unequal to zero only at $t \geq t_0$ because of causality. For very large $t - t_0$ the polarization again tends to zero because of the finite memory time of the medium. This means that the field-strength pulses which have affected the sample a long time before do not contribute to the present value of the polarization. This finite duration of the memory originates from dissipation processes due to which the system tends to equilibrium at any time. Only in systems with finite memory times does the integral in (1.76) converge for all limited field strength functions. In principle, (1.77) can be used to determine the linear response function $\varkappa^{(1)}_{\bullet\bullet}(t)$ by measuring $P_{\bullet}(r_{\bullet}, t)$ after δ-pulse excitation.

1.3.1.2. Nonlinear Polarization. We now consider the first nonlinear term in (1.73). Assuming a local relationship between polarization and field strength, this term, again after a transformation of the integration variables, can be written in the form

$$P^{(2)}_{\bullet}(t) = \varepsilon_0 \int_0^{\infty} d\tau_1 \int_0^{\infty} d\tau_2\, \varkappa^{(2)}_{\bullet\bullet\bullet}(\tau_1, \tau_2) E_{\bullet}(t - \tau_1) E_{\bullet}(t - \tau_2). \tag{1.78}$$

The *second-order susceptibility function* $\varkappa^{(2)}_{\bullet\bullet\bullet}$ is a third-rank tensor. In analogy to the linear case, it can be interpreted as a pulse-response function—in this case, of course, a nonlinear one. If two δ-like field-strength pulses

$$E_{\bullet}(t) = \check{E}_{\bullet}\delta(t - t_0) + \check{E}'_{\bullet}\delta(t - t'_0) \tag{1.79}$$

act on the material, the response is given by

$$\begin{aligned} P^{(2)}_{\bullet}(t) = {} & \varepsilon_0 \varkappa^{(2)}_{\bullet\bullet\bullet}(t - t_0, t - t_0)\check{E}_{\bullet}\check{E}_{\bullet} + \varepsilon_0 \varkappa^{(2)}_{\bullet\bullet\bullet}(t - t'_0, t - t'_0)\check{E}'_{\bullet}\check{E}'_{\bullet} \\ & + \varepsilon_0 \varkappa^{(2)}_{\bullet\bullet\bullet}(t - t_0, t - t'_0)\check{E}_{\bullet}\check{E}'_{\bullet} + \varepsilon_0 \varkappa^{(2)}_{\bullet\bullet\bullet}(t - t'_0, t - t_0)\check{E}'_{\bullet}\check{E}_{\bullet}. \end{aligned} \tag{1.80}$$

This means that the *second-order polarization* depends on products of the field-strength pulses and, via $\varkappa^{(2)}_{\bullet\bullet\bullet}(\tau_1, \tau_2)$, on the time delay between the arrival of the pulses and the instant t at which $P^{(2)}_{\bullet}$ is measured. The nonlinear pulse response can be determined by observing $P^{(2)}_{\bullet}(t)$ after two-pulse excitation of the material in dependence on the delays $t - t_0$ and $t - t'_0$.

Without loss of generality we may require the following symmetry to be satisfied:

$$\varkappa^{(2)}_{ijk}(\tau_1, \tau_2) = \varkappa^{(2)}_{ikj}(\tau_2, \tau_1). \tag{1.81}$$

This means that the second-order susceptibility should be symmetric under a simultaneous change of the tensor indices j, k and the variables τ_1, τ_2. We can prove that it is possible to choose such symmetric susceptibility functions in the following way: Any nonsymmetric susceptibility function $[\varkappa^{(2)}_{ijk}]_{\text{ns}}$ can be represented by a symmetric and an antisymmetric part:

$$\left[\varkappa^{(2)}_{ijk}(\tau_1, \tau_2)\right]_{\text{ns}} = \left[\varkappa^{(2)}_{ijk}(\tau_1, \tau_2)\right]_{\text{s}} + \left[\varkappa^{(2)}_{ijk}(\tau_1, \tau_2)\right]_{\text{as}}, \tag{1.82a}$$

where

$$\left[\varkappa^{(2)}_{ijk}(\tau_1, \tau_2)\right]_{\text{s}} = \tfrac{1}{2}\left[\varkappa^{(2)}_{ijk}(\tau_1, \tau_2)\right]_{\text{ns}} + \tfrac{1}{2}\left[\varkappa^{(2)}_{ikj}(\tau_2, \tau_1)\right]_{\text{ns}} \tag{1.82b}$$

and

$$\left[\varkappa^{(2)}_{ijk}(\tau_1, \tau_2)\right]_{\text{as}} = \tfrac{1}{2}\left[\varkappa^{(2)}_{ijk}(\tau_1, \tau_2)\right]_{\text{ns}} - \tfrac{1}{2}\left[\varkappa^{(2)}_{ikj}(\tau_2, \tau_1)\right]_{\text{ns}}. \tag{1.82c}$$

In this representation the susceptibility tensor is inserted into (1.78). The contribution of $[\varkappa^{(2)}_{ijk}(\tau_1, \tau_2)]_{\text{as}}$ equals

$$\left\{\tfrac{1}{2}\varepsilon_0 \sum_{j,k} \int_0^\infty d\tau_1 \int_0^\infty d\tau_2 \left[\varkappa^{(2)}_{ijk}(\tau_1, \tau_2)\right]_{\text{ns}} E_j(t-\tau_1) E_k(t-\tau_2)\right\}$$

$$-\left\{\tfrac{1}{2}\varepsilon_0 \sum_{j,k} \int_0^\infty d\tau_1 \int_0^\infty d\tau_2 \left[\varkappa^{(2)}_{ikj}(\tau_2, \tau_1)\right]_{\text{ns}} E_j(t-\tau_1) E_k(t-\tau_2)\right\}.$$

This expression is seen to vanish by exchanging the summation indices j, k and the integration variables τ_1, τ_2 in the second term. Hence, the antisymmetric part of the susceptibility function has no physical meaning and may be set equal to zero.

In the *n th order* the local relation between polarization and field strength is given by

$$P^{(n)}_{\bullet}(t) = \varepsilon_0 \int_0^\infty d\tau_1 \cdots \int_0^\infty d\tau_n$$

$$\times \varkappa^{(n)}_{\bullet\bullet\cdots\bullet}(\tau_1, \ldots, \tau_n) E_{\bullet}(t-\tau_1) \cdots E_{\bullet}(t-\tau_n). \tag{1.83}$$

The *susceptibility function* (or *response function*) describes the response of the system to the action of n δ-like field strength pulses. As already explained in the case of $n = 2$, the response function may be chosen so as to exhibit the symmetric form

$$\varkappa^{(n)}_{ij_1 \cdots j_n}(\tau_1, \ldots, \tau_n) = \mathsf{P}\varkappa^{(n)}_{ij_1 \cdots j_n}(\tau_1, \ldots, \tau_n), \tag{1.84}$$

where P is an operation by which the indices $j_1, \ldots, j_n$ are arbitrarily permuted but in the same way as the variables $\tau_1, \ldots, \tau_n$ [e.g. $\varkappa^{(n)}_{ij_1 j_2 \cdots j_n}(\tau_1, \tau_2, \ldots, \tau_n) = \varkappa^{(n)}_{ij_2 j_1 \cdots j_n}(\tau_2, \tau_1, \ldots, \tau_n)$]. The symmetry (1.84) is usually called *intrinsic permutation symmetry*.

1.3.2. Susceptibilities in the Frequency Domain

In Section 1.3.1 the relation between polarization and field strength was described by susceptibility functions depending on *time and space coordinates*. By applying Fourier transformation we may also characterize the relationship between polarization and field strength in the *frequency and wave-vector domain*.

Under the condition that local relations hold, the susceptibilities are dependent only on the frequencies and not on the wave vectors. The Fourier transforms FT are written in the following form [with $\nu = (1/2\pi)\omega$]:

$$E_\bullet(t) = \mathrm{FT}\{\breve{E}_\bullet(\nu)\} = \int_{-\infty}^{\infty} d\nu\, \breve{E}_\bullet(\nu) e^{+i2\pi\nu t}, \tag{1.85a}$$

$$\breve{E}_\bullet(\nu) = \mathrm{FT}\{E_\bullet(t)\} = \int_{-\infty}^{\infty} dt\, E_\bullet(t) e^{-i2\pi\nu t}, \tag{1.85b}$$

$$P_\bullet(t) = \int_{-\infty}^{\infty} d\nu\, \breve{P}_\bullet(\nu) e^{+i2\pi\nu t}, \tag{1.86a}$$

$$\breve{P}_\bullet(\nu) = \int_{-\infty}^{\infty} dt\, P_\bullet(t) e^{-i2\pi\nu t}. \tag{1.86b}$$

For the nth-order susceptibilities an n-dimensional Fourier transformation leads to the relations

$$\varkappa^{(n)}_{\bullet\bullet \cdots \bullet}(\tau_1, \ldots, \tau_n) = \int_{-\infty}^{+\infty} d\nu_1 \cdots \int_{-\infty}^{+\infty} d\nu_n \breve{\varkappa}^{(n)}_{\bullet\bullet \cdots \bullet}(\nu_1, \ldots, \nu_n) \times \exp\left[i2\pi \sum_{j=1}^{n} \nu_j \tau_j\right], \tag{1.87a}$$

$$\breve{\varkappa}^{(n)}_{\bullet\bullet \cdots \bullet}(\nu_1, \ldots, \nu_n) = \int_{-\infty}^{+\infty} d\tau_1 \cdots \int_{-\infty}^{+\infty} d\tau_n\, \varkappa^{(n)}_{\bullet\bullet \cdots \bullet}(\tau_1, \ldots, \tau_n) \times \exp\left[-i2\pi \sum_{j=1}^{n} \nu_j \tau_j\right]. \tag{1.87b}$$

While the susceptibility $\varkappa^{(n)}_{\bullet\bullet\cdots\bullet}$ in the time domain depends on n time coordinates $\tau_1, \ldots, \tau_n$, the susceptibility $\underset{\smile}{\varkappa}^{(n)}_{\bullet\bullet\cdots\bullet}$ in the frequency domain depends on n frequency coordinates $\nu_1, \ldots, \nu_n$. Sometimes it is advantageous to use the form $\underset{\smile}{\varkappa}^{(n)}_{\bullet\bullet\cdots\bullet}(\nu; \nu_1, \ldots, \nu_n)$ with ν equal to $\sum_{j=1}^{n} \nu_j$.

1.3.2.1. First-Order Susceptibility. First we apply these transformations to the linear polarization of (1.76). Using $\mathrm{FT}\{E_{\bullet}(t - \tau_1)\} = \underset{\smile}{E}_{\bullet}(\nu)\exp[i2\pi\nu\tau_1]$ and carrying out the integration over τ_1,

$$\underset{\smile}{P}^{(1)}_{\bullet}(\nu) = \varepsilon_0 \underset{\smile}{\varkappa}^{(1)}_{\bullet\bullet}(\nu; \nu)\underset{\smile}{E}_{\bullet}(\nu) \tag{1.88}$$

is obtained. Compared with the integral relation (1.76), this algebraic relationship is useful in many applications. The linear susceptibility tensor $\underset{\smile}{\varkappa}^{(1)}_{\bullet\bullet}(\nu; \nu)$ is related to the frequency-dependent dielectric permittivity $\varepsilon_{\bullet\bullet}(\nu)$ by

$$\underset{\smile}{\varkappa}^{(1)}_{\bullet\bullet}(\nu; \nu) = \varepsilon_{\bullet\bullet}(\nu) - 1. \tag{1.89}$$

Sometimes it is of advantage to separate $E_{\bullet}(t)$ and $P_{\bullet}(t)$ into contributions with positive and negative frequencies:

$$E_{\bullet}(t) = E^{(+)}_{\bullet}(t) + E^{(-)}_{\bullet}(t), \tag{1.90a}$$

$$P_{\bullet}(t) = P^{(+)}_{\bullet}(t) + P^{(-)}_{\bullet}(t), \tag{1.90b}$$

where

$$E^{(\pm)}_{\bullet}(t) = \int_0^{\infty} d\nu\, \underset{\smile}{E}^{(\pm)}_{\bullet}(\nu) e^{\pm i2\pi\nu t}, \tag{1.90c}$$

$$\underset{\smile}{E}^{(+)}_{\bullet}(+\nu) = \left[\underset{\smile}{E}^{(+)}_{\bullet}(-\nu)\right]^{*} = \underset{\smile}{E}^{(-)}_{\bullet}(-\nu) \equiv \underset{\smile}{E}_{\bullet}(\nu) \tag{1.90d}$$

(for $\nu \geq 0$) and

$$P^{(\pm)}_{\bullet}(t) = \int_0^{\infty} d\nu\, \underset{\smile}{P}^{(\pm)}_{\bullet}(\nu) e^{\pm i2\pi\nu t}, \tag{1.90e}$$

$$\underset{\smile}{P}^{(+)}_{\bullet}(+\nu) = \left[\underset{\smile}{P}^{(+)}_{\bullet}(-\nu)\right]^{*} = \underset{\smile}{P}^{(-)}_{\bullet}(-\nu) \equiv \underset{\smile}{P}_{\bullet}(\nu) \tag{1.90f}$$

(for $\nu \geq 0$). The relations (1.90d) and (1.90f) guarantee that $E_{\bullet}(t)$ and $P_{\bullet}(t)$ are real. The linear relationship between polarization and field strength is

$$\underset{\smile}{P}^{(1)(\pm)}_{\bullet}(\nu) = \varepsilon_0 \underset{\smile}{\varkappa}^{(1)(\pm)}_{\bullet\bullet}(\nu; \nu)\underset{\smile}{E}^{(\pm)}_{\bullet}(\nu), \tag{1.91a}$$

where

$$\underset{\smile}{\varkappa}^{(1)(\pm)}_{\bullet\bullet}(\nu;\nu) = \int_0^\infty d\tau\,\varkappa^{(1)}_{\bullet\bullet}(\tau)e^{\mp i2\pi\nu\tau}, \tag{1.91b}$$

$$\underset{\smile}{\varkappa}^{(1)(+)}_{\bullet\bullet}(+\nu;+\nu) = \left[\underset{\smile}{\varkappa}^{(1)(+)}_{\bullet\bullet}(-\nu;-\nu)\right]^* = \underset{\smile}{\varkappa}^{(1)(-)}_{\bullet\bullet}(-\nu;-\nu) \equiv \underset{\smile}{\varkappa}^{(1)}_{\bullet\bullet}(\nu,\nu)$$

(for $\nu \geq 0$).

In many optical experiments electromagnetic fields are used that contain only frequencies in narrow bands $\Delta\nu_j$ around ν_j, where $\Delta\nu_j \ll \nu_j$. In case only one such frequency band is present, we may eliminate the fast time dependence by setting

$$E^{(\pm)}_\bullet(t) = \tfrac{1}{2}\hat{E}^{(\pm)}_\bullet(\nu_j,t)e^{\pm i2\pi\nu_j t}, \tag{1.92}$$

where $\hat{E}_\bullet(\nu_j,t) \equiv \hat{E}_{j\bullet}(t)$ is a slowly varying amplitude factor, which contains only frequencies of the order of $\Delta\nu_j$. [The factor $\frac{1}{2}$ in (1.92) has been chosen to obtain $E_\bullet(t) = \hat{E}_\bullet(\nu_j;t)\cos 2\pi\nu_j t$ for real $\hat{E}^{(+)}_{j\bullet} = \hat{E}^{(-)}_{j\bullet} = \hat{E}_{j\bullet}$.] Using (1.92) and an analogous expression for $P^{(\pm)}_j$, we get from (1.76) and (1.90)

$$\hat{P}^{(1)(\pm)}_\bullet(\nu_j;t) = \varepsilon_0\int_0^\infty d\tau\,\varkappa^{(1)}_{\bullet\bullet}(\tau)\hat{E}^{(\pm)}_\bullet(\nu_j;t-\tau)e^{\mp i2\pi\nu_j\tau}. \tag{1.93}$$

If the amplitude function $\hat{E}^{(\pm)}(\nu_j;t')$ varies only slightly during the memory time of the medium, it is of use to expand it at $t' = t$ by

$$\hat{E}^{(\pm)}_\bullet(\nu_j,t-\tau) = \hat{E}^{(\pm)}_\bullet(\nu_j;t) - \left[\frac{d}{dt'}\hat{E}^{(\pm)}_\bullet(\nu_j;t')\right]_{t'=t}\tau$$

$$+\frac{1}{2}\left[\frac{d^2}{dt'^2}\hat{E}^{(\pm)}_\bullet(\nu_j;t')\right]_{t'=t}\tau^2 - \cdots. \tag{1.94}$$

Using (1.94) and (1.91b) together with its derivatives

$$\frac{d}{d\nu}\underset{\smile}{\varkappa}^{(1)(\pm)}_{\bullet\bullet}(\nu;\nu) = \mp 2\pi i\int_0^\infty d\tau\,\tau\varkappa^{(1)}_{\bullet\bullet}(\tau)e^{\mp i2\pi\nu\tau},$$

$$\frac{d^2}{d\nu^2}\underset{\smile}{\varkappa}^{(1)(\pm)}_{\bullet\bullet}(\nu;\nu) = -(2\pi)^2\int_0^\infty d\tau\,\tau^2\varkappa^{(1)}_{\bullet\bullet}(\tau)e^{\mp i2\pi\nu\tau},$$

we obtain from (1.93)

$$\hat{P}_{\bullet}^{(1)(\pm)}(\nu_j;t) = \varepsilon_0 \varkappa_{\bullet\bullet}^{(1)(\pm)}(\nu_j;\nu_j)\hat{E}^{(\pm)}(\nu_j;t)$$
$$\mp \frac{i\varepsilon_0}{2\pi}\left[\frac{d}{d\nu}\varkappa_{\bullet\bullet}^{(1)(\pm)}(\nu;\nu)\right]_{\nu=\nu_j}\left[\frac{d}{dt'}\hat{E}_{\bullet}^{(\pm)}(\nu_j,t')\right]_{t'=t}$$
$$-\frac{1}{2}\frac{\varepsilon_0}{(2\pi)^2}\left[\frac{d^2}{d\nu^2}\varkappa_{\bullet\bullet}^{(1)(\pm)}(\nu;\nu)\right]_{\nu=\nu_j}\left[\frac{d^2}{dt'^2}\hat{E}_{\bullet}^{(\pm)}(\nu_j;t')\right]_{t'=t}. \tag{1.95a}$$

If the field and the polarization consist of contributions from several frequency bands, then

$$P_{\bullet}^{(1)}(t) = \sum_j \tfrac{1}{2}\hat{P}_{\bullet}^{(-)}(\nu_j;t)e^{-i2\pi\nu_j t} + \text{c.c.}. \tag{1.95b}$$

In (1.95) the influence of dispersion is represented by the frequency derivatives of the first-order susceptibility. Assuming the dispersion to be negligible, then only the first term in (1.95a) will remain, and the polarization is given by

$$P_{\bullet}^{(1)}(t) = \sum_j \tfrac{1}{2}\varepsilon_0 \varkappa_{\bullet\bullet}^{(1)}(-\nu_j;-\nu_j)\hat{E}_{\bullet}^{(-)}(\nu_j;t)e^{-i2\pi\nu_j t} + \text{c.c.} \tag{1.96}$$

The same relation is obtained for monochromatic waves (i.e., negligible dependence of the field amplitudes on time) without assuming a dispersionless medium.

Up to now we have only considered the local relationship between polarization and field strength. In the following we want to briefly discuss nonlocal influences, assuming weak spatial dispersion.

The positive- and negative-frequency parts of the field strength and polarization are represented by

$$E_{\bullet}^{(\pm)}(r_{\bullet},t) = \frac{1}{(2\pi)^4}\int_0^{\infty} d\omega \int d^3k\, \underset{\smile}{E}_{\bullet}^{(\pm)}(\omega;k_{\bullet})e^{\pm i(\omega t - k_{\bullet}r_{\bullet})} \tag{1.97}$$

and

$$P_{\bullet}^{(1)(\pm)}(r_{\bullet},t) = \frac{1}{(2\pi)^4}\int_0^{\infty} d\omega \int d^3k\, \underset{\smile}{P}_{\bullet}^{(\pm)}(\omega;k_{\bullet})e^{\pm i(\omega t - k_{\bullet}r_{\bullet})} \tag{1.98}$$

—that is, by means of plane waves (cf. Section 1.1).

Using these expressions, the relation (1.74) between $P_\bullet^{(1)}(r_\bullet, t)$ and $E_\bullet(r_\bullet, t)$ is transformed into

$$\breve{P}_\bullet^{(\pm)}(\omega; k_\bullet) = \varepsilon_0 \breve{\varkappa}_{\bullet\bullet}^{(1)(\pm)}(\omega, k_\bullet; \omega, k_\bullet) E^{(\pm)}(\omega, k_\bullet), \tag{1.99}$$

where

$$\breve{\varkappa}_{\bullet\bullet}^{(1)(\pm)}(\omega, k_\bullet; \omega, k_\bullet) = \int_0^\infty d\tau \int d^3R\, \varkappa_{\bullet\bullet}^{(1)}(r_\bullet, t; r_\bullet - R_\bullet, t-\tau) e^{\mp i(\omega\tau - k_\bullet R_\bullet)}. \tag{1.100a}$$

The dependence of $\breve{\varkappa}_{\bullet\bullet}^{(1)(\pm)}(\omega, k_\bullet; \omega, k_\bullet) = \varepsilon_{\bullet\bullet}^{(\pm)}(\omega, k_\bullet) - 1$ on the wave vector $k_\bullet$ is called *spatial dispersion*. In most cases the radius of "nonlocality" R_0 is small compared with the wavelength of the optical field. Then the response function $\varkappa_{\bullet\bullet}^{(1)}(r_\bullet, t; r'_\bullet, t')$ is negligibly small for $R = |r_\bullet - r'_\bullet| \gtrsim R_0 \ll 1/k$. Therefore the expansion

$$\exp\{\pm ik_\bullet R_\bullet\} = 1 \pm ik_\bullet R_\bullet + \cdots$$

converges rapidly, and up to the first order in $k_\bullet R_\bullet$ we obtain from (1.100a)

$$\breve{\varkappa}_{\bullet\bullet}^{(1)(\pm)}(\omega, k_\bullet; \omega, k_\bullet) = \breve{\varkappa}_{\bullet\bullet}^{(1)(\pm)}(\omega; \omega) \pm \breve{\zeta}_{\bullet\bullet\bullet}^{(1)(\pm)}(\omega; \omega) k_\bullet \tag{1.100b}$$

with

$$\breve{\zeta}_{lmn}^{(1)(\pm)}(\omega; \omega) = \int_0^\infty d\tau \int d^3R\, \varkappa_{lm}^{(1)}(r_\bullet, t; r_\bullet - R_\bullet, t-\tau) R_n. \tag{1.100c}$$

For a detailed discussion of nonlocal effects see for instance Ref. 1.7 and the references given there.

1.3.2.2. Nonlinear Optical Susceptibilities. Let us start in our treatment of nonlinear problems with the second-order polarization $\breve{P}_\bullet^{(2)}$ and its susceptibility. Applying the Fourier transformation defined in (1.85)–(1.87) to the local relation (1.78) between $P_\bullet^{(2)}(t)$ and $E_\bullet(t)$ and taking into account

$$\mathrm{FT}\{E_\bullet(t-\tau_1)E_\bullet(t-\tau_2)\}$$

$$= \int_{-\infty}^{\infty} d\nu'\, e^{i2\pi\nu'\tau_1} \breve{E}_\bullet(-\nu') e^{-i2\pi(\nu+\nu')\tau_2} \breve{E}_\bullet(\nu+\nu') \tag{1.101}$$

the integration over τ_1 and τ_2 leads to

$$\underset{\smile}{P}^{(2)}_{\bullet}(\nu) = \varepsilon_0 \int_{-\infty}^{+\infty} d\nu' \, \underset{\smile}{\varkappa}^{(2)}_{\bullet\bullet\bullet}(\nu; \nu', \nu - \nu') \underset{\smile}{E}_{\bullet}(\nu') \underset{\smile}{E}_{\bullet}(\nu - \nu'). \quad (1.102)$$

As in the case of linear optics, the relation between polarization and field strength is simpler in the frequency domain than it is in the time domain; the number of integrations has been reduced. Before giving the general relation in nth order we write (1.102) in the more symmetric form

$$\underset{\smile}{P}^{(2)}_{\bullet}(\nu) = \varepsilon_0 \int_{-\infty}^{\infty} d\nu_1 \, \underset{\smile}{\varkappa}^{(2)}_{\bullet\bullet\bullet}(\nu; \nu_1, \nu_2) \underset{\smile}{E}_{\bullet}(\nu_1) \underset{\smile}{E}_{\bullet}(\nu_2) \quad (1.103)$$

by inserting $\nu_2 = \nu - \nu'$ and $\nu_1 = \nu'$.

In nth order the relation between polarization and field strength is obtained in the same way. Here we have

$$\begin{aligned} \underset{\smile}{\mathrm{P}}^{(n)}_{\bullet}(\nu) &= \varepsilon_0 \int_{-\infty}^{\infty} d\nu_1 \cdots \int_{-\infty}^{\infty} d\nu_{n-1} \\ &\times \underset{\smile}{\varkappa}^{(n)}_{\bullet\bullet \cdots \bullet}(\nu; \nu_1, \ldots, \nu_n) \underset{\smile}{\mathrm{E}}_{\bullet}(\nu_1) \cdots \underset{\smile}{\mathrm{E}}_{\bullet}(\nu_n). \end{aligned} \quad (1.104)$$

Since $E_{\bullet}(t)$ as well as $P^{(n)}_{\bullet}(t)$ and $\varkappa^{(n)}_{\bullet\bullet \cdots \bullet}(\tau_1, \ldots, \tau_n)$ are real functions, the following symmetry relations hold:

$$\underset{\smile}{\varkappa}^{(n)}_{\bullet\bullet \cdots \bullet}(\nu; \nu_1, \ldots, \nu_n) = \underset{\smile}{\varkappa}^{(n)*}_{\bullet\bullet \cdots \bullet}(-\nu; -\nu_1, \ldots, -\nu_n). \quad (1.105a)$$

The symmetry (1.84) of the response function $\varkappa^{(n)}_{\bullet\bullet\bullet \cdots \bullet}(\tau_1, \tau_2, \ldots \tau_n)$ with respect to permutation of indices and time coordinates leads to

$$\underset{\smile}{\varkappa}^{(n)}_{ij_1 \cdots j_n}(\nu; \nu_1, \ldots, \nu_n) = \mathsf{P} \underset{\smile}{\varkappa}^{(n)}_{ij_1 \cdots j_n}(\nu; \nu_1, \ldots, \nu_n), \quad (1.105b)$$

where P is a simultaneous permutation of the indices j_i and the corresponding frequencies ν_i; for instance,

$$\begin{Bmatrix} j_1 j_2 & \cdots & j_n \\ \nu_1 \nu_2 & \cdots & \nu_n \end{Bmatrix} \to \begin{Bmatrix} j_2 j_1 & \cdots & j_n \\ \nu_2 \nu_1 & \cdots & j_n \end{Bmatrix}.$$

These relations are referred to as intrinsic permutation symmetries or index-frequency symmetries (cf. Section 1.3.2.4).

1.3.2.3. Second-Order Susceptibilities for Monochromatic Fields. Often non-linear optical processes are investigated by irradiation of matter by stationary fields within narrow spectral bands (cf. Section 1.3.2.1). An idealization of this situation allows us to discuss monochromatic waves characterized by

$$E_{\bullet}(t) = \sum_{l=1}^{L} \tfrac{1}{2}\hat{E}_{\bullet}^{(-)}(\nu_l)e^{-i2\pi\nu_l t} + \text{c.c.} \tag{1.106}$$

or the Fourier transforms

$$\underset{\smile}{E}_{\bullet}(\nu) = \sum_{l=1}^{L} \tfrac{1}{2}\left[\hat{E}_{\bullet}^{(-)}(\nu_l)\delta(\nu+\nu_l) + \hat{E}_{\bullet}^{(+)}(\nu_l)\delta(\nu-\nu_l)\right]$$

$$= \sum_{l=1}^{L} \tfrac{1}{2}\hat{E}_{\bullet}^{(-)}(\nu_l)\delta(\nu+\nu_l) + \text{m.f.}, \tag{1.107}$$

where m.f. is the Fourier transform of the conjugate complex term c.c. in (1.106). This term m.f. with "minus frequencies" can be obtained by replacing $\nu_l \to -\nu_l$ in the original expression and taking into account $\hat{E}_{\bullet}(-\nu_l) \equiv \hat{E}_{\bullet}^{(-)}(-\nu_l) = \hat{E}_{\bullet}^{(+)}(\nu_l)$. Sometimes it is more useful to introduce "new" frequencies $\nu_{-1} = -\nu_1, \ldots, \nu_{-l} = -\nu_l, \nu_{-L} = -\nu_L$ instead of the negative frequencies $-\nu_1, \ldots, -\nu_l, \ldots, -\nu_L$ and to write

$$E_{\bullet}(t) = \sum_{l=\pm 1}^{\pm L} \tfrac{1}{2}\hat{E}_{\bullet}(\nu_l)e^{-i2\pi\nu_l t}, \tag{1.108a}$$

or

$$\underset{\smile}{E}_{\bullet}(\nu) = \sum_{l=\pm 1}^{\pm L} \tfrac{1}{2}\hat{E}_{\bullet}(\nu_l)\delta(\nu+\nu_l) \tag{1.108b}$$

with

$$\hat{E}_{\bullet}(\nu_l) = \left[\hat{E}_{\bullet}(\nu_{-l})\right]^{*}. \tag{1.108c}$$

Using this representation of the electric field strength, we next calculate the *second-order polarization* $P_{\bullet}^{(2)}$ in analogy to the procedure performed to determine $P_{\bullet}^{(1)}$. We start with (1.102) and by the use of

$$\underset{\smile}{E}_{\bullet}(\nu')\underset{\smile}{E}_{\bullet}(\nu-\nu') = \sum_{m,n} \tfrac{1}{4}\hat{E}_{\bullet}(\nu_m)\hat{E}_{\bullet}(\nu_n)\delta(\nu'+\nu_m)\delta(\nu-\nu'+\nu_n)$$

we obtain the polarization

$$\underset{\smile}{P}_{\bullet}^{(2)}(\nu) = \sum_{m,n} \tfrac{1}{4}\varepsilon_0 \underset{\smile}{\varkappa}_{\bullet\bullet\bullet}^{(2)}(\nu; -\nu_m, \nu+\nu_m)\hat{E}_{\bullet}(\nu_m)\hat{E}_{\bullet}(\nu_n)\delta(\nu+[\nu_m+\nu_n]). \tag{1.109}$$

In deriving (1.109) use has been made of

$$\delta(\nu' + \nu_m)\delta(\nu - \nu' + \nu_n) = \delta(\nu' + \nu_m)\delta(\nu + [\nu_m + \nu_n]).$$

If two monochromatic waves with frequencies ν', ν'' ($\nu', \nu'' > 0$) interact with the medium (i.e., $L = 2$), the nonlinear polarization originates at five frequencies, namely at $\nu = 0, 2\nu', 2\nu'', \nu' + \nu'', |\nu' - \nu''|$. The complex amplitudes of the corresponding polarization terms are given by

$$\begin{aligned}
\hat{P}^{(2)}_{\bullet}(0) &= \tfrac{1}{2}\varepsilon_0 \underset{\smile}{\varkappa}^{(2)}_{\bullet\bullet\bullet}(0; \nu', -\nu')\hat{E}_{\bullet}(\nu')\hat{E}^{*}_{\bullet}(\nu') \\
&\quad + \tfrac{1}{2}\varepsilon_0 \underset{\smile}{\varkappa}^{(2)}_{\bullet\bullet\bullet}(0; \nu'', -\nu'')\hat{E}_{\bullet}(\nu'')\hat{E}^{*}_{\bullet}(\nu''), \\
\hat{P}^{(2)}_{\bullet}(2\nu') &= \tfrac{1}{2}\varepsilon_0 \underset{\smile}{\varkappa}^{(2)}_{\bullet\bullet\bullet}(-2\nu'; -\nu', -\nu')\hat{E}_{\bullet}(\nu')\hat{E}_{\bullet}(\nu'), \\
\hat{P}^{(2)}_{\bullet}(2\nu'') &= \tfrac{1}{2}\varepsilon_0 \underset{\smile}{\varkappa}^{(2)}_{\bullet\bullet\bullet}(-2\nu''; -\nu'', -\nu'')\hat{E}_{\bullet}(\nu'')\hat{E}_{\bullet}(\nu''), \\
\hat{P}^{(2)}_{\bullet}(\nu' + \nu'') &= \varepsilon_0 \underset{\smile}{\varkappa}^{(2)}_{\bullet\bullet\bullet}(-\nu' - \nu''; -\nu', -\nu'')\hat{E}_{\bullet}(\nu')\hat{E}_{\bullet}(\nu''), \\
\hat{P}^{(2)}_{\bullet}(\nu' - \nu'') &= \varepsilon_0 \underset{\smile}{\varkappa}^{(2)}_{\bullet\bullet\bullet}(-\nu' + \nu''; -\nu', +\nu'')\hat{E}_{\bullet}(\nu')\hat{E}^{*}_{\bullet}(\nu''). \qquad (1.110)
\end{aligned}$$

These polarization contributions are associated with *nonlinear optical rectification*, *second-harmonic generation*, and the *generation of sum and difference frequencies*. Measuring the nonlinear polarization amplitude for various frequencies ν', ν'' of the fields, the nonlinear susceptibilities can be determined everywhere in the (ν', ν'') plane. Thereby the symmetry relation (1.105) may be used. The interaction of one optical wave of frequency ν' with an electrostatic field of frequency $\nu'' = 0$ has to be treated separately because the sum and difference frequencies are degenerate. The nonlinear contribution to the polarization amplitude at ν' is given by

$$\hat{P}^{(2)}_{\bullet}(\nu') = 2\varepsilon_0 \underset{\smile}{\varkappa}^{(2)}_{\bullet\bullet\bullet}(-\nu'; -\nu', 0)\hat{E}_{\bullet}(\nu')\hat{E}_{\bullet}(0). \qquad (1.111)$$

This nonlinear term causes a change of the refractive index at ν', which depends on the polarization directions of the interacting fields. This change is called the *Pockels effect*.

Equation (1.110) may be written in brief:

$$\hat{P}^{(2)}_{\bullet}(\nu) = \varepsilon_0 \chi^{(2)}_{\bullet\bullet\bullet}(\nu; \nu_1, \nu_2)\hat{E}_{\bullet}(\nu_1)\hat{E}_{\bullet}(\nu_2), \qquad (1.112)$$

where

$$\chi^{(2)}_{\bullet\bullet\bullet}(\nu;\nu_1,\nu_2)=c^{(2)}\underset{\smile}{\varkappa}^{(2)}_{\bullet\bullet\bullet}(-\nu;-\nu_1,-\nu_2)$$

and $c^{(2)}=1$ and $\frac{1}{2}$ hold for $\nu_1\neq\nu_2$ and $\nu_1=\nu_2$, respectively; the two frequencies ν_1,ν_2 may attain the values $\pm\nu',\pm\nu''$, the value zero being excluded for any of the frequencies ν, ν_1, and ν_2.

In the case of second-harmonic generation ($\nu_1=\nu_2=\nu'$) the intrinsic permutation symmetry (1.105b) leads to an invariance of the susceptibility with permutation of the last two indices. Hence it is convenient to use the matrices $d_{il}(2\nu';\nu',\nu')$ in place of $\chi^{(2)}_{ijk}(2\nu';\nu',\nu')$ or $\underset{\smile}{\varkappa}^{(2)}_{ijk}(-2\nu';-\nu',-\nu')$. The matrix d_{il} is defined by

$$d_{il}(2\nu';\nu',\nu')=\tfrac{1}{2}\underset{\smile}{\varkappa}^{(2)}_{ijk}(-2\nu';-\nu',-\nu')=\chi^{(2)}_{ijk}(2\nu';\nu',\nu') \quad (1.113)$$

with the following correspondence of l with (j,k):

l	1	2	3	4	5	6
jk	xx	yy	zz	zy yz	zx xz	xy yx

The second-harmonic generation is then characterized by a matrix with 18 components, and the polarization amplitude is given in the matrix relation

$$\begin{pmatrix}\hat{P}^{(2)}_x(2\nu')\\ \hat{P}^{(2)}_y(2\nu')\\ \hat{P}^{(2)}_z(2\nu')\end{pmatrix}=\varepsilon_0\begin{pmatrix}d_{11}&\cdots&d_{16}\\ d_{21}&\cdots&d_{26}\\ d_{31}&\cdots&d_{36}\end{pmatrix}\begin{pmatrix}\hat{E}^2_x(\nu')\\ \hat{E}^2_y(\nu')\\ \hat{E}^2_z(\nu')\\ 2\hat{E}_z(\nu')\hat{E}_y(\nu')\\ 2\hat{E}_z(\nu')\hat{E}_x(\nu')\\ 2\hat{E}_x(\nu')\hat{E}_y(\nu')\end{pmatrix}, \quad (1.114a)$$

which may be briefly written in the form (cf. Ref. 46)

$$\hat{P}_i(2\nu')=2\varepsilon_0\sum_{l=1}^{6}d_{il}(2\nu';\nu',\nu')\left(1-\tfrac{1}{2}\delta_{jk}\right)\hat{E}_j(\nu')\hat{E}_k(\nu'). \quad (1.114b)$$

In the case of $\nu_1 \neq \nu_2$ a description of the nonlinear polarization with the help of the d_{il} can be used, assuming the dispersion to be negligible, that is, $\chi^{(2)}_{ijk}(\nu;\, \nu_1, \nu_2) = \chi^{(2)}_{ikj}(\nu;\, \nu_2, \nu_1) \approx \chi^{(2)}_{ikj}(\nu;\, \nu_1, \nu_2)$. Then the relations given above have to be replaced by

$$d_{il}(\nu;\, \nu_1, \nu_2) = \tfrac{1}{2}\chi^{(2)}_{ijk}(-\nu;\, -\nu_1, -\nu_2) = \tfrac{1}{2}\chi^{(2)}_{ijk}(\nu;\, \nu_1, \nu_2), \qquad (1.115)$$

$$\begin{pmatrix} \hat{P}^{(2)}_x(\nu) \\ \hat{P}^{(2)}_y(\nu) \\ \hat{P}^{(2)}_z(\nu) \end{pmatrix} = 2\varepsilon_0 \begin{pmatrix} d_{11} & \cdots & d_{16} \\ d_{21} & \cdots & d_{26} \\ d_{31} & \cdots & d_{36} \end{pmatrix} \begin{pmatrix} \hat{E}_x(\nu_1)\hat{E}_x(\nu_2) \\ \hat{E}_y(\nu_1)\hat{E}_y(\nu_2) \\ \hat{E}_z(\nu_1)\hat{E}_z(\nu_2) \\ \hat{E}_z(\nu_1)\hat{E}_y(\nu_2) + \hat{E}_z(\nu_2)\hat{E}_y(\nu_1) \\ \hat{E}_z(\nu_1)\hat{E}_x(\nu_2) + \hat{E}_z(\nu_2)\hat{E}_x(\nu_1) \\ \hat{E}_x(\nu_1)\hat{E}_y(\nu_2) + \hat{E}_x(\nu_2)\hat{E}_y(\nu_1) \end{pmatrix}, \qquad (1.116a)$$

and

$$\hat{P}^{(2)}_i(\nu) = 2\varepsilon_0 \sum_{l=1}^{6} d_{il}(\nu;\, \nu_1, \nu_2)\left(1 - \tfrac{1}{2}\delta_{jk}\right)\left[\hat{E}_j(\nu_1)\hat{E}_k(\nu_2) + \hat{E}_j(\nu_2)\hat{E}_k(\nu_1)\right]. \qquad (1.116b)$$

Furthermore, on the basis of the matrices $\chi^{(2)}_{ijk}$ or d_{il}, it is possible to define effective nonlinear susceptibilities $\tilde{\chi}^{(2)}$ or $\tilde{d}$ in the following way. The field amplitudes at the frequencies ν_1 and ν_2 are represented in the form

$$\hat{E}_\bullet(\nu_i) = e^{(i)}_\bullet \hat{E}(\nu_i) \qquad (1.117)$$

by the unit vectors $e^{(i)}_\bullet$ and the scalar amplitudes $\hat{E}(\nu_i)$. After inserting these field amplitudes into (1.112) and calculating the projection $\hat{P}^{(2)}(\nu_3)$ of the vector $\hat{P}^{(2)}_\bullet(\nu_3)$ on the direction of the unit vector $e^{(3)}_\bullet$, we obtain

$$\hat{P}^{(2)}(\nu_3) = \varepsilon_0 \tilde{\chi}^{(2)}(\nu_3;\, \nu_1, \nu_2)\hat{E}(\nu_1)\hat{E}(\nu_2) \qquad (1.118)$$

with

$$\tilde{\chi}^{(2)}(\nu_3; \nu_1, \nu_2) = \sum_{i,j,k} e_i^{(3)} \chi_{ijk}^{(2)}(\nu_3; \nu_1, \nu_2) e_j^{(1)} e_k^{(2)},$$

or

$$\widehat{P}^{(2)}(\nu_3) = 2c^{(2)} \varepsilon_0 \tilde{d}(\nu_3; \nu_1, \nu_2) \widehat{E}(\nu_1) \widehat{E}(\nu_2) \tag{1.119}$$

with

$$\tilde{d}(\nu_3; \nu_1, \nu_2) = \sum_{i,l} e_i^{(3)} d_{il}(\nu_3; \nu_1, \nu_2)(1 - \delta_{jk})\left(e_j^{(1)} e_k^{(2)} + e_k^{(1)} e_j^{(2)}\right)$$

and $c^{(2)}$ from (1.112). If all the polarization directions are defined in a given experimental arrangement, the nonlinearity can thus be characterized by only one scalar parameter $\tilde{\chi}^{(2)}$ or $\tilde{d}$.

1.3.2.4. Higher-Order Susceptibilities for Monochromatic Fields. In the nth order of the relation between the polarization amplitude $\widehat{P}_{\bullet}$ and n field strength amplitudes at the frequencies $\nu_l \neq 0$ $(l = 1, \ldots, L)$ we have

$$\widehat{P}_{\bullet}^{(n)}(\nu) = \chi_{\bullet\bullet\cdots\bullet}^{(n)}(\nu; \nu_1, \ldots, \nu_n) \widehat{E}_{\bullet}(\nu_1) \cdots \widehat{E}_{\bullet}(\nu_n), \tag{1.120a}$$

where

$$\chi_{\bullet\bullet\cdots\bullet}^{(n)}(\nu; \nu_1, \ldots, \nu_n) = c^{(n)} \underset{\smile}{\varkappa}_{\bullet\bullet\cdots\bullet}^{(n)}(-\nu; -\nu_1, \ldots, -\nu_n) \tag{1.120b}$$

and

$$c^{(n)} = \frac{1}{2^{n-1}} \frac{n!}{m_1! m_{-1}! \cdots m_l! m_{-l}! \cdots m_L! m_{-L}!}. \tag{1.120c}$$

[m_l, m_{-l} are the numbers of field strength amplitudes at ν_l or $-\nu_l$ occurring in (1.120a). Therefore they fulfil the condition $\sum_{l=1}^{L}(m_l + m_{-l}) = n$, where L is the number of the different absolute values of the frequencies.] We verify this connection between $\chi_{\bullet\bullet\cdots\bullet}^{(n)}(\nu; \nu_1, \ldots, \nu_n)$ and $\underset{\smile}{\varkappa}_{\bullet\bullet\cdots\bullet}^{(n)}(\nu; \nu_1, \ldots, \nu_n)$ only under simplifying conditions, in order to avoid lengthy calculations. Let us consider one component of the field strength and polarization, and assume a material without memory. Then the relation between polarization and field strength can be described by the frequency-independent scalar $\underset{\smile}{\varkappa}^{(n)}$ in the form

$$P^{(n)}(t) = \varepsilon_0 \underset{\smile}{\varkappa}^{(n)} [E(t)]^n. \tag{1.121}$$

Using

$$E(t) = \sum_{l=\pm 1}^{\pm L} \tfrac{1}{2}\hat{E}(\nu_l)e^{-i2\pi\nu_l t}$$

and

$$P(t) = \sum_{r=\pm 1}^{\pm R} \tfrac{1}{2}\hat{P}(\nu_r)e^{-i2\pi\nu_r t},$$

we obtain from (1.120)

$$\sum_{r=\pm 1}^{\pm R} \hat{P}(\nu_r)e^{-i2\pi\nu_r t} = \frac{1}{2^{n-1}}\underset{\smile}{\varkappa}^{(n)}\left[\sum_{l=\pm 1}^{\pm L} \hat{E}(\nu_l)e^{-i2\pi\nu_l t}\right]^n.$$

The use of binomial coefficients leads to

$$\hat{P}(\nu_r) = \frac{1}{2^{n-1}}\sum_{(\nu_r)} \frac{n!}{m_1!m_{-1}!\cdots m_L!m_{-L}!}\underset{\smile}{\varkappa}^{(n)}$$

$$\times\left[\hat{E}(\nu_1)\right]^{m_1}\left[\hat{E}(-\nu_1)\right]^{m_{-1}}\cdots\left[\hat{E}(\nu_L)\right]^{m_L}\left[\hat{E}(-\nu_L)\right]^{m_{-L}}, \quad (1.122)$$

where $\Sigma_{(\nu_r)}$ means a sum over all terms which satisfy the equation

$$\sum_{L=\pm 1}^{\pm L} m_l \nu_l = \nu_r.$$

From (1.122) the validity of (1.112c) is obvious.

In analogy to the calculation of second-order polarization, it is possible to define the effective susceptibilities

$$\tilde{\chi}^{(n)} = \sum_{a,\, j_1,\ldots,\, j_n} e_a^{(n+1)}\chi_{aj_1\cdots j_n}(\nu_a;\nu_1,\ldots,\nu_n)e_{j_1}^{(1)}\cdots e_{j_n}^{(n)}. \quad (1.123)$$

Making use of the general relations, we now discuss the third-order polarization. First we consider the action of one strong monochromatic field at the frequency ν'. In this case the nonlinear polarization contains terms at the frequencies ν' and $3\nu'$. Their amplitudes are given by

$$\hat{P}_\bullet^{(3)}(\nu') = \varepsilon_0\chi_{\bullet\bullet\bullet\bullet}^{(3)}(\nu';\nu',\nu',-\nu')\hat{E}_\bullet(\nu')\hat{E}_\bullet(\nu')\hat{E}_\bullet^*(\nu') \quad (1.124a)$$

with

$$\chi^{(3)}_{\bullet\bullet\bullet\bullet}(\nu';\nu',\nu',-\nu') = \tfrac{3}{4}\underset{\smile}{\kappa}^{(3)}_{\bullet\bullet\bullet\bullet}(-\nu';-\nu',-\nu',+\nu'),$$

and

$$\hat{P}^{(3)}_{\bullet}(3\nu') = \varepsilon_0\chi^{(3)}_{\bullet\bullet\bullet\bullet}(3\nu';\nu',\nu',\nu')\hat{E}_{\bullet}(\nu')\hat{E}_{\bullet}(\nu')\hat{E}_{\bullet}(\nu') \quad (1.124\text{b})$$

with

$$\chi^{(3)}_{\bullet\bullet\bullet\bullet}(3\nu';\nu',\nu',\nu') = \tfrac{1}{4}\underset{\smile}{\kappa}^{(3)}_{\bullet\bullet\bullet\bullet}(-3\nu';-\nu',-\nu',-\nu').$$

$P^{(3)}_{\bullet}(\nu')$ leads to an optically induced change of the dielectric permittivity and the refractive index. *Self-focusing* and *self-defocusing* of optical beams as well as the *optical Kerr effect* are caused by such light-induced changes of the real part of the permittivity ε and the refractive index n (cf. Chapter 4). The light-induced imaginary parts of ε and n describe losses of the exciting field by *two-photon absorption* processes (cf. Chapter 10).

Second, the polarization is calculated by taking into account two absolute values ν', ν'' of the frequency of the exciting field. In addition to the nonlinear contributions at the frequencies ν' and $3\nu'$ as well as at ν'' and $3\nu''$, which have already been given, we obtain third-order terms at the frequencies $2\nu' \pm \nu''$, $2\nu'' \pm \nu'$, ν', and ν''. Their amplitudes are given by

$$\hat{P}^{(3)}_{\bullet}(2\nu'+\nu'') = \varepsilon_0\chi^{(3)}_{\bullet\bullet\bullet\bullet}(2\nu'+\nu'';\nu',\nu',\nu'')\hat{E}_{\bullet}(\nu')\hat{E}_{\bullet}(\nu')\hat{E}_{\bullet}(\nu'') \quad (1.125\text{a})$$

and

$$\hat{P}^{(3)}_{\bullet}(2\nu'-\nu'') = \varepsilon_0\chi^{(3)}_{\bullet\bullet\bullet\bullet}(2\nu'-\nu'';\nu',\nu',-\nu'')\hat{E}_{\bullet}(\nu')\hat{E}_{\bullet}(\nu')\hat{E}^{*}_{\bullet}(\nu''), \quad (1.125\text{b})$$

where

$$\chi^{(3)}_{\bullet\bullet\bullet\bullet}(2\nu'\pm\nu'';\nu',\nu',\pm\nu'') = \tfrac{3}{4}\underset{\smile}{\kappa}^{(3)}_{\bullet\bullet\bullet\bullet}(-(2\nu'\pm\nu'');-\nu',-\nu',\mp\nu''),$$

and

$$\hat{P}^{(3)}_{\bullet}(\nu') = \varepsilon_0\chi^{(3)}_{\bullet\bullet\bullet\bullet}(\nu';\nu',\nu'',-\nu'')\hat{E}_{\bullet}(\nu')\hat{E}_{\bullet}(\nu'')\hat{E}^{*}_{\bullet}(\nu''), \quad (1.125\text{c})$$

where

$$\chi^{(3)}_{\bullet\bullet\bullet\bullet}(\nu'; \nu', \nu'', -\nu'') = \tfrac{3}{2}\varkappa^{(3)}_{\bullet\bullet\bullet\bullet}(-\nu'; -\nu', -\nu'', +\nu''),$$

and by similar expressions, obtained from permutation of ν' and ν'', for $\widehat{P}^{(3)}_{\bullet}(2\nu'' + \nu')$, $\widehat{P}^{(3)}_{\bullet}(2\nu'' - \nu')$, and $\widehat{P}^{(3)}_{\bullet}(\nu'')$. The amplitude $\widehat{P}^{(3)}_{\bullet}(2\nu' \pm \nu'')$ describes the generation of fields at the sum or difference frequency. Through $\widehat{P}^{(3)}_{\bullet}(\nu')$ the light field at ν'' causes a change of the dielectric permittivity and of the refractive index at ν', whose real part gives rise to changes of the propagation properties of the light waves at ν' (cf. Section 4.4). The imaginary part of the permittivity describes losses at frequency ν' induced by the field at ν'' or vice versa. In the language of quantum theory this is a two-photon absorption of photons at ν' and ν''.

Third, we consider three interacting fields at ν', ν'', ν'''. Now additional contributions to the nonlinear polarization occur at $\nu' + \nu'' + \nu'''$, $\nu' + \nu'' - \nu'''$, $\nu' - \nu'' + \nu'''$, $-\nu' + \nu'' + \nu'''$. Their amplitudes are

$$\widehat{P}^{(3)}_{\bullet}(\nu' + \nu'' + \nu''') = \varepsilon_0\chi^{(3)}_{\bullet\bullet\bullet\bullet}(\nu' + \nu'' + \nu'''; \nu', \nu'', \nu''') \times \widehat{E}_{\bullet}(\nu')\widehat{E}_{\bullet}(\nu'')\widehat{E}_{\bullet}(\nu''') \quad (1.126a)$$

and

$$\widehat{P}^{(3)}_{\bullet}(\nu' + \nu'' - \nu''') = \varepsilon_0\chi^{(3)}_{\bullet\bullet\bullet\bullet}(\nu' + \nu'' - \nu'''; \nu', \nu'', -\nu''') \times \widehat{E}_{\bullet}(\nu')\widehat{E}_{\bullet}(\nu'')\widehat{E}^{*}_{\bullet}(\nu'''), \quad (1.126b)$$

where

$$\chi^{(3)}_{\bullet\bullet\bullet\bullet}(\nu' + \nu'' \pm \nu'''; \nu', \nu'', \pm\nu''') = \tfrac{3}{2}\varkappa^{(3)}_{\bullet\bullet\bullet\bullet}(-(\nu' + \nu'' \pm \nu'''); -\nu_1', -\nu_1'', \mp\nu'''),$$

and where $\widehat{P}^{(3)}_{\bullet}(\nu' - \nu'' + \nu''')$ and $\widehat{P}^{(3)}_{\bullet}(-\nu' + \nu'' + \nu''')$ can be obtained by permutation of the frequencies. These nonlinear polarization amplitudes describe the formation of sum and difference frequencies. In general, since a new light wave may be generated at the frequency $\nu_4 = \nu_1 + \nu_2 + \nu_3$ by the polarization $\widehat{P}^{(3)}(\nu_4) = \chi^{(3)}_{\bullet\bullet\bullet\bullet}(\nu_4; \nu_1, \nu_2, \nu_3)\widehat{E}_{\bullet}(\nu_1)\widehat{E}_{\bullet}(\nu_2)\widehat{E}_{\bullet}(\nu_3)$, the third-order terms of nonlinear optics represent a four-wave interaction.

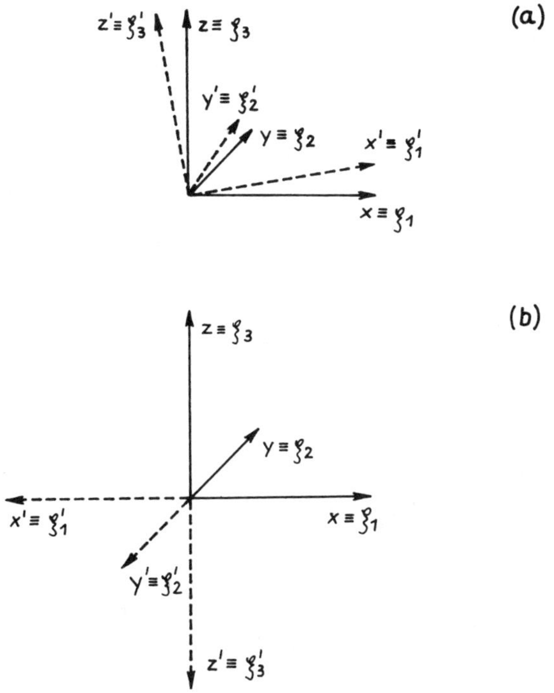

Fig. 1.8. Transformation between the coordinates ξ_i and ξ_i': (a) rotation, (b) inversion.

1.3.2.5. Spatial Symmetry of Susceptibilities. The nth order susceptibility is a rank-$(n+1)$ tensor with 3^{n+1} components, which means that at fixed frequencies $\nu, \nu_1, \ldots, \nu_n$ the susceptibility has in general to be characterized by 3^{n+1} complex numbers. However, symmetry properties of the material may result in not all components being independent of each other; in other words, spatial symmetry of crystals implies relations between the components of the susceptibility tensor.

Let us now transform the Cartesian coordinates $\xi_\bullet = (x, y, z)$, in which the susceptibilities are given by a rotation or inversion, into $\xi'_\bullet = (x', y', z')$. We have

$$\xi'_\bullet = \Theta_{\bullet\bullet}\xi_\bullet \tag{1.127}$$

(see Fig. 1.8). Since the length of the vector remains unchanged under the transformation of the coordinates

$$\sum_a \xi'_a\xi'_a = \sum_{i,j}\delta_{ab}\Theta_{ai}\xi_i\Theta_{bj}\xi_j = \sum_{i,j}\Theta_{ai}\Theta_{aj}\xi_i\xi_j = \sum_i \xi_i\xi_i, \tag{1.128}$$

Table 1.1. Crystal Classes without Inversion Symmetry

Crystal Class: International	Crystal Class: Schönfliess	Crystal System	Number of Optical Axes	$\chi_{ijk}^{(2)}$: Nonvanishing	$\chi_{ijk}^{(2)}$: Independent	d_{il}: Nonvanishing	d_{il}: (Independent)	
1	C_1	Triclinic	2	27	27	18	18 (10)	Iodic acid (HIO_3)
2	C_2	Monoclinic	2	13	13	8	8 (4)	
m	C_{1h} (C_s)	Monoclinic	2	14	14	10	10 (6)	
222	D_2 (V)	Orthorhombic	2	6	6	3	3 (1)	
$mm2$	C_{2v}	Orthorhombic	2	7	7	5	5 (3)	Sodium nitrite ($NaNO_2$)
3	C_3	Trigonal	1	21	9	13	6 (5)	
32	D_3	Trigonal	1	10	4	5	2 (2)	β-quartz (SiO_2), Te
$3m$	C_{3v}	Trigonal	1	11	5	7	3 (2)	Lithium niobate ($LiNbO_3$)
4	C_4	Tetragonal	1	13	7	7	4 (3)	
$\bar{4}$	S_4	Tetragonal	1	12	6	7	4 (2)	
422	D_4	Tetragonal	1	6	3	2	1 (1)	
$4mm$	D_{4v}	Tetragonal	1	7	4	4	2 (1)	Barium titanate ($BaTiO_3$)
$\bar{4}2m$	D_{2d} (V_d)	Tetragonal	1	6	3	3	2 (1)	KDP (KH_2PO_4), ADP
6	C_6	Hexagonal	1	13	7	7	4 (3)	
$\bar{6}$	C_{3h}	Hexagonal	1	8	2	6	2 (2)	
622	D_6	Hexagonal	1	6	3	2	1 (1)	α-quartz (SiO_2)
$6mm$	C_{6v}	Hexagonal	1	7	4	5	3 (2)	Cadmium sulfide (CdS)
$\bar{6}m2$	D_{3h}	Hexagonal	1	4	1	3	1 (1)	
23	T	Cubic	0	6	2	3	1 (1)	Sodium chlorate ($NaClO_3$)
$\bar{4}3m$	T_d	Cubic	0	6	1	3	1 (1)	Gallium arsenide (GaAs)
432	O	Cubic	0	6	1	0	0 (0)	

the components Θ_{ai} must obey the relation

$$\Theta_{aj} = \sum_i \Theta_{ia}^{-1}\delta_{ij}, \qquad \text{or} \quad \sum_a \Theta_{ai}\Theta_{aj} = \delta_{ij}.$$

The susceptibility $\chi^{(n)}_{\bullet\bullet\cdots\bullet}$, as a tensor of rank $n+1$, is transformed like a product of $n+1$ coordinates by the relation

$$\chi^{(n)\prime}_{ab\cdots g} = \sum_{j,k,\ldots,p} \Theta_{aj}\Theta_{bk} \cdots \Theta_{gp}\chi^{(n)}_{jk\cdots p}. \tag{1.129}$$

If the transformation is identified with one of the symmetry transformations of the material under investigation ($\Theta_{ai} \equiv S_{ai}$), the susceptibility as a measurable physical parameter of the system has to remain unchanged, which means that the primed susceptibility in (1.129) equals the unprimed one:

$$\chi^{(n)}_{ab\cdots g} = \sum_{(j,k,\ldots,p)} S_{aj}S_{bk} \cdots S_{gp}\chi^{(n)}_{jk\cdots p}. \tag{1.130}$$

As an example we identify the symmetry transformation of the material S_{aj} with the *inversion*

$$S_{aj} = -\delta_{aj}. \tag{1.131}$$

Then the transformation of the nth-order susceptibility is written as

$$\chi^{(n)}_{ab\cdots g} = (-1)^{n+1}\chi^{(n)}_{ab\cdots g}. \tag{1.132}$$

Obviously, the susceptibility must vanish for all even values of the order n. This means that in materials with inversion symmetry no macroscopic nonlinear polarization of even order n (e.g. $n = 2$) can be generated. Hence in materials with inversion centers, especially in isotropic media, the nonlinear polarization $P^{(2)}_{\bullet}$ vanishes and $P^{(3)}_{\bullet}$ turns out to be the first nonvanishing nonlinear contribution. Among the 32 symmetry classes of crystals there are 21 classes without inversion center. Even-order nonlinear effects can occur only in these classes. It should be noted that the inversion symmetry can be destroyed by spatial inhomogeneities (e.g., near surfaces) or by a static electric field applied to the medium.

Spatial symmetry of nonlinear susceptibilities of lowest order $\chi^{(2)}_{\bullet\bullet\bullet}$ will now be discussed in more detail (see also Table 1.1 and Fig. 1.9). Without spatial symmetry (class 1) there are in general 27 independent components of $\chi^{(2)}_{\bullet\bullet\bullet}$. In the case of second-harmonic generation [or in that of negligible dispersion between the frequencies ν_1 and ν_2, i.e., $\chi^{(2)}_{ijk}(\nu_3; \nu_1, \nu_2) \sim \chi^{(2)}_{ijk}(\nu_3; \nu_2, \nu_1)$] the number of independent elements reduces to 18 (see Section 1.3.2.3). If the Kleinman symmetry (see Section 1.3.2.7)

$$\chi^{(2)}_{ijk}(\nu_3; \nu_1, \nu_2) = \chi^{(2)}_{jik}(\nu_3; \nu_1, \nu_2) = \chi^{(2)}_{kji}(\nu_3; \nu_1, \nu_2) \tag{1.133}$$

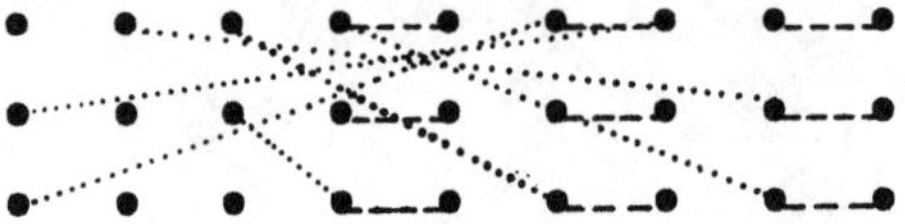

(*a*) Class 1

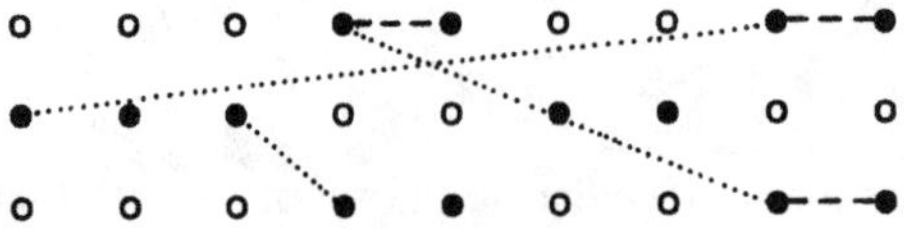

(*b*) Class 2

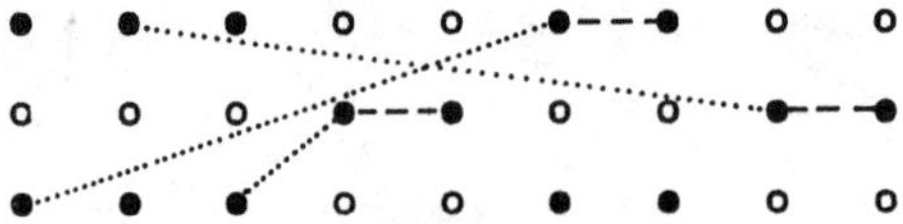

(*c*) Class m

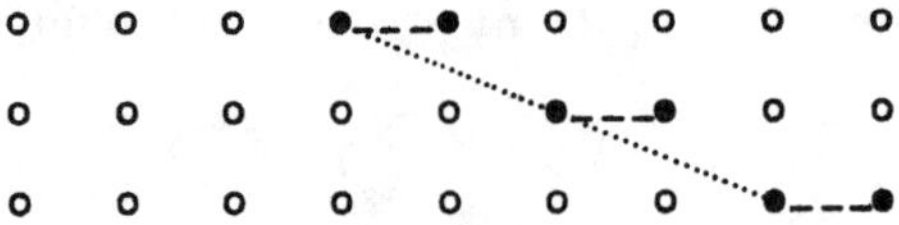

(*d*) Class 222

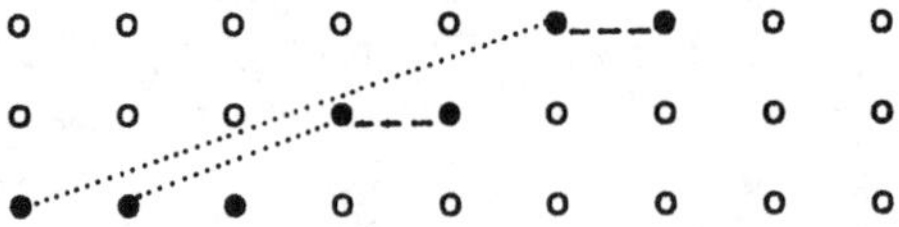

(*e*) Class $mm2$

Fig. 1.9. Second-order susceptibility $\chi_{ijk}^{(2)}$ for crystal classes without inversion symmetry (cf. Refs. 9, 46). The components of the third-rank tensors are given as circles in the following arrangement:

$$\begin{array}{l}(xxx)(xyy)(xzz)(xyz)\cdots(xzy)(xzx)\cdots(xxz)(xxy)\cdots(xyx)\\(yxx)(yyy)(yzz)(yyz)\cdots(yzy)(yzx)\cdots(yxz)(yxy)\cdots(yyx)\\(zxx)(zyy)(zzz)(zyz)\cdots(zzy)(zzx)\cdots(zxz)(zxy)\cdots(zyx)\end{array}$$

Components that are equivalent for second-harmonic generation [see (1.113)] are joined by broken lines. Relations resulting from spatial symmetry are represented by full lines. Components equal in the Kleinman approximation are joined by dotted lines. A bar above a component denotes a negative value.

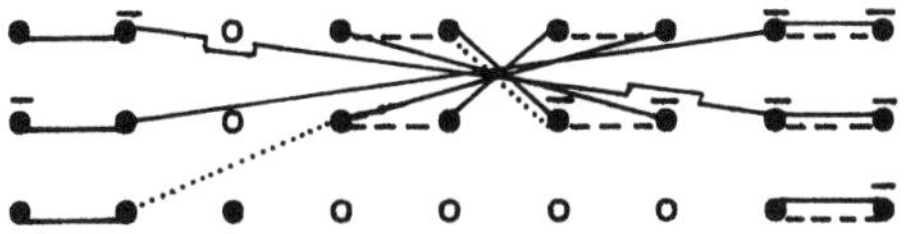

(*f*) Class 3

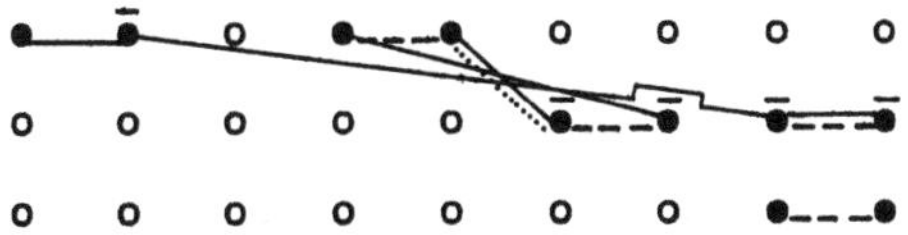

(*g*) Class 32

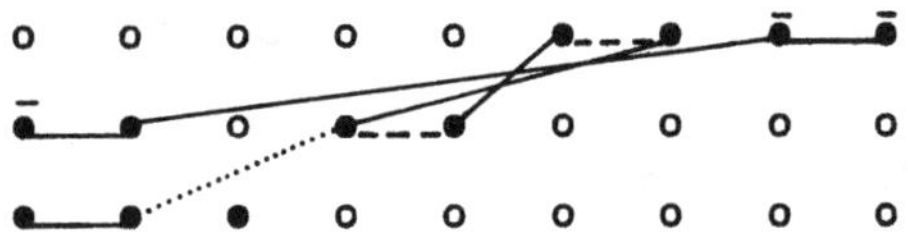

(*h*) Class 3*m*

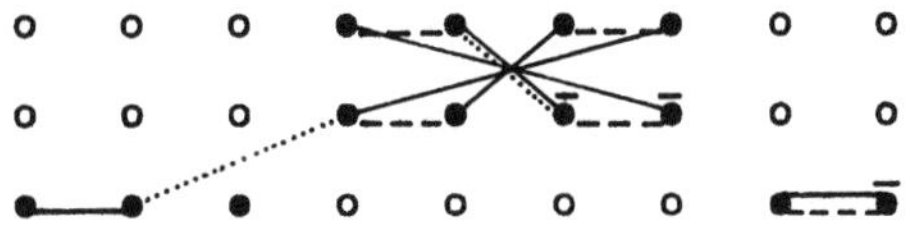

(*i*) Class 4

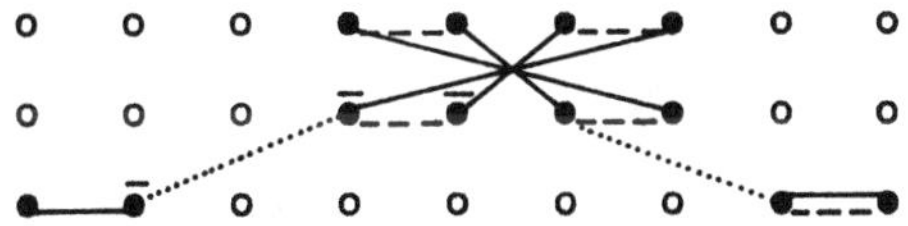

(*j*) Class $\overline{4}$

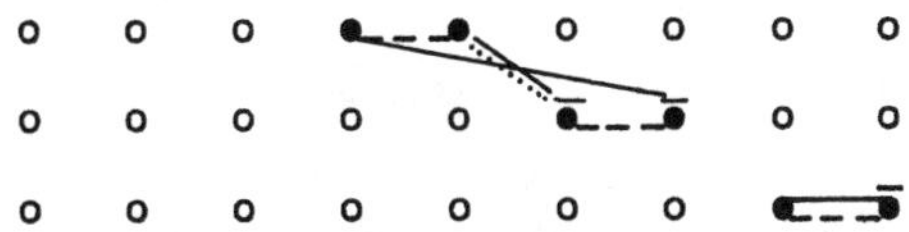

(*k*) Class 422

Fig. 1.9. (*Continued*)

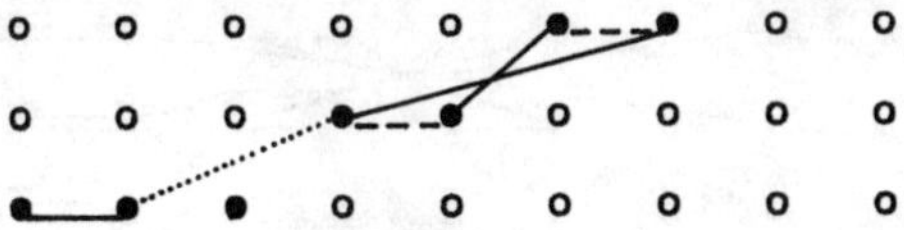

(*l*) Class $4mm$

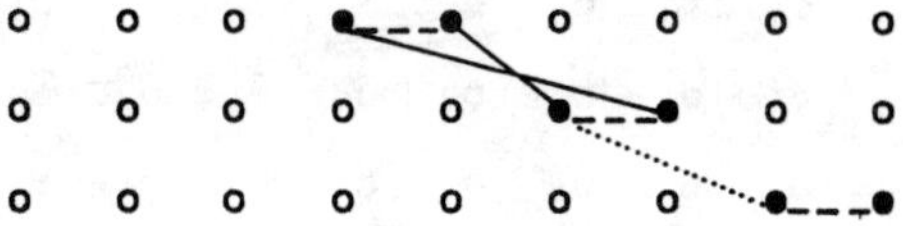

(*m*) Class $\bar{4}2m$

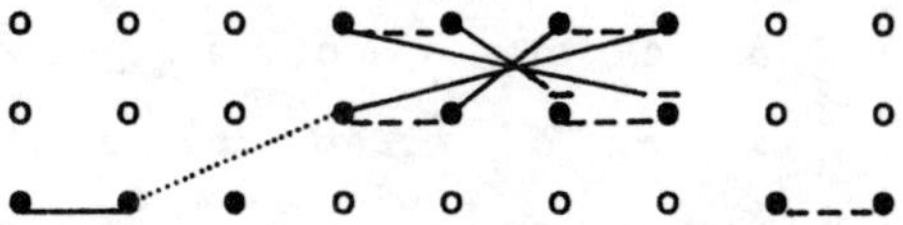

(*n*) Class 6

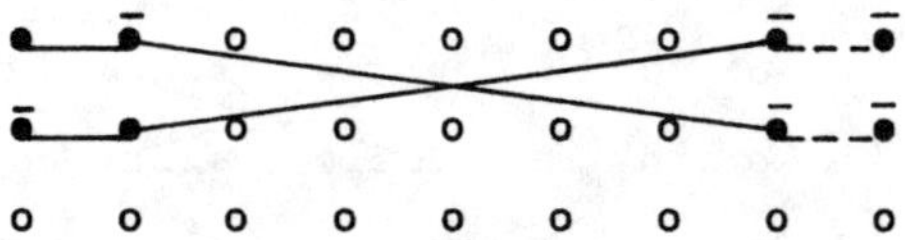

(*o*) Class 6

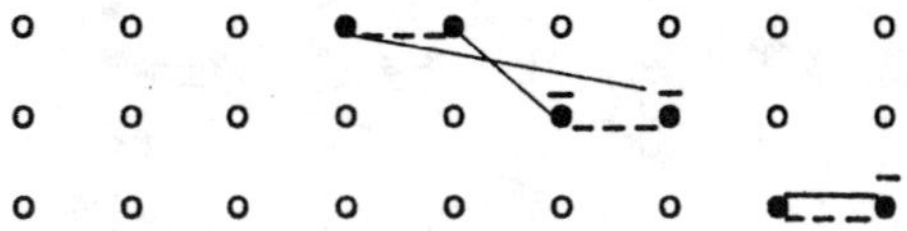

(*p*) Class 622

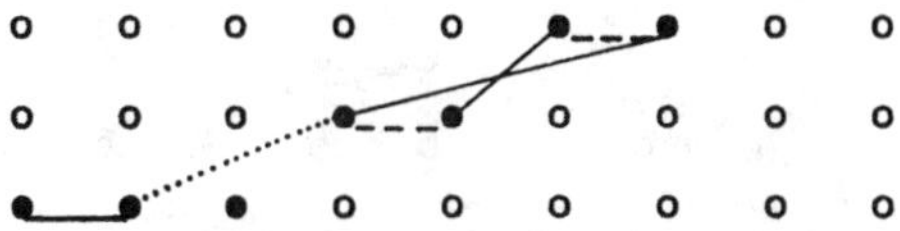

(*q*) Class $6mm$

Fig. 1.9. (*Continued*)

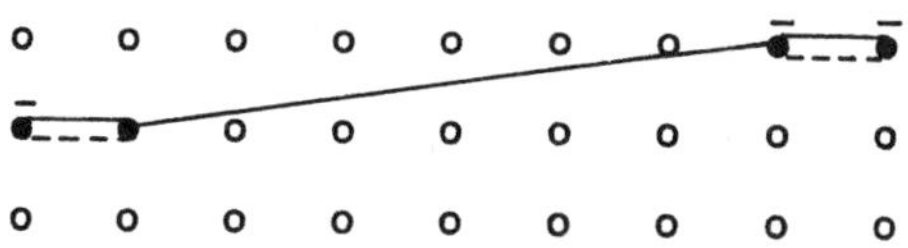

(r) Class $\bar{6}m2$

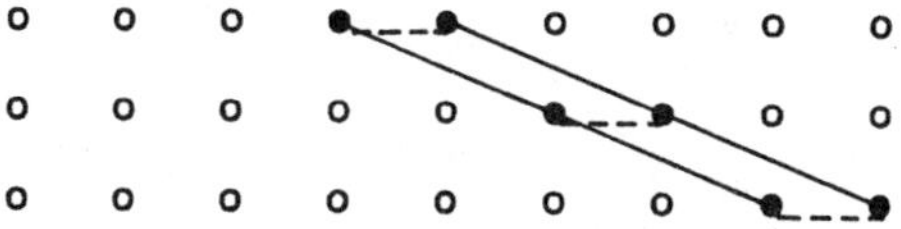

(s) Class 23

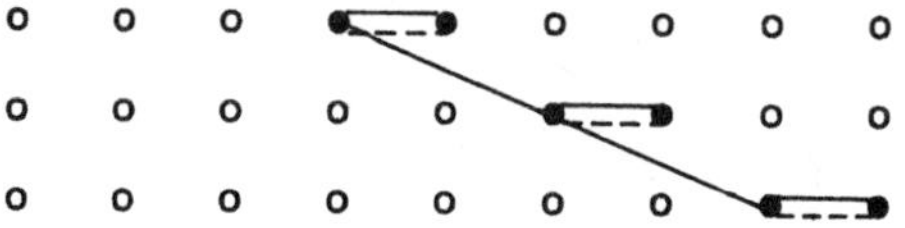

(t) Class $\bar{4}3m$

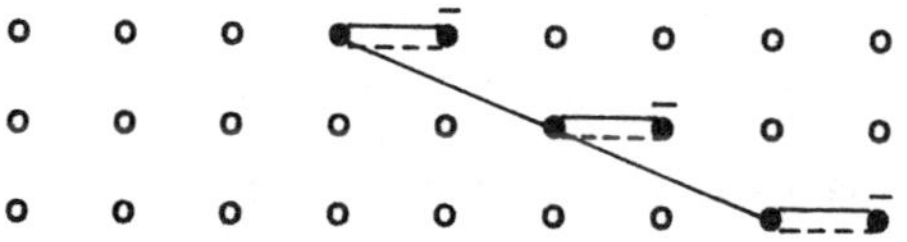

(u) Class 432

Fig. 1.9. (*Continued*)

holds, only 10 independent components will remain in class 1. In all other symmetry classes the number of nonvanishing components and of independent components is lowered by spatial symmetry. This will be demonstrated by some examples.

In class 2 there is only one symmetry element, namely a C_2 axis, here arbitrarily chosen to point in the y direction (see Fig. 1.10a). Under the corresponding symmetry operation the coordinates x, y, z are transformed into $(-x, y, -z)$. Since the tensor components $\chi^{(2)}_{ijk}$ are transformed in the same way as are the coordinate products $\xi_i \cdot \xi_j \cdot \xi_k$, all components with an odd number of indices x and z will vanish:

$$\chi^{(2)}_{xxx} = \chi^{(2)}_{xyy} = \chi^{(2)}_{xzz} = \chi^{(2)}_{xxz} = \chi^{(2)}_{xzx} = \chi^{(2)}_{yzy} = \chi^{(2)}_{yyz}$$

$$= \chi^{(2)}_{yxy} = \chi^{(2)}_{yyx} = \chi^{(2)}_{zxx} = \chi^{(2)}_{zyy} = \chi^{(2)}_{zzz} = \chi^{(2)}_{zxz} = \chi^{(2)}_{zzx} = 0. \quad (1.134)$$

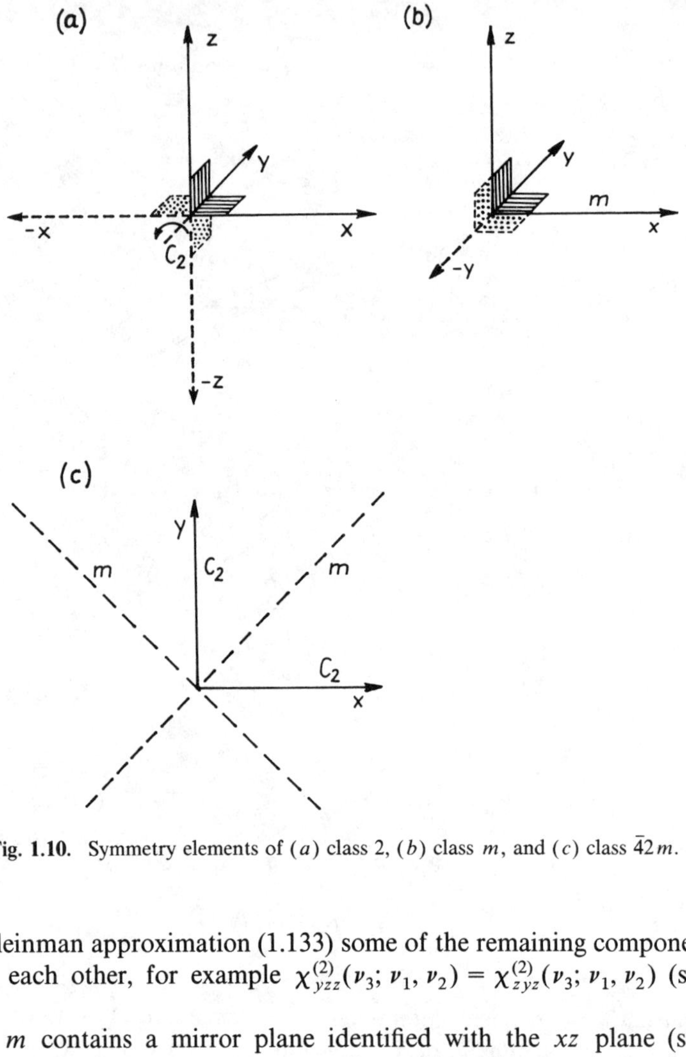

Fig. 1.10. Symmetry elements of (*a*) class 2, (*b*) class *m*, and (*c*) class $\bar{4}2m$.

In the Kleinman approximation (1.133) some of the remaining components are equal to each other, for example $\chi^{(2)}_{yzz}(\nu_3; \nu_1, \nu_2) = \chi^{(2)}_{zyz}(\nu_3; \nu_1, \nu_2)$ (see Fig. 1.9).

Class *m* contains a mirror plane identified with the *xz* plane (see Fig. 1.10*b*). The coordinates *x*, *y*, *z* are transformed into $(x, -y, z)$, that is to say, all components of $\chi^{(2)}_{ijk}$ with an odd number of *y* indices will vanish:

$$\chi^{(2)}_{xzy} = \chi^{(2)}_{xyz} = \chi^{(2)}_{xxy} = \chi^{(2)}_{xyx} = \chi^{(2)}_{yxx} = \chi^{(2)}_{yyy}$$

$$= \chi^{(2)}_{yzz} = \chi^{(2)}_{yzx} = \chi^{(2)}_{yxz} = \chi^{(2)}_{zzy} = \chi^{(2)}_{zyz} = \chi^{(2)}_{zxy} = \chi^{(2)}_{zyx} = 0. \quad (1.135)$$

Next we consider class $\bar{4}2m$, two representatives of which are the crystals KDP and ADP often used in nonlinear optics. The symmetry of this class is characterized by three C_2 axes along the *x*, *y*, *z* directions and by two mirror planes that contain the *z* axis and intersect the *xy* plane at angles of 45° with

Table 1.2. Susceptibilities $d_{il}(2\nu; \nu, \nu)$ of Second-Harmonic Generation[a]

Material	$\lambda = c/\nu$ (μm)	d_{il}
KDP	0.69	$d_{36} = 1$
		$d_{14} = d_{25} = 0.95$
	1.06	$d_{36} = 1$
		$d_{14} = d_{25} = 1.01$
ADP	1.06	$d_{36} = 0.93$
		$d_{14} = d_{25} = 0.89$

[a] Normalized to the component $d_{36}(2\nu; \nu, \nu)$ of KDP.

the x and y axes (see Fig. 1.10c). The action of the symmetry operation $\bar{4}$ can be expressed by the rotations C_2 and reflections m. Using the same arguments as before, we are led to the conclusion that all tensor components with two equal indices must vanish. Then the remaining components are $\chi^{(2)}_{xzy}, \chi^{(2)}_{xyz}, \chi^{(2)}_{yzx}, \chi^{(2)}_{yxz}, \chi^{(2)}_{zxy}, \chi^{(2)}_{zyx}$. Under reflection at one of the mirror planes the coordinate axes x and y are interchanged. This means that all components differing only in the arrangement of x and y indices are equal:

$$\chi^{(2)}_{xzy} = \chi^{(2)}_{yzx}, \qquad \chi^{(2)}_{xyz} = \chi^{(2)}_{yxz}, \qquad \chi^{(2)}_{zxy} = \chi^{(2)}_{zyx}. \tag{1.136}$$

In the case of second-harmonic generation the number of independent components is reduced to two, that is, $d_{14} = d_{25}$ and d_{36} remain. Assuming the validity of the Kleinman symmetry (cf. Section 1.3.2.7), $d_{14} = d_{25} = d_{36}$ holds. Table 1.2 contains components of the susceptibility for second-harmonic generation in the nonlinear crystals KDP and ADP. The Kleinman symmetry is seen to be approximately satisfied for these materials.

For several crystal classes the Kleinman symmetry leads to a reduction in the number of nonvanishing components. If, due to spatial symmetry, a component $\chi^{(2)}_{ijk}$ equals the negative of the value of another component $\chi^{(2)}_{abc}$, and if the two should be equal according to the Kleinman relation, then both components must vanish. The graph for class 3 in Fig. 1.9 serves as an example. Spatial symmetry requires that $\chi^{(2)}_{xzy} = -\chi^{(2)}_{yzx}$; the two components are not only connected by a full line, indicating the spatial symmetry, but also by a dotted line, which represents the Kleinman relation $\chi^{(2)}_{xzy} = \chi^{(2)}_{yzx}$. Therefore it can be concluded that $\chi^{(2)}_{xzy} = \chi^{(2)}_{xyz} = \chi^{(2)}_{yzx} = \chi^{(2)}_{yxz} = 0$.

There is one class, namely class 432, for which the joint validity of spatial symmetry and the Kleinman relations leads to $\chi_{ijk} = 0$ for all values of i, j, k.

Concerning the *third-order susceptibilities* $\chi^{(3)}_{ijkl}(\nu; \nu_1, \nu_2, \nu_3)$, only the special yet very important case of isotropic media will be discussed. In such media, namely, in gases, liquids, and amorphous solids, $\chi^{(3)}_{ijkl}$ is the lowest-order nonlinear susceptibility with nonvanishing components. Because of the isotropy of the material $\chi^{(3)}_{ijkl}$ is invariant under the action of any rotation and reflection.

We first consider the action of a reflection in the yz plane, by which (x, y, z) is transformed into $(-x, y, z)$. As a consequence all components of $\chi^{(3)}_{ijkl}$ with an odd number of x indices must vanish. Analogously, using reflections in the xz and the xy plane, all components with uneven numbers of y and z indices can be shown to be zero. Hence only 21 nonvanishing components remain.

Relations between them will now be derived. We start with a rotation of 90° about the z axis. Then (x, y) is transformed into $(y, -x)$, and the transformation tensor for the (x, y) coordinates S_{ai} is given by

$$S_{ai} = \begin{pmatrix} 0 & 1 \\ -1 & 0 \end{pmatrix}. \tag{1.137}$$

This means that all components $\chi^{(3)}_{ijkl}$ are equal to $\chi^{(3)}_{abcd}$ if *ijkl* is obtained from *abcd* by replacing x with y and y with x. The minus sign in (1.137) has no effect on the transformation of the fourth-rank tensor $\chi^{(3)}_{ijkl}$, because the number of x and y indices of all the nonvanishing components is even. These arguments lead to the relations $\chi^{(3)}_{xxxx} = \chi^{(3)}_{yyyy}$, $\chi^{(3)}_{xxyy} = \chi^{(3)}_{yyxx}$, $\chi^{(3)}_{xyxy} = \chi^{(3)}_{yxyx}$, $\chi^{(3)}_{xyyx} = \chi^{(3)}_{yxxy}$. Equivalent relations are obtained for 90° rotations about the x and y axes. Taken together they are

$$\chi^{(3)}_{xxxx} = \chi^{(3)}_{yyyy} = \chi^{(3)}_{zzzz},$$

$$\chi^{(3)}_{yyzz} = \chi^{(3)}_{zzyy} = \chi^{(3)}_{zzxx} = \chi^{(3)}_{xxzz} = \chi^{(3)}_{xxyy} = \chi^{(3)}_{yyxx} = \chi^{(3)}_{(1)},$$

$$\chi^{(3)}_{yzyz} = \chi^{(3)}_{zyzy} = \chi^{(3)}_{zxzx} = \chi^{(3)}_{xzxz} = \chi^{(3)}_{xyxy} = \chi^{(3)}_{yxyx} = \chi^{(3)}_{(2)},$$

$$\chi^{(3)}_{yzzy} = \chi^{(3)}_{zyyz} = \chi^{(3)}_{zxxz} = \chi^{(3)}_{xzzx} = \chi^{(3)}_{xyyx} = \chi^{(3)}_{yxxy} = \chi^{(3)}_{(3)}, \tag{1.138}$$

where $\chi^{(3)}_{(1)}$, $\chi^{(3)}_{(2)}$, and $\chi^{(3)}_{(3)}$ have been introduced as abbreviations.

Finally, we have to perform rotations through arbitrary angles Φ_x, Φ_y, Φ_z about the coordinate axes x, y, z. The transformation of the x, y coordinates under rotation about z is described by

$$S_{ai} = \begin{pmatrix} \cos\Phi_z & \sin\Phi_z \\ -\sin\Phi_z & \cos\Phi_z \end{pmatrix}. \tag{1.139}$$

By substituting S_{ai} into (1.130) we obtain

$$\chi^{(3)}_{xxxx} = \left(\cos^4\Phi_z + \sin^4\Phi_z\right)\chi^{(3)}_{xxxx} + 2\cos^2\Phi_z \sin^2\Phi_z\left[\chi^{(3)}_{xxyy} + \chi^{(3)}_{xyxy} + \chi^{(3)}_{xyyx}\right]. \tag{1.140}$$

For arbitrary values of Φ_z, (1.140) can only be valid if

$$\chi^{(3)}_{xxxx} = \chi^{(3)}_{xxyy} + \chi^{(3)}_{xyxy} + \chi^{(3)}_{xyyx}. \tag{1.141}$$

If we use the abbreviations $\chi^{(3)}_{(1)}, \chi^{(3)}_{(2)}, \chi^{(3)}_{(3)}$ introduced in (1.138) and the first relation of (1.138), then (1.141) leads to

$$\chi^{(3)}_{xxxx} = \chi^{(3)}_{yyyy} = \chi^{(3)}_{zzzz} = \chi^{(3)}_{(1)} + \chi^{(3)}_{(2)} + \chi^{(3)}_{(3)}. \tag{1.142}$$

Together with (1.138), this constraint implies that there are only three independent tensor components of $\chi^{(3)}_{ijkl}$ in isotropic media. With the help of (1.138) and (1.141) the amplitude of the nonlinear polarization $\hat{P}_i^{(3)}$ ($\nu = \nu_1 + \nu_2 + \nu_3$) can be expressed in the following concise form:

$$\begin{aligned} \varepsilon_0^{-1}\hat{P}^{(3)}_{\bullet}(\nu) &= \chi^{(3)}_{(1)}(\nu; \nu_1, \nu_2, \nu_3)\left\{\hat{E}_{\bullet}(\nu_2)\hat{E}_{\bullet}(\nu_3)\right\}\hat{E}_{\bullet}(\nu_1) \\ &+\chi^{(3)}_{(2)}(\nu; \nu_1, \nu_2, \nu_3)\left\{\hat{E}_{\bullet}(\nu_1)\hat{E}_{\bullet}(\nu_3)\right\}\hat{E}_{\bullet}(\nu_2) \\ &+\chi^{(3)}_{(3)}(\nu; \nu_1, \nu_2, \nu_3)\left\{\hat{E}_{\bullet}(\nu_1)\hat{E}_{\bullet}(\nu_2)\right\}\hat{E}_{\bullet}(\nu_3). \end{aligned} \tag{1.143}$$

This means that $\hat{P}^{(3)}_{\bullet}(\nu)$ is obtained by the addition of three vectors, each of which is the product of one of the field amplitudes $\hat{E}_{\bullet}(\nu_i)$ with a scalar factor. This factor depends on the nonlinear susceptibilities $\chi^{(3)}_{(1)}$ and the scalar product of the two other field amplitudes.

1.3.2.6. Time-Reversal Symmetry. Let us now discuss additional symmetry relations which are only valid in media with negligible losses of the electromagnetic field. For the sake of simplicity the interaction processes are assumed to be stationary, that is, the energy content of a volume element ΔV averaged over several periods of the applied fields in constant due to compensation of input by output. The average electromagnetic energy fed into ΔV per unit time is $-\overline{\nabla_{\bullet} S_{\bullet}}^t$, where $S_{\bullet}$ is the Poynting vector. $\overline{Q}^t$ is the average heat supplied to ΔV per unit time. With respect to the system consisting of the dielectric material and of the electromagnetic field in ΔV, the term $-\overline{Q}^t$ is considered as a *loss*. If there is no other energy exchange, the relation

$$-\overline{\nabla_{\bullet} S_{\bullet}}^t + \overline{Q}^t = 0$$

holds. Then a radiation process with $\overline{Q}^t = 0$, and hence with

$$\overline{\nabla_{\bullet} S_{\bullet}}^t = 0, \tag{1.144}$$

is called a loss-free process. Note that (1.144) does not rule out a weakening or amplification of certain frequency components of the electromagnetic field, but any amplification at a given frequency has to be compensated for by losses at other frequencies and vice versa. [Parametric interactions including the generation of sum and difference frequencies will later be shown to be nearly loss-free if none of the interacting waves is absorbed in linear optical processes, i.e., $\mathrm{Im}\{\underset{\smile}{\varkappa}(\nu;\nu)\} \approx 0$.]

On condition that we deal with a loss-free material, the constitutive relation $P_\bullet[E_\bullet]$ has the following *time-reversal symmetry*: $P_\bullet(-t)$ depends on $E_\bullet(-t)$ in the same way as $P_\bullet(t)$ does on $E_\bullet(t)$.

The proof has to be given by deriving the constitutive equation $P_\bullet[E_\bullet]$ from the equation of motion of the material. This will be done in Section 3.3 on the basis of quantum mechanics. Let us here only treat a simple classical model of the dielectric material. Particles of mass m and charge q are bound to equilibrium position by forces depending nonlinearly on the elongation $R_\bullet$. The charge q of one particle is compensated for by an immovable charge $-q$ at equilibrium position $R_\bullet = 0$. Under the action of an electric field $E_\bullet(t)$ the equation of motion is

$$m\frac{d^2}{dt^2}R_\bullet(t) + \tilde{\Gamma}\frac{d}{dt}R_\bullet(t) + \sum_{j=1}^{\infty} k^{(j)}_{\bullet\bullet\ldots\bullet} \cdot (R_\bullet(t))^j = qE_\bullet(t). \quad (1.145)$$

The second term on the left-hand side of (1.145) describes "friction" processes proportional to the velocity of this particle, $\tilde{\Gamma}$ being a constant parameter. The approximation of a loss-free material implies the neglect of the damping term, that is, $\tilde{\Gamma} = 0$. The third term on the left-hand side represents the backward driving force, which exhibits a nonlinear dependence on the elongation $R_\bullet$. The time reversal-operation T affects the $R_\bullet$-dependent terms of (1.145) in the following way:

$$\mathsf{T}R_\bullet(t) = R_\bullet(\mathsf{T}t) = R_\bullet(-t),$$

$$\mathsf{T}k^{(j)}_{\bullet\ldots\bullet}(R_\bullet(t))^j = k^{(j)}_{\bullet\ldots\bullet} \cdot (R_\bullet(-t))^j,$$

$$\mathsf{T}\frac{d}{dt}R_\bullet(t) = -\frac{d}{dt}R_\bullet(-t),$$

$$\mathsf{T}\frac{d^2}{dt^2}R_\bullet(t) = \frac{d^2}{dt^2}R_\bullet(-t). \quad (1.146)$$

The transformation properties of Maxwell's equations provide

$$\mathsf{T}E_\bullet(t) = E_\bullet(-t). \quad (1.147)$$

Substituting the transformed terms into (1.145), the equation of motion is seen to have the same form as before if the friction term is neglected. This means that $R_\bullet(-t)$ depends on $E_\bullet(-t)$ in the same way as $R_\bullet(t)$ does on $E_\bullet(t)$. The motion of the charged particle yields the time-dependent dipole moment

$$d_\bullet(t) = qR_\bullet(t),$$

and all the dipole moments (assumed to be interaction-free) constitute the polarization of the material,

$$P_\bullet(t) = \gamma q R_\bullet(t), \tag{1.148}$$

where γ is the number of particles per unit volume. The proportionality between $P_\bullet(t)$ and $R_\bullet(t)$ means that $P_\bullet(t)$ exhibits the same behavior as $R_\bullet(t)$ under the action of time reversal.

Now we represent the time-reversal operation T by an equivalent operation $\tilde{\mathsf{T}}$ in the Fourier space, defined by

$$\mathsf{T}F(t) = \mathrm{FT}\{\tilde{\mathsf{T}}\breve{F}(\omega)\}, \tag{1.149}$$

where $F(t)$ is an arbitrary function and $\mathrm{FT}\{\cdots\}$ is the Fourier transform of the function in braces. Because of the relationship

$$F(-t) = \frac{1}{2\pi}\int_{-\infty}^{\infty} d\omega\, \breve{F}(\omega)e^{-i\omega t} = \frac{1}{2\pi}\int_{-\infty}^{\infty} d\omega\, \breve{F}(-\omega)e^{+i\omega t}, \tag{1.150}$$

the operation of time reversal in the Fourier space is given by

$$\tilde{\mathsf{T}}\breve{F}(\omega) = \breve{F}(-\omega). \tag{1.151}$$

This relation is applied to (1.104):

$$\begin{aligned}\tilde{\mathsf{T}}\breve{P}_\bullet^{(n)}(\omega) &= \breve{P}_\bullet^{(n)}(-\omega)\\ &= \tilde{\mathsf{T}}\Bigg\{\frac{\varepsilon_0}{(2\pi)^{n-1}}\int_{-\infty}^{\infty} d\omega_1 \cdots \int_{-\infty}^{\infty} d\omega_{n-1}\\ &\qquad \times \breve{\varkappa}^{(n)}_{\bullet\bullet\cdots\bullet}(\omega;\omega_1,\ldots,\omega_n)\breve{E}_\bullet(\omega_1)\cdots \breve{E}_\bullet(\omega_n)\Bigg\}.\end{aligned} \tag{1.152}$$

Since $P_{\bullet}(-t)$ depends on $E_{\bullet}(-t)$ as does $P_{\bullet}(t)$ on $E_{\bullet}(t)$, the relation

$$\underset{\smile}{P}_{\bullet}^{(n)}(-\omega) = \frac{\varepsilon_0}{(2\pi)^{n-1}} \int_{-\infty}^{\infty} d\omega_1 \cdots \int_{-\infty}^{\infty} d\omega_{n-1}$$

$$\times \underset{\smile}{\varkappa}_{\bullet\bullet\cdots\bullet}^{(n)}(\omega; \omega_1, \ldots, \omega_n) \cdot \underset{\smile}{E}_{\bullet}(-\omega_1) \cdots \underset{\smile}{E}_{\bullet}(-\omega_n) \quad (1.153)$$

is obtained. On the other hand, by taking into account (1.105), the complex conjugation of (1.104) leads to

$$\underset{\smile}{P}_{\bullet}^{(n)}(-\omega) = \frac{\varepsilon_0}{(2\pi)^{n-1}} \int_{-\infty}^{\infty} d\omega_1 \cdots \int_{-\infty}^{\infty} d\omega_{n-1}$$

$$\times \underset{\smile}{\varkappa}_{\bullet\bullet\cdots\bullet}^{(n)}(-\omega; -\omega_1, \ldots, -\omega_n) \cdot \underset{\smile}{E}_{\bullet}(-\omega_1) \cdots \underset{\smile}{E}_{\bullet}(-\omega_n). \quad (1.154)$$

From a comparison of (1.153) and (1.154),

$$\underset{\smile}{\varkappa}_{\bullet\bullet\cdots\bullet}^{(n)}(-\omega; -\omega_1, \ldots, -\omega_n) = [\underset{\smile}{\varkappa}_{\bullet\bullet\cdots\bullet}^{(n)}(\omega; \omega_1, \ldots, \omega_n)]^*$$

$$= \underset{\smile}{\varkappa}_{\bullet\bullet\cdots\bullet}^{(n)}(\omega; \omega_1, \ldots, \omega_n) \quad (1.155)$$

can be deduced. Thus we see that the neglect of losses leads to real susceptibilities not only in linear optics but also in any order of nonlinearity.

1.3.2.7. Overall Permutation Symmetry. In loss-free media the index-frequency permutation symmetry

$$\underset{\smile}{\varkappa}_{i j_1 \cdots j_l \cdots j_n}^{(n)}(\omega; \omega_1, \ldots, \omega_l, \ldots, \omega_n)$$

$$= \underset{\smile}{\varkappa}_{j_l j_1 \cdots i \cdots j_n}^{(n)}(-\omega_l; \omega_1, \ldots, -\omega, \ldots \omega_n) \quad (1.156)$$

holds for $l = 1 \cdots n$ in addition to the intrinsic permutation symmetry of (1.105b). Hence the nonlinear susceptibilities remain invariant under interchange of all tensor indices if the corresponding frequencies are simultaneously interchanged. If the first frequency $\omega = \omega_1 + \cdots + \omega_n$ is interchanged with any other frequency ω_l, we have to change the sign of both these frequencies. The symmetry relations of (1.105b) together with those of (1.156) are called *overall permutation symmetries* or *general index-frequency symmetries*. The validity of (1.156) is connected with the existence of certain state functions describing the interaction between nonlinear material and light.

Let us prove this assertion for the lowest nonlinear order using a simple model of the loss-free dielectric material (cf. Section 1.3.2.3). With respect to a fixed particle of charge $-q$, a movable charge q is displaced to $R_\bullet$ by an electric field $E_\bullet$. If the position of the movable charge is changed by $\delta R_\bullet$, the electric field will perform the work

$$\delta W = qE_\bullet \, \delta R_\bullet. \tag{1.157}$$

Using the relation $P_\bullet = \gamma q R_\bullet$ of (1.148), the work per unit volume, $\delta \mathscr{W} = \gamma \, \delta W$, is calculated to be

$$\delta \mathscr{W} = E_\bullet \, \delta P_\bullet. \tag{1.158}$$

We now consider the interaction of nearly monochromatic waves. The field strength within the element ΔV is given by

$$E_\bullet(t) = \sum_j \tfrac{1}{2} \widehat{E}_\bullet(\omega_j) e^{-i\omega_j t} + \text{c.c.}. \tag{1.159}$$

The polarization contains linear and nonlinear contributions at the same frequencies:

$$P_\bullet(t) = \sum_j \tfrac{1}{2} \widehat{P}_\bullet(\omega_j) e^{-i\omega_j t} + \text{c.c.}. \tag{1.160}$$

The time dependence of the monochromatic waves of field strength and polarization results in rapidly oscillating terms in the work per volume $\delta \mathscr{W}$. On integration over a longer period, these oscillating terms will vanish. If the field-strength amplitudes are adiabatically changed, then the polarization amplitudes will vary as well, and during this process additional work will be done. We are interested only in the work $\overline{\delta \mathscr{W}}^t$ averaged over many periods of the lowest-frequency wave. $\overline{\delta \mathscr{W}}^t$ is calculated to be

$$\overline{\delta \mathscr{W}}^t = \sum_j \tfrac{1}{4} \widehat{E}^*_\bullet(\omega_j) \, \delta \widehat{P}_\bullet(\omega_j) + \text{c.c.}. \tag{1.161}$$

In the time average, the field strength at frequency ω_j performs work only on the polarization of the same frequency.

To give an example we calculate the contribution of the second-order nonlinear polarization of three interacting waves at $\omega_1, \omega_2, \omega_3$ with ω_3

$= \omega_1 + \omega_2$ to $\overline{\delta \mathscr{W}}^t$. The nonlinear polarization is given in the form

$$\hat{P}_c^{(2)}(\omega_3) = \sum_{a,b} \varepsilon_0 \chi_{cab}^{(2)}(\omega_3; \omega_1, \omega_2) \hat{E}_a(\omega_1) \hat{E}_b(\omega_2),$$

$$\hat{P}_a^{(2)}(\omega_1) = \sum_{c,b} \varepsilon_0 \chi_{acb}(\omega_1; \omega_3, -\omega_2) \hat{E}_c(\omega_3) \hat{E}_b(-\omega_2),$$

$$\hat{P}_b^{(2)}(\omega_2) = \sum_{a,c} \varepsilon_0 \chi_{bac}(\omega_2; -\omega_1, \omega_3) \hat{E}_a(-\omega_1) \hat{E}_c(\omega_3). \tag{1.162}$$

Equivalent expressions hold for $P_l(-\omega_j) = P_l^*(\omega_j)$. The use of (1.162) in (1.161) leads to

$$\begin{aligned}\overline{\delta \mathscr{W}}^t = \sum_{a,b,c} \Big\{ & \tfrac{1}{4} \hat{E}_c^*(\omega_3) \varepsilon_0 \chi_{cab}^{(2)}(\omega_3; \omega_1, \omega_2)\, \delta\big[\hat{E}_a(\omega_1) \hat{E}_b(\omega_2)\big] \\ & + \tfrac{1}{4} \hat{E}_a(\omega_1) \varepsilon_0 \chi_{acb}^{(2)}(-\omega_1, -\omega_3, \omega_2)\, \delta\big[\hat{E}_c^*(\omega_3) \hat{E}_b(\omega_2)\big] \\ & + \tfrac{1}{4} \hat{E}_b(\omega_2) \varepsilon_0 \chi_{bac}^{(2)}(-\omega_2; \omega_1, -\omega_3)\, \delta\big[\hat{E}_a(\omega_1) \hat{E}_c^*(\omega_3)\big] + \text{c.c.} \Big\}.\end{aligned} \tag{1.163}$$

Here $\hat{E}_l(-\omega_j) = \hat{E}_l^*(\omega_j)$ has been taken into account, and in the second and third terms the complex conjugate $\hat{E}_\bullet\, \delta P_\bullet^*$ of $\hat{E}_\bullet^*\, \delta P_\bullet$ has been explicitly given to achieve a more symmetric form.

By use of $x\, \delta(yz) = \delta(xyz) - (yz)\, \delta x$ the variation of $\overline{\mathscr{W}}^t$ can be written as

$$\overline{\delta \mathscr{W}}^t = \delta \mathscr{G} + \delta \mathscr{F}, \tag{1.164}$$

where

$$\begin{aligned}\delta \mathscr{G} = \sum_{a,b,c} \Big\{ & \frac{\varepsilon_0}{4} \{ \chi_{cab}^{(2)}(\omega_3; \omega_1, \omega_2) + \chi_{acb}^{(2)}(-\omega_1; -\omega_3, \omega_2) \\ & + \chi_{bac}^{(2)}(-\omega_2; \omega_1, -\omega_3) \}\, \delta\big[\hat{E}_a(\omega_1) \hat{E}_b(\omega_2) \hat{E}_c^*(\omega_3)\big] + \text{c.c.} \Big\}\end{aligned} \tag{1.165a}$$

and

$$\begin{aligned}\delta \mathscr{F} = -\frac{\varepsilon_0}{4} \sum_{a,b,c} \Big\{ & \chi_{cab}^{(2)}(\omega_3; \omega_1, \omega_2) \hat{E}_a(\omega_1) \hat{E}_b(\omega_2)\, \delta \hat{E}_c^*(\omega_3) \\ & + \chi_{acb}^{(2)}(-\omega_1; -\omega_3, \omega_2) \hat{E}_c^*(\omega_3) \hat{E}_b(\omega_2)\, \delta \hat{E}_a(\omega_1) \\ & + \chi_{bac}^{(2)}(-\omega_2; \omega_1, -\omega_3) \hat{E}_a(\omega_1) \hat{E}_c^*(\omega_3)\, \delta \hat{E}_b(\omega_2) + \text{c.c.} \Big\},\end{aligned} \tag{1.165b}$$

which can be written more concisely as

$$\delta\mathscr{G} = \overline{\delta\left[P_{\bullet}^{(2)}(t)E_{\bullet}(t)\right]}^{t} \tag{1.166a}$$

and

$$\delta\mathscr{F} = -\overline{P_{\bullet}^{(2)}(t)\,\delta E_{\bullet}(t)}^{t}. \tag{1.166b}$$

It is obvious that the variation $\delta\mathscr{G}$ represents a total differential, and hence $\mathscr{G}$ is a state function. The term $\delta\mathscr{F}$, however, is only a total differential if the partial derivations of the coefficients in front of the field variations meet certain requirements with respect to the field amplitudes. Concerning the variation $\delta\widehat{E}_{c'}(\omega_3)$ and $\delta\widehat{E}_{a'}(\omega_1)$ these requirements are

$$\frac{\partial}{\partial\widehat{E}_{a'}(\omega_1)}\sum_{a,b}\chi^{(2)}_{c'ab}(\omega_3;\,\omega_1,\omega_2)\widehat{E}_a(\omega_1)\widehat{E}_b(\omega_2)$$

$$= \frac{\partial}{\partial\widehat{E}^*_{c'}(\omega_3)}\sum_{c,b}\chi^{(2)}_{a'cb}(-\omega_1;\,-\omega_3,\omega_2)\widehat{E}^*_c(\omega_3)\widehat{E}_b(\omega_2). \tag{1.167}$$

(a', c' are arbitrary but fixed tensor indices.) On performing the partial differentiation in (1.167), we obtain

$$\sum_b\chi^{(2)}_{c'a'b}(\omega_3;\,\omega_1,\omega_2)\widehat{E}_b(\omega_2) = \sum_b\chi^{(2)}_{a'c'b}(-\omega_1;\,-\omega_3,\omega_2)\widehat{E}_b(\omega_2). \tag{1.168}$$

Since the field amplitudes $\widehat{E}_b(\omega_2)$ can be arbitrarily chosen for each value of b, and since a' and c' can attain any value, it is possible to deduce the relation

$$\chi^{(2)}_{cab}(\omega_3;\,\omega_1,\omega_2) = \chi^{(2)}_{acb}(-\omega_1;\,-\omega_3,\omega_2) \tag{1.169a}$$

from (1.168). In the same way the relations

$$\chi^{(2)}_{cab}(\omega_3;\,\omega_1,\omega_2) = \chi^{(2)}_{bac}(-\omega_2;\,\omega_1,-\omega_3) \tag{1.169b}$$

and

$$\chi^{(2)}_{acb}(-\omega_1;\,-\omega_3,\omega_2) = \chi^{(2)}_{bca}(-\omega_2;\,-\omega_3,\omega_1) \tag{1.169c}$$

can be derived. To satisfy the overall permutation symmetries it is not only a necessary but also a sufficient condition for $\delta\mathscr{F}$ to be a total differential, and for $\mathscr{F}$ to be a state function. This means that the existence of the state functions $\mathscr{F}$ and $\overline{\mathscr{W}}^t$ is equivalent to the fulfilment of the overall permutation symmetry. The property of $\overline{\mathscr{W}}^t$ being a state function implies that the time average of the work per unit volume does not depend on the particular process in which the field amplitudes are switched on adiabatically. $\overline{\mathscr{W}}^t$ depends only on the complex field amplitudes at a given time. Under the conditions mentioned, $\delta\mathscr{F}$ of (1.166b) is the variation of the free energy per unit volume.

Applying the overall permutation symmetry (1.169), the variations $\delta\mathscr{G}$ and $\delta\mathscr{F}$ in (1.165) and (1.166) can be written in a more concise form, and not only $\mathscr{G}$ but also $\mathscr{F}$ and hence $\overline{\mathscr{W}}^t$ can be easily obtained by integration. This gives

$$\mathscr{G} = \sum_{a,b,c} \tfrac{3}{4}\varepsilon_0 \chi^{(2)}_{cab}(\omega_3; \omega_1, \omega_2)\hat{E}_a(\omega_1)\hat{E}_b(\omega_2)\hat{E}^*_c(\omega_3) + \text{c.c.}, \tag{1.170a}$$

$$\mathscr{F} = -\sum_{a,b,c} \frac{\varepsilon_0}{4}\chi^{(2)}_{cab}(\omega_3; \omega_1, \omega_2)\hat{E}_a(\omega_1)\hat{E}_b(\omega_2)\hat{E}^*_c(\omega_3) + \text{c.c.}, \tag{1.170b}$$

$$\overline{\mathscr{W}}^t = \sum_{a,b,c} \frac{\varepsilon_0}{2}\chi^{(2)}_{cab}(\omega_3; \omega_1, \omega_2)\hat{E}_a(\omega_1)\hat{E}_b(\omega_2)\hat{E}^*_c(\omega_3) + \text{c.c.}. \tag{1.170c}$$

By comparison of (1.170b) and (1.170c),

$$\overline{\mathscr{W}}^t = -2\mathscr{F} \tag{1.171}$$

is obtained.

So far we have proved the equivalence of the *overall permutation symmetry* and the existence of the *free energy* of the system as a state function (cf. Refs. 7 and 1.8). Any calculation of susceptibilities by employing a microscopic theory, for example, the solution of (1.145) together with (1.148), yields tensors satisfying the overall permutation symmetry in loss-free media. Hence the existence of $\mathscr{F}$ as a state function can be deduced. In Chapter 4 we shall perform such calculations on the basis of the quantum-mechanical description of the material. Therefore we do not intend to use the microscopic equations of classical mechanics here.

Assuming negligible dispersion of the nonlinear susceptibilities in the region of $\omega_1, \omega_2, \omega_3$, the overall permutation symmetry will lead to an invariance of nonlinear susceptibilities under arbitrary permutation of the tensor indices, for example, to

$$\chi^{(2)}_{ijk}(\omega_3; \omega_1, \omega_2) = \chi^{(2)}_{jik}(\omega_3; \omega_1, \omega_2) = \chi^{(2)}_{kji}(\omega_3; \omega_1, \omega_2). \tag{1.172}$$

Such equations were first derived by Kleinman (1.9) and are therefore called *Kleinman relations*. Examples of susceptibilities that satisfy the Kleinman relations have already been given in Section 1.3.2.5.

If two of the three frequencies in $E_\bullet(t)$ are equal ($\omega_1 = \omega_2 = \omega$, $\omega_3 = 2\omega$) or if one of the frequencies is zero, the overall permutation symmetry in the form of (1.169) remains valid for $\underset{\smile}{\varkappa}^{(2)}_{ijk}(\omega_3;\, \omega_1, \omega_2)$. (The coefficients $\chi^{(2)}/\underset{\smile}{\varkappa}^{(2)}$ given in Section 1.3.2.3 have to be taken into account in order to derive relations for $\chi^{(2)}_{ijk}$.) Thus the following relations hold:

$$\underset{\smile}{\varkappa}^{(2)}_{ijk}(2\omega;\, \omega, \omega) = \underset{\smile}{\varkappa}^{(2)}_{jik}(-\omega;\, -2\omega, \omega) = \underset{\smile}{\varkappa}^{(2)}_{jik}(\omega; 2\omega, -\omega) \qquad (1.173)$$

and

$$\underset{\smile}{\varkappa}^{(2)}_{ijk}(0;\, \omega, -\omega) = \underset{\smile}{\varkappa}^{(2)}_{kji}(\omega;\, \omega, 0). \qquad (1.174)$$

Equation (1.173) represents a relation between the susceptibility $\underset{\smile}{\varkappa}^{(2)}_{ijk}(2\omega;\, \omega, \omega)$ describing the second-harmonic generation and the susceptibility $\underset{\smile}{\varkappa}^{(2)}_{jik}(\omega; 2\omega, -\omega)$ needed to calculate the change of the wave at frequency ω during its interaction with a wave at 2ω (cf. Chapter 11). Equation (1.174) is a relation between susceptibilities of two different processes. The tensors $\underset{\smile}{\varkappa}^{(2)}_{ijk}(0;\, \omega, -\omega)$ and $\underset{\smile}{\varkappa}^{(2)}_{kji}(\omega;\, \omega, 0)$ are associated with optical rectification and the Pockels effect, respectively.

Finally it should be mentioned in this sub-subsection that the overall permutation symmetry as well as the considerations concerning the free energy are also valid if the field strength amplitudes are not constant but are slowly varying in time; this means that the inequality $|(\partial/\partial t)\overset{\frown}{E}_a(\omega_i;\, t)| \ll \omega_i |\overset{\frown}{E}(\omega_i;\, t)|$ should be satisfied.

1.3.2.8. Manley–Rowe Relations. The overall permutation symmetries valid for loss-free media lead to relations between the contributions of several monochromatic waves to the power fed into the material. Since such equations were first derived by Manley and Rowe for microwave problems, they are called Manley–Rowe relations (Ref. 1.10).

Let us derive these relations by considering the energy exchange between matter and electromagnetic field. The time average of the electromagnetic power supplied to the medium per unit volume is $\mathscr{P} = (-\overline{\nabla_\bullet S_\bullet}^{\,t})$, where $S_\bullet$ is the Poynting vector of the radiation. Supposing a real dispersion-free linear permittivity $\varepsilon_{\bullet\bullet}$, the Poynting equation

$$-\nabla_\bullet S_\bullet = E_\bullet \frac{\partial}{\partial t} D_\bullet + H_\bullet \frac{\partial}{\partial t} B_\bullet, \qquad (1.175)$$

which follows immediately from Maxwell's equations, leads, in conjunction with the constitutive relations

$$D_\bullet = \varepsilon_0 \varepsilon_{\bullet\bullet} E_\bullet + P^{\mathrm{NL}}_\bullet$$

and

$$B_\bullet = \mu_0 H_\bullet,$$

to the time average

$$\mathscr{P} = -\overline{\nabla_\bullet S_\bullet}^{\,t} = +\overline{E_\bullet \frac{\partial}{\partial t} P^{\mathrm{NL}}}^{\,t} \tag{1.176}$$

of the electromagnetic power fed into the material per unit volume. (This averaging has to be carried out over a time of at least one period of the electromagnetic wave of lowest frequency.) Note that $\mathscr{P}$ of course equals the variation of work per unit volume and unit time introduced in (1.158). For monochromatic waves of $E_\bullet$ and $P_\bullet$ with frequencies ω_j, the power $\mathscr{P}$ per unit volume is calculated to be

$$\mathscr{P} = \sum_j \mathscr{P}_j,$$

where

$$\mathscr{P}_j = \frac{i}{4}\omega_j\left[\hat{E}_\bullet(\omega_j)\hat{P}_\bullet^{\mathrm{NL}*}(\omega_j) - \hat{E}_\bullet^*(\omega_j)\hat{P}_\bullet^{\mathrm{NL}}(\omega_j)\right]. \tag{1.177}$$

As pointed out in Section 1.3.2.7, the field strength at frequency ω_j performs in the time average work only on the polarization at the same frequency. In loss-free media the total power fed into the material is zero for stationary processes; the interaction of electromagnetic waves due to nonlinear optical processes may however result in an energy exchange between them. That is, some $\mathscr{P}_j$ may be different from zero.

As in linear optics, the power $\mathscr{P}_j$ imparted to the material in ΔV is only different from zero if there is a phase shift between $\hat{P}_\bullet$ and $\hat{E}_\bullet$. The expression (1.177) for $\mathscr{P}_j$ may be written in the form

$$\mathscr{P}_j = \frac{1}{2}\omega_j e_{j\bullet} e'_{j\bullet}\left|\hat{E}(\omega_j)\right|\left|\hat{P}(\omega_j)\right|\sin(\varphi'_j - \varphi_j),$$

where

$$\hat{E}_\bullet(\omega_j) = e_{j\bullet}\left|\hat{E}(\omega_j)\right|e^{i\varphi_j}$$

and

$$\hat{P}_\bullet(\omega_j) = e'_{j\bullet}\left|\hat{P}(\omega_j)\right|e^{i\varphi'_j}$$

has been used. $\mathscr{P}_j$ is positive, negative, and zero for $0 < (\varphi_j' - \varphi_j) \leq \pi$, $0 > (\varphi_j' - \varphi_j) > -\pi$, and $\varphi_j' = \varphi_j$, respectively. In linear optics a real susceptibility means a zero phase delay between field and polarization and therefore the energy change of the radiation is zero. In nonlinear optics a real susceptibility does not imply that $\mathscr{P}_j = 0$. The susceptibility connects $\hat{P}_\bullet(\omega_j)$ with other field amplitudes $\hat{E}_\bullet(\omega_1), \hat{E}_\bullet(\omega_2), \ldots, \hat{E}_\bullet(\omega_n)$ with $\omega_j = \omega_1 + \omega_2 + \cdots + \omega_n$. For real $\chi^{(n)}_{\bullet\bullet\cdots\bullet}$ the phase φ_j' is given by $\varphi_j' = \varphi_1 + \varphi_2 + \cdots + \varphi_n$ and can attain any value.

Let us now calculate the quantities

$$M_j = \frac{1}{\omega_j}\mathscr{P}_j \tag{1.178}$$

with the assumption of overall permutation symmetry. To avoid lengthy calculations we confine ourselves to three interacting waves of frequencies ω_1, ω_2, and $\omega_3 = \omega_1 + \omega_2$ in a material with second-order nonlinearity. Taking into account the relations $\chi^{(2)}_{\bullet\bullet\bullet}(-\omega; -\omega', -\omega'') = \chi^{(2)*}_{\bullet\bullet\bullet}(\omega; \omega', \omega'') = \chi^{(2)}_{\bullet\bullet\bullet}(\omega; \omega', \omega'')$, $\hat{E}_\bullet(-\omega) = \hat{E}^*_\bullet(\omega)$, and $\hat{P}_\bullet(-\omega) = P^*_\bullet(\omega)$, which are valid because of negligible losses and because of the reality of $E_\bullet(t)$, $P_\bullet(t)$, we obtain from (1.177) and (1.178)

$$M_1 = \sum_{a,b,c} \frac{i}{4}\chi^{(2)}_{acb}(-\omega_1; -\omega_3, \omega_2)$$
$$\times\left[\hat{E}_a(\omega_1)\hat{E}^*_c(\omega_3)\hat{E}_b(\omega_2) - \hat{E}^*_a(\omega_1)\hat{E}_c(\omega_3)\hat{E}^*_b(\omega_2)\right],$$

$$M_2 = \sum_{a,b,c} \frac{i}{4}\chi^{(2)}_{bac}(-\omega_2; \omega_1, -\omega_3)$$
$$\times\left[\hat{E}_b(\omega_2)\hat{E}_a(\omega_1)\hat{E}^*_c(\omega_3) - \hat{E}^*_b(\omega_2)\hat{E}^*_a(\omega_1)\hat{E}_c(\omega_3)\right],$$

$$M_3 = \sum_{a,b,c} \frac{i}{4}\chi^{(2)}_{cab}(-\omega_3; -\omega_1, -\omega_2)$$
$$\times\left[\hat{E}_c(\omega_3)\hat{E}^*_a(\omega_1)\hat{E}^*_b(\omega_2) - \hat{E}^*_c(\omega_3)\hat{E}_a(\omega_1)\hat{E}_b(\omega_2)\right]. \tag{1.179}$$

The summation indices a, b, c in each term have been arbitrarily chosen such that $\omega_1, \omega_2, \omega_3$ are connected with a, b, c, respectively. Because of the overall permutation symmetry of (1.169) the three susceptibility tensors in (1.179) are equal. As a consequence the Manley-Rowe relations

$$M_1 = M_2 = -M_3 \tag{1.180a}$$

or

$$\frac{\mathscr{P}_1}{\omega_1} = \frac{\mathscr{P}_2}{\omega_2} = -\frac{\mathscr{P}_3}{\omega_3} \tag{1.180b}$$

are obtained.

Although these equations have been derived entirely on the basis of classical physics, they can be advantageously interpreted in a quantum-mechanical language. To do this we divide (1.180b) by Planck's constant $\hbar$ and obtain

$$\frac{\mathscr{P}_1}{\hbar\omega_1} = \frac{\mathscr{P}_2}{\hbar\omega_2} = -\frac{\mathscr{P}_3}{\hbar\omega_3}. \tag{1.181}$$

Since $\hbar\omega_j$ is the energy of a photon of frequency $\hbar\omega_j$, $\mathscr{N}_j = \mathscr{P}_j/\hbar\omega_j$ is the number of photons per unit volume and unit time created (or annihilated) during the nonlinear optical interaction. In the particular process under consideration the changes of photon numbers at the frequencies ω_1 and ω_2 are the same; the change at frequency ω_3 has the same absolute value but the opposite sign. In other words, one photon at ω_1 and another at ω_2 are annihilated, and simultaneously, one photon is generated at $\omega_3 = \omega_1 + \omega_2$, or vice versa. During such a nonlinear process the total number of photons is not constant in spite of neglected material losses. Only the energy conservation $\Sigma_j \mathscr{P}_j = 0$ holds, as is evident from

$$\hbar\omega_3 = \hbar\omega_1 + \hbar\omega_2$$

in connection with the Manley–Rowe relations in the form of (1.181).

Dropping the special assumption that three waves are interacting in a material of second-order nonlinearity, general Manley–Rowe relations can be derived in a similar manner (Refs. 46, 1.10). If the frequencies of all interacting waves, ω_{mn}, are obtained from two original frequencies ω' and ω'' by means of

$$\omega_{mn} = m\omega' + n\omega'' \tag{1.182}$$

(m, n being positive or negative integers or zero), the Manley–Rowe relations can be represented in the form

$$\sum_{m,n} \frac{n}{\omega_{mn}} \mathscr{P}_{mn} = 0 \tag{1.183}$$

and

$$\sum_{m,n} \frac{m}{\omega_{mn}} \mathscr{P}_{mn} = 0.$$

where the summation is extended over all m, n with $\omega_{mn} > 0$. It is possible to give a quantum-mechanical interpretation which resembles that given before. [It should be mentioned that (1.180b) follows immediately from the more general equation (1.183) with $\omega_1 = 1 \cdot \omega' + 0 \cdot \omega''$, $\omega_2 = 0 \cdot \omega' + 1 \cdot \omega''$, $\omega_3 = 1 \cdot \omega' + 1 \cdot \omega''$; that is, m and n assume only the values 0 and ± 1.]

1.4. WAVE PROPAGATION IN NONLINEAR OPTICAL MEDIA

One of the main tasks of nonlinear optics is the calculation of the propagation and interaction of electromagnetic waves in nonlinear optical media. For this purpose the Maxwell equations (1.54) or (1.56) for the macroscopic fields have to be solved in connection with the nonlinear constitutive relations. In the following some general aspects of this problem will be treated.

We neglect here the magnetization of the medium and the dependence of the polarization on the magnetic field, as was done in the final part of Section 1.2 and in Section 1.3 throughout. Furthermore the density $\rho^{(q)}$ of free charges and their current density $j_{\bullet}^{(q)}$ are assumed to vanish. Then the constitutive relations reduce to the functional interdependence between the polarization $P_{\bullet}(r_{\bullet}, t)$ and the electric field strength $E_{\bullet}(r_{\bullet}, t)$. In many applications the simplification mentioned above is justified. Then the following set of equations has to be solved:

$$\nabla_{\bullet} \times H_{\bullet} = \varepsilon_0 \frac{\partial}{\partial t} E_{\bullet} + \frac{\partial}{\partial t} P_{\bullet}^{\mathrm{L}} + \frac{\partial}{\partial t} P_{\bullet}^{\mathrm{NL}}, \tag{1.184a}$$

$$\nabla_{\bullet} \times E_{\bullet} \equiv -\mu_0 \frac{\partial}{\partial t} H_{\bullet}, \tag{1.184b}$$

$$\nabla_{\bullet}\left(\varepsilon_0 E_{\bullet} + P_{\bullet}^{\mathrm{L}} + P_{\bullet}^{\mathrm{NL}}\right) = 0, \tag{1.184c}$$

$$\nabla_{\bullet} H_{\bullet} = 0, \tag{1.184d}$$

$$P_{\bullet}^{\mathrm{L}} = P_{\bullet}^{\mathrm{L}}\left[E_{\bullet}(r_{\bullet}, t)\right], \qquad P_{\bullet}^{\mathrm{NL}} = P^{\mathrm{NL}}\left[E_{\bullet}(r_{\bullet}, t)\right]. \tag{1.184e}$$

The relations (1.184e) may for instance be given in the form of the integrals of (1.73) or of (1.83).

Using here the same mathematical operations as in linear optics, the wave equation

$$\nabla_\bullet \times (\nabla_\bullet \times E_\bullet) + \frac{1}{c^2}\frac{\partial^2}{\partial t^2}E_\bullet = -\mu_0 \frac{\partial^2}{\partial t^2}(P_\bullet^{\mathrm{L}} + P_\bullet^{\mathrm{NL}}) \tag{1.185}$$

can be obtained from (1.184). Its left-hand side consists of terms that even appear if the electromagnetic field in vacuo is described. The source term on the right-hand side represents the influence of the medium. If we substitute the integral relation (1.83) for $P_\bullet(r_\bullet, t)$ as a function of $E_\bullet(r'_\bullet, t')$, then (1.185) becomes an integrodifferential equation for the electric field. After solving this equation the magnetic field strength and the polarization can be determined from $E_\bullet(r_\bullet, t)$ with the help of (1.184). This means that the solution of (1.185) is the main problem to be tackled. The task to be fulfilled can be generally formulated in the following way: the solution of the wave equation has to be carried out under certain boundary and initial conditions, which are imposed by the physical nature of the problem to be treated.

As an example we consider a nonlinear optical medium occupying the half space with $z \geq 0$. The boundary at $z = 0$ is irradiated with light of field strength $E_\bullet^{(0)}(r_\bullet, t)$. If we neglect the reflection at the boundary and the backscattering of light in the sample, the field strength in the plane $z = 0$ is given by $E_\bullet^{(0)}(r_\bullet, t)$ at any time. If the light signal is not stationary but switched on at a time t_0, the electric field and the polarization at any point $r_\bullet$ of the dielectric medium remain zero until the first part of the signal arrives.

In the following several methods will be discussed that can be employed to facilitate the solution of (1.185). The decision which of these methods should be applied will depend on the particular physical problem to be investigated.

1.4.1. Fourier Transformation

In some cases the treatment of the wave equation (1.185) can be simplified by carrying out a Fourier transformation. To explain the basic idea we first choose a simple geometry for the experiment: only one component (say the x component) of the electric field and of the polarization are different from zero; $E_x = E$ and $P_x = P$ do not depend on the coordinates x and y. Under these conditions $\nabla_\bullet E_\bullet = 0$ and the wave equation reduces to

$$-\frac{\partial^2}{\partial z^2}E(t, z) + \frac{1}{c^2}\frac{\partial^2}{\partial t^2}E(t, z) = -\mu_0 \frac{\partial^2}{\partial t^2}P(t, z). \tag{1.186}$$

As in Section 1.3.2, we represent $E(z, t)$ and $P(z, t)$ by their Fourier transforms $\underset{\smile}{E}(\omega, k)$, $\underset{\smile}{P}(\omega, k)$, which are functions of frequency and wave number k:

$$E(t, z) = \frac{1}{(2\pi)^2} \int_{-\infty}^{\infty} d\omega \int_{-\infty}^{\infty} dk\, \underset{\smile}{E}(\omega, k) e^{+i(\omega t - kz)},$$

$$P(t, z) = \frac{1}{(2\pi)^2} \int_{-\infty}^{\infty} d\omega \int_{-\infty}^{\infty} dk\, \underset{\smile}{P}(\omega, k) e^{+i(\omega t - kz)}, \tag{1.187}$$

The application of the transformation to (1.186) leads to

$$k^2 \underset{\smile}{E}(\omega, k) - \frac{1}{c^2}\omega^2 \underset{\smile}{E}(\omega, k) = \mu_0 \omega^2 \underset{\smile}{P}(\omega, k). \tag{1.188}$$

The source term $P(\omega, k)$ on the right-hand side of (1.188) is given by

$$\begin{aligned} \underset{\smile}{P}(\omega, k) &= \sum_n \underset{\smile}{P}^{(n)}(\omega, k) \\ &= \varepsilon_0 \underset{\smile}{\varkappa}^{(1)}(\omega; \omega) \underset{\smile}{E}(\omega, k) \\ &\quad + \frac{\varepsilon_0}{(2\pi)^2} \int_{-\infty}^{\infty} d\omega' \int_{-\infty}^{\infty} dk'\, \underset{\smile}{\varkappa}^{(2)}(\omega; \omega', \omega - \omega') \\ &\quad \times \underset{\smile}{E}(\omega', k) \underset{\smile}{E}(\omega - \omega', k - k') + \cdots . \end{aligned} \tag{1.189}$$

The equations (1.188) and (1.189) together form an integral equation for $E(\omega, k)$, whereas the corresponding equation (1.186) for $E(z, t)$ is an integro-differential equation.

1.4.2. Monochromatic Plane Waves

In many experiments the nonlinear material experiences stationary irradiation with light whose spectrum is concentrated in narrow frequency bands. Idealization of this situation leads to an interaction of monochromatic fields. Substituting

$$E_\bullet(t, r_\bullet) = \sum_{\mu=1}^{a} \tfrac{1}{2} \hat{E}_\bullet(\omega_\mu; r_\bullet) e^{-i\omega_\mu t} + \text{c.c.}, \tag{1.190a}$$

$$P_\bullet^{\text{L}}(t, r_\bullet) = \varepsilon_0 \sum_{\mu=1}^{a} \tfrac{1}{2} \chi_{\bullet\bullet}^{(\text{L})}(\omega_\mu; \omega_\mu) \hat{E}_\bullet(\omega_\mu, r_\bullet) e^{-i\omega_\mu t} + \text{c.c.}, \tag{1.190b}$$

$$P_\bullet^{\text{NL}}(t, r_\bullet) = \sum_{\nu=1}^{b} \tfrac{1}{2} \hat{P}_\bullet^{\text{NL}}(\omega_\nu, r_\bullet) e^{-i\omega_\nu t} + \text{c.c.}, \tag{1.190c}$$

where $P_{\bullet}^{\mathrm{NL}}$ may contain other frequencies than $E_{\bullet}$ and $P_{\bullet}^{\mathrm{L}}$, into the wave equation (1.185), we have

$$\sum_{\mu=1}^{a} e^{-i\omega_\mu t}\left\{\nabla_{\bullet}\times\left[\nabla_{\bullet}\times\hat{E}_{\bullet}(\omega_\mu, r_{\bullet})\right] - \frac{\omega_\mu^2}{c^2}\varepsilon_{\bullet\bullet}(-\omega_\mu)\hat{E}_{\bullet}(\omega_\mu, r_{\bullet})\right\}$$

$$= \mu_0 \sum_{\nu=1}^{b} \omega_\nu^2 \hat{P}_{\bullet}^{\mathrm{NL}}(\omega_\nu, r_{\bullet})e^{-i\omega_\nu t}, \tag{1.191}$$

where $\varepsilon_{\bullet\bullet}(-\omega_\mu) = 1 + \chi_{\bullet\bullet}^{(\mathrm{L})}(\omega_\mu; \omega_\mu)$, and an analogous equation for the complex conjugate part of the fields and polarizations. In general, a stationary solution of (1.191) requires the relation

$$\nabla_{\bullet}\times\left[\nabla_{\bullet}\times\hat{E}_{\bullet}(\omega_\mu, r_{\bullet})\right] - \frac{\omega_\mu^2}{c^2}\varepsilon_{\bullet\bullet}(-\omega_\mu)\hat{E}_{\bullet}(\omega_\mu, r_{\bullet}) = \mu_0\omega_\mu^2\hat{P}_{\bullet}^{\mathrm{NL}}(\omega_\mu, r_{\bullet}) \tag{1.192}$$

to be satisfied for each frequency ω_μ. This means that only the part of $P_{\bullet}^{\mathrm{NL}}(t; r_{\bullet})$ at the frequency $\omega_\nu = \omega_\mu$ contributes to the variation of the field strength $\hat{E}_{\bullet}(\omega_\mu, r_{\bullet})$ [cf. the arguments leading to (1.161)]. In the nth order the amplitude of nonlinear polarization on the right-hand side of (1.192), in view of (1.120a), is given by

$$\hat{P}_{\bullet}^{(n)}(\omega_\mu; r_{\bullet}) = \varepsilon_0\chi_{\bullet\bullet\cdots\bullet}^{(n)}(\omega_\mu; \omega_1, \dots, \omega_n)\hat{E}_{\bullet}(\omega_1, r_{\bullet})\cdots\hat{E}_{\bullet}(\omega_n, r_{\bullet}). \tag{1.193}$$

[Here one or several of the frequencies ω_β ($\omega_\beta = \omega_1, \dots, \omega_n$) may be negative and/or may appear several times.] The equations (1.192) for $l = 1, 2, \dots, a$ together with (1.193) present a set of partial differential equations for the field-strength amplitudes $\hat{E}_{\bullet}(\omega_\mu, r_{\bullet})$. Compared with (1.185), the difficulties are now reduced because no integrodifferential equations appear.

Wavelike processes are of great importance in experiments with quasimonochromatic fields. This means that the fast spatial variation of the amplitude $\hat{E}_{\bullet}(\omega_\mu, r_{\bullet})$ is given by $\exp\{ik_{\mu\bullet}r_{\bullet}\}$; in other words, the wave amplitudes $\bar{E}_{\bullet}(\omega_\mu, r_{\bullet})$ and their lengths $\bar{E}(\omega_\mu, r_{\bullet})$ defined by

$$\hat{E}_{\bullet}(\omega_\mu, r_{\bullet}) = \bar{E}_{\bullet}(\omega_\mu, r_{\bullet})e^{ik_{\mu\bullet}r_{\bullet}} = e_{\mu\bullet}\bar{E}(\omega_\mu, r_{\bullet})e^{ik_{\mu\bullet}r_{\bullet}} \tag{1.194}$$

are slowly varying functions of the space coordinate $r_{\bullet}$. The separation of fast and slowly varying factors in $\hat{E}$ means, with respect to the Fourier space, that the transform of $\hat{E}(\omega_\mu, r_\bullet)$ contains only components with wave vectors $k_\bullet$ in a small interval $\delta k_\bullet$ around $k_{\mu\bullet}$, and the transform of $\bar{E}(\omega_\mu, r_\bullet)$ has non-vanishing components only in a small region $\delta k_\bullet$ around $k_\bullet = 0$. Based on the assumption

$$|\delta k| \ll |k|, \tag{1.195}$$

the wave amplitudes (i.e. their moduli and phases) vary considerably only over distances large compared with the wavelength $2\pi/k^{(r)}$ (where $k^{(r)}$ is the real part of k). After inserting $\hat{E}_\bullet(\omega_\mu, r_\bullet)$ from (1.194) into (1.192), the second-order derivatives of $\bar{E}(\omega_\mu, r_\bullet)$ can be neglected. More precisely, this is justified if the inequalities

$$\left|1 - \operatorname{Im}\left\{\frac{2}{k_x}\frac{1}{\bar{E}}\frac{\partial}{\partial x}\bar{E}\right\}\right| \gg \left|\operatorname{Re}\left\{\frac{1}{k_x^2}\frac{1}{\bar{E}}\frac{\partial^2}{\partial x^2}\bar{E}\right\}\right|,$$

$$\left|\operatorname{Re}\left\{\frac{2}{k_x}\frac{1}{\bar{E}}\frac{\partial}{\partial x}\bar{E}\right\}\right| \gg \left|\operatorname{Im}\left\{\frac{1}{k_x^2}\frac{1}{\bar{E}}\frac{\partial^2}{\partial x^2}\bar{E}\right\}\right|, \tag{1.196}$$

and analogous expressions for the derivatives with respect to the other coordinates are satisfied. A proof for (1.196) can be provided if (1.195) holds. This connection between (1.195) and (1.196) is obvious from the following consideration. The fastest space variation of $\bar{E}(\omega_\mu, r_\bullet)$ consistent with $\delta k_\bullet$ as the maximum wave vector in the Fourier spectrum $\breve{\bar{E}}(\omega_\mu, k'_\bullet)$ is given by $\bar{E}(\omega_\mu, r_\bullet) = \bar{E}(\omega_\mu)\exp\{i(\delta k_\bullet r_\bullet)\}$, where in general $\delta k_\bullet = \delta k_\bullet^{(r)} + i\,\delta k_\bullet^{(i)}$ is complex. Then the derivatives

$$\frac{1}{\bar{E}(\omega_\mu, r_\bullet)}\frac{\partial}{\partial x}\bar{E}(\omega_\mu, r_\bullet) = i\,\delta k_x = i\,\delta k_x^{(r)} - \delta k_x^{(i)}$$

and

$$\frac{1}{\bar{E}(\omega_\mu, r_\bullet)}\frac{\partial^2}{\partial x^2}\bar{E}(\omega_\mu, r_\bullet) = -(\delta k_x)^2$$

$$= -\left[\delta k_x^{(r)}\right]^2 + \left[\delta k_x^{(i)}\right]^2 - 2i\,\delta k_x^{(i)}\,\delta k_x^{(r)}$$

are obtained, which are seen to obey (1.196) if we assume the validity of (1.195).

Now the expression for $\hat{E}_\bullet(\omega_\mu, r_\bullet)$ of (1.194) is inserted into the time-independent wave equation (1.192). The geometric circumstances are specified in

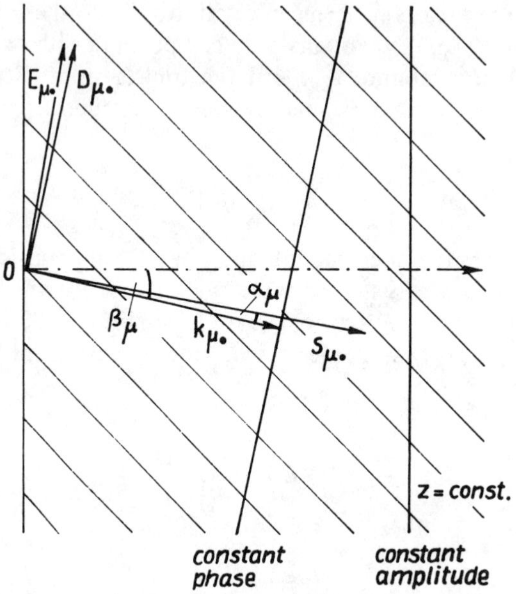

Fig. 1.11. Wave propagation in an anisotropic medium occupying the half space $z \geq 0$ ($k_\bullet$ = wave vector, $S_\bullet$ = Poynting vector, $E_\bullet$ = electric field strength, $D_\bullet$ = dielectric displacement).

Fig. 1.11. We see that the nonlinear material occupies the half space with $z \geq 0$. Within this material the wave vector $k_{\mu\bullet}$ forms an angle β_μ with the z axis. In anisotropic media the direction of $S_{\mu\bullet}$ may differ from that of $k_{\mu\bullet}$. Then the angle between the two vectors is α_μ (cf. e.g. Ref. 8). All waves are infinitely extended in the x and y directions, and the amplitudes $\overline{E}(\omega_\mu, r_\bullet)$ are assumed to be independent of x and y: $\overline{E}(\omega_\mu, r_\bullet) = \overline{E}(\omega_\mu, z)$. As described above, the second derivatives of $\overline{E}(\omega_\mu, r_\bullet)$ are neglected in the evaluation of $\nabla_\bullet \times (\nabla_\bullet \times E_\bullet)$. Here the wave vector $k_{\mu\bullet}$ is chosen to be the same as that of a monochromatic wave in the material with neglect of nonlinearities. Then $e_{\mu\bullet} e^{ik_{\mu\bullet} r_\bullet}$ satisfies the homogeneous wave equation of linear optics,

$$\nabla_\bullet \times \left(\nabla_\bullet \times e_{\mu\bullet} e^{ik_{\mu\bullet} r_\bullet}\right) - \frac{\omega_\mu^2}{c^2} \varepsilon_{\bullet\bullet}(-\omega_\mu) e_{\mu\bullet} e^{ik_{\mu\bullet} r_\bullet} = 0, \qquad (1.197)$$

which is equivalent to

$$k_{\mu\bullet} \times \left(k_{\mu\bullet} \times e_{\mu\bullet}\right) + \frac{\omega_\mu^2}{c^2} \varepsilon_{\bullet\bullet}(-\omega_\mu) e_{\mu\bullet} = 0 \qquad (1.198a)$$

and to

$$k_{\mu\bullet}\left(k_{\mu\bullet} e_{\mu\bullet}\right) + \frac{\omega_\mu^2}{c^2}\left[\varepsilon_{\bullet\bullet}(-\omega_\mu) e_{\mu\bullet} - e_{\mu\bullet}\right] = 0. \qquad (1.198b)$$

Neglecting the term with $(\partial^2/\partial z^2)\bar{E}(\omega_\mu, z)$ we thus obtain

$$e_{\mu\bullet}\left\{\nabla_\bullet \times \left(\nabla_\bullet \times \hat{E}_\bullet(\omega_\mu, z)\right)\right\} = -i2k_\mu \cos\alpha_\mu \cos(\alpha_\mu - \beta_\mu)e^{ik_{\mu\bullet}r_\bullet}\frac{\partial}{\partial z}\bar{E}(\omega_\mu, z).$$

The substitution of this expression into the wave equation (1.192) gives

$$-i2k_\mu \cos\alpha_\mu \cos(\alpha_\mu - \beta_\mu)\frac{\partial}{\partial z}\bar{E}(\omega_\mu, z)$$
$$= \mu_0 \omega_\mu^2 e_{\mu\bullet}\hat{P}_\bullet^{\mathrm{NL}}(\omega_\mu, r_\bullet)e^{-ik_{\mu\bullet}r_\bullet}. \tag{1.199}$$

Since the projection of the nonlinear polarization amplitude of nth order on $e_{\mu\bullet}$ is

$$e_{\mu\bullet}\hat{P}_\bullet^{\mathrm{NL}}(\omega_\mu, z) = \varepsilon_0\tilde{\chi}^{(n)}(\omega_\mu; \omega_1, \ldots, \omega_n)\bar{E}(\omega_1, z) \cdots \bar{E}(\omega_n, z)$$
$$\times \exp\left[i\left(\sum_{\lambda=1}^{n} k_{\lambda\bullet}\right)r_\bullet\right],$$

the relation (1.199) leads to

$$\frac{\partial}{\partial z}\bar{E}(\omega_\mu, z) = i\frac{\omega_\mu^2}{2c^2k_\mu \cos\alpha_\mu \cos(\alpha_\mu - \beta_\mu)}\tilde{\chi}^{(n)}(\omega_\mu; \omega_1, \ldots, \omega_n)$$
$$\times \bar{E}(\omega_1, z) \cdots \bar{E}(\omega_n, z)e^{i\Delta k_\bullet r_\bullet}, \tag{1.200a}$$

where the *momentum* or *wave-vector mismatch* $\Delta k_\bullet$ is given by

$$\Delta k_\bullet = \left(\sum_{\lambda=1}^{n} k_{\lambda\bullet}\right) - k_{\mu\bullet}, \tag{1.200b}$$

and $\tilde{\chi}$ is the effective susceptibility

$$\tilde{\chi}(\omega_\mu; \omega_1, \ldots, \omega_n) = e_{\mu\bullet}\chi^{(n)}_{\bullet\bullet\cdots\bullet}(\omega_\mu; \omega_1, \ldots, \omega_n)e_{1\bullet} \cdots e_{n\bullet} \tag{1.200c}$$

introduced in (1.123). The assumption of wave amplitudes independent of x, y will only be satisfied if the x and y components of the momentum mismatch $\Delta k_\bullet$ obey the relations

$$\Delta k_x = 0, \qquad \Delta k_y = 0. \tag{1.200d}$$

From this it follows that the x and y components of the wave vectors of the fields generating the polarization must precisely compensate the corresponding components of the generated field. In general the relations (1.200b) and (1.200d) are complex, and hence they provide conditions for the real and imaginary parts of $k_{\bullet} = k_{\bullet}^{(r)} + ik_{\bullet}^{(i)}$.

Equations (1.199) and (1.200) in connection with (1.194) show that the field amplitude $\hat{E}(\omega_\mu, z) = \bar{E}(\omega_\mu, z)e^{ik_{\mu\bullet}r_\bullet}$ varies in the nonlinear material for two reasons: First, the amplitude is attenuated by linear optical absorption, and second, it varies in response to the nonlinear polarization P^{NL}, which in the nth order depends on a product of n field-strength amplitudes. In loss-free media only this nonlinear term remains as a source of field-strength variation.

In (1.199) and (1.200) the change of the field amplitude of a given polarization direction $e_{\mu\bullet}$ has been calculated. It should be mentioned that there may occur two waves at one frequency $\omega_\mu \equiv \omega_l$ with different polarization directions $\sigma = 1, 2$. In this case two equations of the form (1.199) or (1.200) are obtained at one frequency, whereby in anisotropic media the wave vector depends also on the polarization direction $e_{\mu\bullet}$, as follows from (1.198).

Whether the nonlinear optical interaction results in attenuation or amplification of the amplitude $\bar{E}(\omega_\mu, z)$ depends on the phase angle Φ of the nonlinear term in (1.200a); Φ in turn depends on the phase angles of both the effective susceptibility and the wave amplitude $\bar{E}(\omega_\lambda, z)$, as well as on $\Delta k_\bullet r_\bullet$. Let us discuss this behavior using the *approximation of given fields*, which will now be explained. If the nonlinearities are small, $P_{\bullet}^{\mathrm{NL}}$ can be obtained approximately from (1.193), where on the right-hand side the field amplitudes $\bar{E}_{\bullet}^{(0)}(\omega_\mu, z)$ of linear optics are inserted. The $\bar{E}_{\bullet}^{(0)}(\omega_\mu, z)$ are calculated from the wave equation by neglecting $P_{\bullet}^{\mathrm{NL}}$; the resulting wave amplitudes $\bar{E}_{\bullet}^{(0)}(\omega_\mu)$ do not depend on the space coordinate. Thus (1.200a) gives

$$\frac{\partial}{\partial z}\bar{E}(\omega_\mu, z) = \frac{i\omega_\mu^2}{2c^2k_\mu \cos\alpha_\mu \cos(\alpha_\mu - \beta_\mu)}\tilde{\chi}^{(n)}(\omega_\mu; \omega_1, \ldots, \omega_n)$$

$$\times \bar{E}^{(0)}(\omega_1) \cdots \bar{E}^{(0)}(\omega_n) e^{i\Delta k_z z}. \tag{1.201}$$

The phase angle Φ of the source term in (1.201) varies with z only via the term $\Delta k_z z$ in space. Hence the solution of (1.201) is

$$\bar{E}(\omega_\mu, z) - \bar{E}^{(0)}(\omega_\mu) = \frac{1}{\Delta k_z}\tilde{\tilde{\chi}}^{(n)}\bar{E}^{(0)}(\omega_1) \cdots \bar{E}^{(0)}(\omega_n)[1 - e^{i\Delta k_z z}] \tag{1.202}$$

with

$$\tilde{\tilde{\chi}}^{(n)} = \frac{\omega_\mu^2}{2c^2k_\mu \cos\alpha_\mu \cos(\alpha_\mu - \beta_\mu)}\tilde{\chi}^{(n)}(\omega_\mu; \omega_1, \ldots, \omega_n).$$

Now suppose we have a material with $\Delta k_z^{(i)} = 0$; this requirement can for example be satisfied if the linear optical losses at ω_μ and $\omega_1, \ldots, \omega_n$ are

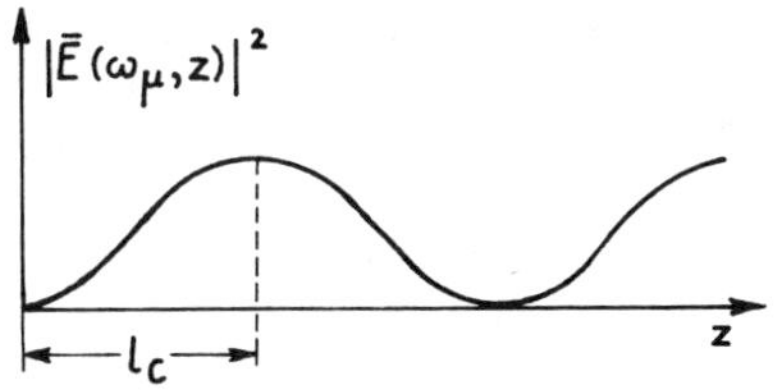

Fig. 1.12. Wave generation by nonlinear interaction. The square of the modulus of the wave amplitude $|\bar{E}(\omega_\mu, z)|$ is plotted versus the space coordinate z; l_c is the phase coherence length.

negligible. Provided the amplitude $\bar{E}^{(0)}(\omega_\mu) = \bar{E}(\omega_\mu, z = 0)$ at the entrance of the nonlinear material is zero, then the square of $\bar{E}(\omega_\mu, z)$, which is proportional to the intensity, is given by

$$|\bar{E}(\omega_\mu, z)|^2 = |\tilde{\tilde{\chi}}^{(n)}|^2 |\bar{E}^{(0)}(\omega_1)|^2 \cdots |\bar{E}^{(0)}(\omega_n)|^2 \left(\frac{\sin \frac{1}{2} \Delta k_z z}{\frac{1}{2} \Delta k_z} \right)^2 . \qquad (1.203)$$

Only if the relation

$$\Delta k_z = \left(\sum_{\lambda=1}^{n} k_{\lambda z} \right) - k_{\mu z} = 0 \qquad (1.204)$$

holds for the z component of the momentum mismatch is there a monotonic intensity increase of the wave generated at $\omega_\mu = \omega_1 + \cdots + \omega_n$. Conversely, for $\Delta k_z \neq 0$ the intensity varies periodically as a function of z (see Fig. 1.12). It increases from zero to its maximum over the distance between $z = 0$ and

$$z = l_c = \frac{1}{\Delta k_z}, \qquad (1.205)$$

which is called the *phase coherence length*.

At $z = 0$ the amplitude $\hat{E}(\omega_\mu, z = 0) = \bar{E}(\omega_\mu, z = 0)$ is generated with the phase angle

$$\varphi_\mu(0) = \varphi_{\rm NL}(0) + \frac{\pi}{2},$$

where $\varphi_{\rm NL}(0)$ is the phase angle of $\hat{P}^{\rm NL}(\omega_\mu, 0)$ according to (1.199). At a phase shift of $\pi/2$ between $\hat{E}(\omega_\mu, 0)$ and $\hat{P}^{\rm NL}(\omega_\mu, 0)$, maximum power is transferred to the electromagnetic field at frequency ω_μ, which is in agreement with (1.177). At $z = l_c$ the phase angles $\varphi_\mu(l_c)$ and $\varphi_{\rm NL}(l_c)$ of $\hat{E}(\omega_\mu, z = l_c)$ and $\hat{P}^{\rm NL}(\omega_\mu, z = l_c)$, respectively, are given by

$$\varphi_\mu(l_c) = \varphi_\mu(0) + k_{\mu z} l_c$$

and

$$\varphi_{\rm NL}(l_c) = \varphi_{\rm NL}(0) + \left(\sum_{\lambda=1}^{n} k_{\lambda z} \right) l_c .$$

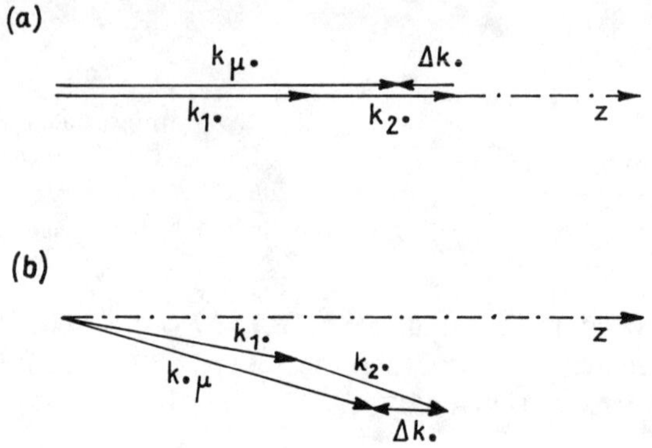

Fig. 1.13. Wave vector mismatch $\Delta k_{\bullet}$: (a) collinear interaction of waves, (b) noncollinear interaction of waves.

Taking into account (1.200b) and (1.205), the equality of the phase angles of $\hat{E}(\omega_\mu, z)$ and $\hat{P}^{\mathrm{NL}}(\omega_\mu, z)$ can be deduced for $z = l_c$. This means that the energy exchange between the wave of frequency ω_μ and the material vanishes. This is again in agreement with (1.177). Beyond $z = l_c$ the phase delay $\varphi_\mu(z) - \varphi_{\mathrm{NL}}(z)$ takes on the opposite sign, and hence energy is pumped back into the other waves with the help of the nonlinear material.

As $\Delta k_{\bullet} \to 0$, the phase angle between $\hat{E}(\omega_\mu, z)$ and $\hat{P}^{\mathrm{NL}}(\omega_\mu, z)$ becomes constant over a long distance. Therefore the equation $\Delta k_{\bullet} = 0$ is referred to as the *phase-matching relation*. In this connection it should be mentioned that for $\Delta k_{\bullet} \to 0$ the monotonic increase of the intensity at ω_μ eventually leads to a pronounced weakening of the generating waves, whose amplitudes have so far been assumed to be constant. Thus if $\Delta k_{\bullet} = 0$, the approximation of constant generating waves has to be dropped at large z.

The *momentum mismatch* $\Delta k_{\bullet}$ (see Fig. 1.13) depends on the lengths of the wave vectors,

$$k_\mu = \frac{\omega_\mu n(k_{\mu\bullet}, \sigma_\mu)}{c}, \tag{1.206}$$

and on their directions; $n(k_{\mu\bullet}, \sigma_\mu)$ stands for the refractive index of a wave of wave vector $k_{\mu\bullet}$ and of polarization direction $\sigma_\mu = 1$ or 2. By proper choice of the propagation direction of incident waves, $\Delta k_{\bullet} = 0$ can always be achieved. But in many experiments a collinear interaction geometry is preferable. Then the phase-matching relation $\Delta k_{\bullet} = 0$ requires that

$$\omega_\mu n(k_{\mu\bullet}, \sigma_\mu) = \sum_{\lambda=1}^{n} \omega_\lambda n(k_{\lambda\bullet}, \sigma_\lambda). \tag{1.207}$$

This equation can only be satisfied if the direction of z with respect to the optical axis of an anisotropic material as well as the polarization directions of the interacting waves are properly chosen. This will be discussed in more detail in the sections of Part II dealing with special effects of nonlinear optics.

1.4.3. Monochromatic Light Beams

If the field amplitude $\hat{E}_\bullet(\omega_\mu, r_\bullet)$ depends not only on z but also on x and y, it is far more difficult to tackle the nonlinear wave equation, in spite of the simplification provided by the assumption of monochromatic fields. The xy dependence of the amplitudes has always to be taken into account if the finite diameter of the interacting light beams and the transverse structure of the nonlinear material or of other parts of the experimental arrangement influence the efficiency of nonlinear processes.

Here we want to discuss how to solve the wave equation for $\hat{E}_\bullet(\omega_\mu, x, y, z) = e_{\mu\bullet}\bar{E}(\omega_\mu, x, y, z)e^{ik_{\mu\bullet}r_\bullet}$ under some simplifying conditions. The directions of the propagation vector $k_{\mu\bullet}$ and of the Poynting vector $S_{\mu\bullet}$ are assumed to coincide and to point into the z direction. Furthermore, we suppose that

$$\nabla_\bullet \hat{E}_\bullet(\omega_\mu, r_\bullet) = 0. \tag{1.208}$$

These conditions are for instance satisfied in homogeneous and isotropic optical materials with small nonlinearity. Using the assumptions mentioned above, we obtain

$$\nabla_\bullet^2 \hat{E}_\bullet(\omega_\mu, r_\bullet) + \frac{\omega_\mu^2}{c^2}\varepsilon_{\bullet\bullet}(\omega_\mu)\hat{E}_\bullet(\omega_\mu, r_\bullet) = -\omega_\mu^2\mu_0 \hat{P}_\bullet^{\mathrm{NL}}(\omega_\mu, r_\bullet) \tag{1.209}$$

from the wave equation (1.192). For wave amplitudes slowly varying with z, the second derivatives with respect to this coordinate are again neglected and

$$2ik_\mu \frac{\partial}{\partial z}\bar{E}(\omega_\mu, r_\bullet) + \Delta_\perp \bar{E}(\omega_\mu, r_\bullet) = -\omega_\mu^2\mu_0 e_{\mu\bullet}\hat{P}_\bullet^{\mathrm{NL}}(\omega_\mu, r_\bullet)e^{-ik_\mu z} \tag{1.210}$$

is obtained, where

$$\Delta_\perp = \frac{\partial^2}{\partial x^2} + \frac{\partial^2}{\partial y^2} \tag{1.211}$$

is the transverse part of the Laplace operator. Neglecting the nonlinearity, (1.210) represents the so-called *quasioptical approximation* for diffraction problems. In nonlinear optics (1.210) may be used for instance to treat self-focusing and self-defocusing effects.

1.4.4. Waves with Slowly Varying Amplitudes

Assuming stationarity, waves with amplitudes slowly varying in space have already been treated in Sections 1.4.2 and 1.4.3. Let us now discuss nonstationary interaction processes in which the dependence of wave amplitudes on space and time coordinates has to be taken into account.

First we consider the following simple geometrical situation: All waves propagate in the z direction; only one component of fields and polarizations—say the x component—is not equal to zero; the amplitudes of the plane waves do not depend on x nor on y. The electric field is supposed to consist of several wavelike parts whose Fourier transforms contain only frequencies near $\omega_1, \ldots, \omega_a$. Then the electric field strength is given by

$$E(t, z) = \sum_{\mu=1}^{a} \tfrac{1}{2}\bar{E}(\omega_\mu; t, z) e^{-i(\omega_\mu t - k_\mu z)} + \text{c.c.}. \tag{1.212}$$

An analogous formula holds for the linear polarization $P^{\mathrm{L}}(t, z)$ and its amplitudes $\bar{P}^{\mathrm{L}}(\omega_\mu; t, z)$. The polarization amplitudes $\bar{P}^{\mathrm{L}}(\omega_\mu; t, z)$ can be expressed by the field amplitudes with the help of the linear optical susceptibility $\breve{\varkappa}^{(1)}(\omega; \omega)$. Since the Fourier transforms of the slowly varying amplitudes contain only frequencies near ω_μ, it is useful to expand $\breve{\varkappa}^{(1)}(\omega, \omega)$ at ω_μ. This yields

$$\begin{aligned}\breve{\varkappa}^{(1)}(\omega; \omega) = \breve{\varkappa}^{(1)}(\omega_\mu; \omega_\mu) &+ \left(\frac{d}{d\omega}\breve{\varkappa}^{(1)}\right)_{\omega=\omega_\mu}(\omega - \omega_\mu)\\ &+ \frac{1}{2}\left(\frac{d^2}{d\omega^2}\breve{\varkappa}^{(1)}\right)_{\omega=\omega_\mu}(\omega - \omega_\mu)^2 + \cdots.\end{aligned} \tag{1.213}$$

The coefficients in this expansion are associated with phenomenological parameters such as the phase and group velocity, the absorption coefficient, and their frequency derivatives at the frequency $\omega = \omega_\mu$. Using (1.213), the slowly varying polarization amplitude can be given in the form

$$\begin{aligned}\bar{P}^{\mathrm{L}}(\omega_\mu; t, z) &= \varepsilon_0 \breve{\varkappa}^{(1)}(-\omega_\mu; -\omega_\mu)\bar{E}(\omega_\mu; t, z)\\ &\quad + i\varepsilon_0\left(\frac{d}{d\omega}\breve{\varkappa}^{(1)}(-\omega; -\omega)\right)_{\omega_\mu}\frac{\partial}{\partial t}\bar{E}(\omega_\mu; t, z)\\ &\quad - \tfrac{1}{2}\varepsilon_0\left(\frac{d^2}{d\omega^2}\breve{\varkappa}^{(1)}(-\omega; -\omega)\right)_{\omega_\mu}\frac{\partial^2}{\partial t^2}\bar{E}(\omega_\mu; t, z) + \cdots,\end{aligned} \tag{1.214}$$

as has been shown in Section 1.3.2. The expansions in (1.213) and (1.214) converge rapidly if the frequencies are restricted to the interval $\delta\omega_\mu$ around ω_μ with

$$\delta\omega_\mu \ll \omega_\mu. \tag{1.214a}$$

[This restriction has its analogue concerning the wave vectors in (1.195).]

Substituting the electric field of (1.212) and the linear polarization term of (1.214) into the wave equation (1.185) or (1.186), one obtains the following set of partial differential equations for the field amplitudes:

$$\left[C_1 + C_2\frac{\partial}{\partial z} + C_3\frac{\partial^2}{\partial z^2} + C_4\frac{\partial}{\partial t} + C_5\frac{\partial^2}{\partial t^2}\right]\overline{E}(\omega_\mu; t, z) = \left[C_4' + C_2'\frac{\partial}{\partial t} + C_3'\frac{\partial^2}{\partial t^2}\right]\widehat{P}^{\mathrm{NL}}(\omega_\mu; t, z)e^{-ik_\mu z} \tag{1.215}$$

where $\mu = 1, \ldots, a$. The constant parameters C_i, C_i' depend on ω_μ, k_μ, $\underset{\smile}{\varkappa}^{(1)}(\omega_\mu; \omega_\mu) = \varepsilon(\omega_\mu) - 1$, and the derivatives of $\underset{\smile}{\varkappa}^{(1)}(\omega, \omega)$ or $\varepsilon(\omega)$. They are (cf. Ref. 38, Vol. I)

$$C_1 = k_\mu^2 - \frac{1}{c^2}\omega_\mu^2\varepsilon(-\omega_\mu)\ (=0),$$

$$C_2 = -i2k_\mu,$$

$$C_3 = -1,$$

$$C_4 = -i\frac{1}{c^2}\omega_\mu\left[2\varepsilon(-\omega_\mu) + \omega_\mu\left(\frac{d}{d\omega}\varepsilon(-\omega)\right)_{\omega_\mu}\right],$$

$$C_5 = \frac{1}{c^2}\left[\varepsilon(-\omega_\mu) + 2\omega_\mu\left(\frac{d}{d\omega}\varepsilon(-\omega)\right)_{\omega_\mu} + \tfrac{1}{2}\omega_\mu^2\left(\frac{d^2}{d\omega^2}\varepsilon\right)_{-\omega_\mu}\right],$$

$$C_1' = \mu_0\omega_\mu^2,$$

$$C_2' = i\mu_0 2\omega_\mu,$$

$$C_3' = -\mu_0. \tag{1.216}$$

If the wave number k_μ is chosen to satisfy the complex dispersion relation of linear optics,

$$k_\mu^2 = \frac{1}{c^2}\omega_\mu^2\varepsilon(-\omega_\mu), \tag{1.217}$$

then C_1 vanishes as indicated in the first line of (1.216). Conversely, sometimes one relates k_μ only to the real part of the linear permittivity

$$k_\mu^2 = \frac{1}{c^2}\omega_\mu^2 \varepsilon^{(r)}(-\omega_\mu). \tag{1.218a}$$

Then using (1.218) in (1.216), C_1 is seen to be given by the imaginary part $\varepsilon^{(i)}(-\omega_\mu)$ of the linear permittivity as

$$C_1 = -i\frac{1}{c^2}\omega_\mu^2 \varepsilon^{(i)}(-\omega_\mu). \tag{1.218b}$$

The solution of the set of coupled partial differential equations is a very difficult mathematical task. Depending on the real physical situation the neglect of specific terms in these equations may be justified. This facilitates solving them. To justify such neglect we use the same arguments as in Section 1.4.2. In particular we take an approach that is called the *slowly-varying-envelope approximation* (SVEA) or the *slowly-varying-amplitude approximation* (SVAA). The equations for the wave amplitudes are Fourier-transformed with respect to time and space coordinates. Since the envelopes (or amplitudes) are slowly varying functions, their Fourier transforms are restricted by (1.195) and (1.214) to the small wave-number and frequency intervals δk_μ and $\delta\omega_\mu$. The partial differential equation (1.215) goes over into a power series in $\delta k_\mu / k_\mu$ and $\delta\omega_\mu/\omega_\mu$ where it is often possible to neglect higher-order terms. In many applications it is justified to neglect for instance the third term $C_3(\partial^2/\partial z^2)\overline{E}$ of (1.215) in comparison with the second one $C_2(\partial/\partial z)\overline{E}$, their ratio being of the order of $\delta k_\mu/k_\mu$. By analogous arguments the fifth term $C_5(\partial^2/\partial t^2)\overline{E}$ gives a negligible contribution in (1.215) in comparison with the fourth term $C_4(\partial/\partial t)\overline{E}$, the ratio of the two terms being of the order of $\delta\omega_\mu/\omega_\mu$. On the right-hand side of (1.215) the main influence results from the first term $C_1'\widehat{P}^{\mathrm{NL}}$, and therefore the other two terms can often be ignored. The remaining equation reads

$$\left\{-i2k_\mu\frac{\partial}{\partial z} - i\frac{1}{c^2}\omega_\mu\left[2\varepsilon(-\omega_\mu) + \omega_\mu\left(\frac{d}{d\omega}\varepsilon(-\omega)\right)_{\omega_\mu}\right]\frac{\partial}{\partial t}\right\}\overline{E}(\omega_\mu; t, z)$$

$$= \mu_0\omega_\mu^2 \widehat{P}^{\mathrm{NL}}(\omega_\mu; t, z)e^{-ik_\mu z}. \tag{1.219}$$

If the linear permittivity ε is real, one obtains

$$\left[\frac{\partial}{\partial z} + \frac{1}{v_\mu}\frac{\partial}{\partial t}\right]\overline{E}(\omega_\mu; t, z) = \frac{i\mu_0\omega_\mu^2}{2k_\mu}\widehat{P}^{\mathrm{NL}}(\omega_\mu; t, z)e^{-ik_\mu z}, \tag{1.220}$$

where the group velocity v_μ is given by

$$\frac{1}{v_\mu} = \frac{\omega_\mu}{2c^2 k_\mu}\left[2\varepsilon(-\omega_\mu) + \omega_\mu\left(\frac{d}{d\omega}\varepsilon(-\omega)\right)_{\omega_\mu}\right]. \tag{1.221}$$

Sometimes it is easier to solve this equation with the help of the new coordinates

$$\eta = t - \frac{1}{v_\mu}z, \qquad \xi = z. \tag{1.222}$$

The derivative of a function with respect to the so-called *retarded time* or *local time*, which is measured in a coordinate frame moving with the light pulse at velocity v_μ, represents its time evolution in that coordinate system. Applying these new coordinates, the relation (1.220) leads to

$$\frac{\partial}{\partial\xi}\bar{E}(\omega_\mu;\eta,\xi) = \frac{i\mu_0\omega_\mu^2}{2k_\mu}\widehat{P}^{\mathrm{NL}}(\omega_\mu;\eta,\xi)e^{-ik_\mu\xi}. \tag{1.223}$$

Similar manipulations can be performed for anisotropic media and arbitrary angles between $k_{\mu\bullet}$ and z. Under the conditions described in Section 1.4.2 (cf. Fig. 1.11) we obtain

$$\left\{\cos\alpha_\mu\cos(\alpha_\mu - \beta_\mu)\frac{\partial}{\partial z} + \frac{1}{v_\mu}\frac{\partial}{\partial t}\right\}\bar{E}(\omega_\mu;t,z) = \frac{i\mu_0\omega_\mu^2}{2k_\mu}\widehat{P}^{\mathrm{NL}}(\omega_\mu;t,z)e^{-ik_\mu z} \tag{1.224a}$$

in place of (1.220).

In deriving (1.220) and (1.224) we did not make use of the nonlinear properties of P^{NL}. Hence we may separate the total polarization in a different way and have only to reinterpret these equations. If there is a one-photon transition near ω_μ, it is of advantage to separate out the nonresonant part $\widehat{P}^{\mathrm{LNR}}$ of the linear polarization around ω_μ, rather than $\widehat{P}^{\mathrm{L}}$ as a whole. Thus we have

$$\widehat{P}(\omega_\mu;t,z) = \widehat{P}^{\mathrm{LNR}}(\omega_\mu;t,z) + \widehat{P}'(\omega_\mu;t,z), \tag{1.224b}$$

where $\widehat{P}'(\omega_\mu;t,z) = \widehat{P}^{\mathrm{LR}}(\omega_\mu,t,z) + \widehat{P}^{\mathrm{NL}}(\omega_\mu;t,z)$. The reinterpretation of the equations given above now means that we have to replace $\widehat{P}^{\mathrm{NL}}$ by $\widehat{P}'$, and the optical parameters $\varepsilon(-\omega_\mu)$, $n(\omega_\mu)$, and $v(\omega_\mu)$ by the values they take on with neglect of the resonant transition at ω_μ or near ω_μ. As an example we consider an atomic ensemble that exhibits only one transition near resonance with $\omega_{10} \approx \omega_\mu$, whereas all other transitions are far from resonance. Thus we may rewrite (1.224) and obtain

$$\left[\cos\alpha_\mu \cos(\alpha_\mu - \beta_\mu)\frac{\partial}{\partial z} + \frac{1}{v_\mu}\frac{\partial}{\partial t}\right]\bar{E}(\omega_\mu, t, z) = \frac{i\mu_0\omega_\mu^2}{2k_\mu}\hat{P}'(\omega_\mu, t, z)e^{-ik_\mu z}. \tag{1.224c}$$

Here the parameters k_μ and v_μ are only given by the nonresonant transitions, whereas $\hat{P}'$ results from the resonant transition $1 \leftrightarrow 0$.

1.4.5. Interaction Processes in Resonators

In Section 1.1 we have treated electromagnetic fields within empty resonators. The field was expanded in terms of the resonator modes according to

$$E_\bullet(r_\bullet, t) = \sum_\mu E_{\mu\bullet}(r_\bullet, t), \tag{1.225}$$

where $E_{\mu\bullet}(r_\bullet, t) = -(1/\sqrt{\varepsilon_0})p_\mu(t)E_{\mu\bullet}(r_\bullet)$. The spatial mode functions $E_{\mu\bullet}(r_\bullet)$ obey the time-independent wave equation of the empty resonator (1.8) and the orthonormality relations (1.9). Maxwell's equations within the empty resonator and the corresponding time-dependent wave equation will be satisfied by the expansion chosen if the time-dependent expansion coefficients $p_\mu(t)$ obey the oscillator equations (1.6) and (1.7).

The spatial modes $E_{\mu\bullet}(r_\bullet)$ form a complete set of functions. Thus arbitrary fields can be expanded in these modes at any time. We are going to perform such an expansion for the electromagnetic field within the resonator filled with matter acting on the field through the polarization $P_\bullet(r_\bullet, t)$. The task consists in replacing (1.6) and (1.7) by adequate equations of motion for the time-dependent expansion coefficients $p_\mu(t)$ and then solving these equations. For this purpose we insert the mode expansion (1.225) into the wave equation (1.185) and obtain

$$\sum_\mu \left\{-\frac{1}{\sqrt{\varepsilon_0}}p_\mu(t)\nabla_\bullet \times [\nabla_\bullet \times E_{\mu\bullet}(r_\bullet)] - \frac{1}{c^2\sqrt{\varepsilon_0}}\ddot{p}_\mu(t)E_{\mu\bullet}(r_\bullet)\right\} = -\mu_0\frac{\partial^2}{\partial t^2}P_\bullet(r_\bullet, t). \tag{1.226}$$

We take into account that according to (1.8) $E_{\mu\bullet}(r_\bullet)$ is an eigenfunction of the operator $\nabla_{\bullet} \times \nabla_{\bullet} \times$ with the eigenvalue ω_μ^2/c^2. Then multiplication of (1.226) by $-\sqrt{\varepsilon_0}\, c^2 E_{\nu\bullet}(r_\bullet)$ and integration over the whole space of the resonator leads to

$$\ddot{p}_\nu(t) + \omega_\nu^2 p_\nu(t) = \frac{1}{\sqrt{\varepsilon_0}} \int dV\, E_{\nu\bullet}(r_\bullet) \frac{\partial^2}{\partial t^2} P_\bullet(r_\bullet, t). \qquad (1.227a)$$

Here the orthonormality relation (1.9) has been used. According to this equation the expansion coefficients $p_\nu(t)$ evolve under the action of the linear and nonlinear polarization of the material.

As it was explained in Section 1.1.3, losses from the resonator due to imperfect reflectivity of the mirrors or due to diffraction can be taken into account approximately by a loss term $(2/T_A)\dot{p}_\nu$ on the left-hand side of (1.227a), where $T_A = 2T_{\mathscr{E}}$ and $T_{\mathscr{E}}$ are the relaxation time of the field amplitude and that of the field energy of the empty resonator, respectively. Thus

$$\ddot{p}_\nu(t) + \frac{2}{T_A}\dot{p}_\nu(t) + \omega_\nu^2 p_\nu(t) = \frac{1}{\sqrt{\varepsilon_0}} \int dV\, E_{\nu\bullet}(r_\bullet) \frac{\partial^2}{\partial t^2} P_\bullet(r_\bullet, t). \qquad (1.227b)$$

It should be mentioned that an equation of the same form is obtained for a cavity surrounded by ideally reflecting and closed walls but filled with a lossy material characterized by its conductivity. In this case the telegraph equation rather than the wave equation has to be used. Thus both kinds of losses can be described by (1.227b) if the parameter $1/T_A$ represents all the attenuation processes.

Sometimes it is useful to expand the field in other modes which, starting with modes of the empty resonator, have been modified by some part of the polarization. The aim is to get modes exhibiting a higher stability under the action of the remaining part of the polarization. Let us give a simple example. Suppose the material to be characterized by a real and dispersion-free linear optical permittivity $\varepsilon_{\bullet\bullet}$ in the frequency region of interest. Then the modified radiation modes of the resonator can be defined by

$$\nabla_\bullet \times \nabla_\bullet \times E_{\mu\bullet}(r_\bullet) = \frac{\omega_\mu}{c^2} \varepsilon_{\bullet\bullet} E_{\mu\bullet}(r_\bullet) \qquad (1.228a)$$

in conjunction with the orthonormality condition

$$\int dV\, E_{\nu\bullet}(r_\bullet) \varepsilon_{\bullet\bullet} E_{\mu\bullet}(r_\bullet) = \delta_{\mu\nu}. \qquad (1.228b)$$

Inserting the expansion of $E_\bullet(r_\bullet, t)$ with respect to these modes into the wave equation, one obtains in the same way as above an equation of motion for the time-dependent expansion coefficients $p_\nu(t)$, where only the nonlinear part of the polarization $P^{NL}(r_\bullet, t)$ acts as a source term. Taking into account resonator losses, we have

$$\ddot{p}_\nu(t) + \frac{2}{T_A}\dot{p}_\nu(t) + \omega_\nu^2 p_\nu(t) = \frac{1}{\sqrt{\varepsilon_0}} \int dV E_{\nu\bullet}(r_\bullet) \frac{\partial^2}{\partial t^2} P_\bullet^{NL}(r_\bullet, t), \quad (1.229)$$

which replaces (1.227b).

Since the part of $P_\bullet$ to be considered in the mode equation can be arbitrarily chosen, one may use another way of splitting $P_\bullet$, which is very useful for materials with linear optical losses and dispersion in certain frequency regions. Then instead of $P_\bullet^{NL}(r_\bullet, t)$ the polarization $P_\bullet'(r_\bullet, t) = P_\bullet(r, t) - \varepsilon_0(\varepsilon_{\bullet\bullet} - 1)E_\bullet(r_\bullet, t)$, which contains $P_\bullet^{NL}(r_\bullet, t)$ as well as $[P_\bullet^{L}(r_\bullet, t) - \varepsilon_0(\varepsilon_{\bullet\bullet} - 1)E_\bullet(r_\bullet, t)]$, has to be inserted on the right-hand side of (1.229), where, as before, $\varepsilon_{\bullet\bullet}$ is the real and dispersion-free part of the linear permittivity.

1.4.6. Excitation of Plane-Wave Modes

In Section 1.4.5 the influence of the polarization on the excitation of resonator modes was discussed. We shall now treat the analogous problem for plane-wave modes of the radiation field. In free space the radiation field has been represented by plane-wave modes in (1.20) together with (1.23). We rewrite these relations in the form

$$E_\bullet(r_\bullet, t) = \sum_\mu E_{\mu\bullet}(r_\bullet, t) = \sum_\mu \left(E_{\mu\bullet}^{(-)}(r_\bullet, t) + E_{\mu\bullet}^{(+)}(r_\bullet, t) \right), \quad (1.230a)$$

where

$$E_{\mu\bullet}^{(-)}(r_\bullet, t) = a_\mu(t) E_{\mu\bullet}(r_\bullet), \quad (1.230b)$$

$$a_\mu(t) = \hat{a}_\mu e^{-i\omega_\mu t}, \quad (1.230c)$$

$$E_{\mu\bullet}(r_\bullet) = i e_{\mu\bullet} \left(\frac{\hbar\omega}{2\varepsilon_0 V} \right) e^{ik_{\mu\bullet} r_\bullet}, \quad (1.230d)$$

and

$$2\varepsilon_0 \int dV E_{\mu\bullet}(r_\bullet) E_{\mu\bullet}^*(r_\bullet) = \hbar\omega. \quad (1.230e)$$

The mode functions $E_{\mu\bullet}(r_\bullet)$ have to obey the time-independent wave equation, and the relation

$$\dot{a}_\mu(t) = -i\omega_\mu a_\mu(t) \quad (1.231)$$

results from the Maxwell equations for the time-dependent expansion coefficients. The contribution of the μth mode to the field energy is given by (1.24).

Let us now define plane-wave modes in an infinitely extended *loss-free linear optical material* characterized by its dielectric permittivity $\varepsilon_{\bullet\bullet}(\omega)$ and magnetic permeability of value 1. According to Ref. 1.11, the contribution of the μth mode to the field energy is then given by

$$H_{\mu}^{\mathrm{L}} = \tfrac{1}{2}\varepsilon_0 \int dV \left\{ \left[\left(\frac{d}{d\omega}[\omega\varepsilon_{\bullet\bullet}(-\omega)] \right)_{\omega_\mu} + \varepsilon_{\bullet\bullet}(-\omega_\mu) \right] \overline{E_{\mu\bullet}(r_\bullet, t) E_{\mu\bullet}(r_\bullet, t)}^{t} \right\}. \tag{1.232}$$

This formula can be derived by starting from the time-averaged Poynting relation

$$-\overline{\nabla_\bullet S_\bullet}^{t} = \overline{E_\bullet \frac{\partial}{\partial t} D_\bullet}^{t} + \overline{H_\bullet \frac{\partial}{\partial t} B_\bullet}^{t} = \frac{d}{dt}\mathscr{H}^{\mathrm{L}}, \tag{1.233}$$

where $\mathscr{H}^{\mathrm{L}}$ is the density of the field energy in the loss-free, dispersive material. From (1.95a) the time derivative of the slowly varying polarization amplitude $\widehat{P}_{\mu\bullet}^{(-)}(r_\bullet, t)$ near the frequency ω_μ is seen to be

$$\begin{aligned} 2\frac{\partial}{\partial t} P_{\mu\bullet}^{(-)}(r_\bullet, t) = & -i\omega_\mu \varepsilon_0 \underset{\smile}{\varkappa}_{\bullet\bullet}^{(1)}(-\omega_\mu; -\omega_\mu) \widehat{E}_{\mu\bullet}^{(-)}(r_\bullet, t) e^{-i\omega_\mu t} \\ & + \omega_\mu \varepsilon_0 \left[\frac{d}{d\omega} \underset{\smile}{\varkappa}^{(1)}(-\omega, -\omega) \right]_{\omega_\mu} e^{-i\omega_\mu t} \frac{\partial}{\partial t} \widehat{E}_{\mu\bullet}^{(-)}(r_\bullet, t) \\ & + \varepsilon_0 \underset{\smile}{\varkappa}_{\bullet\bullet}^{(1)}(-\omega_\mu; -\omega_\mu) e^{-i\omega_\mu t} \frac{\partial}{\partial t} \widehat{E}_{\mu\bullet}^{(-)}(r_\bullet, t). \end{aligned} \tag{1.234}$$

Multiplication of this equation by $E^{(+)}(r_\bullet, t)$ leads to

$$\begin{aligned} & 4E_{\mu\bullet}^{(+)}(r_\bullet, t) \frac{\partial}{\partial t} P_{\mu\bullet}^{(-)}(r_\bullet, t) \\ & = -i\omega_\mu \varepsilon_0 \underset{\smile}{\varkappa}_{\bullet\bullet}^{(1)}(-\omega_\mu; -\omega_\mu) \widehat{E}_{\mu\bullet}^{(+)}(r_\bullet, t) \widehat{E}_{\mu\bullet}^{(-)}(r_\bullet, t) \\ & \quad + \omega_\mu \varepsilon_0 \left[\frac{d}{d\omega} \underset{\smile}{\varkappa}_{\bullet\bullet}^{(1)}(-\omega; -\omega) \right]_{\omega_\mu} \widehat{E}_{\mu\bullet}^{(+)}(r_\bullet, t) \frac{\partial}{\partial t} \widehat{E}_{\mu\bullet}^{(-)}(r_\bullet, t) \\ & \quad + \varepsilon_0 \underset{\smile}{\varkappa}_{\bullet\bullet}^{(1)}(-\omega_\mu; -\omega_\mu) \widehat{E}_{\mu\bullet}^{(+)}(r_\bullet, t) \frac{\partial}{\partial t} \widehat{E}_{\mu\bullet}^{(-)}(r_\bullet, t). \end{aligned} \tag{1.235}$$

Since a corresponding relation holds for $E_{\mu\bullet}^{(-)}(r_\bullet, t)(\partial/\partial t)P_\bullet^{(+)}(r_\bullet, t)$, we obtain by averaging over a small time interval of the order of $2\pi/\omega_\mu$

$$\overline{E_{\mu\bullet}\frac{\partial}{\partial t}D_{\mu\bullet}}^t = \tfrac{1}{4}\varepsilon_0\left\{\left[\frac{d}{d\omega}(\omega\varepsilon_{\bullet\bullet}(-\omega))\right]_{\omega_\mu}\right\}\frac{\partial}{\partial t}\left\{\hat{E}_{\mu\bullet}^{(+)}(r_\bullet, t)\hat{E}_{\mu\bullet}^{(-)}(r_\bullet, t)\right\}, \tag{1.236}$$

where $\varepsilon_{\bullet\bullet}(-\omega) = \varkappa_{\bullet\bullet}^{(1)}(-\omega; -\omega) + 1$ has been used. Substituting

$$\overline{E_{\mu\bullet}\frac{\partial}{\partial t}D_{\mu\bullet}}^t$$

from (1.236) together with

$$\overline{H_{\mu\bullet}\frac{\partial}{\partial t}B_{\mu\bullet}}^t = \tfrac{1}{4}\varepsilon_0\varepsilon_{\bullet\bullet}(-\omega)\frac{\partial}{\partial t}\left\{\hat{E}_{\mu\bullet}^{(+)}(r_\bullet, t)\hat{E}_{\mu\bullet}^{(-)}(r_\bullet, t)\right\}, \tag{1.237}$$

which is derived in the same way, into (1.233), and integrating with respect to time, we obtain the energy density of the μth mode,

$$\mathscr{H}_\mu^{\mathrm{L}} = \tfrac{1}{2}\varepsilon_0\left\{\left[\frac{d}{d\omega}(\omega\varepsilon_{\bullet\bullet}(-\omega))\right]_{\omega_\mu} + \varepsilon_{\bullet\bullet}(-\omega_\mu)\right\}\overline{E_{\mu\bullet}(r_\bullet, t)E_{\mu\bullet}(r_\bullet, t)}^t. \tag{1.238}$$

The integration over the mode volume V leads to (1.232).

To obtain for H_μ^{L} the same expression

$$H_\mu^{\mathrm{L}} = \hbar\omega_\mu a_\mu^* a_\mu \tag{1.239}$$

as given in (1.24), the mode functions $E_{\mu\bullet}(r_\bullet)$ have to obey the normalization condition

$$\varepsilon_0\int dV\left\{\left[\frac{\partial}{\partial\omega}\omega\varepsilon_{\bullet\bullet}(-\omega)\right]_{\omega_\mu} + \varepsilon_{\bullet\bullet}(-\omega_\mu)\right\}E_{\mu\bullet}^*(r_\bullet)E_{\mu\bullet}(r) = \hbar\omega_\mu. \tag{1.240}$$

Besides, the mode functions have to satisfy the homogeneous, time-independent wave equation, which provides—with $E_{\mu\bullet}(r_\bullet) = e_{\mu\bullet}\overline{E}_\mu e^{ik_{\mu\bullet}r_\bullet}$ and (1.198)

—the relation

$$k_{\mu\bullet}(k_{\mu\bullet}e_{\mu\bullet}) + \frac{\omega_\mu^2}{c^2}\left[\varepsilon_{\bullet\bullet}(-\omega_\mu)e_{\mu\bullet} - k_\mu^2 e_{\mu\bullet}\right] = 0. \tag{1.241}$$

As in Section 1.4.5, we substitute here the mode expansion of the electric field into the wave equation of nonlinear optics, (1.185). The resulting relation is multiplied by $E_{\nu\bullet}^*(r_\bullet)$ and integrated over the mode volume V. Thus

$$\dot{a}_\nu(t) = -i\omega_\nu a_\nu(t) - \frac{i}{\hbar\omega_\mu^2}\int dV\, E_{\nu\bullet}^*(r_\bullet)\frac{\partial^2}{\partial t^2}P_\bullet^{\mathrm{NL}}(r_\bullet, t) \tag{1.242}$$

is obtained instead of (1.231). By means of this equation the mode excitation arising from the nonlinear polarization can be calculated. Only the part of $P_\bullet^{\mathrm{NL}}(r_\bullet, t)$ within a small frequency range around ω_ν influences the expansion coefficient $a_\nu(t)$ appreciably. With $P_\bullet^{\mathrm{NL}}(r_\bullet, t) = \Sigma_\mu P_{\mu\bullet}^{(-)}(r_\bullet, t)$ + c.c., the relation

$$\dot{a}_\nu(t) = -i\omega_\nu a_\nu(t) + \frac{i}{\hbar}\int dV\, E_{\nu\bullet}^*(r_\bullet)P_{\nu\bullet}^{(-)}(r_\bullet, t) \tag{1.243}$$

is obtained from (1.242).

With regard to the interaction of quantized fields with matter (cf. Chapter 2) it is of importance that the equation of motion for $a_\nu(t)$ can be written in the form

$$\dot{a}_\nu(t) = -\{a_\nu(t), H\} \tag{1.244}$$

with

$$H = H^{\mathrm{L}} + H^{\mathrm{NL}}$$

where H^{L} and H^{NL} represent the field energy in linear approximation as given in (1.232), and the part of free energy caused by nonlinear polarization, respectively. The braces { } in (1.244) mean the Poisson brackets with respect to the generalized, canonically conjugate coordinates q_ν, p_ν defined by

$$q_\nu(t) = \sqrt{\frac{\hbar}{2\omega_\nu}}\left[a_\nu(t) + a_\nu^*(t)\right],$$

$$p_\nu(t) = i\sqrt{\frac{\hbar\omega_\nu}{2}}\left[a_\nu^*(t) - a_\nu(t)\right]. \tag{1.245}$$

The calculation of the Poisson bracket $\{a_\nu(t), H^{\mathrm{L}}\}$ for instance leads to

$$\begin{aligned}\{a_\nu, H^{\mathrm{L}}\} &= \frac{\partial a_\nu}{\partial p_\nu}\frac{\partial H^{\mathrm{L}}}{\partial q_\nu} - \frac{\partial a_\nu}{\partial q_\nu}\frac{\partial H^{\mathrm{L}}}{\partial p_\nu} \\ &= \frac{\partial a_\nu}{\partial p_\nu}\left(\frac{\partial H^{\mathrm{L}}}{\partial a_\nu}\frac{\partial a_\nu}{\partial q_\nu} + \frac{\partial H^{\mathrm{L}}}{\partial a_\nu^*}\frac{\partial a_\nu^*}{\partial q_\nu}\right) \\ &\quad - \frac{\partial a_\nu}{\partial q_\nu}\left(\frac{\partial H^{\mathrm{L}}}{\partial a_\nu}\frac{\partial a_\nu}{\partial p_\nu} + \frac{\partial H^{\mathrm{L}}}{\partial a_\nu^*}\frac{\partial a_\nu^*}{\partial p_\nu}\right) \\ &= \frac{\partial H^{\mathrm{L}}}{\partial a_\nu^*}\{a_\nu, a_\nu^*\} \end{aligned} \tag{1.246}$$

and, with

$$\frac{\partial H^{\mathrm{L}}}{\partial a_\nu^*} = \hbar\omega_\nu a_\nu$$

and

$$\{a_\nu, a_\nu^*\} = \frac{i}{\hbar},$$

to the relation

$$\dot{a}_\nu = -\{a_\nu, H^{\mathrm{L}}\} = -i\omega_\nu a_\nu \tag{1.247}$$

with neglect of the nonlinear term.

The term H^{NL} has to be calculated by utilizing the variation of the free energy

$$\delta F = -\int dV\, \overline{P_\bullet^{\mathrm{NL}}\, \delta E_\bullet}^{\,t} \tag{1.248}$$

introduced in Section 1.3.2. If the functional dependence of $P_\bullet^{\mathrm{NL}}$ on $E_\bullet$ is known, the free energy can be obtained explicitly. For our purposes it is sufficient to know that the free energy depends on the mode excitations a_ν, a_ν^*:

$$H^{\mathrm{NL}} = F = F(a_1, \ldots, a_\nu, \ldots; a_1^*, \ldots, a_\nu^*, \ldots).$$

Substitution of

$$\{a_\nu, H^{\mathrm{NL}}\} = \frac{\partial F}{\partial a_\nu^*}\{a_\nu, a_\nu^*\}$$

$$= \frac{i}{\hbar}\frac{\partial F}{\partial a_\nu^*} = \frac{i}{\hbar}\int dV E_{\nu\bullet}^*(r_\bullet) P_{\nu\bullet}^{(-)}(r_\bullet, t) \qquad (1.249)$$

into (1.244) provides the equation of motion (1.243).

REFERENCES

1.1. H. Kogelnik and T. Li, *Appl. Optics* **5**, 1550 (1966).

1.2. K. J. T. Böttcher, *Theory of Electric Polarisation*, Vol. 1, 2nd ed., Elsevier, Amsterdam, 1973, and *Dielectrics in Time Dependent Fields*, Vol. 2, 1st ed., Elsevier, Amsterdam, 1978.

1.3. H. Fröhlich, *Theory of Dielectrics*, 2nd ed., Clarendon, Oxford, 1958.

1.4. A. S. Pine, *Phys. Rev. A* **139**, 901 (1965).

1.5. A. Wünsche, *Exp. Techn. Phys.* **23**, 223 (1975).

1.6. W. Fuller Brown, Jr., "Dielectrics," in S. Flügge (Ed.), *Handbuch der Physik*, Vol. 17, Springer, Berlin (West), Göttingen, Heidelberg, 1956.

1.7. V. M. Agranovich and V. L. Ginzburg, *Optics of Crystals with Spatial Dispersion and Exciton Theory*, Nauka, Moscow, 1979 (Russian).

1.8. J. A. Armstrong, N. Bloembergen, J. Ducuing, and P. S. Pershan, *Phys. Rev.* **127**, 1918 (1962).

1.9. D. A. Kleinman, *Phys. Rev.* **126**, 1977 (1962).

1.10. J. M. Manley and H. E. Rowe, *Proc. IRE* **47**, 2115 (1959).

1.11. L. D. Landau and E. M. Lifshits, *Electrodynamics of Continuous Media* (Course of Theoretical Physics, Vol. 8), Addison-Wesley, New York, 1961.

2

The Quantized Free Radiation Field

Owing to mutual interaction between matter and radiation, in linear and nonlinear optical processes the incident electromagnetic radiation undergoes an alteration of its properties or electromagnetic radiation with different properties is generated. A complete and adequate explanation of these basic effects requires the description of the radiation field by quantum-theoretical means, although the classical treatment is capable of describing important effects and of providing their interpretation to a certain extent, as we have seen in Chapter 1. Therefore, in addition to quantities employed in the classical treatment (amplitude, intensity, distribution of partial waves over the frequencies), the quantum state of the radiation field is needed, which involves information about the particles of the radiation field, the photons. It is just these elementary particles whose interaction with the particles (and quasiparticles) of the atomic system underlies the physical process to be discussed. In this chapter we shall describe the free radiation field, which is the field in the absence of charges and conduction currents. After this, we shall continue in Chapter 3 with general aspects of the interaction of the radiation field with atomic systems.

The procedure for the quantization of the radiation field will be discussed in Section 2.1; the general properties of linearly polarized photons and photons with fixed angular momentum and parity are described in Section 2.2. This is followed in Section 2.3 by a discussion of the properties of important field states, where various pure and mixed states are taken into consideration as well as the measurable values of representative observables.

In order to avoid interruptions in our treatment of the physical features of the radiation field that might arise from discussions of general quantum-theoretical laws and terms, we shall refer to the compilation and interpretation of basic quantum-theoretical relations, quantities, and concepts in Appendix A. There, readers who are not familiar with this subject may read it as a whole and/or may use the particular references to Appendix A given in Chapter 2 (as well as in the following chapters).

The measurement of the properties of the quantized radiation field is performed with one or more radiation detectors. To determine the frequency-dependent intensity, radiation receivers placed behind a spectral–dispersive device are used; the detectors have to be characterized by the dependence of the output signal (temperature, photocurrent, photovoltage) on the time-averaged number of incident photons (the radiation intensity). According to the requirements of spectral sensitivity, time resolution, and signal-to-noise ratio, one employs different types of detectors, mainly thermocouples, bolometers, pyroelectric receivers, and receivers operating on the basis of the internal or external photoeffect.

To determine experimentally such fundamental features of the radiation field as photon-statistical and coherence properties, detectors based on the external photoeffect (photoelectric cell, photomultiplier) offer advantages in application, for instance because of their small time constants and their high detectivity. Moreover, from the theoretical point of view, those detectors show a direct connection between the properties of the radiation field to be measured and the output signal, so that a probability relation between the absorption of photons and the emission of photoelectrons can be given. This relation is the basis of the description of photon counting in macrophysical time intervals. In addition, by using several receivers of this type, the measurement of intensity correlation becomes possible. We shall discuss the operating mode of these detectors and their application to the measurement of photon-statistical and coherence properties in Chapter 5, after having treated in Chapter 3 the principles of interaction between radiation and matter (this will enable us to describe the detector atoms interacting with radiation).

2.1. THE QUANTIZATION OF THE FREE RADIATION FIELD

On the one hand, an approach directly based on general field-theoretical aspects [cf. Section A.1.6] will be considered; on the other hand, we shall deal with an approach based on the expansion of the field in plane waves, which is associated with the introduction of creation and annihilation operators for photons.

2.1.1. General Procedure

The free radiation field under consideration in this chapter is an electromagnetic field in the absence of charges and conduction currents. For the sake of preparation for the quantization of this field, let us first display certain aspects of its classical description.

The theory is founded on the system of Maxwell's equations as described in Section 1.1.1. The relations

$$\frac{\partial D_\bullet}{\partial t} = \nabla_\bullet \times H_\bullet, \qquad \frac{\partial B_\bullet}{\partial t} = -\nabla_\bullet \times E_\bullet \tag{2.1}$$

play the role of equations of motion, whereas the relations

$$\nabla_\bullet D_\bullet = 0, \qquad \nabla_\bullet B_\bullet = 0 \tag{2.2}$$

must be regarded as constraints fixing the longitudinal parts of the electric displacement $D_\bullet$ and of the magnetic induction $B_\bullet$. It should be noted that generally the longitudinal and the transverse parts of an arbitrary vector field $C_\bullet$ are defined by the decomposition

$$C_\bullet = C_\bullet^{(\mathrm{lo})} + C_\bullet^{(\mathrm{tr})}, \tag{2.3a}$$

where the longitudinal (irrotational) part $C_\bullet^{(\mathrm{lo})}$ has the properties

$$\nabla_\bullet \times C_\bullet^{(\mathrm{lo})} = 0, \qquad \nabla_\bullet C^{(\mathrm{lo})} = \nabla_\bullet C_\bullet \tag{2.3b}$$

and the transverse (divergenceless) part $C_\bullet^{(\mathrm{tr})}$ has the properties

$$\nabla_\bullet C_\bullet^{(\mathrm{tr})} = 0, \qquad \nabla_\bullet \times C_\bullet^{(\mathrm{tr})} = \nabla_\bullet \times C_\bullet. \tag{2.3c}$$

Under the assumption of square-integrable fields $C_\bullet^{(\mathrm{lo})}$ and $C_\bullet^{(\mathrm{tr})}$, the decomposition (2.3a) is unique.

For subsequent considerations the following statement is of importance: it is possible to formulate the Maxwell theory of the free radiation field with the help of a single, purely transverse vector field. This field, the so-called *vector potential* $A_\bullet(r_\bullet, t)$, obeys the wave equation

$$\nabla_\bullet^2 A_\bullet - \frac{1}{c^2}\frac{\partial^2 A_\bullet}{\partial t^2} = 0 \tag{2.4a}$$

and the equation

$$\nabla_\bullet A_\bullet = 0. \tag{2.4b}$$

The latter relation presents a gauge condition (the so-called *Coulomb* or *radiation gauge*) and guarantees the transversality of the field $A_\bullet$. The vector potential and the field quantities are determined by the relations

$$\nabla_\bullet \times A_\bullet = B_\bullet = \mu_0 H_\bullet \quad \text{and} \quad -\frac{\partial A_\bullet}{\partial t} = E_\bullet = \varepsilon_0^{-1} D_\bullet, \tag{2.5}$$

and hence the equations (2.1) and (2.2) are automatically satisfied. This can be easily shown by applying general rules of vector analysis [as e.g., $\nabla_\bullet(\nabla_\bullet \times A_\bullet) = 0$, $\nabla_\bullet \times (\nabla_\bullet \times A_\bullet) = \nabla_\bullet(\nabla_\bullet A_\bullet) - \nabla_\bullet^2 A_\bullet$)].

Now we must find the quantities that determine the *dynamical state of the free radiation field* at every instant. According to the canonical formalism of the field theory, these quantities are the field $A_\bullet$ and its canonically conjugate field, which can be calculated from the Lagrangian density [cf. (A.84)]. The field $A_\bullet$ has to be assigned the Lagrangian density

$$\tilde{\mathscr{L}} = \frac{\varepsilon_0}{2}\left(\frac{\partial A_\bullet}{\partial t}\right)^2 - \frac{1}{2\mu_0}(\nabla_\bullet \times A_\bullet)^2. \tag{2.6}$$

By doing so, one obtains the true equation of motion (2.4a) for $A_\bullet$, which can be easily verified by applying the general Lagrangian equation [cf. (A.85)]. From the Lagrangian density the lth vector component of the canonically conjugate field turns out to be

$$\frac{\partial \tilde{\mathscr{L}}}{\partial(\partial A_l/\partial t)} = \varepsilon_0 \frac{\partial}{\partial t} A_l. \tag{2.7}$$

Thus the canonically conjugate field of $A_\bullet$ according to (2.5) is identical with $-D_\bullet$.

By means of the field $A_\bullet$, its canonically conjugate field $-D_\bullet$, and the Lagrangian density $\tilde{\mathscr{L}}$, further important quantities can be formed with the help of the field-theoretical formalism [cf. (A.91)–(A.93)]. By spatial integration of the Hamiltonian density the *Hamiltonian function*

$$H[A_\bullet, -D_\bullet] = \int d^3r_\bullet \left[\frac{1}{2\varepsilon_0} D_\bullet^2 + \frac{1}{2\mu_0}(\nabla_\bullet \times A_\bullet)^2\right]$$

$$= \int d^3r_\bullet \left[\frac{\varepsilon_0}{2} E_\bullet^2 + \frac{\mu_0}{2} H^2.\right] \tag{2.8a}$$

is obtained. Furthermore, the *linear momentum* of the field can be represented by

$$G_\bullet[A_\bullet, -D_\bullet] = \int d^3r_\bullet\, [D_\bullet \times (\nabla_\bullet \times A_\bullet)]$$

$$= \int d^3r_\bullet\, c^{-2}[E_\bullet \times H_\bullet]. \tag{2.8b}$$

The density of the linear momentum is given by $[E_\bullet \times H_\bullet]/c^2$, and the density of the angular momentum by $r_\bullet \times [E_\bullet \times H_\bullet]/c^2$. Thus the angular momentum of the field becomes

$$J_\bullet = \int d^3r_\bullet \frac{r_\bullet \times (E_\bullet \times H_\bullet)}{c^2}. \tag{2.8c}$$

The use of the position vector for forming $J_\bullet$ indicates that this quantity depends on the choice of the point of reference (origin).

Taking into account the spatial properties, in particular the transversality of field $A_\bullet$, we may conclude from (2.8c) that $J_\bullet$, in accordance with

$$J_\bullet = J_\bullet^{(\text{or})} + J_\bullet^{(\text{in})}, \tag{2.8d}$$

can be decomposed into two terms in such a way that one term, namely $J_\bullet^{(\text{or})}$, depends on the point of origin, whereas the other, $J_\bullet^{(\text{in})}$, does not. We are mainly interested in the latter term, which is referred to as the *intrinsic part of the angular momentum*; it is given by

$$J_\bullet^{(\text{in})}[A_\bullet, -D_\bullet] = \int d^3r_\bullet (D_\bullet \times A_\bullet) \tag{2.8e}$$

Let us start the discussion about the *quantization* of the free radiation field with a few remarks on the application of the general formalism of quantum field theory (cf. Section A.1.6). After this we shall treat the whole problem in detail on the basis of operators that are associated directly with the existence of light particles in the quantized field; this procedure has the advantage of being in close connection with the usual experimental conditions.

Passing from the classical field theory to the quantum field theory, the field vector $A_\bullet$ and its canonically conjugate field vector $-D_\bullet$ must be replaced by the operators $\hat{A}_\bullet(r_\bullet, t)$ and $-\hat{D}_\bullet(r_\bullet, t)$. These operators act on states of the Hilbert space of the free radiation field. (Hilbert-space operators will generally be marked with a caret. Furthermore, we shall use the convention that an operator with explicitly written time-dependence has to be understood in the Heisenberg picture, while an operator written as a time-independent quantity has to be understood in the Schrödinger picture. An exception is the density operator [cf. (A.77)], which carries the dynamical time dependence in the Schrödinger picture.) By using the operators $\hat{A}_\bullet$ and $\hat{D}_\bullet$, the Hamiltonian function $H[A_\bullet, -D_\bullet]$ goes over into the Hamiltonian operator $\hat{H} = H[\hat{A}_\bullet, -\hat{D}_\bullet]$, the momentum $G_\bullet[A_\bullet, -D_\bullet]$ into the momentum operator $\hat{G}_\bullet = G_\bullet[\hat{A}_\bullet, -\hat{D}_\bullet]$, and the intrinsic part of the angular momentum $J^{(\text{in})}[A_\bullet, -D_\bullet]$ into the operator $\hat{J}^{(\text{in})} = J^{(\text{in})}[\hat{A}_\bullet, -\hat{D}_\bullet]$. There hold analogous relations for the operators of the other field quantities. According to (2.5) the operators of the electric field strength, the electric displacement, the magnetic induction, and the magnetic field strength are connected with the field operator $\hat{A}_\bullet(r_\bullet, t)$ in the following way:

$$-\frac{\partial \hat{A}_\bullet(r_\bullet, t)}{\partial t} = \hat{E}_\bullet(r_\bullet, t) = \varepsilon_0^{-1}\hat{D}_\bullet(r_\bullet, t), \tag{2.9a}$$

$$\nabla_\bullet \times \hat{A}_\bullet(r_\bullet, t) = \hat{B}_\bullet(r_\bullet, t) = \mu_0 \hat{H}_\bullet(r_\bullet, t). \tag{2.9b}$$

Essential properties of the quantized free radiation field are determined by the quantization relations for the field and its canonically conjugate field; in the framework of the general formalism of quantum field theory [cf. Section

A.1.6] the quantization is expressed by means of commutation or anticommutation relations. The choice of the appropriate type of relations is connected with the corpuscular interpretation of the quantized field. The particles of the electromagnetic field, the photons, obey *Bose–Einstein statistics*, in agreement with all experimental results, as will be shown in Chapters 5 and 7 and Section 2.3. The boson character of the photons can only be obtained if commutation relations for $\hat{A}_\bullet$ and $-\hat{D}_\bullet$ are used; we shall prove this fact in Sections 2.2.1 and 2.2.2. The commutation relations that must be chosen for the pth and qth vector component of the fields are

$$[\hat{A}_p(r'_\bullet, t), -\hat{D}_q(r''_\bullet, t)] = i\hbar\delta_{pq}^{(\mathrm{tr}3)}(r'_\bullet - r''_\bullet), \tag{2.10a}$$

$$[\hat{A}_p(r'_\bullet, t), \hat{A}_q(r''_\bullet, t)] = \hat{0}, \qquad [-\hat{D}_p(r_\bullet, t), -\hat{D}_q(r''_\bullet, t)] = \hat{0}. \tag{2.10b}$$

The right-hand side of (2.10a) contains the so-called three-dimensional transverse delta function, which will be explained in Section 2.2.2. The need to use this function and not the ordinary three-dimensional delta function $\delta^{(3)}(r'_\bullet - r''_\bullet) = \delta(x' - x'')\,\delta(y' - y'')\,\delta(z' - z'')$ follows from the assumption of the transversality of the field $\hat{A}_\bullet$ and from the corresponding occurrence of only transverse photons as particles of the field $\hat{A}_\bullet$. It may be added that the quantization of a general electromagnetic field (in the presence of charges) leads to the occurrence of longitudinal and scalar photons. These, however, need not be taken into account in radiative processes where the quantized radiation field interacts with charge carriers, whose attractive and repulsive forces are described by a Coulomb field, which can be regarded as an external unquantized field. This assumption will be made throughout this book. Further assumptions of this kind will be discussed when describing the interaction between radiation and matter in Section 3.1.1.

The vector potential $\hat{A}_\bullet$ is of the type of a dynamical field variable and hence obeys the equation of motion

$$\frac{d}{dt}\hat{A}_\bullet(r_\bullet, t) = \frac{1}{i\hbar}[\hat{A}_\bullet(r_\bullet, t), \hat{H}]. \tag{2.11}$$

A deeper insight into the properties of photons and states of the radiation field can be gained by analyzing the transverse field $\hat{A}_\bullet$ by means of a complete set $\{u_{\mu\bullet}(r_\bullet)\}$ of orthonormal transverse fields [cf. (A.103)]. For the photon field the corresponding expansion is given by the formula

$$\hat{A}_\bullet(r_\bullet, t) = \sum_\mu \{\hat{a}_\mu(t)u_{\mu\bullet}(r_\bullet) + \hat{a}^\dagger_\mu(t)u^*_{\mu\bullet}(r_\bullet)\}. \tag{2.12}$$

The index μ represents a certain mode, for instance a mode with a certain linear and/or angular momentum (this will be discussed in detail in Section

2.2). The *operator* character of the field $\hat{A}_\bullet(r_\bullet, t)$ is expressed by the *operators* $\hat{a}_\mu(t)$ and $\hat{a}_\mu^\dagger(t)$ of the right-hand side of (2.12). Later on we shall demonstrate that the operator $\hat{N}_\mu = \hat{a}_\mu^\dagger \hat{a}_\mu$ represents the number of photons in the mode μ and yields the occupation-number representation. With this representation a relatively simple survey of the Hilbert-space properties is possible, which can be used to predict the measurable values of representative observables under various experimental conditions. Within this framework the expansion in plane waves will be treated in detail in Section 2.1.2.

2.1.2. Quantization of the Field Expanded in Plane Waves

Let us first consider the classical field function of the vector potential $A_\bullet$ that satisfies the wave equation (2.4a) and the Coulomb-gauge condition (2.4b). The expansion of $A_\bullet$ in plane traveling waves follows the same pattern as for the field strength $E_\bullet$ which is described in Section 1.1.2.2. The entire space is subdivided into cubes with the volume $V = L^3$, where the nth cube is characterized by the triple (n_x, n_y, n_z) of integers. The periodicity condition for $A_\bullet(r_\bullet, t)$ reads

$$A_\bullet(r_\bullet, t) = A_\bullet(r_\bullet + R_{n\bullet}, t), \tag{2.13}$$

where $R_{n\bullet} = (n_x L, n_y L, n_z L)$. In the following $A_\bullet$ will be considered only in the basic cube ($n_x = n_y = n_z = 0$). The continuation into other periodicity volumes is physically irrelevant. The basic cube may be supposed sufficiently big so that it encloses all experimental arrangements. The superposition of plane traveling waves as fundamental solutions of (2.4a) leads to

$$A_\bullet(r_\bullet, t) = \sum_\mu e_{\mu\bullet} \tilde{a}_\mu e^{i(k_{\mu\bullet} r_\bullet - \omega_\mu t)} + \text{c.c.}, \tag{2.14}$$

where the dispersion relation

$$\omega_\mu^2 = c^2 k_\mu^2 \tag{2.15}$$

holds. The $\tilde{a}_\mu$ are complex numbers that play the role of integration constants. The right-hand side of (2.14) is a superposition of transverse plane traveling waves; from the Coulomb-gauge condition the transversality condition

$$e_{\mu\bullet} \cdot k_{\mu\bullet} = 0 \tag{2.16}$$

for the polarization unit vector $e_{\mu\bullet}$ and the wave number vector $k_{\mu\bullet}$ follows. The mode index characterizes the wave vector $k_{\mu\bullet} = (2\pi l_x/L, 2\pi l_y/L, 2\pi l_z/L)$ with the integers (l_x, l_y, l_z) and one of two polarization directions ($\sigma = 1, 2$) belonging to one wave vector $k_{l\bullet}$. For convenience, $e_{l,\sigma=1\,\bullet}$ will be chosen perpendicular to $e_{l,\sigma=2\,\bullet}$, so that $e_{l1\bullet} \cdot e_{l2\bullet} = 0$. The quantity $\tilde{a}_\mu$ is split into a dimensionless factor a_μ and a factor $(\hbar/2\varepsilon_0 V\omega_\mu)^{1/2}$, which has the same dimension as $A_\bullet$. (The introduction of Planck's constant into $A_\bullet$ will lead to a simpler expression in the quantization to be performed later.) Altogether we have

$$A_\bullet(r_\bullet, t) = \sum_l \sum_{\sigma=1,2} \left(\frac{\hbar}{2\varepsilon_0 V\omega_l}\right)^{1/2} e_{l\sigma\bullet} a_{l\sigma} e^{i(k_{l\bullet} r_\bullet - \omega_l t)} + \text{c.c.}. \tag{2.17}$$

Obviously, the field $A_\bullet$ is represented by summands that are characterized by discrete indices $\mu = (l, \sigma)$. This means that $A_\bullet$, because of the application of the periodicity condition (2.13), can be described at every instant by a countable set of variables (the quantities a_μ). The formulas for the mode number density are the same as in the case of the electric field [cf. (1.26) and (1.27)].

Now we take up the *quantization*; to do this we compare (2.17) with (2.12). The function $u_{\mu\bullet}(r_\bullet)$ is found to be

$$u_{\mu\bullet}(r_\bullet) = \left(\frac{\hbar}{2\varepsilon_0 V\omega_\mu}\right)^{1/2} e_{\mu\bullet} e^{ik_{\mu\bullet} r_\bullet}. \tag{2.18}$$

In passing from the classical quantity $A_\bullet$ to the quantum-theoretical operator $\hat{A}_\bullet$, the classical quantity $a_\mu \exp(-i\omega_\mu t)$ on the right-hand side changes into the operator $\hat{a}_\mu \exp(-i\omega_\mu t) \equiv \hat{a}_\mu(t)$. The fact that $A_\bullet$ is a real field leads in quantization to a hermitian field operator $\hat{A}_\bullet$. Therefore a_μ^* has to be identified with $\hat{a}_\mu^\dagger$. The conjugate complex part (c.c.) in $A_\bullet$ goes over into a hermitian conjugate (h.c.) in $\hat{A}_\bullet$. Altogether we obtain

$$\hat{A}_\bullet(r_\bullet, t) = \sum_\mu \hat{a}_\mu e^{-i\omega_\mu t} u_{\mu\bullet}(r_\bullet) + \text{h.c.}. \tag{2.19}$$

The $u_{\mu\bullet}(r_\bullet)$ satisfy the orthonormality relations

$$\int_V d^3 r_\bullet [u_{l\sigma\bullet}^*(r_\bullet) u_{l'\sigma'\bullet}(r_\bullet)] = \frac{\hbar}{2\varepsilon_0 \omega_l} \delta_{ll'} \delta_{\sigma\sigma'}. \tag{2.19a}$$

Corresponding to our general convention, the operator $\hat{A}_\bullet(r_\bullet, t)$ in (2.19) has to be understood in the Heisenberg picture. It should be mentioned that the field operator $\hat{A}_\bullet(r_\bullet, t)$, exhibits a dependence of the same kind as the classical quantity $A_\bullet(r_\bullet, t)$ does on position and time coordinates in the Minkowski space. Thus a fundamental requirement of special relativity is satisfied. Here, in the free radiation field, the position vector as well as the time coordinate plays the role of a parameter and is not transformed into an operator. Note that also the field operator $\hat{A}_\bullet$ is represented by summands characterized by discrete indices $\mu = (l, \sigma)$.

We have now to turn to the commutation relations for the operators $\hat{a}_\mu$ and $\hat{a}_\mu^\dagger$. Under the assumption that the commutation relations

$$[\hat{a}_\mu, \hat{a}_{\mu'}^\dagger] = \delta_{\mu\mu'}\hat{I}, \tag{2.20a}$$

$$[\hat{a}_\mu, \hat{a}_{\mu'}] = \hat{0}, \qquad [\hat{a}_\mu^\dagger, \hat{a}_{\mu'}^\dagger] = \hat{0} \tag{2.20b}$$

hold for $\hat{a}_\mu$ and $\hat{a}_\mu^\dagger$, the following two essential facts can be derived. First, the so-called particle-number operator $\hat{N}_\mu = \hat{a}_\mu^\dagger \hat{a}_\mu$ leads to the experimentally confirmed *boson* properties of the particles of the radiation field, the photons (see Section 2.2.1). Second, commutation relations constructed by fields $\hat{A}_\bullet$ and $\hat{D}_\bullet$ that are expanded in traveling plane waves [cf. (2.19) and (2.20)] pass over to the general "free-space" commutation relations (2.10) if the length of the normalization cube V becomes infinitely large (see Section 2.2.2). It should be mentioned that it is also possible to derive reversely the relations (2.20) from the general relations (2.10) that contain the fields in the free (infinite) space. However, in this case a Fourier integral instead of the Fourier series (2.19) must be used; thereby one would be deprived of the advantage to represent the field operators with the help of a countable set of terms and to have a finite metric of the Hilbert space.

In what follows the knowledge of the *Hamiltonian* $\hat{H}$ is needed. To achieve this we have to replace $A_\bullet$ and $D_\bullet$ on the right-hand side of (2.8a) by $\hat{A}_\bullet$ and by $\hat{D}_\bullet = -\varepsilon_0\, \partial \hat{A}_\bullet / \partial t$, which can be derived from (2.19). After carrying out the spatial integration over the volume V of the basic region, we obtain, taking into account the orthonormality (2.19a) of $u_{\mu\bullet}(r_\bullet)$, the expression

$$\hat{H} = \sum_\mu \frac{\hbar\omega_\mu}{2}\left(\hat{a}_\mu^\dagger \hat{a}_\mu + \hat{a}_\mu \hat{a}_\mu^\dagger\right).$$

From (2.20a) $\hat{a}_\mu \hat{a}_\mu^\dagger = \hat{a}_\mu^\dagger \hat{a}_\mu + \hat{I}$ follows, and hence we obtain

$$\hat{H} = \sum_\mu \hat{H}_\mu \qquad \text{with} \quad \hat{H}_\mu = \hbar\omega_\mu(\hat{a}_\mu^\dagger \hat{a}_\mu + \tfrac{1}{2}\hat{I}). \tag{2.21}$$

Note that the Hamiltonian of the total free radiation field is a superposition of energy parts $\hat{H}_\mu$ of the single modes; interaction terms between the single modes do not appear. If we additionally take into account that according to (2.20) all operators $\hat{a}_\mu$, $\hat{a}_{\mu'}^\dagger$ belonging to different modes commute, we arrive at the statement that the total field consists of single modes (transverse waves with given wave vectors and polarization directions) independent of each other. Formally, there is an analogy with a problem of mechanics: The operator $\hat{a}$, which represents the complex normal amplitude of the one-dimensional harmonic oscillator of mechanics (cf. Section A.2.5.1) satisfies with $\hat{a}^\dagger$ the same commutation relations as do the operators $\hat{a}_\mu$ and $\hat{a}_\mu^\dagger$ of the radiation field.

Also the form $\hbar\omega(\hat{a}^\dagger\hat{a} + \hat{I}/2)$ of the Hamiltonian operator of the mechanical system is in conformity with $\hat{H}_\mu$. Therefore the single modes of the radiation field are occasionally called *radiation oscillators*.

Let us now calculate the *temporal* dependence of $\hat{a}_\mu(t)$. According to the general equation of motion [cf. (A.70)] and according to (2.21), we have

$$\frac{d\hat{a}_\mu(t)}{dt} = \frac{1}{i\hbar}\left[\hat{a}_\mu(t), \sum_{\mu'} \hbar\omega_{\mu'}\left(\hat{a}^\dagger_{\mu'}(t)\hat{a}_{\mu'}(t) + \frac{\hat{I}}{2}\right)\right]. \tag{2.22}$$

Since according to general rules commutators are picture-invariant, the commutator in (2.22) may be calculated in the Schrödinger picture rather than in the Heisenberg picture. Since $\hat{a}_\mu$ commutes with $\hat{H}_{\mu' \neq \mu}$, we obtain the commutator of (2.22) in the form

$$\begin{aligned}\left[\hat{a}_\mu, \hbar\omega_\mu\left(\hat{a}^\dagger_\mu\hat{a}_\mu + \frac{\hat{I}}{2}\right)\right] &= \hbar\omega_\mu\left(\hat{a}_\mu\hat{a}^\dagger_\mu\hat{a}_\mu - \hat{a}^\dagger_\mu a^2_\mu\right) \\ &= \hbar\omega_\mu\hat{a}_\mu.\end{aligned} \tag{2.23}$$

Returning to the Heisenberg picture we have

$$\frac{d\hat{a}_\mu(t)}{dt} = -i\omega_\mu\hat{a}_\mu(t). \tag{2.24}$$

From this we obtain, with $\hat{a}_\mu(0) = \hat{a}_\mu$, the solution

$$\hat{a}_\mu(t) = \hat{a}_\mu e^{-i\omega_\mu t}. \tag{2.25}$$

Here the convention is used that operators in the Heisenberg picture coincide with those in the Schrödinger picture at $t = 0$. Inspection of (2.19) shows that the time dependence of $\hat{A}_\bullet(r_\bullet, t)$ is carried by the operators $\hat{a}_\mu(t)$ and $\hat{a}^\dagger_\mu(t)$.

Following the same scheme according to which we calculated the Hamiltonian $\hat{H}$ in (2.21), the momentum operator $\hat{G}_\bullet$ can be calculated; this gives

$$\hat{G}_\bullet = \sum_\mu \hbar k_{\mu\bullet}\left(\hat{a}^\dagger_\mu\hat{a}_\mu + \frac{\hat{I}}{2}\right). \tag{2.26}$$

Since the sum $\Sigma_\mu k_{\mu\bullet}$ contains $k_{l\bullet}$ and $k_{-l\bullet}$, with $k_{-l\bullet}$ equal to $-k_{l\bullet}$, the sum $\Sigma_\mu \hbar k_{\mu\bullet}\hat{I}/2$ vanishes.

2.2. THE PHOTON FIELD

A deeper insight into the nature of the particles of the free radiation field, the photons, will be gained from solving the eigenvalue problem of the Hamiltonian $\hat{H}$. In accordance with the remarks on (2.21), we may regard the contributions of the individual modes as independent subspaces of the total Hilbert space. For this reason we start in Section 2.2.1 with the eigenvalue problem of a one-mode field; these results can be used to describe the total system in a rather simple manner.

2.2.1. The One-Mode Field of Linearly Polarized Photons

The eigenvalue problem of one mode with the index μ [see (2.19)–(2.21)] can be solved on the basis of the relations

$$\hat{H}_\mu = \hbar\omega_\mu\left(\hat{a}_\mu^\dagger\hat{a}_\mu + \frac{\hat{I}}{2}\right), \qquad [\hat{a}_\mu, \hat{a}_\mu^\dagger] = \hat{I},$$
$$[\hat{a}_\mu, \hat{a}_\mu] = \hat{0}, \quad [\hat{a}_\mu^\dagger, \hat{a}_\mu^\dagger] = \hat{0}. \tag{2.27}$$

One masters the eigenvalue problem of $\hat{H}_\mu$ if one masters that of the operator

$$\hat{N}_\mu = \hat{a}_\mu^\dagger\hat{a}_\mu = \frac{\hat{H}_\mu}{\hbar\omega_\mu} - \frac{\hat{I}}{2}. \tag{2.28}$$

Obviously $\hat{N}_\mu$ is a hermitian operator, and it may be regarded as an observable, since it coincides with the energy observable apart from an additive constant and a real, constant factor. Moreover, we shall see that a fundamental physical meaning can be assigned to $\hat{N}_\mu$.

The assumption of $\hat{N}_\mu$ to be an observable implies the existence of a system of eigenvectors $\{|n_\mu\rangle\}$ of $\hat{N}_\mu$; they obey the eigenvalue equation

$$\hat{N}_\mu|n_\mu\rangle = n_\mu|n_\mu\rangle, \tag{2.29}$$

where n_μ is the eigenvalue belonging to the eigenstate $|n_\mu\rangle$, which must in general have a positive norm $\langle n_\mu|n_\mu\rangle \equiv \| |n_\mu\rangle\|^2$. The following statement is of fundamental importance: The operator $\hat{N}_\mu$ has a discrete spectrum with the eigenvalues

$$n_\mu = 0, 1, 2, 3, \ldots . \tag{2.30}$$

To prove this important assertion we may proceed as follows. From the commutation relations in (2.27) we can obtain by means of complete induction on the nonnegative integers α and β the more general commutation relations

$$\left[\hat{N}_\mu, (\hat{a}_\mu)^\alpha\right] = -\alpha(\hat{a}_\mu)^\alpha, \qquad \left[\hat{N}_\mu, (\hat{a}_\mu^\dagger)^\beta\right] = \beta(\hat{a}_\mu^\dagger)^\beta.$$

If we operate with both sides of these equations on an eigenket $|n_\mu\rangle$ (positive norm!) with the eigenvalue n_μ, we arrive at

$$\hat{N}_\mu\left[(\hat{a}_\mu)^\alpha|n_\mu\rangle\right] = (n_\mu - \alpha)\left[(\hat{a}_\mu)^\alpha|n_\mu\rangle\right], \tag{2.31a}$$

$$\hat{N}_\mu\left[(\hat{a}_\mu^\dagger)^\beta|n_\mu\rangle\right] = (n_\mu + \beta)\left[(\hat{a}_\mu^\dagger)^\beta|n_\mu\rangle\right]. \tag{2.31b}$$

The action of the operators $(\hat{a}_\mu)^\alpha$ and $(\hat{a}_\mu^\dagger)^\beta$ on $|n_\mu\rangle$ has to result in a vector of the Hilbert space; this is required by general rules of quantum theory (cf. Section A.1.3). Hence, the norm of $(\hat{a}_\mu)^\alpha|n_\mu\rangle$ and $(\hat{a}_\mu^\dagger)^\beta|n_\mu\rangle$ must not be negative. To study the norm of the vectors in (2.31a) we take the product with $\langle n_\mu|(\hat{a}_\mu^\dagger)^\alpha$, which leads to

$$\left\|(\hat{a}_\mu)^{\alpha+1}|n_\mu\rangle\right\|^2 = (n_\mu - \alpha)\left\|(\hat{a}_\mu)^\alpha|n_\mu\rangle\right\|^2. \tag{2.32}$$

First we will consider (2.32) under the assumption $\alpha = 0$. Because of $\|\,|n_\mu\rangle\|^2 > 0$, the norm of $\hat{a}_\mu|n_\mu\rangle$ is nonnegative only if $n_\mu \geq 0$. This means that negative eigenvalues n_μ of $\hat{N}_\mu$ do not exist. For $n_\mu = 0$ the vector $\hat{a}_\mu|n_\mu\rangle$ is equal to the zero vector $|0_v\rangle$ of the Hilbert space. Now we will check the case of noninteger n_μ values; assume that the integer α takes a positive value and that $\alpha > n_\mu > \alpha - 1$ holds for n_μ. From a stepwise application of (2.32)—starting with the vector $|n_\mu\rangle$ on the right-hand side—it becomes obvious that the vectors $|n_\mu\rangle, \hat{a}_\mu|n_\mu\rangle, \ldots, (\hat{a}_\mu)^\alpha|n_\mu\rangle, (\hat{a}_\mu)^{\alpha+1}|n_\mu\rangle$ possess a positive norm only up to $(\hat{a}_\mu)^\alpha|n_\mu\rangle$ inclusive, while $(\hat{a}_\mu)^{\alpha+1}|n_\mu\rangle$ has a negative norm. This means that the assumption of noninteger positive eigenvalues n_μ leads to a contradiction; noninteger eigenvalues n_μ do not exist. Now, what remains to be done is to consider the integer, nonnegative n_μ values. Inspection of (2.31a) and (2.31b) shows that the vectors

$$\ldots, (\hat{a}_\mu^\dagger)^\beta|n_\mu\rangle, \ldots, \hat{a}_\mu^\dagger|n_\mu\rangle, |n_\mu\rangle, \hat{a}_\mu|n_\mu\rangle, \ldots, (\hat{a}_\mu)^\alpha|n_\mu\rangle, \ldots \tag{2.33a}$$

can be regarded as eigenkets of $\hat{N}_\mu$ with the eigenvalues

$$\ldots, n_\mu + \beta, \ldots, n_\mu + 1, n_\mu, n_\mu - 1, \ldots, n_\mu - \alpha, \ldots, \tag{2.33b}$$

if these vectors possess a positive norm. Inspection of (2.32) reveals that the sequence of eigenkets in (2.33a) breaks off on the right at $\alpha = n_\mu$. The norm of vectors with $\alpha > n_\mu$ is zero. Although they belong to the Hilbert space according to

$$(\hat{a}_\mu)^\alpha |n_\mu\rangle = |0_v\rangle, \tag{2.34a}$$

they are not eigenkets. All the other vectors in (2.33a) are eigenkets.

The eigenvectors of the observable $\hat{N}_\mu$ are orthonormal according to

$$\langle n_\mu | n'_\mu \rangle = \delta_{n_\mu n'_\mu} \tag{2.34b}$$

and form a complete set, which leads to

$$\sum_{n_\mu = 0} |n_\mu\rangle\langle n_\mu| = \hat{I}. \tag{2.35}$$

As expected, the eigenstates $|n_\mu\rangle$ of $\hat{N}_\mu$ are also eigenstates of the energy operators $\hat{H}_\mu$

$$\hat{H}_\mu |n_\mu\rangle = \hbar\omega_\mu \left(\hat{N}_\mu + \frac{\hat{I}}{2} \right) |n_\mu\rangle = \hbar\omega_\mu \left(n_\mu + \tfrac{1}{2} \right) |n_\mu\rangle. \tag{2.36}$$

As a result the energy eigenvalue of the state $|n_\mu\rangle$ is

$$\mathscr{E}_{n_\mu} = \hbar\omega_\mu (n_\mu + \tfrac{1}{2}). \tag{2.37}$$

In an analogous way the contribution $G_{n_\mu \bullet}$ of the *momentum* of the μth mode in the state $|n_\mu\rangle$ can be calculated; from (2.26) it can be concluded that

$$G_{n_\mu \bullet} = \hbar k_{\mu \bullet} (n_\mu + \tfrac{1}{2}). \tag{2.38}$$

The results (2.30), (2.37), and (2.38) are to be interpreted in the following way. The eigenket $|n_\mu\rangle$ represents a state in which n_μ particles of the radiation field exist. The integer n_μ is the *photon number* in the μth mode, the corresponding observable $\hat{N}_\mu$ is the *photon-number operator*, and $|n_\mu\rangle$ is the *photon-number state*. Inspection of (2.37) and (2.38) shows that every photon contributes $\hbar\omega_\mu$ to the energy $\mathscr{E}_{n_\mu}$ and $\hbar k_{\mu \bullet}$ to the momentum $G_{n_\mu \bullet}$; the term $\frac{1}{2}$ is the photon vacuum part (which will be discussed later on in Section 2.3.1). Since, by (2.30), n_μ may take on any arbitrary nonnegative integer, it is possible that an arbitrary number of photons may be in the same dynamical state characterized by the energy $\hbar\omega_\mu$, the momentum $\hbar k_{\mu \bullet}$, and the polarization direction $e_{\mu \bullet}$. This result follows directly from the commutation relations (2.27) for $\hat{a}_\mu^\dagger$ and $\hat{a}_\mu$. Note that the derivation of the eigenvalue spectrum (2.30) of the photon-number operator did not depend on the application of a special representation, but only made use of basic properties of the Hilbert space. The possibility that one dynamical state can be occupied by an arbitrary number of particles characterizes the photons as *bosons*. If we had employed anticommutators instead of commutators, then, in accordance with (A.152) and (A.153), this approach would have led to only two eigenvalues, namely to $n_\mu = 0$ and $n_\mu = 1$. In this case a state would be occupied by not more than one particle, which is true of *fermions* (particles obeying the Fermi–Dirac statistics). In Section 2.2.3 another important boson property of photons, namely their integer spin, will be discussed.

The set of states $\{|n_\mu\rangle\}$ with $n_\mu = 0, 1, 2, 3, \ldots$ as eigenkets of the photon-number operators $\hat{N}_\mu$ form the *occupation-number representation*. A general (pure) state of the one-mode field can be represented by

$$|\psi_\mu\rangle = \sum_{n_\mu=0} c(n_\mu)|n_\mu\rangle \qquad \text{with} \quad \langle\psi_\mu|\psi_\mu\rangle = 1, \tag{2.39}$$

where $|c(n_\mu)|^2 = |\langle n_\mu|\psi_\mu\rangle|^2$ means the probability of finding n_μ photons in the field. The occupation-number representation exhibits definitively the physical significance of the operators $\hat{a}_\mu^\dagger$ and $\hat{a}_\mu$. A comparison of (2.33a) with (2.33b) indicates that $\hat{a}_\mu^\dagger|n_\mu\rangle$ characterizes an eigenstate with the eigenvalue $n_\mu + 1$, and $\hat{a}_\mu|n_\mu\rangle$ another one with the eigenvalue $n_\mu - 1$. Thus

$$\hat{a}_\mu^\dagger|n_\mu\rangle = c_+|n_\mu + 1\rangle \quad \text{and} \quad \hat{a}_\mu|n_\mu\rangle = c_-|n_\mu - 1\rangle. \tag{2.40}$$

The complex numbers c_+ and c_- are normalization factors; $|c_+|^2$ can be obtained from the relations

$$\left(\langle n_\mu|\hat{a}_\mu\right)\left(\hat{a}_\mu^\dagger|n_\mu\rangle\right) = \langle n_\mu|\hat{N}_\mu + \hat{I}|n_\mu\rangle = n_\mu + 1 = |c_+|^2\langle n_\mu + 1|n_\mu + 1\rangle.$$

In an analogous way $|c_-|^2$ has to be determined. Altogether this leads to

$$\hat{a}_\mu^\dagger|n_\mu\rangle = \sqrt{n_\mu + 1}\,|n_\mu + 1\rangle \qquad \text{for} \quad n_\mu = 0, 1, 2, 3, \ldots, \tag{2.41a}$$

$$\hat{a}_\mu|n_\mu\rangle = \begin{cases} \sqrt{n_\mu}\,|n_\mu - 1\rangle & \text{for} \quad n_\mu = 1, 2, 3, \ldots, \\ |0_v\rangle & \text{for} \quad n_\mu = 0. \end{cases} \tag{2.41b}$$

By applying the operator $\hat{a}_\mu^\dagger$, the system is transferred into a state that contains $n_\mu + 1$ instead of n_μ photons. Because of this, $\hat{a}_\mu^\dagger$ is called the *photon creation operator* for the μth mode. Since the application of $\hat{a}_\mu$ to $|n_\mu\rangle$ results in a state with $n_\mu - 1$ photons, the operator $\hat{a}_\mu$ is called the *photon annihilation operator*. Note that the effect of $\hat{a}_\mu$ on the eigenstate $|n_\mu = 0\rangle$ (the so-called vacuum state) leads to the zero vector $|0_v\rangle$. Photon creation or annihilation can be caused by interaction of the free radiation field with another system, for instance, an atomic system. The operators $\hat{a}_\mu^\dagger$ and $\hat{a}_\mu$ describe the creation and annihilation, respectively, of one photon. This will be treated in detail in Section 3.3. The operators $(\hat{a}_\mu^\dagger)^\beta$ and $(\hat{a}_\mu)^\beta$ describe for the integer $\beta > 1$ the simultaneous creation and simultaneous annihilation, respectively, of several namely β photons; it is these multiphoton processes that will form mainly the topic of Part II of this book.

The fact that the $\hat{a}_\mu^\dagger$ and $\hat{a}_\mu$ correspond to transition processes and do not describe the occupation of a state as $\hat{N}_\mu$ does, becomes also apparent from considering their matrix elements, which can be obtained directly from (2.41), and characterize the $\hat{a}_\mu^\dagger$ and $\hat{a}_\mu$ as non-Hermitian operators. We want to emphasize that by successive application of $\hat{a}_\mu^\dagger$ to the eigenket $|n_\mu = 0\rangle$ of the lowest eigenvalue (vacuum state), any photon-number state can be gained in the form

$$|n_\mu\rangle = \frac{1}{\sqrt{n_\mu!}}(\hat{a}_\mu^\dagger)^{n_\mu}|n_\mu = 0\rangle, \tag{2.42}$$

which can be easily shown by means of complete induction; $(n_\mu!)^{-1/2}$ is a normalization factor.

2.2.2. The Total Field

In Section 2.1.2 we stated that the Hilbert space of the total field can be regarded as a set of independent subspaces assigned to the different modes. According to general rules of quantum theory (cf. Section A.1.1.2), an eigenket of the total Hamiltonian operator $\hat{H}$ is therefore given as a direct product of the eigenkets $|n_\mu\rangle$ of the one-mode fields, and the eigenvalue of $\hat{H}$ as a sum of the eigenvalues $\mathscr{E}_{n_\mu}$ of the one-mode field. With regard to the physical content nothing new is being added for the present, although the formulas for the total field have a somewhat voluminous appearance. The eigenstate of the total field characterized by n_μ photons in the μth mode with the photon energy $\hbar\omega_\mu$, the photon linear momentum $\hbar k_{\mu\bullet}$, and the photon polarization direction $e_{\mu\bullet}$ is given by

$$|n_1, \ldots, n_\mu, \ldots\rangle \equiv |n_1\rangle \cdots |n_\mu\rangle \cdots. \tag{2.43}$$

The corresponding energy eigenvalue has the form

$$\mathscr{E}_{n_1, \ldots, n_\mu, \ldots} = \sum_{n_1, \ldots, n_\mu, \ldots} \hbar\omega_\mu(n_\mu + \tfrac{1}{2}). \tag{2.44}$$

The orthonormality and completeness of the eigenstates $|n_1, \ldots, n_\mu, \ldots\rangle$ are described by

$$\langle n_1, \ldots, n_\mu, \ldots |n_1', \ldots, n_\mu', \ldots\rangle = \delta_{n_1 n_1'} \cdots \delta_{n_\mu n_\mu'} \cdots \tag{2.45}$$

$$\sum_{n_1, \ldots, n_\mu, \ldots} |n_1, \ldots n_\mu, \ldots\rangle\langle n_1, \ldots, n_\mu, \ldots | = \hat{I}. \tag{2.46}$$

From (2.43) and the results of the one-mode consideration the following relations are obtained:

$$\hat{N}_\mu|n_1,\ldots,n_\mu,\ldots\rangle = n_\mu|n_1,\ldots,n_\mu,\ldots\rangle, \tag{2.47}$$

$$\hat{a}_\mu^\dagger|n_1,\ldots,n_\mu,\ldots\rangle = \sqrt{n_\mu + 1}\,|n_1,\ldots,n_\mu + 1,\ldots\rangle, \tag{2.48}$$

$$\hat{a}_\mu|n_1,\ldots,n_\mu,\ldots\rangle = \sqrt{n_\mu}\,|n_1,\ldots,n_\mu - 1,\ldots\rangle; \tag{2.49}$$

in particular,

$$\hat{a}_\mu|n_1,\ldots,(n_\mu = 0),\ldots\rangle = |0_v\rangle. \tag{2.49a}$$

A general (pure) photon state of the free radiation field is represented by a linear combination of eigenstates (2.43), and hence by

$$|\psi\rangle = \sum_{n_1,\ldots,n_\mu,\ldots} c(n_1,\ldots,n_\mu,\ldots)|n_1,\ldots,n_\mu,\ldots\rangle, \tag{2.50}$$

where $|c(n_1,\ldots n_\mu,\ldots)|^2$ is the probability of finding n_1 photons in the first mode, $\ldots$, n_μ photons in the μth mode, and so on. In Section 2.3 we shall discuss the properties of important field states.

Special attention shall be drawn to the fact that according to (2.48) and (2.49) the result of applying $\hat{a}_\mu^\dagger$ and $\hat{a}_\mu$ to the eigenstate $|n_1,\ldots,n_\mu,\ldots\rangle$ is only dependent on the occupation number n_μ of the μth mode and not on the occupation numbers of the other modes of the total field. This is a boson property resting on the commutation relations (2.20). It should be noted that creation and annihilation operators of fermions with the anticommutation relations (A.155) exhibit qualitatively different behavior under the action on eigenstates of the total system.

Let us now sketch the special features that characterize the photons among the bosons. The representation (2.19) of the field operator $\hat{A}_\bullet(r_\bullet, t)$ can be formally related to the so-called second quantization of the one-component Schrödinger field, where the field operator $\hat{\psi}$ is expanded in terms of eigenfunctions (wave functions) of the one-particle Hamiltonian of the time-independent Schrödinger equation (cf. Section A.1.6). In the expansion of the field operator $\hat{A}_\bullet(r_\bullet, t)$ the factors $u_{\mu\bullet}(r_\bullet)$ of the annihilation operators $\hat{a}_\mu(t)$ can be thought of as eigenfunctions (wave functions) of photons; by action of problem-matched operators on these vector wave functions $u_{\mu\bullet}(r_\bullet)$ the corresponding eigenvalues can be reproduced. For example, the momentum operator $(\hbar/i)\nabla_\bullet$ yields the momentum eigenvalue $\hbar k_{\mu\bullet}$; further examples will be discussed in Section 2.2.3. However, it has to be noted that the wave function $u_{\mu\bullet}(r_\bullet)$ must not be interpreted as probability amplitude for the spatial density, since a corresponding continuity equation for the conservation of the particle number cannot be established in connection with the transversality.

The expansion (2.19) for the electromagnetic field $\hat{A}_\bullet$ contains a second term with the index μ, namely $\hat{a}_\mu^\dagger(t)u_{\mu\bullet}^*(r_\bullet)$. To this term—containing the creation operator $\hat{a}_\mu^\dagger(t)$—photons with just the same properties $(e_{\mu\bullet}, \hbar k_{\mu\bullet})$ have to be assigned, as they are represented by the term $\hat{a}_\mu(t)u_{\mu\bullet}(r_\bullet)$ with the annihilation operator $\hat{a}_\mu(t)$. Here a characteristic feature of photons becomes obvious. In the expansion of a *general nonhermitian* boson field $\hat{\psi}$ in eigenfunctions of states of free particles, the expression $[\hat{a}_\mu(t)u_{\mu\bullet}(r_\bullet) + \hat{b}_\mu^\dagger(t)u_{\mu\bullet}^*(r_\bullet)]$ belongs to the index μ [cf. (A.103)]; here $\hat{a}_\mu(t)$ has to be interpreted as the annihilation operator of one species of particles and $\hat{b}_\mu^\dagger(t)$ as the creation operator of a different species of particles. The two species of particles are referred to as particles and antiparticles.

In the case of the electromagnetic field the particles coincide with their antiparticles: there are only identical particles, the photons. This behavior is associated with the hermiticity of $\hat{A}$; the operator $\hat{b}_\mu^\dagger(t)u_{\mu\bullet}^*(r_\bullet)$ must agree with $\hat{a}_\mu^\dagger(t)u_{\mu\bullet}^*(r_\bullet)$. In contrast to those electrically neutral particles that possess antiparticles, the photons are called *strictly neutral particles*. While the former particles can only be annihilated (created) in pairs, the photons can also be annihilated (created) as single particles, as will be seen in Section 3.3 and Part II.

Let us now consider the *field operators and their space-time dependence*. From the operator $\hat{A}_\bullet(r_\bullet, t)$ in (2.19) we obtain the operator $-\partial\hat{A}_\bullet/\partial t$ of the electric field strength:

$$\hat{E}_\bullet(r_\bullet, t) = \sum_\mu \hat{E}_{\mu\bullet}(r_\bullet, t), \tag{2.51}$$

where

$$\hat{E}_{\mu\bullet}(r_\bullet, t) = e_{\mu\bullet}\left(\frac{\hbar\omega_\mu}{2\varepsilon_0 V}\right)^{1/2} i\hat{a}_\mu e^{i(k_{\mu\bullet}r_\bullet - \omega_\mu t)} + \text{h.c.}.$$

The magnetic induction $\nabla_\bullet \times A$ is given by

$$\hat{B}_\bullet(r_\bullet, t) = \sum_\mu \hat{B}_{\mu\bullet}(r_\bullet, t), \tag{2.52}$$

where

$$\hat{B}_{\mu\bullet}(r_\bullet, t) = \left(\frac{k_{\mu\bullet}}{k_\mu} \times e_{\mu\bullet}\right)\left(\frac{\hbar\omega_\mu}{2\varepsilon_0 V c^2}\right)^{1/2} i\hat{a}_\mu e^{i(k_{\mu\bullet}r_\bullet - \omega_\mu t)} + \text{h.c.}.$$

The field operator $\hat{A}_\bullet$ satisfies the wave equation. The equation (2.19) leads directly to the expressions

$$\nabla_\bullet^2 \hat{A}_\bullet = -\sum_\mu \left(\frac{\hbar}{2\varepsilon_0 V \omega_\mu}\right)^{1/2} e_{\mu\bullet} k_\mu^2 \left[\hat{a}_\mu e^{i(k_{\mu\bullet} r_\bullet - \omega_\mu t)} + \text{h.c.}\right],$$

$$\frac{\partial^2}{\partial t^2}\hat{A}_\bullet = -\sum_\mu \left(\frac{\hbar}{2\varepsilon_0 V \omega_\mu}\right)^{1/2} e_{\mu\bullet} \omega_\mu^2 \left[\hat{a}_\mu e^{i(k_{\mu\bullet} r_\bullet - \omega_\mu t)} + \text{h.c.}\right].$$

Taking into account the dispersion relation $\omega_\mu^2 = c^2 k_{\mu\bullet}^2$, we immediately arrive at the wave equation

$$\nabla_\bullet^2 \hat{A}_\bullet - \frac{1}{c^2}\frac{\partial^2}{\partial t^2}\hat{A}_\bullet = 0. \tag{2.53}$$

Since the differential operators $\nabla_\bullet$ and $-\partial/\partial t$ commute, the same wave equation holds for the electric field strength $\hat{E}_\bullet$ as well.

In the classical description the real field strength $E_\bullet(r_\bullet, t)$ can be represented by the sum of two complex conjugate parts, $E_\bullet^{(-)}(r_\bullet, t)$ and $E_\bullet^{(+)}(r, t)$. The quantity $E_\bullet^{(-)}$ is called the *complex analytic signal*; it contains all the information about the field. The quantity $E_\bullet^{(-)}(r_\bullet, t)$ is connected with the negative-frequency part of the temporal Fourier transform $\breve{E}_\bullet(r_\bullet, \omega)$ of the field strength $E_\bullet(r_\bullet, t)$ by the relation

$$E_\bullet^{(-)}(r_\bullet, t) = \frac{1}{2\pi}\int_{-\infty}^{0} d\omega\, \breve{E}_\bullet(r_\bullet, \omega) e^{i\omega t}. \tag{2.54}$$

The operator corresponding to $E_\bullet^{(-)}$ in a quantized free radiation field that is expanded in plane waves is

$$\hat{E}_\bullet^{(-)}(r_\bullet, t) = \sum_\mu \hat{E}_{\mu\bullet}^{(-)}(r_\bullet, t), \tag{2.55}$$

where

$$\hat{E}_{\mu\bullet}^{(-)}(r_\bullet, t) = e_{\mu\bullet}\left(\frac{\hbar\omega_\mu}{2\varepsilon_0 V}\right)^{1/2} i\hat{a}_\mu e^{i(k_{\mu\bullet} r_\bullet - \omega_\mu t)}.$$

Since the Coulomb gauge $\nabla_\bullet \hat{A}_\bullet = 0$ is assumed, $\nabla_\bullet^2 \hat{A}_\bullet = -\nabla_\bullet \times (\nabla_\bullet \times \hat{A}_\bullet)$ follows according to general rules of vector analysis. This leads, by (2.53), to

$$\frac{\partial}{\partial t}\varepsilon_0\left(-\frac{\partial \hat{A}_\bullet}{\partial t}\right) = \nabla_\bullet \times \left(\frac{1}{\mu_0}\nabla_\bullet \times \hat{A}_\bullet\right),$$

showing that

$$\frac{\partial}{\partial t}\hat{D}_{\bullet}(r_{\bullet}, t) = \nabla_{\bullet} \times \hat{H}_{\bullet}(r_{\bullet}, t). \tag{2.56a}$$

In an analogous way we may obtain

$$\frac{\partial}{\partial t}\hat{B}_{\bullet}(r_{\bullet}, t) = -\nabla_{\bullet} \times \hat{E}_{\bullet}(r_{\bullet}, t). \tag{2.56b}$$

The quantum-theoretical relations (2.56a, b) for the field operators correspond to the classical Maxwell equations (2.1).

Having introduced the commutation relations (2.20) for the creation and annihilation operators of the single modes without making use of the basic commutation relations (2.10) of the field variables, we have now to discuss the connection between these relations.

Let us start from the left-hand side commutation relation (2.10a) for the operators $\hat{A}_{\bullet}$ and $-\hat{D}_{\bullet}$ at equal times. Since the form of commutation relations is generally independent of the picture used, we may pass to the Schrödinger picture. The time t can be taken as $t = 0$, where we use the convention that quantities in the Heisenberg picture coincide with those in the Schrödinger picture at $t = 0$. Substituting the components $\hat{A}_p(r'_{\bullet}, 0)$ and $-\hat{D}_q(r''_{\bullet}, 0)$ from (2.19) and (2.51) into the left-hand side of (2.10a), we get

$$\left[\hat{A}_p(r'_{\bullet}, 0), -\hat{D}_q(r''_{\bullet}, 0)\right] = i\frac{\hbar}{2V}\sum_{l,\sigma}(e_{l\sigma})_p(e_{l\sigma})_q(e^{ik_{l\bullet}s_{\bullet}} + \text{c.c.}),$$

where $s_{\bullet} = r'_{\bullet} - r''_{\bullet}$. The subscript $l = (l_x, l_y, l_z)$ characterizes the components of the wave vectors, and σ the polarization directions. The summation over the index σ will be carried out. From (2.16) we know that the vectors $e_{l1\bullet}, e_{l2\bullet}, k_{l\bullet}$ are perpendicular to each other; by using the well-known relations of the direction cosines, the relation

$$\sum_{\sigma=1}^{2}(e_{l\sigma})_p(e_{l\sigma})_q + \frac{(k_l)_p(k_l)_q}{k_{l\bullet}^2} = \delta_{pq} \tag{2.57}$$

is obtained. Thus we have

$$\left[\hat{A}_p(r'_{\bullet}, 0), -\hat{D}_q(r''_{\bullet}, 0)\right] = i\frac{\hbar}{V}\sum_{l}\left[\delta_{pq} - \frac{(k_l)_p(k_l)_q}{k_{l\bullet}^2}\right]e^{ik_{l\bullet}s_{\bullet}}, \tag{2.58}$$

where the relation $k_{l\bullet} = -k_{l\bullet}$ has been used to eliminate the factor $e^{-ik_{l\bullet}s_{\bullet}}$.

The commutation relations (2.10b) can be treated analogously; thus we have

$$\left[\hat{A}_p(r',0),\, \hat{A}_q(r'',0)\right]$$

$$= i\frac{\hbar}{\varepsilon_0 V}\sum_l \frac{1}{\omega_l}\left[\delta_{pq} - \frac{(k_l)_p(k_l)_q}{k_{l\bullet}^2}\right]\sin(k_{l\bullet}s_\bullet). \tag{2.59}$$

Because of the properties of $k_{l\bullet}$ and $\omega_l^2 = c^2k_{l\bullet}^2$, the terms in the sum change their sign if $l = (l_x, l_y, l_z)$ is replaced by $-l = (-l_x, -l_y, -l_z)$. Since the summation must be taken over all positive and negative integers l_x, l_y, l_z, the sum on the right-hand side is zero.

In the expression on the right-hand side of (2.58) two important physical features are incorporated: namely, on the one hand, via (2.16), the transversality of the field, and on the other hand, via the periodicity condition (2.13), the existence of an independent field only in a cube with a finite volume $V = L^3$. The latter feature leads to the advantage of discrete indices (l, σ) of the basis set, and it is connected with a finite metric in the Hilbert space. If we let the length L of the normalization volume V become infinitely large, the components of the wave vector [cf. (1.21)] will tend to continuous quantities. Therefore, we can replace the operation $(1/V)\Sigma_l$ in (2.58) by $(2\pi)^{-3}\int d^3k_\bullet$, and then we get the relation

$$\left[\hat{A}_p(r'_\bullet,0),\, -\hat{D}_q(r''_\bullet,0)\right] = i\hbar\delta_{pq}^{(\mathrm{tr}\,3)}(r'_\bullet - r''_\bullet), \tag{2.60}$$

where the so-called threedimensional transverse delta function $\delta_{qp}^{(\mathrm{tr}\,3)}(r'_\bullet - r''_\bullet)$ is defined by

$$\delta_{qp}^{(\mathrm{tr}\,3)}(r'_\bullet - r''_\bullet) \equiv \frac{1}{(2\pi)^3}\int d^3k_\bullet\left(\delta_{qp} - \frac{(k)_p(k)_q}{k_\bullet^2}\right)e^{ik_\bullet(r'_\bullet - r''_\bullet)} \tag{2.61}$$

Physically, the transition from (2.58) to (2.60) means the consideration of a field in infinite space instead of a field within a cube with finite volume. The application of (2.60) is associated with an infinite metric of the Hilbert space; the orthonormality of the basis waves is then described by $\delta(k_\bullet - k'_\bullet)$ rather than by the corresponding Kronecker symbol $\delta_{ll'}$ as used in (2.19a). From the commutation relations in the Schrödinger picture we may again pass to those in the Heisenberg picture and finally arrive at

$$[\hat{A}_p(r'_\bullet, t),\, -\hat{D}_q(r''_\bullet, t)] = i\hbar\delta_{qp}^{(\mathrm{tr}\,3)}(r'_\bullet - r''_\bullet); \tag{2.62}$$

this relation exactly coincides with (2.10a).

2.2.3. Photons of Fixed Angular Momentum and Parity

Previously, we have calculated the energy (2.21) and the linear momentum (2.26) of the quantized radiation field. In this subsection we concentrate on the angular momentum of the radiation field. We may proceed as we did when discussing the quantities mentioned above and replace the classical field strengths in the expression (2.8d) by the corresponding field operators. Thus we obtain for the operator of the angular momentum

$$\hat{J}_{\bullet} = \hat{J}_{\bullet}^{(\mathrm{or})} + \hat{J}_{\bullet}^{(\mathrm{in})}. \tag{2.63}$$

From (2.8e) the operator $\hat{J}_{\bullet}^{(\mathrm{in})}$ of the intrinsic part of the angular momentum (which does not depend on the origin of the coordinates) results in the form

$$\hat{J}_{\bullet}^{(\mathrm{in})} = \int_V d^3r_{\bullet}\,(\hat{D}_{\bullet} \times \hat{A}_{\bullet}). \tag{2.64}$$

As will be shown, this quantity can be associated with the spin of the photons. But we cannot expect that in connection with the photon the spin refers to a property of a particle at rest (as is possible, for example, with the electron), since for a photon, which moves at the velocity of light, the rest mass is zero and a reference system at rest does not exist.

According to Section 2.2.1, two modes with the polarization directions $e_{l1\bullet}$ and $e_{l2\bullet}$ belong to one wave vector $k_{l\bullet}$. For the present let us consider only these two modes; for simplicity we omit the index l and form, according to

$$e_{1\bullet} \times e_{2\bullet} = e_{\bullet}, \tag{2.65}$$

a right-handed Cartesian triad with $e_{\bullet}$ as unit vector in the direction of $k_{\bullet}$. We consider the operators in the Schrödinger picture at $t = 0$. Thus we obtain for the vector potential and electric displacement

$$\hat{A}_{\bullet}(r_{\bullet}) = \hat{a}_1 e_{1\bullet} v + \hat{a}_1^{\dagger} e_{1\bullet} v^* + \hat{a}_2 e_{2\bullet} v + \hat{a}_2^{\dagger} e_{2\bullet} v^*, \tag{2.66a}$$

$$\frac{\hat{D}(r_{\bullet})}{i\omega\varepsilon_0} = \hat{a}_1 e_{1\bullet} v - \hat{a}_1^{\dagger} e_{1\bullet} v^* + \hat{a}_2 e_{2\bullet} v - \hat{a}_2^{\dagger} e_{2\bullet} v^*, \tag{2.66b}$$

where $v = v(r_{\bullet}) = (\hbar/2\varepsilon_0 V\omega)^{1/2}\exp(ik_{\bullet}r_{\bullet})$. From this, with (2.64), we obtain directly the intrinsic part of the angular momentum:

$$\hat{J}_{\bullet}^{(\mathrm{in})} = i\hbar(\hat{a}_2^{\dagger}\hat{a}_1 - \hat{a}_1^{\dagger}\hat{a}_2)e_{\bullet}. \tag{2.67}$$

Let us now try to find eigenstates of the operator $\hat{J}^{(\mathrm{in})} \equiv \hat{J}_{\bullet}^{(\mathrm{in})} e_{\bullet}$ that can be interpreted as one-photon states. A general one-photon state at a given $k_{\bullet}$ value can be represented as a linear combination of the kets $|1, 0\rangle$ and $|0, 1\rangle$;

the ket $|1,0\rangle$ means the state with one photon of the polarization direction $e_{1\bullet}$, and the ket $|0,1\rangle$ that of one photon with the polarization direction $e_{2\bullet}$. Hence the standard procedure [cf. (A.42)] for the determination of the eigenvalues $J^{(\text{in})}$ and the eigenstates leads to the secular equation

$$\begin{vmatrix} -J^{(\text{in})} & -i\hbar \\ i\hbar & -J^{(\text{in})} \end{vmatrix} = 0, \tag{2.68}$$

from which the eigenvalues $J^{(\text{in})} = \pm\hbar$ with the eigenkets

$$|1_\pm\rangle = \alpha_\pm(|1,0\rangle \pm i|0,1\rangle) \tag{2.69}$$

can be inferred. In normalizing to unity, $|\alpha_\pm| = 1/\sqrt{2}$ holds for the magnitudes of the normalization factors; to achieve a fully consistent interpretation the phase factor $\alpha_\pm$ is fixed at $+1$. From the eigenvalue equations

$$\hat{J}^{(\text{in})}|1_+\rangle = +\hbar|1_+\rangle, \qquad \hat{J}^{(\text{in})}|1_-\rangle = -\hbar|1_-\rangle \tag{2.70}$$

the significance of the states $|1_+\rangle$ and $|1_-\rangle$ becomes clear. Each of the photons has a fixed value of the intrinsic part of the angular momentum along its direction $e_\bullet$ of propagation. If this value is $+\hbar$, we speak of a *clockwise circularly polarized photon*; if it is $-\hbar$, of a *counterclockwise circularly polarized photon*. In this sense we can assign the spin 1 (in units of $\hbar$) to the photon; since this value proves to be an integral multiple of $\hbar$, the *boson character* is again explicitly exhibited. It may be mentioned that the action of $\hat{J}^{(\text{or})}_\bullet e_\bullet$ on the states $|1_\pm\rangle$ can also be considered in a manner analogous to that used above in detail for $\hat{J}^{(\text{in})}$; as a result one gets the zero vector. Therefore the results contained in (2.70) with respect to the action of $\hat{J}^{(\text{in})}$ on the states $|1_+\rangle$ and $|1_-\rangle$ hold for the total angular momentum as well.

Let us now determine the creation operators $\hat{a}_+^\dagger$ and $\hat{a}_-^\dagger$ for the clockwise and counterclockwise polarized photons. To do this we recall the fact described at the end of Section 2.2.1 that eigenkets of a photon can be generated by applying the creation operator to the ground state of the photon. The treatment for linearly polarized photons presented in (2.42) is now employed for circularly polarized photons. Let $|0,0\rangle$ be the photon ground state (no photon in the modes with the polarization directions $e_{1\bullet}, e_{2\bullet}$). Then the creation operators $\hat{a}_+^\dagger$ and $\hat{a}_-^\dagger$ can be derived from the equations

$$\hat{a}_+^\dagger|0,0\rangle = |1_+\rangle, \qquad \hat{a}_-^\dagger|0,0\rangle = |1_-\rangle. \tag{2.71}$$

Let $\hat{a}_\pm^\dagger$ be constructed by the linear combination $a_\pm^\dagger = c_\pm \hat{a}_1^\dagger + d_\pm \hat{a}_2^\dagger$. From (2.71) the constants $c_\pm$ and $d_\pm$ can be derived; thus we have

$$\hat{a}_\pm^\dagger = \frac{1}{\sqrt{2}}\left(\hat{a}_1^\dagger \pm i\hat{a}_2^\dagger\right), \qquad \hat{a}_\pm = \frac{1}{\sqrt{2}}(\hat{a}_1 \mp i\hat{a}_2). \tag{2.72}$$

It can be easily shown that the operators $\hat{a}_{\pm}$, $\hat{a}^{\dagger}_{\pm}$ satisfy the same commutation relations as do $\hat{a}_1$, $\hat{a}^{\dagger}_1$, $\hat{a}_2$, $\hat{a}^{\dagger}_2$ in (2.20); the operators $\hat{a}^{\dagger}_{+}$, $\hat{a}_{+}$ commutate with $\hat{a}^{\dagger}_{-}$ and $\hat{a}_{-}$. The relations

$$[\hat{a}_{\pm}, \hat{a}^{\dagger}_{\pm}] = \hat{I}, \qquad [\hat{a}_{\pm}, \hat{a}_{\pm}] = [\hat{a}^{\dagger}_{\pm}, \hat{a}^{\dagger}_{\pm}] = \hat{0} \tag{2.72a}$$

hold. This leads to the possibility of forming the particle-number operators

$$\hat{N}_{+} = \hat{a}^{\dagger}_{+}\hat{a}_{+}, \qquad \hat{N}_{-} = \hat{a}^{\dagger}_{-}\hat{a}_{-}, \tag{2.73}$$

which, for the same reasons as were used for verifying (2.30), have the nonnegative integers as their eigenvalues. By (2.72), we can rewrite the expression for the intrinsic part of the angular momentum and obtain

$$\hat{J}^{(\mathrm{in})}_{\bullet} = \hbar(\hat{N}_{+} - \hat{N}_{-})e_{\bullet}. \tag{2.74}$$

From this it becomes obvious that in general the eigenvalue of $J^{(\mathrm{in})}_{\bullet}e_{\bullet}$ is an integral multiple of $\hbar$ and is proportional to the difference between the number of clockwise and counterclockwise circularly polarized photons.

According to Section 2.2.2, the factors of the annihilation operators in (2.66a) can be regarded as eigenfunctions (wave functions) of photons with the polarization directions $e_{1\bullet}$, $e_{2\bullet}$ and the wave vectors $k_{\bullet}$. Let us now find the analogous eigenfunctions for circularly polarized photons. To do this, we substitute the relations $\hat{a}_1 = (1/\sqrt{2})(\hat{a}_{-} + \hat{a}_{+})$ and $\hat{a}_2 = (+i/\sqrt{2})(\hat{a}_{+} - \hat{a}_{-})$, which can be directly calculated from (2.72), into (2.66a). Then we have

$$\hat{A}_{\bullet}(r_{\bullet}) = \hat{a}_{+}g_{+\bullet}(r_{\bullet}) + \hat{a}^{\dagger}_{+}g^{*}_{+\bullet}(r_{\bullet}) + \hat{a}_{-}g_{-\bullet}(r_{\bullet}) + \hat{a}^{\dagger}_{-}g^{*}_{-\bullet}(r_{\bullet}), \tag{2.75}$$

where

$$g_{\pm\bullet}(r_{\bullet}) = \frac{1}{\sqrt{2}}(e_{1\bullet} \pm ie_{2\bullet})v(r_{\bullet}) \equiv \gamma_{\pm\bullet}v(r_{\bullet}). \tag{2.75a}$$

In the sense of Section 2.2.2, the functions $g_{\pm\bullet}(r_{\bullet})$ can be regarded as the eigenfunctions (wave functions) that belong to photons with the wave vector $k_{\bullet}$ in the + state or − state, respectively; by the action of the operator $(\hbar/i)\nabla_{\bullet}$ the factor $v(r_{\bullet})$ reproduces the momentum $\hbar k_{\bullet}$. The factor $\gamma_{\pm\bullet}$ in $g_{\pm\bullet}(r_{\bullet})$ can be interpreted as the spin part of the wave function. To illustrate this, the operation of the spin matrix $\hat{\hat{S}}$ for one particle with the spin quantum number 1 on $\gamma_{\pm\bullet}$ will be considered. The direction of propagation (z direction) is characterized by $e_{\bullet}$, while $e_{1\bullet}$ and $e_{2\bullet}$ coincide with the x and the y direction, respectively. The z component of the spin matrix $\hat{\hat{S}}$ is known to be

$$\hat{\hat{S}}_z = \hbar\begin{pmatrix} 1 & 0 & 0 \\ 0 & 0 & 0 \\ 0 & 0 & -1 \end{pmatrix}. \tag{2.76}$$

It is easily seen that

$$\hat{\hat{S}}_z\gamma_{+\bullet} = +\hbar\gamma_{+\bullet} \quad \text{and} \quad \hat{\hat{S}}_z\gamma_{-\bullet} = -\hbar\gamma_{-\bullet}. \tag{2.76a}$$

It must be concluded from (2.74) that in addition to the spin parts γ_+ and γ_- a spin part belonging to the spin value zero in the z direction does not exist.

Thus, the meaning of the spin function $\gamma_{\pm\bullet}$ becomes clear. It can be described as follows; according to (2.75a), the photon spin is determined by a fixed combination of polarization vectors of plane traveling waves and cannot be interpreted as a property of a particle at rest.

Up to now we have only considered the two polarization states belonging to a single $k_\bullet$ value. The complete expansion of $\hat{A}_\bullet$ in circularly polarized plane waves can be obtained by summing over all the independent $k_\bullet$ values:

$$\hat{A}_\bullet(r_\bullet) = \sum_{k_\bullet}[\hat{a}_+(k_\bullet)g_{+\bullet}(r_\bullet; k_\bullet) + \hat{a}_-(k_\bullet)g_{-\bullet}(r_\bullet; k_\bullet)] + \text{h.c.} \tag{2.77}$$

The components of $k_\bullet$ are, of course, given by $l = (l_x, l_y, l_z)$.

In (2.19) and (2.77) we became acquainted with the expansion in linearly and circularly polarized plane waves. The expansion in plane waves is doubtless the most frequently used. Sometimes also the expansion in spherical waves is of some importance, since in such an expansion the properties of the angular momentum and of the parity of photons become apparent, by which also atoms and molecules may be characterized while emitting or absorbing photons.

In principle, the procedure of the expansion in spherical waves is the same as that in plane waves; therefore it is sufficient to give only a brief discussion here. Before dealing with the wave function of the photon, let us make some remarks on those wave functions that are eigenfunctions of the angular-momentum operators $\hat{j}_\bullet^2$ and $\hat{j}_z$, as they are generally needed for describing a particle with spin 1. These eigenfunctions $Y_{jm\bullet}$ are called vector spherical harmonics; they must satisfy the eigenvalue equations

$$\hat{j}_\bullet^2 Y_{jm\bullet} = \hbar^2 j(j+1)Y_{jm\bullet}, \qquad \hat{j}_z Y_{jm\bullet} = \hbar m Y_{jm\bullet}, \tag{2.78}$$

where j and m are quantum numbers or combinations of them. The $Y_{jm\bullet}$ are linear combinations formed by the familiar scalar spherical functions (used e.g. in the description of the H atom) containing the unit vector $k_\bullet/k$, and by the spin functions $\gamma_{\eta\bullet}$. The spin functions agree for $\eta = +1$ and $\eta = -1$ with the functions $\gamma_{+\bullet}$ and $\gamma_{-\bullet}$ from (2.75a), where $e_{1\bullet} = e_{x\bullet}$ and $e_{2\bullet} = e_{y\bullet}$. The unit vector $e_{z\bullet}$ can formally be written as spin function $\gamma_{\eta\bullet}$ with $\eta = 0$.

The vector spherical functions or combinations of them can be tested for their parity. We recall that the parity is connected with the behavior of a wave function if an inversion of the coordinate system takes place. While only the sign of $r_\bullet$ is changed on applying the parity operator $\hat{\hat{P}}$ to a scalar function $v(r_\bullet)$, the operation of $\hat{\hat{P}}$ on a vector function additionally implies the inversion of the directions of the axes; thus

$$\hat{\hat{P}}A_\bullet(r_\bullet) = -A_\bullet(-r_\bullet). \tag{2.79a}$$

The eigenvalue P of the operator $\hat{\hat{P}}$ is given by

$$\hat{\hat{P}}A_\bullet(r_\bullet) = PA_\bullet(r_\bullet). \tag{2.79b}$$

By linear combination of vector spherical harmonics the eigenfunctions (wave functions) $g_{el\bullet}$ and $g_{ma\bullet}$ of photons possessing certain properties can be obtained. Thus we obtain the expansion

$$\begin{aligned}\hat{A}_\bullet(r_\bullet) = \sum_{k,j,m} [&\hat{a}_{el}(k,j,m)g_{el\bullet}(r_\bullet;k,j,m) + \\ &+\hat{a}_{ma}(k,j,m)g_{ma}(r_\bullet;k,j,m)] + \text{h.c.},\end{aligned} \tag{2.80}$$

which is analogous to (2.77). The eigenfunctions $g_{el\bullet}$ characterize the so-called electric 2^j-pole contribution with energy $\hbar ck$, quantum number j of the angular momentum, quantum number m of the z component of the angular momentum, and parity $(-1)^j$. The eigenfunctions $g_{ma\bullet}$ characterize the magnetic 2^j-pole contribution with $\hbar ck$, j, m, and parity $(-1)^{j+1}$. The operators $\hat{a}_{el}$ and $\hat{a}_{ma}$ are the annihilation operators of the photons in these states. Also the creation operators $\hat{a}_{el}^\dagger$ and $\hat{a}_{ma}^\dagger$ as well as the particle number operators $\hat{N}_{el} = \hat{a}_{el}^\dagger\hat{a}_{el}$ and $\hat{N}_{ma} = \hat{a}_{ma}^\dagger\hat{a}_{ma}$ have the meanings corresponding to the previously introduced analogous quantities. Since the creation and annihilation operators indexed by el and ma satisfy the boson commutation relations (2.20), $\hat{N}_{el}$ and $\hat{N}_{ma}$ also have the nonnegative integers as their eigenvalues. The radiation field is represented by a multipole expansion. We assign a state with $j = 1$ to an electric (el) dipole photon, a state with $j = 2$ to an electric (el) quadrupole photon, and so on; analogous statements hold for magnetic (ma) multipole photons, where however the parity takes the opposite value to that of an el photon of the same j value. The sum (2.80) represents an expansion in a complete set of basis functions; photons with zero angular momentum do not appear.

2.3. PROPERTIES OF TYPICAL FIELD STATES

An insight into the features of the quantized free radiation field can be gained from the discussion of the expectation values of basic operators. A comparison of these physically relevant values with their corresponding classical quantities allows us also to set the limits between quantum-theoretical and classical description. We have to consider pure field states as well as mixed field states.

2.3.1. Pure States of the Field

Any pure field state can be represented by a specification of the expression for the general photon state given in (2.50). In this subsection we deal with photon-number states, coherent states, and eigenstates of the operator of the electric field strength. Later on, in Section 10.6.3 and Section 14.3, we shall discuss states that exhibit extraordinary coherence behavior, namely two-photon coherent states and squeezed states, respectively.

2.3.1.1. Photon-Number States. First we consider photon-number states $|n_\mu\rangle$ belonging to a single mode with the index μ.

Since $|n_\mu\rangle$ is an eigenvector of the photon-number operator $\hat{N}_\mu$ [cf. (2.29)], the mean-square deviation of the particle number for photon number states is zero. We have

$$\langle n_\mu|(\widehat{\Delta N_\mu})^2|n_\mu\rangle = \langle n_\mu|(\hat{N}_\mu - \langle n_\mu|\hat{N}_\mu|n_\mu\rangle)^2|n_\mu\rangle = 0. \tag{2.81}$$

Let us now calculate the expectation value of the field strength $\hat{E}_{\mu\bullet}$, which, by (2.51), is that part of the total field strength which belongs to the μth mode. Since $\hat{E}_{\mu\bullet}$ depends linearly on $\hat{a}_\mu$ and $\hat{a}_\mu^\dagger$, the expectation value vanishes because of (2.41):

$$\langle n_\mu|\hat{E}_{\mu\bullet}(r_\bullet, t)|n_\mu\rangle = 0. \tag{2.82}$$

Since the mean value of the field strength vanishes for all $r_\bullet$ and t, a state with fixed photon number is not appropriate for describing a wavelike phenomenon.

We now determine the expectation value of the square $\hat{E}_{\mu\bullet}^2$ of the field strength. According to (2.51), $\hat{E}_{\mu\bullet}^2$ consists of four summands containing the operator products $\hat{a}_\mu\hat{a}_\mu^\dagger$, $\hat{a}_\mu^\dagger\hat{a}_\mu$, $\hat{a}_\mu^2$, and $(a_\mu^\dagger)^2$. By virtue of (2.41) the expectation values $\langle n_\mu|\hat{a}_\mu^2|n_\mu\rangle$ and $\langle n_\mu|(a_\mu^\dagger)^2|n_\mu\rangle$ vanish, while the expectation values of $\hat{a}_\mu\hat{a}_\mu^\dagger$ and $\hat{a}_\mu^\dagger\hat{a}_\mu$ yield $n_\mu + 1$ and n_μ, respectively. Thus we arrive at

$$\langle n_\mu|\hat{E}_{\mu\bullet}^2|n_\mu\rangle = \frac{1}{\varepsilon_0 V}\hbar\omega_\mu(n_\mu + \tfrac{1}{2}). \tag{2.83}$$

This result can be easily related to the energy density. To satisfy (2.8a), the operator $\varepsilon_0\hat{E}_{\mu\bullet}^2/2$ should correspond to the electric part of the classical energy density, and $\varepsilon_0\hat{E}_{\mu\bullet}^2$ should correspond to that of the total energy density of the μth mode, because in the radiation field the electric and magnetic parts are equal in the time average. According to (2.83) and (2.36) we obtain

$$\langle n_\mu|\varepsilon_0\hat{E}_{\mu\bullet}^2|n_\mu\rangle = \frac{\langle n_\mu|\hat{H}_\mu|n_\mu\rangle}{V} = \frac{\mathscr{E}_{n_\mu}}{V}. \tag{2.84}$$

This result shows that the interpretation of $\varepsilon_0 E_{\mu\bullet}^2$ as an operator of the spatial energy density is appropriate.

Let us now turn to the calculation of expectation values with the multimode state $|n_1, \ldots n_\mu, \ldots\rangle$ (with a fixed number of photons in each mode).

We first calculate the expectation value of the total field strength $\hat{E}_\bullet(r_\bullet, t)$. Since the total field strength is additively made up of the contributions $\hat{E}_{\mu\bullet}$ of the single modes, and $|n_1, \ldots, n_\mu, \ldots\rangle$ is a direct product of the eigenkets, it follows immediately that

$$\langle n_1, \ldots, n_\mu, \ldots |\hat{E}_\bullet(r_\bullet, t)| n_1, \ldots, n_\mu, \ldots\rangle = 0. \tag{2.85}$$

This result shows that it is impossible to describe a superposition of waves of different modes with a multimode state $|n_1, \ldots, n_\mu, \ldots\rangle$.

In the mean square of the total field strength the contributions of the single modes add because of their independence of each other; thus one has

$$\begin{aligned}\langle n_1, \ldots, n_\mu, \ldots |E_\bullet^2(r_\bullet, t)| n_1, \ldots, n_\mu, \ldots\rangle \\ = \frac{1}{\varepsilon_0 V} \sum_\mu \hbar\omega_\mu (n_\mu + \tfrac{1}{2}).\end{aligned} \tag{2.86}$$

Since generally $\langle[\widehat{\Delta E}_\bullet(r_\bullet, t)]^2\rangle = \langle \hat{E}_\bullet^2(r_\bullet, t)\rangle - \langle \hat{E}_\bullet(r_\bullet, t)\rangle^2$ holds for the mean-square fluctuations, we find because of (2.85) that the right-hand side of (2.86) agrees with the mean squared deviation

$$\langle n_1, \ldots, n_\mu, \ldots | \left[\widehat{\Delta E}_\bullet(r_\bullet, t)\right]^2 | n_1, \ldots, n_n, \ldots\rangle$$

of the field strength.

Particular reference has to be made to the situation where there exists a photon vacuum ($n_\mu = 0$ for any μ). Obviously, every mode contributes the value $\hbar\omega_\mu / 2\varepsilon_0 V$ to the mean-square deviation of the total field strength, which becomes

$$\frac{1}{\varepsilon_0 V} \sum_\mu \frac{\hbar\omega_\mu}{2}. \tag{2.87}$$

This expression diverges, since the number of modes per unit circular frequency increases with increasing ω_μ according to (1.27). This divergence is due to the fact that $\langle n_1, \ldots n_\mu, \ldots | [\widehat{\Delta E}_\bullet(r_\bullet, t)]^2 | n_1, \ldots n_\mu, \ldots\rangle$ was calculated for a pointlike space–time region. From such a procedure divergences arise even in the quantum-theoretical treatment of basic problems. For instance, the Heisenberg uncertainty relation

$$\langle(\widehat{\Delta p_x})^2\rangle\langle(\widehat{\Delta x})^2\rangle \geq \frac{\hbar^2}{4} \tag{2.88}$$

for a particle with the momentum operator $\hat{p}_x$ and the position operator $\hat{x}$ reveals that the average deviation of the momentum diverges if $(\Delta x)^2 = 0$ (i.e., a precise position value) occurs. In measuring the electric field strength with a realistic measuring device, the measured values have to be related to a nonvanishing volume (measurement by an atomic system or by elementary particles) and a nonvanishing time interval (measurement with finite bandwidth of the measuring device). We shall sketch how this concept can be mathematically expressed. Instead of the operator $\hat{E}_\bullet(r_\bullet, t)$, an "average operator" $\bar{\hat{E}}(\Delta V, \Delta t)$ related to the space–time region $\Delta V \Delta t$ around $r_\bullet$ and t is introduced. Then we have

$$\bar{\hat{E}}_\bullet(\Delta V, \Delta t) = \frac{1}{\Delta V \Delta t} \int_{\Delta V} d^3 r_\bullet \int_{\Delta t} dt\, \hat{E}_\bullet(r_\bullet, t). \tag{2.89}$$

Calculating the expectation value with the vacuum state, we obtain

$$\begin{aligned}
&\langle \ldots, 0, \ldots | \bar{\hat{E}}_\bullet^2(\Delta V, \Delta t) | \ldots, 0, \ldots \rangle \\
&= \frac{1}{(\Delta V \Delta t)^2} \int_{\Delta V} d^3 r_\bullet \int_{\Delta V} d^3 r'_\bullet \int_{\Delta t} dt \int_{\Delta t} dt \\
&\quad \times \langle \ldots, 0, \ldots | \hat{E}_\bullet(r_\bullet, t) \hat{E}_\bullet(r'_\bullet, t') | \ldots, 0, \ldots \rangle.
\end{aligned}$$

The integrand can be explicitly formed by (2.51); it is a function of $r_\bullet - r'_\bullet$ and $t - t'$. It can be shown that $\langle \ldots, 0, \ldots | \bar{\hat{E}}_\bullet^2(\Delta V, \Delta t) | \ldots, 0, \ldots \rangle$ takes finite values for nonvanishing ΔV and Δt, but it diverges if ΔV or Δt tend to zero.

2.3.1.2. Coherent States. In contrast to the photon-number state, the coherent state may serve for describing wavelike phenomena. These states are of fundamental importance in quantum electronics and quantum optics.

Let us first consider the coherent state of a single mode. The coherent state $|\alpha_\mu\rangle$ of the μth mode can be defined as a right-hand eigenstate of the annihilation operator $\hat{a}_\mu$; the corresponding eigenvalue equation is

$$\hat{a}_\mu |\alpha_\mu\rangle = \alpha_\mu |\alpha_\mu\rangle, \tag{2.91}$$

where α_μ is supposed to be a complex number. Note that $\hat{a}_\mu$ is a nonhermitian operator with *complex* eigenvalues $\{\alpha_\mu\}$; this means that here the situation essentially differs from that in the case of the hermetian operator $\hat{N}_\mu$ with its *real* eigenvalues $\{n_\mu\}$. The consideration of the limiting case $\alpha_\mu = 0$ shows that also the photon ground state $|n_\mu = 0\rangle$ satisfies the equation (2.91). According to (2.39), $|\alpha_\mu\rangle$ can be represented by the expansion

$$|\alpha_\mu\rangle = \sum_{n_\mu = 0} \langle n_\mu | \alpha_\mu \rangle | n_\mu \rangle. \tag{2.92}$$

Inserting this expression into (2.91) and taking into account the action of $\hat{a}_\mu$ on $|n_\mu\rangle$ according to (2.41), we obtain the recursion formula

$$\langle n_\mu + 1|\alpha_\mu\rangle\sqrt{n_\mu + 1} = \alpha_\mu\langle n_\mu|\alpha_\mu\rangle, \tag{2.93}$$

which leads to

$$\langle n_\mu|\alpha_\mu\rangle = \alpha_\mu^{n_\mu}(n_\mu!)^{-1/2}\langle n_\mu = 0|\alpha_\mu\rangle. \tag{2.94}$$

The constant $\langle n_\mu = 0|\alpha_\mu\rangle$ is determined from the normalization condition $\langle\alpha_\mu|\alpha_\mu\rangle = 1$; one obtains $\langle n_\mu = 0|\alpha_\mu\rangle = \exp(-|\alpha_\mu|^2/2)$, where the arbitrary phase factor has been fixed. Hence the expansion of the coherent state in terms of photon-number states is

$$|\alpha_\mu\rangle = \exp\left(-\frac{|\alpha_\mu|^2}{2}\right)\sum_{n_\mu=0}\alpha_\mu^{n_\mu}(n_\mu!)^{-1/2}|n_\mu\rangle. \tag{2.95}$$

The relations (2.91) and (2.95) can be interpreted as equivalent definitions of the coherent state $|\alpha_\mu\rangle$. In (2.42) it has been shown that an arbitrary photon-number state $|n_\mu\rangle$ can be generated from the ground state $|n_\mu = 0\rangle$. An analogous procedure can be given also for the coherent state; we have

$$|\alpha_\mu\rangle = \hat{D}(\alpha_\mu)|n_\mu = 0\rangle, \tag{2.96}$$

where $\hat{D}(\alpha_\mu)$ is the so-called displacement operator

$$\hat{D}(\alpha_\mu) = \exp\left(\alpha_\mu\hat{a}_\mu^\dagger - \alpha_\mu^*\hat{a}_\mu\right) \tag{2.96a}$$

with the property of unitarity: $\hat{D}^{-1}(\alpha_\mu) = \hat{D}^\dagger(\alpha_\mu)$.

In the case of photon-number states the orthonormality and completeness relations (2.34a) and (2.35) are important means for practical calculations. Analogous relations are useful in the case of coherent states. For two coherent states $|\alpha_\mu\rangle$ and $|\alpha'_\mu\rangle$ of the μth mode, (2.95) yields the scalar product

$$\langle\alpha_\mu|\alpha'_\mu\rangle = \exp(-|\alpha_\mu - \alpha'_\mu|^2). \tag{2.97}$$

This relation reveals that coherent states are indeed normalized to unity (for $\alpha_\mu = \alpha'_\mu$ the right-hand side becomes equal to unity). But they are not orthogonal for $\alpha_\mu \neq \alpha'_\mu$; only for large $|\alpha_\mu - \alpha'_\mu|$ does the right-hand side of (2.97) tend to zero. The system of the kets $\{|\alpha_\mu\rangle\}$ is a so-called *overcomplete system*; the $\{|\alpha_\mu\rangle\}$ are interconnected by certain relations. It is possible to expand general photon states $|\psi_\mu\rangle$ in terms of coherent states. We next expand the photon-number state $|n_\mu\rangle$. The procedure is as follows: we multiply (2.95) by $\pi^{-1}(n_\mu!)^{-1/2}\exp(-|\alpha_\mu|^2/2)(\alpha_\mu^*)^{n_\mu}$ and integrate over the α_μ plane; on the right-hand side this leads to $|n_\mu\rangle$. Finally we obtain

$$|n_\mu\rangle = \frac{1}{\pi}\int d^2\alpha_\mu \exp\left(-\frac{|\alpha_\mu|^2}{2}\right)(\alpha_\mu^*)^{n_\mu}(n_\mu!)^{-1/2}|\alpha_\mu\rangle, \tag{2.97a}$$

where $\int d^2\alpha_\mu$ is an abbreviation of $\int d\,\mathrm{Re}\{\alpha_\mu\}\int d\,\mathrm{Im}\{\alpha_\mu\}$. Using this result, one obtains with the help of the completeness relation $\hat{I} = \Sigma_{n_\mu}|n_\mu\rangle\langle n_\mu|$ the relation

$$\hat{I} = \frac{1}{\pi}\int d^2\alpha_\mu\, |\alpha_\mu\rangle\langle\alpha_\mu|. \tag{2.98}$$

The application of this formula allows us to express general states $|\psi_\mu\rangle$ and operators $\hat{C}$ in terms of coherent states by using the relations $|\psi_\mu\rangle = \hat{I}|\psi_\mu\rangle$ and $\hat{C} = \hat{I}\hat{C}\hat{I}$.

So far we have dealt with some mathematical properties of the coherent states. Now we are going to discuss their physical properties by deriving measurable expectation values. From (2.91) we obtain at once

$$\langle\alpha_\mu|\hat{N}_\mu|\alpha_\mu\rangle = \langle\alpha_\mu|\hat{a}_\mu^\dagger\hat{a}_\mu|\alpha_\mu\rangle = |\alpha_\mu|^2. \tag{2.99}$$

Thus, the square of the absolute value of α_μ characterizes the mean photon number of the coherent state. The expectation value of $\hat{N}_\mu^2$ is given by

$$\langle\alpha_\mu|\hat{N}_\mu^2|\alpha_\mu\rangle = |\alpha_\mu|^4 + |\alpha_\mu|^2. \tag{2.100}$$

Since generally $\langle(\widehat{\Delta N_\mu})^2\rangle = \langle\hat{N}_\mu^2\rangle - \langle\hat{N}_\mu\rangle^2$, the mean-square deviation of the photon number is given, in the case of the coherent state, by

$$\langle\alpha_\mu|(\widehat{\Delta N_\mu})^2|\alpha_\mu\rangle = \langle\alpha_\mu|\hat{N}_\mu|\alpha_\mu\rangle. \tag{2.101}$$

Hence the ratio of the mean-square deviation of the photon number to the mean photon number, the so-called reduced variance, is equal to unity.

Now we turn to the electric field strength. Since $\langle\alpha_\mu|\hat{a}_\mu|\alpha_\mu\rangle = \alpha_\mu$ and $\langle\alpha_\mu|\hat{a}_\mu^\dagger|\alpha_\mu\rangle = \alpha_\mu^*$, (2.51) directly gives

$$\langle\alpha_\mu|\hat{E}_{\mu\bullet}(r_\bullet, t)|\alpha_\mu\rangle = \left(\frac{\hbar\omega_\mu}{2\varepsilon_0 V}\right)^{1/2} e_{\mu\bullet}[i\alpha_\mu e^{i(k_{\mu\bullet}r_\bullet - \omega_\mu t)} + \text{c.c.}]. \tag{2.102}$$

This equation involves an important statement: the expectation value of the field strength represents a plane traveling wave whose complex amplitude factor is determined by α_μ. The quantity α_μ thus determines the phase and the absolute value $(2\hbar\omega_\mu/\varepsilon_0 V)^{1/2}\langle\hat{N}_\mu\rangle^{1/2}$ of the wave amplitude, the latter being proportional to the square root of the mean photon number because of (2.99).

The right-hand side of (2.102) indicates a correspondence to the classical description of a wave. But we have to remember that the quantum-theoretical expectation value $\langle \alpha_\mu | \hat{E}_{\mu\bullet}(r_\bullet, t) | \alpha_\mu \rangle$ represents the arithmetic mean. The single measurements reveal, according to the quantum-theoretical foundations, unavoidable fluctuations of the measured field strength $E_{\mu\bullet}$ about the mean value with a probability distribution $w(E_{\mu\bullet}; |\alpha_\mu\rangle)$. Only if these fluctuations become sufficiently small may we speak of a good approximation to classical behavior. We shall return to this fact in Section 2.3.1.4.

The probability $w(n_\mu; |\alpha_\mu\rangle)$ of measuring the photon number n_μ in a coherent state $|\alpha_\mu\rangle$ is given by the expectation value of the projection operator $|n_\mu\rangle\langle n_\mu|$; we have

$$w(n_\mu; |\alpha_\mu\rangle) = \left|\langle \alpha_\mu | n_\mu \rangle\right|^2 = (n_\mu!)^{-1} \langle \hat{N}_\mu \rangle^{n_\mu} e^{-\langle \hat{N}_\mu \rangle}, \tag{2.103}$$

which is just the *Poisson probability distribution.*

In analogy to the procedure used to form the multimode photon-number state (2.43), the multimode coherent state (the so-called *global coherent state*) can also be constructed. Since the respective operators $\hat{a}_\mu$ of the different modes commute with each other, the global coherent state can be defined by the direct product

$$|\{\alpha_\mu\}\rangle = \prod_\mu |\alpha_\mu\rangle. \tag{2.104}$$

As shown above, the one-mode coherent state represents a state belonging to a plane traveling wave with fixed frequency, whereas the global coherent state characterizes a superposition of plane traveling waves with different frequencies. The state $|\{\alpha_\mu\}\rangle$ is the right-hand eigenstate of the operator $\hat{E}_\bullet^{(-)}(r_\bullet, t)$ of the complex analytic signal [cf. (2.55)]. There holds the eigenvalue equation

$$\hat{E}_\bullet^{(-)}(r_\bullet, t) |\{\alpha_\mu\}\rangle = E_\bullet^{(-)}(r_\bullet, t; \{\alpha_\mu\}) |\{\alpha_\mu\}\rangle, \tag{2.105}$$

where

$$E_\bullet^{(-)}(r_\bullet, t; \{\alpha_\mu\}) = \sum_\mu \left(\frac{\hbar \omega_\mu}{2\varepsilon_0 V} \right)^{1/2} i \alpha_\mu e^{i(k_{\mu\bullet} r_\bullet - \omega_\mu t)} e_{\mu\bullet} \tag{2.106}$$

is the complex eigenvalue, which coincides with the classical expression for the complex analytic signal (cf. Section 1.3.2). Global coherent states play an important role with respect to the coherence properties of the radiation (cf. Section 5.2.2).

2.3.1.3. Eigenstates of the Electric Field Strength. The classical description of linear and nonlinear optical properties (Sections 1.3 and 1.4) revealed that the relationship between the polarization and the electric field strength is of fundamental importance. Introducing this relation into the wave equation (1.85), one obtains a differential equation for the electric field strength, whose solution allows the description of the linear and nonlinear optical phenomena. Because of this central position we shall now treat the eigenvalue problem of the corresponding quantum-theoretical quantity, namely of the operator $\hat{E}_\bullet$ of the electric field strength (see Ref. 2.1). Before doing this, let us mention that there exists also another approach to get a relationship between the traditional classical description and the quantum-theoretical representation; this attempt rests on the introduction of Hilbert-space operators corresponding to the modulus and the phase of the complex field-strength amplitude (cf. Refs. 2.2, 2.3, 2.4).

It is obvious from (2.51) and (2.20) that the total field-strength operator $\hat{E}_\bullet$ is additively composed of the parts $\hat{E}_{\mu\bullet}$ of the single modes commuting with each other. Thus the eigenvalue of $\hat{E}_\bullet$ can be represented as the sum of the parts of the single modes and the eigenket as direct product of the eigenkets of the single modes [cf. (A.49)]. Therefore essential knowledge can be gained by considering one single mode. Thus we can use a scalar approach and denote the field-strength operator in the direction of the unit polarization vector by $\hat{E}_\mu(r)$, where r is the space coordinate in the direction of $k_{\mu\bullet}$; we use the Schrödinger picture, which is assumed to coincide with the Heisenberg picture at $t = 0$.

The eigenvalue equation of the field strength observable $\hat{E}_\mu(r)$ is

$$\hat{E}_\mu(r)\big|E_\mu(r)\big\rangle = E_\mu(r)\big|E_\mu(r)\big\rangle. \tag{2.107}$$

The solution of this eigenvalue problem leads to a continuous eigenvalue spectrum; the eigenket belonging to $E_\mu(r)$ is

$$\big|E_\mu(r)\big\rangle = \sum_{n_\mu=0} \big\langle n_\mu|E_\mu(r)\big\rangle|n_\mu\rangle, \tag{2.108}$$

where an expansion in terms of the photon-number states is used; the expansion coefficients are given by

$$\begin{aligned}\big\langle n_\mu|E_\mu(r)\big\rangle &= [-is_\mu(r)]^{n_\mu}(2\pi)^{-1/4}(2^{n_\mu}n_\mu!)^{-1/2}\left(\frac{\hbar\omega_\mu}{2\varepsilon_0 V}\right)^{-1/4}\\ &\times\exp\left[-\frac{1}{4}\left(\frac{\hbar\omega_\mu}{2\varepsilon_0 V}\right)^{-1}E_\mu^2(r)\right]\\ &\times H_{n_\mu}\left(\frac{1}{\sqrt{2}}\left(\frac{\hbar\omega_\mu}{2\varepsilon_0 V}\right)^{-1/2}E_\mu(r)\right).\end{aligned} \tag{2.108a}$$

Here $H_{n_\mu}(\xi)$ is the Hermite polynomial of n_μth order, and $s_\mu(r)$ is defined by

$s_\mu(r) \equiv \exp(ik_\mu r)$. The eigenstates $|E_\mu(r)\rangle$ satisfy the orthonormality and completeness relation

$$\langle E_\mu(r)|E'_\mu(r)\rangle = \delta\left[E_\mu(r) - E'_\mu(r)\right],$$

$$\int dE_\mu(r)\,|E_\mu(r)\rangle\langle E_\mu(r)| = \hat{I}. \tag{2.109}$$

Let us outline the way to derive the solution (2.108a) from the eigenvalue equation (2.107). First the commutation relations of the operators $\hat{E}_\mu(r)$ and $\hat{A}_\mu(r)$ are established. From (2.19) and (2.51), by taking into account the commutation relations (2.27) for $\hat{a}_\mu$ and $\hat{a}^\dagger_\mu$, we obtain

$$\left[\hat{E}_\mu(r), \hat{A}_\mu(r)\right] = i\frac{\hbar}{\varepsilon_0 V}\hat{I},$$

$$\left[\hat{E}_\mu(r), \hat{E}_\mu(r)\right] = \hat{0}, \qquad \left[\hat{A}_\mu(r), \hat{A}_\mu(r)\right] = \hat{0}. \tag{2.110}$$

Furthermore we refer to the hermiticity relations of the field quantities,

$$\hat{E}_\mu(r) = \hat{E}^\dagger_\mu(r), \qquad \hat{A}_\mu(r) = \hat{A}^\dagger_\mu(r) \tag{2.110a}$$

As to their *mathematical structure* the five equations in (2.110) and (2.110a) are in exact agreement with the relations (A.126) for fundamental quantities in quantum mechanics, if we identify $\hat{E}_\mu(r)$ with the position operator $\hat{x}$ and $\hat{A}_\mu(r)$ with the operator of the linear momentum $\hat{p}$; note that the constant $\hbar/\varepsilon_0 V$ in (2.110) corresponds to the constant $\hbar$ in (A.126). For the eigenvalue spectrum of $\hat{x}$ and for matrix elements of general operators $\hat{F} = F(\hat{x}, \hat{p})$ *mathematical conclusions* are drawn in Section A.2.4 which, because of the mathematical structure of (2.110) and (2.110a), are also valid for the operator $\hat{E}_\mu(r)$ and its eigenkets $|E_\mu(r)\rangle$. That means that the eigenvalue equation (2.107) has a continuous eigenvalue spectrum $E_\mu(r)$, and the matrix element $\langle E_\mu(r)|F\{\hat{E}_\mu(r), \hat{A}_\mu(r)\}|\psi_\mu\rangle$ is given by

$$\langle E_\mu(r)|F\{\hat{E}_\mu(r), \hat{A}_\mu(r)\}|\psi_\mu\rangle = F\left\{E_\mu(r), \frac{\hbar}{i\varepsilon_0 V}\frac{d}{dE_\mu(r)}\right\}\langle E_\mu(r)|\psi_\mu\rangle. \tag{2.111}$$

Here $\hat{F} = F\{\hat{E}_\mu(r), \hat{A}_\mu(r)\}$ is an operator function, and $|\psi_\mu\rangle$ is an arbitrary state of the Hilbert space of the μth mode. Now we identify $|\psi_\mu\rangle$ in particular with the photon-number state $|n_\mu\rangle$, and $\hat{F}$ with the photon-number operator $\hat{N}_\mu = \hat{a}^\dagger_\mu \hat{a}_\mu$, which via

$$\hat{a}_\mu = -i\left(\frac{\varepsilon_0 V}{2\hbar\omega_\mu}\right)^{1/2} s^*_\mu(r)\left[\hat{E}_\mu(r) + i\omega_\mu \hat{A}_\mu(r)\right] \tag{2.112}$$

can be represented in the form $\hat{N}_\mu = N_\mu\{\hat{E}_\mu(r), \hat{A}_\mu(r)\}$. If we equate the right-hand sides of the relations

$$\langle E_\mu(r)|\hat{N}_\mu|n_\mu\rangle = n_\mu\langle E_\mu(r)|n_\mu\rangle, \tag{2.113a}$$

$$\langle E_\mu(r)|\hat{N}_\mu|n_\mu\rangle = N_\mu\left\{E(r), \frac{\hbar}{i\varepsilon_0 V}\frac{d}{dE_\mu(r)}\right\}\langle E_\mu(r)|n_\mu\rangle, \tag{2.113b}$$

we obtain the following differential equation for $\langle E_\mu(r)|n_\mu\rangle$:

$$\left\{-\frac{\hbar^2}{2\left(\varepsilon_0 V/\omega_\mu^2\right)}\frac{d^2}{dE_\mu^2(r)} + \frac{\varepsilon_0 V}{2}E_\mu^2(r) - \hbar\omega_\mu\left(n_\mu + \tfrac{1}{2}\right)\right\}\langle E_\mu(r)|n_\mu\rangle = 0. \tag{2.114}$$

Formally, this relation corresponds to the time-independent Schrödinger equation (A.145a) of the one-dimensional harmonic oscillator of point mechanics, where $\varepsilon_0 V/\omega_\mu^2$ is identified with the "mass" of the mass point. The well-known procedure for solving the oscillator problem leads to (2.108a). Here an n_μ-dependent normalizing factor has been fixed with respect to phase and magnitude so as to make the right-hand side of (2.108) satisfy the eigenvalue equation (2.107).

Analogously, the eigenket $|E_\mu(r)\rangle$ can be expanded in terms of coherent states $|\alpha_\mu\rangle$. The treatment of the relation $\langle E_\mu(r)|\hat{a}_\mu|\alpha_\mu\rangle = \alpha_\mu\langle E_\mu(r)|\alpha_\mu\rangle$ instead of (2.113a) leads to a differential equation for $\langle E_\mu(r)|\alpha_\mu\rangle$ whose solution allows the representation

$$|E_\mu(r)\rangle = \frac{1}{\pi}\int d^2\alpha_\mu \langle\alpha_\mu|E_\mu(r)\rangle|\alpha_\mu\rangle, \tag{2.115}$$

where

$$\langle\alpha_\mu|E_\mu(r)\rangle = (2\pi)^{-1/4}\left(\frac{\hbar\omega_\mu}{2\varepsilon_0 V}\right)^{-1/4}$$
$$\times\exp\left\{-\frac{1}{4}\left(\frac{\hbar\omega_\mu}{2\varepsilon_0 V}\right)^{-1}E_\mu^2(r) - i\left(\frac{\hbar\omega_\mu}{2\varepsilon_0 V}\right)^{-1/2}E_\mu(r)\left[\alpha_\mu s_\mu(r)\right]^*\right.$$
$$\left.-\operatorname{Im}^2\left[\alpha_\mu s_\mu(r)\right] - \frac{i}{2}\operatorname{Im}\left[\alpha_\mu^2 s_\mu^2(r)\right]\right\}.$$

The relations (2.108a) and (2.115) reveal that $|E_\mu(r)\rangle$ explicitly depends on the position coordinate via $s_\mu(r)$.

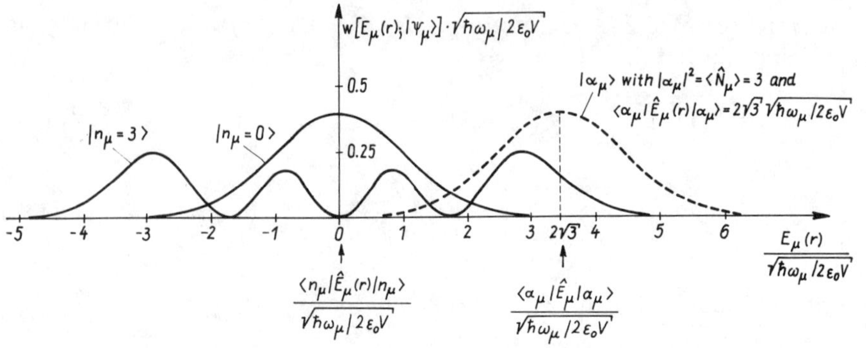

Fig. 2.1. Field distributions for the photon-number states $|n_\mu = 0\rangle$ and $|n_\mu = 3\rangle$ (solid curves) and for the coherent state $|\alpha_\mu\rangle$ (dashed curve) with $\langle \hat{N}_\mu \rangle = 3$.

2.3.1.4. Summary of the Physical Interpretation. The results from Section 2.3.1.3 can be used to determine the probability distribution of the field strength at the point r for an arbitrary state $|\psi_\mu\rangle$; we have

$$w\left[E_\mu(r);\, |\psi_\mu\rangle\right] = \left|\left\langle \psi_\mu | E_\mu(r)\right\rangle\right|^2. \tag{2.116}$$

If $|\psi_\mu\rangle$ equals the photon-number state $|n_\mu\rangle$, a probability distribution of the field strength arises [cf. (2.108b)] that is essentially determined by the square of the Hermite polynomial of n_μth order. Examples of these relations are plotted in Fig. 2.1 for $n_\mu = 0$ and 3. Since$|\langle n_\mu | E_\mu(r)\rangle|^2$ symmetrically depends on E_μ, a vanishing mean value of the field strength values results, which is in agreement with (2.82). In the photon-number state the probability distribution proves to be independent of the position r. The r.m.s. deviation is given by

$$\left\{\int dE_\mu(r)\, E_\mu^2(r) w\left[E_\mu(r);\, |n_\mu\rangle\right]\right\}^{1/2} = \left[\frac{\hbar\omega_\mu}{\varepsilon_0 V}\left(n_\mu + \tfrac{1}{2}\right)\right]^{1/2}. \tag{2.117}$$

Generally, we obtain the result that in the photon number state $|n_\mu\rangle$ the expectation value of the field strength $E_\mu(r_\bullet, t)$ vanishes at all $r_\bullet$ and t, whereas nonvanishing mean-square fluctuations of the field strength exist. These field fluctuations are not caused by imperfect measuring instruments, but must be regarded as a fundamental property of the quantized electromagnetic field. They increase with increasing photon number. Note that these fluctuations also occur in the case $n_\mu = 0$, that is, if no photon exists in the μth mode.

Now we identify the state $|\psi_\mu\rangle$ in (2.116) with the coherent state $|\alpha_\mu\rangle$. Then we have the probability distribution of the field strength:

$$w\left[E_\mu(r);\,|\alpha_\mu\rangle\right] = \left(2\pi\frac{\hbar\omega_\mu}{2\varepsilon_0 V}\right)^{-1/2} \times \exp\left\{-\frac{1}{2}\left(\frac{\hbar\omega_\mu}{2\varepsilon_0 V}\right)^{-1}\left[E_\mu(r) + 2\left(\frac{\hbar\omega_\mu}{2\varepsilon_0 V}\right)^{1/2}\operatorname{Im}\{\alpha_\mu s_\mu(r)\}\right]^2\right\}. \tag{2.118}$$

This is a Gaussian distribution over $E_\mu(r)$ with the following characteristic quantities:

$$\text{mean value} = -2\left(\frac{\hbar\omega_\mu}{2\varepsilon_0 V}\right)^{1/2}\operatorname{Im}\{\alpha_\mu s_\mu(r)\}, \tag{2.118a}$$

$$\text{r.m.s. deviation} = \left(\frac{\hbar\omega_\mu}{2\varepsilon_0 V}\right)^{1/2}. \tag{2.118b}$$

The mean value represents a sine wave depending on $k_\mu r$; the expression (2.118a) corresponds to (2.102). The r.m.s. deviation is independent of the coordinate r and of the mean value of the photon number; its value coincides

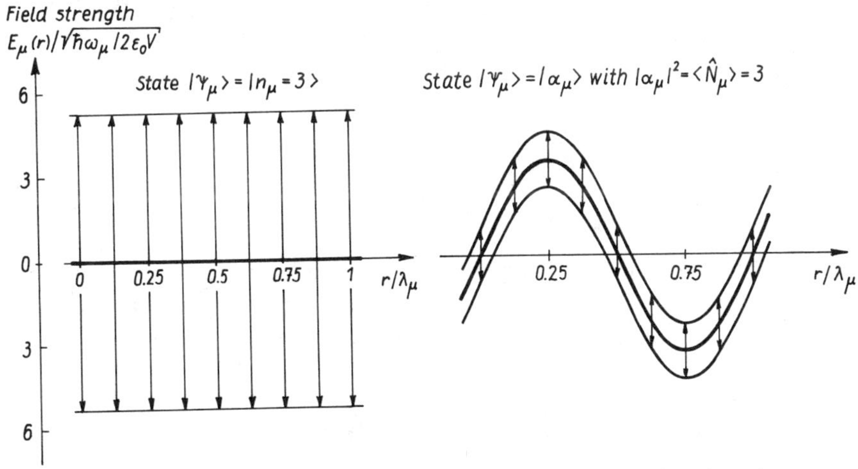

Fig. 2.2. Field-strength distributions for the photon-number state $|n_\mu = 3\rangle$ and a coherent state with $\langle N_\mu\rangle = 3$. Bold curve: expectation value $\langle\psi_\mu|\hat{E}_\mu(r)|\psi_\mu\rangle$; ↕ : range of fluctuation, $2\langle\psi_\mu|[\widehat{\Delta E}_\mu(r)]^2|\psi_\mu\rangle^{1/2}$.

Table 2.1. Important Expectation Values

Expectation Value	State $\|\psi_\mu\rangle$ $\|n_\mu\rangle$	$\|\alpha_\mu\rangle$
$\langle \hat{N}_\mu \rangle$	n_μ	$\|\alpha_\mu\|^2$
$\langle [\widehat{\Delta N_\mu}]^2 \rangle$	0	$\|\alpha_\mu\|^2 = \langle \hat{N}_\mu \rangle$
$\langle \hat{E}_{\mu\bullet}(r_\bullet, t) \rangle$	0	$e_{\mu\bullet}\sqrt{\dfrac{2\hbar\omega_\mu}{\varepsilon_0 V}}\sqrt{\langle \hat{N}_\mu \rangle}\sin\left(k_{\mu\bullet}r_\bullet - \omega_\mu t + \varphi_{\alpha_\mu} + \dfrac{\pi}{2}\right)$
$\langle [\widehat{\Delta E_{\mu\bullet}}(r_\bullet, t)]^2 \rangle$	$\dfrac{\hbar\omega_\mu}{\varepsilon_0 V}(n_\mu + \frac{1}{2})$	$\dfrac{\hbar\omega_\mu}{2\varepsilon_0 V}$
$w[n'_\mu; \|\psi_\mu\rangle]$	$\delta_{n_\mu n'_\mu}$	$(n'_\mu!)^{-1} e^{-\langle \hat{N}_\mu \rangle} \langle N_\mu \rangle^{n'_\mu}$

with that of the vacuum state. The spatial dependence of the mean value and the r.m.s. deviation are given in Fig. 2.2 at fixed time t for a photon-number state and a coherent state.

The photon number state $|n_\mu\rangle$ has to be regarded as a photon state corresponding to the particle aspect, whereas the coherent state $|\alpha_\mu\rangle$ is a photon state that (under certain circumstances) corresponds to the classical wave behavior. To illustrate this fact the foregoing results on the photon number and coherent state are compiled in Table 2.1.

For states with a fixed photon number the mean-square deviation of the particle number is zero, which indicates the connection with the particle aspect. In the coherent state the relative fluctuation of the values of the field strength reveals to what extent a classical wave is approximated. Because of the sinusoidal dependence of the mean value $\langle \alpha_\mu | \hat{E}_\mu(r) | \alpha_\mu \rangle$ on r we relate the mean-square deviations $\langle \alpha_\mu | [\widehat{\Delta E}_\mu]^2 | \alpha_\mu \rangle$ to the spatially averaged value $\overline{\overline{\langle \alpha_\mu | \hat{E}_\mu(r) | \alpha_\mu \rangle^2}}$. The results from Table 2.1 lead to the following equation for the coherent state:

$$\frac{\overline{\overline{\langle \alpha_\mu | [\widehat{\Delta E_\mu}]^2 | \alpha_\mu \rangle}}}{\overline{\overline{\langle \alpha_\mu | \hat{E}_\mu(r) | \alpha_\mu \rangle^2}}} \langle \alpha_\mu | [\widehat{\Delta N_\mu}]^2 | \alpha_\mu \rangle = \frac{1}{2}. \tag{2.119}$$

The field-strength quotient can be regarded as a measure of the wave character. If sufficiently small relative fluctuations occur, an approach to a classical wave with its well-defined values of the field strength at the space–time points exists. The relation (2.119) shows that the relative field fluctuations tend to zero with increasing mean squared photon-number deviation. Thus it may be said that the relation (2.119) expresses the complementarity of the wave and the particle interpretation.

By means of a single-mode laser operating far above threshold, radiation fields can be generated that exhibit in good approximation the properties of a coherent state (this will be shown in detail in Section 5.1.1). Therefore light in the coherent state is called ideal laser light. It is far more difficult to generate light fields in a photon-number state.

2.3.2. Mixed States of the Field

So far we have dealt with pure states of the field. In addition to these states, mixed states are of great importance, as for example, the states describing *thermal radiation* or, more generally, the so-called *chaotic radiation*. In the case of a mixed state the physical system under consideration is subjected to a weak experimental preparation, so that only an incomplete knowledge of the system exists [cf. Section A.1.5]. According to the quantum-theoretical foundations (A.73) the radiation field in a mixed state can be adequately described by the *density operator*

$$\hat{\rho} = \sum_{|\psi\rangle} p_{|\psi\rangle} |\psi\rangle\langle\psi|, \tag{2.120a}$$

where the states $|\psi\rangle$ are normalized pure photon states and $p_{|\psi\rangle}$ is the probability of finding the state $|\psi\rangle$ in the ensemble that represents the field. Of course, in a consideration like this pure states such as the photon-number state or the coherent state are included as special cases. For instance, the density operator of a global coherent state $\{|\alpha_\mu\rangle\}$ [cf. Section 2.3.1] takes the form

$$\hat{\rho} = |\{\alpha_\mu\}\rangle\langle\{\alpha_\mu\}|. \tag{2.120b}$$

Since the properties of a coherent state are closely connected with those of a classical wave, it is advantageous to represent the general density operator (2.120a) of the radiation field as a superposition of the operators $|\{\alpha_\mu\}\rangle\langle\{\alpha_\mu\}|$; this leads to the so-called *Glauber-Sudarshan P-representation* of $\hat{\rho}$.

Its characteristic features can be demonstrated by a one-mode consideration. The P-representation (of the μth mode) is defined by

$$\hat{\rho}_\mu = \int d^2\alpha_\mu\, P(\alpha_\mu) |\alpha_\mu\rangle\langle\alpha_\mu|, \tag{2.121}$$

where $P(\alpha_\mu)$ is a c-number. The quantity $P(\alpha_\mu)$ has the character of a weight function. Because $\hat{\rho}_\mu = \hat{\rho}_\mu^\dagger$, $P(\alpha_\mu)$ is a real function of the complex variable α_μ; since generally $\operatorname{Tr}\hat{\rho}_\mu = 1$, the function $P(\alpha_\mu)$ is normalized. So we have the properties

$$P(\alpha_\mu) = P^*(\alpha_\mu), \qquad \int d^2\alpha_\mu\, P(\alpha_\mu) = 1. \tag{2.122}$$

The general representation of $\hat{\rho}_\mu$ in terms of coherent states leads, by (2.98), to

$$\hat{\rho}_\mu = \frac{1}{\pi^2} \int d^2\alpha_\mu \int d^2\alpha'_\mu \langle\alpha_\mu|\hat{\rho}_\mu|\alpha'_\mu\rangle |\alpha_\mu\rangle\langle\alpha'_\mu|, \tag{2.123}$$

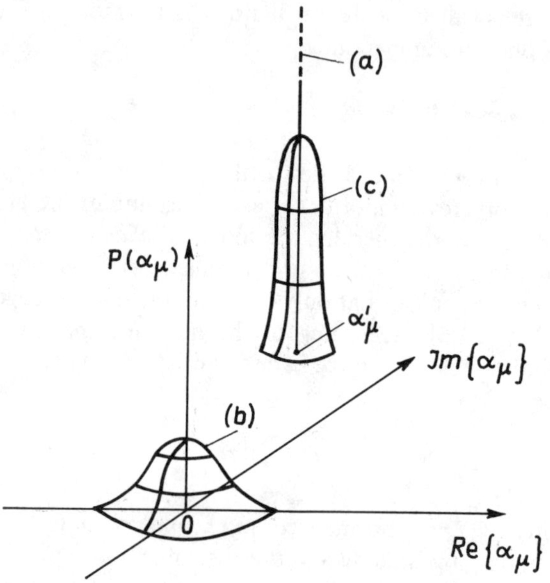

Fig. 2.3. *P*-representation. Function $P(\alpha_\mu)$ for (a) a coherent state, (b) chaotic radiation, (c) a superposition model.

whereas the *P*-representation has the simpler diagonal form (2.121). However, it must be emphasized that $P(\alpha_\mu)$, in spite of the properties expressed in (2.122), cannot generally be interpreted as a proper probability distribution, since $P(\alpha_\mu)$ takes on negative values under certain conditions.

Let us now discuss some important examples of the *P*-representation of mixed states.

Obviously, the *P* representation for a coherent state $|\alpha'_\mu\rangle$ that has the density operator $\hat{\rho}_\mu = |\alpha'_\mu\rangle\langle\alpha'_\mu|$ is given by the two-dimensional delta function

$$P(\alpha_\mu) = \delta^{(2)}(\alpha_\mu - \alpha'_\mu). \tag{2.124}$$

Only at $\alpha_\mu = \alpha'_\mu$ does the function $P(\alpha_\mu)$ not vanish. This means that the absolute value and the phase of the complex amplitude are fixed; these properties characterize the so-called ideal laser radiation. An illustration of $P(\alpha_\mu)$ in the α_μ-plane is given in Fig. 2.3(a).

We now turn to the so-called chaotic light. It is physically characterized by the requirement for maximum entropy of the radiation at a given mean photon number $\langle\hat{N}_\mu\rangle$. According to general quantum-theoretical principles this means that $-k_B\mathrm{Tr}\{\hat{\rho}_\mu \ln \hat{\rho}_\mu\}$ is a maximum under the secondary condition of a given mean value $\langle\hat{N}_\mu\rangle = \mathrm{Tr}\{\hat{\rho}_\mu\hat{N}_\mu\}$, where of course the general condition $\mathrm{Tr}\{\rho_\mu\} = 1$ has to be satisfied [cf. Section A.1.5]. Under these conditions the variational calculus provides the solution

$$\hat{\rho}_\mu = \frac{e^{-\lambda \hat{N}_\mu}}{\mathrm{Tr}\{e^{-\lambda \hat{N}_\mu}\}}, \tag{2.125a}$$

where the Lagrange multiplier λ has to be determined from the secondary condition for $\langle \hat{N}_\mu \rangle$ [cf. (A.81)]. This leads to

$$\hat{\rho}_\mu = \frac{\langle \hat{N}_\mu \rangle^{\hat{N}_\mu}}{\left(1 + \langle \hat{N}_\mu \rangle\right)^{\hat{N}_\mu + \hat{I}}} = \sum_{n_\mu} \frac{\langle \hat{N}_\mu \rangle^{n_\mu}}{\left(1 + \langle \hat{N}_\mu \rangle\right)^{n_\mu + 1}} |n_\mu\rangle\langle n_\mu|. \tag{2.125b}$$

Making use of (2.98), one has

$$\hat{\rho}_\mu = \frac{1}{\pi} \int d^2\alpha_\mu \, |\alpha_\mu\rangle\langle\alpha_\mu| \sum_{n_\mu} \frac{\langle \hat{N}_\mu \rangle^{n_\mu}}{\left(1 + \langle \hat{N}_\mu \rangle\right)^{n_\mu + 1}} |n_\mu\rangle\langle n_\mu|; \tag{2.125c}$$

with the help of (2.95) and (2.97a) one finally obtains

$$\hat{\rho}_\mu = \int d^2\alpha_\mu \left[\pi \langle \hat{N}_\mu \rangle\right]^{-1} \exp\left[-\frac{|\alpha_\mu|^2}{\langle \hat{N}_\mu \rangle}\right] |\alpha_\mu\rangle\langle\alpha_\mu|. \tag{2.125d}$$

Thus the chaotic light has the *P*-representation

$$P(\alpha_\mu) = [\pi \langle \hat{N}_\mu \rangle]^{-1} \exp\left[-\frac{|\alpha_\mu|^2}{\langle \hat{N}_\mu \rangle}\right]. \tag{2.126}$$

This case is illustrated in Fig. 2.3(b). There is an equipartition of phases and a Gaussian distribution of the absolute values of the amplitudes of the waves involved in $\hat{\rho}_\mu$; this is referred to as a *chaotic radiation* state. Light from thermal sources is chaotic; however, it can also be generated artificially by randomizing laser radiation—for example, by passing it through a rotating ground glass.

Often the properties of real radiation sources lie between the abovementioned two limiting cases of ideal laser radiation and completely chaotic radiation. Then, a so-called superposition model can be applied. The corresponding *P*-representation

$$P(\alpha_\mu) = [\pi \langle \hat{N}_\mu \rangle]^{-1} \exp\left[-\frac{|\alpha_\mu - \alpha'_\mu|^2}{\langle \hat{N}_\mu \rangle}\right] \tag{2.127}$$

is illustrated in Fig. 2.3(c). The mean value of the amplitude factors is α'_μ. The quantity $\langle \hat{N}_\mu \rangle$ can obviously be regarded as a measure of the amplitude fluctuations with respect to their phases and absolute values.

In particular, when dealing with problems of nonlinear optics and multiphoton processes, an important point is the calculation of expectation values of products of field operators. Because of (2.51), this means that the expectation values of the products of creation and annihilation operators have to be determined. To give an example, consider the product $(\hat{a}_\mu^\dagger)^\sigma(\hat{a}_\mu)^\tau$, where σ and τ are integers; note that this is a so-called normally ordered product, where the annihilation operators stand to the right of the creation operators. This means that annihilation operators act on a state vector before creation operators do. For antinormally ordered products the succession is interchanged.

The expectation value of the product operator $(\hat{a}_\mu^\dagger)^\sigma(\hat{a}_\mu)^\tau$ can be represented with the aid of $P(\alpha_\mu)$:

$$\operatorname{Tr}\left\{\hat{\rho}_\mu\left(\hat{a}_\mu^\dagger\right)^\sigma\left(\hat{a}_\mu\right)^\tau\right\} = \int d^2\alpha_\mu\, P(\alpha_\mu)(\alpha_\mu^*)^\sigma(\alpha_\mu)^\tau. \tag{2.128}$$

It is also possible to achieve a simple form of this expectation value by using the two-dimensional Fourier transform $\varkappa(\lambda_\mu)$ of $P(\alpha_\mu)$; the Fourier transformation and the respective inverse transformation are given by

$$\varkappa(\lambda_\mu) = \int d^2\alpha_\mu\, P(\alpha_\mu)\exp\left[\lambda_\mu\alpha_\mu^* - \lambda_\mu^*\alpha_\mu\right], \tag{2.129a}$$

$$P(\alpha_\mu) = \frac{1}{\pi^2}\int d^2\lambda_\mu\, \varkappa(\lambda_\mu)\exp\left[-\lambda_\mu\alpha_\mu^* + \lambda_\mu^*\alpha_\mu\right]. \tag{2.129b}$$

The first relation reveals that $\varkappa(\lambda_\mu)$ can be written as

$$\varkappa(\lambda_\mu) = \operatorname{Tr}\left\{\hat{\rho}_\mu\hat{D}_N(\lambda_\mu)\right\}, \qquad \text{where} \quad \hat{D}_N(\lambda_\mu) = \exp\left[\lambda_\mu\hat{a}_\mu^\dagger\right]\exp\left[-\lambda_\mu^*\hat{a}_\mu\right]. \tag{2.129c}$$

The so-called characteristic function $\varkappa(\lambda_\mu)$ yields the expectation value of the operator product by differentiating:

$$\operatorname{Tr}\left\{\hat{\rho}_\mu\left(\hat{a}_\mu^\dagger\right)^\sigma\left(\hat{a}_\mu\right)^\tau\right\} = \left.\frac{\partial^\sigma}{\partial\lambda_\mu^\sigma}\frac{\partial^\tau}{\partial\left(-\lambda_\mu^*\right)^\tau}\varkappa(\lambda_\mu)\right|_{\lambda_\mu=0}. \tag{2.130}$$

Let us add some remarks concerning the mathematical properties of $P(\alpha_\mu)$ and $\varkappa(\lambda_\mu)$. In the case of chaotic radiation and of the superposition model the functions $P(\alpha_\mu)$ proved to be mathematically well behaved and nonnegative; they could be clearly interpreted as probability distributions of the wave amplitudes. In other cases the situation is more complicated. So, for instance, the P-representation of the photon number state $|n_\mu\rangle$ is proportional to the n_μth derivative of the delta function. This shows that generalized functions may also occur. Then special investigations of the existence problem for these functions and of the interpretation of the relations given above have to be performed.

Because of these "impractical" properties it is often advantageous to use better-behaved functions for the description of radiation field states (Ref. 2.5). In this regard a useful tool is the function $P_A(\alpha_\mu) = \pi^{-1}\langle\alpha_\mu|\hat{\rho}_\mu|\alpha_\mu\rangle$, which is associated with its two-dimensional Fourier transform $\varkappa_A(\lambda_\mu)$ by an analogeous relation to that given in (2.129a), namely

$$\varkappa_A(\lambda_\mu) = \int d^2\alpha_\mu\, P_A(\alpha_\mu)\exp\left[\lambda_\mu\alpha_\mu^* - \lambda_\mu^*\alpha_\mu\right]. \tag{2.131}$$

The functions $\varkappa(\lambda_\mu)$ and $\varkappa_A(\lambda_\mu)$ are connected in a simple way with the mean value of the displacement operator $\hat{D}(\lambda_\mu)$ from (2.96a); we have

$$\varkappa(\lambda_\mu) = \mathrm{Tr}\left\{\hat{\rho}_\mu e^{\lambda_\mu\hat{a}_\mu^\dagger}e^{-\lambda_\mu^*\hat{a}_\mu}\right\} = \mathrm{Tr}\left\{\hat{\rho}_\mu\hat{D}(\lambda_\mu)\right\}e^{\frac{1}{2}|\lambda_\mu|^2}, \tag{2.132a}$$

$$\varkappa_A(\lambda_\mu) = \mathrm{Tr}\left\{\hat{\rho}_\mu e^{-\lambda_\mu^*\hat{a}_\mu}e^{\lambda_\mu\hat{a}_\mu^\dagger}\right\} = \mathrm{Tr}\left\{\hat{\rho}_\mu\hat{D}(\lambda_\mu)\right\}e^{-\frac{1}{2}|\lambda_\mu|^2}. \tag{2.132b}$$

Because of the unitarity of $\hat{D}$, the inequality $|\mathrm{Tr}\{\hat{\rho}_\mu\hat{D}(\lambda_\mu)\}| \le 1$ holds; therefore we have

$$\left|\varkappa(\lambda_\mu)\right| \le e^{\frac{1}{2}|\lambda_\mu|^2}, \qquad \left|\varkappa_A(\lambda_\mu)\right| \le e^{-\frac{1}{2}|\lambda_\mu|^2}. \tag{2.132c}$$

Obviously, in certain cases $\varkappa(\lambda_\mu)$ may diverge as $\exp(|\lambda_\mu|^2/2)$ for large $|\lambda_\mu|$. However, the function $\varkappa_A(\lambda_\mu)$ always tends to zero as $\exp(-|\lambda_\mu|^2/2)$ for large $|\lambda_\mu|$; the function $\varkappa_A(\lambda_\mu)$ and its Fourier transform $P_A(\alpha_\mu)$ are always well behaved, with the properties $|\varkappa_A(\lambda_\mu)| \le 1$ and $|P_A(\alpha_\mu)| \le 1$. As can be easily seen, the equations (2.132) contain the simple relation $\varkappa_A(\lambda_\mu) = \varkappa(\lambda_\mu)e^{-|\lambda_\mu|^2}$ between the functions $\varkappa(\lambda_\mu)$ and $\varkappa_A(\lambda_\mu)$; it can be rewritten as a relation between $P(\alpha_\mu)$ and $P_A(\alpha_\mu)$ by using the Fourier transformation described in (2.129). The functions $P_A(\alpha_\mu)$ and $\varkappa_A(\lambda_\mu)$ have the same significance with respect to the expectation values of antinormally ordered operator products as $P(\alpha_\mu)$ and $\varkappa(\lambda_\mu)$ have for normally ordered ones:

$$\begin{aligned}\mathrm{Tr}\left\{\hat{\rho}_\mu(\hat{a}_\mu)^\sigma(\hat{a}_\mu^\dagger)^\tau\right\} &= \int d^2\alpha_\mu\, P_A(\alpha_\mu)(\alpha_\mu)^\sigma(\alpha_\mu^*)^\tau \\ &= \frac{\partial^\tau}{\partial\lambda_\mu^\tau}\frac{\partial^\sigma}{\partial(-\lambda_\mu^*)^\sigma}\varkappa_A(\lambda_\mu)\bigg|_{\lambda_\mu=0}\end{aligned} \tag{2.133}$$

The functions $\varkappa(\lambda_\mu)$ and $\varkappa_A(\lambda_\mu)$ are called the *normally ordered* and the *antinormally ordered characteristic function*, respectively.

Now, we pass to the *multimode P-representation* of a general density operator $\hat{\rho}$. For this purpose the one-mode P-representation (2.121) of $\hat{\rho}_\mu$ has to be replaced by

$$\hat{\rho} = \int d^2\{\alpha_\mu\}\, P(\{\alpha_\mu\})|\{\alpha_\mu\}\rangle\langle\{\alpha_\mu\}|. \tag{2.134}$$

The real function $P(\{\alpha_\mu\})$ now depends on a set of amplitude factors α_μ for all modes. In analogy to the one-mode case, the corresponding functions $P_A(\{\alpha_\mu\})$, $\varkappa(\{\lambda_\mu\})$, and $\varkappa_A(\{\lambda_\mu\})$ may also be introduced for the multimode case.

The function $P(\{\alpha_\mu\})$ characterizes the state of the radiation field, just as $\hat{\rho}$ does. With the help of $P(\{\alpha_\mu\})$, we shall represent an important expectation value, namely the photon-number probability distribution $w(n)$ in the normalization volume V. Let us recall the result (2.103), where the photon-number distribution for a one-mode coherent state was derived, and turned out to be the Poisson distribution. Following the same pattern, the probability of finding n photons in the general multimode case can be calculated; one obtains

$$w(n) = \int_0^\infty d\tilde{M}\, \mathscr{W}(\tilde{M})(n!)^{-1}\tilde{M}^n e^{-\tilde{M}}, \tag{2.135}$$

where $\tilde{M} = \Sigma_\mu|\alpha_\mu|^2$ is the expectation value of the total photon number operator $\hat{N} = \Sigma_\mu \hat{N}_\mu$ in the state $|\{\alpha_\mu\}\rangle$. The function $\mathscr{W}(\tilde{M})$ expresses the connection with the P-representation $P(\{\alpha_\mu\})$ of the state by

$$\mathscr{W}(\tilde{M}) = \int d^2\{\alpha'_\mu\}\, P(\{\alpha'_\mu\})\delta\Big(\tilde{M} - \sum_\mu|\alpha'_\mu|^2\Big). \tag{2.136}$$

Here, in the general case, $w(n)$ may be regarded as an averaging of Poisson distributions with the weight function $\mathscr{W}(\tilde{M})$.

In Section 5.1.1 we shall discuss in detail the close relationship between $w(n)$ and the photon-counting distribution, which governs the explanation of photon-counting experiments on the basis of the external photoeffect. However, already at this point we want to consider conditions and quantities occurring in the discussion of the counting process with a photodetector. The normalization volume V can be regarded as the volume AcT in front of a detector upon the receiving area A onto which the photons are perpendicularly incident. Suppose cT to be the length of the path the photons travel in a (macrophysical) time interval of duration T, given, for instance, by the measuring time in the experiment considered. Then $\tilde{M}$ is given by

$$\tilde{M} = \int_t^{t+T} dt' \, S(t'), \tag{2.137}$$

where $S(t')$ is the number of incident photons per unit time.

In the multimode case the chaotic radiation is characterized by the requirement for maximum entropy with fixed mean photon number $\langle \hat{N}_\mu \rangle$ in each of the modes; this leads to

$$P(\{\alpha_\mu\}) = \prod_\mu \langle \hat{N}_\mu \rangle^{-1} \exp\left[-\frac{|\alpha_\mu|^2}{\langle \hat{N}_\mu \rangle} \right], \tag{2.138}$$

which is the product of the single-mode expressions (2.126). Thermal radiation is a special case of chaotic radiation; here $\langle \hat{N}_\mu \rangle$ from (2.138) coincides with the ensemble mean value of the photon number in thermal equilibrium, which is

$$\langle \hat{N}_\mu \rangle_{\text{th}} = \left[\exp\left(\frac{\hbar \omega_\mu}{k_B T} \right) - 1 \right]^{-1}. \tag{2.139}$$

The quantity $\hbar \omega_\mu \langle \hat{N}_\mu \rangle_{\text{th}}$ describes the mean energy of the photons in the μth mode; multiplying this quantity by the mode density [cf. (1.26)], we obtain Planck's radiation formula for the radiation energy per unit frequency per unit volume which has been found to be in excellent agreement with the experimental results.

Let us finally sketch the experimental verification of certain photon distributions. While laser light well above threshold (ideal laser light) exhibits a Poisson distribution [cf. (2.103)], chaotic light [cf. (2.125)] has a super-Poisson distribution. In the measurement of Franck-Hertz light under certain conditions also light with sub-Poisson distribution has been detected (cf. Ref. 2.6).

REFERENCES

2.1. M. Schubert and W. Vogel, *Wiss. Z. Univ. Jena, Math. Nat. Reihe* **27**, 179 (1978).
2.2. W. H. Louisell, *Phys. Lett.* **7**, 49 (1963).
2.3. M. Schubert, Phys. Lett. **27A**, 698 (1968).
2.4. H. Paul, *Fortschr. d. Phys.* **22**, 657 (1975).
2.5. J. Peřina, *Coherence of Light*, Van Nostrand Reinhold, London, 1971, Sections 13.2 and 13.3.
2.6. M. C. Teich, B. E. A. Saleh, J. Peřina, *J. Opt. Soc. Amer.* **B1**, 366 (1984).

3

Interaction Between Radiation and Matter

All the nonlinear optical effects and multiphoton processes dealt with so far in this book and also those to be still treated in it rest on the mutual interaction of the electromagnetic radiation with matter in the form of single particles, atoms, ions, molecules, crystals, and liquids. In this sense the study of this interaction can be regarded as the main purpose of the general theoretical concepts presented in this book.

For describing the interaction between light and matter the classical treatment of Chapter 1 makes use of the susceptibilities, which involve phenomenological parameters and forms a link between electric field strength and electric polarization. In Chapter 3 the interaction is described by a special term in the total Hamiltonian, the interaction or coupling Hamiltonian; this corresponds to the quantum-theoretical concept of the interaction between quasiisolated systems, as sketched in connection with the interaction picture (Section A.2.1). It will become apparent that a relationship exists between these two procedures. So in the semiclassical treatment it is possible to interpret the parameters of the susceptibilities as intrinsic properties (transition frequencies, dipole moments) of the atomic systems; this will be shown in Chapter 4. Furthermore, in cases where there are no real transitions of the atomic system, an effective interaction operator can be formulated in terms of phenomenologically introduced susceptibilities (Section 3.2.3).

With the help of the interaction Hamiltonian the physically relevant expectation value of an arbitrary observable can in principle be evaluated; the general foundation of this procedure is given in Section 3.1. To gain results from general equations that can be compared with experiments, it is mostly necessary to employ specific concepts or approximation methods. Some of these procedures are of great importance and have wide application; they will be presented in Section 3.2. For the description of specific nonlinear optical processes and multiphoton processes as given in Part II, we need terms, laws, and order-of-magnitude evaluations for linear one-photon processes, namely spontaneous and stimulated emission and absorption. These aspects will be discussed in Section 3.3.

3.1. FOUNDATIONS OF THE INTERACTION BETWEEN RADIATION AND MATTER

The interaction of electromagnetic radiation with matter is based on the behavior of electrically charged particles interacting with the radiation field. In the treatment of Section A.2.1, our consideration begins with two uncoupled ("free") subsystems, the radiation field and the atomic system. If a coupling between these subsystems is introduced, a combined system will result in which the properties of the two subsystems are modified, since the subsystems affect each other. The coupling is represented by a coupling term in the total-energy operator. The derivation of this term will be discussed in Section 3.1.1. The general procedure for the derivation of physically relevant relations and of expectation values for the given interaction operator will be dealt with in Section 3.1.2.

3.1.1. The Interaction Operator

The derivation of the interaction part in the total Hamiltonian can be performed analogously to the classical approach. Therefore we shall consider the classical case before treating the semiclassical and the fully quantum-theoretical case.

Let us begin the classical description with the consideration of a single pointlike particle, which may be characterized by its electric charge q and its mass m. We assume that the particle is acted on by the electric field

$$E_{\bullet}(r_{\bullet}, t) = -\frac{\partial}{\partial t}A_{\bullet}(r_{\bullet}, t) - \nabla_{\bullet}V(r_{\bullet}) \tag{3.1}$$

and by the magnetic field

$$B_{\bullet}(r_{\bullet}, t) = \nabla_{\bullet} \times A_{\bullet}(r_{\bullet}, t), \tag{3.2}$$

where $A_{\bullet}(r_{\bullet}, t)$ is the vector potential of a radiation field with the properties described in (2.4a) and (2.4b). The quantity $V(r_{\bullet})$ represents a scalar electric potential; if, for instance, we identify the particle with an electron in an outer atomic shell, $V(r_{\bullet})$ is the potential of the atomic core. Taking into account the electrostatic force and the Lorentzian force, the (nonrelativistic) equation of motion

$$m\frac{d^2 r_{\bullet}}{dt^2} = qE_{\bullet} + q\left(\frac{dr_{\bullet}}{dt} \times B_{\bullet}\right) \tag{3.3}$$

is obtained; thus the Lagrangian function

$$L\left(r_\bullet, \frac{dr_\bullet}{dt}\right) = \frac{m}{2}\left(\frac{dr_\bullet}{dt}\right)^2 - qV + q\frac{dr_\bullet}{dt}A_\bullet \tag{3.4}$$

must be assigned to the system under consideration. From this expression the equation of motion

$$\frac{d}{dt}\frac{\partial L}{\partial(dr_j/dt)} - \frac{\partial L}{\partial r_j} = 0 \tag{3.5}$$

for the jth component of the vector $r_\bullet$ can be derived. This result agrees with the differential equation (3.3). The canonically conjugate momentum $p_\bullet$ of the position coordinate $r_\bullet$ is

$$p_\bullet = \nabla_{(\dot{r}_\bullet)_\bullet} L = m\frac{dr_\bullet}{dt} + qA_\bullet. \tag{3.6}$$

Thus the Hamiltonian function is given by

$$H(r_\bullet, p_\bullet) = p_\bullet\frac{dr_\bullet}{dt} - L = \frac{1}{2m}(p_\bullet - qA_\bullet)^2 + qV. \tag{3.7}$$

This expression can be written in the form

$$H = \left[\frac{1}{2m}p_\bullet^2 + qV\right] + H^C, \tag{3.7a}$$

where the interaction term H^C is given by

$$H^C = -\frac{q}{m}A_\bullet p_\bullet + \frac{q^2}{2m}A_\bullet^2. \tag{3.8}$$

The first term on the right-hand side of (3.7a) does not contain the vector potential representing the radiation field; therefore the expression in brackets can be interpreted as the Hamiltonian function of the isolated atomic subsystem. The term H^C represents the coupling between the atomic and the radiation subsystem; it contains atomic as well as radiation quantities. H is the Hamiltonian function of an atomic system in the presence of a given external field $A_\bullet$. In reality, the atomic system exerts a reactive force on the radiation field. If we want to obtain the Hamiltonian function of the total matter–radiation system, then on the right-hand side of (3.7) the term (2.8a) representing the isolated radiation system must be added.

The interaction term H^C consists of two terms, the first of which (denoted by $H^{(1)}$) is linearly dependent on the vector potential of the radiation field, whereas the second (denoted by $H^{(2)}$) exhibits a quadratic dependence on it. Let us compare these two terms under simple conditions. To this end we consider an electron (with charge $-e$) in the field of an atomic core (with charge $+e$); we may think here of a hydrogenlike structure. Suppose the radiation field to consist only of one mode (frequency ω, amplitude $\hat{A}_\bullet$); this means that the amplitude of the electric field strength is $|\hat{E}_\bullet| = \omega|\hat{A}_\bullet|$. Thus the ratio of the quadratic to the linear term has the following order of magnitude:

$$\frac{|H^{(2)}|}{|H^{(1)}|} \sim \frac{e|\hat{E}_\bullet|}{\omega|p_\bullet|}. \tag{3.9}$$

For an electron in a hydrogenlike structure the quantity $e/\omega|p_\bullet|$ can be evaluated from quantum mechanics. The mean momentum value $|p_\bullet|$ is of order $\hbar/r$, where r is the mean value of the distance from the nucleus. Moreover, for optical frequencies in the visible region the value $\hbar\omega$ is of the order $er|E_{\text{atom}}|$, where $|E_{\text{atom}}|$ is the (averaged) field strength of the atomic core acting on the electron, having a value of some 10^{11} V/m. Altogether we arrive at

$$\frac{|H^{(2)}|}{|H^{(1)}|} \sim \frac{|\hat{E}_\bullet|}{|E_{\text{atom}}|}. \tag{3.10}$$

Consequently, for an electron in an outer atomic shell the quadratic term in H^C becomes comparable with the linear one when the field strength $|E_\bullet|$ of the incident radiation approaches $|E_{\text{atom}}|$.

Let us now pass over to the consideration of the case of an atomic system consisting of several pointlike charges (electrons, nuclei). Let the charge carrier α have the charge q_α and the mass m_α; its position and momentum vectors will be denoted by $r_{\alpha\bullet}$ and $p_{\alpha\bullet}$. Then the same formalism as has just been given for the treatment of one point charge leads to the following Hamiltonian function for an atomic system in an external field described by the vector potential $A_\bullet$:

$$H = H^A + H^C, \tag{3.11}$$

where

$$H^A = \sum_\alpha \frac{1}{2m_\alpha} p_{\alpha\bullet}^2 + U(\ldots, r_{\alpha\bullet}, \ldots) \tag{3.12}$$

and

$$H^C = \sum_\alpha -\frac{q_\alpha}{m_\alpha} A_\bullet(r_{\alpha\bullet}, t) p_{\alpha\bullet} + \sum_\alpha \frac{q_\alpha^2}{2m_\alpha} A_\bullet^2(r_{\alpha\bullet}, t). \tag{3.13}$$

H^A represents the unperturbed (isolated) atomic system with the potential energy U. The coupling term H^C arises from the superposition of the respective parts of all the charge carriers of the atomic system.

Let us now give the *semiclassical description* of an atomic system in an external electromagnetic field $A_\bullet$. We assume that the dependence of $A_\bullet$ on the position and time coordinates is given and that it can be described by a classical function; the reaction of the charge carriers on the electromagnetic field is neglected. The atomic system is quantized, and the coordinates $r_{\alpha\bullet}$ and the momenta $p_{\alpha\bullet}$ of the charge carriers go over into the Hilbert-space operators $\hat{r}_{\alpha\bullet}$ and $\hat{p}_{\alpha\bullet}$. The components of these vector operators obey the commutation relations (A.8). Replacing the quantities $r_{\alpha\bullet}$, $p_{\alpha\bullet}$ by $\hat{r}_{\alpha\bullet}$, $\hat{p}_{\alpha\bullet}$ on the right-hand side of (3.12), one obtains the Hamiltonian operator $\hat{H}^A$ of the atomic system. To allow the description of the action of the electromagnetic field on the pointlike charge carriers, their space coordinates $r_{\alpha\bullet}$ have to be replaced by the operators $\hat{r}_{\alpha\bullet}$, in particular in the field function $A_\bullet(r_{\alpha\bullet}, t)$. But because of the noncommutability of the components of $\hat{r}_{\alpha\bullet}$ and $\hat{p}_{\alpha\bullet}$, it makes a difference whether the classical expression $A_\bullet(r_{\alpha\bullet}, t)p_{\alpha\bullet}$, which is equal to $p_{\alpha\bullet}A_\bullet(r_{\alpha\bullet}, t)$, in the quantization is transformed into the operator $A_\bullet(\hat{r}_{\alpha\bullet}, t)\hat{p}_{\alpha\bullet}$ or into the operator $\hat{p}_{\alpha\bullet}A_\bullet(\hat{r}_{\alpha\bullet}, t)$. As in similar cases, its representation as hermitized product

$$\tfrac{1}{2}\left[A_\bullet(\hat{r}_{\alpha\bullet}, t)\hat{p}_{\alpha\bullet} + \hat{p}_{\alpha\bullet}A_\bullet(\hat{r}_{\alpha\bullet}, t)\right] \tag{3.14a}$$

leads to the correct expression. Using (A.29), this formula can be rewritten as

$$\tfrac{1}{2}\left[2A_\bullet(\hat{r}_{\alpha\bullet}, t)\hat{p}_{\alpha\bullet} - i\hbar\nabla_{(\hat{r}_{\alpha\bullet})\bullet}A_\bullet(\hat{r}_{\alpha\bullet}, t)\right]. \tag{3.14b}$$

Because of the Coulomb gauge condition (2.4b), the second term in the bracket vanishes. Thus the linear term of the interaction Hamiltonian must be formulated by the expression $A_\bullet(\hat{r}_{\alpha\bullet}, t)\hat{p}_{\alpha\bullet}$, where the momentum operator $\hat{p}_{\alpha\bullet}$ is to the right of the operator function $A_\bullet(\hat{r}_{\alpha\bullet}, t)$. Altogether, the Hamiltonian of an atomic system in a radiation field is

$$\hat{H} = \hat{H}^A + \hat{H}^C, \tag{3.15}$$

where the interaction Hamiltonian $\hat{H}^C$ is represented by

$$\hat{H}^C = \sum_\alpha -\frac{q_\alpha}{m_\alpha}A_\bullet(\hat{r}_{\alpha\bullet}, t)\hat{p}_{\alpha\bullet} + \sum_\alpha \frac{q_\alpha^2}{2m_\alpha}A_\bullet^2(\hat{r}_{\alpha\bullet}, t). \tag{3.16}$$

In the *fully quantum-theoretical description* of the interaction we start from two subsystems assumed to be independent of each other. These are on the one hand the atomic system with the Hamiltonian $\hat{H}^A$, and on the other hand the radiation field with the Hamiltonian $\hat{H}^R$. In contrast to the semiclassical treatment, also the radiation field will now be described as a quantized system. To do this, we may use the relations for the quantized free radiation field represented in Chapter 2. For an expansion of the field in linearly polarized plane waves the Hamiltonian

$$\hat{H}^R = \sum_{\mu} \hbar\omega_{\mu}\left(\hat{a}_{\mu}^{\dagger}\hat{a}_{\mu} + \tfrac{1}{2}\hat{I}\right) \tag{3.17}$$

must be used, which coincides with the expression derived in (2.21). The operators $\hat{a}_{\mu}^{\dagger}$ and $\hat{a}_{\mu}$ represent the creation and annihilation operators of the μth mode, and ω_{μ} is the frequency. The Hamiltonian $\hat{H}$ of the total system is

$$\hat{H} = \hat{H}^A + \hat{H}^R + \hat{H}^C, \tag{3.18}$$

where the interaction operator is given by

$$\hat{H}^C = \sum_{\alpha} -\frac{q_{\alpha}}{m_{\alpha}}\hat{A}_{\bullet}(\hat{r}_{\alpha\bullet})\hat{p}_{\alpha\bullet} + \sum_{\alpha}\frac{q_{\alpha}^2}{2m_{\alpha}}\hat{A}_{\bullet}^2(\hat{r}_{\alpha\bullet}). \tag{3.19}$$

The vector potential $\hat{A}_{\bullet}(\hat{r}_{\alpha\bullet})$ has the form

$$\hat{A}_{\bullet}(\hat{r}_{\alpha\bullet}) = \sum_{\mu}\left(\frac{\hbar}{2\varepsilon_0 V\omega_{\mu}}\right)^{1/2} e_{\mu\bullet}\left\{\hat{a}_{\mu}\exp\left(ik_{\mu\bullet}\hat{r}_{\alpha\bullet}\right) + \text{h.c.}\right\}. \tag{3.20}$$

All the operators used in (3.18)–(3.20) are written in the Schrödinger picture. Note that the expression (3.20) agrees with the right-hand side of (2.19) at $t = 0$ except that now the operator $\hat{r}_{\alpha\bullet}$ has been inserted, which is assigned to the position of the αth charge carrier in the case of the quantum-theoretical description. The linear term on the right-hand side of (3.19) is the interaction term which is of particular importance for the applications to be dealt with in the following sections.

The Hamiltonian (3.18) of the total system satisfies the assumptions of the interaction picture as sketched in Section A.2.1. The operator $\hat{H}^A + \hat{H}^R$ must be identified with the Hamiltonian $\hat{H}^F$ of the "total free system." The operators $\hat{H}^A$ and $\hat{H}^R$ can be assigned to independent subspaces in the total Hilbert space. Therefore, according to (A.49), the eigenvalues of $\hat{H}^F$ are given by the sum of the eigenvalues of $\hat{H}^A$ and $\hat{H}^R$, whereas, the eigenkets of $\hat{H}^F$ are given by the direct product. Thus we have

$$\hat{H}^F|\mathscr{E}^F\rangle = \mathscr{E}^F|\mathscr{E}^F\rangle, \tag{3.21}$$

where

$$\mathscr{E}^F = \mathscr{E}^A + \sum_{\mu}\hbar\omega_{\mu}\left(n_{\mu} + \tfrac{1}{2}\right), \qquad |\mathscr{E}^F\rangle = |\mathscr{E}^A\rangle|\ldots, n_{\mu}, \ldots\rangle. \tag{3.22}$$

The Hamiltonians assigned to the subsystems satisfy the eigenvalue equations

$$\hat{H}^A|\mathscr{E}^A\rangle = \mathscr{E}^A|\mathscr{E}^A\rangle,$$

$$\hat{H}^R|\ldots, n_\mu, \ldots\rangle = \left[\sum_\mu \hbar\omega_\mu\left(n_\mu + \tfrac{1}{2}\right)\right]|\ldots, n_\mu, \ldots\rangle. \tag{3.23}$$

We conclude this subsection with some remarks on the connection between classical, semiclassical, and fully quantum-theoretical descriptions as well as on the approximations and the range of validity of the expressions derived.

The above expressions for the fully quantum-theoretical description, especially the interaction operator (3.19), can be regarded as the basis of the treatment of linear and nonlinear optical processes. They have proved to be a useful tool in the description of interaction effects of electromagnetic radiation with atoms, molecules, and solids. These basic relations were obtained by inductive derivation in correspondence with the classical description, starting from the nonrelativistic equation of motion of the atomic system in the presence of a radiation field. It may be remarked here that the interaction operators given above can also be obtained in the very same form (3.19) by deduction with the help of the formalism of quantum field theory (see Refs. 3.1, 3.2). To achieve this, the Dirac field has to be treated in interaction with the electromagnetic field. By means of an appropriate transformation (the Foldy–Wouthuysen transformation), the contribution of positrons is eliminated and a Hamiltonian for a fixed number of electrons is formed by the consistent expansion in terms of the elementary charge and the reciprocal velocity of light. In such an approach relativistic correction terms arise automatically that are connected with the properties of the discussed pointlike particles (mass, orbital momentum, spin). But in the description of most of the important nonlinear optical effects they may be neglected or globally incorporated in $\hat{H}^A$. In addition, let us assume that the energies involved are sufficiently small so that no relativistic effects occur, such as the creation or annihilation of the charge carriers of which the atomic system consists [cf. the explanation of the equations (2.10a, b)].

Since the relations (3.18) and (3.19) can be considered to describe adequately the interaction of radiation and matter, we may infer from them under what conditions the semiclassical description can be regarded as being valid. This will be so if the quantized radiation field has approximately the properties of a classical wave with its well-defined field-strength values at each space–time point (these conditions were explained in Section 2.3.1.2). Spontaneous processes and the photon noise cannot be expressed by the semiclassical theory without additional assumptions. However, the fully quantum-theoretical formalism allows a comprehensive treatment, which in the limiting case leads to the semiclassical description. The classical description may be applied if the quantization of the atomic system can be ignored or single particles are considered (for instance, in the case of the interaction of radiation with free charge carriers).

3.1.2. The Determination of Physically Relevant Quantities

By means of the Hamiltonian representing the interaction between radiation and matter as given in (3.16) or (3.19), the expectation value of an arbitrary observable can in principle be evaluated. The general methods for the derivation of time-dependent expectation values are described in Appendix A (see Sections A.1.1.4 and A.1.2.1). In practice, one has to distinguish mainly between two procedures.

Firstly, the *density operator* $\hat{\rho}(t)$ at time t, which characterizes the system in a comprehensive manner, will be determined under the assumption that its initial value $\hat{\rho}(t_0)$ is known. In the semiclassical description $\hat{\rho}(t_0)$ only refers to the atomic system, whereas in the fully quantum-theoretical description $\hat{\rho}(t_0)$ refers to the total system composed of the atomic system and the radiation field. With the help of the density operator the expectation value of an observable $\hat{L}$ at time t can be determined. We may assume that the total Hamiltonian $\hat{H}$ consists of a free part $\hat{H}^F$ and an interaction part $\hat{H}^C$ [cf. (3.15) or (3.18)]. Therefore, the application of the interaction picture is advantageous. The expectation value $\langle\hat{L}(t)\rangle$ is given by

$$\langle\hat{L}(t)\rangle = \mathrm{Tr}\{\hat{L}_I(t)\hat{\rho}_I(t)\}, \tag{3.24}$$

where

$$\hat{L}_I(t) = \left[\hat{U}^F(t,t_0)\right]^{-1}\hat{L}_S\hat{U}^F(t,t_0) \tag{3.25}$$

is the representative of the observable in the interaction picture and

$$\hat{\rho}_I(t) = \hat{\rho}(t_0) + \sum_{n=1}^{\infty}(i\hbar)^{-n}\int_{t_0}^{t}dt_1\cdots\int_{t_0}^{t_{n-1}}dt_n\left[\hat{H}_I^C(t_1),\cdots\left[\hat{H}_I^C(t_n),\hat{\rho}(t_0)\right]\cdots\right] \tag{3.26}$$

is the density operator in the interaction picture. According to (A.107) the unitary operator $\hat{U}^F$ contains the "free" operator $\hat{H}^F$, which in our case is identical with $\hat{H}^A$ in the semiclassical case and identical with $\hat{H}^A + \hat{H}^R$ in the quantum-theoretical case. The expression (3.26) displays the advantage that the expectation value $\langle\hat{L}(t)\rangle$ is, as we may say, automatically decomposed into terms containing the interaction operator (i.e., the field quantity) in different powers. This feature facilitates the explanation of the experimental data, as they are mostly expressed as effects of different orders with respect to the field strength.

At this point an exact definition of the term "process of nth order" must be given. Equation (A.115) clearly shows the dependence of a general state vector

$|\psi\rangle_I$ on the interaction operator. That part $|\psi^{(n)}\rangle_I$ of the state vector which contains the interaction operator of the nth power describes a *process of the nth order*. The corresponding expectation value to be formed by $\langle\psi^{(n)}|_I$ and $|\psi^{(n)}\rangle_I$ contains the interaction operator to the $2n$th power. Let us discuss the significant example of the transition from the initial state $|\mathscr{E}_i\rangle$ at time t_0 to the final state $|\mathscr{E}_f\rangle$ at time $t > t_0$, where $|\mathscr{E}_i\rangle$ and $|\mathscr{E}_f\rangle$ are assumed to be different eigenstates of the free system. For an nth-order process the transition probability is given by

$$p_{i\to f}^{(2n)}(t) = \left|\left\langle \mathscr{E}_f \middle| \psi^{(n)}(t)\right\rangle_I\right|^2. \tag{3.27}$$

For simplicity we assume t_0 to be zero. Thus we have

$$\left\langle \mathscr{E}_f \middle| \psi^{(n)}(t)\right\rangle_I = \langle \mathscr{E}_f|(i\hbar)^{-n}\int_0^t dt_1 \cdots \int_0^{t_{n-1}} dt_n\, \hat{H}_I^C(t_1)\cdots \hat{H}_I^C(t_n)|\mathscr{E}_i\rangle. \tag{3.28}$$

In the case of the fully quantum-theoretical treatment the interaction operator $\hat{H}^C$ given in (3.19) must be used. Passing from $\hat{H}_I^C$ to the time-independent operator $\hat{H}^C$ and inserting the identity operator $\hat{I} = \Sigma_j|\mathscr{E}_j\rangle\langle\mathscr{E}_j|$ between the operator factors, we finally arrive at

$$\left\langle \mathscr{E}_f \middle| \psi^{(n)}(t)\right\rangle_I = (i\hbar)^{-1}\sum_{j_1,\ldots,j_{n-1}} \frac{\langle \mathscr{E}_f|\hat{H}^C|\mathscr{E}_{j_1}\rangle\cdots\langle \mathscr{E}_{j_{n-1}}|\hat{H}^C|\mathscr{E}_i\rangle}{\left(\mathscr{E}_i - \mathscr{E}_{j_1}\right)\cdots\left(\mathscr{E}_i - \mathscr{E}_{j_{n-1}}\right)} G_n(t), \tag{3.29}$$

where the function $G_n(t)$ carries the time dependence according to

$$\begin{aligned} G_n(t) &= (i\hbar)^{n-1}\left(\mathscr{E}_i - \mathscr{E}_{j_1}\right)\cdots\left(\mathscr{E}_i - \mathscr{E}_{j_{n-1}}\right) \\ &\quad\times \int_0^t dt_1 \exp\left(i\omega_{fj_1}t_1\right)\cdots\int_0^{t_{n-1}} dt_n \exp\left(i\omega_{j_{n-1}i}t_n\right). \end{aligned} \tag{3.30}$$

Here the transition frequencies $\omega_{j'j''}$ are given by $\hbar^{-1}(\mathscr{E}_{j'} - \mathscr{E}_{j''})$. In (3.29) we have to sum $n-1$ times over the eigenstates $|\mathscr{E}_j\rangle$ of the free system. The transition probability $p^{(2n)}(t)$ consists of a $2n$-fold product of matrix elements of the operator $\hat{H}^C$.

Secondly, for the determination of physically relevant quantities one may start from the *equations of motion* for the respective operators. With the help of the total Hamiltonian $\hat{H}$ as given in Section 3.1.1, the equation of motion for each of the observables and dynamical variables needed for the description of the matter–radiation system can be formulated by (A.70) or (A.110). To derive the explicit time dependence of a certain variable, in general a system of differential equations must be solved simultaneously. If one succeeds in doing this (in most cases approximations as described in Section 3.2 have to be applied), then this procedure will prove to be advantageous in that it yields general relations between the time-dependent expectation values of different observables that can often be compared directly with the corresponding classical relations.

3.2. CONCEPTS AND REPRESENTATIVE APPROXIMATION METHODS

For a qualitative and quantitative interpretation of experimental results, the fundamental quantities and relations, as they are given in Section 3.1, will be used. To perform this with an acceptable amount of calculation it is necessary that from the very beginning some particular features of the atomic system and of the radiation field under consideration, as well as of the interaction process under study, be utilized for simplifying the theoretical description.

One of these simplifying approximations, which have a wide application, is the dipole approximation. It can be employed if in the atomic system the charge carriers under investigation are in a volume whose linear dimension is less than the wavelengths of the light waves involved. This will be discussed in Section 3.2.1. Another such basic approximation is the rotating-wave approximation (neglect of nonresonant terms), which will be dealt with in Section 3.2.2. In this approximation only those terms will be used in the interaction operator in which the initial and final energy values are nearly equal; as to the time behavior, this corresponds to the neglect of terms with fast time dependence. Another useful basic concept is the description of the interaction Hamiltonian with the help of phenomenologically introduced susceptibility functions (see Section 3.2.3). To describe the interaction between a radiation field and an atomic system under realistic conditions, it is necessary to take into account the interaction with dissipative systems. Because of this interaction damping, relaxation, and fluctuation phenomena occur, which will be dealt with in Section 3.2.4.

3.2.1. The Dipole Approximation

That term of the interaction operators $\hat{H}^C$ in (3.13) and (3.19), which contains the field strength linearly, is of special importance for the description of the main nonlinear effects. This term (denoted by $\hat{H}^{(1)}$) can be introduced into the calculation in a considerably simplified form if the following conditions are satisfied: The charge carriers interacting with the field are in a finite space region with diameter L much less than the wavelengths of those waves that constitute the radiation field. This means

$$|k_{\mu\bullet}|L \ll 1, \tag{3.31}$$

where $k_{\mu\bullet}$ is the wave number of the μth mode present in the radiation field. This condition is for instance fulfilled for an electron in an atom or small molecule if it is acted on by radiation of the ultraviolet, visible, or infrared region. In this case the product $|k_{\mu\bullet}|L$ is less than $(2\pi/10^{-7}\text{ m}) \times 10^{-9}\text{ m} \approx 0.06$. We shall see that the detailed discussion of the dipole approximation not only yields a mathematically easy-to-handle form of $\hat{H}^{(1)}$ but also facilitates the interpretation of empirical findings.

Let us now discuss the dipole approximation for the fully quantum-theoretical case. The equation (3.19) shows that the linear interaction operator $\hat{H}^{(1)}$ contains additively the contributions of the different charge carriers. Therefore we start with considering only one of them. Let both $\hat{H}^A$ and $\hat{H}^{(1)}$ in (3.19) refer to a single particle, whose particle index α can be dropped.

For representing physically relevant quantities the matrix elements with the eigenkets $|\mathscr{E}^F\rangle$ of the "free" Hamiltonian $\hat{H}^F = \hat{H}^A + \hat{H}^R$ are needed, where

$$|\mathscr{E}^F\rangle = |\mathscr{E}^A\rangle|\mathscr{E}^R\rangle. \tag{3.32}$$

If $|\mathscr{E}_0^F\rangle$ and $|\mathscr{E}_1^F\rangle$ are two eigenkets of $\hat{H}^F$, we obtain from (3.19) and (3.20) the matrix element

$$\begin{aligned}\langle\mathscr{E}_0^F|\hat{H}^{(1)}|\mathscr{E}_1^F\rangle = \sum_\mu -\frac{q}{m}\left(\frac{\hbar}{2\varepsilon_0 V\omega_\mu}\right)^{1/2} e_{\mu\bullet} \\ \times\Big[\langle\mathscr{E}_0^R|\hat{a}_\mu|\mathscr{E}_1^R\rangle\langle\mathscr{E}_0^A|e^{ik_{\mu\bullet}\hat{r}_\bullet}\hat{p}_\bullet|\mathscr{E}_1^A\rangle \\ +\langle\mathscr{E}_0^R|\hat{a}^\dagger|\mathscr{E}_1^R\rangle\langle\mathscr{E}_0^A|e^{-ik_{\mu\bullet}\hat{r}_\bullet}\hat{p}_\bullet|\mathscr{E}_1^A\rangle\Big]\end{aligned} \tag{3.33}$$

of the operator $\hat{H}^{(1)}$. The right-hand side contains the matrix elements

$$\langle\mathscr{E}_0^R|\hat{a}_\mu|\mathscr{E}_1^R\rangle, \qquad \langle\mathscr{E}_0^R|\hat{a}_\mu^\dagger|\mathscr{E}_1^R\rangle, \tag{3.34a}$$

$$\langle\mathscr{E}_0^A|e^{ik_{\mu\bullet}\hat{r}_\bullet}\hat{p}_\bullet|\mathscr{E}_1^A\rangle, \qquad \langle\mathscr{E}_0^A|e^{-ik_{\mu\bullet}\hat{r}_\bullet}\hat{p}_\bullet|\mathscr{E}_1^A\rangle \tag{3.34b}$$

The elements with the photon annihilation and creation operator can be easily evaluated by means of the relations of the quantized radiation field as given in (2.48) and (2.49). The matrix elements in (3.34b) depend on the operators and eigenkets of the atomic system. Without loss of generality we may assume that the charge center of the particle under consideration coincides with the origin.

Let us consider the expansion

$$\langle \mathscr{E}_0^A | e^{\pm i k_{\mu\bullet} \hat{r}_\bullet} \hat{p}_\bullet | \mathscr{E}_1^A \rangle = \langle \mathscr{E}_0^A | \left(\hat{I} \pm i k_{\mu\bullet} \hat{r}_\bullet + \cdots \right) \hat{p}_\bullet | \mathscr{E}_1^A \rangle. \tag{3.35}$$

If the condition (3.31) holds, the right-hand side can obviously be approximated by the matrix element $\langle \mathscr{E}_0^A | \hat{p}_\bullet | \mathscr{E}_1^A \rangle$. This approximation is called the dipole approximation; it means physically that the exponential factor describing the space dependence of the field is approximately constant within the region of the nonvanishing probability of finding the particle at a given position, and that only interaction processes associated with dipole radiation are described. The terms of higher order in the expansion (3.35) would lead to multipole radiation of a higher order than dipole [cf. (2.80)].

If the matrix element $\langle \mathscr{E}_0^A | \hat{p}_\bullet | \mathscr{E}_1^A \rangle$ obtained in dipole approximation is subjected to an appropriate transformation, an instructive interpretation becomes possible. According to (A.29) the momentum operator can be written as

$$\hat{p}_\bullet = \frac{m}{i\hbar} [\hat{r}_\bullet, \hat{H}^A], \tag{3.36}$$

if the kinetic energy of the particle in $\hat{H}^A$ is assumed to be the nonrelativistic expression $\hat{p}_\bullet^2/2m$. This assumption seems to be justified over a wide range; otherwise, as a rule only small additional energy contributions will occur. Equation (3.36) directly leads to

$$\langle \mathscr{E}_0^A | \hat{p}_\bullet | \mathscr{E}_1^A \rangle = \frac{im}{\hbar} \left(\mathscr{E}_0^A - \mathscr{E}_1^A \right) \langle \mathscr{E}_0^A | \hat{r}_\bullet | \mathscr{E}_1^A \rangle. \tag{3.37}$$

If a charge carrier with the charge $-q$ is fixed at the origin, we can introduce the operator $\hat{d}_\bullet = q\hat{r}_\bullet$ as the operator of an electric dipole moment assigned to the charge carrier under consideration. Thus we arrive at

$$\langle \mathscr{E}_0^A | \hat{p}_\bullet | \mathscr{E}_1^A \rangle = i \frac{m}{q} \omega_{01} \langle \mathscr{E}_0^A | \hat{d}_\bullet | \mathscr{E}_1^A \rangle, \tag{3.38}$$

where the matrix element of the momentum depends on the transition frequency $\omega_{01} = \hbar^{-1} \ (\mathscr{E}_0^A - \mathscr{E}_1^A)$ and on the matrix element of the electric dipole moment.

We started from the matrix element $\langle \mathscr{E}_0^F | \hat{H}^{(1)} | \mathscr{E}_1^F \rangle$ in (3.33) and shall now return to it. By inserting the result of the dipole approximation on the right-hand side of (3.38) instead of $\langle \mathscr{E}_0^A | e^{\pm i k_{\mu\bullet} \hat{r}_\bullet} \hat{p}_\bullet | \mathscr{E}_1^A \rangle$, one obtains

$$\langle \mathscr{E}_0^A | \hat{H}^{(1)} | \mathscr{E}_1^A \rangle = \sum_\mu \frac{\omega_{01}}{\omega_\mu} \langle \mathscr{E}_0^F | - \hat{d}_\bullet \left[\left(\frac{\hbar \omega_\mu}{2\varepsilon_0 V} \right)^{1/2} e_{\mu\bullet} i \hat{a}_\mu + \text{h.c.} \right] | \mathscr{E}_1^F \rangle. \tag{3.39}$$

The expression in the brackets is the contribution $\hat{E}_{\mu\bullet}$ of the μth mode to the operator of the total electric field strength in the Schrödinger picture (taken at the point of origin), as can immediately be seen from (2.51). A simplification of the right-hand side of (3.39) can be obtained if certain relations between the atomic transition frequency ω_{01} and the frequencies ω_μ of the modes incorporated in the field exist. On conditions that $\omega_{01} \approx \omega_\mu$, summing would lead to $\langle \mathscr{E}_0^F| - \hat{d}_\bullet \hat{E}_\bullet |\mathscr{E}_1^F\rangle$, where $\hat{E}_\bullet$ is the operator of the total field strength, $\hat{E}_\bullet = \Sigma_\mu \hat{E}_{\mu\bullet}$.

Since there actually exist resonance conditions (see Section 3.3.1) of the form $\omega_{01} \approx \omega_\mu$ in a great number of linear and nonlinear processes, (3.39) suggests that the exact interaction operator $-(q/m)\hat{A}_\bullet \hat{p}_\bullet$ may be replaced by the operator $-\hat{d}_\bullet \hat{E}_\bullet$, which represents the energy of an electric dipole moment in an electric field. In the following we shall compare the exact linear interaction operator $-(q/m)\hat{A}_\bullet \hat{p}_\bullet$ with the interaction operator in the dipole approximation

$$\hat{H}_{\text{di}}^C = -\hat{d}_\bullet \hat{E}_\bullet. \tag{3.40}$$

Under the assumption of the dipole approximation the operators $\hat{A}_{\bullet I}(t)$ and $\hat{E}_{\bullet I}(t)$ do not depend on the time-dependent position operator $\hat{r}_{\bullet I}(t)$ of the particle. Thus the relation

$$\begin{aligned} \hat{H}_I^{(1)}(t) &= -\frac{q}{m}\hat{A}_{\bullet I}(t)\hat{p}_{\bullet I}(t) \\ &= -\hat{d}_{\bullet I}(t)\hat{E}_{\bullet I}(t) + \frac{d}{dt}\{\hat{A}_{\bullet I}(t)\hat{d}_{\bullet I}(t)\} \end{aligned} \tag{3.41}$$

holds, as can be easily verified by differentiating. For the determination of physically relevant quantities, matrix elements of operators with the form

$$\int_{t_0}^{t_{j-1}} dt'\, \hat{H}_I^{(1)}(t') \tag{3.42}$$

must be constructed [cf. (3.32)]. The time integration of the first term $-\hat{d}_{\bullet I}\hat{E}_{\bullet I}$ in (3.41) yields an expression with the denominator $\omega_{01} \pm \omega_\mu$; thus, the resulting contribution to the matrix element is proportional to $(\omega_\mu)^{1/2}/(\omega_{01} \pm \omega_\mu)$. The second term on the right-hand side of (3.41) obviously leads to a contribution that is proportional to $(\omega_\mu)^{-1/2}$. As a result the ratio of the first to the second term is of the order $\omega_\mu/(\omega_{01} \pm \omega_\mu)$. Substantial contributions to the matrix element of

$$\int_{t_0}^{t_{j-1}} dt'\, \hat{H}_I^{(1)}(t') \tag{3.42}$$

will come only from those modes where the denominator becomes sufficiently small. This means that the contribution of $\hat{H}_{\mathrm{di}}^{C}$ dominates the second term

$$\frac{d}{dt}\{\hat{A}_{\bullet I}\hat{d}_{\bullet I}\}. \tag{3.43}$$

Consequently, in the dipole approximation the interaction operator $\hat{H}_{\mathrm{di}}^{C}$ may be used instead of $\hat{H}^{(1)}$.

Finally, we will drop the simplifying assumption that the point of origin coincides with the center of charge. If the center of charge has the position coordinate $\bar{r}_{\bullet}$, in the dipole approximation the matrix element $\langle \mathscr{E}_0^A|\exp[\pm ik_{\mu\bullet}\hat{r}_{\bullet}]\cdot\hat{p}_{\bullet}|\mathscr{E}_1^A\rangle$ must be replaced by $\exp[\pm ik_{\mu\bullet}\bar{r}_{\bullet}]\langle\mathscr{E}_0^A|\hat{p}_{\bullet}|\mathscr{E}_1^A\rangle$, and the interaction operator $\hat{H}^{(1)}$ by $-\hat{d}_{\bullet}\hat{E}_{\bullet}(\bar{r}_{\bullet})$, where the operator of the field strength—as in the case of the free radiation field—now depends on the vector $\bar{r}_{\bullet}$ consisting of c-number components. It should be noted that the operators $\hat{d}_{\bullet}$ and $\hat{E}_{\bullet}(\bar{r}_{\bullet})$ are commuting quantities, whereas the operators $\hat{A}_{\bullet}(\hat{r}_{\bullet})$ and $\hat{p}_{\bullet}$ in $\hat{H}^{(1)}$ are noncommuting ones.

Now we will deal with the case of several charge carriers. It will be assumed that the dipole approximation can be applied to each of the charge carriers. Then the total interaction Hamiltonian (in the dipole approximation) is

$$\hat{H}_{\mathrm{di}}^{C} = -\sum_{\alpha}\hat{d}_{\alpha\bullet}\hat{E}_{\bullet}(\bar{r}_{\alpha\bullet}), \tag{3.44}$$

where $\hat{d}_{\alpha\bullet}$ is the electric-dipole-moment operator and $\bar{r}_{\alpha\bullet}$ the coordinate of the center of charge of the αth particle. The formula (3.44) includes the case of a single atom or a small molecule, in which the electric field is approximately constant. Therefore all the $\hat{r}_{\alpha\bullet}$ can be identified with an arbitrary atomic coordinate, say the center of mass $\bar{r}_{\bullet}$. So we arrive at

$$\hat{H}_{\mathrm{di}}^{C} = -\hat{d}'_{\bullet}\hat{E}_{\bullet}(\bar{r}_{\bullet}) \tag{3.45}$$

where $\hat{d}'_{\bullet} = \Sigma_{\alpha}\hat{d}_{\alpha\bullet}$ is the operator of the atomic or molecular dipole moment. The matrix elements of $\hat{d}'_{\bullet}$ mainly determine the quantitative features of emission and absorption processes of atoms and molecules, while the operator $\Sigma\hat{p}_{\alpha\bullet}$, which would result directly from $\hat{H}^{(1)}$ under the same condition of a quasiconstant field in the atomic system, cannot be interpreted in such an instructive way.

So far we have treated the fully quantum-theoretical case. The result (3.44) can be easily transferred to the semiclassical case. In the dipole approximation the interaction Hamiltonian is given by

$$\hat{H}_{\mathrm{di}}^{C}(t) = -\sum_{\alpha}\hat{d}_{\alpha\bullet}E(\bar{r}_{\alpha\bullet}, t), \tag{3.46}$$

where $E_{\bullet}(\bar{r}_{\alpha\bullet}, t)$ is the classical electric field strength at the position $\bar{r}_{\alpha\bullet}$ at time t.

3.2.2. The Rotating-Wave Approximation

In Section A.2.3 it is shown that a considerable transition probability between an initial state and a final state of a closed quantized system can in general result only if the energies of the two states are approximately equal. The knowledge of this fact enables us to simplify lengthy calculations in nonlinear optics by considering from the very beginning only those terms in the interaction operator that represent a process with approximate energy conservation. This approach is equivalent to the so-called rotating-wave approximation, which was originally introduced in connection with the semiclassical theory of spin-resonance spectroscopy (see Section 4.8). Under certain conditions this approximation corresponds to the neglect of rapidly varying terms in the classical theory of the interaction between light and matter (see Section 7.1.2).

Let us first explain the main features of this approximation under simple conditions for the case of one-photon processes. Then we shall consider its application to two-photon processes.

3.2.2.1. Basic Concept. A one-photon process will be discussed in which the field of the μth radiation mode causes the transition between the atomic states 0 and 1 (see Fig. 3.1). With respect to what follows it is very useful to describe the atomic system by means of the atomic flip operators $\hat{B}_{kl}$, which, when applied to the eigenkets $|\mathscr{E}_j^A\rangle$ of the free atomic system, satisfy the relations

$$\hat{B}_{kl}|\mathscr{E}_j^A\rangle = \begin{cases} |0_v\rangle & \text{for} \quad l \neq j, \\ |\mathscr{E}_k^A\rangle & \text{for} \quad l = j \end{cases} \tag{3.47}$$

and $\hat{B}_{kl} = \hat{B}_{lk}^{\dagger}$ [cf. Section A.2.5]. For a two-level system the general state $|\psi\rangle$

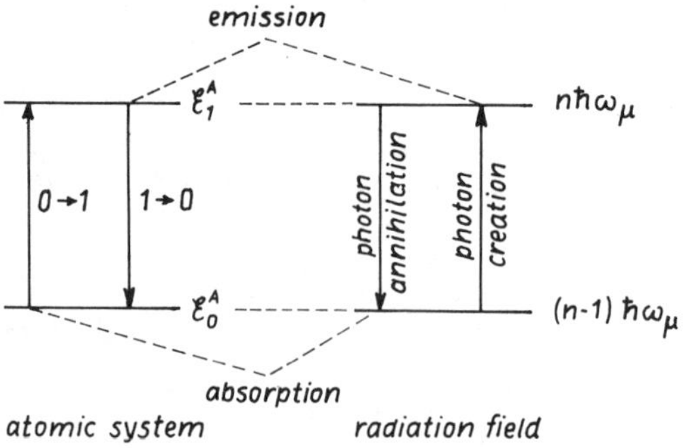

Fig. 3.1. One-photon absorption and emission: the transitions in the atomic system (left) and in the μth mode of the radiation field (right).

of the atomic system is represented as

$$|\psi\rangle = c_0|\mathscr{E}_0^A\rangle + c_1|\mathscr{E}_1^A\rangle, \tag{3.48}$$

and we have only to consider the flip operators $\hat{B}_{01}$ and $\hat{B}_{10} = \hat{B}_{01}^{\dagger}$.

In dipole approximation the interaction Hamiltonian $\hat{H}^C$ is given as $\hat{H}^C = -\hat{d}_{\bullet}\hat{E}_{\bullet}$, and after expressing the dipole operator $\hat{d}_{\bullet}$ in terms of the atomic flip operators [cf. (A.174)] we obtain

$$\hat{H}^C = -\left(d_{01\bullet}\hat{B}_{01} + d_{10\bullet}\hat{B}_{10}\right)\hat{E}_{\bullet}, \tag{3.49}$$

where $d_{10\bullet} = \langle\mathscr{E}_1^A|\hat{d}_{\bullet}|\mathscr{E}_0^A\rangle$. The first term in $\hat{H}^C$ represents the transition of the atomic system from the upper state to the lower one; we may say that the atom flips from $|\mathscr{E}_1^A\rangle$ into state $|\mathscr{E}_0^A\rangle$. The second term is responsible for the reversed process.

Let us now consider the field-strength operator in the interaction Hamiltonian (3.49). To simplify our argumentation, at the beginning we assume that only one mode of the radiation field takes part in the interaction. According to Section 2.2.2 a one-mode field may be described by the field-strength operator

$$\hat{E}_{\mu\bullet} = \hat{E}_{\mu\bullet}^{(-)} + \hat{E}_{\mu\bullet}^{(+)}, \tag{3.50}$$

where $\hat{E}_{\mu\bullet}^{(-)}$ and $\hat{E}_{\mu\bullet}^{(+)}$ contain the annihilation and creation operator of the μth mode, respectively. It should be mentioned that here the creation and annihilation operators are chosen to be time-independent operators in the Schrödinger picture.

Using (3.50) in (3.49), we obtain

$$\begin{aligned}\hat{H}^C = -\hat{d}_{\bullet}\hat{E}_{\mu\bullet} = &-d_{01\bullet}\hat{B}_{01}\hat{E}_{\mu\bullet}^{(+)} - d_{10\bullet}\hat{B}_{10}\hat{E}_{\mu\bullet}^{(-)}\\ &-d_{01\bullet}\hat{B}_{01}\hat{E}_{\mu\bullet}^{(-)} - d_{10\bullet}\hat{B}_{10}\hat{E}_{\mu\bullet}^{(+)}.\end{aligned} \tag{3.51}$$

The first term on the right-hand side of (3.51) is responsible for an atomic transition from the upper to the lower level and for the simultaneous creation of one photon (see Fig. 3.1). The reversed process in which the atomic system gains energy by annihilation of a photon is described by the second term. These two processes approximately satisfy the condition of energy conservation because $\omega_{10} \approx \omega_{\mu}$. In contrast to this, the third and the fourth term of (3.51) are connected with processes that violate energy conservation seriously. Thus, the third term represents the atomic transition from the upper to the lower level together with the annihilation of a photon; this process would have to be represented in Fig. 3.1 by two downward arrows. As a consequence, the third and fourth term of (3.51) make no considerable contribution to the transition probability from the initial state to the final one. [Here the time for which the transition probability is calculated has to meet the requirements of coarse graining (see Section A.2.3)]. Therefore, when dealing with one-photon processes it is justified to neglect the last two terms of (3.51) from the very beginning. This neglect is called the *rotating-wave approximation* for one-photon processes (an explanation of the origin of this term is given in Section 4.8). This approximation, which has been explained so far for one mode, can analogously be applied to the total Hamiltonian, which is composed of terms from all modes of the radiation field.

The great advantage of the rotating-wave approximation becomes obvious when multiphoton processes are treated. As an example we apply this concept to multiphoton processes of lowest order.

3.2.2.2. Application to Two-Photon Processes. We start with the transition probability from the initial state $|\mathscr{E}_i\rangle$ into the final state $|\mathscr{E}_f\rangle$ by two-photon processes, which is given, according to the general perturbation theory (3.27)–(3.30), by

$$p_{i\to f}(t) = \left| \frac{1}{(i\hbar)^2} \sum_j \frac{\langle \mathscr{E}_f | \hat{H}^C | \mathscr{E}_j \rangle \langle \mathscr{E}_j | \hat{H}^C | \mathscr{E}_i \rangle}{\mathscr{E}_i - \mathscr{E}_j} G_2(t) \right|^2, \tag{3.52}$$

where

$$G_2(t) = (\mathscr{E}_i - \mathscr{E}_j) \int_0^t dt_1\, e^{i\omega_{fj}t_1} \int_0^{t_1} dt_2\, e^{i\omega_{ji}t_2}$$

$$= \frac{\hbar}{\omega_{fi}}[e^{i\omega_{fi}t} - 1] - \frac{\hbar}{\omega_{fj}}[e^{i\omega_{fj}t} - 1].$$

Here ω_{fi} and ω_{fj} represent the transition frequencies between the final state $|\mathscr{E}_f\rangle$ and the initial state $|\mathscr{E}_i\rangle$ and between the final state $|\mathscr{E}_f\rangle$ and the intermediate state $|\mathscr{E}_j\rangle$. In agreement with the consideration in Section 3.2.2.1, energy resonance is assumed to exist between the initial and final states of the whole system; this implies $\omega_{fi} \approx 0$. The transition frequencies ω_{fj} however are supposed to attain only nonvanishing values, which means that resonance between any intermediate state and the final state is excluded. Under these conditions the second term in $G_2(t)$ is small in comparison with the first one and can thus be neglected, provided that the transition probability is calculated by using again the concept of time coarse-graining. Thus we have

$$\frac{\Delta p_{i\to f}(t)}{\Delta t} = \frac{2\pi}{\hbar} \left| \sum_j \frac{\langle \mathscr{E}_f | \hat{H}^C | \mathscr{E}_j \rangle \langle \mathscr{E}_j | \hat{H}^C | \mathscr{E}_i \rangle}{\mathscr{E}_i - \mathscr{E}_j} \right|^2 \delta(\mathscr{E}_f - \mathscr{E}_i). \tag{3.53}$$

Let us now utilize specific features of the two-photon absorption (TPA) of an atomic system. This process is schematically shown in Fig. 3.2. The

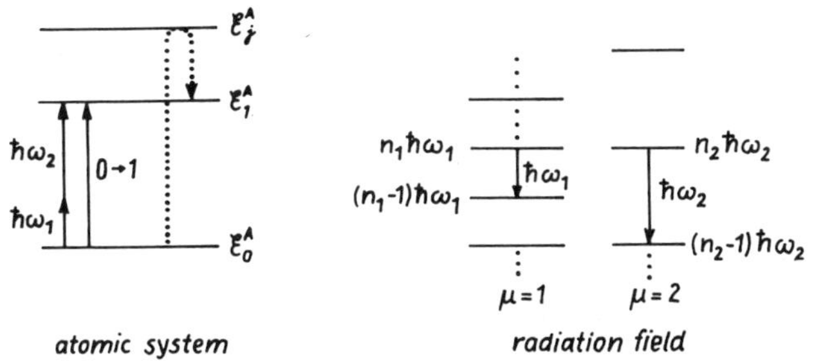

Fig. 3.2. Two-photon absorption: (*left*) Transition between the initial atomic level 0 and the final level 1. One intermediate level j, which is reached only by virtual transitions (dotted arrow), has been plotted. (*right*) annihilation of one photon in mode 1 and of another one in mode 2.

transition $|\mathcal{E}_0^A\rangle \to |\mathcal{E}_1^A\rangle$ of the atom is connected with the absorption of one photon of energy $\hbar\omega_1$ in mode $\mu = 1$ and one of energy $\hbar\omega_2$ in mode $\mu = 2$. The initial and final states of the whole system may be represented by

$$|\mathcal{E}_i\rangle = |\mathcal{E}_0^A\rangle|n_1, n_2\rangle \tag{3.54a}$$

and

$$|\mathcal{E}_f\rangle = |\mathcal{E}_1^A\rangle|n_1 - 1, n_2 - 1\rangle \tag{3.54b}$$

provided that the radiation is in eigenstates of the photon number operator with respect to mode 1 and 2. Other modes are assumed to be not relevant in this interaction. The energy conservation ($\mathcal{E}_f = \mathcal{E}_i$) can be expressed in the form

$$\mathcal{E}_1^A - \mathcal{E}_0^A = \hbar\omega_1 + \hbar\omega_2. \tag{3.55}$$

In accordance with the rotating-wave approximation, only the intermediate states

$$|\mathcal{E}_j\rangle = \begin{cases} |\mathcal{E}_j^A\rangle|n_1 - 1, n_2\rangle, \\ |\mathcal{E}_j^A\rangle|n_1, n_2 - 1\rangle \end{cases} \tag{3.56}$$

can provide substantial contributions to the product of matrix elements in (3.53). Therefore, applying again the dipole approximation, we arrive at

$$\frac{\Delta p_{1\to f}(t)}{\Delta t} = \frac{2\pi}{\hbar}\left|\sum_j \frac{(d_{fj})_2\langle n_2 - 1|\hat{E}_2^{(-)}|n_2\rangle(d_{ji})_1\langle n_1 - 1|\hat{E}_1^{(-)}|n_1\rangle}{\mathcal{E}_0^A - \mathcal{E}_j^A + \hbar\omega_1}\right.$$
$$\left. + \frac{(d_{fj})_1\langle n_1 - 1|\hat{E}_1^{(-)}|n_1\rangle(d_{ji})_2\langle n_2 - 1|\hat{E}_2^{(-)}|n_2\rangle}{\mathcal{E}_0^A - \mathcal{E}_j^A + \hbar\omega_2}\right|^2 \delta(\mathcal{E}_f - \mathcal{E}_i), \tag{3.57a}$$

where

$$(d_{kl})_\mu = \langle \mathscr{E}_k^A | \hat{d}_\bullet | \mathscr{E}_l^A \rangle e_\bullet^{(\mu)} \tag{3.57b}$$

is the projection of the dipole matrix element on the unit vector $e_\bullet^{(\mu)}$ in the direction of polarization of mode μ.

We next describe the two-photon absorption and emission processes by introducing the specific operator

$$\hat{H}_{\text{TP}}^C = \hat{H}_{\text{TPA}}^C + \hat{H}_{\text{TPE}}^C \tag{3.58a}$$

where

$$\hat{H}_{\text{TPA}}^C = \eta \hat{B}_{10} \hat{E}_1^{(-)} \hat{E}_2^{(-)}, \tag{3.58b}$$

$$\hat{H}_{\text{TPE}}^C = \left(\hat{H}_{\text{TPA}}^C\right)^\dagger, \tag{3.58c}$$

and

$$\eta = \sum_j \frac{(d_{fj})_2 (d_{ji})_1}{\mathscr{E}_0^A - \mathscr{E}_j^A + \hbar\omega_1} + \frac{(d_{fj})_1 (d_{ji})_2}{\mathscr{E}_0^A - E_j^A + \hbar\omega_2}. \tag{3.58d}$$

$\hat{B}_{10}$ and $\hat{B}_{01}$ are the flip operators of (3.47). Using the first term of (3.58a) in place of the general interaction operator in perturbation theory, the state vector calculated in first order will already yield the transition probability of the two-photon absorption process. To this end the matrix element

$$\langle \mathscr{E}_f | \hat{H}_{\text{TPA}}^C | \mathscr{E}_i \rangle = \langle n_1 - 1, n_2 - 1 | \langle \mathscr{E}_1^A | \hat{H}_{\text{TPA}}^C | \mathscr{E}_0^A \rangle | n_1, n_2 \rangle \tag{3.59}$$

has to be calculated. Taking into account the normalization $\langle \mathscr{E}_1^A | \hat{B}_{10} | \hat{E}_0^A \rangle = 1$, we obtain

$$\langle \mathscr{E}_f | \hat{H}_{\text{TPA}}^C | \mathscr{E}_i \rangle = \eta \langle n_1 - 1 | \hat{E}_1^{(-)} | n_1 \rangle \langle n_2 - 1 | \hat{E}_2^{(-)} | n_2 \rangle. \tag{3.60}$$

The substitution of this matrix element into the first-order transition probability [cf. (A.123)] gives a relation identical with (3.57a), which was obtained by means of second-order perturbation theory with the "basic" interaction Hamiltonian $\hat{H}^C = -\hat{d}_\bullet \hat{E}_\bullet$.

Not only can the transition probability be calculated, but also other important parameters of the system, such as

$$\frac{d}{dt}\langle \hat{a}_\mu^\dagger \hat{a}_\mu \rangle \quad \text{and} \quad \frac{d}{dt}\langle \hat{a}_\mu^\dagger \rangle,$$

representing the change of the expectation values of the photon number and of the creation operator in mode.

The first term of the effective interaction Hamiltonian (3.58a) clearly shows the features of two-photon absorption: it contains the annihilation operators of the photons in the modes 1 and 2 and the flip operator that flips the atom from $|\mathscr{E}_0^A\rangle$ into $|\mathscr{E}_1^A\rangle$. The second term, which is the hermitian conjugate of the first one, is connected with two-photon emission. For this reason the second term does not contribute to the transition probability calculated with the assumption of an atom initially in the ground state. This term comes into play if we consider atomic systems initially in the excited state $|\mathscr{E}_1^A\rangle$. Then, by the use of the same procedure, we obtain the transition probability for two-photon emission in first-order perturbation theory.

The procedures used here for a concise description of two-photon absorption and emission processes can also be successfully applied to higher multiphoton processes. Then, of course, other specific multiphoton interaction Hamiltonians must be used. Similar methods are also applicable to processes in which the radiation field does not resonantly exchange energy with the atomic system; examples of such processes are the nonresonant generation of harmonics and the parametric effects.

3.2.3. The Use of Susceptibilities in the Interaction Hamiltonian

As already mentioned, specific features of the interaction between light and matter, such as spontaneous emission and fluctuation processes, can only be fully understood on the basis of the quantization of the radiation field. In treating such effects it is often advantageous to describe the material by phenomenologically introduced susceptibilities or to use susceptibilities calculated for the interaction of classical fields with classically described atomic systems (Section 1.3) or with quantized atomic systems (Chapter 4). This procedure proves to be successful for loss-free materials in which, during the interaction process, energy is exchanged between several radiation modes but not between field and matter. In our treatment we use interaction Hamiltonians containing the operators of the radiation field and susceptibilities that represent experimental results or calculated expectation values of the atomic observables.

Depending on whether interaction processes in resonators or those in free space are dealt with, the classical considerations of Section 1.4.5 or 1.4.6 can be used. We demonstrate the method for the case of radiation processes in free space. According to (1.244) the mode expansion coefficient $a_\nu(t)$ of mode ν varies under the influence of the Hamiltonian function H, which consists of the two parts H^{L} and H^{NL}. The part H^{L} represents the energy of the field including the contribution that arises from the linear polarization within the material [cf. (1.238)] and H^{NL} represents the nonlinear polarization [cf. (1.248) and (1.249)].

The classical equation of motion (1.244) can be "translated" into the corresponding quantum-mechanical relation by replacing $a_\nu(t)$, $H = H^{\mathrm{L}} + H^{\mathrm{NL}}$, and the Poisson bracket $\{A, B\} = C$ by the annihilation operator $\hat{a}_\nu(t)$, the

Hamiltonian operator $\hat{H} = \hat{H}^{L} + \hat{H}^{NL}$, and the commutation relation $[\hat{A}, \hat{B}] = i\hbar\hat{C}$, respectively. Thus we obtain

$$\frac{d}{dt}\hat{a}_{\nu}(t) = \frac{1}{i\hbar}[\hat{a}_{\nu}(t), \hat{H}^{L}(t) + \hat{H}^{NL}(t)]. \tag{3.61}$$

Using the Hamiltonians introduced in Section 1.4.6, this can be written as

$$\frac{d}{dt}\hat{a}_{\nu}(t) = -i\omega_{\nu}\hat{a}_{\nu}(t) + \frac{i}{\hbar}\int dV\,\hat{E}^{(+)}_{\nu\bullet}(r_{\bullet})\hat{P}^{(-)}_{\nu\bullet}(r_{\bullet}, t), \tag{3.62}$$

where $\hat{P}^{(-)}_{\nu\bullet}(r_{\bullet}, t)$, the operator of nonlinear polarization at the frequency ω_{ν}, depends on the field-strength operators of the exciting field (which may be expressed in terms of the annihilation and creation operators) and on the nonlinear susceptibility in the same way as the classical nonlinear polarization depends on the field strength and susceptibility. Equation (3.61) will be used in Section 11.2.2 in treating the influence of quantum noise on parametric interactions.

3.2.4. The Influence of Dissipative Systems on Optical Phenomena

The adequate description of linear and nonlinear optical effects requires the treatment of the interaction between a radiation field and an atomic system that acts under realistic conditions. In the determination of the properties of such a "realistic atomic system" the fact must be taken into account that every system of charge carriers (atoms, molecules, solids) is more or less influenced by dissipative systems to which it is coupled. A *dissipative system* stands for a reservoir with a large number of degrees of freedom, whereas the atomic system, which is called a *dynamical system*, has a relatively small number of degrees of freedom. For instance, the field fluctuations of the photon vacuum (see Section 2.3.1) and the magnetic or electric fields arising from a system of stochastically moving atoms may be regarded as dissipative systems acting on an atomic system. Due to such interactions, phenomena such as damping, spontaneous transitions, and fluctuation forces occur. The foundations of these effects will be discussed in a fully quantum-theoretical manner in Section 3.2.4.1 by studying simple models; the more practical introduction of phenomenological relaxation parameters on the basis of a semiclassical master-equation approach is dealt with in Section 3.2.4.2, and the basic equations for the interaction between a radiation field and a realistic atomic system (that is to say an atomic system acted on by a dissipative system) are given in Section 3.2.4.3.

3.2.4.1. Atomic System Interacting with a Dissipative System. Although in nature a great diversity of interactions between dissipative and dynamical systems exist, the basic and general aspects can be explained with the help of a few simple models. We assume the dissipative system to be a boson system interacting with a harmonic oscillator and a two-level system, respectively.

The dissipative system is characterized by the Hamiltonian

$$\hat{H}^D = \sum_\mu \hbar\omega_\mu \hat{a}_\mu^\dagger \hat{a}_\mu \tag{3.63}$$

with

$$[\hat{a}_\mu, \hat{a}_{\mu'}] = 0, \qquad [\hat{a}_\mu^\dagger, \hat{a}_{\mu'}^\dagger] = 0, \qquad [\hat{a}_\mu, \hat{a}_{\mu'}^\dagger] = \hat{I},$$

where the zero-point energy in $\hat{H}^D$ has been appropriately fixed. The operators $\hat{a}_\mu^\dagger$ and $\hat{a}_\mu$ are the creation and annihilation operators of the μth mode, which satisfy the general commutation relations for bosons [see (A.149)]. The energy of the μth state is $\hbar\omega_\mu$. Let us assume that the dissipative system has a quasicontinuous energy spectrum with the state density $\sigma(\omega_\mu)$ per unit angular frequency and that, arising from the dissipative system, the force

$$\hat{M} = \sum_\mu g_\mu \hat{a}_\mu \tag{3.64}$$

acts on the dynamical system. This "driving force" is linear in the annihilation operator of the dissipative system; the g_μ are the coupling constants between the dissipative and the dynamical system. The initial value of the density operator at $t = 0$ is assumed to be

$$\hat{\rho}^D(0) = \prod_\mu \hat{\rho}_\mu^D(0). \tag{3.65}$$

In the case of thermal equilibrium $\hat{\rho}_\mu^D$ will be expressed by

$$\hat{\rho}_\mu^D = \frac{\exp\left(-\hbar\omega_\mu \hat{a}_\mu^\dagger \hat{a}_\mu / k_B \mathscr{T}\right)}{\mathrm{Tr}\left[\exp\left(-\hbar\omega_\mu \hat{a}_\mu^\dagger \hat{a}_\mu / k_B \mathscr{T}\right)\right]} \tag{3.66}$$

[cf. (A.81)].

We start our discussion by treating a dynamical system that may be regarded as a one-dimensional harmonic oscillator; the operator of the complex normal amplitude will be denoted by $\hat{C}$. Then, according to the general description of a harmonic oscillator [cf. Section A.2.5.1], the Hamiltonian of the dynamical system is given by

$$\hat{H}^A = \hbar\omega \hat{C}^\dagger \hat{C} \qquad \text{with} \quad [\hat{C}^\dagger, \hat{C}^\dagger] = [\hat{C}, \hat{C}] = 0, \quad [\hat{C}, \hat{C}^\dagger] = \hat{I}, \tag{3.67}$$

where the zero-point energy has been appropriately fixed.

We set the interaction operator equal to

$$\hat{H}^C = \hbar(\hat{C}^\dagger \hat{M} + \hat{M}^\dagger \hat{C}); \tag{3.68}$$

it is bilinearly in the operators of the dynamical and dissipative systems. Furthermore, $\hat{H}^C$ depends linearly on the coupling constants g_μ and g_μ^*, respectively. Let us remark that the form of $\hat{H}^C$ is quite reasonable, and it covers the important case of dipole interaction of an atomic system with a stochastic electromagnetic field under the condition of neglecting nonresonant processes [cf. Sections 3.2.1 and 3.2.2].

The Hamiltonian of the total system is

$$\hat{H} = \hat{H}^A + \hat{H}^D + \hat{H}^C. \tag{3.69}$$

Furthermore, it is assumed that initially, at $t = 0$, the operators of the dynamical and dissipative systems commute. At this time both the systems may be taken to be uncoupled, so that

$$\hat{\rho}(0) = \hat{\rho}^A(0)\hat{\rho}^D(0); \tag{3.69a}$$

here $\hat{\rho}(0)$, $\hat{\rho}^A(0)$, and $\hat{\rho}^D(0)$ are the density operators of the total, dynamical, and dissipative systems at $t = 0$.

The foregoing characteristics are in principle sufficient to carry out an exact calculation of the time evolution of the operator $\hat{C}^\dagger(t)$. From this the most important statements about the dynamical system (e.g., about the energy, the dipole moment, and the statistics of fluctuations) can be derived. For the exact determination of $\hat{C}^\dagger(t)$ it would be necessary to include in the consideration the equations of motions for all operators $\hat{a}_\mu^\dagger(t)$, $\hat{a}_\mu(t)$ of the dissipative system. Such detailed treatment would, however, not provide any physical knowledge of importance for our purposes, but it would require a great deal of work. It is in fact possible to set up, in a good approximation, a relatively simple differential equation solely for operators of the dynamical system and to incorporate the influence of the dissipative system globally. We begin our discussion of this procedure by applying it to the operator $\hat{C}^\dagger(t)$.

The following exact equation of motion for $\hat{C}^\dagger(t)$ directly results from the Hamiltonian (3.69):

$$i\hbar\frac{d}{dt}\hat{C}^\dagger(t) = [\hat{C}^\dagger(t), \hat{H}^A(t)] + [\hat{C}^\dagger(t), \hat{H}^D(t)] + [\hat{C}^\dagger(t), \hat{H}^C(t)]. \tag{3.70}$$

Because of the given expressions for $\hat{H}^A$, $\hat{H}^D$, $\hat{H}^C$ and the commutation relations of the operators involved, this differential equation may be rewritten in the form

$$\frac{d}{dt}\hat{C}^\dagger(t) - i\omega\hat{C}^\dagger(t) = i\hat{M}^\dagger(t). \tag{3.70a}$$

The first commutator on the right-hand side of (3.70) leads to $i\omega\hat{C}^{\dagger}(t)$, the second commutator vanishes; the third commutator leads to $i\hat{M}^{\dagger}(t)$. To achieve an expansion in terms of the coupling constants we rewrite the right-hand side and get

$$i\hat{M}^{\dagger}(t) = i\left[\hat{U}^{C}(t,0)\right]^{-1}\hat{M}_{I}^{\dagger}\hat{U}^{C}(t,0) \tag{3.71}$$

where $\hat{U}^{C}$ is the unitary operator characteristic of the interaction picture [$\hat{U}^{C}$ is connected with the interaction operator $\hat{H}_{I}^{C}$ via the differential equation (A.114)]. The expansion of $\hat{U}^{C}$ and $\hat{M}_{I}^{\dagger}$ in terms of the coupling constants defined in (3.64) leads to

$$\frac{d}{dt}\hat{C}^{\dagger}(t) - i\omega\hat{C}^{\dagger}(t) = i\hat{M}_{I}^{\dagger}(t)$$
$$- \hat{C}_{I}^{\dagger}(t)\int_{0}^{t} dt' \sum_{\mu}|g_{\mu}|^{2}e^{i(\omega_{\mu}-\omega)(t-t')} + \hat{I}\cdot O(1)|g|^{3}. \tag{3.72}$$

The first term on the right-hand side contains the coupling constants in the first order, the second term contains them in the second order; the term with the order symbol represents the expressions in the third and higher orders. The relation (3.72) shows a somewhat undesirable property: the quantity to be determined appears in two different pictures, namely in the Heisenberg picture —$\hat{C}^{\dagger}(t)$—and in the interaction picture—$\hat{C}_{I}^{\dagger}(t)$. Let us correct this situation by introducing the operator

$$\underline{\hat{C}}^{\dagger}(t) \equiv \hat{C}^{\dagger}(t)e^{-i\omega t}, \tag{3.73}$$

which is constructed by splitting off the fast main time dependence from the operator in the Heisenberg picture. The operator $\underline{C}^{\dagger}(t)$ is slowly varying in time; if the coupling constants tend to zero, the operator $\underline{\hat{C}}^{\dagger}(t)$ tends to $\underline{\hat{C}}^{\dagger}(0)$. The connection of $\underline{\hat{C}}^{\dagger}(t)$ with the representation in the interaction picture is given by

$$\hat{C}_{I}^{\dagger}(t) = \underline{\hat{C}}^{\dagger}(0)e^{i\omega t}. \tag{3.74}$$

Finally, the equation of motion (3.70) may be written in the form

$$\frac{d}{dt}\underline{\hat{C}}^{\dagger}(t) = ie^{-i\omega t}\hat{M}_{I}^{\dagger}(t) - \underline{\hat{C}}^{\dagger}(0)\int_{0}^{t} dt'\, Q(t';t) + \hat{I}\cdot O(1)|g|^{3}, \tag{3.75}$$

where $Q(t';t)$ is the integrand of the integral in (3.72). Up to now we have only carried out a mathematical reformulation; now we shall perform physical approximations in the sense of the coarse-graining in the time domain (cf. Section A.2.3). First we make use of the quasicontinuous spectrum of the dissipative system; because of this assumption the quantity $Q(t'; t)$ may be written as

$$Q(t'; t) \to \int_0^\infty d\omega' \, \sigma(\omega') |g_{\omega'}|^2 e^{i(\omega' - \omega)(t - t')}$$

$$\to \sigma(\omega) |g_\omega|^2 2\pi\delta(t - t'), \tag{3.76}$$

where $g_{\omega'} \equiv g_\mu$ with $\omega_\mu = \omega'$; if $\sigma(\omega')$ and $|g_{\omega'}|^2$ can be regarded as slowly varying functions and $t \gg \omega^{-1}$ holds, $Q(t'; t)$ becomes deltafunction-like. After integrating with respect to time, the second term on the right-hand side of (3.75) reads

$$-\underline{C}^\dagger(0)\beta, \qquad \text{where} \quad \beta = \sigma(\omega)|g_\omega|^2. \tag{3.77}$$

Until now we have assumed that the interval $(0, t)$ is sufficiently large so that the integral on the right-hand side can be replaced by the constant β. On the other hand we shall assume that, when taken macrophysically, the interval $(0, t)$ is sufficiently small, so that the variation of the quantity $\hat{\underline{C}}^\dagger(t')$ in the interval $0 \leq t' \leq t$ is relatively small (this means that the coupling constants are assumed to be small); therefore in the second term on the right-hand side of (3.75) $\hat{\underline{C}}^\dagger(0)$ can be replaced by $\hat{\underline{C}}^\dagger(t)$. Since small coupling constants are assumed, the term with the order symbol is omitted, so that the relation (3.75) finally goes over into

$$\frac{d}{dt}\hat{\underline{C}}^\dagger(t) + \beta\hat{\underline{C}}^\dagger(t) = ie^{-i\omega t}\hat{M}_I^\dagger(t). \tag{3.78}$$

In deriving (3.78) obviously different assumptions with respect to the length of the interval $(0, t)$ are used; but they coexist in many important cases. It turns out in practice that the differential equation (3.78) is a good approximation for determining, via $\hat{\underline{C}}^\dagger(t)$, the basic operator $\hat{C}^\dagger(t)$ of a harmonic oscillator under the influence of a dissipative system. The quantity β has the character of a damping constant. The quantity on the right-hand side of (3.78) has the nature of a *fluctuation force* with the general property that on averaging over the variables of the dissipative system it becomes zero. The fluctuation force associated with $\hat{\underline{C}}^\dagger$, which will be denoted by $\hat{\Gamma}^\dagger(t)$, has the following properties:

$$\hat{\Gamma}^\dagger(t) = ie^{-i\omega t}\hat{M}_I^\dagger(t) = ie^{-i\omega t}\sum_\mu g_\mu^* \hat{a}_\mu^\dagger e^{i\omega_\mu t}, \tag{3.79}$$

$$\mathrm{Tr}\{\hat{\rho}^D \hat{\Gamma}^\dagger(t)\} = 0, \qquad \text{since} \quad \mathrm{Tr}\{\hat{\rho}^D \hat{a}_\mu^\dagger\} = 0.$$

Generally, the expectation values of odd-order products of $\hat{\Gamma}^\dagger$ and $\hat{\Gamma} = (\hat{\Gamma}^\dagger)^\dagger$ vanish. The expectation values of the second-order products either vanish [as the average values of $\hat{\Gamma}^\dagger(t_1)\hat{\Gamma}^\dagger(t_2)$ and $\hat{\Gamma}(t_1)\hat{\Gamma}(t_2)$] or are delta functions, namely

$$\mathrm{Tr}\{\hat{\rho}^D\hat{\Gamma}^\dagger(t_1)\hat{\Gamma}(t_2)\} = 2\beta\zeta_\omega\delta(t_1 - t_2),$$

$$\mathrm{Tr}\{\hat{\rho}^D\hat{\Gamma}(t_1)\hat{\Gamma}^\dagger(t_2)\} = 2\beta(\zeta_\omega + 1)\delta(t_1 - t_2), \tag{3.80}$$

where

$$\zeta_\omega = \frac{1}{\exp(\hbar\omega/k_B\mathcal{T}) - 1} \tag{3.81}$$

means the average occupation number of the quanta of the dissipative system in the mode $\omega_\mu = \omega$. The general solution of the differential equation (3.78) is given by

$$\underline{\hat{C}}^\dagger(t) = e^{-\beta(t-t_0)}\underline{\hat{C}}^\dagger(t_0) + \int_{t_0}^{t} dt'\, e^{-\beta(t-t')}\hat{\Gamma}^\dagger(t'). \tag{3.82}$$

If the fluctuation force is ignored, an exactly determined time evolution for $\underline{\hat{C}}^\dagger(t)$ results. From the operator $\underline{\hat{C}}^\dagger(t)$ determined by (3.78) we may now revert to an approximate operator with fast time dependence by multiplying by the factor $e^{i\omega t}$; it must be emphasized that the resulting operator $\hat{C}^\dagger(t) = e^{i\omega t}\underline{\hat{C}}^\dagger(t)$ only approximately agrees with the exact operator in the Heisenberg picture because of the approximations used in deriving (3.78). The equation for this approximated Heisenberg operator is given by

$$\frac{d\hat{C}^\dagger(t)}{dt} - i\omega\hat{C}^\dagger(t) + \beta\hat{C}^\dagger(t) = \hat{\Gamma}^\dagger_{C^\dagger}(t), \tag{3.83}$$

where $\hat{\Gamma}^\dagger_{C^\dagger}$ is the fluctuation force belonging to $\hat{C}^\dagger(t)$.

Later on a physical interpretation of this equation will be given. But before doing so, let us first treat the two-level system influenced by a dissipative system. According to its general description [cf. (A.173a)] we can use the Hamiltonian

$$\hat{H}^A = \hbar\omega\hat{B}^\dagger\hat{B} \tag{3.84}$$

and the dipole moment operator

$$\hat{d} = d_{21}\hat{B}^\dagger + d_{12}\hat{B}, \tag{3.85}$$

where d_{21}, d_{12} are the transition matrix elements between the ground state $|1\rangle$

and the upper state $|2\rangle$; $\hbar\omega$ is the transition energy. The flip operators $\hat{B}^\dagger$, $\hat{B}$ obey the anticommutation relations

$$[\hat{B}, \hat{B}]_+ = [\hat{B}^\dagger, \hat{B}^\dagger]_+ = 0, \qquad [\hat{B}, \hat{B}^\dagger]_+ = \hat{I}. \tag{3.86}$$

The interaction operator is assumed to be

$$\hat{H}^C = \hbar(\hat{B}^\dagger\hat{M} + \hat{M}^\dagger\hat{B}), \tag{3.87}$$

where $\hat{M}$ again denotes the driving force arising from the dissipative system. This helps us to set up the exact equation of motion for $\hat{B}^\dagger(t)$ in analogy to (3.70). We apply the same approximations to it as we used for $\hat{C}^\dagger(t)$ (neglect of coupling constants of third and higher orders; coarse-graining in the time domain). One finally ends up with

$$\frac{d}{dt}\hat{B}^\dagger(t) - i\omega\hat{B}^\dagger(t) + \beta(1 + 2\zeta_\omega)\hat{B}^\dagger(t) = \hat{\Gamma}^\dagger_{\hat{B}^\dagger}(t), \tag{3.88}$$

where $\hat{B}^\dagger(t)$ is the approximated Heisenberg operator of the atomic system in the sense of the discussion preceding (3.83). The left-hand side contains a damping term with the temperature-dependent damping parameter $\beta(1 + 2\zeta_\omega)$. The term on the right-hand side represents the fluctuation force $\hat{\Gamma}^\dagger_{\hat{B}^\dagger}(t)$ belonging to $\hat{B}^\dagger(t)$; averaging of $\hat{\Gamma}^\dagger_{\hat{B}^\dagger}(t)$ with respect to the variables of the dissipative system leads to zero.

Also, from other operators [as for instance $\hat{C}^\dagger(t)\hat{C}(t)$ and $\hat{B}^\dagger(t)\hat{B}(t)$] which correspond to the energy operator of the particular system except for the factor $\hbar\omega$, equations of motion can be derived which take globally into account the influence of the dissipative system. To gain them we start from the exact equations of motion

$$i\hbar\frac{d}{dt}\hat{C}^\dagger(t)\hat{C}(t) = [\hat{C}^\dagger(t)\hat{C}(t), \hat{H}(t)]$$

and

$$i\hbar\frac{d}{dt}\hat{B}^\dagger(t)\hat{B}(t) = [\hat{B}^\dagger(t)\hat{B}(t), \hat{H}(t)]$$

and apply the abovementioned approximations (neglect of coupling constants of third and higher orders; time coarse-graining). In the case of the two-level system we obtain the equation

$$\frac{d}{dt}\hat{B}^\dagger(t)\hat{B}(t) = -2\beta(1 + 2\zeta_\omega)\left\{\hat{B}^\dagger(t)\hat{B}(t) - \hat{I}\frac{\zeta_\omega}{1 + 2\zeta_\omega}\right\} + \hat{\Gamma}_{\hat{B}^\dagger\hat{B}}(t), \tag{3.89}$$

where $\hbar\omega\hat{B}^\dagger(t)\hat{B}(t)$ is the approximated Heisenberg operator for the energy and $\hat{\Gamma}_{\hat{B}^\dagger\hat{B}}(t)$ is the corresponding fluctuation force, which vanishes if the trace over the variables of the dissipative system is taken.

Having discussed how the basic equations are formed, let us now turn to some physical interpretations.

We begin with the consideration of *fluctuation forces*. The equations (3.79) and (3.83), which pertain to the harmonic oscillator, reveal a simple feature of the fluctuation force $\hat{\Gamma}^\dagger_{\hat{C}^\dagger}(t)$, namely, that it does not depend on atomic operators. It represents the driving force (multiplied by the imaginary unit) in the interaction picture and may therefore be interpreted as a force that is caused by a dissipative system which is not affected by the atomic system. In the case of the two-level system the situation is complicated, as a more detailed calculation shows: the fluctuation operator $\hat{\Gamma}^\dagger_{\hat{B}^\dagger}(t)$ [cf. (3.88)] contains the operators $\hat{B}^\dagger(t)$ and $\hat{B}(t)$, so that an interpretation of it as being due to a force of a free dissipative system fails. It may be mentioned here that for describing significant nonlinear optical phenomena we need to include in our consideration the fluctuation forces—e.g., for explaining the generation of relatively phase-fixed electromagnetic vibrations originating from spontaneous processes. Examples of this are the radiation of a laser, which will be discussed in Chapter 7, and that of the stimulated Raman effect, which will be treated in Chapter 11.

We now proceed to deal with *relaxation times*. To obtain the time-dependent expectation value of $\hat{C}^\dagger(t)$ or $\hat{B}^\dagger(t)$, the expression $\mathrm{Tr}\{\hat{\rho}(0)\hat{C}^\dagger(t)\}$ or $\mathrm{Tr}\{\hat{\rho}(0)\hat{B}^\dagger(t)\}$ must be evaluated, where $\hat{\rho}(0)$ is the density operator of the total system at $t = 0$. This can be performed in two steps. The procedure will here be demonstrated for the operator $\hat{B}^\dagger(t)$ of the two-level system. We first take the trace over all the variables of the dissipative system. The resulting quantity $\hat{B}^{\frown\dagger}(t) = \mathrm{Tr}\{\hat{\rho}^D(0)\hat{B}^\dagger(t)\}$ has the character of an operator in the Hilbert space of the dynamical system. The effect of this operation on the quantities in Equation (3.88) leads to the differential equation

$$\frac{d}{dt}\overset{\frown}{\underset{\smile}{\hat{B}}}{}^\dagger(t) - i\omega\overset{\frown}{\underset{\smile}{\hat{B}}}{}^\dagger(t) + \beta(1 + 2\zeta_\omega)\overset{\frown}{\underset{\smile}{\hat{B}}}{}^\dagger(t) = 0, \tag{3.90}$$

where the term representing the fluctuation force disappears. If the solution of (3.90), namely

$$\underset{\smile}{\hat{B}}{}^\dagger(t) = e^{-\beta(1+2\zeta_\omega)t}e^{i\omega t}\underset{\smile}{\hat{B}}{}^\dagger(0), \tag{3.91}$$

is inserted into (3.85), we obtain the dipole-moment operator

$$\overset{\frown}{\underset{\smile}{\hat{d}}}\,(t) = e^{-\beta(1+2\zeta_\omega)t}\left\{d_{21}e^{i\omega t}\overset{\frown}{\underset{\smile}{\hat{B}}}{}^\dagger(0) + \text{h.c.}\right\} \tag{3.92}$$

in the Hilbert space of the dynamical system. A further averaging by taking the trace with respect to the atomic part $\hat{\rho}^A(0)$ of the density operator leads to the expectation value $\langle \hat{d}(t) \rangle$ of the dipole moment. From the time dependence given in (3.92) it can be easily inferred that the dipole moment oscillates with a *fixed phase* at frequency ω within the time interval $0 < t \lesssim \beta^{-1}(1 + 2\zeta_\omega)^{-1}$. Let us assume that an ensemble of uncoupled uniform particles have all attained the same value of the dipole moment at $t = 0$ due to experimental manipulations. Then we may conclude from (3.92) that for $t \gtrsim 1/\beta(1 + 2\zeta_\omega) = T_2$ the polarization, which is known to be the sum of the dipole moments, tends to zero. Consequently, T_2 is called the *phase relaxation time* (or phase destroying time, or phase decay time). In an analogous way the phase relaxation time $1/\beta$ for the harmonic oscillator results from (3.83). The equation (3.89) for the energy $\hbar\omega\hat{B}^\dagger(t)\hat{B}(t)$ of the two-level system can be treated in a similar way to the equation (3.88) for $\hat{B}^\dagger(t)$. This immediately leads to the following result: the energy tends to its equilibrium value $\hbar\omega\zeta_\mu/(1 + 2\zeta_\mu)$ by an exponential law with the relaxation time $T_1 = 1/2\beta(1 + 2\zeta_\omega)$. Thus T_1 is called the *energy relaxation time*.

In addition to the relaxation times, the *transition probabilities* in the rate equations, which we shall turn to now, are of importance. The general description of the two-level system (Section A.2.5.2) shows that $\hat{B}^\dagger\hat{B}$ is the occupation-number operator $\hat{B}_{22}$ of the upper level, and $\hat{B}\hat{B}^\dagger = B_{11}$ is that of the ground state. Utilizing this connection and (3.86), we obtain (3.89) in the form

$$\frac{d}{dt}\hat{B}_{22}(t) = -2\beta(1 + \zeta_\omega)B_{22}(t) + 2\beta\zeta_\omega B_{11}(t) + \hat{\Gamma}_{\hat{B}_{22}}(t), \quad (3.93)$$

where $\hat{\Gamma}_{\hat{B}_{22}}(t) \equiv \hat{\Gamma}_{\hat{B}^\dagger\hat{B}}(t)$. By averaging with respect to the variables of the dissipative and the dynamical system one obtains the rate equation for the mean occupation number of the upper level,

$$\frac{d}{dt}\langle \hat{B}_{22}(t) \rangle = -p_{2\to 1}\langle \hat{B}_{22}(t) \rangle + p_{1\to 2}\langle \hat{B}_{11}(t) \rangle. \quad (3.94)$$

The first term on the right-hand side represents the contribution of the transition from $|2\rangle$ to $|1\rangle$ with the transition probability $p_{2\to 1} = 2\beta(1 + \zeta_\omega)$, while the second term characterizes the transition from $|1\rangle$ to $|2\rangle$ with $p_{1\to 2} = 2\beta\zeta_\omega$. Correspondingly, the transition probabilities depend on the properties of the atomic and the dissipative system, on the coupling constants, and via ζ_ω on the temperature. Obviously, the transition probability from the upper to the lower level exceeds the transition probability from the lower to the upper level by the factor

$$\frac{p_{2\to 1}}{p_{1\to 2}} = \frac{1 + \zeta_\omega}{\zeta_\omega} = \exp\left(\frac{\hbar\omega}{k_B\mathcal{T}}\right); \quad (3.94a)$$

this important fact is illustrated in Fig. 3.3.

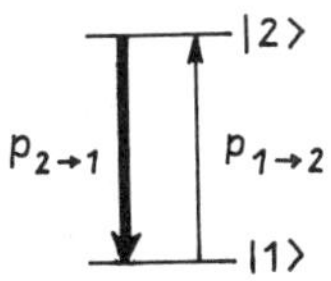

Fig. 3.3. Scheme of the transition probabilities between the lower atomic state $|1\rangle$ and the upper atomic state $|2\rangle$ at finite temperature.

Finally, let us sketch the connection of the *structure of the interaction operator* with the relaxation processes. As it can be concluded from (3.88) and (3.89), for the two-level system the relation

$$\frac{1}{T_2} = \frac{1}{2T_1} \tag{3.95}$$

results between the phase relaxation time and the energy relaxation time. This result is implied by the (special) structure of the employed interaction operator on the right-hand side of (3.87), which consists of summands with the operator products $\hat{B}^\dagger \hat{a}_\mu$ and $\hat{a}_\mu^\dagger \hat{B}$. The first product characterizes a physical process in which the atomic system changes from the lower to the upper state and a quantum of the dissipative system is simultaneously annihilated. In the second product the transitions are reversed. As considerations of transition probabilities reveal (Section A.2.3), such processes can only occur with a sufficiently high probability if the energy absorbed (emitted) by the atomic system is approximately equal to the energy emitted (absorbed) by the dissipative system. It is obvious that such processes are associated with both phase and energy relaxation. There are other types of interaction mechanisms. So the additional introduction of terms with operator products of the form $\hat{B}^\dagger \hat{B} \hat{a}_\mu^\dagger \hat{a}_{\mu'}$ into the interaction operator may represent further important features of the real interaction process (see e.g. Ref. 3.3). These are processes in which one quantum of the μth mode of the dissipative system is created and another one of the μ'th mode is annihilated. The operator $\hat{B}^\dagger \hat{B}$ indicates that no effective change in energy of the atomic system occurs. Hence, because of the required quasiresonance of the entire process, $\hbar\omega_\mu \approx \hbar\omega_{\mu'}$ must hold. The overall energy of the atomic system is conserved but the phase is not. When in the interaction operator one also includes phase-decay terms of the form $\hat{B}^\dagger \hat{B} \hat{a}_\mu^\dagger \hat{a}_{\mu'}$, the equation (3.95) goes over into

$$\frac{1}{T_2} = \frac{1}{2T_1} + \varkappa . \tag{3.96}$$

The relation (3.95) is empirically confirmed only in exceptional (though important) cases, as for instance in the case of radiative damping by spontaneous emission (which will be described in Section 3.3.1). The relation (3.96) has a wider range of application.

In summary, the following statements may be made regarding the applicability of the foregoing results: The resulting differential equations (3.83), (3.88), (3.89) for the approximated atomic operators $\hat{C}^\dagger(t)$, $\hat{B}^\dagger(t)$, $\hat{B}^\dagger(t)\hat{B}(t)$ of the harmonic oscillator and of the two-level system can also be applied, in view of their structure (occurrence of fluctuation forces and of relaxation times), to general atomic systems influenced by dissipative systems. The calculations on the basis of the above models reveal the principal way to calculate the relaxation constants from quantities of the atomic and dissipative system and from the coupling constants. In practice, this is very difficult because of the complexity of the interaction mechanisms, so that phenomenological relaxation times are frequently used.

3.2.4.2. Description of Dissipation by Ensemble Averages. Now we describe the time development of atomic ensemble averages under the influence of dissipative systems by *semiclassical* means. This yields the possibility of presenting a relatively simple treatment of atomic systems, even with more than two levels taking part in the interaction, and of showing how coupling parameters can be phenomenologically introduced. In doing so, certain results of the fully quantum-theoretical description will be incorporated. The Hamiltonian of the dynamical system under the action of perturbations is given by

$$\hat{H} = \hat{H}^A + \hat{H}^{CD} + \hat{H}^{CR} \tag{3.97}$$

where $\hat{H}^A$ is the operator of the uncoupled atomic systems, whereas $\hat{H}^{CD}$ and $\hat{H}^{CR}$ represent the action of dissipative systems and of coherent radiation fields on the atom. At the beginning let us assume that no radiation field acts on the atom, this implies $\hat{H}^{CR} = 0$. In the semiclassical approximation $\hat{H}^{CD}$ may be written as

$$\hat{H}^{CD} = \hat{G}F(t) \tag{3.98}$$

with $\hat{G}$ an operator of the atomic system (e.g., the dipole moment operator) and $F(t)$ a (generalized) classical force caused by the dissipative system at the position of the atom. $F(t)$ is a stochastic function. For example, we may think of $F(t)$ as being the electric field strength E_{stoch} generated by charged particles of the dissipative system moving stochastically around the atom under investigation (see Fig. 3.4). Important properties of $F(t)$ can be represented by its lowest-order correlation function, which is characterized by a finite but very small correlation time τ_c.

Let us start with the equation of motion of the rth ensemble member in the interaction picture,

$$\frac{d}{dt}\hat{\rho}_I^{(r)}(t) = -\frac{i}{\hbar}\left[\hat{H}_I^{CD(r)}(t), \hat{\rho}_I^{(r)}(t)\right]. \tag{3.99}$$

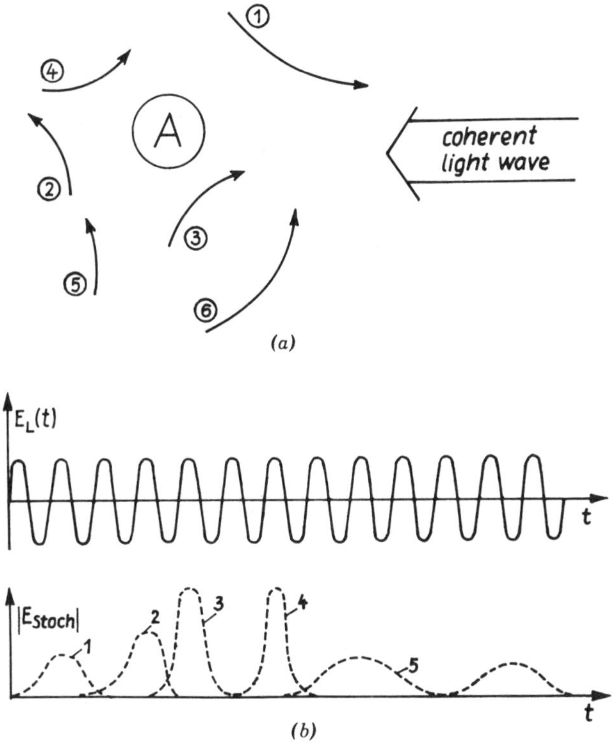

Fig. 3.4. Perturbation of the atom by coherent light and stochastically moving particles. (*a*) Schematic: *A* is the atom; $1, \ldots, 6$ are moving particles. (*b*) Electric fields acting on the atom: E_L results from the coherent light wave; E_{stoch} is composed of the contributions of the moving particles.

By formal integration we obtain

$$\hat{\rho}_I^{(r)}(t) = \hat{\rho}_I^{(r)}(0) - \frac{i}{\hbar}\int_0^t dt' \left[\hat{H}_I^{CD(r)}(t'), \hat{\rho}_I^{(r)}(t')\right]. \tag{3.100}$$

Using this expression on the right-hand side of (3.99), we arrive at

$$\frac{d}{dt}\hat{\rho}_I^{(r)}(t) = -\frac{i}{\hbar}\left[\hat{H}_I^{CD(r)}(t), \hat{\rho}_I^{(r)}(0)\right]$$
$$-\frac{1}{\hbar^2}\int_0^t d\tau \left[\hat{H}_I^{CD(r)}(t), \left[\hat{\Pi}_I^{CD(r)}(\iota - \tau), \hat{\rho}_I^{(r)}(t-\tau)\right]\right]. \tag{3.101}$$

Now the ensemble average is taken, and the average $\overline{\hat{\rho}_I^{(r)}(t)}$ is denoted by $\hat{\rho}_I(t)$. We may assume that $\hat{\rho}_I^{(r)}(0)$ and $\hat{H}^{CD(r)}(t)$ are uncorrelated, which is justified if $|t-0| > \tau_c$, and that the ensemble average of all matrix elements $\overline{H_{I\,kl}^{CD(r)}}$ (constructed with the kth and lth atomic eigenstate) vanishes because $\overline{F(t)} = 0$. Nonzero contributions may be taken into account in $\hat{H}^A$. Hence the ensemble average of the first term on the right-hand side of (3.101) vanishes. Furthermore, the dynamical system is supposed to develop, under the action of the dissipative system, slowly compared with the fast variations within the dissipative system; in other words, the changes of $\hat{\rho}_I(t)$ during τ_c are negligible $[\hat{\rho}_I(t) \approx \hat{\rho}_I(t \pm \tau_c)]$, which requires $\hbar^{-2}(H_{kl}^{CD})^2\tau_c^2 \ll 1$. Under this condition the correlation of $\hat{\rho}_I^{(r)}(t-\tau)$ with the Hamiltonians $\hat{H}^{CD(r)}(t)$ and $\hat{H}^{CD(r)}(t-\tau)$ and with their products can be neglected. Since the time range $\tau > \tau_c$ gives no substantial contribution to the integral in (3.101), the upper integration limit t may be replaced by ∞ for $t > \tau_c$. Within the time range $\tau < \tau_c$ the density operator $\hat{\rho}_I(t-\tau)$ is nearly equal to $\hat{\rho}_I(t)$ and therefore does not depend on the integration variable τ. Thus we obtain

$$\begin{aligned}\frac{d}{dt}\rho_{I\,kl}(t) = -\frac{1}{\hbar^2}\sum_{mn}\int_0^\infty d\tau\,\Big\{&\overline{H_{I\,km}^{CD}(t)H_{mn}^{CD}(t-\tau)}\rho_{I\,nl}(t)\\ &-\overline{H_{I\,km}^{CD}(t)H_{I\,nl}^{CD}(t-\tau)}\rho_{I\,mn}(t)\\ &-\overline{H_{I\,nl}^{CD}(t)H_{I\,km}^{CD}(t-\tau)}\rho_{I\,mn}(t)\\ &+\overline{H_{I\,nl}^{CD}(t)H_{I\,mn}^{CD}(t-\tau)}\rho_{I\,km}(t)\Big\}. \end{aligned} \quad (3.102)$$

The substitution of the matrix elements

$$H_{mn}^{CD}(t) = e^{-i\omega_{mn}t}H_{I\,mn}^{CD}(t)$$

which are given here in the Schrödinger picture, into (3.102) gives

$$\frac{d}{dt}\rho_{I\,kl}(t) = \sum_{mn} R_{kl,\,mn}\rho_{I\,mn}(t)e^{i(\omega_{km}+\omega_{nl})t}, \quad (3.103a)$$

where

$$\begin{aligned}R_{kl,\,mn} = -\frac{1}{\hbar^2}\sum_{m,n}\int_0^\infty d\tau\,\Bigg\{&\left[\sum_p \overline{H_{kp}^{CD}(t)H_{pm}^{CD}(t-\tau)e^{-i\omega_{mp}\tau}}\right]\delta_{nl}\\ &-\overline{H_{km}^{CD}(t)H_{nl}^{CD}(t-\tau)e^{-i\omega_{nl}\tau}}\\ &-\overline{H_{nl}^{CD}(t)H_{km}^{CD}(t-\tau)e^{-i\omega_{km}\tau}}\\ &+\left[\sum_p \overline{H_{pl}^{CD}(t)H_{np}^{CD}(t-\tau)e^{-i\omega_{np}\tau}}\right]\delta_{km}\Bigg\} \end{aligned} \quad (3.103b)$$

are the elements of the so-called *relaxation matrix*. [In deriving (3.103) the summation indices have been appropriately changed.] Because of the stationarity of the perturbation of the dissipative system, the averaged products like $\overline{H_{km}^{CD}(t)H_{nl}^{CD}(t-\tau)}$ and hence, the elements of the relaxation matrix $R_{kl,\,mn}$ are independent of time. In the following we shall not try to calculate the relaxation matrix $R_{kl,\,mn}$ from the matrix elements of the interaction operator, but we shall use them as parameters to be determined phenomenologically.

In (3.103) we can neglect the rapidly oscillating terms because they do not give substantial contributions to the evolution of the density matrix $\rho_{I\,kl}(t)$. Let us assume that there are no equally or nearly equally spaced energy levels in the dynamical system (this means that an accidental degeneracy is excluded). Then the relation $\omega_{km} + \omega_{nl} \approx 0$ can be satisfied only in the following two cases:

1. $m = k$ and $n = l$,
2. $k = l$ and $m = n$,

where the relations

$$\frac{d}{dt}\rho_{I\,kl}(t) = R_{kl,\,kl}\rho_{I\,kl}(t) \qquad (k \neq l) \tag{3.104a}$$

and

$$\frac{d}{dt}\rho_{I\,kk}(t) = \sum_{m} R_{kk,\,mm}\rho_{I\,mm}(t) \tag{3.104b}$$

can be derived from (3.103). Note that the derivation of these equations requires weak coupling between the dynamical and the dissipative system ($|R|^2\tau_c^2 \ll 1$). The two equations of (3.104) allow the calculation of $\rho_I(t)$ for $t \gg \tau_c$ if $\rho_I(0)$ is known.

To facilitate the interpretation, (3.104) may be written as

$$\frac{d}{dt}\rho_{kl}(t) = -i\omega_{kl}\rho_{kl} - \frac{1}{\tau_{kl}}\rho_{kl}(t) \tag{3.105a}$$

and

$$\frac{d}{dt}\rho_{kk}(t) = \sum_{m} w_{mk}\rho_{mm}(t) - \left(\sum_{n} w_{kn}\right)\rho_{kk}(t), \tag{3.105b}$$

where

$$\frac{1}{\tau_{kl}} = -R_{kl,\,kl}$$

with

$$R_{kk,mm} = w_{mk} \quad \text{and} \quad R_{kk,kk} = w_{kk} - \sum_n w_{kn}.$$

Here use has been made of

$$\rho_{I\,km}(t) = e^{i\omega_{km}t}\rho_{km}(t).$$

The dying away of any off-diagonal excitation $\rho_{kl}(t)$ is described by the relaxation time τ_{kl}, which is called the *phase decay time* because the off-diagonal matrix element carries the phase information (cf. Section 3.2.4.1). For two-level systems τ_{kl} is also called the *transverse relaxation time*, because in the special case of magnetic-resonance spectroscopy τ_{kl} describes the decay of the magnetization components transverse to the magnetic field. The diagonal element ρ_{kk} is the probability of finding the system in the kth state, and w_{mk} is the *transition probability* $m \to k$. Hence (3.105b) is a system of rate equations connecting the occupation probabilities of the molecular levels. Multiplying (3.105) by the whole number N of molecules taking part in the process or by the molecular number density γ, rate equations for the population numbers N_k or the population number densities γ_k are obtained. The relaxation rates w_{mk} and τ_{kl}^{-1} can be related to the matrix elements of the interaction operator $\hat{H}^{CD}$ by the use of (3.103)–(3.105). Such calculation reveals some shortcomings of the semiclassical theory. Thus, we obtain $w_{mk} = w_{km}$, which is in contradiction to the results of the fully quantum-mechanical description of (3.94a):

$$\frac{p_{k\to m}}{p_{m\to k}} = \frac{w_{km}}{w_{mk}} = e^{\hbar\omega_{km}/k_B\mathcal{T}}. \tag{3.106}$$

This unsymmetric relation is in agreement with experiment; the decrease of the population of higher levels by relaxation is preferred (see Fig. 3.1). Only for temperatures $\mathcal{T} \to \infty$ does the relation become symmetric ($w_{mk} = w_{km}$). The semiclassical equations (3.105) correctly describe molecules acted on by dissipative systems if the right rate parameters are inserted, that is, rate parameters determined experimentally or calculated by the use of quantum theory. Since the calculation of relaxation parameters is rather difficult, experimental values are used in most cases.

Instead of the transition rates w_{km} one often uses relaxation times T_{km} defined by

$$T_{mk} = T_{km} = \frac{\rho_{mm}^e}{w_{km}} = \frac{\rho_{kk}^e}{w_{mk}}, \tag{3.107}$$

where ρ_{mm}^e is the population probability of state m in thermal equilibrium (attained, e.g., at $t \to \infty$ after an excitation at $t = 0$). Using (3.107) in

(3.105b), we obtain

$$\frac{d}{dt}\rho_{kk}(t) = -\sum_m \left(\frac{\rho^e_{mm}}{T_{km}}\rho_{kk}(t) - \frac{\rho^e_{kk}}{T_{km}}\rho_{mm}(t) \right). \qquad (3.108)$$

Assuming a two-level system (or assuming all T_{km} to be equal, $T_{km} = T$) one obtains, because $\Sigma_n \rho_{mm} = \Sigma_m \rho^e_{mm} = 1$,

$$\frac{d}{dt}\rho_{kk}(t) = -\frac{1}{T}\left[\rho_{kk}(t) - \rho^e_{kk}\right]. \qquad (3.109)$$

Here T is referred to as the (energy) *lifetime* of the system or, because of its meaning in magnetic-resonance spectroscopy, as the *longitudinal relaxation* time.

3.2.4.3. Atomic Systems Interacting with Dissipative Systems and Radiation. Now let us discuss an atom being simultaneously influenced by a dissipative system and by a light field representing a coherent perturbation. The term *coherent perturbation* is used here with the following meaning. The correlation time τ_c' of this perturbation is assumed to be large in comparison with τ_c of the dissipative system. In this case we have to employ the full Hamiltonian of (3.97). The two perturbations are supposed to act independently of one another on the time development of the atom. Then we obtain

$$\frac{d}{dt}\hat{\rho}_I(t) = R\hat{\rho}_I(t) + \frac{1}{i\hbar}\left[\hat{H}_I^{CR}, \hat{\rho}_I\right], \qquad (3.110)$$

where the first term on the right-hand side represents the influence of the dissipative system [cf. (3.104)] with R the "dissipation matrix," and the second term that of the coherent perturbation. The action of these two perturbations is often considered in the following way: Since in most cases the same dissipative perturbation is present at all times, we regard the atom influenced by it as a given system that is now perturbed only by the radiation field. For this purpose we write (3.110) as

$$\frac{d}{dt}\rho_{kk} + \frac{1}{T}\left(\rho_{kk} - \rho^e_{kk}\right) = \frac{1}{i\hbar}\left[\hat{H}^{CR}, \hat{\rho}\right]_{kk} \qquad (3.111)$$

and

$$\frac{d}{dt}\rho_{kl} + i\omega_{kl}\rho_{kl} + \frac{1}{\tau_{kl}}\rho_{kl} = \frac{1}{i\hbar}\left[\hat{H}^{CR}, \hat{\rho}\right]_{kl}$$

with

$$\left[\hat{H}^{CR}, \hat{\rho}\right]_{kl} = \sum_m \left(H^{CR}_{km}\rho_{ml} - \rho_{km}H^{CR}_{ml} \right).$$

If the terms on the right-hand sides of these differential equations are considered as perturbations, then the unperturbed system is identical with (3.105a), and (3.109) and gives $\rho^{(0)}$. Hence, the nth-order contributions $\hat{\rho}^{(n)}$ to $\hat{\rho}$ $(n \geq 1)$ are obtained from

$$\frac{d}{dt}\rho_{kk}^{(n)}(t) + \frac{1}{T}\left(\rho_{kk}^{(n)} - \rho_{kk}^{e}\right) = \frac{1}{i\hbar}\left[\hat{H}^{CR}, \hat{\rho}^{(n-1)}\right]_{kk}$$

and

$$\frac{d}{dt}\rho_{kl}^{(n)}(t) + i\omega_{kl}\rho_{kl}^{(n)} + \frac{1}{\tau_{kl}}\rho_{kl}^{(n)} = \frac{1}{i\hbar}\left[\hat{H}^{CR}, \rho^{(n-1)}\right]_{kl}.$$

3.3. BASIC ONE-PHOTON PROCESSES

Important features of nonlinear optical processes and multiphoton processes, which are the main subject of this book, are described by terms and relations that have been developed in connection with basic one-photon processes. Therefore we shall deal with the spontaneous and stimulated emission and absorption of one photon by an atomic system in Section 3.3.1. This is followed by the discussion of line-broadening effects in Section 3.3.2.

3.3.1. Emission and Absorption of One Photon by an Atomic System

Let us consider the interaction of an atomic system, say an atom, with electromagnetic radiation. The atomic subsystem is to be characterized by the Hamiltonian $\hat{H}^A$, and the radiation subsystem by the Hamiltonian $\hat{H}^R$ as presented in (2.21). For simplicity we assume that the center of mass of the atom coincides with the origin $r_\bullet = 0$. The conditions of the dipole approximation (Section 3.2.1) are assumed to be satisfied. Then we can start from the interaction operator

$$\hat{H}^C = -\hat{d}_\bullet \hat{E}_\bullet, \tag{3.112}$$

where $\hat{d}_\bullet$ denotes the atomic dipole-moment operator and $\hat{E}_\bullet$ the operator of the electric field strength. The "free system"—in the sense of the interaction picture (Section A.2.1)—is represented by the Hamiltonian

$$\hat{H}^F = \hat{H}^A + \hat{H}^R. \tag{3.113}$$

It has the eigenvalues

$$\mathscr{E}^F = \mathscr{E}^A + \sum_\mu \hbar\omega_\mu\left(n_\mu + \tfrac{1}{2}\right) \tag{3.114}$$

and the eigenstates

$$|\mathscr{E}^F\rangle = |\mathscr{E}^A\rangle|n_1,\ldots,n_\mu,\ldots\rangle. \tag{3.115}$$

$\mathscr{E}^A$ and $|\mathscr{E}^A\rangle$ belong to the atom; ω_μ and n_μ are the frequency and the photon numbers of the μth mode of the radiation field.

We start by discussing the *emission process*. An excited atom is supposed to change from the initial state $|\mathscr{E}_i^A\rangle$ to the lower final state $|\mathscr{E}_f^A\rangle$, accompanied by the emission of one photon of the μth mode. Thus the initial and final states of the total system are given by

$$|\mathscr{E}_i^F\rangle = |\mathscr{E}_i^A\rangle|n_1,\ldots,n_\mu,\ldots\rangle, \tag{3.116a}$$

$$|\mathscr{E}_f^F\rangle = |\mathscr{E}_f^A\rangle|n_1,\ldots,n_\mu+1,\ldots\rangle; \tag{3.116b}$$

only in the μth mode the photon number increases by one after the atomic transition. We want to evaluate the rate of the total transition probability; therefore, according to Fermi's "golden rule" (A.125), we must next calculate the matrix element $\langle\mathscr{E}_f^F|\hat{H}^C|\mathscr{E}_i^F\rangle$. Since the operator $\hat{d}_\bullet$ acts only on atomic states and the operator $\hat{E}_\bullet$ only on field states, the matrix element takes the form of a product, namely

$$\langle\mathscr{E}_f^F|\hat{H}^C|\mathscr{E}_i^F\rangle = -\langle\mathscr{E}_f^A|\hat{d}_\bullet|\mathscr{E}_i^A\rangle\langle n_1,\ldots,n_\mu+1,\ldots|\hat{E}_\bullet|n_1,\ldots,n_\mu,\ldots\rangle. \tag{3.117}$$

Let the factor containing the dipole moment be designated by $d_{fi\bullet}$; according to basic properties of the radiation field as represented in (2.51) and (2.48), the second factor is

$$e_{\mu\bullet}\left(\frac{\hbar\omega_\mu}{2\varepsilon_0 V}\right)^{1/2}(n_\mu+1)^{1/2}.$$

Thus we get the expression

$$\left|\langle\mathscr{E}_f^F|\hat{H}^C|\mathscr{E}_i^F\rangle\right|^2 = |d_{fi\bullet}e_{\mu\bullet}|^2\frac{\hbar\omega_\mu n_\mu}{2\varepsilon_0 V} + |d_{fi\bullet}e_{\mu\bullet}|^2\frac{\hbar\omega_\mu}{2\varepsilon_0 V}, \tag{3.118}$$

which is directly proportional to the rate of the total transition probability.

While the first term on the right-hand side of (3.118) yields a nonvanishing contribution only in the presence of radiation ($n_\mu > 0$), the second term characterizes an emission process, which takes place in the absence of radiation; the latter is called *spontaneous emission* and will be treated now. The energy of the final state is

$$\mathscr{E}_f^F = \mathscr{E}_i^F + \hbar\omega_\mu - \left(\mathscr{E}_i^A - \mathscr{E}_f^A\right) = \mathscr{E}_i^F + \hbar\omega_\mu - \hbar\omega_{if}, \tag{3.119}$$

where ω_{if} denotes the atomic transition frequency. Generally, appreciable transition probabilities can only occur if the final energy $\mathscr{E}_f^F$ is *approximately* equal to the initial energy $\mathscr{E}_i^F$ [cf. Section A.2.3]. Therefore (3.119) leads to the following condition for the energy of the emitted photon:

$$\hbar\omega_\mu \approx \hbar\omega_{if}. \tag{3.120}$$

This means that there is a spread of the photon energy $\hbar\omega_\mu$ within a certain interval around the atomic transition energy. To obtain the total transition probability, we have to sum over all the final states that are involved. The application of Fermi's "golden rule" requires the insertion of the number of states $\sigma(\mathscr{E}_f^F)$ per unit energy of the radiation field, which is given [cf. (1.27)] by

$$\sigma\left(\mathscr{E}_f^F\right) = \frac{V}{\pi^2 c^3 \hbar}\omega_{if}^2, \tag{3.121}$$

where the integration over the entire solid angle and the summation over the directions of polarization is included. The factor $|d_{fi\bullet} e_{\mu\bullet}|^2$ in (3.118) can be expressed by $|d_{fi\bullet}|^2 \cos^2\vartheta_\mu$, where $|d_{fi\bullet}|$ is the length of the vector $|d_{fi\bullet}|$ and ϑ_μ is the angle between $d_{fi\bullet}$ and $e_{\mu\bullet}$. Since emitted light will be considered independently of its direction of propagation and polarization, an averaging over all angles ϑ_μ must be performed, while the direction of the vector $d_{fi\bullet}$ is fixed. This orientation averaging yields

$$\overline{|d_{fi\bullet} e_{\mu\bullet}|^2}^{\text{orient}} = |d_{fi\bullet}|^2 \overline{\cos^2\vartheta_\mu} = \frac{|d_{fi\bullet}|^2}{3}. \tag{3.122}$$

Altogether, because of the second term on the right-hand side of (3.118), we arrive at the total transition probability per unit time

$$\frac{d}{dt}P_{\text{sp em}} = \frac{1}{3\varepsilon_0 \pi c^3 \hbar}|d_{ul\bullet}|^2 \omega_{ul}^3 = A_{ul} \tag{3.123}$$

for the spontaneous emission of unpolarized radiation into the entire solid angle. The expression on the right-hand side is called the *Einstein transition coefficient* A_{ul} of the atomic transition from the upper state (index u) to the lower state (index l). It depends (apart from universal constants) on the atomic transition moment $|d_{ul\bullet}|$ and the atomic transition frequency ω_{ul}, which stands for a mean value of the frequencies emitted. Note that in the case of emission the initial state has to be identified with the upper one, and the final state with the lower one.

The first term on the right-hand side of (3.118) represents (in the case of $n_\mu > 0$) a contribution to the one-photon emission, which is stimulated (induced) by the radiation present. This process is called the *stimulated emission*. In deriving (3.118) we only considered the emission of radiation in the μth mode. In general, the incident radiation shows a spread in frequency. Therefore, in order to obtain the total transition probability we must take into account all the modes for which the condition $\omega_\mu \approx \omega_{if}$ is fulfilled. If $\sigma(\hbar\omega_\mu)$ is the state density around ω_{if}, the relations (3.118), (3.122) together with Fermi's "golden rule" (A.125) lead to the expression

$$\frac{d}{dt}P_{\text{st em}} = \frac{\pi}{3\hbar\varepsilon_0}|d_{fi\bullet}|^2\left\{\frac{\hbar\omega_\mu n_\mu}{V}\sigma(\hbar\omega_\mu)\right\} \tag{3.124}$$

for the transition probability per unit time for stimulated emission. It is obvious that the ratio $\hbar\omega_\mu n_\mu/V$ is the energy density per unit volume for a single mode and that therefore the expression in the braces represents the energy density per unit volume and per unit energy. Introducing the energy density ζ per unit volume and per unit angular frequency, we finally obtain

$$\frac{d}{dt}P_{\text{st em}} = \frac{\pi}{3\hbar^2\varepsilon_0}|d_{ul\bullet}|^2\zeta(\omega_{ul}) = B_{ul}\zeta(\omega_{ul}). \tag{3.125}$$

The transition rate per unit angular frequency divided by the energy density ζ is called the *Einstein transition coefficient* B_{ul} for the transition from the upper to the lower state (if we relate the transition coefficient to the frequency, we get the coefficient $B_{ul}/2\pi$). Apart from universal constants, the Einstein transition coefficient B_{ul} depends on the atomic transition moment.

In analogy to the stimulated one-photon emission, the transition rate due to one-photon *absorption* of radiation can also be evaluated. An atom assumed to be initially in a lower state $|\mathscr{E}_i^A\rangle$ absorbs one photon and flips into the upper state $|\mathscr{E}_f^A\rangle$. Also the radiation that induces the absorption has a certain spread in frequencies around the atomic transition frequency $\omega_{fi} = \hbar^{-1}(\mathscr{E}_f^A - \mathscr{E}_i^A)$. If the corresponding energy density per unit volume per unit angular frequency is $\zeta(\omega_{fi})$, one obtains the total transition rate

$$\frac{d}{dt}P_{\text{abs}} = \frac{\pi}{3\hbar^2\varepsilon_0}|d_{ul\bullet}|^2\zeta(\omega_{ul}) = B_{lu}\zeta(\omega_{ul}). \tag{3.126}$$

for absorption from the lower (initial) state to the upper (final) state.

From (3.125) and (3.126) it can be immediately seen that (under our tacit assumption of nondegenerate atomic levels) the Einstein transition coefficients B_{lu} for the transition from the lower to the upper state is identical with the coefficient B_{ul} for the reversed transition. Thus, the expressions for the rate of

Table 3.1. Transition Frequencies ω_{ul} and Transition Dipole Moments $|d_{ul}|$

Transition Frequencies ω_{ul} (s^{-1})	
10^{16}	Electronic transitions in atoms, molecules, and solids
10^{14}	Optical vibrational transitions in molecules and solid-state material
10^{12}	Rotational transitions in molecules
Transition Dipole Moments $\lvert d_{ul}\rvert$ (A s m)	
10^{-29}	Electronic transitions in atoms and molecules
10^{-31}	Vibrational transitions in molecules

stimulated emission and absorption agree with each other. The ratio A_{ul}/B_{ul} is proportional to ω_{ul}^3; this means that the spontaneous emission dominates the stimulated emission for sufficiently large values of ω_{ul}. Let us estimate the value of A_{ul}. The matrix element of the dipole moment, in the one-electron approximation, is

$$|d_{ul\bullet}| = e\left|\int d^3r_\bullet\, \psi_u^*(r_\bullet) r_\bullet \psi_l(r_\bullet)\right| = e|r_{ul}|. \tag{3.127}$$

As a rule, for an atom the functions $\psi_u^*(r_\bullet)$ and $\psi_l(r_\bullet)$ attain appreciable values only within a region of the order of 10^{-10} m. It follows $|d_{ul\bullet}| \lesssim 10^{-29}$ A s m. For radiation in the visible region the angular frequency ω_{ul} is taken to be 4×10^{15} s^{-1}. For "strong" atomic spectral lines—that is, large values of $dP_{\text{sp em}}/dt$—one obtains $A_{ul} \sim 10^8$ s^{-1}. We may conclude that for atomic transitions generally $A_{ul} \ll \omega_{ul}$ is satisfied. Table 3.1 displays typical values of transition frequencies and transition moments of atoms and molecules.

From the equations for spontaneous and stimulated emission and for absorption [see (3.123), (3.125), (3.126)]; one can calculate how many atomic systems per unit time make a transition between the two levels. If the atomic systems are assumed to be independent of each other, the rate for the transition from the upper to the lower level is

$$\frac{dN_{u\to l}}{dt} = N_u\left[A_{ul} + B_{ul}\zeta(\omega_{ul})\right], \tag{3.128}$$

whereas the rate for the transition from the lower to the upper level is

$$\frac{dN_{l\to u}}{dt} = N_l B_{lu}\zeta(\omega_{ul}). \tag{3.129}$$

N_u and N_l represent the number of atoms in the upper and lower level, respectively. The validity of these two equations is independent of the particular conditions from which the occupation numbers N_u, N_l and the radiation density $\zeta(\omega_{ul})$ arise.

Important conclusions can be drawn from the consideration of the special case that there is thermal equilibrium between the emitting and absorbing atoms on the one hand and the radiation field of a cavity at the temperature $\mathscr{T}$ on the other hand. This means that for the balance of the transitions and the occupation numbers the following two relations

$$\frac{dN_{u\to l}}{dt} = \frac{dN_{l\to u}}{dt}, \qquad \frac{N_u}{N_l} = \exp\left(-\frac{\hbar\omega_{ul}}{k_B\mathscr{T}}\right) \tag{3.130}$$

hold and that $\zeta(\omega_{ul})$ must be identified with the energy density of the cavity radiation $\zeta(\omega_{ul}, \mathscr{T})$, which is given by

$$\zeta(\omega_{ul}, \mathscr{T}) = \frac{\hbar\pi^{-2}c^{-3}\omega_{ul}^3}{e^{\hbar\omega_{ul}/k_B\mathscr{T}} - 1}. \tag{3.131}$$

From the relations (3.128)–(3.130) we find

$$\zeta(\omega_{ul}, \mathscr{T}) = \frac{A_{ul}/B_{ul}}{(B_{lu}/B_{ul})[\exp(\hbar\omega_{ul}/k_B\mathscr{T} - 1)]}. \tag{3.132}$$

From this expression Einstein's relations

$$B_{ul} = B_{lu}, \tag{3.133a}$$

$$A_{ul} = \frac{\hbar}{\pi^2c^3}\omega_{ul}^3 B_{ul} \tag{3.133b}$$

immediately result. The relation (3.133b) reveals the following important result: Although from the above equilibrium considerations the quantities A_{ul} and B_{ul} cannot be derived singly, it is possible to derive a unique interrelationship between them (if A_{ul} is known, B_{ul} can be calculated, and vice versa). This knowledge will be made use of in the discussion on the line-broadening effects.

3.3.2. Line-Broadening Effects

Instead of monochromatic radiation at the exact atomic transition frequency, a certain spread in frequencies is generally found when emission and absorption are observed. The width of the frequency interval depends on specific properties of the medium investigated as well as on external experimental conditions (temperature, pressure, and so on). One of the line-broadening mechanisms is a fundamental one that is connected with the so-called natural line width. It will be treated in Section 3.3.2.1, before dealing with other features of line-broadening effects in Section 3.4.2.2.

3.3.2.1. Natural Line Width. Under the conditions of spontaneous emission the energy of the emitted photons spreads in a certain interval around the atomic transition energy, which is expressed by the relation (3.120). V. Weisskopf and E. Wigner (Ref. 3.4) have evaluated the energy distribution of spontaneously emitted photons. We shall sketch the procedure applied by them, where the notation of the foregoing subsection is used. For simplicity, only the first excited state and the ground state of the atomic system (say an atom) are taken into consideration. Assuming that at $t = 0$ the atom is in its first excited state $|\mathscr{E}_1^A\rangle$ and the radiation field in the vacuum state ($n_\mu = 0$ for all modes), we have the initial state

$$|\mathscr{E}_i^F\rangle = |\mathscr{E}_1^A\rangle|0,0,0,\dots\rangle. \tag{3.134}$$

It is assumed that the expression for the dipole interaction (3.112) between the atomic and the radiation system holds.

We shall attempt to calculate at a time $t > 0$ the probability of finding the total system in one of the final states defined by

$$|\mathscr{E}_\mu^F\rangle \equiv |\mathscr{E}_0^A\rangle|0,\dots 0,\ n_\mu = 1,0,\dots\rangle, \tag{3.135}$$

where the atom is in its ground state and the radiation field is in a certain one-photon state (one photon in the μth mode). We aim at the result for a long interaction duration, $(t - 0) \to \infty$. Therefore the time-dependent perturbation theory (see Section A.2.2) is inadequate for solving our problem, and we must start from the exact equation of motion for the state vector, (A.109); from it the system of differential equations

$$i\hbar\frac{d}{dt}c_j(t) = \sum_{j'}\langle\mathscr{E}_j^F| - \hat{d}_\bullet\hat{E}|\mathscr{E}_{j'}^F\rangle e^{i\omega_{jj'}t}c_{j'}(t) \qquad \text{for all } j \tag{3.136}$$

directly follows for the time-dependent coefficients $c_j(t) = \langle\mathscr{E}_j^F|\psi(t)\rangle_I$ of a general state $|\psi(t)\rangle_I$ in the interaction picture. The states $|\mathscr{E}_j^F\rangle$ and $|\mathscr{E}_{j'}^F\rangle$ are arbitrary states of the free system characterized by (3.114). It should be emphasized that the sum must be taken over all terms containing nonzero matrix elements $\langle\mathscr{E}_j^F| - \hat{d}_\bullet\hat{E}_\bullet|\mathscr{E}_{j'}^F\rangle$ if the exactness of equation (3.136) is to be maintained.

Let us first consider the differential equation for $c_i(t) = \langle\mathscr{E}_i^F|\psi(t)\rangle_I$, which concerns the coefficient for the first excited atomic state and the photon vacuum state. The respective matrix element $\langle\mathscr{E}_i^F| - \hat{d}_\bullet\hat{E}_\bullet|\mathscr{E}_{j'}^F\rangle$ in this equation are nonzero only if $|\mathscr{E}_{j'}^F\rangle$ contains a one-photon state as presented in

(3.135). Thus we have

$$i\hbar\frac{d}{dt}c_i(t) = \sum_{\mu}\langle\mathscr{E}_i^F| - \hat{d}_\bullet\hat{E}_\bullet|\mathscr{E}_\mu^F\rangle e^{i\omega_{i\mu}t}c_\mu(t), \tag{3.137}$$

where $\hbar\omega_{i\mu} = \mathscr{E}_i^F - \mathscr{E}_\mu^F$. Furthermore we need the equations for the coefficients of the one-photon states $|\mathscr{E}_\mu^F\rangle$, that is, for $c_\mu(t) = \langle\mathscr{E}_\mu^F|\psi(t)\rangle_I$. The respective matrix elements $\langle\mathscr{E}_\mu^F| - \hat{d}_\bullet\hat{E}_\bullet|\mathscr{E}_{j'}^F\rangle$ are nonzero only if $|\mathscr{E}_{j'}^F\rangle$ contains the photon vacuum state or a two-photon state that can be generated from the photon state $|\mathscr{E}_\mu^F\rangle$ by the additional creation of one photon. Because of these two-photon states, further coefficients must be introduced into the system of differential equations, where also three-photon states, four-photon states, and so on come finally into play. Weisskopf and Wigner omitted all terms containing multiphoton radiation states; this means physically that multiphoton emission is neglected. Thus the equations for the coefficients $c_\mu(t)$ become

$$i\hbar\frac{d}{dt}c_\mu(t) = \langle\mathscr{E}_i^F| - \hat{d}_\bullet\hat{E}_\bullet|\mathscr{E}_\mu^F\rangle^* e^{-i\omega_{i\mu}t}c_i(t) \qquad \text{for all } \mu. \tag{3.138}$$

To obtain a mathematical solution, the procedure is as follows: The integration $\int_0^t dt'$ is performed on both sides of (3.138). The resulting expression $c_\mu(t)$, which among other quantities contains $c_i(t')$, is inserted into (3.137). This leads to an integrodifferential equation for $c_i(t)$, which can be solved approximately. The solution $c_i(t)$ is inserted into (3.138), which will lead to the solution for $c_\mu(t)$. The quantity $|c_\mu(t)|^2$ is the probability of finding at time t the state $|\mathscr{E}_\mu^F\rangle$, which is characterized by a certain wave vector and polarization direction. We want to find the probability density that one photon —independently of polarization and direction of emission—is spontaneously emitted with an energy lying in the interval $(\mathscr{E}, \mathscr{E} + d\mathscr{E})$. This probability density can be calculated from $|c_\mu(t)|^2$ by multiplying it by the mode density (in analogy with the approach in Section 3.3.1), and for large times t it becomes approximately

$$G(\mathscr{E}; \mathscr{E}_m, \Delta\mathscr{E}) = \frac{\hbar A_{10}/2\pi}{[\mathscr{E} - (\mathscr{E}_{10} + \mathscr{E}_{LR})]^2 + (\hbar A_{10}/2)^2}. \tag{3.139}$$

In this result the Einstein transition coefficient A_{10} from (3.123), which represents the transition between the atomic states $|\mathscr{E}_1^A\rangle$ and $|\mathscr{E}_0^A\rangle$, occurs. The result for the given probability density holds for $t \gg A_{10}^{-1}$. The function $G(\mathscr{E}; \mathscr{E}_m, \Delta\mathscr{E})$ represents a Lorentzian shape with the width (fwhm) $\Delta\mathscr{E} = \hbar A_{10}$ and the center energy $\mathscr{E}_m = \mathscr{E}_{10} + \mathscr{E}_{\text{LR}}$.

The Lamb–Retherford shift $\mathscr{E}_{\text{LR}}$ can be regarded as small compared with the atomic transition energy $\mathscr{E}_{10} = \mathscr{E}_1^A - \mathscr{E}_0^A$; therefore the center energy $\mathscr{E}_m$ agrees in good approximation with $\mathscr{E}_{10}$. It can be inferred from the estimation

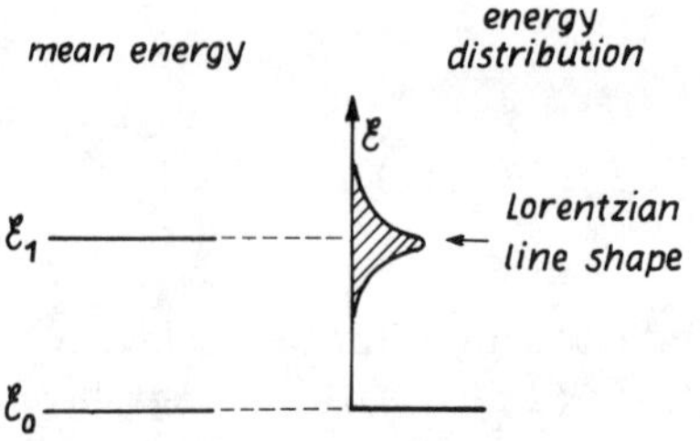

Fig. 3.5. Mean energy and energy distribution for radiative damping.

of A_{10} given in Section 3.3.1 that the relative line width $\Delta\mathcal{E}/\mathcal{E}_{10}$ is $\lesssim 10^{-8}$ for atoms. The integration of $G(\mathcal{E};\, \mathcal{E}_m,\, \Delta\mathcal{E})$ with respect to all photon energies $\mathcal{E}$ yields approximately the value 1 (exactly if the lower integral limit is extended to $-\infty$, which may be done because of the small relative line width). Since the line width of the spontaneous emission is independent of external conditions and arises only from the coupling of the atomic system to the radiation field in the vacuum state, it is called the *natural line width*. This width is said to be caused by radiation damping.

If the physical content of (3.139) is interpreted in the atomic energy-level scheme, the natural line width must be regarded as the result of the transition from a broadened energy level with the Lorentzian shape $G(\mathcal{E};\, \mathcal{E}_1^A,\, \hbar A_{10})$ around the mean energy $\mathcal{E}_1^A$ into the sharp ground state with the energy $\mathcal{E}_0^A$; this is illustrated in Fig. 3.5. For a general atomic system the interpretation is as follows: A higher excited atomic level j can be depleted by spontaneous emission of a photon with the decay of the excited state of the atom to a lower level; the process is quantitatively characterized by (3.123). The probability that the atom jumps from level j either into level $j-1$ or into level $(j-2)\ldots$ or into level 0 is the sum of all the transition probabilities. Therefore the energy width of level j (with mean energy $\mathcal{E}_j^A$) is

$$\Delta\mathcal{E}_j^A = \hbar \sum_{j'<j} A_{jj'}. \tag{3.140}$$

The energy line width of the transition from level j to j' is the sum of the two line widths, namely

$$\Delta\mathcal{E}_{jj'}^A = \Delta\mathcal{E}_j^A + \Delta\mathcal{E}_{j'}^A. \tag{3.141}$$

The line shape for the transition $j \to j'$ is Lorentzian, and is the convolution of the two Lorentzian shapes of the levels taking part. The theoretical results of (3.139) and (3.141) are in good agreement with experiment.

At this point let us briefly consider the time evolution of the occupation probability of the excited atomic state. We again refer to the two-level system. The calculation presented at the beginning of this sub-subsection yields

$$|c_i(t)|^2 = \exp(-A_{10}t); \tag{3.142}$$

this means that the occupation of the excited state of the atom decreases exponentially, where A_{10}^{-1} is the lifetime; the equilibrium occupation value of the upper level tends to zero. The conditions of spontaneous emission can be interpreted as interaction of the atom (two-level system) with a dissipative system (cf. Section 3.2.4.1). To this end we regard the radiation field in the vacuum state with its field fluctuations as the dissipative system. The temperature is assumed to tend to zero, since for the two-level system the equilibrium occupation value of the upper state is assumed to be zero. The relations that govern the interaction between the atomic and the dissipative system [see (3.88) and (3.89)] contain the damping parameter β, which is proportional to the product of the mode density of the radiation field and the coupling constant squared [see (3.77)]. The coupling constant can be obtained by a comparison between the interaction operator (3.87) and the dipole interaction operator given in (3.49). The damping constant $\beta = \omega^3 |d_{10}|^2 / 6\pi c^3 \hbar \varepsilon_0$ results, and thus the spontaneous emission has the phase relaxation time $T_2 = \beta^{-1} = 2A_{10}^{-1}$. As a consequence of (3.89) the energy relaxation time T_1 is A_{10}^{-1}, and this corresponds to the result expressed by (3.142).

3.3.2.2. Homogeneous and Inhomogeneous Line Broadening. First we shall modify the form of the result on the natural line width so as to improve its usefulness both for practical application and for extension to other line-broadening mechanisms. To achieve this, let us start from the transition rate (A.123) as derived in the framework of the concept of time coarse-graining:

$$\frac{d}{dt} p_{i \to f} = \frac{2\pi}{\hbar} \left| \langle \mathscr{E}_f | \hat{H}^C | \mathscr{E}_i \rangle \right|^2 \delta\left(\mathscr{E}_f^F - \mathscr{E}_i^F \right). \tag{3.143}$$

Because of the delta function this expression represents a relation between sharp energy levels. As one may conclude from the results of Section 3.3.2.1, in reality continuous energy distributions around sharp energy values occur. Therefore the delta function in (3.143) has to be replaced by an appropriate weight function $\bar{g}_{if}(\mathscr{E})$, so that the transition rate per unit energy is

$$\left(\frac{d}{dt} p_{i \to f} \right)_{\mathscr{E}} = \frac{2\pi}{\hbar} \left| \langle \mathscr{E}_f | \hat{H}^C | \mathscr{E}_i \rangle \right|^2 \bar{g}_{if}(\mathscr{E}). \tag{3.144}$$

The integration of $(dp_{i \to f}/dt)_{\mathscr{E}}$ with respect to $\mathscr{E}$ yields the transition rate for the total line,

$$\frac{dp'_{i \to f}}{dt} = \int_{\text{line}} d\mathscr{E} \left(\frac{dp_{i \to f}}{dt} \right)_{\mathscr{E}} = \int_{\text{line}} d\omega \left(\frac{dp_{i \to f}}{dt} \right)_{\omega}, \tag{3.145}$$

where

$$\left(\frac{dp_{i \to f}}{dt} \right)_{\omega} = \hbar \left(\frac{dp_{i \to f}}{dt} \right)_{\mathscr{E}}.$$

For spontaneous emission the expressions (3.123) and (3.139) lead to

$$\left(\frac{d}{dt}p_{u\to l}^{\text{sp em}}\right)_{\omega} = A_{ul}g_{ul}^{\text{sp em}}(\omega - \omega_{ul}), \tag{3.146}$$

where

$$g_{ul}^{\text{sp em}}(\omega - \omega_{ul}) = \frac{A_{ul}/2\pi}{(\omega - \omega_{ul})^2 + (A_{ul}/2)^2}$$

and

$$\int d\omega\, g_{ul}^{\text{sp em}}(\omega - \omega_{ul}) = 1.$$

So far we have considered only one atom. But for practical reasons the treatment of an ensemble of atomic systems is of interest. Let us therefore pass over to the description of an ensemble of N_u excited, uniform atoms, which are assumed to interact neither with each other nor with an external field. Then the number $N_{u\to l}(\omega)$ of atoms that return to the lower state by emitting a photon in the frequency interval $(\omega, \omega + d\omega)$ per unit time is

$$\frac{dN_{u\to l}(\omega)}{dt} = N_u A_{ul} g_{ul}^{\text{sp em}}(\omega - \omega_{ul}). \tag{3.147}$$

In deriving this equation we have tacitly assumed that all the atoms are fixed in space. This means that an observer finds exactly the same conditions for all the radiating particles. If such conditions exist, we generally speak of *homogeneous line broadening*. Supposing that the ensemble of particles is a real gas at a finite temperature $\mathscr{T}$, then all the particles will move because of thermal agitation. We consider in particular the particle with the index α. If this particle moves with the velocity $v_{\alpha\bullet}$, the experimentalist who is observing photons emitted in the direction of the unit vector $e_{\bullet}$ will find a particle with the mean frequency

$$\omega_{m,\alpha} = \omega_{ul} + \Delta\omega_{\alpha}, \qquad \text{where} \quad \Delta\omega_{\alpha} = \frac{\omega_{ul}}{c} v_{\alpha\bullet} e_{\bullet}. \tag{3.148}$$

The frequency shift $\Delta\omega_{\alpha}$ is caused by the Doppler effect. For the ensemble of atoms, which have different mean frequencies, the transition rate is

$$\begin{aligned}\frac{d}{dt}N_{u\to l}(\omega) &= \sum_{\alpha} A_{ul} g_{ul}^{\text{sp em}}(\omega - \omega_{m,\alpha}) \\ &= N_u A_{ul} g_{ul}(\omega - \omega_{ul}),\end{aligned} \tag{3.149}$$

where

$$g_{ul}(\omega - \omega_{ul}) \equiv N_u^{-1} \sum_{\alpha} g_{ul}^{\text{sp em}}(\omega - \omega_{m,\alpha}). \tag{3.149a}$$

The total line-shape function $g_{ul}(\omega - \omega_{ul})$ results from superposition of Lorentzian line shapes [cf. (3.146)] of individual atoms with *different* mean frequencies $\omega_{m,\alpha}$. Their distribution over the frequency scale arises from the distribution of the velocities $v_{\alpha\bullet}$ of the gas particles. According to the Maxwell velocity distribution, the probability of finding a particle with the velocity component $v_\bullet e_\bullet$ is, as is well known, proportional to $\exp[-M(v_\bullet e_\bullet)^2/2k_B\mathcal{T}]$, where M is the mass of the particle. Thus, by (3.148), the probability density of finding a particle with the mean frequency ω_m is

$$p(\omega_m) = \left(\frac{2\pi\omega_{ul}^2 k_B\mathcal{T}}{Mc^2}\right)^{-1/2} \exp\left[-\frac{(\omega_m - \omega_{ul})^2}{2\omega_{ul}^2 k_B\mathcal{T}/Mc^2}\right]. \tag{3.150}$$

This is a distribution with Gaussian character possessing the mean frequency ω_{ul} and the width (fwhm)

$$\Delta\omega_{\text{Do}} = \omega_{ul}\sqrt{(2\ln 2)\frac{k_B\mathcal{T}}{Mc^2}}\,. \tag{3.151}$$

Since $N_u p(\omega_m)\, d\omega_m$ particles have a mean frequency within the interval $(\omega_m, \omega_m + d\omega_m)$, the equation (3.149a) leads to the total line shape

$$g_{ul}(\omega - \omega_{ul}) = \int d\omega_m\, p(\omega_m) g_{ul}^{\text{sp em}}(\omega - \omega_m), \tag{3.152}$$

which is illustrated in Fig. 3.6. Obviously, this is a convolution integral containing the Doppler line shape $p(\omega_m)$ and the Lorentzian line shape of the

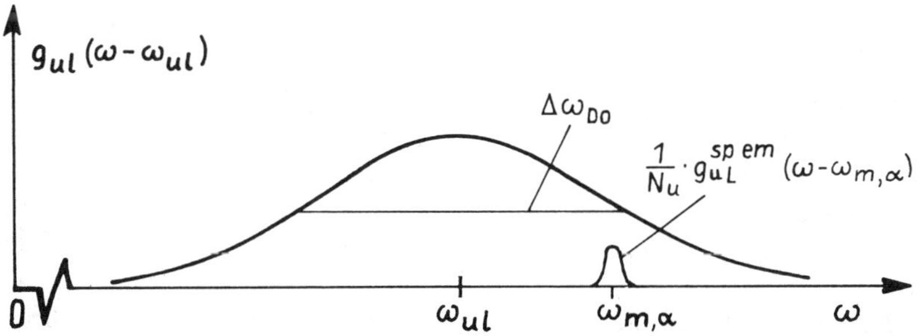

Fig. 3.6. Total line shape $g_{ul}(\omega - \omega_{ul})$ in the case of $\Delta\omega_{\text{Do}}$ being much greater than the natural line width.

Table 3.2. Line-Broadening Mechanisms

Mechanism	Type	Line-Width Formula	Relative Broadening $\Delta\omega_{ul}/\omega_{ul}$
Radiation damping of isolated atoms and molecules	Hom.	$\Delta\omega_{ul} = \hbar^{-1}(\Delta\mathscr{E}_u + \Delta\mathscr{E}_l)$	10^{-7}–10^{-12}
Collisional broadening	Hom.		$\sim 10^{-10}\dfrac{p}{m\text{ Torr}}$ (a)
Power broadening	Hom.	$\Delta\omega_{ul} \sim \hbar^{-1}\lvert d_{ul}\rvert \cdot \lvert E\rvert$	
Finite interaction duration	Hom.	$\Delta\omega_{ul} \sim v/L$	
Doppler broadening	Inhom.	$\Delta\omega_{ul} \sim (\sqrt{v^2}/c)\omega_{ul}$	$\sim 10^{-6}$ (b)
Inhom. fields and vibrations in solid	Inhom. (hom.)		$\gtrsim 10^{-5}$
Inhomogeneities in microphysically disordered material	Inhom.		$\gtrsim 10^{-3}$

[a] For $p \lesssim 1$ Torr.
[b] For $\mathscr{T} = 300$ K.

spontaneous emission. Under normal conditions the quotient $k_B\mathscr{T}/Mc^2$ in $\Delta\omega_{\text{Do}}$ is of the order of 10^{-12}; therefore the Doppler line width $\Delta\omega_{\text{Do}}$ exceeds the natural line width by about the factor 10^2; in this case the total line shape resembles a Gaussian line with $\Delta\omega_{\text{tot}} \approx \Delta\omega_{\text{inh}} = \Delta\omega_{\text{Do}}$. Doppler broadening rests on the fact that the radiating systems are in different conditions with respect to the observer (observing apparatus). If such conditions exist, generally the phenomenon is called *inhomogeneous line broadening*. Therefore $p(\omega_m)$ may be designated by $g_{\text{inh}}(\omega_m)$.

Radiation damping and Doppler broadening may be regarded as prototypes for homogeneous and inhomogeneous line broadening. For other important line-broadening mechanisms the type of line broadening, the dependence of the line width on internal and external parameters, and a quantitative estimate of the relative line width are given in Table 3.2. The value of the line width of radiation damping becomes apparent from (3.141). The line width of collision broadening is proportional to the pressure in the case of small pressures p. In power broadening the line width is proportional to the strength $|E|$ of the external field acting on the atom. In the case of finite interaction duration in a gas, the line width depends on the velocity $|v|$ of the particle moving through electromagnetic beam with an interaction length L.

Having discussed the line-broadening effects with respect to spontaneous emission only, we now include the induced one-photon processes (stimulated emission and absorption). The discussion at the end of Section 3.3.1 revealed that, given the knowledge of the Einstein coefficient A for spontaneous emission, the Einstein coefficient B for stimulated emission can be calculated. The arguments given at that place [see (3.133b)] referred to quantities concerning the total transition of a line. Since the equilibrium conditions given at the end of Section 3.3.1 must be valid for any frequency interval $(\omega, \omega + d\omega)$, the conclusions drawn there for the relationships between the quantities of spontaneous and stimulated emission can be transferred to the frequency-dependent line shape. Let

$$\left(\frac{dp_{u\to l}^{\mathrm{sp\,em}}}{dt}\right)_\omega = A_{ul} g_{ul}(\omega - \omega_{ul}) \tag{3.153}$$

be the transition rate per unit frequency for one atom that is spontaneously emitting one photon with the frequency ω; here the function $g_{ul}(\omega - \omega_{ul})$, the so-called *fluorescence line shape*, is supposed to contain all homogeneous and inhomogeneous line-broadening mechanisms involved. Then the transition rates for the stimulated processes (stimulated emission and absorption) are given by

$$\left(\frac{dp_{u\to l}^{\mathrm{st\,em}}}{dt}\right)_\omega = B_{ul}\zeta(\omega) g_{ul}(\omega - \omega_{ul}), \tag{3.154}$$

$$\left(\frac{dp_{l\to u}^{\mathrm{abs}}}{dt}\right)_\omega = B_{ul}\frac{g_u}{g_l}\zeta(\omega) g_{ul}(\omega - \omega_{ul}). \tag{3.155}$$

The quantities g_u and g_l are the weight factors, if degeneracy of the upper and the lower level exists. For practical reasons it is sometimes advantageous to replace the energy density $\zeta(\omega)$ in (3.154)–(3.155) by the quantity $S(\omega)/c$, with $S(\omega)$ the irradiance per unit angular frequency.

REFERENCES

3.1. T. Itoh, *Rev. Mod. Phys.* **37**, 159 (1965).

3.2. G. Weber, *Wiss. Z. Univ. Jena, Math.-Nat. Reihe* **18**, 213 (1969).

3.3. H.-E. Ponath and M. Schubert, *Ann. d. Phys.* **34**, 456 (1977).

3.4. V. F. Weisskopf and E. Wigner, *Z. Phys.* **63**, 54 (1930).

4

Semiclassical Description of Nonlinear Optics

As already mentioned in Section 3.1, complete agreement between calculations and all types of experimental results can only be obtained by applying quantum theory to the entire system of light and matter. But important phenomena can be described with sufficient accuracy by using a semiclassical treatment where the interaction between classical electromagnetic fields and quantized atomic systems is considered. The success of such a treatment is based on the possibility of describing strong electromagnetic fields with low fluctuations by means of classical physics (cf. Section 2.3.1). Indeed, nonlinear optics often deals with the interaction of such strong electromagnetic fields of negligible fluctuations.

Compared with the full quantum-theoretical treatment of light–matter interaction, the semiclassical description has two main advantages. First, by its use calculations become simpler and often more concise. Second, the analogy with the classical theory becomes more pronounced (see Sections 1.3 and 1.4), because the evolution of the electromagnetic fields, which are the measured quantities, is calculated in both cases from the classical Maxwell equations, where either the classical polarization or the corresponding quantum-mechanical (quantum-statistical) expectation value has to be inserted.

In this chapter the expectation value of the polarization will be shown to display the same functional dependence on field strength as classical polarization does. This means that the structure of the constitutive relations (or material equations) is the same in both cases. Thus susceptibilities of analogous meaning may be defined within the framework of semiclassical theory. Their values may be compared with classically calculated susceptibilities. Therefore it is possible to pass over from classical relations to semiclassical ones by exchanging the susceptibilities. In this way, measured quantities, such as alterations of the intensity of classically described fields, are related to quantum-mechanical parameters of the atomic systems under investigation.

4.1. RELATIONSHIP BETWEEN POLARIZATION AND FIELD STRENGTH

The aim of this section is to establish relations between polarization and field strength by means of a quantum-mechanical treatment of the material under consideration. The resultant *constitutive relations* have the same form as those of Section 1.3 (see, e.g., Refs. 2, 9, 29, 38, 4.1–4.6).

For clarity, we suppose simple conditions that allow the consideration of local relations between field strength and material response as well as the application of the dipole approximation (cf. Section 3.2.1). On the one hand, the part of the sample under investigation is assumed to be of small linear extent in comparison with the wavelengths of the macroscopic electric fields. But on the other hand this part, with volume ΔV, should be so large that it contains a great number of particles.

All properties of the sample, which may be regarded as a manifold of many atoms, molecules, or elementary cells, can be evaluated if the time dependence of its density operator $\hat{\rho}(t)$ is known. The operator $\hat{\rho}(t)$ evolves in time under the action of the time-independent Hamiltonian $\hat{H}^0$, which describes the material without an external influence, and under the action of the time-dependent operator $\hat{H}^C(t)$, which is switched on at $t = t_0$ and represents the effect of given external fields on the material. As long as the perturbation is not applied, we assume the density operator to be time-independent. This condition is for example fulfilled in thermal equilibrium, where $\hat{\rho}^{(0)}$ is given by

$$\hat{\rho}^{(0)} \equiv \hat{\rho}^e = \frac{\exp\left(-\hat{H}^0/k_B\mathcal{T}\right)}{\mathrm{Tr}\left\{\exp\left(-\hat{H}^0/k_B\mathcal{T}\right)\right\}}.$$

Then the density operator in the interaction picture,

$$\hat{\rho}_I(t) = \left[\hat{U}^0(t, t_0')\right]^{-1}\hat{\rho}(t)\hat{U}^0(t, t_0'), \tag{4.1}$$

can be expressed by

$$\hat{\rho}_I(t) = \hat{\rho}^{(0)} + \sum_{n=1} (i\hbar)^{-n}\int_{t_0}^{t} dt_1 \cdots \int_{t_0}^{t_{n-1}} dt_n$$

$$\times\left[\hat{H}_I^C(t_1), \left[\hat{H}_I^C(t_2), \cdots \left[\hat{H}_I^C(t_n), \hat{\rho}^{(0)}\right] \cdots\right]\right] \tag{4.2}$$

[cf. (A.116)]. Employing this relation, $\hat{\rho}_I(t)$ can be calculated at any time if the time-independent operator $\hat{\rho}^{(0)}$ is known. The time t_0' in (4.1) may be chosen arbitrarily, because $\hat{\rho}^{(0)}$ does not depend on time, and it may differ from the time t_0 when the perturbation starts. We set $t_0' = 0$. Hence the unitary transformation operator is expressed as

$$\hat{U}^{(0)}(t) = \hat{I}\exp\left(-\frac{i}{\hbar}t\hat{H}^0\right). \tag{4.3}$$

Having calculated $\hat{\rho}_I(t)$ and from it $\hat{\rho}(t)$, we may evaluate the expectation value of the dipole moment of the sample,

$$\langle\hat{d}_\bullet\rangle = \mathrm{Tr}\{\hat{\rho}(t)\hat{d}_\bullet\}, \tag{4.4}$$

and the polarization

$$P_\bullet(t) = \frac{1}{\Delta V}\langle\hat{d}_\bullet\rangle. \tag{4.5}$$

In this calculation we use the interaction operator

$$\hat{H}^C(t) = -\hat{d}_\bullet E_\bullet(t), \tag{4.6}$$

which is written as

$$\begin{aligned}\hat{H}_I^C(t) &= [\hat{U}^0(t)]^{-1}[-\hat{d}_\bullet E_\bullet(t)]\hat{U}^0(t)\\ &= -\hat{d}_{I\bullet}(t)E_\bullet(t)\end{aligned} \tag{4.7}$$

in the interaction picture. According to the semiclassical concept the electric field strength $E_\bullet(t)$ is a classical quantity, which is given as function of time. Because of (4.6) the expansion in increasing powers of $\hat{H}^C$ in (4.2) is equivalent to an expansion in terms of $E_\bullet(t)$ (cf. Section 1.3).

In the first order of perturbation theory we obtain the polarization $P_\bullet^{(1)}(t)$ from (4.5) in conjunction with the term $\hat{\rho}^{(1)}$ of (4.2) and with (4.4) and (4.7):

$$P_i^{(1)}(t) = \left(\frac{-1}{i\hbar}\right)\left(\frac{1}{\Delta V}\right)$$

$$\times\mathrm{Tr}\left\{\sum_j \hat{U}^0(t)\int_{t_0}^{t} dt_1\, E_j(t_1)\left[\hat{d}_{Ij}(t_1), \hat{\rho}^{(0)}\right]\left(\hat{U}^0(t)\right)^{-1}\hat{d}_i\right\}. \tag{4.8}$$

Applying the transformation $t_1 = t - \tau_1$, setting $t_0 = -\infty$, and using the relation

$$\hat{U}^0(t)\hat{d}_{I\bullet}(t-\tau_1)\left(\hat{U}^0(t)\right)^{-1} = \hat{d}_{I\bullet}(-\tau_1), \tag{4.9}$$

which is valid for any operator that is time-independent in the Schrödinger picture, the polarization is obtained as

$$P_i^{(1)}(t) = \sum_j \varepsilon_0 \int_0^\infty d\tau_1 \left[\left(\frac{-1}{i\hbar} \right) \frac{1}{\varepsilon_0 \, \Delta V} \mathrm{Tr}\left\{ \left[\hat{d}_{Ij}(-\tau_1), \hat{\rho}^{(0)} \right] \hat{d}_i \right\} \right] E_j(t - \tau_1) \tag{4.10}$$

This equation is equivalent to (1.76) with respect to the functional dependence of $P_\bullet(t)$ on $E_\bullet(t)$. By comparison, the linear response function $\varkappa_{\bullet\bullet}^{(1)}(\tau_1)$ introduced in Section 1.3 is found to be

$$\varkappa_{ij}^{(1)}(\tau_1) = \left(\frac{-1}{i\hbar} \right) \frac{1}{\varepsilon_0 \, \Delta V} \mathrm{Tr}\left\{ \left[\hat{d}_{Ij}(-\tau_1), \hat{\rho}^{(0)} \right] \hat{d}_i \right\}. \tag{4.11}$$

For calculations to be performed later it is convenient to change the arrangement of operators in (4.10). By utilizing the invariance of the trace of operator products under cyclic rearrangement of their factors (e.g. $\mathrm{Tr}\{\hat{A}\hat{B}\hat{C}\} = \mathrm{Tr}\{\hat{B}\hat{C}\hat{A}\} = \mathrm{Tr}\{\hat{C}\hat{A}\hat{B}\}$), the response function $\varkappa_{ij}^{(1)}(\tau_1)$ may be written as

$$\varkappa_{ij}^{(1)}(\tau_1) = \left(\frac{-1}{i\hbar} \right)^1 \frac{1}{\varepsilon_0 \, \Delta V} \mathrm{Tr}\{ \hat{\rho}^{(0)} [\hat{d}_i, \hat{d}_{Ij}(-\tau_1)] \}. \tag{4.12}$$

The correspondence between the classical and semiclassical constitutive relations can be regarded as a further justification of the classical treatment given in Section 1.3. Furthermore, it is thus possible to substitute quantum-mechanically calculated susceptibilities for classical ones in the relations derived in Section 1.3 within the framework of classical physics.

In an analogous way the higher-order terms $P_\bullet^{(n)}(t)$ are calculated by using the contributions $\hat{\rho}^{(n)}(t)$ of the density operator expansion in (4.1). Then we have

$$P_i^{(n)}(t) = \sum_{a_1, \ldots, a_n} \varepsilon_0 \int_0^\infty d\tau_1 \cdots \int_0^\infty d\tau_n \varkappa_{ia_1 \cdots a_n}^{(n)}(\tau_1, \ldots, \tau_n) \cdot E_{a_1}(t - \tau_1) \cdots E_{a_n}(t - \tau_n), \tag{4.13}$$

whose form is equivalent to the classical relation (1.83); the nonlinear response function is given by

$$\varkappa_{ia_1 \cdots a_n}^{(n)}(\tau_1, \ldots, \tau_n) = \left(\frac{-1}{i\hbar} \right)^n \frac{1}{\varepsilon_0 \, \Delta V} \times \mathrm{Tr}\left\{ \hat{\rho}^{(0)} \left[\cdots \left[\hat{d}_i, \hat{d}_{Ia_1}(-\tau_1) \right], \cdots \hat{d}_{Ia_n}(-\tau_n) \right] \right\}. \tag{4.14}$$

Obviously, these response functions do not satisfy the symmetry relations (1.84). But a symmetrized form can be achieved by summing all possible time-index transformed expressions and dividing the sum by the number of terms. This procedure will be discussed in detail with respect to frequency variables and tensor indices in Section 4.3.

4.2. THE TIME-DEPENDENT MATERIAL RESPONSE

We describe the time-dependent polarization of a material with electric fields acting on it in terms of the linear and nonlinear response functions given in Section 4.1. Their dependence on the time coordinates will now be discussed in more detail. With this aim in mind the response functions will be calculated by means of the energy eigenstates $|\mathscr{E}_\alpha\rangle$ of the Hamiltonian $\hat{H}^0$ of the unperturbed material. The density operator $\hat{\rho}^{(0)}$ is assumed to be diagonal within this basic system; this means

$$\langle \mathscr{E}_\alpha | \hat{\rho}^{(0)} | \mathscr{E}_\beta \rangle = \rho^{(0)}_{\alpha\alpha} \delta_{\alpha\beta}. \tag{4.15}$$

This assumption is for instance justified if the material is in thermodynamic equilibrium with a reservoir of given temperature. It should be mentioned that the interaction between sample and reservoir has not been taken into account explicitly in Section 4.1; such a coupling will be discussed later, in Section 4.5. Using (4.15),

$$\langle \mathscr{E}_\alpha | \hat{d}_\bullet | \mathscr{E}_\beta \rangle = d_{\alpha\beta\bullet}, \tag{4.16}$$

and

$$\mathscr{E}_\alpha - \mathscr{E}_\beta = \hbar \omega_{\alpha\beta}, \tag{4.17}$$

we obtain from (4.12)

$$\varkappa^{(1)}_{ij}(\tau_1) = \left(\frac{-1}{i\hbar} \right)^1 \frac{1}{\varepsilon_0 \, \Delta V} \sum_{\alpha, \beta} \rho^{(0)}_{\alpha\alpha} \left(d_{\alpha\beta i} d_{\beta\alpha j} e^{i\omega_{\alpha\beta}\tau_1} - d_{\alpha\beta j} d_{\beta\alpha i} e^{-i\omega_{\alpha\beta}\tau_1} \right). \tag{4.18}$$

Note that the density operator and the energy levels are those of the whole sample. The equation (4.18) expresses the fact that the response function $\varkappa^{(1)}_{ij}$ as a function of the time coordinate τ_1 is given by a superposition of harmonic contributions. Providing that $\rho^{(0)}_{\alpha\alpha} = \delta_{\alpha 0}$ (i.e., all particles of the unperturbed sample are in the ground state) and that only one vanishing vector component $d_{\alpha\beta i} = d_{\alpha\beta}$ exists, (4.18) yields

$$\varkappa^{(1)}(\tau_1) = \left(\frac{2}{\hbar} \right) \left(\frac{1}{\varepsilon_0 \, \Delta V} \right) \sum_\beta |d_{0\beta}|^2 \sin \omega_{\beta 0} \tau_1. \tag{4.19}$$

At first glance (4.19) seems to describe only entirely undamped processes, as the susceptibility is a sum of terms that depend sinusoidally on time. Let us imagine that the excited states are represented by rather sharp and discretely arranged energy levels. Then the response function $\varkappa^{(1)}(\tau_1)$ would only consist of harmonic components with well-separated frequencies. Such discrete harmonic components may lead to beats in the time behavior of $\varkappa^{(1)}(\tau_1)$, but they are incapable of bringing about any dying away of the response function at $\tau_1 \to \infty$. Thus the memory of the sample, which is characterized by the decay time of $\varkappa^{(1)}(\tau_1)$, would last infinitely long.

In contrast to this, any real sample is characterized by a finite decay time, as already mentioned in Sections 1.3 and 3.3. Such decay results from important processes like radiation damping, collisions, inhomogeneities of the sample, and interactions between the particles under consideration. The coupling of the atoms to dissipative systems will be treated in Section 4.5 on the basis of the consideration in Section 3.3.2.

The influence of the interaction between all the individual particles of the sample on the response function can be accounted for in the following way by using a simple model. We assume the sample to contain M identical atoms whose transition $0 \to 1$ is nearly resonant with the radiation field for infinite distances between the atoms. If the atoms are brought closer together, their mutual interactions will increase, and consequently, the transition will split into M sublevels arranged in a band around the unperturbed atomic level (see Fig. 4.1). The bandwidth increases with increasing interaction between the atoms. The transition moment $d_{0\beta}$ is assumed to be equal for the M sublevels, and therefore it may be written in front of the sum in (4.19). For a quasicontinuous distribution of the sublevels the remaining sum can be converted into an integral $[\Sigma_\beta \to \int_{\text{band}} d\Omega\, \sigma(\Omega)]$ by introducing a density function $\sigma(\omega - \omega_{10})$, which represents the distribution of sublevels around ω_{10}. Thus we obtain

$$\varkappa^{(1)}(\tau_1) = \frac{2}{\hbar\varepsilon_0\, \Delta V}|d_{10}|^2 \int_{\text{band}} d\Omega\, \sigma(\Omega) \sin(\omega_{10} + \Omega)\tau_1. \tag{4.20}$$

If $\sigma(\Omega)$ is an even function and if it decreases rather strongly as $\Omega \to \infty$, the

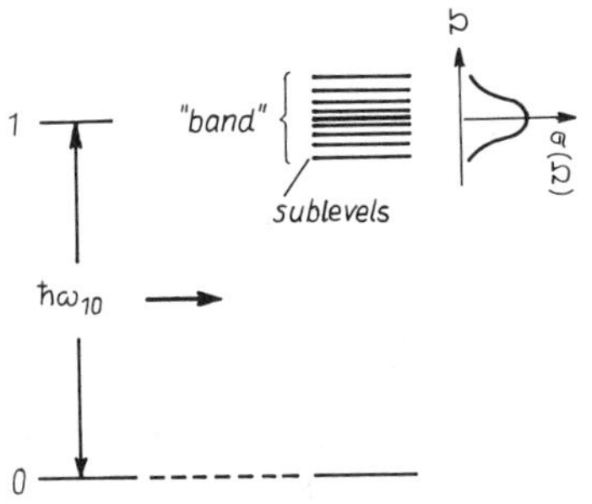

Fig. 4.1. Splitting-off of the transition $0 \to 1$ caused by interaction between atoms $[\sigma(\Omega) = \text{density of sublevels}]$.

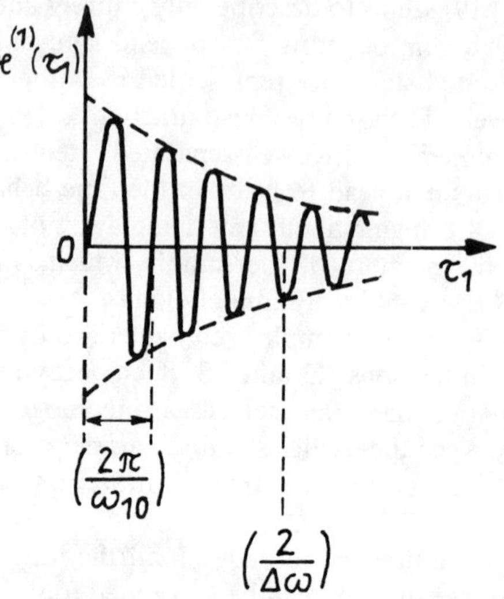

Fig. 4.2. Linear response function $\varkappa^{(1)}(\tau_1)$ versus time τ_1.

relation

$$\varkappa^{(1)}(\tau_1) = \frac{4}{\hbar\varepsilon_0\,\Delta V}|d_{10}|^2 \sin\omega_{10}\tau_1 \int_0^\infty d\Omega\,\sigma(\Omega)\cos\Omega\tau_1 \tag{4.21}$$

follows from (4.20). The particular shape of $\sigma(\omega - \omega_{10})$ depends not only on the specific atom–atom interaction but also on the coupling of the atoms to dissipative systems.

Let us assume an overall Lorentzian band shape

$$\sigma(\omega - \omega_{10}) = \frac{1}{\pi}\frac{\Delta\omega/2}{(\omega - \omega_{10})^2 + (\Delta\omega/2)^2}, \tag{4.22}$$

which is consistent with many experimental situations. Substitution of (4.22) into (4.21) leads to

$$\varkappa^{(1)}(\tau_1) = \frac{1}{\hbar\varepsilon_0\,\Delta V}|d_{10}|^2 e^{-(\Delta\omega/2)\tau_1}\sin\omega_{10}\tau_1 \tag{4.23}$$

(see Fig. 4.2). Again the response function exhibits oscillations, but their amplitudes decrease with increasing τ_1, and thus the decay time of $\varkappa^{(1)}(\tau_1)$ in correspondence with the memory time of the system is no longer infinite. The same functional dependence of the response function is obtained for one classical oscillator if friction is taken into account. Note that the exponential decrease of the response function results from the Lorentzian line shape. In fact, other line shapes provide modified response functions, but the general behavior characterized by a finite memory time is preserved.

The equations (4.20) and (4.21) can be used even if the band shape is not only caused by the mutual interaction of the particles under consideration but also by their coupling to other systems—as, for instance, an ensemble of phonons, solvent molecules, or radiation-field oscillators. Since such a coupling has been neglected so far in our theory, we must replace the level density $\sigma(\Omega)$, describing only atom–atom interaction, by measured band profiles, which of course contain all coupling effects.

An analogous consideration can be applied to the higher-order response functions of (4.14). As a result the $\varkappa^{(n)}(\tau_1, \dots, \tau_n)$ are seen to approach zero if at least one of the τ_i approaches infinity. Hence, in agreement with experiment the higher-order response functions possess finite correlation times as well.

4.3. SUSCEPTIBILITIES IN THE FREQUENCY DOMAIN

In applications, we have often to make use of linear and nonlinear susceptibilities in the frequency domain (Section 1.3). We may get them by applying the Fourier transformation to the response functions obtained in the previous subsection. Here however we follow another line, starting from the results of nth-order perturbation theory for the polarization $P_\bullet^{(n)}(t)$ as a function of the field strength $E_\bullet(t)$, which is replaced by its Fourier transform

$$\underset{\smile}{E}_\bullet(\omega) = \int_{-\infty}^{\infty} dt\, E_\bullet(t) e^{-i\omega t}. \tag{4.24}$$

In *first order* ($n = 1$) we obtain from (4.10) and (4.12)

$$P_i^{(1)}(t) = \left(\frac{-1}{i\hbar}\right)^1 \frac{1}{\Delta V} \sum_j \int_0^\infty d\tau_1 \operatorname{Tr}\{\hat{\rho}^{(0)}[\hat{d}_i, \hat{d}_{Ij}(-\tau_1)]\}$$

$$\times \frac{1}{2\pi} \int_{-\infty}^{\infty} d\omega\, e^{+i\omega(t-\tau_1)} \underset{\smile}{E}_j(\omega). \tag{4.25}$$

After interchanging the order of integration we can write

$$P_i^{(1)}(t) = \frac{\varepsilon_0}{2\pi} \sum_j \int_{-\infty}^{\infty} d\omega\, \underset{\smile}{\varkappa}_{ij}^{(1)}(\omega; \omega) \underset{\smile}{E}_j(\omega) e^{+i\omega t}, \tag{4.26}$$

where

$$\underset{\smile}{\varkappa}_{ij}^{(1)}(\omega; \omega) = \left(\frac{-1}{i\hbar}\right)^1 \frac{1}{\varepsilon_0 \Delta V} \int_0^\infty d\tau_1\, e^{-i\omega\tau_1} \operatorname{Tr}\{\hat{\rho}^{(0)}[\hat{d}_i, \hat{d}_{Ij}(-\tau_1)]\}.$$

This first-order susceptibility $\underset{\smile}{\varkappa}_{ij}^{(1)}(\omega;\omega)$, which in Section 1.3.2 was defined by

$$\underset{\smile}{P}_{\bullet}(\omega) = \varepsilon_0 \underset{\smile}{\varkappa}_{\bullet\bullet}(\omega;\omega)\underset{\smile}{E}_{\bullet}(\omega),$$

can be easily shown to be the Fourier transform of the response function $\varkappa_{ij}^{(1)}(\tau_1)$ given in (4.11), where $\varkappa_{ij}^{(1)}(\tau_1) = 0$ for $\tau_1 < 0$ has to be taken into account.

In *second order* we start from (4.13) and (4.14) with $n = 2$. This gives

$$P_i^{(2)}(t) = \left(\frac{-1}{i\hbar}\right)^2 \frac{1}{\Delta V} \sum_{j,k} \int_0^{\infty} d\tau_1 \int_{\tau_1}^{\infty} d\tau_2 \, \mathrm{Tr}\left\{ \hat{\rho}^{(0)} \left[\left[\hat{d}_i, \hat{d}_{Ij}(-\tau_1) \right], \hat{d}_{Ik}(-\tau_2) \right] \right\}$$

$$\times E_j(t-\tau_1) E_k(t-\tau_2). \tag{4.27}$$

Substituting again the Fourier transforms of the electric field and changing the order of integration, we have

$$P_i^{(2)}(t) = \frac{\varepsilon_0}{(2\pi)^2} \sum_{j,k} \int_{-\infty}^{+\infty} d\omega_1$$

$$\times \int_{-\infty}^{\infty} d\omega_2 \, \underset{\smile}{\tilde{\varkappa}}_{ijk}^{(2)}(\omega;\omega_1,\omega_2) \underset{\smile}{E}_j(\omega_1) \underset{\smile}{E}_k(\omega_2) e^{+i(\omega_1+\omega_2)t}, \tag{4.28}$$

where

$$\underset{\smile}{\tilde{\varkappa}}_{ijk}^{(2)}(\omega;\omega_1,\omega_2) = \left(\frac{-1}{i\hbar}\right)^2 \left(\frac{1}{\varepsilon_0 \Delta V}\right) \int_0^{\infty} d\tau_1 \int_0^{\infty} d\tau_2$$

$$\times \mathrm{Tr}\left\{ \rho^{(0)} \left[\left[\hat{d}_i, \hat{d}_{Ij}(-\tau_1) \right], \hat{d}_{Ik}(-\tau_2) \right] \right\} e^{-(i\omega_1\tau_1 + i\omega_2\tau_2)}. \tag{4.29}$$

The second-order polarization $P_i^{(2)}(t)$ in (4.28) remains of course invariant if the summation indices j, k and the integration variables ω_1, ω_2 are exchanged. A simultaneous exchange of summation indices and integration variables requires that we replace $\underset{\smile}{\tilde{\varkappa}}_{ijk}^{(2)}(\omega;\omega_1,\omega_2)$ by $\underset{\smile}{\tilde{\varkappa}}_{ikj}^{(2)}(\omega;\omega_2,\omega_1)$. According to (4.29) the susceptibility is not invariant under such transformations. But as explained in Section 1.3, invariance may be enforced by introducing the *second-order susceptibility*

$$\underset{\smile}{\varkappa}_{ijk}^{(2)}(\omega;\omega_1,\omega_2) = \frac{1}{2!} \mathsf{P}_2 \underset{\smile}{\tilde{\varkappa}}_{ijk}^{(2)}(\omega;\omega_1,\omega_2)$$

$$\equiv \frac{1}{2!} \left[\underset{\smile}{\tilde{\varkappa}}_{ijk}^{(2)}(\omega;\omega_1,\omega_2) + \underset{\smile}{\tilde{\varkappa}}_{ikj}(\omega;\omega_2,\omega_1) \right]. \tag{4.30}$$

The application of P_2 to an arbitrary expression of the kind of $\tilde{\underset{\smile}{\varkappa}}^{(2)}_{ijk}(\omega;\ \omega_1, \omega_2)$ means that its possible 2! frequency-index permutations are formed and summed. Such susceptibilities, which remain invariant under the frequency-index transformations, may be used in (4.28).

In the *n th order* we obtain

$$P_i^{(n)}(t) = \frac{\varepsilon_0}{(2\pi)^n} \sum_{a_1, \ldots, a_n} \int_{-\infty}^{\infty} d\omega_1 \cdots \int_{-\infty}^{\infty} d\omega_n \, \underset{\smile}{\varkappa}^{(n)}_{ia_1 \cdots a_n}(\omega;\ \omega_1, \ldots, \omega_n)$$

$$\times \underset{\smile}{E}_{a_1}(\omega_1) \cdots \underset{\smile}{E}_{a_n}(\omega_n) e^{+it\sum_{j=1}^{n}\omega_j} \qquad (4.31)$$

where the susceptibility may be chosen to be in the symmetric form

$$\underset{\smile}{\varkappa}^{(n)}_{ia_1 \cdots a_n}(\omega;\ \omega_1, \ldots, \omega_n) = \left(\frac{-1}{i\hbar}\right)^n \left(\frac{1}{\varepsilon_0\, \Delta V}\right) \frac{1}{n!} \mathsf{P}_n \int_0^{\infty} d\tau_1 \cdots \int_0^{\infty} d\tau_n$$

$$\times \mathrm{Tr}\left\{ \hat{\rho}^{(0)} \left[\cdots \left[\hat{d}_i, \hat{d}_{Ia_1}(-\tau_1) \right], \cdots \hat{d}_{Ia_n}(-\tau_n) \right] \right\}$$

$$\times e^{-i\sum_{j=1}^{n}\omega_j\tau_j}. \qquad (4.32)$$

The permutation operation P_n is defined in an analogous way to P_2.

4.4. SUSCEPTIBILITIES OF LOSS-FREE ATOMIC SYSTEMS

In the preceding two subsections nonlinear response functions and susceptibilities have been calculated for a sample that is specified only by its size and particle content. Now the sample will be assumed to be composed of a large number M of identical atomic systems (see Fig. 4.3) that are oriented identically in space and interact neither with each other nor with other systems, so that they do not experience any loss. The condition of identical orientation will be dropped later. There are two possibilities for calculating the linear and nonlinear susceptibilities of such an ensemble.

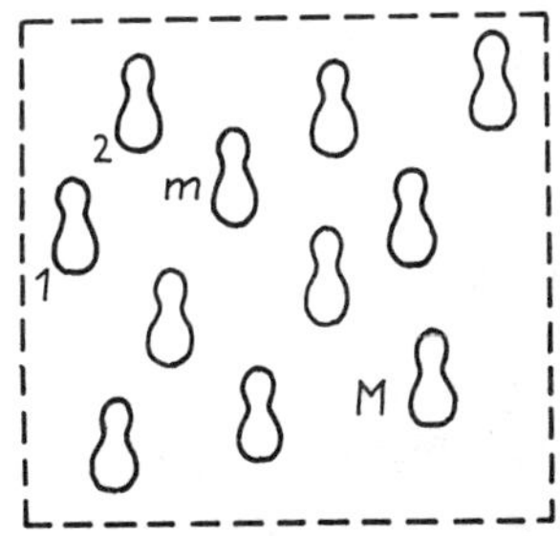

Fig. 4.3. Ensemble of M molecules oriented identically in space.

The first one is to start with the calculation of the dipole moment of one atom as a function of the electric field strength. To do this, the relations of Section 4.2 may be used if the Hamiltonian $\hat{H} = \hat{H}^0 + \hat{H}^C$ of the entire system is replaced by the Hamiltonian $\hat{H}_m = \hat{H}_m^0 + \hat{H}_m^C$ of the mth atom and if, accordingly, $\hat{\rho}^{(j)}(t)$ is replaced by $\hat{\rho}_m^{(j)}(t)$, $\hat{U}^0(t)$ by $\hat{U}_m^0(t)$, and so on. In this way the first-order contribution to the dipole moment expectation value is obtained as

$$\left(d_i^{(1)}(t)\right)_m = \left(\frac{-1}{i\hbar}\right)\sum_j \int_0^\infty d\tau_1 \operatorname{Tr}\left\{\hat{\rho}_m^{(0)}\left[(\hat{d}_i)_m, \left(\hat{d}_{Ij}(-\tau_1)\right)_m\right]\right\} E_j(t-\tau_1). \tag{4.33}$$

The polarization is now obtained by summing all dipole moments of the sample. This gives

$$P_i(t) = \frac{1}{\Delta V}\sum_{m=1}^{M} (d_i)_m. \tag{4.34}$$

Since all atoms give equal contributions, the summation in (4.34) can be easily carried out by multiplying the dipole moment of one arbitrarily chosen atom by the number M or by the number density $\gamma = M/\Delta V$, so that we have

$$P_i^{(1)}(t) = \varepsilon_0 \sum_j \int_0^\infty d\tau_1\, \varkappa_{ij}^{(1)}(\tau_1) E_j(t-\tau_1), \tag{4.35a}$$

where

$$\varkappa_{ij}^{(1)}(\tau_1) = \left(\frac{-1}{i\hbar}\right)^1 \frac{1}{\varepsilon_0}\gamma \operatorname{Tr}\left\{\hat{\rho}_m^{(0)}\left[(\hat{d}_i)_m, \left(\hat{d}_{Ij}(-\tau_1)\right)_m\right]\right\}. \tag{4.35b}$$

The corresponding first-order susceptibility in the frequency domain is

$$\breve{\varkappa}_{ij}^{(1)}(\omega;\omega) = \left(\frac{-1}{i\hbar}\right)^1 \frac{1}{\varepsilon_0}\gamma \times \int_0^\infty d\tau_1 \operatorname{Tr}\left\{\hat{\rho}_m^{(0)}\left[(\hat{d}_i)_m, \left(\hat{d}_{Ij}(-\tau_1)\right)_m\right]\right\} e^{-i\omega\tau_1}. \tag{4.35c}$$

Higher-order response functions and susceptibilities can be obtained in an analogous way.

This means that the final results of Sections 4.2 and 4.3 need to be modified only in the following way: the ensemble operators and the factor $1/\Delta V$ have to be replaced by one-atom operators and by the number density $\gamma = M/\Delta V$, respectively. To avoid an overloading of the symbols, the index m indicating the individual atom will be omitted. Mistaking single-atom operators for ensemble operators is ruled out by the following convention: operators will respectively refer to the whole ensemble and to single atoms in equations that contain the factors $1/\Delta V$ and γ. Where confusion might arise, an additional explanation will be given. For instance, we obtain

$$\varkappa^{(n)}_{ia_1 \cdots a_n}(\tau_1, \ldots, \tau_n) = \frac{1}{n!}\mathsf{P}_n\left(\frac{-1}{i\hbar}\right)^n \gamma$$
$$\times \operatorname{Tr}\left\{\hat{\rho}^{(0)}\left[\cdots\left[\hat{d}_i, \hat{d}_{Ia_1}(-\tau_1)\right], \cdots \hat{d}_{Ia_n}(-\tau_n)\right]\right. \tag{4.36a}$$

and

$$\varkappa^{(1)}_{ia_1 \cdots a_n}(\omega; \omega_1, \ldots, \omega_n) = \frac{1}{n!}\mathsf{P}_n\left(\frac{-1}{i\hbar}\right)^n \gamma$$
$$\times \int_0^\infty d\tau_1 \cdots \int_0^\infty d\tau_n$$
$$\times \operatorname{Tr}\left\{\hat{\rho}^{(0)}\left[\cdots\left[\hat{d}_i, \hat{d}_{Ia_1}(-\tau_1)\right], \cdots \hat{d}_{Ia_n}(-\tau_n)\right]\right\} e^{-i\Sigma_{j=1}^{n}\omega_j\tau_j} \tag{4.36b}$$

for the nth-order response function and susceptibility of an ensemble of noninteracting atoms.

The second way to perform the calculation is to express the ensemble operators by single-atom operators,

$$\hat{H}^0 + \hat{H}^C = \sum_{m=1}^{M}\left(\hat{H}^0_m + \hat{H}^C_m\right),$$

$$\hat{U}^0 = \prod_{m=1}^{M}\hat{U}^0_m,$$

$$\hat{\rho}^{(0)} = \prod_{m=1}^{M}\hat{\rho}^{(0)}_m,$$

and to substitute them into the relations of Sections 4.2 and 4.3. This approach which for example is used in Ref. 9, of course leads to the same results as the first one.

We can now proceed to evaluate the *linear and nonlinear susceptibilities* in the basic system of the energy eigenstates $|\mathscr{E}_\alpha\rangle$ of the unperturbed single atoms. (In the following all operators and states refer to one individual atom without any special notation.) Being unaffected by electromagnetic fields, the atomic systems are assumed to be in equilibrium with a reservoir, and hence $\rho_{\alpha\beta}^{(0)} = \rho_{\alpha\alpha}^{e}\delta_{\alpha\beta}$. With the help of

$$\hat{U}_{\alpha\beta}^{0}(t) = \delta_{\alpha\beta}e^{-(i/\hbar)\mathscr{E}_\alpha t}$$

and

$$\mathscr{E}_\alpha - \mathscr{E}_\beta = \hbar\omega_{\alpha\beta},$$

the dipole-moment matrix element in the interaction picture becomes

$$d_{I\alpha\beta\bullet}(t) = \left(\left[\hat{U}^0(t)\right]^{-1}\hat{d}_\bullet\hat{U}^0(t)\right)_{\alpha\beta}$$

$$= e^{i\omega_{\alpha\beta}t}d_{\alpha\beta\bullet}, \tag{4.37}$$

where $d_{\alpha\beta\bullet} = \langle\mathscr{E}_\alpha|\hat{d}_\bullet|\mathscr{E}_\beta\rangle$.

In first order the substitution of the matrix element of (4.37) into (4.35c) yields

$$\varkappa_{ij}^{(1)}(\omega;\omega) = \left(\frac{-1}{i\hbar}\right)\frac{1}{\varepsilon_0}\gamma\sum_{\alpha,\beta}\int_0^\infty d\tau_1\,\rho_{\alpha\alpha}^{e}$$

$$\times\left(d_{\alpha\beta i}d_{\beta\alpha j}e^{i\omega_{\alpha\beta}\tau_1} - d_{\alpha\beta j}d_{\beta\alpha i}e^{-i\omega_{\alpha\beta}\tau_1}\right)e^{-i\omega\tau_1}. \tag{4.38}$$

In evaluating this integral we are confronted with the following difficulty, which originates from the assumption of a loss-free material. Since the energy eigenvalues $\mathscr{E}_\alpha$ and their differences $\hbar\omega_{\alpha\beta}$ are real, the integral converges only if the frequency ω lies in the lower half of the complex plane and not on the real axis $\omega = \omega_r$. In order to guarantee convergence, we set $\omega = \omega_r - i\delta$ ($\delta = \text{const} > 0$). The factor $e^{-\delta\tau_1}$ then leads to a sufficiently strong decrease of the integrand for $\tau_1 \to \infty$, and the integration yields

$$\varkappa_{ij}^{(1)}(\omega;\omega) = \left(\frac{-1}{\hbar}\right)\frac{1}{\varepsilon_0}\gamma\sum_{\alpha,\beta}\rho_{\alpha\alpha}^{e}$$

$$\times\left(\frac{d_{\alpha\beta i}d_{\beta\alpha j}}{\omega_{\alpha\beta} - \omega_r + i\delta} - \frac{d_{\alpha\beta j}d_{\beta\alpha i}}{-\omega_{\alpha\beta} - \omega_r + i\delta}\right). \tag{4.39}$$

Upon integration the limiting process $\delta \to 0$ may be performed (in other words, the real axis may be approached) if the singularities $\omega = \pm\omega_{\alpha\beta}$ are excluded. Thus the first-order susceptibility on the real axis is

$$\varkappa_{ij}^{(1)}(\omega;\omega) = \left(\frac{-1}{\hbar}\right)\frac{\gamma}{\varepsilon_0}\sum_{\alpha,\beta}\rho_{\alpha\alpha}^{e}\left(\frac{d_{\alpha\beta i}d_{\beta\alpha j}}{\omega_{\alpha\beta} - \omega} + \frac{d_{\alpha\beta j}d_{\beta\alpha i}}{\omega_{\alpha\beta} + \omega}\right). \tag{4.40}$$

Our calculation, which neglects relaxation processes, fails in the vicinity of the points $\omega = \pm\omega_{\alpha\beta}$. The region to be excluded around a singularity depends on the strength of the relaxation process and is comparable with the half-width of the absorption coefficient, which may be determined experimentally. Susceptibilities without any singularities will be obtained by taking into account relaxation processes from the very beginning (cf. Section 4.5).

The *first-order susceptibility* of (4.40) obviously satisfies the overall permutation symmetry; that is, that $\varkappa_{ij}^{(1)}(\omega; \omega_1)$ is invariant under the simultaneous exchange of indices $i \leftrightarrow j$ and of frequencies $-\omega \leftrightarrow \omega_1$. Thus we have

$$\underset{\smile}{\varkappa}_{ij}^{(1)}(\omega; \omega) = \underset{\smile}{\varkappa}_{ji}^{(1)}(-\omega; -\omega).$$

(The first index has to be related to the frequency $\omega = -\omega_1$ if the second one is associated with ω_1.) This symmetry is a consequence of the neglect of losses (cf. Section 1.3.2.7). Bearing in mind this symmetry, the full expression for $\underset{\smile}{\varkappa}_{ij}^{(1)}(\omega; \omega)$ can be generated by applying the operation $\tilde{\mathsf{P}}_1$, which interchanges indices and frequencies simultaneously ($i \leftrightarrow j$, $\omega \leftrightarrow -\omega$) and sums the resultant terms, to only one of the terms of (4.40). This yields

$$\underset{\smile}{\varkappa}_{ij}^{(1)}(\omega; \omega) = \left(\frac{-1}{\hbar}\right)^1 \frac{\gamma}{\varepsilon_0} \frac{1}{1!} \tilde{\mathsf{P}}_1 \sum_{\alpha, \beta} \rho_{\alpha\alpha}^e \frac{d_{\alpha\beta i} d_{\beta\alpha j}}{\omega_{\alpha\beta} + \omega}. \tag{4.41}$$

With the assumption that all unperturbed molecules are in the ground state ($\rho_{\alpha\alpha}^e = \delta_{\alpha 0}$), we estimate the contribution $\Delta\underset{\smile}{\varkappa}^{(1)}$ of a very strong transition $0 \to 1$ to one of the susceptibility components $\underset{\smile}{\varkappa}^{(1)}(\omega; \omega)$ at low frequencies ($\omega \ll \omega_{10}$). From (4.41) we obtain

$$\Delta\underset{\smile}{\varkappa}^{(1)} \approx \frac{2}{\varepsilon_0 \hbar \omega_{10}} \gamma |d_{10}|^2.$$

With the parameters $\omega_{10} \sim 10^{16}\ \mathrm{s}^{-1}$, $d_{10} \sim 10^{-29}$ A s m, and $\gamma \sim 10^{28}\ \mathrm{m}^{-3}$, which are typical of strong electronic transitions in dense molecular ensembles, we get

$$\Delta\varkappa^{(1)} \sim 10^{-1}.$$

(In its order of magnitude the value of d_{10} corresponds to a change of the distance between positive and negative elementary charges by about one atomic diameter.) At higher frequencies but still far from resonance, $\Delta\underset{\smile}{\varkappa}^{(1)}(\omega; \omega)$ and together with it the contribution $\Delta\varepsilon(\omega)$ to the dielectric permittivity increases by a factor $\omega_{10}/(\omega_{10} - \omega)$ and can attain values in the order of 10^0–10^1.

Next we calculate the *second-order susceptibility*

$$\varkappa_{ijk}^{(2)}(\omega;\omega_1,\omega_2)=\frac{1}{\hbar^2}\frac{\gamma}{\varepsilon_0}\frac{1}{2!}\mathsf{P}_2\int_0^\infty d\tau_1\int_{\tau_1}^\infty d\tau_2$$

$$\times\operatorname{Tr}\left\{\hat{\rho}^e\left[\left[\hat{d}_i,\hat{d}_{Ij}(-\tau_1)\right],\hat{d}_{Ik}(-\tau_2)\right]\right\}e^{-i\omega_1\tau_1-i\omega_2\tau_2} \quad (4.42)$$

in terms of one-atom operators, upon rewriting (4.29) by use of (4.30). Again the trace is evaluated in the basic system of the eigenstates of the unperturbed atomic system. Taking into account similar considerations concerning convergence, integration leads to

$$\varkappa_{ijk}^{(2)}(\omega;\omega_1,\omega_2)=\frac{1}{\hbar^2}\frac{\gamma}{\varepsilon_0}\frac{1}{2!}\mathsf{P}_2$$

$$\times\sum_{\alpha,\beta,\gamma}\rho_{\alpha\alpha}^e\left(\frac{d_{\alpha\beta i}d_{\beta\gamma j}d_{\gamma\alpha k}}{(\omega_{\gamma\alpha}+\omega_2)(\omega_{\beta\alpha}+\omega_1+\omega_2)}\right.$$

$$-\frac{d_{\alpha\beta j}d_{\beta\gamma i}d_{\gamma\alpha k}}{(\omega_{\gamma\alpha}+\omega_2)(\omega_{\gamma\beta}+\omega_1+\omega_2)}$$

$$-\frac{d_{\alpha\beta k}d_{\beta\gamma i}d_{\gamma\alpha j}}{(\omega_{\alpha\beta}+\omega_2)(\omega_{\gamma\beta}+\omega_1+\omega_2)}$$

$$\left.+\frac{d_{\alpha\beta k}d_{\beta\gamma j}d_{\gamma\alpha i}}{(\omega_{\alpha\beta}+\omega_2)(\omega_{\alpha\gamma}+\omega_1+\omega_2)}\right), \quad (4.43)$$

where the points $\omega_2=-\omega_{\alpha\beta}$ and $\omega_1+\omega_2=-\omega_{\alpha\beta}$ have to be excluded on the real frequency axis for all combinations of the states $|\mathscr{E}_\alpha\rangle$ and $|\mathscr{E}_\beta\rangle$. Because of the specific action of P_2, which simultaneously interchanges indices (except the first one) and frequencies and sums the resultant terms, we may perform arbitrary substitutions $(i,\omega_i)\leftrightarrow(j,\omega_j)$ in one of the terms of (4.43) without changing the result. Thus we interchange $(j,\omega_1)\leftrightarrow(k,\omega_2)$ in the third term of (4.43) in order to add this term to the second one, and get

$$\frac{-d_{\alpha\beta j}d_{\beta\gamma i}d_{\gamma\alpha k}}{(\omega_{\gamma\alpha}+\omega_2)(\omega_{\gamma\beta}+\omega_1+\omega_2)}-\frac{d_{\alpha\beta j}d_{\beta\gamma i}d_{\gamma\alpha k}}{(\omega_{\alpha\beta}+\omega_1)(\omega_{\alpha\beta}+\omega_1+\omega_2)}$$

$$=-\frac{d_{\alpha\beta j}d_{\beta\gamma i}d_{\gamma\alpha k}}{(\omega_{\alpha\beta}+\omega_1)(\omega_{\gamma\alpha}+\omega_2)}$$

After substituting this expression into (4.43) and interchanging $(j, \omega_1) \leftrightarrow (k, \omega_2)$ in the last term, we obtain

$$\underset{\smile}{\varkappa}_{ijk}^{(2)}(\omega; \omega_1, \omega_2) = \left(\frac{1}{\hbar}\right)^2 \left(\frac{\gamma}{\varepsilon_0}\right)\left(\frac{1}{2!}\right)\mathsf{P}_2 \times \sum_{\alpha,\beta,\gamma} \rho_{\alpha\alpha}^e \left(\frac{d_{\alpha\beta i} d_{\beta\gamma j} d_{\gamma\alpha k}}{(\omega_{\alpha\gamma} - \omega_2)(\omega_{\alpha\beta} - \omega_1 - \omega_2)} + \frac{d_{\alpha\beta j} d_{\beta\gamma i} d_{\gamma\alpha k}}{(\omega_{\alpha\beta} + \omega_1)(\omega_{\alpha\gamma} - \omega_2)} + \frac{d_{\alpha\beta j} d_{\beta\gamma k} d_{\gamma\alpha i}}{(\omega_{\alpha\beta} + \omega_1)(\omega_{\alpha\gamma} + \omega_1 + \omega_2)}\right). \quad (4.44)$$

Like the first-order susceptibility, the second-order susceptibility $\underset{\smile}{\varkappa}_{ijk}^{(2)}(\omega; \omega_1, \omega_2)$ is seen to satisfy the overall permutation symmetry, where the first index has to be related to the frequency $-\omega = -(\omega_1 + \omega_2)$. Thus

$$\underset{\smile}{\varkappa}_{ijk}^{(2)}(\omega; \omega_1, \omega_2) = \underset{\smile}{\varkappa}_{jik}^{(2)}(-\omega_1; -\omega_1 - \omega_2, \omega_2) = \underset{\smile}{\varkappa}_{kji}^{(2)}(-\omega_2; \omega_1, -\omega_1 - \omega_2).$$

Hence the three terms in (4.44) can be generated by starting from one term. If we replace P_2 by the overall permutation operation $\tilde{\mathsf{P}}_2$, we may write

$$\underset{\smile}{\varkappa}_{ijk}^{(2)}(\omega; \omega_1, \omega_2) = \left(\frac{1}{\hbar}\right)^2 \frac{\gamma}{\varepsilon_0} \frac{1}{2!} \tilde{\mathsf{P}}_2 \sum_{\alpha,\beta,\gamma} \rho_{\alpha\alpha}^e \frac{d_{\alpha\beta i} d_{\beta\gamma j} d_{\gamma\alpha k}}{(\omega_{\alpha\gamma} - \omega_2)(\omega_{\alpha\beta} - \omega_1 - \omega_2)}. \quad (4.45)$$

[$\tilde{\mathsf{P}}_2$ generates all possible permutations of indices and related frequencies, including $(i, -\omega = -\omega_1 - \omega_2)$, and sums the resultant terms.]

The magnitude of one component (e.g. $\underset{\smile}{\varkappa}_{111}^{(2)} \equiv \underset{\smile}{\varkappa}^{(2)}$) of the second-order susceptibility will now be estimated under simple conditions. The atomic systems are described by a three-level scheme (see Fig. 4.4) with the level energies $\hbar\omega_{10}$ and $\hbar\omega_{20}$ large compared with the photon energies $\hbar\omega_1$ and $\hbar\omega_2$. All atomic systems are supposed to be in the ground state ($\rho_{\alpha\alpha}^e = \delta_{\alpha 0}$). Under these conditions we obtain from (4.45)

$$\underset{\smile}{\varkappa}^{(2)}(\omega; \omega_1, \omega_2) \approx \left(\frac{6}{\hbar^2}\right)\left(\frac{\gamma}{\varepsilon_0}\right)\frac{d_{01} d_{12} d_{20}}{\omega_{10}\omega_{20}}.$$

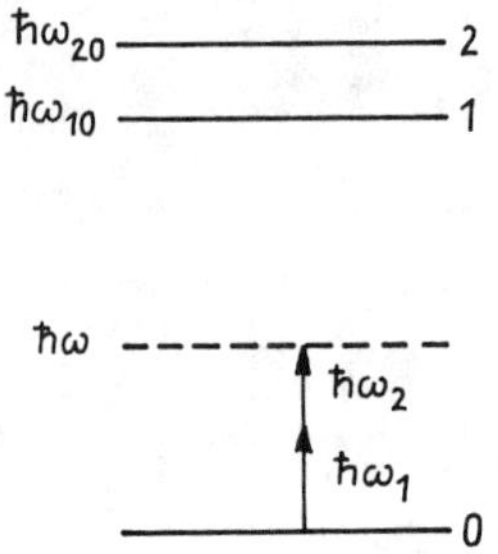

Fig. 4.4. Atomic three-level scheme used for the calculation of sum frequency susceptibilities ($\hbar\omega_{10}$ and $\hbar\omega_{20}$ are energies of excited states; $\hbar\omega_1$ and $\hbar\omega_2$ are photon energies).

Inserting values typical of strong electronic transitions of a dense sample in the ultraviolet region ($d_{01}, d_{12}, d_{20} \sim 10^{-29}$ A s m, $\omega_{10}, \omega_{20} \sim 10^{16}\ \mathrm{s}^{-1}$, $\gamma \sim 10^{28}\ \mathrm{m}^{-3}$), the maximum component of the second-order susceptibility is estimated to attain values of

$$\underset{\smile}{\varkappa}^{(2)}(\omega; \omega_1, \omega_2) \sim 10^{-11}\ \mathrm{m/V}.$$

To take into account a field correction as described in Section 1.2.3, $\underset{\smile}{\varkappa}^{(2)}$ has to be multiplied by a factor

$$C^{(2)}_{\text{field}} = \left(\frac{\varepsilon_{\text{NR}} + 2}{3}\right)^3,$$

and thus effective susceptibilities are obtained.

The susceptibility $\underset{\smile}{\varkappa}^{(2)}$ can be enhanced in the case of resonance. If for instance $\omega_1 + \omega_2$ is near ω_{10}, but still far away from regions where strong absorption occurs, then one term in (4.45) increases substantially and the susceptibility is given by

$$\underset{\smile}{\varkappa}^{(2)}(\omega; \omega_1, \omega_2) \approx \left(\frac{1}{\hbar^2}\right)\left(\frac{\gamma}{\varepsilon_0}\right)\frac{d_{01}d_{12}d_{20}}{(\omega_{20} - \omega_1)(\omega_{10} - \omega_1 - \omega_2)}.$$

Compared with the previous relation, the enhancement factor $\omega_{10}\omega_{20}/[6(\omega_{20} - \omega_1)(\omega_{10} - \omega_1 - \omega_2)]$ may here increase the susceptibility by some orders of magnitude.

Our rough evaluation of first- and second-order susceptibilities may be used to estimate the ratio

$$\left|\frac{\underset{\smile}{\varkappa}^{(1)}}{\underset{\smile}{\varkappa}^{(2)}}\right| \sim \left|\frac{F_1}{F_2}\frac{\hbar\omega_{10}}{d_{10}}\right|$$

where F_1 and F_2 are numerical factors that depend on the specific atomic model with its energy levels and transition moments as well as on the frequencies ω, ω_1, and ω_2. Substituting the same values as above for the transition frequencies and transition moments we get $F_1/F_2 \sim 1$, and thus this ratio is found to be about 10^{11} V/m, which is of the order of magnitude of the inner atomic fields. (An elementary charge generates for instance an electric field of this order of magnitude at a distance of about one atomic diameter.) This estimation suggests that we obtain an expansion of the polarization in powers of the ratio between the field strengths of the light field and of the atomic field. As already mentioned, nonlinearities become important whenever the strength of the radiation field is not negligible compared with that of the atomic field.

For the *third-order susceptibility* we obtain

$$\breve{\varkappa}^{(3)}_{ijkl}(\omega;\omega_1,\omega_2,\omega_3) = \left(\frac{-1}{\hbar}\right)^3\left(\frac{\gamma}{\varepsilon_0}\right)\frac{1}{3!}\tilde{\mathsf{P}}_3 \sum_{\alpha,\beta,\gamma,\delta} \rho^e_{\alpha\alpha} \times \frac{d_{\alpha\beta i}d_{\beta\gamma j}d_{\gamma\delta k}d_{\delta\alpha l}}{(\omega_{\alpha\beta}-\omega_1-\omega_2-\omega_3)(\omega_{\alpha\gamma}-\omega_2-\omega_3)(\omega_{\alpha\delta}-\omega_3)} \tag{4.46}$$

by analogous calculations. To estimate the off-resonance susceptibility $\breve{\varkappa}^{(3)}$ and the ratio $|\breve{\varkappa}^{(2)}/\breve{\varkappa}^{(3)}|$, the expression (4.46) can be used in the condensed, rough form

$$\breve{\varkappa}^{(3)} = F_3\frac{1}{\hbar^3}\left(\frac{\gamma}{\varepsilon_0}\right)\frac{|d_{10}|^4}{\omega_{10}^3}.$$

A comparison with $\breve{\varkappa}^{(2)}$ yields the ratio

$$\left|\frac{\breve{\varkappa}^{(2)}}{\breve{\varkappa}^{(3)}}\right| \sim \left|\frac{F_2}{F_3}\frac{\hbar\omega_{10}}{d_{10}}\right|,$$

which is again of the order of magnitude of the inner atomic field, 10^{11} V/m. Hence $\breve{\varkappa}^{(3)}$ may attain values of about 10^{-22} m^2/V^2.

The *n th-order susceptibility* is evaluated to be

$$\breve{\varkappa}^{(n)}_{ia_1\cdots a_n}(\omega;\omega_1,\ldots,\omega_n) = \left(\frac{-1}{\hbar}\right)^n\frac{\gamma}{\varepsilon_0}\frac{1}{n!}\tilde{\mathsf{P}}_n \sum_{\alpha,\beta_1,\ldots,\beta_n} \times \frac{\rho^e_{\alpha\alpha}d_{\alpha\beta_1 i}\cdots d_{\beta_n\alpha a_n}}{(\omega_{\alpha\beta_1}-\omega_1-\cdots-\omega_n)\cdots(\omega_{\alpha\beta_n}-\omega_n)}, \tag{4.47}$$

and the order of magnitude of the ratio $\underset{\smile}{\varkappa}^{(n-1)}/\underset{\smile}{\varkappa}^{(n)}$ is estimated to be

$$\left|\frac{\underset{\smile}{\varkappa}^{(n-1)}}{\underset{\smile}{\varkappa}^{(n)}}\right| \sim \left|\frac{F_{n-1}}{F_n}\frac{\hbar\omega_{10}}{d_{10}}\right| \sim \left|\frac{\hbar\omega_{10}}{d_{10}}\right| \sim 10^{11}\ \mathrm{V/m}.$$

To explain the physical meaning of such ratios we calculate the change $\Delta\varepsilon(\omega)$ of the dielectric permittivity at frequency ω that is caused by a strong radiation field at frequency ω', and then we compare it with the noninfluenced permittivity. The strong field leads, via the nonlinear polarization

$$\widehat{P}^{(3)}(-\omega) = \tfrac{3}{2}\underset{\smile}{\varkappa}^{(3)}(\omega;\,\omega,\omega',-\omega')\widehat{E}(-\omega)\left|\widehat{E}(\omega')\right|^2$$

[cf. (1.125)], to

$$\Delta\varepsilon(\omega) = \tfrac{3}{2}\underset{\smile}{\varkappa}^{(3)}(\omega;\,\omega,\omega',-\omega')\left|\widehat{E}(\omega')\right|^2.$$

The ratio

$$\frac{\Delta\varepsilon(\omega)}{\varepsilon(\omega)} = \frac{\tfrac{3}{2}\underset{\smile}{\varkappa}^{(3)}(\omega;\,\omega,\omega',-\omega')\left|\widehat{E}(\omega')\right|^2}{1+\underset{\smile}{\varkappa}^{(1)}(\omega;\,\omega)}$$

obviously has the order of magnitude of $|\widehat{E}(\omega')|^2/|E_A|^2$, where E_A is the strength of the inner atomic field.

4.5. SUSCEPTIBILITIES OF ATOMIC SYSTEMS WITH LOSSES

In the previous section we considered an ensemble of atoms that were assumed to interact neither with each other nor with any other system except the classical radiation field. Such an ensemble of atomic systems cannot dissipate energy, and therefore it possesses an infinite memory time. With these assumptions specific difficulties arose in calculating the linear and nonlinear susceptibilities, and as a result, certain frequency regions around the (sharp) transition frequencies had to be excluded from our consideration. Thus the susceptibilities calculated so far are not applicable if resonance effects—such as for instance, one-photon or multiphoton absorption—have to be described.

In this section losses of the atomic ensemble will therefore be taken into account. Each single atom is assumed to be coupled to a reservoir, that is, to a dissipative system possessing an infinite number of degrees of freedom and being in thermodynamic equilibrium characterized by a certain temperature. This concept is described in detail in Section 3.2.4. Any nonequilibrium state of the atomic ensemble will be destroyed by relaxation processes due to the interaction of the atoms with the reservoir. As to its physical nature, this reservoir may be thought of as being a buffer gas surrounding the atoms under consideration, or as consisting of the phonons of a crystal. The interaction of the atoms with each other may also be roughly described by the coupling of each atom to a reservoir.

To calculate the response functions or susceptibilities of atoms under the action of a classical radiation field and a reservoir, we may start from the results of Section 3.2.4. The matrix elements of the atomic density operator calculated by means of the eigenstates of the unperturbed atom obey the relations

$$\frac{d}{dt}\rho_{\alpha\beta} + i\omega_{\alpha\beta}\rho_{\alpha\beta} + \frac{1}{\tau_{\alpha\beta}}\rho_{\alpha\beta} = \frac{1}{i\hbar}\left[\hat{H}^C, \hat{\rho}\right]_{\alpha\beta},$$

$$\frac{d}{dt}\rho_{\alpha\alpha} + \frac{1}{T_\alpha}\left(\rho_{\alpha\alpha} - \rho^e_{\alpha\alpha}\right) = \frac{1}{i\hbar}\left[\hat{H}^C, \hat{\rho}\right]_{\alpha\alpha}. \tag{4.48}$$

The *relaxation times* $\tau_{\alpha\beta}$ and T_α describe the dying away of nonequilibrium values of the off-diagonal and diagonal density matrix elements, respectively. As in the previous section, the operator $\hat{H}^C$ stands for the action of the classical radiation field on an atom and is given by $\hat{H}^C(t) = -\hat{d}_\bullet E_\bullet(t)$ in the dipole approximation. In the following, the diagonal elements $\rho_{\alpha\alpha}$ of the density matrix are assumed to preserve throughout the interaction process their equilibrium values $\rho^e_{\alpha\alpha}$, which are only determined by the influence of the reservoir on the atom. Changes of the atomic level populations caused by the radiation field are neglected. The treatment based on these assumptions is called the *irreversible approximation*; its applicability depends on the strength of the interaction between atom and reservoir as well as on that between atom and radiation field and has to be checked in any case. When making use of the irreversible approximation, we only need to treat the first equation of (4.48), which can be solved with the help of a perturbation procedure, in which the atom under the action of the dissipative system is considered as the unperturbed system and the radiation field as the perturbation.

In first order the equation

$$\frac{d}{dt}\rho^{(1)}_{\alpha\beta}(t) + i\omega_{\alpha\beta}\rho^{(1)}_{\alpha\beta}(t) + \frac{1}{\tau_{\alpha\beta}}\rho^{(1)}_{\alpha\beta}(t) = \frac{1}{i\hbar}\left(\rho^e_{\alpha\alpha} - \rho^e_{\beta\beta}\right)d_{\alpha\beta\bullet}E_\bullet(t) \tag{4.49}$$

has to be solved. We get thc solution

$$\rho^{(1)}_{\alpha\beta}(t) = \frac{1}{i\hbar}\left(\rho^e_{\alpha\alpha} - \rho^e_{\beta\beta}\right)d_{\alpha\beta\bullet}\int_{-\infty}^{t} dt_1\, e^{i(\omega_{\alpha\beta} - i/\tau_{\alpha\beta})(t_1 - t)}E_\bullet(t_1). \tag{4.50}$$

After setting $\tau_1 = t - t_1$ and replacing $E_\bullet(t_1)$ by its Fourier transform, one obtains by integration over τ_1

$$\rho_{\alpha\beta}^{(1)}(t) = \frac{1}{2\pi}\int_{-\infty}^{\infty} d\omega\, \underset{\smile}{E}_\bullet(\omega)e^{+i\omega t}$$

$$\times\left\{\left(\frac{-1}{\hbar}\right)\left(\rho_{\alpha\alpha}^{e} - \rho_{\beta\beta}^{e}\right)d_{\alpha\beta\bullet}\frac{1}{\omega_{\alpha\beta} - \left(i/\tau_{\alpha\beta}\right) + \omega}\right\}. \tag{4.51}$$

We see that, in contrast to the previous section, all difficulties concerning the convergence of the integral have been overcome as a result of taking into account relaxation parameters $1/\tau_{\alpha\beta}$, which mathematically have the same effect as the artificially introduced parameter δ.

With the help of the density matrix we calculate the expectation value of the polarization $P_\bullet^{(1)}(t) = \gamma\Sigma_{\alpha,\beta}\rho_{\alpha\beta}^{(1)}(t)d_{\beta\alpha\bullet}$ in first order. By comparison with

$$P_\bullet^{(1)}(t) = \frac{\varepsilon_0}{2\pi}\int_{-\infty}^{\infty} d\omega\, \underset{\smile}{\varkappa}_{\bullet\bullet}^{(1)}(\omega;\omega)\, \underset{\smile}{E}_\bullet(\omega)e^{+i\omega t},$$

we find the susceptibility

$$\underset{\smile}{\varkappa}_{ij}^{(1)}(\omega;\omega) = \left(\frac{-1}{\hbar}\right)\frac{\gamma}{\varepsilon_0}\sum_{\alpha,\beta}\left(\rho_{\alpha\alpha}^{e} - \rho_{\beta\beta}^{e}\right)\frac{d_{\alpha\beta j}d_{\beta\alpha i}}{\omega_{\alpha\beta} + \omega - i/\tau_{\alpha\beta}}. \tag{4.52}$$

Keeping in mind that $\omega_{\beta\alpha} = -\omega_{\alpha\beta}$ and $\tau_{\beta\alpha} = \tau_{\alpha\beta}$ we interchange the summation indices α and β in the term with $\rho_{\beta\beta}^{e}$. This leads to

$$\underset{\smile}{\varkappa}_{ij}^{(1)}(\omega;\omega) = \left(\frac{-1}{\hbar}\right)\frac{\gamma}{\varepsilon_0}\sum_{\alpha,\beta}\rho_{\alpha\alpha}^{e}\left\{\frac{d_{\alpha\beta i}d_{\beta\alpha j}}{\omega_{\alpha\beta} - \omega + i/\tau_{\alpha\beta}} + \frac{d_{\alpha\beta j}d_{\beta\alpha i}}{\omega_{\alpha\beta} + \omega - i/\tau_{\alpha\beta}}\right\}. \tag{4.53}$$

The susceptibility is seen to possess singularities only in the upper half of the complex frequency plane at $-\omega_{\alpha\beta} + i/\tau_{\alpha\beta}$ and $\omega_{\alpha\beta} + i/\tau_{\alpha\beta}$, but not on the real axis. This means that we are allowed to use this susceptibility at any real frequency with the transition frequencies included. Another consequence of the losses that result in the occurrence of relaxation parameters in the susceptibility is the violation of the overall permutation symmetry; the operation $(i, -\omega) \leftrightarrow (j, \omega)$ now changes the value of the susceptibility in (4.53). Only far away from resonance frequencies, where the losses are small, is the overall permutation symmetry approximately valid. We can thus estimate the application field of the loss-free model discussed in Section 4.4 by utilizing the results of calculations that take into account the interaction between atom and reservoir. Since it is very difficult to calculate the relaxation parameters $1/\tau_{\alpha\beta}$ from theory, they are mostly determined by comparing calculated and measured first-order susceptibilities or response functions. In the neighbourhood of a particular transition frequency, say ω_{10}, the dominant term in the frequency dependence of the first-order susceptibility is given by

$$\Delta \underset{\smile}{\varkappa}_{ij}^{(1)}(\omega; \omega) = \frac{\gamma}{\hbar\varepsilon_0}(\rho_{00}^e - \rho_{11}^e)\frac{d_{01j}d_{10i}}{\omega_{10} - \omega + i/\tau_{10}} \tag{4.54a}$$

with the real and imaginary parts

$$\mathrm{Re}\left\{\Delta \underset{\smile}{\varkappa}_{ij}^{(1)}(\omega; \omega)\right\} = \frac{\gamma}{\hbar\varepsilon_0}(\rho_{00}^e - \rho_{11}^e)d_{01j}d_{10i}\frac{\omega_{10} - \omega}{(\omega_{10} - \omega)^2 + (1/\tau_{10})^2} \tag{4.54b}$$

and

$$\mathrm{Im}\left\{\Delta \underset{\smile}{\varkappa}_{ij}^{(1)}(\omega; \omega)\right\} = \frac{-\gamma\pi}{\hbar\varepsilon_0}(\rho_{00}^e - \rho_{11}^e)d_{01j}d_{10i}\frac{1/\pi\tau_{10}}{(\omega_{10} - \omega)^2 + (1/\tau_{10})^2}, \tag{4.54c}$$

which are shown schematically in Fig. 4.5 for an isotropic medium. The imaginary part is represented by a Lorentzian profile of half-width $\Delta\omega = 2/\tau_{10}$. Since the imaginary part of the linear susceptibility is proportional to the one-photon absorption coefficient, the relaxation parameter $1/\tau_{10}$ can be achieved by measuring the profile of the absorption line. Of course, the atomic ensemble under investigation has to be checked to see whether it satisfies the

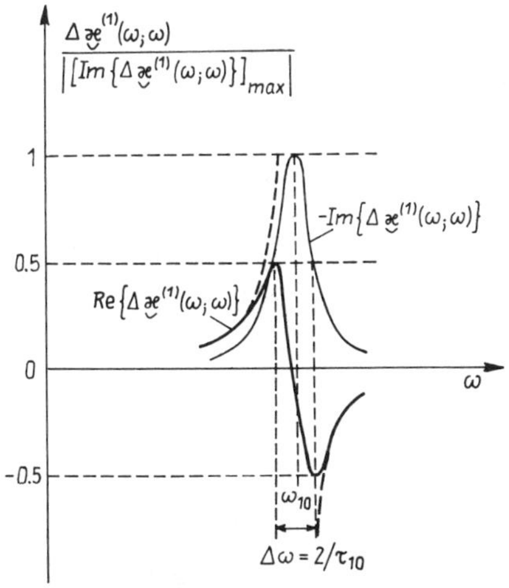

Fig. 4.5. Real and imaginary parts of the linear susceptibility versus frequency (full line: with attenuation, dashed line; without attenuation).

conditions assumed, under which all atomic systems behave identically and hence possess equal resonance frequencies.

Ensembles containing atoms with different properties (e.g., different resonance frequencies) have to be divided into subensembles which fulfil the conditions mentioned. The susceptibility of the whole ensemble will then be obtained by averaging the susceptibilities calculated for all the subensembles. Inhomogeneously broadened systems may be treated in this way. Furthermore, this method allows the calculation of susceptibilities for an ensemble of differently oriented atoms or molecules. In the final expressions for the susceptibilities obtained under the condition of identical orientation we have only to replace products of dipole matrix elements by their orientation averages. To this end the component $d_{\alpha\beta j}$ taken in a laboratory-fixed frame (x, y, z) is expressed by the components of the same vector $\tilde{d}_{\alpha\beta\bullet}$ taken in a coordinate system (X, Y, Z) which is fixed in the molecule. Hence

$$d_{\alpha\beta j} = \sum_J a_{jJ} \tilde{d}_{\alpha\beta J}, \tag{4.55}$$

where a_{jJ} are direction cosines. The orientation averages of the dipole matrix elements and their products are given by

$$\overline{d_{\alpha\beta j}}^{\,0} = \sum_J \overline{a_{jJ}}^{\,0}\, \tilde{d}_{\alpha\beta J}, \tag{4.56a}$$

$$\overline{d_{\alpha\beta j} d_{\beta\alpha k}}^{\,0} = \sum_{JK} \overline{a_{jJ} a_{kK}}^{\,0}\, \tilde{d}_{\alpha\beta J} \tilde{d}_{\beta\alpha K}, \tag{4.56b}$$

and so on. The average over the direction cosines can be easily calculated if all molecular directions have equal probability—in other words, if the macroscopic system behaves isotropically. Then, for example, the averages

$$\overline{a_{jJ}}^{\,0} = 0,$$

$$\overline{a_{jJ} a_{kK}}^{\,0} = \tfrac{1}{3}\delta_{jk}\delta_{JK}$$

are obtained. Substituting such averages into (4.56) and using the resulting relations in (4.53), the first-order susceptibility of isotropic systems takes on the form

$$\underset{\smile}{\varkappa}_{ii}^{(1)}(\omega; \omega) = \left(\frac{-1}{\hbar}\right) \frac{\gamma}{\varepsilon_0} \sum_{\alpha, \beta} \rho_{\alpha\alpha}^{e} \frac{|\tilde{d}_{\alpha\beta}|^2}{3}$$

$$\times \left\{ \frac{1}{\omega_{\alpha\beta} - \omega + i/\tau_{\alpha\beta}} + \frac{1}{\omega_{\alpha\beta} + \omega - i/\tau_{\alpha\beta}} \right\}. \tag{4.57}$$

In the higher orders of perturbation theory, nonlinear optical susceptibilities can be calculated in an analogous way without additional physical assumptions. In the second and third order we obtain

$$
\begin{aligned}
\underset{\smile}{\varkappa}^{(2)}_{ijk}(\omega;\omega_1,\omega_2) = \left(\frac{-1}{\hbar}\right)^2 \frac{\gamma}{\varepsilon_0}\frac{1}{2!}\sum_{\alpha,\beta,\gamma}\rho^e_{\alpha\alpha}
\times \mathsf{P}_2 \Bigg\{ & \frac{d_{\alpha\beta i}d_{\beta\gamma j}d_{\gamma\alpha k}}{\left(\omega_{\alpha\gamma}+\frac{i}{\tau_{\alpha\gamma}}-\omega_2\right)\left(\omega_{\alpha\beta}+\frac{i}{\tau_{\alpha\beta}}-\omega_1-\omega_2\right)} \\
&+ \frac{d_{\alpha\beta j}d_{\beta\gamma i}d_{\gamma\alpha k}}{\left(\omega_{\alpha\gamma}+\frac{i}{\tau_{\alpha\gamma}}-\omega_2\right)\left(\omega_{\alpha\beta}-\frac{i}{\tau_{\alpha\beta}}+\omega_1\right)} \\
&+ \frac{d_{\alpha\beta j}d_{\beta\gamma k}d_{\gamma\alpha i}}{\left(\omega_{\alpha\beta}-\frac{i}{\tau_{\alpha\beta}}+\omega_1\right)\left(\omega_{\alpha\gamma}-\frac{i}{\tau_{\alpha\gamma}}+\omega_1+\omega_2\right)}\Bigg\}
\end{aligned}
\tag{4.58}
$$

and

$$
\begin{aligned}
\underset{\smile}{\varkappa}^{(3)}_{ijkl}(\omega;\omega_1,\omega_2,\omega_3) = \left(\frac{-1}{\hbar}\right)^3 \frac{\gamma}{\varepsilon_0}\frac{1}{3!}\sum_{\alpha,\beta,\gamma,\delta}\rho^e_{\alpha\alpha} \\
\times \mathsf{P}_3 \Bigg\{ & \frac{d_{\alpha\beta i}d_{\beta\gamma j}d_{\gamma\delta k}d_{\delta\alpha l}}{\left(\omega_{\alpha\beta}+\frac{i}{\tau_{\alpha\beta}}-\omega_1-\omega_2-\omega_3\right)\left(\omega_{\alpha\gamma}+\frac{i}{\tau_{\alpha\gamma}}-\omega_2-\omega_3\right)\left(\omega_{\alpha\delta}+\frac{i}{\tau_{\alpha\delta}}-\omega_3\right)} \\
&+ \frac{d_{\alpha\beta j}d_{\beta\gamma i}d_{\gamma\delta k}d_{\delta\alpha l}}{\left(\omega_{\alpha\beta}-\frac{i}{\tau_{\alpha\beta}}+\omega_1\right)\left(\omega_{\alpha\gamma}+\frac{i}{\tau_{\alpha\gamma}}-\omega_2-\omega_3\right)\left(\omega_{\alpha\delta}+\frac{i}{\tau_{\alpha\delta}}-\omega_3\right)} \\
&+ \frac{d_{\alpha\beta j}d_{\beta\gamma k}d_{\gamma\delta i}d_{\delta\alpha l}}{\left(\omega_{\alpha\beta}-\frac{i}{\tau_{\alpha\beta}}+\omega_1\right)\left(\omega_{\alpha\gamma}-\frac{i}{\tau_{\alpha\gamma}}+\omega_1+\omega_2\right)\left(\omega_{\alpha\delta}+\frac{i}{\tau_{\alpha\delta}}-\omega_3\right)} \\
&+ \frac{d_{\alpha\beta j}d_{\beta\gamma k}d_{\gamma\delta l}d_{\delta\alpha i}}{\left(\omega_{\alpha\beta}-\frac{i}{\tau_{\alpha\beta}}+\omega_1\right)\left(\omega_{\alpha\gamma}-\frac{i}{\tau_{\alpha\gamma}}+\omega_1+\omega_2\right)\left(\omega_{\alpha\delta}-\frac{i}{\tau_{\alpha\delta}}+\omega_1+\omega_2+\omega_3\right)}\Bigg\}.
\end{aligned}
\tag{4.59}
$$

The linear as well as the nonlinear optical susceptibilities evaluated here are obviously free from singularities on the real frequency axes. Thus we can use them for any combination of the frequencies $\omega_1, \omega_2, \omega_3$. Far from resonances, there is no important difference between the susceptibilities of interaction-free and relaxing atomic ensembles. In those regions we may neglect the relaxation parameters in (4.57)–(4.59), and by doing so, the susceptibilities of Section 4.4 are obtained. On the other hand, though only in a formal manner, we may achieve the susceptibilities of relaxing ensembles from those of nonrelaxing ones by substituting the complex parameters $(\omega_{\alpha\beta} \pm i/\tau_{\alpha\beta})$ for the real transition frequencies $\omega_{\alpha\beta}$ in such a way that all singularities are shifted to the upper half of the frequency plane. Interesting phenomena associated with relaxation processes can only be expected to occur near resonance. Concerning linear susceptibilities the resonant energy exchange between radiation field and matter has already been discussed. In second order, one- and two-photon resonances are possible, and both may provide a substantial enhancement of the susceptibilities of second-harmonic and sum-frequency generation as well as of parametric effects, which can now be calculated for any value of ω_1 and ω_2. But compared with off-resonance regions, no specific new effects will occur. The situation changes in the third order (and in higher orders). In addition to an enhancement of susceptibilities responsible for processes that arise also in off-resonance regions, such as third-harmonic generation, frequency mixing, higher-order parametric effects, and light-induced refractive-index change, the resonances induce two specific effects namely, two-photon absorption and Raman processes. Let us consider the susceptibilities of both effects in detail.

We start with evaluating the *susceptibility* of *two-photon absorption*. This evaluation is based on the following assumptions. Let the susceptibility describe the attenuation of a weak signal at the frequency ω' induced by a strong pump wave at ω'', where $\omega' + \omega'' \approx \omega_{10}$; the atomic ensemble is supposed to be in the ground state before it is acted on by electromagnetic fields. Any one-photon resonance is excluded. To calculate the signal attenuation, the nonlinear polarization amplitude $\hat{P}^{(3)}_{\bullet}(-\omega')$, which is proportional to $\underset{\smile}{\varkappa}^{(3)}_{\bullet\bullet\bullet\bullet}(\omega'; \omega', \omega'', -\omega'')$, must be known. This susceptibility will be calculated for $\omega', \omega'' > 0$, starting from (4.59). Only terms enhanced by resonance denominators of the form $\omega_{10} + (i/\tau_{10}) - \omega' - \omega''$ make substantial contributions and are therefore taken into account. Such denominators occur only in particular permutations of the third and the fourth terms in (4.59) for $\alpha = 0$ and $\gamma = 1$. Thus we obtain

$$\underset{\smile}{\varkappa}^{(3)}_{ijkl}(\omega'; \omega', \omega'', -\omega'') = \frac{\gamma}{6\hbar^3\varepsilon_0} \frac{1}{\left(\omega_{10} - \omega' - \omega'' + \dfrac{i}{\tau_{10}}\right)}$$

$$\times \sum_{\beta,\delta} \left\{ \frac{d_{0\beta j} d_{\beta 1 k} d_{1\delta i} d_{\delta 0 l}}{(\omega_{\beta 0} - \omega')(\omega_{\delta 0} - \omega'')} \right.$$

$$+\frac{d_{0\beta j}d_{\beta 1k}d_{1\delta l}d_{\delta 0i}}{(\omega_{\beta 0}-\omega')(\omega_{\delta 0}-\omega')}$$

$$+\frac{d_{0\beta k}d_{\beta 1j}d_{1\delta i}d_{\delta 0l}}{(\omega_{\beta 0}-\omega'')(\omega_{\delta 0}-\omega'')}$$

$$\left.+\frac{d_{0\beta k}d_{\beta 1j}d_{1\delta l}d_{\delta 0i}}{(\omega_{\beta 0}-\omega'')(\omega_{\delta 0}-\omega')}\right\}. \quad (4.60)$$

The expression in the braces can be represented by a product of two factors, of which the first contains only the summation index β and the second contains only δ. We divide these factors by $\hbar$ and obtain

$$\alpha_{01jk}(\omega',\omega'')=\frac{1}{\hbar}\sum_{\beta}\left\{\frac{d_{0\beta j}d_{\beta 1k}}{\omega_{\beta 0}-\omega'}+\frac{d_{0\beta k}d_{\beta 1j}}{\omega_{\beta 0}-\omega''}\right\} \quad (4.61a)$$

and

$$\alpha^{*}_{01il}(\omega',\omega'')=\frac{1}{\hbar}\sum_{\delta}\left\{\frac{d_{\delta 0i}d_{1\delta l}}{\omega_{\delta 0}-\omega'}+\frac{d_{\delta 0l}d_{1\delta i}}{\omega_{\delta 0}-\omega''}\right\}. \quad (4.61b)$$

The parameter α_{01jk} has the dimension of a polarizability and may be considered as the specific matrix element of the *generalized polarizability tensor*. In this notation the polarizability tensor of linear optics

$$\alpha_{ij}(\omega)=\frac{1}{\gamma}\varkappa^{(1)}(\omega;\omega)\equiv\alpha_{00ij}(-\omega;\omega)$$

reads

$$\alpha_{ij}(\omega)=\frac{1}{\hbar}\sum_{\beta}\left\{\frac{d_{0\beta i}d_{\beta 0j}}{\omega_{\beta 0}+\omega}+\frac{d_{0\beta j}d_{\beta 0i}}{\omega_{\beta 0}-\omega}\right\}. \quad (4.62)$$

Substituting (4.61) into (4.60), we obtain

$$\varkappa^{(3)}_{ijkl}(\omega';\omega',\omega'',-\omega'')=\frac{\gamma}{6\hbar\varepsilon_0}\frac{\omega_{10}-\omega'-\omega''-i/\tau_{10}}{(\omega_{10}-\omega'-\omega'')^2+(1/\tau_{10})^2}$$

$$\times\alpha_{01jk}(\omega',\omega'')\alpha^{*}_{01il}(\omega',\omega''). \quad (4.63)$$

Far from one-photon resonances the generalized polarizabilities do not strongly depend on the frequencies ω', ω''. Hence, the frequency dependence of $\varkappa^{(3)}_{ijkl}$ is mainly determined by the two-photon resonance, which implies a Lorentzian

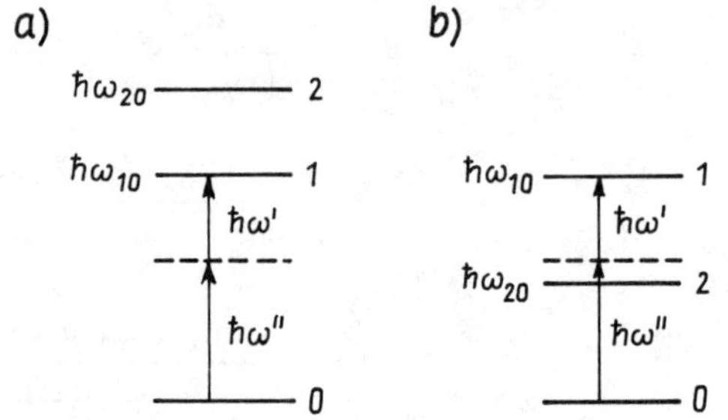

Fig. 4.6. Atomic three-level scheme used for the calculation of two-photon absorption ($\hbar\omega_{10}$ and $\hbar\omega_{20}$ are energies of excited states; $\hbar\omega'$ and $\hbar\omega''$ are energies of absorbed photons). In (a) the intermediate level $\hbar\omega_{20}$ is far from resonance; in (b) it is near resonance.

resonance line and the corresponding dispersion profile for the imaginary and the real part of $\underset{\smile}{\varkappa}^{(3)}$, respectively.

Let us now estimate the order of magnitude of the real and the imaginary part of $\underset{\smile}{\varkappa}^{(3)}(\omega'; \omega', \omega'', -\omega'')$ for electronic transitions. A three-level system is assumed where the single intermediate level $\beta = \delta = 2$ is located high above the final state 1 (see Fig. 4.6a). With the transition moments $d_{02}, d_{21} \sim 10^{-29}$ A s m and the transition frequency $\omega_{20} \sim 10^{16}$ s^{-1}, the generalized susceptibility is estimated as

$$\alpha_{01} \approx \frac{2}{\hbar} \frac{d_{02} d_{21}}{\omega_{20}} \sim 10^{-40} \text{ A s m}^2/\text{V}.$$

Assuming for dense media a relaxation time $\tau_{10} \sim 10^{-13}$ s and the high population density of $\gamma \sim 10^{28}$ m^{-3}, the modulus of the imaginary part attains a maximum value

$$\left|\text{Im}\left\{\underset{\smile}{\varkappa}^{(3)}(\omega'; \omega', \omega'', -\omega'')\right\}\right|_{\max} \approx \frac{\gamma\tau_{10}}{6\hbar\varepsilon_0}|\alpha_{01}|^2 \sim 10^{-21} \text{ m}^2/\text{V}^2.$$

The generalized polarizability α_{01} and (in conjunction with it) the susceptibility are strongly enhanced by an intermediate level between the levels 0 and 1 near the one-photon energy $\hbar\omega'$ or $\hbar\omega''$ (see Fig. 4.6b).

The nonlinear optical susceptibility may be substituted into the relations of classical electrodynamics (Section 1.4), and thus the attenuation of the signal wave caused by the strong pump wave may be calculated. Such processes will be discussed in detail in Section 10.3.2.

Let us now consider the *Stokes susceptibility* $\underset{\smile}{\varkappa}^{(3)}_{ijkl}(\omega_S; \omega_S, \omega_L, -\omega_L)$ of the *stimulated Raman effect*. In contrast to two-photon absorption, the frequencies ω_S and ω_L are assumed to approximately satisfy the resonance condition $\omega_L - \omega_S = \omega_{10}$ (see Fig. 4.7). That susceptibility allows us, for example, to

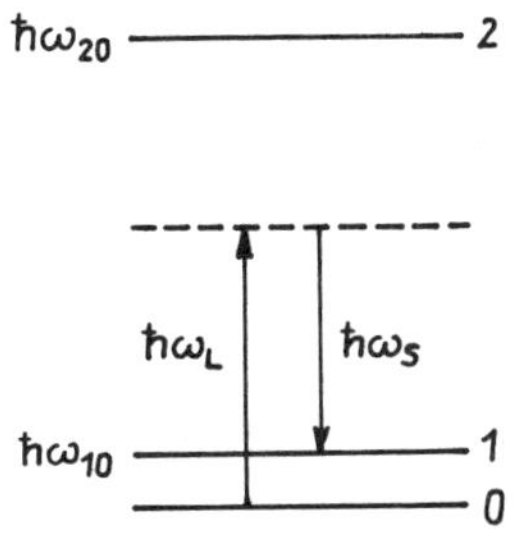

Fig. 4.7. Atomic three-level scheme for the calculation of Raman susceptibilities ($\hbar\omega_{10}$ and $\hbar\omega_{20}$ are energies of excited states, $\hbar\omega_L$ and $\hbar\omega_S$ are energies of laser and Stokes photons, respectively).

describe the amplification of a signal wave at the Stokes frequency ω_S at the cost of a strong laser field at ω_L. As with two-photon absorption, we again start from (4.59) and suppose all atomic systems to be initially in the ground state, which means $\rho^e = \delta_{\alpha 0}$. Resonance denominators of the form $(\omega_{10} + i\tau_{10}^{-1} + \omega_S - \omega_L)$ occur only for particular frequency-index permutations of the first and the second term in (4.59) where the summation indices α and γ take the values $\alpha = 0$ and $\gamma = 1$. By exclusively taking into account such resonant terms, we obtain the Raman susceptibility as

$$\varkappa^{(3)}_{ijkl}(\omega_S; \omega_S, \omega_L, -\omega_L) = \frac{1}{6\hbar^3\varepsilon_0(\omega_{10} - \omega_L + \omega_S - i/\tau_{10})}$$

$$\times \sum_{\beta,\delta}\left\{\frac{d_{0\beta i}d_{\beta 1k}d_{1\delta j}d_{\delta 0l}}{(\omega_{\beta 0} + \omega_S)(\omega_{\delta 0} - \omega_L)}\right.$$

$$+ \frac{d_{0\beta i}d_{\beta 1k}d_{1\delta l}d_{\delta 0j}}{(\omega_{\beta 0} + \omega_S)(\omega_{\delta 0} + \omega_S)}$$

$$+ \frac{d_{0\beta k}d_{\beta 1i}d_{1\delta j}d_{\delta 0l}}{(\omega_{\beta 0} - \omega_L)(\omega_{\delta 0} - \omega_L)}$$

$$\left. + \frac{d_{0\beta k}d_{\beta 1i}d_{1\delta l}d_{\delta 0j}}{(\omega_{\beta 0} - \omega_L)(\omega_{\delta 0} + \omega_S)}\right\}. \quad (4.64)$$

Utilizing, in analogy to the previous treatment of two-photon absorption, the Raman polarizability tensor

$$\alpha_{01ik}(-\omega_S, \omega_L) = \frac{1}{\hbar}\sum_\beta\left\{\frac{d_{0\beta i}d_{\beta 1k}}{\omega_{\beta 0} + \omega_S} + \frac{d_{0\beta k}d_{\beta 1i}}{\omega_{\beta 0} - \omega_L}\right\},$$

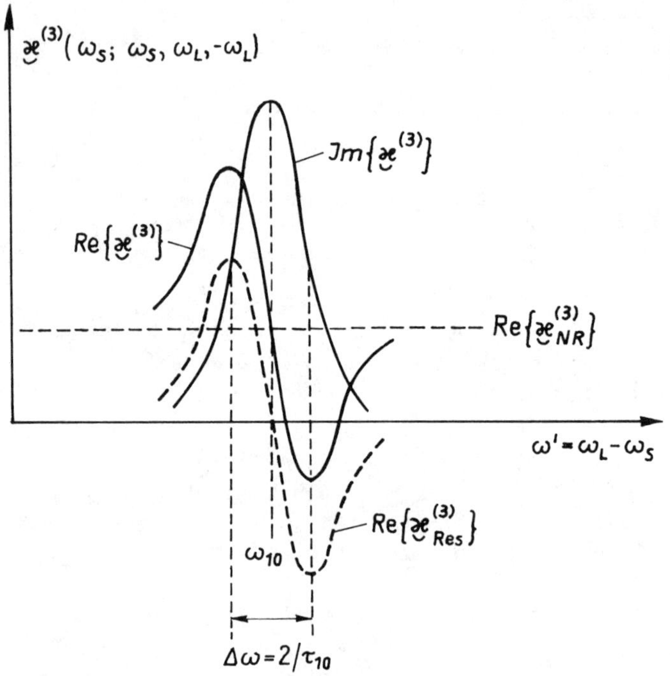

Fig. 4.8. Real and imaginary parts of the Stokes susceptibility versus frequency. (The real part of $\chi^{(3)}$ is composed of the resonant term $\mathrm{Re}\{\chi^{(3)}_{\mathrm{Res}}\}$ and the nonresonant term $\mathrm{Re}\{\chi^{(3)}_{\mathrm{NR}}\}$).

the Raman susceptibility of (4.64) can be written more comprehensively as

$$\chi^{(3)}_{ijkl}(\omega_S; \omega_S, \omega_L, -\omega_L) = \frac{\gamma}{6\hbar\varepsilon_0} \frac{\omega_{10} - \omega_L + \omega_S + i/\tau_{10}}{(\omega_{10} - \omega_L + \omega_S)^2 + (1/\tau_{10})^2}$$

$$\times \alpha_{01ik}(-\omega_S, \omega_L)\alpha^*_{01jl}(-\omega_S, \omega_L). \qquad (4.65)$$

In Fig. 4.8 the real and the imaginary part of $\chi^{(3)}(\omega_S; \omega_S, \omega_L; -\omega_L)$ are shown in the vicinity of the resonance frequency ω_{10}. Note the positive sign of the imaginary part, which indicates amplification of the Stokes wave in the presence of the strong pump field at the laser frequency. In addition to the resonant part of the susceptibility, the nonresonant contribution $\chi^{(3)}_{\mathrm{NR}}$ caused by all transitions except $0 \to 1$ has been plotted. It is real and nearly constant in the vicinity of ω_{10} if all other resonances lie far off.

Now we estimate the values of the resonant and nonresonant contributions to $\chi^{(3)}$. We assume only one intermediate level $\beta = \delta = 2$, the energy of which is far from one-photon resonance (see Fig. 4.7). If $0 \to 2$ and $1 \to 2$ are strong electronic transitions in the ultraviolet region—say atomic transitions—then the transition moments and together with them the polarizability matrix element have the same order of magnitude as they have in the case of two-photon absorption (d_{02}, $d_{12} \sim 10^{-29}$ A s m, $\alpha_{01} \sim 10^{-40}$ A s m^2/V). Hence, setting again $\gamma \sim 10^{28}$ m^{-3} and $\tau_{10} \sim 10^{-13}$ s for dense media, the maximum of the imaginary part of the Raman susceptibility attains values of about

$$\left(\text{Im}\{\underset{\smile}{\varkappa}^{(3)}(\omega_S; \omega_S, \omega_L, -\omega_L)\}\right)_{\max} \sim 10^{-21} \text{ m}^2/\text{V}^2.$$

In gases and vapors the susceptibility may approach the same order of magnitude in spite of the much lower density because the relaxation time τ_{10} is higher. In many cases the product $\gamma\tau_{10}$ is nearly constant over a certain pressure range and achieves values of about 10^{14} m^{-3} s. Thus $(\text{Im}\{\underset{\smile}{\varkappa}^{(3)}\})_{\max} \sim 10^{-20}\text{m}^2/\text{V}^2$ is attainable. The maximum of $\text{Re}\{\underset{\smile}{\varkappa}^{(3)}\}$ is of the same order of magnitude as that of the imaginary part. Far from resonance the real part of the contribution of the transition $0 \to 1$ to the susceptibility is decreased by a factor of about $(\omega_{10}\tau_{10})^{-1} \lesssim 10^{-2}$ from its maximum. Usually several nonresonant transitions contribute to $\underset{\smile}{\varkappa}^{(3)}_{\text{NR}}$ at a given frequency, say at ω_{10}. Hence the ratio $\text{Re}\{\underset{\smile}{\varkappa}^{(3)}_{\text{NR}}\}/(\text{Re}\{\underset{\smile}{\varkappa}^{(3)}_{\text{Res}}\})_{\max}$ can attain values of $\sim 10^{-1}$.

Of special interest is the vibrational Raman effect, where the states 0 and 1 belong, as vibrational substates $v = 0, 1$, to one electronic level e of the molecule under investigation, for instance, to the electronic ground level $e = 0$. Within the adiabatic approximation of molecular states the transition moment $d_{(e_1 v_1)(e_2 v_2)\bullet}$ between the vibronic states ($e = e_1$, $v = v_1$) and ($e = e_2$, $v = v_2$) is given by

$$d_{(e_1 v_1)(e_2 v_2)\bullet} \approx (v_1|v_2)\, d_{e_1 e_2 \bullet}(R^e), \tag{4.66}$$

where $d_{e_1 e_2 \bullet}(R^e)$ represents the electronic transition moment calculated by the use of the electronic eigenfunctions of the molecule for all nuclei at equilibrium positions R^e, and $(v_1|v_2)$ is the so-called Franck–Condon factor and represents the overlap integral of the two vibrational eigenfunctions. The electronic transition moment may attain values up to 10^{-29} A s m. The Franck–Condon factor attains its maximum value $(v_1|v_2) = 1$ only for complete overlapping of the vibrational eigenfunctions; this means that the two vibrational eigenfunctions must be equal. In the Raman polarizability, products of transition moments occur in the form

$$d_{(e_1 v_1)(e_2 v_2)} d_{(e_2, v_2)(e_1, v_1 + 1)} \approx (v_1|v_2)(v_2|v_1 + 1)\left|d_{e_1 e_2}(R^e)\right|^2. \tag{4.67}$$

If one of the Franck–Condon factors—say $(v_1|v_2)$—reaches the maximum value of 1, the other can attain only values below 1, because the relation $(v_1|v_2) \approx 1$ implies the same functional structure of the vibrational eigenfunctions ψ_{v_1} and ψ_{v_2}, and consequently the functions ψ_{v_2} and ψ_{v_1+1} behave differently. The product of the Franck–Condon factors, η, can be estimated as

$$(v_1|v_2)(v_2|v_1+1) = \eta \sim 10^{-1}.$$

Hence, the product of transition moments in (4.67) can attain values of about 10^{-59} $(\mathrm{A\,s\,m})^2$, and the polarizability $\alpha_{01}(-\omega_S, \omega_L)$ of strong vibrational Raman transitions is evaluated to be

$$\alpha_{01}(-\omega_S, \omega_L) \sim 10^{-41}\ \mathrm{A\,s\,m^2/V}.$$

The Raman susceptibility $\underset{\smile}{\varkappa}^{(3)}(\omega_S; \omega_S, \omega_L, -\omega_L)$ is calculated for a dense medium ($\gamma \sim 10^{28}$ m^{-3}), supposing a relaxation time τ_{10} of the vibrational transition of about 10^{-12} s. Substitution of these values into (4.65) yields

$$\left(\mathrm{Im}\left\{\underset{\smile}{\varkappa}^{(3)}(\omega_S; \omega_S, \omega_L, -\omega_L)\right\}\right)_{\max} \sim 10^{-22}\ \mathrm{m^2/V^2}.$$

To conclude this section we would like to draw attention once more to the very restrictive assumptions of noninteracting molecules. We have neglected the influence of any overlap between the wave functions of different atomic systems as well as the microscopic electric fields at the space coordinate of one atom that arise from the charges, dipole, and multipole moments of all other atoms. In particular, the second neglect leads to a local electric field strength that is determined by the external sources and is independent of the interaction process. Thus our results seem to be applicable only to dilute ensembles. But the area of applicability can be extended by using the appropriate relation between microscopic and macroscopic electric fields and introducing effective susceptibilities (cf. Section 1.2).

If the assumption of identical atomic systems is dropped, ensemble averages of the susceptibility expressions have to be calculated as explained for the first order. For instance, if the molecules differ in orientation, the products of dipole-moment matrix elements have to be replaced by ensemble averages that are taken over all possible orientations.

4.6. DIRECT EVALUATION OF TRANSITION PROBABILITIES

The semiclassical theory enabled us to calculate the efficiency of radiative processes, including those with resonant interactions, by utilizing the nonlinear optical susceptibilities already calculated in order to achieve higher-order contributions to the polarization. The polarization acts as the source term in Maxwell's equations or in the wave equation, and hence allows the evaluation of the field-strength variation at particular frequencies in the course of interaction (Section 1.4). In this section we shall use a different and more direct method applicable to resonant interaction between matter and radiation. The time derivative $((d/dt)\rho_{\alpha\alpha}(t))_R$ of the molecular level population probability resulting from the action of the radiation field is calculated. On the assumption that initially the molecules are in the ground state ($\rho_{\alpha\beta} = \delta_{\alpha 0}\delta_{\beta 0}$) and go to state 1, the transition probability per unit time due to radiation is given by

$$\left(\frac{\Delta W_{0\to 1}}{\Delta t}\right)_R = \overline{\left(\frac{d}{dt}\rho_{11}\right)_R}^{t}, \tag{4.68}$$

where an appropriate time average (e.g. over one period of a harmonic perturbation) has to be taken. The power per unit volume, $\Delta\mathscr{P}_L/\Delta V$, that is exchanged between radiation and matter is associated with the transition probability by summing the contributions of all individual molecules:

$$\frac{\Delta\mathscr{P}_L}{\Delta V} = \frac{\hbar\omega_{10}}{\Delta V}\sum_m \overline{\left(\frac{d}{dt}\rho_{11(m)}\right)_R}^{t}. \tag{4.69}$$

In the case of interaction-free, identical molecules the summation gives

$$\frac{\Delta\mathscr{P}_L}{\Delta V} = \hbar\omega_{10}\gamma\overline{\left(\frac{d}{dt}\rho_{11}\right)_R}^{t} \tag{4.70}$$

The one-particle *interaction cross section* σ_{01} is defined by dividing the power per unit volume by the number of particles per unit volume and by the incident radiation power per area A. It is

$$\sigma_{01} = \frac{\Delta\mathscr{P}_L}{\Delta V}\frac{A}{\mathscr{P}_L}\frac{1}{\gamma} = \hbar\omega_{10}\frac{A}{\mathscr{P}_L}\left(\frac{\Delta W_{0\to 1}}{\Delta t}\right)_R. \tag{4.71}$$

At this point in our discussion we may introduce the photon concept into the semiclassical theory. If the excitation of one molecule is for instance accompanied by the absorption of j photons of equal energy $\hbar\omega = j^{-1}\hbar\omega_{10}$, the number of photons annihilated per unit time and unit volume of matter is

$$\frac{\Delta N_{\text{Ph}}}{\Delta V\Delta t} = \frac{1}{\hbar\omega}\left(\frac{\Delta\mathscr{P}_L}{\Delta V}\right) = j\gamma\left(\frac{\Delta W_{0\to 1}}{\Delta t}\right)_R.$$

In the case of multiphoton absorption ($j > 1$) the cross section $\sigma_{01}^{(j)}$ is not independent of the strength of the radiation field. To get, in spite of this fact, a pure molecular parameter, we define a *normalized* cross section

$$\tilde{\sigma}_{01}^{(j)} = \sigma_{01}^{(j)}\frac{1}{I^{j-1}}, \tag{4.72}$$

where $I = \mathscr{P}_L/(A\hbar\omega)$ is the photon flux per unit area.

The density matrix will be calculated by a perturbation procedure in the irreversible approximation; this means that the diagonal components are assumed to preserve their equilibrium values [e.g. $\rho_{\alpha\alpha}(t) = \rho^e_{\alpha\alpha} = \delta_{\alpha 0}$] because the radiation-induced changes such as

$$\left(\frac{d}{dt}\rho_{11}\right)_R = \frac{1}{i\hbar}[\hat{H}^C, \hat{\rho}]_{11} \tag{4.73}$$

are compensated by relaxation transitions (cf. Section 3.2.4). The radiation field is supposed to be monochromatic, and we write the interaction Hamiltonian as

$$\hat{H}^C = -\hat{d}_\bullet \widehat{E}_\bullet \cos \omega t. \tag{4.74}$$

Here the particular choice of the phase of the electric field has no influence on the results. To avoid the occurrence of many indices we restrict ourselves to transition moments aligned in the field direction. The interaction Hamiltonian of (4.74) has to be substituted into the perturbation equations of the density matrix. In mth order we have

$$\left(\frac{d}{dt} + i\omega_{\alpha\beta} + \frac{1}{\tau_{\alpha\beta}}\right)\rho^{(m)}_{\alpha\beta} = \frac{1}{i\hbar}\left[\hat{H}^C, \hat{\rho}^{(m-1)}\right]_{\alpha\beta} \tag{4.75a}$$

and

$$\left(\frac{d}{dt}\rho^{(m)}_{\alpha\alpha}\right)_R = \frac{1}{i\hbar}\left[\hat{H}^C, \hat{\rho}^{(m-1)}\right]_{\alpha\alpha} \tag{4.75b}$$

[cf. (4.48)].

In first-order perturbation theory the differential equation for $\rho^{(1)}_{\alpha 0}(t)$ has the stationary solution

$$\rho^{(1)}_{\alpha 0}(t) = -\frac{d_{\alpha 0}}{2\hbar}\widehat{E}\left\{\frac{e^{-i\omega t}}{\omega - \omega_{\alpha 0} + i/\tau_{\alpha 0}} - \frac{e^{i\omega t}}{\omega + \omega_{\alpha 0} - i/\tau_{\alpha 0}}\right\}. \tag{4.76}$$

All the other off-diagonal matrix elements $\rho^{(1)}_{\alpha\beta}(t)$ and all the diagonal elements $\rho^{(1)}_{\alpha\alpha}(t)$ vanish. The latter fact means that there are no changes of population probability in first order.

In the second-order calculation the first-order off-diagonal elements of (4.76) are substituted into (4.75b). After taking the time average, we obtain the solution

$$\overline{\left(\frac{d}{dt}\rho_{11}^{(2)}(t)\right)_R}^t = \frac{1}{2\hbar^2}|d_{10}|^2\hat{E}^2$$
$$\times\left\{\frac{1/\tau_{10}}{(\omega-\omega_{10})^2+(1/\tau_{10})^2}+\frac{1/\tau_{10}}{(\omega+\omega_{10})^2+(1/\tau_{10})^2}\right\}, \tag{4.77}$$

which, according to (4.68), is the second-order transition probability per unit time of molecular excitations from the ground level 0 into level 1. High values of the transition probability are attained near $\omega = \omega_{10}$, which indicates a one-photon absorption process. Thus, the second-order density-matrix perturbation theory provides the one-photon transition probability. Near resonance ($\omega \approx \omega_{10}$) the frequency dependence of the transition probability is given by a Lorentzian profile. The power absorbed per unit volume may be compared with the result

$$\frac{\Delta\mathscr{P}_L}{\Delta V} = \overline{E\frac{d}{dt}P}^t = -\tfrac{1}{2}\omega\hat{E}^2\operatorname{Im}\{\underset{\smile}{\varkappa}^{(1)}(\omega;\omega)\}$$

of the calculation in which the polarization as a function of the field strength is used as the source term in Maxwell's equations. The comparison gives the same expression for the imaginary part of the susceptibility, $\operatorname{Im}\{\underset{\smile}{\varkappa}^{(1)}(\omega;\omega)\}$, as does (4.54c).

The one-photon absorption cross section $\sigma_{01}^{(1)}$ can be obtained by substituting the resonant part of the transition probability from (4.77) and the intensity $n^{(L)}c\varepsilon_0\hat{E}^2/2$ into (4.71). The result is

$$\sigma_{01}^{(1)}(\omega) = \frac{|d_{01}|^2\pi\omega_{10}}{n^{(L)}c\varepsilon_0\hbar}g_L(\omega-\omega_{10}), \tag{4.78}$$

where

$$g_L(\omega-\omega_{10}) = \frac{1}{\pi}\frac{1/\tau_{10}}{(\omega-\omega_{10})^2+(1/\tau_{10})^2}.$$

If other line-broadening processes than the phase decay process that we have considered occur, the Lorentzian line-shape function has to be replaced by a general line-shape function $g(\omega - \omega_{10})$, the integral over which is also normalized. In most cases the line-shape functions are obtained from experimental results because *ab initio* calculations are rather tedious (cf. Section 3.2).

In the fourth-order perturbation theory we obtain with analogous calculations the two-photon transition probability

$$\frac{\Delta W_{0\to 1}^{(4)}}{\Delta t} = \frac{\hat{E}^4}{32\hbar^2}\left|\alpha_{01}(\omega,\omega)\right|^2 \pi g_L(2\omega-\omega_{10}), \tag{4.79}$$

where

$$\alpha_{01}(\omega,\omega) = \frac{2}{\hbar}\sum_{\beta}\frac{d_{0\beta}d_{\beta 1}}{\omega_{\beta 0}-\omega}$$

is the generalized polarizability matrix element for two-photon absorption [cf. (4.61)]. Note that in deriving (4.79) one has to calculate certain off-diagonal elements of the second and third orders, namely $\rho_{01}^{(2)}$ and $\rho_{\beta 1}^{(3)}$. A comparison of (4.79) with the power absorbed per unit volume,

$$\overline{E(t)\frac{d}{dt}P(t)}^{\,t} = -\tfrac{3}{8}\omega\hat{E}^4 \operatorname{Im}\{\underset{\smile}{\varkappa}^{(3)}(\omega;\omega,\omega,-\omega)\},$$

yields

$$\operatorname{Im}\{\underset{\smile}{\varkappa}^{(3)}(\omega;\omega,\omega,-\omega)\} = \frac{-\gamma}{6\hbar}|\alpha_{01}|^2 \pi g_L(2\omega-\omega_{10}). \tag{4.80}$$

This relation is in agreement with (4.63). Now the unnormalized and normalized cross sections for the two-photon transition $0 \to 1$ defined in (4.71) and (4.72) are obtained as

$$\sigma_{01}^{(2)}(\omega) = \frac{\pi\omega_{10}}{16\hbar n^{(L)}\varepsilon_0 c}g_L(2\omega-\omega_{10})|\alpha_{01}|^2|\hat{E}|^2 \tag{4.81}$$

and

$$\tilde{\sigma}_{01}^{(2)}(\omega) = \frac{\pi\omega_{10}^2}{8\left(n^{(L)}\right)^2\varepsilon_0^2 c^2}g_L(2\omega-\omega_{10})|\alpha_{01}|^2. \tag{4.82}$$

Let us estimate typical values of the molecular parameter $\tilde{\sigma}^{(2)}$. With the assumption that there is no intermediate level between levels 0 and 1 and that only one intermediate state above level 1 with $\beta = 2$ takes effective part in the interaction (see Fig. 4.6*a*), the polarizability matrix element α_{01} of (4.79) is in good approximation given by

$$\alpha_{01} \approx \frac{2}{\hbar}\frac{d_{02}d_{21}}{\omega_{20}}.$$

Assuming transition moments d_{02} and d_{21} of $\sim 10^{-29}$ A s m and transition frequencies ω_{20} of $\sim 10^{16}$ s^{-1}, which are typical of strong electronic transitions in the ultraviolet, the polarizability matrix element α_{01} is estimated to be $\sim 10^{-40}$ A s m^2/V. Substituting these values together with $\omega_{10} \sim 10^{16}$ s^{-1}, $n^{(L)} \sim 1$, and $\tau_{01} \sim 10^{-13}$ s into (4.82), we obtain

$$\left(\tilde{\sigma}_{01}^{(2)}\right)_{\max} \sim 10^{-57} \text{ m}^4 \text{ s}.$$

To attain cross sections $\tilde{\sigma}_{\max}^{(2)}$ that are of the same order of magnitude as the one-photon cross sections $\sigma_{\max}^{(1)} \sim 10^{-20}$ m^2 of strong electronic transitions, photon fluxes I of about 10^{37} m^{-2} s^{-1} are needed. This corresponds to an intensity of about 5×10^{18} W/m^2. But at such high intensities, which are equivalent to fields of the order of the inner molecular fields, the assumptions of the perturbation treatment are no longer justified and the atoms and molecules are destroyed. At a photon flux I of $\sim 10^{30}$ m^{-2} s^{-1} (equivalent to intensities of about 5×10^{11} W/m^2), which is rather high but not destructive for typical molecules and atoms, the cross section $\sigma_{\max}^{(2)} = \tilde{\sigma}_{\max}^{(2)} \cdot I$ together with the highest possible population density γ of about 10^{28} m^{-3} gives the two-photon absorption coefficient as

$$\left(k_a^{(2)}\right)_{\max} = \gamma \tilde{\sigma}_{\max}^{(2)} \cdot I \sim 10 \text{ m}^{-1}.$$

Thus a rather thick sample is needed to observe large absorption. In most experiments performed to determine $\tilde{\sigma}^{(2)}$, a smaller population density has to be chosen in order to avoid interactions between the active molecules, and hence very sensitive methods for measuring small losses are required (see Chapter 10).

It should be mentioned that the two-photon absorption cross section can be strongly enhanced if one of the intermediate levels lies between levels 0 and 1 near the one-photon resonance energy $\hbar\omega$ (see Fig. 4.6*b*). The enhancement factor $\omega_{20}^2/(\omega_{20} - \omega)^2$ may attain values as high as $\omega_{20}^2 \tau_{20}^2$. For $|\omega_{20} - \omega| \lesssim 1/\tau_{20}$ the calculations performed so far fail because in this case a two-step excitation has to be considered. Two-photon and multiphoton absorption will be treated in more detail in Chapter 10.

4.7. EQUATION OF MOTION FOR MEASURABLE QUANTITIES

In the previous sections expectation values of observables of atomic systems under the influence of electromagnetic radiation and relaxation have been calculated in the following way. We first solved the set of differential equations

$$\frac{d}{dt}\rho_{\alpha\beta} = -R_{\alpha\beta}\left(\rho_{\alpha\beta} - \rho_{\alpha\beta}^e\right) + \frac{1}{i\hbar}[\hat{H}, \hat{\rho}]_{\alpha\beta}, \tag{4.83}$$

where

$$\hat{H} = \hat{H}^0 + \hat{H}^C$$

and

$$R_{\alpha\beta} = \frac{1}{T_\alpha}\delta_{\alpha\beta} + \frac{1}{\tau_{\alpha\beta}}(1 - \delta_{\alpha\beta}),$$

which describes the atomic density matrix $\rho_{\alpha\beta}$ calculated with the energy eigenstates of $\hat{H}^0$ [cf. (4.48)]. In (4.83) the relaxation processes are characterized by the lifetimes T_α and the phase-decay times $\tau_{\alpha\beta}$. In the dipole approximation the coherent perturbation Hamiltonian $\hat{H}^C$ can be represented by $\hat{H}^C = -\hat{d}_\bullet E_\bullet(t)$. After solving (4.83), the expectation value $\langle \hat{X} \rangle$ of an observable $\hat{X}$, for instance the dipole moment, the polarization, or the population of a certain level, was calculated by substituting the result for $\rho_{\alpha\beta}(t)$ into

$$X(t) = \langle \hat{X} \rangle = \text{Tr}\{\hat{\rho}(t)\hat{X}\} \tag{4.84}$$

[cf. (A.75)].

In this section we shall take another path, following Refs. 29, 38, 41. The matrix elements of the density matrix and their time derivatives in the density matrix equation (4.83) can be replaced by certain expectation values. Thus we obtain equations of motion for measurable quantities. This procedure has the following advantages. The equations describing the motion of atomic systems contain only quantities of immediate experimental significance, and often an analogy with classical equations becomes apparent. Another advantage arises if in particular we replace the matrix elements of the density operator by atomic expectation values that occur in Maxwell's equations. Then we gain a consistent set of differential equations for the field strengths and the atomic expectation values, which allows a complete semiclassical description of the interaction between light and matter.

Let us explain the procedure under simple conditions. The atomic systems are supposed to be identical and equally oriented. With respect to the interaction with the radiation field applied, they can be described by two-level systems. The vector character of field and polarization is ignored, and so is the tensor character of the susceptibilities. Thus we may choose a one-component description. Later, some results that are valid under less restrictive assumptions will be given without proof.

If the observable $\hat{X}$ does not explicitly depend on time, the time derivative of its expectation value is given by

$$\frac{d}{dt}\langle \hat{X} \rangle = \text{Tr}\left\{\left(\frac{d}{dt}\hat{\rho}(t)\right)\hat{X}\right\}. \tag{4.85}$$

The trace may be evaluated by using the time derivatives of the density matrix elements in (4.83), where α and β are restricted to the values 0 and 1 because of the assumption of two-level systems. Thus the relation

$$\frac{d}{dt}\langle \hat{X} \rangle = \sum_{\alpha,\beta}\left(-R_{\alpha\beta}\left(\rho_{\alpha\beta} - \rho^{e}_{\alpha\beta}\right) + \frac{1}{i\hbar}[\hat{H},\hat{\rho}]_{\alpha\beta}\right)X_{\beta\alpha} \tag{4.86}$$

is obtained. We may write the second term more comprehensively as

$$\sum_{\alpha,\beta}[\hat{H},\hat{\rho}]_{\alpha\beta}X_{\beta\alpha} = \mathrm{Tr}\{[\hat{H},\hat{\rho}]\hat{X}\}.$$

Utilizing the invariance of the trace under cyclic permutation of the operators and employing (4.84), we get

$$\mathrm{Tr}\{[\hat{H},\hat{\rho}]\hat{X}\} = \mathrm{Tr}\{\hat{\rho}[\hat{X},\hat{H}]\} = \langle[\hat{X},\hat{H}]\rangle.$$

Hence the equation (4.86) may be rewritten as

$$\frac{d}{dt}\langle \hat{X} \rangle = \sum_{\alpha,\beta} - R_{\alpha\beta}\left(\rho_{\alpha\beta} - \rho^{e}_{\alpha\beta}\right)X_{\beta\alpha} + \frac{1}{i\hbar}\langle[\hat{X},\hat{H}]\rangle. \tag{4.87}$$

We use this relation in two special cases.

First, the matrix $X_{\beta\alpha}$ is assumed to be diagonal, that is, $X_{\alpha\beta} = X_{\alpha\alpha}\delta_{\alpha\beta}$. Substituting the definition of $R_{\alpha\beta}$ as well as $\langle \hat{X} \rangle = \sum_{\alpha,\beta}\rho_{\alpha\beta}X_{\alpha\beta}$ and $\langle \hat{X} \rangle^{e} = \sum_{\alpha}\rho^{e}_{\alpha\alpha}X_{\alpha\alpha}$ into (4.87), we obtain

$$\frac{d}{dt}\langle \hat{X} \rangle = -\frac{1}{T_{10}}(\langle \tilde{X} \rangle - \langle \hat{X} \rangle^{e}) + \frac{1}{i\hbar}\langle[\hat{X},\hat{H}]\rangle. \tag{4.88a}$$

Second, the matrix $X_{\alpha\beta}$ is assumed to contain only off-diagonal elements (i.e. $X_{\alpha\alpha} = 0$), and hence (4.87) implies

$$\frac{d}{dt}\langle \hat{X} \rangle = -\frac{1}{\tau_{10}}\langle \hat{X} \rangle + \frac{1}{i\hbar}\langle[\hat{X},\hat{H}]\rangle. \tag{4.88b}$$

In this case we also derive an equation for $(d^2/dt^2)\langle \hat{X} \rangle$, which is needed for several applications. The operator $\hat{X}$ is now assumed to commute with $\hat{H}^{C}$; the dipole moment may serve as an example. Differentiation of (4.88b) gives

$$\frac{d^2}{dt^2}\langle \hat{X} \rangle = -\frac{1}{\tau_{10}}\frac{d}{dt}\langle \hat{X} \rangle + \frac{1}{i\hbar}\frac{d}{dt}\langle[\hat{X},\hat{H}^{0}]\rangle. \tag{4.89}$$

The operator $\hat{X}' = [\hat{X},\hat{H}^{0}]$ does not depend on time and contains only off-diagonal elements. Hence it satisfies the requirements for (4.88b) to be applied; this leads to

$$\frac{d}{dt}\langle[\hat{X}, \hat{H}^0]\rangle = -\frac{1}{\tau_{10}}\langle[\hat{X}, \hat{H}^0]\rangle + \frac{1}{i\hbar}\langle[[\hat{X}, \hat{H}^0], \hat{H}]\rangle. \qquad (4.90)$$

Inserting (4.90) and (4.88b) into (4.89), we obtain

$$\left(\frac{d^2}{dt^2} + \frac{2}{\tau_{10}}\frac{d}{dt} + \frac{1}{\tau_{10}^2}\right)\langle\hat{X}\rangle = -\frac{1}{\hbar^2}\langle[[\hat{X}, \hat{H}^0], \hat{H}]\rangle. \qquad (4.91)$$

Occupation-number inversion and polarization of one-photon processes will now be treated as examples of such measurable quantities. In the case of homogeneously broadened transitions the expectation values of the occupation-number inversion and the polarization of the ensemble of two-level systems are given as

$$\gamma_I = \mathrm{Tr}\{\hat{\rho}\hat{\gamma}_I\} = \gamma(\rho_{11} - \rho_{00}) \qquad (4.92)$$

and

$$P = \gamma\,\mathrm{Tr}\{\hat{\rho}\hat{d}\} = \gamma(\rho_{01}d_{10} + \rho_{10}d_{01}), \qquad (4.93)$$

and hence they meet the requirements for the application of (4.88a) and (4.88b), respectively.

For the occupation-number inversion we obtain

$$\frac{d}{dt}\gamma_I + \frac{1}{T_{10}}(\gamma_I - \gamma_I^e) = \frac{1}{i\hbar}\langle[\gamma_I, \hat{H}^C]\rangle. \qquad (4.94)$$

The expression on the right-hand side is evaluated by using the operator equation

$$[\hat{\gamma}_I, \hat{H}^C] = \frac{2\gamma}{\hbar\omega_{10}}E(t)[\hat{d}, \hat{H}^0], \qquad (4.95)$$

the validity of which can be easily proved in the matrix representation. The expectation value of $[\hat{d}, \hat{H}^0]$ satisfies the relation

$$\frac{\gamma}{i\hbar}\langle[\hat{d}, \hat{H}^0]\rangle = \frac{d}{dt}P + \frac{1}{\tau_{10}}P. \qquad (4.96)$$

In its derivation, (4.88b) and $P = \gamma\langle d\rangle$ have been used. Substituting (4.94) and (4.95) into (4.93), we obtain

$$\frac{d}{dt}\gamma_I + \frac{1}{T_{10}}(\gamma_I - \gamma_I^e) = \frac{2}{\hbar\omega_{10}} E\left(\frac{d}{dt}P + \frac{1}{\tau_{10}}P\right). \tag{4.97}$$

Using the photon concept, this relation can be easily interpreted. The time average of the product $E(d/dt)P$ is nothing but the power per unit volume supplied to the medium by the radiation field (cf. Section 1.3). Hence the power per unit volume divided by the photon energy $\hbar\omega_{10}$ gives the number of photons lost in the medium per unit volume and unit time. The absorption of one photon changes the population inversion by $+2$. Thus (4.97) represents the energy conservation of the whole system if the relaxation term in it is neglected.

Concerning the polarization, (4.91) implies

$$\left(\frac{d^2}{dt^2} + \frac{2}{\tau_{10}}\frac{d}{dt} + \frac{1}{\tau_{10}^2}\right)P = -\frac{1}{\hbar^2}\gamma\left(\left\langle[[\hat{d}, \hat{H}^0], \hat{H}^0]\right\rangle - E\left\langle[[\hat{d}, \hat{H}^0], \hat{d}]\right\rangle\right). \tag{4.98}$$

Evaluation of the expectation values on the right-hand side gives

$$\gamma\left\langle[[\hat{d}, \hat{H}^0], \hat{H}^0]\right\rangle = \gamma\hbar^2\omega_{10}^2\langle\hat{d}\rangle = \hbar^2\omega_{10}^2 P$$

and

$$-\gamma E\left\langle[[\hat{d}, \hat{H}^0], \hat{d}]\right\rangle = 2\hbar\omega_{10}E\gamma(\rho_{11} - \rho_{00})|d_{10}|^2 = 2\hbar\omega_{10}|d_{10}|^2\gamma_I E.$$

Thus (4.98) yields

$$\left(\frac{d^2}{dt^2} + \frac{2}{\tau_{10}}\frac{d}{dt} + \omega_{10}^2 + \frac{1}{\tau_{10}^2}\right)P = -\frac{2\omega_{10}}{\hbar}|d_{10}|^2\gamma_I E. \tag{4.99}$$

On the assumption that $\omega_{10}\tau_{10} \gg 1$, the term $\tau_{10}^{-2}P$ can be neglected on the left-hand side of (4.99). Taking into account the vector character of field and polarization, we obtain the equation of motion for $P_i(t)$ as

$$\left(\frac{d^2}{dt^2} + \frac{2}{\tau_{10}}\frac{d}{dt} + \omega_{10}^2\right)P_i(t) = -\frac{\omega_{10}}{\hbar}\sum_j \gamma_I(t)(d_{i01}d_{j10} + d_{i10}d_{j01})E_j(t). \tag{4.100}$$

This equation represents a relation between $P_{\bullet}(t)$ and $E_{\bullet}(t)$ similar to that found for an ensemble of classical oscillators with the resonance frequency ω_{10} and the friction coefficient $2/\tau_{10}$ and with $2\hbar^{-1}\omega_{10}|d_{10}|^2$ in place of q^2/m where q and m are the oscillating charge and mass. The essential difference consists in the appearance of $-\gamma_I = \gamma(\rho_{00} - \rho_{11})$ in the quantum-mechanical relation in place of only γ in the classical one. This means that full analogy will only be obtained if all particles are in the ground state and $-\gamma_I = \gamma$.

The equations (4.97) and (4.100) provide a complete description of the ensemble under consideration. In conjunction with the wave equation of classical electrodynamics they allow a consistent treatment of the interaction between light and matter. If the density of active atomic systems is high, one has of course to take into account differences between local and macroscopic electromagnetic fields. The simplest procedure is to substitute $[(\varepsilon^{(L)} + 2)/3]E_{\bullet}$ for the field strength $E_{\bullet}$ in the atomic equations of motion and to substitute the effective polarization $[(\varepsilon^{(L)} + 2)/3]P_{\bullet}$ for $P_{\bullet}$ in the wave equation (cf. Section 1.2).

One might doubt whether it is useful to replace the first-order differential equations for the off-diagonal density matrix elements by a second-order differential equation for the polarization. The mathematical problems in these approaches are nearly the same, because we have to solve either two first-order differential equations for $\text{Re}\{\rho_{01}\}$ and $\text{Im}\{\rho_{01}\}$ or one second-order differential equation for the real physical polarization P. Hence the advantage gained by employing the procedure described is not a mathematical one but concerns the physical insight into the problem. The fact that both treatments are mathematically nearly equivalent becomes more obvious if we separate the fast-varying factors $\exp\{\mp i\omega t\}$ in the field strength

$$E(t) = \tfrac{1}{2}\widehat{E}(t)e^{-i\omega t} + \text{c.c.}$$

and the polarization

$$P(t) = \tfrac{1}{2}\widehat{P}(t)e^{-i\omega t} + \text{c.c.}$$

If the second derivatives of the slowly varying amplitudes $\widehat{P}(t)$ are neglected and if we assume $\omega_{10}\tau_{10} \gg 1$ and $|\omega_{10} - \omega| \ll \omega_{10}$, the real second-order differential equation (4.100) can be replaced by the complex first-order differential equation

$$\frac{d}{dt}\widehat{P}(t) + \left(\frac{1}{\tau_{10}} + i(\omega_{10} - \omega)\right)\widehat{P}(t) = -\frac{i}{\hbar}|d_{10}|^2\gamma_I(t)\widehat{E}(t) \quad (4.101\text{a})$$

or by two real first-order differential equations for the real and imaginary parts

$\hat{P}_1$ and $\hat{P}_2$ of the polarization amplitude $\hat{P} = \hat{P}_1 + i\hat{P}_2$, which are

$$\frac{d}{dt}\hat{P}_1 + \frac{1}{\tau_{10}}\hat{P}_1 - (\omega_{10} - \omega)\hat{P}_2 = \frac{1}{\hbar}|d_{10}|^2\gamma_I\hat{E}_2 \tag{4.101b}$$

and

$$\frac{d}{dt}\hat{P}_2 + \frac{1}{\tau_{10}}\hat{P}_2 + (\omega_{10} - \omega)\hat{P}_1 = -\frac{1}{\hbar}|d_{10}|^2\gamma_I\hat{E}_1. \tag{4.101c}$$

The source term in the differential equation (4.97) for γ_I may also be expressed by the amplitudes $\hat{P}$ and $\hat{E}$. Then this equation becomes

$$\frac{d}{dt}\gamma_I + \frac{1}{T_{10}}(\gamma_I - \gamma_I^e) = \frac{i}{2\hbar}(\hat{E}\hat{P}^* - \hat{E}^*\hat{P}). \tag{4.102}$$

Resonant multiphoton processes can be treated in the same way if the essential atomic properties can be described by the two-level approach, which is of course justified only if all intermediate states are far from resonance. Under this condition the multiphoton interaction between the effective two-level atoms and the radiation field may be described by effective interaction operators that contain terms being nonlinear in the field strength. Let us explain this procedure for two-photon resonant processes such as two-photon absorption and stimulated Raman scattering. Then the specific interaction Hamiltonian is quadratic in $E_\bullet$ and has the structure

$$\hat{H}^C = -\frac{1}{2}\sum_{i,j}\hat{\alpha}_{ij}E_iE_j, \tag{4.103}$$

where the generalized polarizability $\hat{\alpha}_{\bullet\bullet}$ acts as an interaction operator. For *two-photon absorption and emission*, $\hat{H}^C$ is represented by

$$\hat{H}^C_{\text{TPA}} = -\frac{1}{2}\sum_{i,j}\hat{\alpha}^{\text{TPA}}_{ij}E_i^{(-)}E_j^{(-)} \tag{4.104a}$$

and

$$\hat{H}^C_{\text{TPE}} = \left(\hat{H}^C_{\text{TPA}}\right)^\dagger = -\frac{1}{2}\sum_{i,j}\left(\hat{\alpha}^{\text{TPA}}_{ij}\right)^\dagger E_i^{(+)}E_j^{(+)}. \tag{4.104b}$$

For the *Raman effect*, $\hat{H}^C$ is given by

$$\hat{H}^C_R = \sum_{i,j}\left[-\tfrac{1}{2}\hat{\alpha}^R_{ij}E_i^{(-)}E_j^{(+)} - \tfrac{1}{2}\left(\hat{\alpha}^R_{ij}\right)^\dagger E_i^{(+)}E_j^{(-)}\right]. \tag{4.104c}$$

The expectation value of $\hat{\alpha}$ can be obtained by applying (4.91). The result is

$$\left(\frac{d^2}{dt^2} + \frac{2}{\tau_{10}}\frac{d}{dt} + \omega_{10}^2\right)\langle\hat{\alpha}_{ij}\rangle$$

$$= -\frac{1}{2}\frac{\omega_{10}}{\hbar}\sum_{k,l}\left[(\alpha_{ij})_{01}(\alpha_{kl})_{10}\right.$$

$$\left.+(\alpha_{ij})_{10}(\alpha_{kl})_{01}\right]E_kE_l(\rho_{11}-\rho_{00}). \quad (4.105)$$

In place of $\langle\hat{\alpha}_{ij}\rangle$ we may use a macroscopic expectation value $Q_{ij} = \gamma\langle\hat{\alpha}_{ij}\rangle$ together with $\gamma_I = \gamma(\rho_{11} - \rho_{00})$. These two macroscopic parameters must satisfy the equations

$$\left(\frac{d^2}{dt^2} + \frac{2}{\tau_{10}}\frac{d}{dt} + \omega_{10}^2\right)Q_{ij}$$

$$= -\frac{1}{2}\frac{\omega_{10}}{\hbar}\gamma_I\sum_{k,l}\left((\alpha_{ij})_{10}(\alpha_{kl})_{01} + (\alpha_{ij})_{01}(\alpha_{kl})_{10}\right)E_kE_l \quad (4.106a)$$

and

$$\frac{d}{dt}\gamma_I + \frac{1}{T_{10}}(\gamma_I - \gamma_I^e) = \frac{1}{\hbar\omega_{10}}\sum_{i,j}\left(\frac{d}{dt}Q_{ij}\right)E_iE_j. \quad (4.106b)$$

If all molecules are in the ground state ($\gamma_I = -\gamma$), full analogy with the classical oscillator equation of polarizable molecules is achieved. The analogy requires that $|\alpha_{10}|^2\hbar^{-1}\omega_{10} = \alpha_1^2/2$, where α_1 is the first derivative of the polarizability with respect to the normal coordinate of the oscillator vibration.

The procedure described may also be applied if the molecular transition $0 \to 1$ is acted upon by one- and two-photon transitions simultaneously. Of great interest are transitions that are infrared- and Raman-active. Phonon polaritons are connected with such transitions. Utilizing the interaction operator

$$\hat{H}^C = -\hat{d}_{\bullet}E_{\bullet} - \tfrac{1}{2}\hat{\alpha}_{\bullet\bullet}E_{\bullet}E_{\bullet},$$

we obtain the equations of motion as

$$\left(\frac{d^2}{dt^2} + \frac{2}{\tau_{10}}\frac{d}{dt} + \omega_{10}^2\right)\tilde{P}_i$$
$$= -\frac{\omega_{10}}{\hbar}\gamma_I\left\{\sum_j \left(d_{i01}d_{j10} + d_{i10}d_{j01}\right)E_j + \frac{1}{2}\sum_{j,k}\left(d_{i01}\alpha_{jk10} + d_{i10}\alpha_{jk01}\right)E_jE_k\right\},$$

(4.107a)

$$\left(\frac{d^2}{dt^2} + \frac{2}{\tau_{10}}\frac{d}{dt} + \omega_{10}^2\right)Q_{ij}$$
$$= -\frac{\omega_{10}}{\hbar}\gamma_I\left\{\sum_k \left((\alpha_{ij})_{01}d_{k10} + (\alpha_{ij})_{10}d_{k01}\right)E_k + \frac{1}{2}\sum_{k,l}\left((\alpha_{ij})_{01}(\alpha_{kl})_{10} + (\alpha_{ij})_{10}(\alpha_{kl})_{01}\right)E_kE_l\right\},$$

(4.107b)

$$\frac{d}{dt}\gamma_I + \frac{1}{T_{10}}(\gamma_I - \gamma_I^e) = \frac{2}{\hbar\omega_{10}}\left\{\sum_i E_i\frac{d}{dt}\tilde{P}_i + \frac{1}{2}\sum_{i,j}\left(\frac{d}{dt}Q_{ij}\right)E_iE_j\right\}, \qquad (4.107c)$$

and

$$P_i = \tilde{P}_i + \sum_j Q_{ij}E_j. \qquad (4.107d)$$

If we want to drop the assumption that all molecules are identically oriented, we simply have to take orientation averages of the products of transition moments and polarization matrix elements in the final equations for γ_I, $P_\bullet$, and $Q_{\bullet\bullet}$.

So far we have treated a homogeneously broadened transition, which means that all molecules have been assumed to possess the same transition frequency ω_{10}. To deal with inhomogeneously broadened transitions we divide the ensemble of molecules into subensembles with transition frequencies in narrow intervals $(\omega_{10}', \omega_{10}' + d\omega_{10}')$. These subensembles can be considered as homogeneously broadened and hence can be treated in the way described. For instance, using for the one-photon transitions a one-component description as before, we obtain

$$\left(\frac{d^2}{dt^2} + \frac{2}{\tau_{10}}\frac{d}{dt} + (\omega'_{10})^2\right)P'(\omega'_{10}, t)$$

$$= -\frac{2\omega_{10}}{\hbar}\gamma'_I(\omega'_{10}, t)|d_{10}|^2 E(t) \qquad (4.108a)$$

and

$$\frac{d}{dt}\gamma'_I(\omega'_{10}, t) + \frac{1}{T_{10}}[\gamma_I(\omega'_{10}, t) - \gamma^e_I(\omega'_{10})] = \frac{2}{\hbar\omega_{10}}E\frac{d}{dt}P'(\omega'_{10}, t), \qquad (4.108b)$$

where $P'(\omega'_{10}, t)$ and $\gamma'_I(\omega'_{10}, t)$ are the contributions of all molecules with transition frequencies between ω'_{10} and $\omega'_{10} + d\omega'_{10}$ to polarization and population inversion, respectively. The total polarization and population inversion are obtained by summing all contributions of atomic subgroups or by integrating them over the distribution of transition frequencies $g_{\text{inh}}(\omega'_{10} - \omega_{10})$ around the center frequency ω_{10}. Then we have

$$P(t) = \int_0^\infty d\omega'_{10}\, g_{\text{inh}}(\omega'_{10} - \omega_{10})P'(\omega'_{10}, t) \qquad (4.108c)$$

and

$$\gamma_I(t) = \int_0^\infty d\omega'_{10}\, g_{\text{inh}}(\omega'_{10} - \omega_{10})\gamma'_I(\omega'_{10}, t). \qquad (4.108d)$$

Inhomogeneously broadened multiphoton transitions may be treated analogously.

4.8. DESCRIPTION OF TWO-LEVEL SYSTEMS IN ANALOGY TO THE BLOCH EQUATIONS

As outlined before, a full description of a two-level system is possible by making use of its density matrix elements ρ_{00}, ρ_{11}, ρ_{10}, and ρ_{01} as functions of time. Since $\rho_{00} + \rho_{11} = 1$ and $\rho_{01} = \rho^*_{10}$, three independent real quantities are required. In Section 4.7 equivalent descriptions utilizing expectation values of certain observables were given. Now we shall choose three real quantities, the use of which allows an interesting insight into the physical features of the equations of motion, though from the standpoint of mathematics there is nothing new compared with the treatment in Section 4.7. They are

$$R_1 = \rho_{01} + \rho_{10},$$

$$R_2 = -i(\rho_{01} - \rho_{10}),$$

$$R_3 = -\rho_{00} + \rho_{11}. \tag{4.109}$$

Introducing the abbreviations

$$\Omega_1 = \frac{1}{\hbar}(H_{01} + H_{10}),$$

$$\Omega_2 = \frac{-i}{\hbar}(H_{01} - H_{10}),$$

$$\Omega_3 = \frac{1}{\hbar}(H_{11} - H_{00}) = \omega_{10}, \tag{4.110}$$

where $H_{\alpha\beta} = \langle \mathscr{E}_\alpha | \hat{H} | \mathscr{E}_\beta \rangle$ is the matrix element of the Hamiltonian of (4.83), and considering the R_i and Ω_i as "vector" components, we obtain from (4.83) the "vector" equation

$$\frac{dR_\bullet}{dt} = \Omega_\bullet \times R_\bullet + \left(\frac{d}{dt} R_\bullet \right)_{\text{Rel}}, \tag{4.111}$$

where

$$\left(\frac{d}{dt} R_{1,2} \right)_{\text{Rel}} = -\frac{1}{\tau_{10}} R_{1,2}$$

and

$$\left(\frac{d}{dt} R_3 \right)_{\text{Rel}} = -\frac{1}{T_{10}} (R_3 - R_3^e).$$

Obviously, (4.111) is formally equivalent to the Bloch equations, which describe the interaction of time-dependent magnetic fields with the two-level system that results from one atomic level by splitting in a permanent magnetic field (see e.g. Refs. 4.7, 12, 42). Hence, with neglect of the relaxation terms, a simple geometrical interpretation of (4.111) may be given: $R_\bullet(t)$ rotates about the "vector" $\Omega_\bullet(t)$ with the angular velocity $\Omega(t)$, $R_\bullet(t)R_\bullet(t)$ remaining constant (see Fig. 4.9). This conservation law of the "vector" length is equivalent to $\rho_{00} + \rho_{11} = \text{const}$. [The transformation of the equations of motion into the form of the Bloch equations is not unique. Some authors use another choice of signs in defining the R_i and the Ω_i (see, e.g., Ref. 42).]

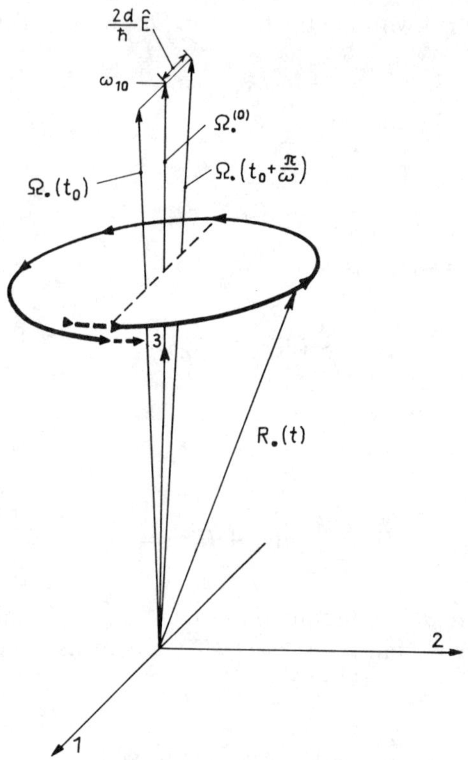

Fig. 4.9. Precession of the "vector" $R_\bullet(t)$ about $\Omega_\bullet(t)$ (for comparison $\Omega_\bullet^0$, which is parallel to axis 3, is also shown, for $t_0 \le t \le t_0 + 2\pi/\omega$ the path of $R_\bullet(t)$ is represented by the full line.)

Under the condition of an irradiation with linearly polarized light, we shall prove that (4.111) is equivalent to the equations of motion derived in the previous section. Without loss of generality d_{01} is assumed to be real. Writing $H_{01} = H_{10} = -d_{01}E$, we obtain from (4.111)

$$\frac{d}{dt}R_1 = -\omega_{10}R_2 - \frac{1}{\tau_{10}}R_1, \tag{4.112a}$$

$$\frac{d}{dt}R_2 = +\omega_{10}R_1 + \frac{2d_{01}}{\hbar}R_3E - \frac{1}{\tau_{10}}R_2, \tag{4.112b}$$

$$\frac{d}{dt}R_3 = -\frac{2d_{01}}{\hbar}ER_2 - \frac{1}{T_{10}}(R_3 - R_3^e). \tag{4.112c}$$

In the case of homogeneously broadened transitions the macroscopic expectation values P and γ_I are connected with the R_i by

$$P = \gamma d_{01} R_1 \tag{4.113}$$

and

$$\gamma_I = \gamma R_3.$$

The parameter R_2 has no direct macroscopic equivalent, and we may eliminate it in (4.112). For this purpose (4.112a) is differentiated and $(d/dt)R_2$ from the second equation is inserted. Under the condition that $\omega_{10}\tau_{10} \gg 1$, the term $\omega_{10}\tau_{10}^{-1}R_2$ is small compared with others in the resulting equation. Therefore it can be replaced by utilizing the approximation

$$\frac{d}{dt}R_1 \approx -\omega_{10}R_2, \tag{4.114}$$

which follows from (4.112a) under the condition mentioned. In (4.112c), the parameter R_2 can also be replaced by employing (4.114). Thus we have

$$\frac{d^2}{dt^2}R_1 + \frac{2}{\tau_{10}}\frac{d}{dt}R_1 + \omega_{10}^2 R_1 = -\frac{2d_{01}\omega_{10}}{\hbar}ER_3 \tag{4.115a}$$

and

$$\frac{d}{dt}R_3 + \frac{1}{T_{10}}(R_3 - R_3^e) = -\frac{2d_{01}}{\hbar\omega_{10}}E\frac{d}{dt}R_1. \tag{4.115b}$$

Substituting P and γ_I for R_1 and R_3 in (4.115), the relations (4.97) and (4.99) result.

The results obtained so far in this subsection are equivalent to the operator relations discussed at the end of Section A.1.2.5.

Let us now consider the case of inhomogeneously broadened transitions. In (4.109)–(4.115) the $R_i(t)$ are replaced by $R_i'(\omega_{10}', t)$, where the primed components refer to the atoms with transition frequencies between ω_{10}' and $\omega_{10}' + d\omega_{10}'$. As in the previous section, the total macroscopic occupation-number inversion and polarization are now calculated by summing the contributions of all the atomic subgroups with different ω_{10}'. Thus the occupation-number inversion

$$\gamma_I = \gamma\int_0^\infty d\omega_{10}'\, g_{\text{inh}}(\omega_{10}' - \omega_{10})R_3'(\omega_{10}') \tag{4.116}$$

is obtained, where $g_{\text{inh}}(\omega_{10}' - \omega_{10})$ is the line-shape function of the inhomogeneously broadened transition.

Because of their comparatively simple geometrical interpretation Bloch's equations are very suitable for discussing the rotating-wave approximation already introduced in Section 3.2.2. If the radiation field is absent, the "vector" R_i rotates about axis 3 with the angular frequency ω_{10} (see Fig. 4.9).

Now we consider the influence of a linearly polarized electric field of strength

$$E(t) = \widehat{E}(t)\cos[\omega t + \varphi(t)], \tag{4.117}$$

using here the real amplitude $\widehat{E}(t)$ and the phase $\varphi(t)$, both slowly varying with time and space coordinates. [According to the notation used by us in other places, we should denote the real amplitude function in (4.117) by $|\widehat{E}(t)|$. Since however no confusion can arise in this section, we omit the modulus sign.] Following the procedure of high-frequency magnetic spectroscopy, this field may be expanded in two fields E^{c} and E^{cc} of equal amplitudes $\widehat{E}/2$ rotating with instantaneous angular frequencies of absolute value $|\omega + d\varphi/dt|$ clockwise and counterclockwise about axis 3 if one looks in the -3 direction onto the 12 plane of Fig. 4.9. This means that the "vector" $\Omega_\bullet$ is split into

$$\Omega_\bullet = \Omega_\bullet^0 + \Omega_\bullet^{cc} + \Omega_\bullet^{c}, \tag{4.118}$$

where

$$\Omega_\bullet^0 = (0, 0, \omega_{10}),$$

$$\Omega_\bullet^{cc} = \left(-\frac{d_{10}}{\hbar}\widehat{E}\cos[\omega t + \varphi(t)], -\frac{d_{10}}{\hbar}\widehat{E}\sin[\omega t + \varphi(t)], 0\right),$$

$$\Omega_\bullet^{c} = \left(-\frac{d_{10}}{\hbar}\widehat{E}\cos[\omega t + \varphi(t)], +\frac{d_{10}}{\hbar}\widehat{E}\sin[\omega t + \varphi(t)], 0\right).$$

The field strength is supposed to be not too high; more precisely, the relation $d_{10}\widehat{E}(t) \ll \hbar\omega_{10}$ should be satisfied. Thus the fast motion of the "vector" $R_\bullet$ is only slightly affected by the field. If the radiation field is nearly in resonance with the atomic transition ($\omega \approx \omega_{10}$), the component E^{cc} rotates with nearly the same angular velocity about axis 3 as does $R_\bullet$, whereas E^{c} propagates in the opposite direction and thus rotates relative to $R_\bullet$ with a very high angular velocity of about $\omega_{10} + \omega \approx 2\omega$. Hence the action of E^{c} on the "vector" components R_i changes their sign very rapidly, and as a consequence they have no appreciable influence on the time average. [For example, in the dipole approximation the energy exchange between field and the atoms is characterized by the term $\overline{d_{10}R_1(t)E(t)}^{\,t}$ averaged over some periods of the radiation. This averaging procedure leads to vanishingly small contributions of the field component E^{c} because of the high angular frequency of $d_{10}R_1E^{c}$.]

For these reasons we neglect the field component E^{c} from the very beginning. This procedure is known as the rotating-wave approximation. In this approximation we may neglect the action of E^{c} in the equation of motion

(4.111), and hence we obtain

$$\frac{d}{dt}R_1(t) = -\omega_{10}R_2(t) - \frac{d_{10}}{\hbar}\hat{E}(t)R_3(t)\sin[\omega t + \varphi(t)] - \frac{1}{\tau_{10}}R_1(t), \tag{4.119a}$$

$$\frac{d}{dt}R_2(t) = \omega_{10}R_1(t) + \frac{d_{10}}{\hbar}\hat{E}(t)R_3(t)\cos[\omega t + \varphi(t)] - \frac{1}{\tau_{10}}R_2(t), \tag{4.119b}$$

$$\frac{d}{dt}R_3(t) = -\frac{d_{10}}{\hbar}\hat{E}(t)\{R_2(t)\cos[\omega t + \varphi(t)] - R_1(t)\sin[\omega t + \varphi(t)]\} - \frac{1}{T_{10}}[R_3(t) - R_3^e]. \tag{4.119c}$$

The small influence of the field component E^c on the resonance frequency, which was first calculated by Bloch and Siegert (Ref. 4.8), is in the frequency domain only of the order of $(d_{10}\hat{E})^2/\hbar^2\omega_{10}$ and can thus be neglected in comparison with ω_{10} in most cases of interest.

In the rotating-wave approximation it is useful to describe the interaction between field and atom in the new coordinate system $(1', 2', 3')$ rotating at the angular velocity $\omega + (d/dt)\varphi(t)$ about the axis 3 of the old system and being aligned with its axis $1'$ along the direction of the field component E^{cc} (see Fig. 4.9). Thus the transformation between the two coordinate systems is

$$R_1 = R_{1'}\cos[\omega t + \varphi(t)] - R_{2'}\sin[\omega t + \varphi(t)], \tag{4.120a}$$

$$R_2 = +R_{1'}\sin[\omega t + \varphi(t)] + R_{2'}\cos[\omega t + \varphi(t)], \tag{4.120b}$$

$$R_3 = R_{3'}. \tag{4.120c}$$

The main advantage of the new system $(1', 2', 3')$ is that $R_{1'}$ and $R_{2'}$ are the components in phase and 90° out of phase with the electric field, respectively. The same holds for the corresponding components of the transition moment $d_{12}R_{1'}$ and $d_{12}R_{2'}$.

The Bloch equations in the new coordinate system may be derived by using (4.120) in (4.119). This gives

$$\frac{d}{dt}R_{1'} = \left(\Delta + \frac{d}{dt}\varphi\right)R_{2'} - \frac{R_{1'}}{\tau_{10}}, \tag{4.121a}$$

$$\frac{d}{dt}R_{2'} = -\left(\Delta + \frac{d}{dt}\varphi\right)R_{1'} + \frac{d_{10}}{\hbar}\hat{E}R_{3'} - \frac{R_{2'}}{\tau_{10}}, \tag{4.121b}$$

$$\frac{d}{dt}R_{3'} = -\frac{d_{10}}{\hbar}\hat{E}R_{2'} - \frac{R_{3'} + 1}{T_{10}}, \tag{4.121c}$$

where $\Delta = \omega - \omega_{10}$ is the detuning from resonance and where R_3^e has been set equal to -1. Compared with (4.119) the differential equations (4.121) contain only terms varying slowly with time. This proves to be useful in many applications.

Instead of the microscopic equations of motion (4.121) we may derive relations for the macroscopic polarization and population inversion. In the case of homogeneously broadened transitions, multiplication of (4.121a, b) and (4.121c) by the factors γd_{10} and γ, respectively, gives

$$\frac{d}{dt}\hat{P}_1(t) = -\left(\Delta + \frac{d}{dt}\varphi(t)\right)\hat{P}_2(t) - \frac{1}{\tau_{10}}\hat{P}_1(t), \tag{4.122a}$$

$$\frac{d}{dt}\hat{P}_2(t) = \left(\Delta + \frac{d}{dt}\varphi(t)\right)\hat{P}_1(t)$$

$$-\frac{1}{\hbar}d_{10}^2\hat{E}(t)\gamma_I(t) - \frac{1}{\tau_{10}}\hat{P}_2(t), \tag{4.122b}$$

$$\frac{d}{dt}\gamma_I(t) = \frac{1}{\hbar}\hat{E}(t)\hat{P}_2(t) - \frac{1}{T_{10}}(\gamma_I(t) + \gamma), \tag{4.122c}$$

where the polarization amplitudes

$$\hat{P}_1(t) = \gamma d_{10}R_{1'}(t), \tag{4.122d}$$

$$\hat{P}_2(t) = -\gamma d_{10}R_{2'}(t) \tag{4.122e}$$

have been introduced, and the total polarization is

$$P(t) = \gamma d_{10}R_1(t) = P_1(t) + P_2(t), \tag{4.122f}$$

where

$$P_1(t) = \hat{P}_1(t)\cos[\omega t + \varphi(t)], \tag{4.122g}$$

$$P_2(t) = \hat{P}_2(t)\sin[\omega t + \varphi(t)]. \tag{4.122h}$$

$P_1(t)$ and $P_2(t)$ are the polarization components in phase and out of phase by 90° with the electric field.

In the case of inhomogeneously broadened transitions we get by an analogous procedure

$$\frac{d}{dt}\hat{P}'_1(\omega'_{10},t) = -\left(\Delta' + \frac{d}{dt}\varphi(t)\right)\hat{P}'_2(\omega'_{10},t) - \frac{1}{\tau_{10}}\hat{P}'_1(\omega'_{10},t), \tag{4.123a}$$

$$\frac{d}{dt}\hat{P}'_2(\omega'_{10},t) = \left(\Delta' + \frac{d}{dt}\varphi(t)\right)\hat{P}'_1(\omega'_{10},t) - \frac{1}{\tau_{10}}d_{10}^2\hat{E}(t)\gamma'_I(\omega'_{10},t) - \frac{1}{\tau_{10}}\hat{P}'_2(\omega'_{10},t), \tag{4.123b}$$

$$\frac{d}{dt}\gamma'_I(\omega'_{10},t) = \frac{1}{\hbar}\hat{E}(t)\hat{P}'_2(\omega'_{10},t) - \frac{1}{T_{10}}\left[\gamma'_I(\omega'_{10},t) - (\gamma_I^e)'\right], \tag{4.123c}$$

where

$$\gamma_I(t) = \int d\omega'_{10}\, g_{\text{inh}}(\omega'_{10} - \omega_{10})\gamma'_I(\omega'_{10},t), \tag{4.123d}$$

$$\hat{P}_1(t) = \int d\omega'_{10}\, g_{\text{inh}}(\omega'_{10} - \omega_{10})\hat{P}'_1(\omega'_{10},t), \tag{4.123e}$$

$$\hat{P}_2(t) = \int d\omega'_{10}\, g_{\text{inh}}(\omega'_{10} - \omega_{10})\hat{P}'_2(\omega'_{10},t), \tag{4.123f}$$

$$\Delta' = \omega - \omega'_{10}, \tag{4.123g}$$

and

$$P(t) = \hat{P}_1(t)\cos[\omega t + \varphi(t)] + \hat{P}_2(t)\sin[\omega t + \varphi(t)]. \tag{4.123h}$$

REFERENCES

4.1. J. A. Armstrong, N. Bloembergen, J. Ducuing, and P. S. Pershan, *Phys. Rev. A* **127**, 1918 (1962).

4.2. N. Bloembergen and Y. R. Shen, *Phys. Rev. A* **133**, 37 (1964).

4.3. H. E. Puthoff, M. L. Report No. 1547, Stanford Univ., 1967.

4.4. R. W. Terhune and P. D. Maker, *Phys. Rev. A* **137**, 801 (1965).

4.5. P. S. Pershan, "Nonlinear Optics," in E. Wolf (Ed.), *Progress in Optics*, North Holland, Amsterdam, 1966.

4.6. D. A. Kleinman, *Phys. Rev.* **126**, 1977 (1962).

4.7. A. Abragam, *The Principles of Nuclear Magnetism*, Clarendon, Oxford, 1961.

4.8. F. Bloch and A. J. F. Siegert, *Phys. Rev.* **57**, 522 (1940).

5

Statistical and Coherence Properties of the Radiation Field and Their Measurement

There is a mutual relationship between nonlinear optical effects and the statistical and coherence properties of radiation. This fact will be made clear in the discussion of nonlinear one-photon and multiphoton processes in Part II. To this end we need the characterization of the field by means of correlation functions and the coherence properties associated with them. The analysis of radiation detection will show that the properties of the radiation field enter into the measurement data of the photodetector via correlation functions. Photodetection may be regarded as the interaction of radiation with a special atomic system, the photodetector. This interaction changes the state of the atomic system, and from the measurement of this change conclusions on the state of the radiation field can be drawn. Various types of radiation receivers are used. To measure light, in particular its statistical features, photodetectors working on the basis of the external photoeffect have gained great importance in practice. On the one hand, they possess a small time constant and a high detectivity. On the other hand, from the macrophysical measurement results they allow one to draw clear, direct conclusions on the state of the radiation field. This aspect will be dealt with in Section 5.1, and we shall find out how the state of the radiation field determines the measurement results by means of special correlation functions. The general description of the statistical and coherence properties of the radiation field will be given in Section 5.2. There we shall also discuss experimental aspects of measurement and significant experimental results.

5.1. PHOTODETECTION ON THE BASIS OF THE EXTERNAL PHOTOEFFECT

On the basis of the photoelectric detection process in one photodetector the photon-counting distribution is now derived, which is associated with measurement data. Moreover, photoelectric counting by several photodetectors is described.

5.1.1. Photoelectric Counting by a Photocell. The Photon-Counting Distribution

Light releases electrons, the so-called photoelectrons, from the cathode of the detector. The photocathode is composed of atoms, which are assumed to be initially in their ground state. The release of the photoelectrons is caused by the absorption of light quanta by the atoms. We have already discussed the one-photon absorption process connected with transitions between discrete atomic energy levels (Section 3.3.1). Now, in the case of the external photoeffect, we have to treat the transition from the ground level to a level in the continuous part of the atomic spectrum. From these considerations an insight can be gained into how the parameters characterizing the radiation field are associated with the values measured by photon counting.

We model the detector as an ensemble of numerous atoms of the same kind. First we seek the probability that one of the detector atoms ejects a photoelectron due to photon absorption. Above the energy value $\mathscr{E}_l$ an atom at the position $r_\bullet$ may possess a continuous spectrum of one-electron states $|\mathscr{E}_c\rangle$ with the state density $\sigma(\mathscr{E}_c)$ per unit energy. If the atom takes one of the state $|\mathscr{E}_c\rangle$, the electron can be thought of as being released from the atom; it can then be extracted by an external field, so that it yields a contribution to the photoelectric current. Let us consider the two-level system consisting of the ground state $|\mathscr{E}_g\rangle$ and an arbitrary state $|\mathscr{E}_c\rangle$ of the continuous part. According to the general description of a two-level system (Section A.2.5), its dipole-moment operator is given by

$$\hat{d} = d_{cg}\hat{B}_{cg} + \text{h.c.}, \tag{5.1}$$

where d_{cg} is the transition dipole moment and $\hat{B}_{cg}$ the flip operator for the transition from the lower to the upper level. The atom is assumed to be acted on by the field $\hat{E}(r_\bullet)$. Then, under the dipole approximation, the interaction operator $\hat{H}^c$ becomes $-\hat{d}\hat{E}(r_\bullet)$. We want to evaluate the transition probability and take into account only contributions that become appreciable in times much greater than $\hbar/(\mathscr{E}_c - \mathscr{E}_g)$. Then the interaction operator can be written as

$$\hat{H}^C = -d_{cg}\hat{B}_{cg}\hat{E}^{(-)}(r_\bullet) - d_{cg}^*\hat{B}_{cg}^\dagger\hat{E}^{(+)}(r_\bullet), \tag{5.2}$$

where terms giving rise to a fast time dependence are omitted, which corresponds to the rotating-wave approximation (cf. Section 3.2.2). $\hat{E}^{(-)}(r_\bullet)$ is the operator of the complex analytic field signal, and $\hat{E}^{(+)}(r_\bullet)$ its hermitian conjugate. So far, we have tacitly ignored the vector character of the dipole moment and the field strength, and we suppose $\hat{E}(r_\bullet)$ to represent a one-component field in a certain polarization direction; $\hat{d}$ is the vector component of the dipole moment in this direction. It is assumed that at $t = 0$ the (normalized) state $|\Phi(0)\rangle$ of the total system factorizes into the state $|\mathscr{E}_g\rangle$ of the atom and the photon state $|\psi(0)\rangle$; consequently

$$|\Phi(0)\rangle = |\mathscr{E}_g\rangle|\psi(0)\rangle. \tag{5.3}$$

We next aim at the probability $p(|\mathscr{E}_c\rangle, t)$ of finding the atom at $t > 0$ in the state $|\mathscr{E}_c\rangle$, independent of the state $|\mathscr{E}_R\rangle$ taken by the radiation field after one photon has been absorbed. According to general rules we have

$$\begin{aligned} p(|\mathscr{E}_c\rangle, t) &= \sum_{|\mathscr{E}_R\rangle} {}_I\langle\Phi(t)|\mathscr{E}_c\rangle|\mathscr{E}_R\rangle\langle\mathscr{E}_R|\langle\mathscr{E}_c|\Phi(t)\rangle_I \\ &= {}_I\langle\Phi(t)|\mathscr{E}_c\rangle\langle\mathscr{E}_c|\Phi(t)\rangle_I, \end{aligned} \tag{5.4}$$

where $|\Phi(t)\rangle_I$ is the ket vector of the total system at time $t \geq 0$ in the interaction picture. Inserting the unitary operator $\hat{U}^C(t,0)$, which characterizes the time evolution in the interaction picture [cf. (A.114a)], we have

$$p(|\mathscr{E}_c\rangle, t) = \langle\psi(0)|\langle\mathscr{E}_g|[\hat{U}^C(t,0)]^{-1}|\mathscr{E}_c\rangle\langle\mathscr{E}_c|\hat{U}^C(t,0)|\mathscr{E}_g\rangle|\psi(0)\rangle. \tag{5.4a}$$

This expression will be evaluated by means of time-dependent perturbation theory up to the second order in $\hat{H}_I^C$. Taking into account that $\langle\mathscr{E}_g|\mathscr{E}_c\rangle = 0$, we arrive at

$$\begin{aligned} p(|\mathscr{E}_c\rangle, t) &= \hbar^{-2}|d_{cg}|^2 \int_0^t dt' \int_0^t dt'' \, e^{i\omega_{cg}(t'-t'')} \\ &\quad \times \langle\psi(0)|\hat{E}_I^{(+)}(\mathbf{r}_\bullet, t'')\hat{E}_I^{(-)}(\mathbf{r}_\bullet, t')|\psi(0)\rangle, \end{aligned} \tag{5.5}$$

where $\hbar\omega_{cg}$ is the transition energy $(\mathscr{E}_c - \mathscr{E}_g)$.

Let us generalize the assumption on the initial radiation state; instead of the pure photon state $|\psi(0)\rangle$ we now assume the field to be initially in a mixed state described by the density operator

$$\hat{\rho}(0) = \sum_{|\psi(0)\rangle} p_{|\psi(0)\rangle}|\psi(0)\rangle\langle\psi(0)|. \tag{5.6}$$

Correspondingly, we have to average the probability $p(|\mathscr{E}_c\rangle, t)$. Thus we obtain the mean value

$$\begin{aligned} \overline{\overline{p(|\mathscr{E}_c\rangle, t)}} &= \sum_{|\psi(0)\rangle} p_{|\psi(0)\rangle} p(|\mathscr{E}_c\rangle, t) \\ &= \hbar^{-2}|d_{cg}|^2 \int_0^t dt' \int_0^t dt'' \, e^{i\omega_{cg}(t'-t'')} \\ &\quad \times \mathrm{Tr}\{\hat{\rho}(0)\hat{E}_I^{(+)}(\mathbf{r}_\bullet, t'')\hat{E}_I^{(-)}(\mathbf{r}_\bullet, t')\}. \end{aligned} \tag{5.7}$$

The transition into the state $|\mathscr{E}_c\rangle$, which belongs to the continuous spectrum, means physically that a photoelectron has been contributed to the photocurrent. With respect to the photocurrent it is irrelevant from which state $|\mathscr{E}_c\rangle$ the photoelectron arises. Therefore we evaluate the total probability $P(t)$ by summing over all the states $|\mathscr{E}_c\rangle$; this operation may be replaced by the integration over the states of the continuous part of the spectrum:

$$P(t) = \sum_{|\mathscr{E}_c\rangle} \overline{\overline{p(|\mathscr{E}_c\rangle, t)}} = \int_{\mathscr{E}_I}^{\infty} d\mathscr{E}_c\, \sigma(\mathscr{E}_c) \overline{\overline{p(|\mathscr{E}_c\rangle, t)}}. \tag{5.8}$$

Obviously, this expression can be put into the form

$$P(t) = \int_0^t dt' \int_0^t dt''\, S(t'' - t')$$
$$\times \mathrm{Tr}\{\hat{\rho}(0)\hat{E}_I^{(+)}(r_\bullet, t'')\hat{E}_I^{(-)}(r_\bullet, t')\}. \tag{5.9}$$

The function $S(t'' - t')$ plays the role of a characteristic function in the time domain, that is, the Fourier transform of the frequency response. The second factor $\langle \hat{E}_I^{(+)}(r_\bullet, t'')\hat{E}_I^{(-)}(r_\bullet, t')\rangle$ of the integrand may be interpreted as the correlation function of the field variable. The probability $P(t)$ depends on the time evolution of both the factors.

A relatively simple result can be obtained if the frequency spectrum is assumed to be flat in the entire range. This physically characterizes the limiting case of a detector whose response function is exactly independent of frequency (ideal broadband detector). In this case $S(t'' - t')$ has the character of a delta function and we may insert the expression

$$S(t'' - t') = \frac{2\pi}{\hbar}\sigma|d|^2\delta(t'' - t') \tag{5.10}$$

into (5.9), where σ and $|d|^2$ are effective values of $\sigma(\hbar\omega_{cg})$ and $|d_{cg}|^2$. The integration with respect to t'' leads to

$$P(t) = \frac{2\pi}{\hbar}\sigma|d|^2 \int_0^t dt' \left\langle \hat{E}_I^{(+)}(r_\bullet, t')\hat{E}_I^{(-)}(r_\bullet, t')\right\rangle. \tag{5.11}$$

This result shows that under the assumption of an ideal broadband frequency response the probability $P(t)$ depends only on the value of the field correlation function at pairs of equal time arguments. In the general case of an appreciably varying frequency response function, the function $S(t'' - t')$ differs from a delta function, and thus also values of the correlation function at pairs of unequal time arguments are involved. Equation (5.9) reveals that in the general case $P(t)$ depends on the time differences $t'' - t'$ for which the correlation function $\langle \hat{E}^{(+)}(r_\bullet, t'')\hat{E}^{(-)}(r_\bullet, t')\rangle$ yields considerable contributions to the double integral; the maximum of these time differences is called the field correlation time.

In the following we shall assume that the conditions for an ideal broadband detector are approximately fulfilled, and therefore we shall utilize (5.11). The photocurrent—that is, the number of photoelectrons per unit time—is not determined by $P(t)$ itself but by the rate of $P(t)$; thus, in view of (5.11), we have

$$\frac{\Delta P}{\Delta t} = \frac{2\pi}{\hbar}\sigma|d|^2 \mathrm{Tr}\{\hat{\rho}(0)\hat{E}_I^{(+)}(r_\bullet, t)\hat{E}_I^{(-)}(r_\bullet, t)\}. \tag{5.12}$$

In deriving $\Delta P/\Delta t$ the statements on transition probabilities and time coarse-graining in Section A.2.3 have implicitly been used. To satisfy these conditions, Δt must be taken greater than the duration of microphysical "switching" (transient) processes. The right-hand side is interpreted as a time average over the interval Δt around t. Since operators in the interaction picture generally have the time dependence of the free system (cf. Section A.2.1), we may interpret the operator $\hat{E}_I^{(+)}(r_\bullet, t)\hat{E}_I^{(-)}(r_\bullet, t)$ as a characteristic of a radiation field which is free, that is to say, not affected by interaction with the detector atoms. Keeping in mind this fact, we may drop the interaction-picture index of the field quantities in the following.

So far we have only discussed the contribution of a single atom to the photoelectric current. To determine the effect of the total detector, we have to introduce a factor η, which comprises the joint action of all the detector atoms and depends on their number and positions. In this connection we imagine a pointlike detector at the position $r_\bullet$. If we let $p(t)\,\Delta t$ be the probability that any atom of the detector ejects a photoelectron as a result of photon absorption in the interval $(t, t + \Delta t)$, then we have

$$\begin{aligned} p(t)\,\Delta t &= \eta\,\mathrm{Tr}\{\hat{\rho}(0)\hat{E}^{(+)}(r_\bullet, t)\hat{E}^{(-)}(r_\bullet, t)\}\,\Delta t \\ &= \eta\langle\hat{I}(x)\rangle\,\Delta t. \end{aligned} \tag{5.13}$$

The interval duration Δt is assumed to be sufficiently small so that the probability may be neglected that more than one atom releases a photoelectron due to photon absorption within Δt; under physically reasonable conditions this requirement for an upper bound does not contradict the existence of the lower bound mentioned when discussing (5.12). The constant η is called the sensitivity factor of the photoelectric detector. The photoelectric current is proportional to the rate $p(t)$, and therefore, by (5.13), it is proportional to $\mathrm{Tr}\{\hat{\rho}(0)\hat{E}^{(+)}(r_\bullet, t)\hat{E}^{(-)}(r_\bullet, t)\}$. Hence this quantity may be interpreted as the radiation intensity $\langle\hat{I}(x)\rangle$ at the space–time point $x = x(r_\bullet, t)$. This interpretation agrees with the fact that the rate $p(t)$ vanishes if the density operator $\hat{\rho}(0)$ represents the photon vacuum, which is a state yielding no contribution to photodetection on the basis of the external photoeffect. Altogether, it turns out to be justifiable to regard the operator $\hat{E}^{(+)}(x)\hat{E}^{(-)}(x)$ as the operator $\hat{I}(x)$ for the radiation intensity at the space–time point x.

Note that $\hat{E}^{(+)}\hat{E}^{(-)}$ does not coincide with the operator $\hat{E}\hat{E}/2$, which must be regarded as the direct formal analogue of the corresponding classical quantity. The expectation value of $\hat{E}\hat{E}/2$ contains the zero-point field fluctuations (as can be concluded from Section 2.3.1), whereas the expectation value of $\hat{E}^{(+)}\hat{E}^{(-)}$ does not. Making use of the explicit expressions for $\hat{E}^{(-)}$ and $\hat{E}^{(+)}$ [cf. (2.35)], the expectation value of $\hat{E}^{(+)}\hat{E}^{(-)}$ can be connected with the radiation energy ζ (minus vacuum energy) per unit volume, or the normally incident energy $\tilde{\zeta}$ per unit time per unit area, by the formula

$$\langle \hat{E}^{(+)}\hat{E}^{(-)} \rangle = \frac{\zeta}{2\varepsilon_0} = \frac{\tilde{\zeta}}{2\varepsilon_0 c}. \tag{5.14}$$

For one-mode radiation of frequency ω_μ, the mean photon number $\langle \hat{N}_\mu \rangle$ is associated with $\zeta = (\hbar\omega_\mu/V)\langle \hat{N}_\mu \rangle$, where V is the normalization volume.

In the remainder of this subsection we shall discuss more practical problems of *photon counting*. The basic equation (5.13) for the probability of registering one photoelectron (photon) within the small interval $(t, t + \Delta t)$ by the detector can be exploited directly for the description of the statistics of photon counting. From (5.13) the probability $P_n(t, t + T)$ can be derived that in an *arbitrarily macrophysical* interval $(t, t + T)$ a certain number n of photoelectrons (photons) are counted by the detector. With respect to the derivation of $P_n(t, t + T)$ the position of the detector is irrelevant; thus, for brevity, we shall write $I(t)$ instead of $\langle \hat{I}(x) \rangle$. In general, the function $I(t)$ is represented in different intervals of the duration T by different realizations of a stochastic process; at the beginning we shall ignore these fluctuations.

Following Mandel (Ref. 5.1), we subdivide the interval $(t, t + T)$ into $T/\Delta t \equiv b$ small subintervals by introducing the points $t_j = t + j\Delta t$, where j takes on the integer values $0, 1, \ldots, b$. The probability of finding a count in the subinterval at $t_{j'}$ is, by (5.13),

$$W(t_{j'}) = \eta I(t_{j'})\,\Delta t;$$

the probability of registering no count in this interval is the counterprobability, namely

$$\overline{W}(t_{j'}) = 1 - \eta I(t_{j'})\,\Delta t,$$

since two or more counts can be neglected. Thus the probability of a fixed sequence s of n counts, say one count in each of the subintervals characterized by $t_{j_1}, t_{j_2}, \ldots t_{j_n}$ and no count in each of the subintervals $t_{k_1}, t_{k_2}, \ldots, t_{k_{b-n}}$, is proportional to the product

$$U_s = \prod_{j_\alpha = j_1}^{j_n} W(t_{j_\alpha}) \prod_{k_\beta = k_1}^{k_{b-n}} \overline{W}(t_{k_\beta}). \tag{5.15}$$

To obtain $P_n(t, t+T)$ one has to sum over the values U_s of all possible sequences s of n counts in the interval $(t, t+T)$. The resulting expression depends on Δt (among other quantities); taking into account the relative smallness of Δt, we may pass to the limiting expression for $\Delta t \to 0$. In summary, one obtains the probability

$$P_n(t, t+T) = \frac{1}{n!} M^n e^{-M} \tag{5.16}$$

of finding n counts in the interval $(t, t+T)$ independently of their temporal distribution; here the quantity M is the integrated number of counts,

$$M = \int_t^{t+T} dt' \, \eta I(t'). \tag{5.16a}$$

Formula (5.16) is a Poisson distribution, where M plays the role of the mean number of counts. Like classical light sources that obey the Poisson distribution, ideal laser light [which can be described by a one-mode coherent state (cf. Section 2.3.1.2)] does not fluctuate; it displays the Poisson distribution (5.16) of counts. If fluctuations of the radiation intensity occur, the integrated number of counts M must also be subjected to a probability distribution, which will be designated by $\mathscr{W}(M)$. Then the right-hand side of (5.16) must be averaged with respect to $\mathscr{W}(M)$; this leads to the photon-counting distribution

$$P_n(t, t+T) = \int_0^\infty dM \, \mathscr{W}(M) \frac{1}{n!} M^n e^{-M}. \tag{5.17}$$

For a general function $\mathscr{W}(M)$ this expression does not coincide with the Poisson distribution. For a stationary field the probability $P_n(t, t+T)$ becomes independent of t, that is, $P_n(t, t+T) \equiv P_n(T)$.

It must be emphasized that the relation (5.17), which governs the counting of photoelectrons, exhibits the same structure as does the relation (2.135) for the photon distribution in a normalization volume that is formed by the surface A of a radiation receiver and the path cT traversed by the photons in the time T. Note that M in (5.16a) refers to the photoelectrons ejected due to photon absorption, and $\tilde{M}$ in (2.137) refers to the incident photons. For sufficiently large values of T, the quantity $M/\tilde{M}$ is equal to the ratio of the mean number of emitted photoelectrons to the number of incident photons. If

$M/\tilde{M}$ is equal to unity, the distributions for the photoelectrons and the photons in the normalization volume AcT are exactly the same. The relation (5.17) is based on (5.13), which in turn is based on the mechanism of the simultaneous absorption of one photon and the generation of one photoelectron. Thus, altogether one may say that the counting of photoelectrons reflects the properties of the light itself.

It is an important problem in physical practice to deduce the probability distribution $\mathscr{W}(M)$, which reflects the intrinsic statistical behavior of the radiation, from the measurable photon-counting distribution. To this end the relation (5.17) must be inverted. In the stationary case this leads to the formula

$$\mathscr{W}(M) = \frac{1}{2\pi} e^{M} \int_{-\infty}^{+\infty} d\xi \, e^{-\xi M} \sum_{n=0}^{\infty} (i\xi)^n P_n(T). \tag{5.18}$$

From $P_n(t, t+T)$ important statistical numerical measures can be derived, for example, the lth moment $\Sigma_{n=0} P_n(t, t+T) n^l = \langle n^l \rangle$ and the variance $\langle (\Delta n)^2 \rangle$. Straightforward calculations lead to the results

$$\langle n \rangle = \langle M \rangle, \tag{5.19a}$$

$$\langle n^2 \rangle = \langle M \rangle + \langle M^2 \rangle, \tag{5.19b}$$

$$\left\langle (\Delta n)^2 \right\rangle = \left\langle (n - \langle n \rangle)^2 \right\rangle = \langle M \rangle + \left\langle (\Delta M)^2 \right\rangle. \tag{5.19c}$$

In the special case of a Poisson distribution the variance $\langle (\Delta n)^2 \rangle$ agrees with $\langle M \rangle$. However, in general the variance $\langle (\Delta M)^2 \rangle$ of the integrated number of counts is nonzero, and it appears added to $\langle M \rangle$ in the formula for $\langle (\Delta n)^2 \rangle$, which codetermines the dispersion of the photon-counting distribution $P_n(t, t+T)$ and hence also its character. It is obvious that the variance $\langle (\Delta M)^2 \rangle$ can be expressed by

$$\eta^2 \int_t^{t+T} dt' \int_t^{t+T} dt'' \langle [I(t') - \langle I \rangle][I(t'') - \langle I \rangle] \rangle.$$

Thus, on the one hand, it depends on the intrinsic radiation behavior, in particular on the intensity correlation essentially characterized by its correlation time (which will be denoted by τ_c). On the other hand, it depends on the measuring time T, which has to be chosen according to the experimental conditions. We do not intend to go into details here, but want to mention only that for large T values the photon-counting distribution turns out to be not sensitive to the special kind of the radiation distribution; it tends to the Poisson distribution if $\tau_c \ll T$.

Let us now discuss the more interesting case of small T. Under the condition $T/\tau_c \to 0$ we can write $M = \eta T I$, and we see that $\mathscr{W}(M)$ gives the probability distribution of the quasiinstantaneous intensity I. Two important examples of this limiting case may be presented, based on the application of (5.17) and (5.18). From these equations it follows by a straightforward calculation that the quantities

$$\mathscr{W}(M) = \delta(M - \langle M \rangle) \quad \text{and} \quad P_n(T) = \frac{\langle M \rangle^n}{n!} e^{-\langle M \rangle} \tag{5.20}$$

refer to the same type of radiation. $P_n(T)$ represents the Poisson distribution. A comparison with the expression (2.103) reveals that the quantities in (5.20) refer to ideal laser light. Its probability distribution for the intensity I is a delta function. Similarly, the quantities

$$\mathscr{W}(M) = \langle M \rangle^{-1} e^{-M/\langle M \rangle} \quad \text{and} \quad P_n(T) = \frac{\langle M \rangle^n}{(1 + \langle M \rangle)^{n+1}} \tag{5.21}$$

refer to the same type of radiation. Here, $P_n(T)$ represents a Bose–Einstein distribution, and a comparison with (2.125) shows that here chaotic light is described whose intensity distribution obeys an exponential law.

In principle, the experimental proof of the photon-counting distribution requires the detection of single photons. This subject has for instance been reviewed by Arecchi and Degiorgio (Ref. 5.2) and by Akhmanov, Djakov, and Chirkin (Ref. 5.3). In the optical frequency range, the detection is mainly performed by means of photomultiplier tubes consisting of a photocathode (the basic element of a photocell) and dynodes for the multiplication of the number of photoelectrons. The photoefficiency of the photocathode, that is, the ratio of the mean number of emitted photoelectrons to that of the incident photons, takes on values up to 0.8. The multiplication factor of photomultipliers in practical use is of the order of 10^7.

The efficiency of photocathodes is subject to certain limitations. There exists a statistical influence on the multiplication process. It can often be neglected or corrected and so the desired distribution of photoelectrons emitted by the photocathode can be inferred from the photomultiplier output. Noise pulses (pulses not caused by the radiation under investigation, but by other sources such as thermionic emission, cosmic rays, etc.), can mostly be discriminated from the standardized signal pulses, since they have a different pulse height and different correlation behavior. The finite duration of a pulse at the photomultiplier output leads to a lower limit for resolving different photon counts, whereby a resolving time T_r is defined (typical experimental value: $T_r \sim 10^{-9}$ s). The device as a whole (photomultiplier plus electronic equipment) has a dead time T_d: after one count is registered, no further event can be processed during the interval T_d. A lower bound on T_d is T_r. However, the output current of the multiplier is sent directly through a gate with opening time T to an integrating capacitor whose charge is a measure of the number of

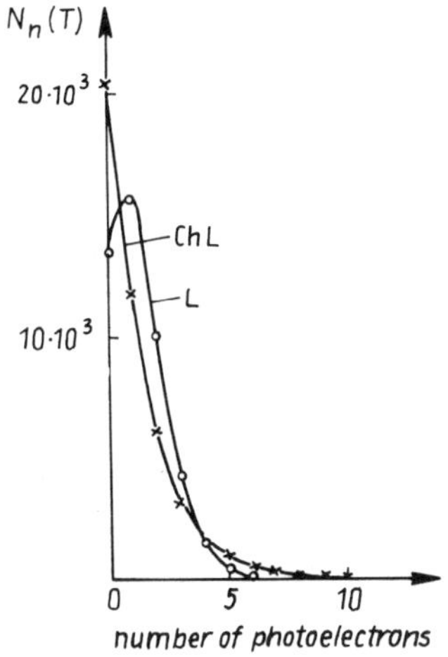

Fig. 5.1. Measurement of the photon-counting distribution for several light sources (L, laser light; ChL, thermal (chaotic) light). $N_n(T)$ is the number of events characterized by n counts during time interval T. [After (Ref. 5.4).]

count events, even two partially overlapping photoelectron responses can be resolved. In this way the dead time is overcome; however, a certain noise arising from the first dynodes occurs (which can be diminished by using special photomultipliers). These perturbing influences can be eliminated (to a certain extent) by applying correction formulas to the experimental results.

We have mentioned above that the region for which the measuring time T is much less than the correlation time τ_c of the radiation deserves greater interest in principle than the other limiting case $T \gg \tau_c$. Since τ_c is the order of the reciprocal line width, even in the case of thermal light emitted in a narrow line we have $\tau_c \sim 10^{-9}$ s. In such a case, the fulfilment of the condition $T \ll \tau_c$ may become difficult. For a short measuring time T the accuracy is not high because of statistical error due to the small number of signal pulses, which is limited by the intensity of the light source or by the dead time (resolving time) of the device. Some of these difficulties can be compensated for by performing a large number of independent measurements; this, however, requires high stability of all experimental parameters.

Arecchi (Ref. 5.4) has measured the photon-counting distribution $P_n(T)$ for several light sources; in Fig. 5.1 the measured number of events $N_n(T)$ characterized by n counts during the time interval T is plotted versus n. The calculation of the first three moments reveals that the distribution assigned to the laser light (L) is Poissonian, whereas the thermal (chaotic) light (ChL) exhibits a Bose–Einstein distribution. These experimental results are in good agreement with the theoretical distribution as given in (5.20) and (5.21).

5.1.2. Photoelectric Counting by Several Photocells. Joint Probabilities

As shown in Section 5.1.1, the measurement of the radiation intensity is achieved by the operation of a single detector. However, in order to explain the intensity correlation, which plays an important role among the correlation and coherence phenomena associated with nonlinear effects, it is not sufficient to consider the operation of only one photon detector; it is necessary to consider the operation of several photodetectors with respect to a certain radiation field. Let each of m detectors be of the same type as discussed in Section 5.1.1, and let them be located at the positions $r_{1\bullet}, r_{2\bullet}, \ldots, r_{m\bullet}$. The intensity correlation experiments to be considered are described with the help of the *joint probability* $p(x_1, \ldots, x_m)\,\Delta t_1 \cdots \Delta t_m$ of observing a count of a photoelectron due to photon absorption in the interval Δt_1 around t_1 by the first detector *and* a count in the interval Δt_2 around t_2 by the second detector *and*... *and* a count in the interval Δt_m around t_m by the mth detector; $x_j = (r_{j\bullet}, t_j)$ is the space–time point assigned to the jth detector.

To derive the joint probability one may proceed along the same lines as for deriving the probability for a single detector given in (5.13). In Section 5.1.1 we started with the treatment of a single atom acted on by a radiation field, from which we passed over to the discussion of the operation of a photodetector. Analogously, we now start with the consideration of m atoms (at the positions $r_{1\bullet}, \ldots, r_{m\bullet}$), which are supposed to represent m photodetectors. Instead of the probability (5.9) for a single atom, the result

$$P(t_1, \ldots, t_m) = \int_0^{t_1} dt_1' \int_0^{t_1} dt_1'' \cdots \int_0^{t_m} dt_m' \int_0^{t_m} dt_m''\, S(t_1' - t_1'') \cdots S(t_m' - t_m'') \times \operatorname{Tr}\{\hat{\rho}(0)\hat{E}_I^{(+)}(r_{1\bullet}, t_1'') \cdots \hat{E}_I^{(+)}(r_{m\bullet}, t_m'')\hat{E}_I^{(-)}(r_{m\bullet}, t_m') \times \cdots \hat{E}_I^{(-)}(r_{1\bullet}, t_1')\} \quad (5.22)$$

is obtained by applying mth-order perturbation theory (see e.g. Ref. 5.5). $P(t_1, \ldots, t_m)$ is the probability that the first atom undergoes a counted photon absorption in the interval $(0, t_1)$, the second atom undergoes one in the interval $(0, t_2)$, and so on. No further measurements on the state of the field are made between the individual absorption processes. The function $S(t_j' - t_j'')$ represents the characteristic function of the jth atom (detector) in the time domain. If the same assumptions (pointlike, broadband detectors operating on the basis of the external photoeffect) as in Section 5.1.1 are made and assuming the time ordering $t_1 \le t_2 \le \cdots \le t_m$ to take into account free fields as well as interacting fields (see e.g., Ref. 32b, Sect. 3.7), it can be concluded from (5.22) that the joint probability we seek is given by

$$p(x_1, \ldots, x_m)\,\Delta t_1 \cdots \Delta t_m = \eta^m \operatorname{Tr}\{\hat{\rho}(0)\hat{E}^{(+)}(x_1) \cdots \hat{E}^{(+)}(x_m)\hat{E}^{(-)}(x_m) \cdots \hat{E}^{(-)}(x_1)\}, \quad (5.23)$$

where η is the sensitivity factor of a single photodetector. The expression on the right-hand side of (5.23) agrees, apart from the factor $\eta^m \Delta t_1 \cdots \Delta t_m$, with a special kind of the $2m$th-order field correlation function whose significance with respect to coherence properties will be illustrated in the next section. Note that the correlation function appearing in (5.23) is normally ordered, this means that the creation operators contained in $\hat{E}^{(+)}$ stand to the left of the annihilation operators contained in $\hat{E}^{(-)}$. As has already been explained in Section 5.1.1, this feature is connected with the fact that the detection process is based on photon absorption and not on photon emission.

5.2. CORRELATION FUNCTIONS AND COHERENCE PROPERTIES

Now we shall deal with the relationship between coherence properties and correlation functions. Theoretical and experimental aspects of classical as well as of quantum correlation functions will be discussed.

5.2.1. Classical Correlation Functions

The modern classical concept of coherence is based on the stochastic behavior of the radiation field. From the discussion of (2.54) we may conclude that the radiation field at the space–time point $x = (r_\bullet, t)$ can be described by the complex analytic signal $E_\bullet^{(-)}(x)$. It may be assumed that the stochastic properties of $E_\bullet^{(-)}(x)$ can be completely expressed by the following set of *differential joint probabilities*:

$$dp_1 = P_1\left[E_\bullet^{(-)}(x_1)\right] d^2E_\bullet^{(-)}(x_1),$$

$$dp_2 = P_2\left[E_\bullet^{(-)}(x_1), E_\bullet^{(-)}(x_2)\right] d^2E_\bullet^{(-)}(x_1)\, d^2E_\bullet^{(-)}(x_2),$$

$$\vdots$$

$$dp_g = P_g\left[E_\bullet^{(-)}(x_1), \dots, E_\bullet^{(-)}(x_j), \dots, E_\bullet^{(-)}(x_g)\right] \times d^2E_\bullet^{(-)}(x_1) \cdots d^2E_\bullet^{(-)}(x_g) \tag{5.24}$$

$$\vdots$$

for different values of g. Here $d^2E_\bullet^{(-)}$ stands for $d(\operatorname{Re} E_\bullet^{(-)})\, d(\operatorname{Im} E_\bullet^{(-)})$, where both the factors are infinitesimal volume elements in the $E_\bullet^{(-)}$ vector space. The probability dp_g is proportional to the infinitesimal volume $d^2E_\bullet^{(-)}(x_1) \cdots d^2E_\bullet^{(-)}(x_g)$ and to the joint probability density P_g of finding the complex analytic signal $E_\bullet^{(-)}(x_1)$ at the space–time point x_1 *and* the signal $E_\bullet^{(-)}(x_2)$ at $x_2 \dots$ *and* the signal $E_\bullet^{(-)}(x_g)$ at x_g.

By means of the joint probabilities dp_g the ensemble average of a function depending on the complex analytic signal at g space–time points can be constructed. In particular, those functions $F^{m,m'}_{\bullet\cdots\bullet}$ are of importance that consist of m' factors $E^{(-)}_{\bullet}(x_l)$ and m factors $E^{(+)}_{\bullet}(x_l) \equiv [E^{(-)}_{\bullet}(x_l)]^*$, namely

$$F^{m,m'}_{\bullet\cdots\bullet}(x_1,\ldots,x_g) = \prod_{l=1}^{m} E^{(+)}_{\bullet}(x_l) \prod_{l=m+1}^{m+m'} E^{(-)}_{\bullet}(x_l), \tag{5.25}$$

where $m + m' = g$. The ensemble average of $F^{m,m'}_{\bullet\cdots\bullet}$,

$$\Gamma^{m,m'}_{\bullet\cdots\bullet}(x_1,\ldots,x_g) = \int dp_g \prod_{l=1}^{m} E^{(+)}_{\bullet}(x_l) \prod_{l=m+1}^{m+m'} E^{(-)}_{\bullet}(x_l), \tag{5.26}$$

is called the *field correlation function of the* $(m + m')$*th order*. Since the electromagnetic field may have different polarization directions at the different space–time points, in general $F^{m,m'}_{\bullet\cdots\bullet}$ and $\Gamma^{m,m'}_{\bullet\cdots\bullet}$ have the character of a tensor of gth rank. Writing the signal vector $E^{(\pm)}_{\bullet}(x_l)$ in the form of $e_{l\bullet}E^{(\pm)}(x_l)$, we obtain

$$F^{m,m'}_{\bullet\cdots\bullet} = e_{1\bullet} \cdots e_{g\bullet} \prod_{l=1}^{m} E^{(+)}(x_l) \prod_{l=m+1}^{m+m'} E^{(-)}(x_l), \tag{5.27}$$

where $g = m + m'$. The factor consisting of the polarization unit vectors $e_{l\bullet}$ carries the tensor character of $F^{m,m'}_{\bullet\cdots\bullet}$, whereas the remaining factors are scalar quantities. In the following discussion the special tensor notation may be omitted, since only one-component fields are displayed.

Let us now sketch some physically relevant properties of these correlation functions.

Mathematical statistics asserts that the knowledge of the correlation functions allows (via the statistical moments) the determination of the probability densities as defined in (5.24).

In general, the correlation function is a complex-valued function whose complex conjugate can be constructed by a certain interchange of factors and arguments:

$$\Gamma^{m,m'}(x_1,\ldots,x_{m+m'}) = \left[\Gamma^{m',m}(x_{m+m'},\ldots,x_1)\right]^*. \tag{5.28}$$

This implies that $\Gamma^{(m,m)}(x_1,\ldots,x_m,x_m,\ldots,x_1)$ is a real (nonnegative) function. From the discussion of (5.13) we may conclude that if zero-point fluctuations can be neglected, the product $E^{(+)}(x_1)E^{(-)}(x_1)$ can be interpreted as the classical expression for the radiation intensity $I(x_1)$ at the space–time point x_1. Hence $\Gamma^{1,1}(x_1, x_1)$ is the corresponding ensemble average. For the same reason the function $\Gamma^{m,m}(x_1,\ldots,x_m,x_m,\ldots,x_1)$ can be thought of as the ensemble average of the product $I(x_1)\cdots I(x_m)$.

In the case of a *stationary field* the probability density as well as the correlation functions are independent of a change of the time origin. In the case of an *ergodic field* the ensemble average can be replaced by the time average. Thus for stationary and ergodic fields the correlation function can be rewritten in the form

$$\begin{aligned}\Gamma^{m,m'}(x_1,\dots,x_g) &\equiv \Gamma^{m,m'}(r_{1\bullet},t_1;\cdots;r_{g\bullet},t_g)\\ &= \lim_{2T\to\infty}\frac{1}{2T}\int_{-T}^{+T}dt\prod_{l=1}^{m}E^{(+)}(r_{l\bullet},t+\tau_l)\\ &\quad\times\prod_{l=m+1}^{m+m'}E^{(-)}(r_{l\bullet},t+\tau_l),\end{aligned}\tag{5.29}$$

where $\tau_l = t_l - t_1$ and $m + m' = g$.

With the help of the correlation functions (5.26), a logically satisfactory and applicable *definition of coherence* can be introduced. A radiation field is said to be *coherent in Kth order* if the correlation functions $\Gamma^{m,m'}$ satisfy the condition

$$\Gamma^{m,m'}(x_1,\dots,x_g) = \prod_{l=1}^{m}V^{(+)}(x_l)\prod_{l=m+1}^{m+m'}V^{(-)}(x_l)\tag{5.30}$$

for all $m, m' \leq K$. The functions $V^{(-)}(x)$ and $V^{(+)}(x) = [V^{(-)}(x)]^*$ are assumed to be independent of the index l of x and must be solutions of Maxwell's equations with the boundary conditions imposed on the problem; thus $V^{(-)}(x)$ corresponds to the complex analytic signal $E^{(-)}(x)$. Because of the form of the product on its right-hand side, the relation (5.30) is called the *factorization condition.* Note that the assertion of coherence depends on the choice of the g space–time points, which are the arguments in the correlation function. Therefore, according to the experimental conditions the coherence of a certain order can in general be asserted only for a limited space–time region, which is called the *coherence volume*. The number of factorizable correlation functions increases with increasing order K. As K tends to infinity, the field is said to be *fully coherent*.

The connection of the definition of coherence given above with stochastic properties can be easily seen from the following consideration. Let us suppose that no field fluctuations occur; then the probability densities P_g for all g have to be written in the form

$$P_g = \prod_{l=1}^{g}\delta^{(2)}\left[E^{(-)}(x_l) - U^{(-)}(x_l)\right],\tag{5.30a}$$

where $E^{(-)}(x_l)$ is the statistical field variable and $U^{(-)}(x_l)$ is an exactly fixed predicted field function. Using the probability densities (5.30a), the correlation functions $\Gamma^{m,m'}$ with arbitrary m and m' obviously become factorized because of the delta function, and consequently the field, which does not deviate from a predicted field, possesses full coherence.

So far, we have discussed in this section explicitly correlation functions $\Gamma^{m,m'}(r_{1\bullet}, t_1; \ldots; r_{(m+m')\bullet}, t_{m+m'})$ depending on space and time coordinates. By application of a Fourier transformation with respect to the time variables [as explicitly represented in (1.87a) and (1.87b)] one may pass over from the functions $\Gamma^{m,m'}$ to the *spectral correlation functions* $\underset{\smile}{\Gamma}^{m,m'}$, which, besides the space coordinates, depend on $m + m'$ frequency variables.

Important theoretical and practical aspects of spectral properties can already be demonstrated by considering the second-order correlation function

$$\Gamma^{1,1}(t_1, t_2) = \langle E^{(+)}(t_1) E^{(-)}(t_2) \rangle,$$

where the space coordinates are ignored. This function will be rewritten in the form

$$\Gamma(t, t-\tau) = \langle E^{(+)}(t) E^{(-)}(t-\tau) \rangle \tag{5.31}$$

(where $t_1 = t$ and $t_2 = t - \tau$) in order to display nonstationarity (if present) and correlation properties. In the stationary case Γ is a function of τ only and possesses a constant correlation time, which term has the same meaning as in the discussion of (5.9). In the nonstationary case the correlation time is, in general, a function of t (Ref. 5.6). The one-to-one Fourier transformation yields the so-called *time-dependent spectrum*

$$\underset{\smile}{\Gamma}(t, \omega) = \int_0^\infty d\tau\, \Gamma(t, t-\tau) e^{-i\omega\tau} + \text{c.c.}, \tag{5.31a}$$

which is a (nonstationary) spectral correlation function. In connection with experiments in spectroscopy it is advantageous to deduce from $\underset{\smile}{\Gamma}(t, \omega)$ the following quantities:

total energy spectrum

$$W(\omega) = \int dt\, \underset{\smile}{\Gamma}(t, \omega)$$

and the *instantaneous intensity*

$$I(t) = \frac{1}{2\pi} \int d\omega\, \underset{\smile}{\Gamma}(t, \omega).$$

The spectral analysis of electromagnetic radiation is of great importance in the investigation of fundamental and practical problems. In addition to the traditional stationary investigations, spectral analysis of nonstationary radiation connected with short-time processes and short-time radiation sources (see, e.g., Chapter 6) plays an increasing role. The use of a spectrometer requires the knowledge of the relationship between the output data and the input radiation $E(t)$ (in the entrance slit) which is to be analyzed. Measurement can be performed by a spectrochronograph (dynamical spectrometer), which may consist of a grating spectrograph and a streak camera. The streak camera records the square Q of the field strength in the exit plane, which is averaged at least over some optical cycles. The output spectrum Q is a function of the time t and of the spectral coordinate ω_s; Q can be represented as a functional of the dynamical instrumental function and of the time-dependent spectrum $\underset{\smile}{\Gamma}(t, \omega)$ of the input radiation (Refs. 5.7, 5.8a, b). In general the output spectrum exhibits another shape than the input spectrum $\underset{\smile}{\Gamma}$, because of the influence of the instrumental function. However, under experimentally important conditions (duration of the nonstationary radiation shorter than $T_{sp}/2$, where T_{sp} is the maximum traveling-time difference between interfering beams) a determination of the input spectrum from the output spectrum by differentiating is possible:

$$\underset{\smile}{\Gamma}(t, \omega) = T_{sp}^2 \frac{\partial}{\partial t} Q(t, \omega_s = \omega). \tag{5.31b}$$

Consequently, one may say that the spectrochronograph allows one to determine the time-dependent spectrum and thus the correlation function of the (nonstationary) radiation to be analyzed. It should be mentioned that for a Fabry–Perot interferometer with a fixed frequency setting the output spectrum as a function of the input radiation is given in Ref. 5.9. However, in contrast to the spectrochronograph, if a single light phenomenon of finite duration is to be analyzed, a large number of Fabry–Perot interferometers with different frequency settings must be used.

Finally, we discuss the aspect of real time in determining the input spectrum (Ref. 5.8b). It should be mentioned that second-order radiation functionals differing from $\underset{\smile}{\Gamma}(t, \omega)$ have been proposed, for instance the Wigner distribution function (Ref. 5.10). However, only the abovementioned time-dependent spectrum $\underset{\smile}{\Gamma}$ has the property that $\underset{\smile}{\Gamma}(t', \omega)$ can be fully determined in the region $t' \leq t$ if a measurement is carried out for times $t' \leq t$. In the case of the other proposed radiation functionals, information on the radiation from the future $(t' \geq t)$ is needed in order to calculate the spectrum at time t.

In previous sections (e.g., Section 2.1) we referred to the correspondence between the classical and the quantum-theoretical description of the electromagnetic field. An analogous relationship exists between the classical correlation functions discussed here and the quantum correlation functions, which will be introduced in the next subsection.

5.2.2. Quantum Correlation Functions and Their Application

As shown in Chapter 2, the quantized radiation field is described by the field operator $\hat{E}_\bullet(r_\bullet, t)$ (in the Heisenberg picture), which can be decomposed into the operator of the complex analytic signal $\hat{E}_\bullet^{(-)}(x)$ and its hermitian conjugate $\hat{E}_\bullet^{(+)}(x)$, where x denotes the space–time point $x = (r_\bullet, t)$. We recall that $\hat{E}_\bullet^{(-)}(x)$ contains the photon annihilation operators of the different modes [cf. (2.55)]. In formal analogy to the classical correlation function, the quantum correlation function of $(m + m')$th order is constructed by

$$\Gamma^{m,m'}_{\bullet\cdots\bullet\mathrm{norm}}(x_1,\ldots,x_g) = \mathrm{Tr}\left\{\hat{\rho}(0)\prod_{l=1}^{m}\hat{E}_\bullet^{(+)}(x_l)\prod_{l=m+1}^{m+m'}\hat{E}_\bullet^{(-)}(x_l)\right\}. \quad (5.32)$$

This function is the quantum mean value of an operator product consisting of $m + m'$ field factors, where the averaging is performed by means of the density operator $\hat{\rho}(0)$, which characterizes the initial (pure or mixed) state of the radiation field. It must be emphasized that the ordering of the field factors in (5.32) has now, in contrast to the classical description, a physical meaning because of the noncommutability of the operators $\hat{E}_\bullet^{(+)}$ and $\hat{E}_\bullet^{(-)}$. The correlation function given in (5.32) is normally ordered; all operators of the operator product containing photon creation operators stand to the left of all operators containing photon annihilation operators. (It may be added that a correlation function with a different ordering can be decomposed into a sum of normally ordered correlation functions by utilizing the commutation relations.) Since in the present section only normally ordered correlation functions will be considered, we shall drop the index "norm."

The quantum correlation function formulated in (5.32) has the same space–time and polarization behavior as the respective classical correlation function; therefore we may adopt the corresponding statements concerning stationarity, ergodicity, and tensor character from Section 5.1.1. In the following we shall again only deal with one-component fields, so that the quantum correlation function is given by

$$\Gamma^{m,m'}(x_1,\ldots,x_g) = \mathrm{Tr}\left\{\hat{\rho}(0)\prod_{l=1}^{m}\hat{E}^{(+)}(x_l)\prod_{l=m+1}^{m+m'}\hat{E}^{(-)}(x_l)\right\}, \quad (5.32\mathrm{a})$$

which contains the scalar quantities $\hat{E}^{(\pm)}(x_l)$.

It has to be pointed out that the quantity $\hat{\rho}(0)$ characterizing the initial state of the radiation field is sufficient for the evaluation of the quantum correlation functions of arbitrary order; on the other hand, in the classical case the joint probability densities P_g from (5.24) for all g are needed for the evaluation of the classical correlation functions. It can be shown that a connection between $\hat{\rho}(0)$ and joint probability densities, now termed $\tilde{P}_g[\ldots, E^{(-)}(x_j),\ldots]$, exists also in the quantum-theoretical case. If $\hat{\rho}(0)$ can be represented by the P-representation $P(\{\alpha_\mu\})$ [see e.g. (2.134)], then with the aid of $P(\{\alpha_\mu\})$ the joint probability densities

$$\tilde{P}_g = \int d^2\{\alpha_\mu\}\, P(\{\alpha_\mu\}) \prod_{j=1}^{g} \delta^{(2)}\left[E^{(-)}(x_j) - E^{(-)}(x_j; \{\alpha_\mu\})\right] \quad (5.33)$$

can be constructed, where $E^{(-)}(x_j; \{\alpha_\mu\})$ is the right-hand eigenvalue of the field operator $\hat{E}^{(-)}(x_j)$ [cf. (2.105)]. By means of the $\tilde{P}_g$ given in (5.33), the normally ordered quantum correlation functions can be derived formally in the same manner as used for deriving the classical correlation functions in (5.26) with the help of P_g. This allows us to take over the classical procedure—at least in the case of normal ordering.

Also, the definition of coherence follows the same pattern as in the classical case. The quantum correlation functions $\Gamma^{m,m'}$ from (5.32a) must satisfy the factorization conditions

$$\Gamma^{m,m'}(x_1, \dots x_g) = \prod_{l=1}^{m} V^{(+)}(x_l) \prod_{l=m+1}^{m+m'} V^{(-)}(x_l) \quad (5.34)$$

for all $m, m' \leq K$ if coherence of Kth order exists in the radiation field. The functions $V^{(-)}(x)$ and $V^{(+)}(x)$ must fulfil the same conditions as were mentioned below (5.30). If K tends to infinity, the field possesses *full coherence*.

In Section 5.1.1 it was shown that a classical radiation field without fluctuations possesses full coherence. We shall now see that coherent states play an analogous role in the case of a quantized field. From (2.105) we know that a global coherent state $|\{\alpha_\mu\}\rangle$ is a right-hand eigenstate of the operator $\hat{E}^{(-)}(x)$ with the eigenvalue $E^{(-)}(x; \{\alpha_\mu\})$. This immediately leads to the following instructive expression for the correlation function of arbitrary order:

$$\langle\{\alpha_\mu\}| \prod_{l=1}^{m} \hat{E}^{(+)}(x_l) \prod_{l=m+1}^{m+m'} \hat{E}^{(-)}(x_l) |\{\alpha_\mu\}\rangle$$

$$= \prod_{l=1}^{m} E^{(+)}(x_l; \{\alpha_\mu\}) \prod_{l=m+1}^{m+m'} E^{(-)}(x_l; \{\alpha_\mu\}), \quad (5.35a)$$

where $E^{(+)}(x_l; \{\alpha_\mu\})$ is the complex conjugate of $E^{(-)}(x_l; \{\alpha_\mu\})$. Equation (5.35a) demonstrates that a correlation function evaluated with a global coherent state is factorized independently of its order. This reveals that global coherent states—which, apart from the unavoidable vacuum fluctuations [cf. the discussion of (2.102)], correspond to a classical wave with spatially and temporally predicted field-strength values—possess full coherence.

Besides the coherent state, another characteristic radiation type, namely chaotic light [see (2.138)], possesses a very simple expression for the correlation functions of arbitrary order. Straightforward calculation yields the property of

Gaussian distribution functions, namely

$$\Gamma^{m,m'}(x_1, x_2, \ldots, x_{m+m'}) = 0 \qquad \text{for} \quad m \neq m',$$

$$\Gamma^{m,m}(x_1, x_2, \ldots, x_{m+m}) = \sum_s \Gamma^{1,1}(x_1, x_{m+1}) \cdots \Gamma^{1,1}(x_m, x_{m+m}), \quad (5.35b)$$

where the sum is taken over all $m!$ permutations s of the indices $1, 2, \ldots, m$. This result says that for chaotic light the $(m + m')$th-order correlation functions is completely determined by the second-order correlation functions of the radiation.

In the modern concept of coherence the terms "monochromaticity" and "coherence" must not be mixed up, since they express different facts. This can be concluded from the following statement: monochromatic radiation always shows coherence of first order, whereas monochromaticity is not sufficient to imply coherence of higher order. To illustrate this, let us consider the special field state $|R\rangle = |0, \ldots, 0, \psi_\mu, 0, \ldots\rangle$, which is supposed to consists of the μth mode of the arbitrary photon state $|\psi_\mu\rangle$ with a nonzero mean photon number $\langle \hat{N}_\mu \rangle$, whereas in all the other modes there exists a photon vacuum. The quantum correlation function $\Gamma^{1,1}(x_1, x_2)$ evaluated with the state $|R\rangle$ is

$$\Gamma^{1,1}(x_1, x_2) = V^{(+)}(x_1) V^{(-)}(x_2), \quad (5.36)$$

where

$$V^{(-)}(x) = i\left(\frac{\hbar\omega_\mu}{2\varepsilon_0 V}\right)^{1/2} \langle \hat{N}_\mu \rangle^{1/2} e^{i(q_\mu \cdot r_\bullet - \omega_\mu t)}.$$

This means that certainly the monochromatic state $|R\rangle$ is coherent in the first order. Let us additionally suppose that $n_{\max}$ is the maximum photon number in the state $|\psi_\mu\rangle$. Then we have to make the ansatz

$$|\psi_\mu\rangle = \sum_{n_\mu = 0}^{n_{\max}} c(n_\mu) |n_\mu\rangle. \quad (5.37)$$

Under the assumption $m' > n_{\max}$ the quantum correlation function $\Gamma^{m,m'}(x_1, x_2)$ turns out to be

$$\Gamma^{m,m'}(x_1, x_2) \equiv 0, \quad (5.37a)$$

since the m'-fold action of the annihilation operator on $|\psi_\mu\rangle$ gives zero. This leads to the conclusion that, although the state $|R\rangle$ is monochromatic, coherence of higher order than $K = n_{\max}$ cannot be maintained.

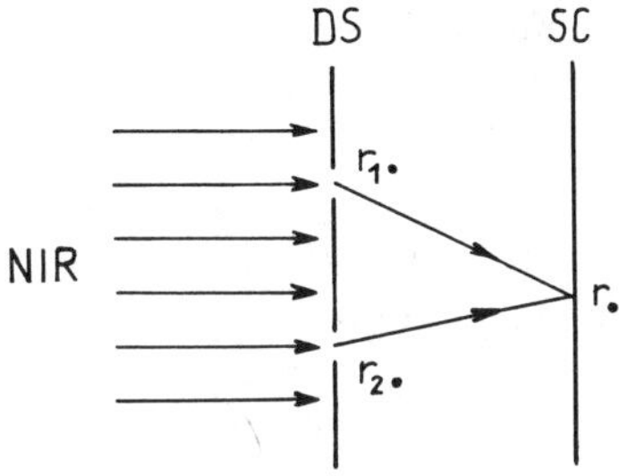

Fig. 5.2. Scheme of Young's interference experiment (NIR normally incident radiation; DS, plane of the double slit; SC, screen).

We now discuss the connection of the quantum correlation functions with the explanation of experimental results. Let us start with the correlation functions $\Gamma^{m,m'}$ with $m = 1$ and $m' = 1$.

As already shown in Section 5.1.1, the correlation function $\Gamma^{1,1}(x, x)$ with equal arguments has to be assigned to the radiation intensity at the space–time point x. Equations (5.14) and (5.13) reveal that the quantity considered is exactly related to the radiation energy density and the counting probability of a detector operating on the basis of the external photoeffect.

The correlation function $\Gamma^{1,1}(x_1, x_2)$ with unequal arguments is called the *mutual coherence function*. Its physical significance clearly can be illustrated by considering Young's interference experiment. We may suppose stationarity of the field and simple geometrical conditions as shown in Fig. 5.2, namely, pointlike pinholes (at the positions $r_{1\bullet}$, $r_{2\bullet}$) and normally incident radiation with unit wave vector $k_\bullet/k$, so that $(k_\bullet/k)(r_{2\bullet} - r_{1\bullet})$ vanishes. From the discussion of (2.55) we know that if the time dependence of the field operators (in the Heisenberg picture) arises only from the dynamical time dependence, then the space–time relations between the operators will be the same as between the corresponding classical quantities. This means that the field operator $\hat{E}(r_\bullet, t)$ satisfies the wave equation. Therefore the total field strength $\hat{E}(r_\bullet, t)$ at a point $r_\bullet$ on the screen turns out to be the superposition of two terms representing the contributions from the two pinholes. We have

$$\hat{E}(r_\bullet, t) = \hat{E}\left(r_{1\bullet}, t - \frac{s_1}{c}\right) + \hat{E}\left(r_{2\bullet}, t - \frac{s_2}{c}\right). \tag{5.38}$$

The quantities s_1/c and s_2/c are the travel times.

Equation (5.38) leads, through the relation

$$\hat{E}^{(\pm)}(x) = \hat{E}^{(\pm)}(x_1) + \hat{E}^{(\pm)}(x_2), \tag{5.38a}$$

directly to

$$\begin{aligned}\Gamma^{1,1}(x, x) &= \Gamma^{1,1}(x_1, x_1) + \Gamma^{1,1}(x_2, x_2) + \left[\Gamma^{1,1}(x_1, x_2) + \Gamma^{1,1}(x_2, x_1)\right] \\ &= \Gamma^{1,1}(x_1, x_1) + \Gamma^{1,1}(x_2, x_2) \\ &\quad + \left|\Gamma^{1,1}(x_1, x_2)\right|\left(e^{i\varphi(x_1, x_2)} + e^{-i\varphi(x_1, x_2)}\right),\end{aligned} \tag{5.39}$$

where $x_j = (r_{j\bullet}, t - s_j/c)$. While the term on the left-hand side represents the intensity on the screen, the first and the second term on the right-hand side represent the intensity values at the two pinholes, which are assumed to be equal. The third term on the right-hand side is the interference term, containing unequal arguments x_1 and x_2. The phase $\varphi(x_1, x_2)$ of $\Gamma^{1,1}(x_1, x_2)$ determines the position of the intensity maximum $I_{\max}$ [at $\varphi(x_1, x_2) = m\,2\pi$ with $m = 0, \pm 1, \pm 2, \ldots$] and of the intensity minimum $I_{\min}$ [at $\varphi(x_1, x_2) = \pi + m\,2\pi$] on the screen. The absolute value $|\Gamma^{1,1}(x_1, x_2)|$ is associated with the visibility M' (as introduced by Michelson) of the interference fringes by

$$M' = \frac{I_{\max} - I_{\min}}{I_{\max} + I_{\min}} = \left| \frac{\Gamma^{1,1}(x_1, x_2)}{\left[\Gamma^{1,1}(x_1, x_1)\right]^{1/2}\left[\Gamma^{1,1}(x_2, x_2)\right]^{1/2}} \right| \equiv \left|\gamma^{1,1}(x_1, x_2)\right|. \tag{5.40}$$

Hence the value of Michelson's visibility depends on the modulus of the normalized correlation function $\gamma^{1,1}(x_1, x_2)$, and thus on the arguments x_1 and x_2, as well as on the quantum state of the radiation field.

We add that the quantity $\gamma^{1,1}(x_1, x_2)$ may be regarded as a *degree of coherence*. In the sense of (5.34), first-order coherence exists if $|\gamma^{1,1}(x_1, x_2)| = 1$; partial coherence is characterized by $0 < |\gamma^{1,1}(x_1, x_2)| < 1$, and full incoherence by $|\gamma^{1,1}(x_1, x_2)| = 0$.

For the sake of illustration, we now discuss the values of M' and $|\gamma^{1,1}(x_1, x_2)|$, respectively, for two field states with an instructive physical meaning. The one-mode photon state $|R\rangle = |0, \ldots, 0, \psi_\mu, 0, \ldots\rangle$ already introduced above is connected—as can be easily seen from (5.36)—with the visibility

$$M'_{|R\rangle} = |e^{i\omega\tau}| = 1. \tag{5.41}$$

Here $\tau = (s_2/c) - (s_1/c)$ is the travel-time difference [the scalar product $k_{\mu\bullet}(r_{2\bullet} - r_{1\bullet}) = (\omega_\mu/c)(k_{\mu\bullet}/k_\mu)(r_{2\bullet} - r_{1\bullet})$ vanishes because of geometrical conditions]. This result expresses that the visibility of an exactly monochromatic state is unity, independent of the photon distribution in the state and of the travel-time difference.

Now let us discuss chaotic light, whose nature has already been characterized [(2.138) and (5.21)]. For this purpose we consider a spectral line where the photon energy $\hbar\omega_\mu\langle\hat{N}_\mu\rangle$ of the μth mode has a Gaussian distribution with the r.m.s. angular-frequency line width $\varkappa^{-1}$ around the mean frequency ω_m. The line width $\varkappa^{-1}$ is assumed to be much greater than the frequency distance of neighboring modes. Thus the shape of the line becomes proportional to $\exp[-(\omega - \omega_m)^2/2\varkappa^{-2}]$. In view of the mode representation (2.109) of the field operators, the correlation function $\Gamma^{1,1}(x_1, x_2)$ reads

$$\Gamma^{1,1}(x_1, x_2) = \frac{1}{2\varepsilon_0 V} \sum_{\mu,\mu'} \sqrt{\hbar\omega_\mu}\sqrt{\hbar\omega_{\mu'}}\langle \hat{a}^\dagger_\mu \hat{a}_{\mu'}\rangle e^{-i\omega_\mu s_1/c}\, e^{i\omega_{\mu'} s_2/c}$$

$$\times e^{i[(\omega_\mu - \omega_{\mu'})t - k_{\mu\bullet} r_{1\bullet} + k_{\mu'\bullet} r_{2\bullet}]}. \tag{5.42a}$$

The quantum average $\langle \hat{a}^\dagger_\mu \hat{a}_{\mu'}\rangle$ is evaluated for chaotic light with the help of (2.138). It becomes $\langle \hat{N}_\mu \rangle$ when $\mu = \mu'$, and vanishes if $\mu \neq \mu'$. Thus the double sum reduces to the sum $\Sigma_\mu \cdots$. Representing this sum as an integral, we get

$$\Gamma^{1,1}(x_1, x_2) \propto \int d\omega\, e^{-(\omega-\omega_m)^2/2\varkappa^{-2}} e^{i\omega\tau}. \tag{5.42b}$$

The dependence on $r_{2\bullet} - r_{1\bullet}$ does not appear, because of the geometrical conditions chosen. The Fourier integral (5.42b) is proportional to $\exp[i\omega_\mu\tau]\exp[-\tau^2/2\varkappa^2]$. Thus the modulus of the normalized correlation function $\gamma^{1,1}_{\text{ChL}}(\tau)$ for chaotic light can be expressed by

$$M'_{\text{ChL}} = \exp\left(-\frac{1}{2}\frac{\tau^2}{\varkappa^2}\right). \tag{5.42c}$$

Sufficiently small time differences $\tau = s_2/c - s_1/c$ result in a high visibility. On the other hand, the visibility decreases exponentially to zero with increasing $\tau^2/2\varkappa^2$. Therefore, since M' coincides with the degree of coherence $|\gamma^{1,1}|$, the reciprocal bandwidth may be regarded as the *coherence time*. The result contained in (5.41) and (5.42c), namely the connection between visibility, travel-time difference, and bandwidth of the radiation, has been confirmed in numerous interference experiments for laser light as well as for conventional light sources. It should be mentioned that for chaotic light the quantity $|\gamma^{1,1}(x_1, x_2)|$ is uniquely associated with the normalized higher-order quantity $\Gamma^{2,2}(x_1, x_2, x_2, x_1)/[\Gamma^{1,1}(x_1, x_1)\Gamma^{1,1}(x_2, x_2)]$, which can be easily deduced from (5.35b). Therefore, an experimental measurement of this function, to be discussed later on [see (5.48)], may also be regarded as an experimental proof of $|\gamma^{1,1}(x_1, x_2)|$.

To conclude our discussion of lowest-order effects, it should be mentioned that besides Young's interference experiment, other important phenomena can be described by the mutual-coherence function. In this manner Wolf (Ref. 5.11) clarified the relationship between the coherence behavior of light sources and the directionality of light beams.

We now pass over to the properties and applications of *higher-order correlation functions* of a radiation field.

A natural extension of the quantum correlation function $\Gamma^{1,1}(x_1, x_1)$, which is connected with photon counting by a single detector, is the quantum correlation function $\Gamma^{m,m}(x_1,\ldots,x_m,x_m,\ldots,x_1)$ for $m>1$. It is connected with the photon counting by m detectors [cf. (5.23)]. The correlation function under consideration is associated with a joint probability; in general, the intensities measured at different space–time points are not independent of one another. However, in the case of full coherence the correlation function factorizes according to

$$\Gamma^{m,m}(x_1,\ldots,x_m,x_m,\ldots,x_1)=\langle\hat{I}(x_1)\rangle\cdots\langle\hat{I}(x_m)\rangle, \qquad (5.43a)$$

which can be easily derived from (5.35a). This result leads, by comparison of (5.23) with (5.13), to

$$p(x_1,\ldots,x_m)=p(x_1)\cdots p(x_m). \qquad (5.43b)$$

This means that under the condition of full coherence each of the detectors behaves independently of the action of the $m-1$ other detectors.

We will now discuss in more detail the function $\Gamma^{2,2}(x_1,x_2,x_2,x_1)$ taken at the same space point $r_{1\bullet}=r_{2\bullet}=r_\bullet$, but at different time arguments t_1, t_2. This function carries important information about the statistical properties of the radiation field. Its physical significance can be elucidated by studying the intensity correlation experiment performed by Hanbury Brown and Twiss (Ref. 5.12). A somewhat simplified experimental scheme—see Fig. 5.3—reveals the following features. Radiation from the stationary source S to be investigated is incident on a half-transparent mirror M, where the beam is split into two beams impinging on two uniform photodetectors D_1 and D_2; we suppose D_2 to be movable, so that the travel-time difference $\tau=t_2-t_1$ can be adjusted. The mirror M may be regarded as the fictitious joint position $r_\bullet$ of both the detectors. The counting takes place during two short time intervals T_1 and T_2 at both the detectors, where both intervals may be chosen to have the same length T. We suppose the detectors to yield n_1 and n_2 counts. These values are sent to the correlator C, where the averaging of the product is performed. Thus the correlator yields $\langle n_1n_2\rangle$. In addition, the mean value $\langle n\rangle$ of counts within the interval T is measured, which is the same at both the detectors. If T is smaller than the time difference τ and the correlation time τ_c

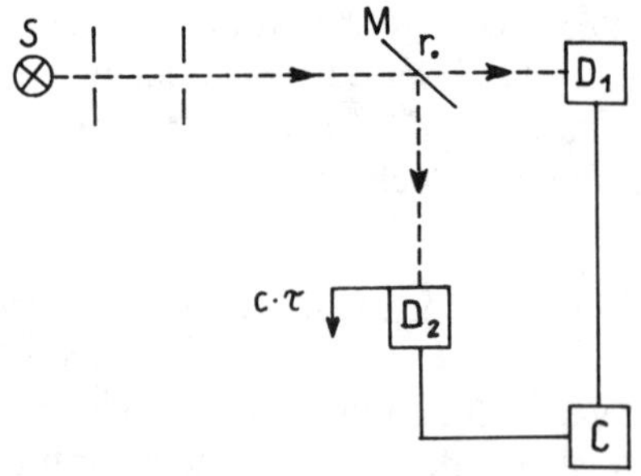

Fig. 5.3. Arrangement for the measurement of intensity correlation (schematic): S, radiation source; D_1, D_2, photon detectors; C, correlator; M, half-transparent mirror.

of the radiation, the quotient $\langle n_1 n_2 \rangle / \langle n \rangle^2$ equals the ratio of the corresponding correlation functions, so that we obtain

$$\frac{\langle n_1 n_2 \rangle}{\langle n \rangle^2} = \frac{\Gamma^{2,2}(x_1, x_2, x_2, x_1)}{\Gamma^{1,1}(x_1, x_1)\Gamma^{1,1}(x_2, x_2)} \equiv \gamma^{2,2}(x_1, x_2). \tag{5.44}$$

On the other hand, a comparison of (5.43) with the relations (5.13) and (5.25) for the counting probabilities yields

$$\gamma^{2,2}(x_1, x_2) = \frac{p(x_1, x_2)}{p(x_1)p(x_2)}. \tag{5.45a}$$

In the following, we shall ignore the dependence on $r_{\bullet}$ and take into account only the time arguments; then, assuming stationarity, we have

$$\gamma^{2,2}(\tau = t_2 - t_1) = \frac{p(t_1, t_2)}{p(t_1)p(t_2)}. \tag{5.45b}$$

This relation implies a clear physical meaning for the normalized correlation function $\gamma^{2,2}(\tau)$. To prove this, we exploit the multiplication theorem of probabilities,

$$p_c(t_2|t_1) = \frac{p(t_1, t_2)}{p(t_1)}, \tag{5.46a}$$

where $p_c(t_2|t_1)$ is the conditional probability for a count at time t_2, when already at the time $t_1 = t_2 - \tau$ a count has been recorded. Inserting (5.45b), one obtains

$$p_c(t_2|t_2 - \tau) = \gamma^{2,2}(\tau)p(t_2). \tag{5.46b}$$

This means: If $\gamma^{2,2}(\tau) = 1$, then $p_c(t_2|t_2 - \tau)$ is equal to $p(t_2)$; in this case, the counts are statistically independent of one another. If $\gamma^{2,2}(\tau) > 1$, the conditional probability p_c surpasses the value of pure random counting; then we speak of photons that are *bunched*; a *bunching effect* appears. If $\gamma^{2,2}(\tau) < 1$, the conditional probability p_c falls below the value of pure random counting; we then speak of an *antibunching effect*.

To illustrate these statements, we consider some field states with an instructive physical meaning. The one-mode coherent state $|C\rangle = |0, \ldots, 0, \alpha_\mu, 0, \ldots\rangle$ leads to

$$\gamma_{|C\rangle}^{2,2}(\tau) = \frac{\langle \alpha_\mu | (\hat{a}_\mu^\dagger)^2 (\hat{a}_\mu)^2 | \alpha_\mu \rangle}{\langle \alpha_\mu | \hat{a}_\mu^\dagger \hat{a}_\mu | \alpha_\mu \rangle^2} = \frac{(\alpha_\mu^*)^2 \alpha_\mu^2}{(\alpha_\mu^* \alpha_\mu)^2} = 1. \tag{5.47}$$

Furthermore, we again consider chaotic light with a Gaussian photon energy distribution as characterized in connection with the derivation of (5.42c). Applying (5.35b), we arrive at

$$\gamma^{2,2}_{\mathrm{ChL}}(\tau) = 1 + \left|\gamma^{1,1}_{\mathrm{ChL}}(\tau)\right|^2 = 1 + e^{-\tau^2/\varkappa^2}, \tag{5.48}$$

where $\varkappa^{-1}$ again is the r.m.s. angular-frequency line width. The one-mode state $|\mathrm{F}\rangle = |0,\ldots 0, n_\mu, 0,\ldots\rangle$ with a fixed photon number $n_\mu \geq 2$ in the μth mode is associated with the function

$$\gamma^{2,2}_{|\mathrm{F}\rangle} = 1 - \frac{1}{n_\mu}. \tag{5.49}$$

In Fig. 5.4 the quantity $\gamma^{2,2}(\tau)$ is plotted versus τ for three states: for the coherent state (C); for chaotic light (ChL), where $\varkappa$ is taken to be 0.65 ms; and for a state of fixed photon number (F), where n_μ is taken to be 4. In the case of the monochromatic states $|\mathrm{C}\rangle$ and $|\mathrm{F}\rangle$, $\gamma^{2,2}(\tau)$ does not depend on the time difference τ. The experimental results (Ref. 5.8) concerning one-mode laser radiation show good agreement with the theoretical values of the coherent state. Thus, for laser light sufficiently above threshold, the photon counts can be regarded as independent of one another.

In the mid fifties Hanbury Brown and Twiss measured the quantity $\gamma^{2,2}_{\mathrm{ChL}}(\tau)$ for thermal light. However, they failed in obtaining $\varkappa$ values larger than some 10^{-8} s even with the narrow line of a mercury isotope. Therefore accurate counting of photons in the range $\tau \lesssim \varkappa$, which was of interest, became very difficult, since the measuring time T could not be chosen to be small compared with τ. Later Arecchi, Gatti, and Sona (Ref. 5.13) obtained the measured values for chaotic light given in Fig. 5.4. This light was generated artificially by sending laser light through a rotating ground-glass disc, whereby a randomiza-

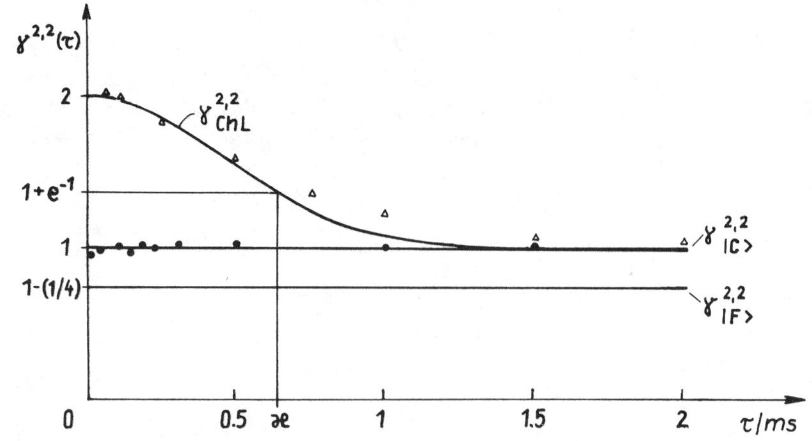

Fig. 5.4. Normalized second-order correlation function $\gamma^{2,2}(\tau)$ versus time difference. Theoretical curves: $\gamma^{2,2}_{|\mathrm{C}\rangle}$, coherent light; $\gamma^{2,2}_{\mathrm{ChL}}$, chaotic light ($\varkappa = 0.65$ ms); $\gamma^{2,2}_{|\mathrm{F}\rangle}$, fixed photon number. Experimental results (after Arecchi, Gatti, and Sona, Ref. 5.13): △, chaotic light; ●, laser light.

tion of the laser light resulted. For this radiation (ChL) the values measured are in good agreement with the theoretical curve for $\gamma^{2,2}(\tau)$, at least in the region $\tau \leq \varkappa$, if $\varkappa = 0.65$ ms is chosen. In this region the $\gamma^{2,2}(\tau)$ values markedly exceed unity; this means that the randomized laser light clearly displays a bunching effect (in contrast to the unrandomized laser light). It should be mentioned that in Ref. 5.13 the information about the (normalized) correlation function $\gamma^{2,2}$ was obtained by measuring the time distributions of photoelectrons from a single-photon counter. The experimental setup consisted of a variable-delay generator triggered by a photoelectron pulse operating a gate of fixed duration Δt_2 at a time τ after the occurrence of another pulse at $\tau = 0$. The number of pulses in Δt_2 was counted and recorded. This method is equivalent to that used by Hanbury Brown and Twiss (Ref. 5.12).

Since $\gamma^{2,2}_{|F\rangle}(\tau)$ is less than unity, an antibunching effect can be expected for the state $|F\rangle$. Experiments on the detection of antibunched radiation have been mainly performed in connection with those on resonance fluorescence (Ref. 5.14). During these experiments difficulties arose because of the fluctuations of the number of emitting atoms within the atomic beam. From the analyses of the experimental results given in Refs. 5.14, 5.15, 5.16, and 5.17, it may be concluded that the radiation of a single atom indeed reveals antibunching with $\gamma^{2,2}(\tau \to 0) < 1$. (Compare the discussion on transient resonance fluorescence in Chapter 9 and the treatment of the change of the correlation by multiphoton effects in Chapters 10, 11, and 12.)

Note that measuring the time-interval distribution of two successive pulses also yields information about the second-order intensity correlation. Time intervals can be measured electronically with high accuracy, and in this way the measurement of the distance of two successive pulses at t_1 and t_2 is performed. One pulse serves as the start signal of a clock, the other as the stop signal.

Correlation functions of the type $\Gamma^{m,m}(x_1, \ldots x_m, x_m, \ldots x_1)$, where $m > 2$, also contain important information about radiation. To give an example let us consider the case $m = 3$, that is, a third-order intensity correlation function. Corti and Degiorgio (Ref. 5.18) studied the radiation of a single-mode intensity-stabilized He–Ne laser. They investigated the (stationary) normalized correlation functions

$$\gamma^{3,3}(\tau = t_2 - t_1, \tau' = t_3 - t_1) = \frac{\Gamma^{3,3}(t_1, t_2, t_3, t_3, t_2, t_1)}{\Gamma^{1,1}(t_1, t_1)\Gamma^{1,1}(t_2, t_2)\Gamma^{1,1}(t_3, t_3)}. \tag{5.50}$$

By means of the quantities $\gamma^{3,3}$ and $\gamma^{2,2}$, an intrinsic third-order intensity correlation function $A^{3,3}(\tau, \tau')$ can be constructed, which is not predictable from the knowledge of second-order intensity correlations. The function $A^{3,3}(\tau, \tau')$ is normalized in such a way that it vanishes in the case of chaotic light [with the Gaussian distribution (2.138) of the field fluctuations]. By varying the value of the so-called laser pump parameter $\tilde{p}$ which will be introduced in Section 7.2.2, it could be shown that in a narrow region around the threshold ($\tilde{p} = 0$) the radiation source reveals a characteristic non-Gaussianity. The procedure can also be applied to the measurement of quantities that yield an additional information beyond the correlation effects of lower order. The experimental method follows, in principle, the same pattern as employed for measuring $\gamma^{2,2}(\tau)$. But now all the difficulties concerning the measuring time T, the correlation time τ_c, and the time differences of the t_j are greater, since now a third-order intensity correlation with the two time differences τ and τ' must be measured. Here the conditions are as follows. The correlation time of the radiation was about 40 μs. For a single fixed τ value, 107 different τ' values were measured. A relatively small statistical error of the measured correlation function, about 0.3% at each point (τ, τ'), could be achieved.

We shall treat other higher-order correlation functions in connection with the discussion of special nonlinear effects in Part II.

REFERENCES

5.1. Leonard Mandel, "Fluctuations of Light Beams," in E. Wolf (Ed.), *Progress in Optics*, Vol. 2, North-Holland, Amsterdam, 1963, p. 181.

5.2. F. T. Arecchi and V. Degiorgio, "Measurement of the statistical properties of optical fields," in F. T. Arecchi and E. O. Schulz-Dubois (Eds.), *Laser Handbook*, Vol. 1, North-Holland, Amsterdam, 1972, p. 191.

5.3. S. A. Akhmanov, Ju. E. Djakov, and A. S. Chirkin, "Introduction into statistical radiophysics and optics", Nauka, Moscow, 1981 (Russian), Chapter 2, §9.

5.4. F. T. Arecchi, *Phys. Rev. Lett.* **15**, 912 (1965).

5.5. R. J. Glauber, in C. De Witt, A. Blandin, and C. Cohen-Tannoudji (Eds.), *Quantum Optics and Electronics*, Gordon and Breach, New York, 1965, p. 84.

5.6. H.-E. Ponath and M. Schubert, *Ann. d. Phys.* **37**, 109 (1980).

5.7. R. Gase and M. Schubert, *Opt. Acta* **29**, 1331 (1982).

5.8. (a) R. Gase and M. Schubert, *Opt. Acta* **30**, 1125 (1983); (b) R. Gase and M. Schubert, *Opt. Acta* **32**, 433 (1985).

5.9. J. H. Eberly and K. Wodkiewicz, *J. Opt. Soc. Amer.* **67**, 1252 (1977).

5.10. K.-H. Brenner and K. Wodkiewicz, *Opt. Comm.* **43**, 103 (1982).

5.11. E. Wolf, *J. Opt. Soc. Amer.* **68**, 6 (1978).

5.12. R. Hanbury Brown and R. Q. Twiss, *Nature (Lond.)* **177**, 27 (1956).

5.13. F. T. Arecchi, E. Gatti, and A. Sona, *Phys. Lett.* **20**, 27 (1966).

5.14. N. Dagenais and L. Mandel, *Phys. Rev. A* **18**, 2217 (1978).

5.15. M. Schubert and B. Wilhelmi, "The mutual dependence between coherence properties of light and nonlinear optical processes," in E. Wolf (Ed.), *Progress in Optics*, Vol. 17, North-Holland, Amsterdam, 1980, p. 163.

5.16. H. J. Carmichael, P. Drummond, P. Meystre, and D. F. Walls, *Opt. Acta* **27**, 581 (1980).

5.17. M. Schubert, K.-E. Süsse, W. Vogel, D.-G. Welsch, and B. Wilhelmi, *Kvant. Elektr.* **9**, 495 (1982).

5.18. M. Corti and V. Degiorgio, in B. Havelka and J. Blabla (Eds.), *Proceedings of ICO-10*, Prague, 1975, p. 59.

6

Nonstationary Processes

Nonstationary processes are of importance in nonlinear optics for several reasons (see, e.g., Refs. 6.1 and 6.2). First, the generation of short laser pulses allows the application of higher field strengths than can be obtained in the continuous-wave regime, so that the nonlinearities of the irradiated materials become more obvious. By means of intense light pulses substantial changes in the populations of the atomic levels involved as well as strong coherent excitations of the whole system can be temporarily achieved. Second, by nonlinear optical interactions the incident light pulses can be changed in their parameters. Such effects can be exploited for changing pulse duration, maximum intensity, and pulse shape. In such a way, pulses can be shortened or stabilized in their intensity in a controlled manner. Third, the generation of ultrashort light pulses in lasers is a probe of the nonlinearity of the nonstationary interaction processes involved.

On the one hand, nonstationary processes are influenced by the noninstantaneous response of the atomic systems in building up a new charge distribution after the external field has been changed. Since this influence can be observed at a fixed place, it is called a *local* effect.

On the other hand, the dispersive properties of the sample result in transit-time spread between pulses of different midfrequencies as well as in distortions of single pulses, and this may affect the efficiency of nonlinear optical interactions considerably. Since the measurement of this second phenomenon requires longer paths of light in the materials under study, these effects are called *nonlocal.*

In many experiments the local and nonlocal effects appear coupled. To measure them singly requires appropriate measuring and evaluation methods.

The local nonstationary effects, which strongly depend on the relaxation parameters of the resonantly excited individual atomic transitions, will be treated in the chapters that deal with specific one-photon and multiphoton interaction processes. In this chapter we restrict our consideration to nonresonant optical interactions, in which only nonlocal effects appear, because the inertia of the atomic response will have no influence on the nonstationary process.

First of all we want to discuss the kind of influence exerted by the linear optical dispersion on nonlinear interaction processes and to study the conditions under which this influence can be neglected. To do this, we assume in the first and the second section that the nonlinearities are so weak that their influence on quantities such as refractive index and propagation velocity can be neglected. Thus we may imagine the propagation properties of all light pulses to be mainly determined by the optical constants of linear optics, whereas the weak nonlinear coupling only gives contributions to the amplification or attenuation of incident pulses or induces the generation of new pulses and influences their further development. In Section 6.3 we treat the possible compensation of the influence of linear optical dispersion by nonlinear processes.

6.1. PULSE PROPAGATION THROUGH DISPERSIVE LINEAR OPTICAL MEDIA

The effect of linear optical dispersion on nonlinear interactions between light and matter mainly results from two phenomena. First, two pulses with different midfrequencies ω_1 and ω_2 and group velocities $v_1 \equiv (d\omega/dk)_{\omega_1}$ and $v_2 \equiv (d\omega/dk)_{\omega_1}$, which enter a sample at $z = 0$ at the same time, move apart; and second, each single pulse of midfrequency ω_L changes its shape due to the dispersion of the group velocity at ω_L.

In discussing the propagation of two given pulses that are moving apart, we neglect the group-velocity dispersion within the frequency band of each of the two pulses. The arrival of the pulse maximum of the pulse with the higher group velocity v_1 at the point z differs from that with v_2 in that it arrives there earlier by

$$\Delta t = D\frac{z}{c}, \tag{6.1}$$

where the dimensionless dispersion parameter is given by

$$D = c\left(\frac{1}{v_2} - \frac{1}{v_1}\right). \tag{6.2}$$

Provided both entrance pulses have the same duration τ_L, this means that after a distance of $z = L_D$, where

$$L_D = \frac{1}{D}c\tau_L, \tag{6.3}$$

only a small amount of the overlap of the two pulses remains, and thus their interaction is greatly reduced in nonlinear optical processes. The tendency of the pulses to move apart can only be neglected as long as $z \ll L_D$. As an example we consider pulses at the Nd-laser wavelength $\lambda_1 = 1.06\ \mu$m and at the wavelength $\lambda_2 = 0.53\ \mu$m of the second harmonic of this laser radiation, which respectively propagate with ordinary and extraordinary directions of polarization in the crystal $LiIO_3$, the direction of propagation being chosen such that the two refractive indices coincide $[n^{(o)}(\omega_1) = n^{(e)}(\omega_2)]$. Then from the dispersion data on $LiIO_3$ the value $D = 0.084$ can be obtained for the dispersion parameter (Ref. 6.2). For input pulses of 3-ps duration this means that the overlapping and hence the nonlinear interaction can change significantly over the length $L_D \approx 10$ mm, whereas $L_D \approx 0.1$ mm results for extremely short pulses of 30-fs duration. Hence, for pulses of such a short duration very short interaction lengths or materials with a very small dispersion parameter D must be chosen.

Now we are going on to calculate the change in the shape of single pulses when passing the dispersive medium. Any single input pulse of given mid-frequency ω_L and pulse duration τ_{L0} changes its maximum power, pulse shape, and pulse duration in its passage through long enough linear dispersive materials. Since in any nonlinear optical effect the field strength of the incident radiation enters into the nonlinear polarization (see Section 1.3) in high powers, such a pulse shaping affects the efficiency of the nonlinear processes considerably. If the pulse shaping is to be calculated, one cannot start from the abbreviated wave equation (1.220) for the slowly varying amplitudes, but has to take into account additional terms of the general wave equation (1.215). If we further assume infinite plane waves propagating in the sample in the z direction and if we for the present neglect the nonlinear polarization as a driving force in the field equation, then from (1.215) we have

$$\frac{\partial^2}{\partial \eta^2}\bar{E}_L(\xi, \eta) - \frac{2i}{k''}\frac{\partial}{\partial \xi}\bar{E}_L(\xi, \eta) = 0, \tag{6.4a}$$

where $\eta = t - (1/v)z$ is the *retarded time* or *local time*, $\xi = z$ is the space coordinate, and

$$k'' \equiv \left(\frac{d^2k}{d\omega^2}\right)_{\omega=\omega_L} = \frac{\lambda_L^3}{2\pi c^2}\left(\frac{d^2n}{d\lambda^2}\right)_{\lambda=\lambda_L} \tag{6.4b}$$

is the second derivative of the dispersion curve at the frequency $\omega = \omega_L$. Instead of k'' the material parameter

$$M \equiv \frac{\omega_L^2}{2\pi c}k'' = \frac{\lambda_L}{c}\left(\frac{d^2n}{d\lambda^2}\right)_{\lambda=\lambda_L}$$

is frequently used. In loss-free spectral regions the magnitude and the wavelength dependence of k'' and M can be easily calculated if we start from a simple dispersion formula, for instance from a three-parameter Sellmeier

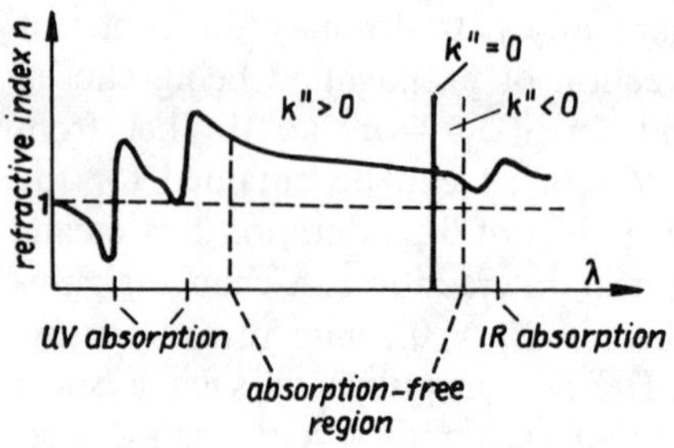

Fig. 6.1. Refractive index versus wavelength.

formula of the form

$$n^2 - 1 = \frac{\omega_d \omega_0}{\omega_0^2 - \omega^2} - \frac{\omega_l^2}{\omega^2}.$$

Depending on the values of the parameters ω_d, ω_0, and ω_l and of the light frequency under consideration, the quantities M and k'' may attain positive and negative values. Figure 6.1 displays the regions with $k'' > 0$ and $k'' < 0$ for a typical optical material, which is assumed to be loss-free in the visible region and to exhibit strong absorption in the ultraviolet and infrared spectral regions. It can be clearly seen that in the major part of the absorption-free region $k'' > 0$, whereas it is only in the transition zone towards the infrared region that k'' becomes negative.

The general solution of the linear optical equation (6.4) is

$$\overline{E}_L(\xi, \eta) = \frac{1 + ik''/|k''|}{2\sqrt{\pi |k''| \xi}} \int_{-\infty}^{\infty} d\eta' \, \overline{E}_{L0}(\eta') \exp\left(-\frac{i(\eta - \eta')^2}{2k''\xi} \right), \qquad \xi > 0, \tag{6.5}$$

where $\overline{E}_{L0}(\eta) = \overline{E}_L(\xi = 0, \eta)$ is the given field-strength amplitude at the boundary. By differentiation this solution can be easily seen to satisfy (6.4), and by use of the representation

$$\delta(\eta - \eta') = \frac{1}{\sqrt{\pi}} \lim_{a \to 0} \frac{1}{\sqrt{a}} e^{-(\eta - \eta')^2/a}$$

of Dirac's delta function we verify that (6.5) satisfies the boundary condition at $\xi = 0$.

We now suppose that a Gaussian-shaped input pulse with

$$\overline{E}_{L0}(\eta) = |\overline{E}_{Lm}| e^{-(2 \ln 2)(\eta/\tau_{L0})^2} e^{-i[\beta_0 \eta^2 + \varphi_{L0}]} \tag{6.6}$$

enters the sample at $\xi = 0$, where phase modulation that might occur is taken into account by the contribution to the phase rising quadratically with time; τ_{L0}, β_0 and φ_{L0} are free parameters. This expression for the time-dependent phase corresponds to an expansion of the phase in the neighborhood of the pulse maximum, where any contribution linear in η is already contained in the midfrequency ω_L. Such phase modulations $\Delta\varphi(\eta)$—which cause, according to

$$\Delta\omega(\eta) = \frac{d}{d\eta}\Delta\varphi(\eta),$$

an approximately linear variation of the instantaneous frequency with time during the action of the main part of the pulse—frequently occur with laser pulses, which then, of course, will not be bandwidth-limited. Such a frequency change is called *pulse chirp*, and we distinguish pulses with *up chirp* and those with *down chirp* according as the instantaneous frequency increases or decreases with time. The cause of such phase modulations, which are due to linear and nonlinear optical processes during pulse generation, will be discussed later. Assuming the input pulse is given by (6.6), we calculate the change in the pulse shape while passing through the medium by using (6.5), and so we obtain the dependence of the pulse duration τ_L on the path length z in the dispersive medium as

$$\frac{\tau_L(z)}{\tau_{L0}} = \sqrt{1 + 2\beta_0 k'' L_\beta \left[1 - \left(1 - \frac{z}{L_\beta}\right)^2\right]} \tag{6.7a}$$

where

$$L_\beta = -\frac{2\beta_0 \tau_{L0}^4 k''}{(4\ln 2\, k'')^2 + (2\beta_0 \tau_{L0}^2 k'')^2}. \tag{6.7b}$$

This relation reveals that the pulse for $L_\beta > 0$, which requires that $k''\beta_0 < 0$, is shortened on its path from $z = 0$ to $z = L_\beta$ (see Fig. 6.2). At this point the pulse is *bandwidth-limited*. For $z > L_\beta$ the pulse duration increases in any case, so that L_β gives the optimum thickness of a dispersive sample to be used for shortening pulses with given input parameters, the length L_β being critically dependent on the duration of the input pulses. For bandwidth-limited input pulses we have $\beta_0 = 0$ and, as a consequence, $L_\beta = 0$, and hence the pulses experience a lengthening from the entrance at $z = 0$ according to

$$\frac{\tau_L(z)}{\tau_{L0}} = \sqrt{1 + (4\ln 2\, k'')^2 \tau_{L0}^{-4} z^2}.$$

From (6.7b) it follows that the compression of pulses with down chirp requires

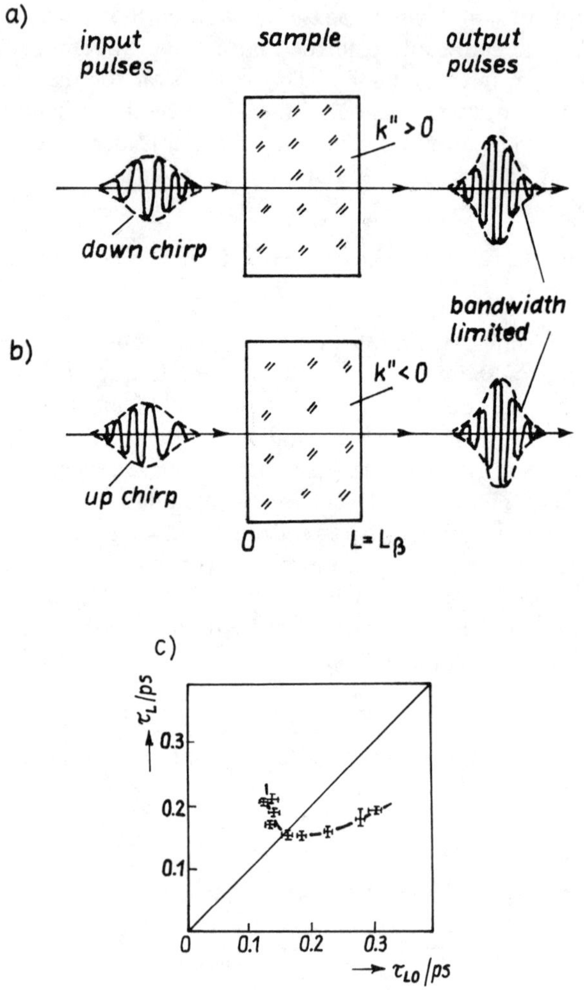

Fig. 6.2. Pulse compression: (*a*) compression of down-chirped pulses in a sample with $k'' > 0$ and with length L_β, (*b*) Compression of up-chirped pulses in a sample with $k'' < 0$ and with length L_β, (*c*) Shaping of down-chirped femtosecond pulses ($\lambda = 0.61\ \mu$m) in a BK5 glass sample of 17-cm length (after Ref. 6.12). The long entrance pulses with $\tau_{L0} > 0.17$ ps experience compression, whereas very short pulses increase in duration because of $L > L_\beta$).

dispersive elements with $k'' > 0$; that of pulses with up chirp, elements with $k'' < 0$ (see Fig. 6.1). The region with $k'' < 0$ is of great importance for pulse propagation in optical fibers, since for pulses of such wavelengths and of appropriate power, distortionless propagation over large distances may occur, which results from the fact that the action of the linear optical effect is compensated for by that of a nonlinear effect (cf. Section 6.3). Between the regions with positive and negative curvature of the $k(\omega)$ and $n(\lambda)$ curves there is the point with $k'' = 0$ and $d^2n/d\lambda^2 = 0$. At this wavelength the pulse passes the material with hardly any distortion. This means that an almost distortionless propagation is possible even without any action of nonlinear processes at low input powers.

6.2. GENERATION OF LIGHT PULSES IN DISPERSIVE NONLINEAR OPTICAL MEDIA

In this section we study the influence of different group velocities of signal pulse and pump pulses on the efficiency of nonlinear conversion processes. The conditions are chosen to be as simple as possible. Let all light pulses propagate in the z direction, and at each pulse midfrequency ω_i let only one value of the wave number $k(\omega_i)$ occur in the sample. We next suppose the frequencies of all light pulses involved to be far from resonance frequencies of the sample, and the sample to be so thin that the attenuation of the pump pulses, which may arise from the nonlinear conversion, can be neglected.

To describe such interaction processes let us start from the results of Section 1.3. The "abbreviated" wave equation derived here, which describes the buildup of the slowly varying amplitude $\overline{E}_S(z,t)$ of the signal at frequency ω_S in the field of pump pulses at frequencies $\omega_1, \ldots, \omega_n$, reads

$$\left(\frac{\partial}{\partial z} + \frac{1}{v_S}\frac{\partial}{\partial t}\right)\overline{E}_S(z,t) = i\tilde{\tilde{\chi}}^{(n)}\overline{E}_1(z,t)\cdots\overline{E}_n(z,t)e^{+i\Delta k z} \qquad (6.8)$$

where

$$\tilde{\tilde{\chi}}^{(n)} = \frac{\omega_S^2}{2c^2 k_S}\sum_{\alpha,\beta,\ldots,\mu} e_\alpha^{(S)}\chi_{\alpha\beta\cdots\mu}^{(n)}(\omega_S;\omega_1,\ldots,\omega_n)e_\beta^{(1)}\cdots e_\mu^{(n)}$$

is calculated from the nonlinear susceptibility $\chi^{(n)}_{\bullet\bullet\cdots\bullet}$ and the unit vectors $e^{(S)}_\bullet, e^{(1)}_\bullet, \ldots, e^{(n)}_\bullet$ in the direction of the field-strength vectors, and where $\Delta k = k_1 + \cdots + k_n - k_S$. We now pass over from the coordinates z and t to the retarded (local) time $\eta \equiv \eta_S = t - z/v$ with respect to the signal pulse and to $\xi = z$, where $v \equiv v_S$ is the group velocity of the signal pulse. Given the assumption that, in spite of linear dispersion and nonlinear coupling, the pump pulses propagate within the sample without being significantly distorted, we conclude that the slowly varying amplitude $\overline{E}_i$ of the ith pulse in a coordinate system $\eta_i = t - z/v_i$ and $\xi = z$ is independent of ξ. Hence the amplitude only depends on $\eta_i = \eta + D_i\xi/c$, where $D_i \equiv D_{Si} = c(v^{-1} - v_i^{-1})$. From (6.8) we thus obtain the equation

$$\frac{\partial}{\partial\xi}\overline{E}_S(\xi,\eta) = i\tilde{\tilde{\chi}}^{(n)}\overline{E}_1\left(\eta + D_1\frac{1}{c}\xi\right)\cdots\overline{E}_n\left(\eta + D_n\frac{1}{c}\xi\right)e^{i\Delta k\xi}, \qquad (6.9)$$

the solutions of which will be discussed as a function of the dispersion parameters of the medium.

The solution for the special case of all the pulses having the same propagation velocity ($v_i = v_S = v$, $i = 1, \ldots, n$) is the simplest, because in it the pump pulses in the local coordinate system of the signal pulse are independent of ξ. In this case the treatment is reduced to the solution of the stationary problem, which has been given in Section 1.4.2, especially in (1.202); but now the local time η appears as an additional parameter.

Provided the pump pulses propagate with different group velocities, the product of amplitudes on the right-hand side of (6.9) changes with increasing ξ for $\eta = \text{const}$. As an example let us consider a situation in which two square pump pulses of different group velocities v_1 and v_2 but of equal duration $\tau_{L1} = \tau_{L2} = \tau_L$ simultaneously enter the sample at $\xi = 0$. The pulse with the higher group velocity runs ahead of the slower one, and for $\xi > L_D = c\tau_L/D$ the pulses no longer overlap, which makes the influence of the nonlinear coupling on the signal pulse disappear. With different pulse shapes the argument given remains valid and qualitatively yields the same results.

In the following we want to discuss the consequences arising from the differences between the group velocity of the signal pulse and the group velocities of the pump pulses. Also in this treatment we assume the conditions to be as simple as possible. Let all pump pulses be of the same duration τ_L, simultaneously arrive at $\xi = 0$, and propagate with the same group velocity v_1. We further assume either that all pump pulses have the same midfrequency or that the midfrequencies differ only slightly from each other. [As an example we may consider second-harmonic generation (SHG) at the frequency $\omega = 2\omega_1$ in the field of one or two pump pulses with midfrequency ω_1.] In addition we assume that $\Delta k = 0$ and that the pump pulses are Gaussian-shaped, where

$$\bar{E}_i(\eta_i) = \bar{E}_{im} e^{-(2\ln 2)(\eta_i/\tau_L)^2}.$$

If we assume all the conditions mentioned to be satisfied, then we obtain from (6.9)

$$\frac{\partial}{\partial \xi} \bar{E}_S(\xi, \eta) = i\tilde{\bar{\chi}}^{(n)} \bar{E}_{1m} \cdots \bar{E}_{nm} \exp\left[-\frac{(2\ln 2)n}{\tau_L^2} \left(\eta + \frac{D_1 \xi}{c} \right)^2 \right], \tag{6.10}$$

where

$$D_1 = c\left[\frac{1}{v_S} - \frac{1}{v_1} \right].$$

The integration yields

$$\bar{E}_S(\xi, \eta) - \bar{E}_S(0, \eta) = i\tilde{\bar{\chi}}^{(n)} \bar{E}_{1m} \cdots \bar{E}_{nm} \frac{\sqrt{\pi}\, \tau_L c}{2\sqrt{2\ln 2}\, \sqrt{n}\, D_1}$$
$$\times \left\{ -\Phi\left[\frac{\sqrt{2\ln 2}\, \sqrt{n}}{\tau_L} \eta \right] + \Phi\left[\frac{\sqrt{2\ln 2}\, \sqrt{n}}{\tau_L} \left(\eta + \frac{D_1}{c} \xi \right) \right] \right\}, \tag{6.11}$$

where the Gaussian error integral is defined as

$$\Phi(x) = \frac{2}{\sqrt{\pi}} \int_0^x dt\, e^{-t^2}$$

(see Refs. 1, 6.2).

At this point it should be briefly mentioned that for the more general case where $\Delta k \neq 0$, error integrals of complex argument are obtained as solutions, since then the factor $e^{i\Delta k \xi}$ additionally occurs in the differential equation that is analogous to (6.10) (cf. Ref. 1). For $\eta = 0$ and short distances ξ in the sample, the first term of an expansion of the Gaussian error function gives the stationary solution, where the dispersion parameter D_1 does not appear. The other terms of this expansion cause the signal pulse to increase more slowly than does the stationary solution.

In Fig. 6.3 the slowly varying signal amplitude $\bar{E}_S$ is plotted as a function of the normalized local time $\tilde{\eta} = \sqrt{2 \ln 2}\sqrt{n}\,\eta/\tau_L$ and the normalized path length $\tilde{\xi} = \sqrt{2 \ln 2}\sqrt{n}\, D_1 \xi/(\tau_L c)$, setting $\bar{E}_S(0, \tilde{\eta}) = 0$. The figure shows that the field-

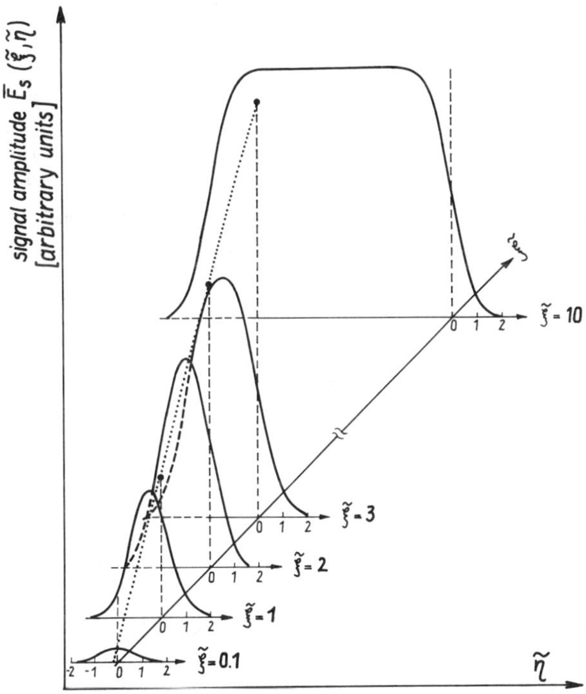

Fig. 6.3. Generation of the signal pulse S in the field of given pump pulses. The dependence of the field-strength amplitude $\bar{E}_S(\tilde{\xi}, \tilde{\eta})$ of the signal pulse on the normalized time coordinate $\tilde{\eta}$ is shown at several positions $\tilde{\xi}$ in the sample. For comparison the dotted line gives the increase of the stationary solution versus $\tilde{\xi}$.

strength pulses increase for large ξ more slowly than they do in the stationary case, where there is a linear dependence on ξ. With increasing ξ the maximum of the pulses generated in this process shifts forward. This can be explained from the fact that the pump pulses travel faster through the sample (with $D_1 > 0$) than does the signal pulse. The consequence is that the conditions for signal-pulse generation are more advantageous at the leading edge of the pulses than they are at the trailing edge. Finally, because of the different group velocities the pulse duration increases with increasing ξ. At very large ξ the maximum amplitude does not continue to increase with ξ, whereas the pulse duration increases approximately linearly with $D_1\xi$. As a result, the signal pulse assumes an almost rectangular shape. This is due to the fact that the pump pulses continually generate new signal light, which, after traveling a certain distance in the sample, will not be amplified any longer because it propagates more slowly than the pump pulses.

6.3. PULSE PROPAGATION THROUGH NONRESONANT NONLINEAR OPTICAL MEDIA

The aim of this section is the study of the action of nonresonant nonlinearities on the change of the complex field-strength amplitude of a single pulse of midfrequency ω_L during its propagation. In Section 6.3.1 we neglect the linear optical dispersion of the sample, whose influence on the pulse parameters has been dealt with in Section 6.1. In Section 6.3.2 we consider the simultaneous action of linear optical dispersion and nonlinearities on the pulse shape. Special attention is paid to the possibility of pulse narrowing and the propagation of solitary pulses.

6.3.1. Pulse Propagation Through Nondispersive Nonlinear Optical Media

Let us consider a situation that is representative for a large class of actual experiments. We suppose a very short pulse to propagate as a plane wave of given direction of polarization in a nonlinear dispersive material in the z direction. The nonresonant nonlinearity of the material is assumed to be given by the following dependence of the refractive index n on the slowly varying field strength amplitude $\bar{E}(z, t)$ of the pulse:

$$n(z, t) = n^{(0)} + \tilde{n}^{(2)}|\bar{E}(z, t)|^2, \tag{6.12}$$

where $\tilde{n}^{(2)}|\bar{E}(z, t)|^2 \ll n^{(0)}$ is assumed. This relation is equivalent to the expression

$$\bar{P}^{(3)}(z, t) = 2\varepsilon_0 n^{(0)}\tilde{n}^{(2)}|\bar{E}(z, t)|^2\bar{E}(z, t) \tag{6.13}$$

for the nonlinear polarization $P^{(3)}$ in the direction of the field strength of the incident pulse (cf. Sections 1.3.2 and 4.4). If we compare (6.13) with the general relations between polarization and field strength in the third order, we see that here the nonlinear susceptibility $\varkappa^{(3)}$ is given by the field-strength-independent part of the refractive index and the part of the refractive index that depends on the squared modulus of the field strength. From the discussion in Section 4.4 it follows that the electronic transitions in the ultraviolet spectral region may give rise to such nonresonant nonlinearities in the visible and near infrared regions. By means of the relations given there we estimate this electronic contribution to be of the order of $\tilde{n}^{(2)} \sim 10^{-22}$ m^2/V^2. Other contributions, for instance those caused by the Kerr effect, will not be discussed, since we deal here with solid samples in the visible and near infrared regions, where the electronic transitions give the main contribution. Regarding the value of $\tilde{n}^{(2)}$, the requirement for small nonlinearity implies that the field-strength amplitude $\bar{E}$ must be less than 10^8 V/cm and the photon flux density of visible radiation less than about 10^{31} cm^{-2} s^{-1}.

The polarization amplitude of (6.13) is substituted into the wave equation for the slowly varying field-strength amplitude (6.8), and thus we obtain

$$\frac{\partial}{\partial \xi}\bar{E}(\xi, \eta) = +i\tilde{\tilde{\chi}}^{(3)}|\bar{E}(\xi, \eta)|^2\bar{E}(\xi, \eta), \tag{6.14}$$

where

$$\tilde{\tilde{\chi}}^{(3)} = \frac{k\tilde{n}^{(2)}}{n^{(0)}}$$

is a nonlinear optical coupling parameter. The solution of (6.14) for an entrance pulse $\bar{E}(\xi = 0, \eta) = \bar{E}_0(\eta)$ reads

$$\bar{E}(\xi, \eta) = \bar{E}_0(\eta)e^{i\tilde{\tilde{\chi}}^{(3)}|\bar{E}_0(\eta)|^2\xi}. \tag{6.15}$$

Since $\tilde{\tilde{\chi}}^{(3)}$ is real in loss-free media, the modulus of the field-strength amplitude remains constant, whereas the phase of the pulse changes in passing through the medium. Let us consider this phase modulation around the maximum of the pulse at $\eta = 0$, where the squared modulus of the field strength amplitude can be represented as

$$|\bar{E}_0|^2 = |\bar{E}_{\max}|^2 + \frac{1}{2}\left(\frac{\partial^2}{\partial\eta^2}|\bar{E}_0|^2\right)_{\eta=0}\eta^2.$$

Substituting this expression into (6.15), we have

$$\bar{E}(\xi, \eta) = |\bar{E}_0(\eta)|e^{-i\varphi_0(\eta)}e^{-i\alpha}e^{-i\beta\eta^2}, \tag{6.16a}$$

where

$$\alpha = -\frac{\omega}{c}\tilde{n}^{(2)}|\bar{E}_{\max}|^2\xi \tag{6.16b}$$

and

$$\beta = -\frac{1}{2}\tilde{n}^{(2)}\left(\frac{\partial^2}{\partial\eta^2}|\bar{E}_0(\eta)|^2\right)_{\eta=0}\xi. \tag{6.16c}$$

This means that the nonlinearity gives rise to a frequency chirp, which increases linearly with the distance ξ in the material. In most cases, $\tilde{n}^{(2)} > 0$ for the electronic contribution to nonlinearity; that is, the refractive index increases with intensity. Then positive values for β result, which corresponds to an up chirp or, in other words, to an increase of frequency with increasing time. From the considerations in Section 6.1 it can be inferred that it is possible to eliminate such a chirp and to achieve a shortening when allowing the pulse to pass through a material with $k'' < 0$. We see that in this way pulse compressions can be achieved in arrangements of separate nonlinear and linear optical elements (see, e.g., Refs. 6.2–6.5).

Because of its importance let us here insert the description of an experiment of this kind, by means of which Shank, Fork, and Yen (Ref. 6.6) succeeded in generating the shortest pulses ever produced until that time (see Fig. 6.4). The experimental approach is based on using pulses of a mode-synchronized dye ring laser of about 80-fs duration (cf. Ref. 6.7). Behind the oscillator the pulses are amplified, wherefrom a slight pulse lengthening may result. These pulses are focused into a short piece of glass fiber. As a result, according to what was stated above, an up chirp is generated, which is associated with a broadening of the spectral width from about 6 nm to 20 nm. Subsequently, in an arrangement of two diffraction gratings, which is equivalent to an optical material with $k'' < 0$, the pulses can be compressed until they become bandwidth-limited. In this manner a pulse duration of 30 fs was achieved. Weiner and Ippen (Ref. 6.13), by using the same method in a refined version, were successful in generating light pulses of only 16-fs duration, which consisted of only eight wave periods. (See Section 8.6.3 for further results and references.)

In typical experiments the conditions are somewhat different from those assumed in our simplified consideration. In addition to the central part of the

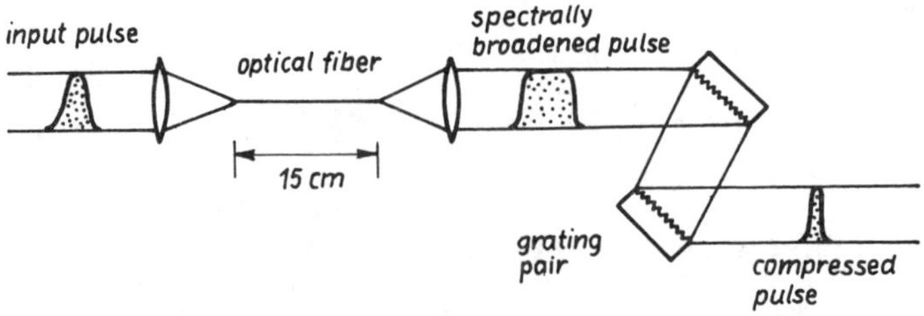

Fig. 6.4. An arrangement for pulse compression (after Ref. 6.6).

pulse, where the chirp can be approximated by one linear function of η, the pulse has wings where the chirp exhibits another functional dependence on η. Therefore the wing parts of the pulse will not be compressed together with the central part, but during the compression of the latter, small satellite pulses can occur before and after the main pulse. In Refs. 6.6 and 6.13 care was taken to minimize the power content of the satellites and to achieve maximum compression. For this purpose the combined action of linear dispersion and nonlinearity in the optical glass fiber had to be taken into account (cf. Refs. 6.8 and 6.14). We shall come back to this point at the end of the next subsection.

6.3.2. Pulse Propagation Through Dispersive Nonlinear Optical Media. Solitary Pulses

We have already pointed out that if we want to investigate the distortion of extremely short pulses, we are not allowed to start from the "abbreviated" wave equation but rather have to take into account in (1.215) the term that contains the second derivative with respect to the retarded time introduced in (1.222). This leads to a relation that corresponds to (6.4a) upon adding to the latter the term with the nonlinear polarization. It reads

$$i\frac{\partial}{\partial\xi}\bar{E}(\xi,\eta) = +\frac{k''}{2}\frac{\partial^2}{\partial\eta^2}\bar{E}(\xi,\eta) - \tilde{\tilde{\chi}}^{(3)}|\bar{E}(\xi,\eta)|^2\bar{E}(\xi,\eta). \quad (6.17)$$

For certain ranges of the parameters of the material this nonlinear partial differential equation provides very interesting solutions. In particular, in appropriate materials bandwidth-limited input pulses can be shortened under certain conditions. Such a shortening, as already mentioned, cannot be accomplished by linear optical means. Moreover, pulses can even propagate over long path lengths without changing their shape. These two phenomena are due to the simultaneous action of linear optical dispersion and nonlinearity.

According to the statements made above and the results presented in Section 6.3.1, it is obvious that at $\tilde{n}^{(2)} > 0$ a shortening of bandwidth-limited input pulses can be expected only in spectral regions where $k'' < 0$. As is apparent from Fig. 6.1, typical optical media possess negative k'' in loss-free wavelength regions only where in addition to the ultraviolet absorption bands the infrared bands have a considerable influence on the dispersion. At such wavelengths chirp generation and pulse compression may occur simultaneously, and pulse shortening as well as the propagation of distortionless pulses may be observed.

In Ref. 6.9 the propagation of sech-shaped laser pulses has been investigated. The pulse envelope $\bar{E}(0,\eta)$ at the entrance is given as

$$\bar{E}(0,\eta) = \frac{a}{\tau_L'}\sqrt{\frac{k''}{\tilde{\tilde{\chi}}^{(3)}}}\,\mathrm{sech}\frac{\eta}{\tau_L'}, \quad (6.18)$$

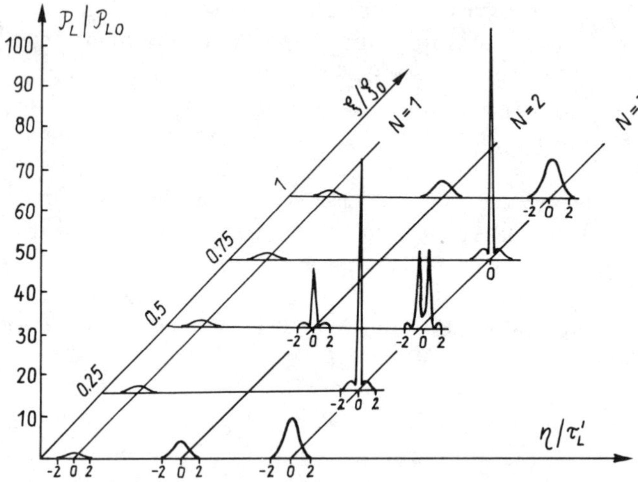

Fig. 6.5. Light power $\mathcal{P}_L$ of N soliton pulses ($N = 1, 2, 3$) versus space coordinate ξ and time η (after Ref. 6.9). $\mathcal{P}_{L0}$ is the peak power of the soliton ($N = 1$), ξ_0 is the soliton period, $\tau_L = 1.76\tau_L'$ is the pulse width at $\xi = 0$.

where $\tau_L = 1.76\tau_L'$ is the pulse duration (fwhm) and a is a dimensionless amplitude factor. It should be noted here that this sech pulse shape can be used for an approximate description of pulses of passively mode-locked dye lasers and color-center lasers.

The numerical evaluation of (6.17) reveals the interesting phenomenon that input pulses with $a = 1$ experience distortionless propagation in the sample. Pulses exhibiting this property are called *solitary pulses* (see e.g. Ref. 6.10). Other types of solitary waves, and especially *solitons*, will be given attention when discussing resonant interactions in Chapter 8. When a is a natural number > 1 there appear solitary solutions of a more complicated type. The passage of these pulses through the material is not distortionless throughout, but the pulses reproduce their shape only after the *periodicity length*

$$\xi_0 = \frac{\pi}{2} \frac{\tau_L'^2}{k''}. \tag{6.19}$$

Within the *periodicity interval* these pulses are split up, which is shown in Fig. 6.5.

These calculations were confirmed experimentally by Mollenauer, Stolen, and Gordon (Ref. 6.9). The schematic in Fig. 6.6 shows the setup and procedure employed by them. The actively mode-locked ion gas laser pumps the color-center laser with F_2^+ centers in $NaCl$ synchronously. This laser is

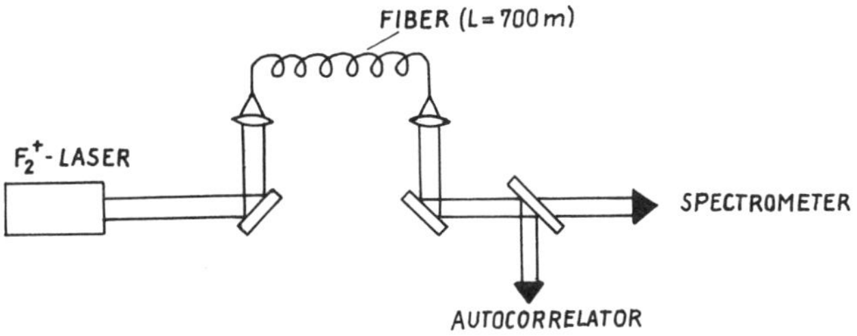

Fig. 6.6. Experimental setup for the observation of soliton propagation (after Ref. 6.9).

tunable in the wavelength region from 1.35 to 1.75 μm, and it emits ultrashort light pulses with a duration of $\tau_L = 7$ ps, a pulse repetition frequency of 100 MHz, and a pulse-duration–bandwidth product $\tau_L \Delta\nu = 0.18$. By means of focusing, the pulse train is coupled into a 700-m-long optical glass fiber. At the end of the fiber the duration of the pulses is measured with an autocorrelator, and the bandwidth with a spectrometer. The experiments of Ref. 6.9 were carried out at $\lambda = 1.55$ μm, where the dispersion curve of glass fiber has a negative curvature with $M = -16$ ps/nm km. At low time-averaged laser powers ($\overline{\mathscr{P}}_{L0} \lesssim 0.3$ W) the pulse duration increases to 16 ps during the passage through the fiber. Such a prolongation corresponds to the linear optical properties of the sample. At $\overline{\mathscr{P}}_{L0} = 1.2$ W the duration of the pulse at the entrance and exit of the fiber is the same. The duration of the output pulse decreases for increasing average power, and a minimum of 2 ps is reached at $\overline{\mathscr{P}}_{L0} = 5$ W. At still higher power splittings can be observed, which are in qualitative agreement with the higher-order solitary solutions.

In an exact comparison between measured and calculated results, attention has to be paid to the fact that the experiment is not based on the propagation of extended plane waves but on guided modes in a waveguide. (The propagation of pulses in nonlinear optical fibers is treated, e.g., in Ref. 6.11.) Nevertheless, from this experiment and from our simple calculations we recognize the strong influence of the interplay of linear optical dispersion and nonlinearity of the sample material on the evolution of the pulses.

Finally let us come back to the pulse-compression experiment described at the end of Section 6.3.1. In our first discussion we did not take into account the linear optical dispersion of the nonlinear medium, which was represented in that case by a piece of optical monomode fiber. At the pulse wavelength $\lambda = 0.6$ μm the dispersion parameter k'' is positive, and since $\tilde{n}^{(2)}$ is positive, this means that the pulse becomes longer. After the pulse has traveled an appropriate distance in the medium, it attains a nearly rectangular temporal shape and a frequency that increases linearly with time over nearly the whole pulse length. For this reason most of the pulse energy can afterwards be compressed into one very short single pulse (cf. Refs. 6.8 and 6.14).

REFERENCES

6.1. S. L. Shapiro (Ed.), *Ultrashort Light Pulses*, Springer, Berlin (West), Göttingen, Heidelberg, New York, 1977.

6.2. J. Herrmann and B. Wilhelmi, *Laser for Ultrashort Light Pulses* Akademie, Berlin, 1984 (German); Elsevier, Amsterdam, 1986 (English).

6.3. E. B. Treacy, *Phys. Lett.* **28A**, 34 (1968).

6.4. T. K. Gustafson, J. P. Taran, H. A. Haus, J. R. Lifschitz, and P. L. Kelley, *Phys. Rev.* **177**, 1196 (1969).

6.5. A. Laubereau, *Phys. Lett.* **29A**, 539 (1969).

6.6. C. V. Shank, R. L. Fork, R. T. Yen, R. H. Stolen, and W. J. Tomlinson, *Appl. Phys. Lett.* **40**, 761 (1982).

6.7. R. L. Fork, B. I. Greene, and C. V. Shank, *Appl. Phys. Lett.* **38**, 671 (1981).

6.8. H. Nakatsuka, D. Grischkowsky, and A. C. Balant, *Phys. Rev. Lett.* **47**, 1910 (1981).

6.9. L. F. Mollenauer, R. H. Stolen, and J. P. Gordon, *Phys. Rev. Lett.* **45**, 1095 (1980).

6.10. R. K. Bullough and P. J. Caudrey (Eds.), *Solitons*, Springer, Berlin (West), Heidelberg, New York, 1980.

6.11. A. B. Shvartsburg, *Optical and Quantum Electronics* **14**, 475 (1982).

6.12. E. Döpel, W. Dietel, D. Kühlke, and B. Wilhelmi, *Opt. Comm.* **43**, 433 (1982).

6.13. A. M. Weiner and E. P. Ippen, *Opt. Lett.* 9, 53 (1984).

6.14. B. Wilhelmi, in *Proceedings of the International School of Laser Applications ISLA III*, Vilnjus, 1984.

Part II

Effects and Processes of Nonlinear Optics

On the basis of the general concepts and methods given in Part I, both the foundations and the applications of significant nonlinear optical effects and processes will now be discussed. In Chapter 7 the basic one-photon laser equations as well as important laser parameters and empirical data are presented. Nonlinearities in transient one-photon processes in the case of resonant excitation of matter are discussed in Chapter 8. Chapter 9 is devoted to nonlinear quantum phenomena in transient one-photon processes, including spontaneous emission processes. Multiphoton absorption and multiphoton emission will be treated in Chapter 10, where also the corresponding spectral investigations of molecules and solids as well as two-photon lasers will be dealt with. Second-harmonic generation, the generation of sum and difference frequencies, and parametric amplification are important processes, especially with respect to application, since a transformation of radiation into other frequency ranges becomes possible; those effects are described in Chapter 11. Chapter 12 is devoted to spontaneous and stimulated Raman scattering; the latter effect provides the basis of modern spectroscopical methods and important tunable lasers. Optical bistability and phase conjugation are relatively young branches of nonlinear optics. They promise wide application and are described in Chapter 13 and Chapter 14, respectively.

7

Nonlinear One-Photon Processes in Lasers

After the publication of basic papers on the maser and laser principle [Basov and Prokhorov (Ref. 7.1), Schawlow and Townes (Ref. 7.2)] and on the observation of the first laser action [Maiman (Ref. 7.3)], a description of laser radiation including statistical properties was given in Ref. 7.4. This treatment was performed on the basis of a linear theory; thus these results are only valid for laser action below threshold and for light amplifiers, but they cannot be applied to laser oscillation above threshold, for which nonlinear relations are needed. Nonlinear as well as statistical mechanisms essentially determine the properties of the output radiation of a real laser, as will be shown in this chapter.

In describing laser mechanisms one may distinguish between continuously running processes and discontinuously running ones. The former allow one to describe the mean values of the electromagnetic energy inside the laser resonator and of the polarization of the active medium and its occupation inversion as functions with a predictable, continuous time dependence. These aspects will be dealt with in Section 7.1. Discontinuously running processes connected with fluctuation forces are caused by certain quantum processes (e.g. photon-number loss in the resonator, spontaneous emission of activated atomic systems). They lead, together with the continuous processes, to the exact or sufficiently complete basic equations for laser radiation. In Section 7.2 we shall derive them in the form of equations of motions including fluctuation (Langevin) forces for the basic field and matter variables, as well as in the form of a master equation for the field density operator and its Glauber-Sudarshan P-representation, respectively. In Section 7.3 we discuss main properties of the laser output radiation (e.g. phase and amplitude noise, line width, photon distribution, correlation functions) and some consequences for application.

7.1. CONTINUOUSLY RUNNING LASER PROCESSES

In this section we shall study laser action by rather simple means. The radiation field strength will be treated as a classical quantity, whereas the ensemble of atomic systems inside the laser resonator will be described quantum-mechanically. If, as has been confirmed by measurement for typical cases, a sufficiently large number of photons is contained inside the laser cavity, the relative error due to disregarding the operator character of the radiation variables is very small. Therefore, the semiclassical laser theory is capable of explaining important aspects of laser processes. Later on we shall compare the semiclassical theory with the results of a fully quantum-theoretical description that will be presented in Section 7.2. Then we shall complete the semiclassical equations with fluctuation forces describing discontinuous (stochastic) processes, which are ignored in the present section.

To display the typical features of the constitutive equations (in Section 7.1.1) and of the interaction between the active medium and the electric field (in Section 7.1.2), we use a simple laser model. The active medium is thought of as an ensemble of uniform two-level atomic systems that do not interact with one another. The electric field strength is assumed to arise from a single mode in a ring resonator with linearly polarized light in one direction of propagation, so that we have essentially a spatially homogeneous behavior. Later, in Section 7.3, we shall see that the results of this discussion can be applied to more complicated and more general conditions (such as standing waves and inhomogeneous broadening).

7.1.1. Nonlinear Constitutive Equations of the Active Laser Medium

The general semiclassical description of the interaction of matter with fields was given in Section 3.1 and in Chapter 4. Now we will specify this connection for the active medium in a laser resonator. Due to the interaction with the radiation field E, a deviation from the equilibrium value of the occupation-number inversion in the active laser medium can occur. This change of occupation leads to a nonlinear dependence of the polarization on the field strength. To explain this quantitatively, we consider a volume element of the active medium which is so small that the electric field may be assumed to be constant over it at a certain time. On the other hand it is supposed to be so large that it contains many of the atomic systems constituting the active medium.

The Hamiltonian of one atomic system interacting with the field E is given by

$$\hat{H} = \hat{H}_A - E\hat{d}, \tag{7.1}$$

where $\hat{H}_A$ is the two-level Hamiltonian of a free atomic system, $\hat{d}$ is its dipole-moment operator, and $-E\hat{d}$ is the interaction operator. It is supposed that the dipole moment and the electric field are perpendicularly polarized with respect to the resonator axis; thus both quantities can be regarded as scalar. Without loss of generality the transition matrix elements of $\hat{d}$ are chosen to be real, and they are denoted by d; permanent dipole moments are ignored. The spectral representation of $\hat{H}_A$ is

$$\hat{H}_A = \mathscr{E}_1|1\rangle\langle 1| + \mathscr{E}_2|2\rangle\langle 2|, \tag{7.2}$$

with $|1\rangle, |2\rangle$ being the eigenkets and $\mathscr{E}_1, \mathscr{E}_2$ being the energy eigenvalues; $\hbar\omega_{21} = \mathscr{E}_2 - \mathscr{E}_1$ is the transition energy. On the basis of the general equation of motion (A.77) for the density operator $\hat{\rho}$ of one atomic system, one can readily derive the following equations for the matrix elements of $\hat{\rho}$:

$$\frac{d}{dt}\rho_{12} = i\omega_{21}\rho_{12} + \frac{i}{\hbar}Ed(\rho_{22} - \rho_{11}), \tag{7.3a}$$

$$\frac{d}{dt}(\rho_{22} - \rho_{11}) = \frac{i2Ed}{\hbar}(\rho_{12} - \rho_{21}). \tag{7.3b}$$

We assume the atomic systems to be acted on not only by the radiation field E, but also by dissipative systems, so that energy and phase relaxations take place. Consequently we have to complete (7.3a, b) with relaxation terms [cf. Section 3.2.4, (3.111), (4.83)], so that we obtain

$$\frac{d}{dt}\rho_{12} = i\omega_{21}\rho_{12} + \frac{i}{\hbar}Ed(\rho_{22} - \rho_{11}) - \frac{1}{\tau}\rho_{12}, \tag{7.4a}$$

$$\frac{d}{dt}(\rho_{22} - \rho_{11}) = \frac{i2Ed}{\hbar}(\rho_{12} - \rho_{21}) - \frac{1}{T}\left[(\rho_{22} - \rho_{11}) - \frac{\gamma_I^e}{\gamma}\right]. \tag{7.4b}$$

Here τ and T are the phase and the energy relaxation time, respectively; γ is the number density of the atomic systems. The term γ_I^e/γ has to be interpreted as the contribution of one atomic system to the equilibrium value of the inversion density γ_I^e without influence of radiation. Note that the assumption of equal transition frequencies ω_{21} and relaxation constants τ and T for all atomic systems expresses the case of a homogeneous line (cf. Section 3.3.2.2).

From the microscopic relations (7.4a, b) we are now going to derive equations for macroscopic quantities, namely for the expectation value

$$P = \gamma\langle\hat{d}\rangle = \gamma d(\rho_{12} + \rho_{21}) \tag{7.5}$$

of the polarization and the expectation value

$$\Delta\gamma_I = \gamma(\rho_{22} - \rho_{11}) - \gamma_I^e \tag{7.6}$$

of the difference between the inversion density and its equilibrium value γ_I^e. The quantities P and $\Delta\gamma_I$ are related to small volume elements whose linear dimensions are assumed to be smaller than the wavelength. Making use of (7.5) and (7.6), the relations (7.4) go over into

$$\frac{d^2P}{dt^2} + \frac{2}{\tau}\frac{dP}{dt} + \left(\omega_{21}^2 + \frac{1}{\tau^2}\right)P$$

$$= -2\omega_{21}d^2\hbar^{-1}E\gamma_I^e - 2\omega_{21}d^2\hbar^{-1}E\,\Delta\gamma_I, \tag{7.7}$$

$$\frac{d}{dt}\,\Delta\gamma_I + \frac{1}{T}\,\Delta\gamma_I = \frac{2}{\hbar\omega_{21}}E\left(\frac{d}{dt}P + \frac{1}{\tau}P\right). \tag{7.8}$$

The second equation shows that $\Delta\gamma_I$ tends to zero (i.e., the inversion density tends to its equilibrium value) if the field strength becomes zero. On the other hand, the first equation represents an equation of motion for the polarization, which may be regarded as the differential equation of a damped harmonic oscillator, where the driving force is proportional to the product of the field strength and the inversion density.

The set of coupled differential equations (7.7) and (7.8) can yield the wanted connection between P and E; for this purpose the quantity $\Delta\gamma_I$ has to be eliminated. The coupled differential equations directly reveal that there must exist a nonlinear dependence of P on E. To describe it, we may employ the scheme presented in Section 1.3.2.4. The problem under study requires the harmonic ansatz

$$E(t) = \frac{1}{2}\widehat{E}(\omega_0)e^{-i\omega_0 t} + \text{c.c.} \tag{7.9a}$$

for the field strength with frequency ω_0, and the ansatz

$$P = P^{(1)}(t) + P^{(3)}(t), \tag{7.9b}$$

where

$$P^{(n)}(t) = \frac{1}{2}\widehat{P}^{(n)}(\omega_0)e^{-i\omega_0 t} + \text{c.c.}, \tag{7.9c}$$

for the polarization. We proceed step by step and calculate the polarization in increasing orders of the amplitude $\widehat{E}(\omega_0)$. First we consider the differential equation (7.7) with the second term on the right-hand side excluded. This truncated differential equation is a linear one that readily yields the solution

$$\widehat{P}^{(1)}(\omega_0) = -\frac{2\omega_{21}d^2\hbar^{-1}\gamma_I^e}{-\omega_0^2 + \left(\omega_{21}^2 + \tau^{-2}\right) - i2\tau^{-1}\omega_0}\widehat{E}(\omega_0), \tag{7.10}$$

so we have $P^{(1)}(t)$. Next, P in (7.8) is replaced by $P^{(1)}(t)$. Now the right-hand side of (7.8) depends quadratically on $\hat{E}(\omega_0)$. The solution for $\Delta\gamma_I$, which can easily be obtained, is inserted into the second term on the right-hand side of (7.7), which then will contain the field amplitude in the third order and we can derive $\hat{P}^{(3)}(\omega_0)$ from (7.7).

We shall now apply to approximations justified by practice with respect to the frequencies and times involved. For example, for a ruby laser $\omega_{21} \approx 3 \times 10^{15}\ \mathrm{s}^{-1}$, $\tau^{-1} \sim 10^{12}\ \mathrm{s}^{-1}$, $T^{-1} \sim 10^{3}\ \mathrm{s}^{-1}$. In the following the connection between polarization and electric field strength is of interest only for those frequencies ω_0 of the electric field that are very close to the transition frequency ω_{21} ($|\omega_0 - \omega_{21}| \ll \Delta\omega_{\mathrm{fl}}$, where $\Delta\omega_{\mathrm{fl}}$ is the fluorescence line width). Therefore we make use of the relation

$$\omega_0 \approx \omega_{21} \gg \tau^{-1} \gg T^{-1}. \tag{7.11}$$

Thus we end up with the following expression for the polarization amplitude, which contains the terms up to the third order in the field-strength amplitude:

$$\hat{P}(\omega_0) = \varepsilon_0 \chi^{(1)} \hat{E}(\omega_0) + \varepsilon_0 \chi^{(3)} \left| \hat{E}(\omega_0) \right|^2 \hat{E}(\omega_0) \overset{\mathrm{def}}{=} \varepsilon_0 \chi \hat{E}(\omega_0), \tag{7.12}$$

where

$$\chi^{(1)} = -i d^2 \tau \gamma_I^e \hbar^{-1} \varepsilon_0^{-1}, \qquad \chi^{(3)} = -\chi^{(1)} d^2 T \tau \hbar^{-2}. \tag{7.12a}$$

The estimation (7.11) allows the neglect of the real parts of $\chi^{(1)}$ and $\chi^{(3)}$. The derivation of (7.12) reveals that the nonlinear term in $\hat{P}(\omega_0)$ arises from the change in occupation of the atomic levels of the active medium under the influence of the field, as was stated at the beginning.

Now we want to demonstrate that the relations (7.12) and (7.12a) for the active medium lead to a property that is of special importance for the laser process, namely, that under certain conditions radiant energy can be generated.

To demonstrate this, we employ the Poynting theorem [see (1.175), (1.176)], which enables us to readily calculate the generated radiation energy as the time averaged divergence of the Poynting vector $S_\bullet$. At the frequency ω_0 the radiant energy per unit volume per unit time

$$\begin{aligned}\overline{\nabla_\bullet S_\bullet} &= -\overline{E_\bullet \frac{dP_\bullet}{dt}} = -\tfrac{1}{2}\omega_0 \varepsilon_0 \operatorname{Im}\{\chi\} \left| \hat{E}(\omega_0) \right|^2 \\ &= \tfrac{1}{2}\omega_0 d^2 \tau \hbar^{-1} \left\{ \gamma_I^e - \gamma_I^e d^2 T \tau \hbar^{-2} \left| \hat{E}(\omega_0) \right|^2 \right\} \left| \hat{E}(\omega_0) \right|^2 \end{aligned} \tag{7.13}$$

is produced by the active medium. The time averaging in this equation covers several periods of the length $2\pi/\omega_0$. $P(t)$ is replaced by

$$\varepsilon_0 \chi \frac{1}{2} \widehat{E}(\omega_0) \exp(-i\omega_0 t) + \text{c.c.},$$

so that contributions of the susceptibility up to the third order are included, and the relation (7.13) is therefore valid in a wide range for field amplitudes being not too high. The expression in braces represents the density of the inversion induced by the field. The first term, the equilibrium inversion γ_I^e, is called the *undepleted* (field-independent) inversion; it is determined by the pumping process acting on the atomic systems and by other incoherent relaxation processes. In the presence of the electric field a lowering (saturation) of the inversion occurs, which is described by the second term, the so-called *saturation* term, which arises from the third-order susceptibility $\chi^{(3)}$. The quantity γ_I^e increases with increasing pumping rate. Pumping causes the population of the upper atomic levels to become higher than that of the lower ones, so that γ_I^e attains positive values and thus $\overline{\nabla \cdot S} > 0$, which means that radiation energy is generated by the active medium. The relation between pumping rate and the unsaturated inversion γ_I^e will be dealt with at the end of the present section.

The generation rate of radiation energy increases as the saturation term decreases. Furthermore, (7.13) exhibits the interrelation between generation rate and the parameters of the atomic systems. For small values of the saturation term the generation rate is proportional to the square of the dipole moment and proportional to the phase relaxation time of the atomic system.

7.1.2. Interaction of the Active Medium with the Resonator Field

Let us first discuss the electric field in an *ideal resonator* without any loss or gain of radiation energy. The wave equation that governs the one-dimensional ring resonator is

$$\frac{\partial^2 E}{\partial t^2} - c^2 \frac{\partial^2 E}{\partial z^2} = 0, \tag{7.14}$$

where E is the radiation field inside the resonator and z the coordinate in the direction of propagation. The solution for the field strength of a single mode at frequency ω_0 turns out to be

$$E(t; z) = \frac{\overline{E}}{2} e^{-i\omega_0(t - z/c)} + \text{c.c.}, \tag{7.15}$$

with $\overline{E}$ a constant complex wave amplitude of the field (for simplicity, in this subsection the parameter index ω_0 assigned to the field amplitude $\widehat{E}$ in Section 7.1.1 is omitted). We consider the tuned case, where the field is periodic with the resonator length L, so that the frequency ω_0 is equal to $m\,2\pi c/L$ with m an integer. We are mainly interested in the time behavior of the spatially homogeneous ring resonator; thus we will regard z as a parameter and write $E = E(t)$, with the time t being the variable. At every point z the field strength obeys the temporal differential equation

$$\frac{d^2}{dt^2}E + \omega_0^2 E = 0. \tag{7.15a}$$

We now turn to the *realistic laser resonator*, taking into account the active medium in it as well as the losses of radiation energy in the passive resonator. This requires the wave equation (7.14) of the ideal resonator to be changed, and according to (1.186), we obtain

$$\frac{\partial^2 E}{\partial t^2} + v'\frac{\partial E}{\partial t} - c^2\frac{\partial^2 E}{\partial z^2} = -\frac{1}{\varepsilon_0}\frac{\partial^2 P}{\partial t^2}. \tag{7.16}$$

The term inserted on the right-hand side contains the polarization P of the active medium, which is induced by the radiation field. P contains the first-order and third-order parts of the susceptibility, which are given in (7.12). Moreover, a phenomenological second term on the left-hand side has been introduced; it represents the losses from the mode energy, which are caused by absorption, diffraction, and the output of laser radiation. The positive constant $v'/2$ corresponds to the reciprocal decay time of the field as a result of the losses mentioned above (if P is set equal to zero).

In the following we have to carry out estimations, and therefore the data preceding (7.11) must be supplemented by $v' \sim 10^9\ \mathrm{s}^{-1}$, a typical value for the resonator losses. Thus we may set

$$\omega_0 \approx \omega_{21} \gg \tau^{-1} \gg v' \gg T^{-1}. \tag{7.17}$$

The solution (7.15) for the ideal resonator with constant wave amplitude $\bar{E}$ cannot be employed as solution of (7.16): we have to take into account the losses and the interaction with the active medium, resulting in a variation of the wave amplitude $\bar{E}$. Therefore we now set as a solution of (7.16):

$$E(t; z) = \frac{\bar{E}(t)}{2}e^{-i\omega_0(t-z/c)} + \text{c.c.}. \tag{7.18}$$

Supposing relatively weak resonator losses and a relatively weak interaction of the field with the active medium, $\bar{E}$ can be assumed to vary slowly with time. The field will be restricted to a single mode, so that $\bar{E}$ does not depend on the coordinate z. With the ansatz (7.18) the third summand of the left-hand side of (7.16) becomes $\omega_0^2 E$. In the following consideration we again assume z to be a parameter, and both the field strength and the polarization to be functions only of the time variable t. We can exploit the spatially homogeneous behavior: at every point z the field strength $E(t)$ obeys the differential equation

$$\frac{d^2E(t)}{dt^2} + v'\frac{dE(t)}{dt} + \omega_0^2 E(t) = -\frac{1}{\varepsilon_0}\frac{d^2P(t)}{dt^2}, \tag{7.19}$$

where

$$E(t) = \frac{\widehat{E}(t)}{2}e^{-i\omega_0 t} + \text{c.c.}. \tag{7.19a}$$

The field amplitude $\widehat{E}(t)$ is supposed to vary only relative slightly in times of the order of $2\pi\omega_0^{-1}$, so that we still may speak of an oscillatorlike behavior of the field strength $E(t)$.

The concept of slowly varying field amplitudes will now be exploited to simplify the basic equations. We shall apply the more general concept represented in Section 1.4.4 and in Ref. 7.5 to the case of the laser field determined by the differential equation (7.19). The fact that the amplitude $\widehat{E}(t)$ in $E(t)$ varies slowly with time means that the Fourier transform of $\widehat{E}(t)$—denoted by $\breve{\widehat{E}}(\omega)$—takes on noticeable values only for small Fourier frequencies ω, say for $|\omega| \lesssim \delta\omega$, where $\delta\omega/\omega_0 \ll 1$. We subject the differential equation (7.19) to a Fourier transformation (Ref. 7.6); there results an *algebraic* relation for the Fourier transform $\breve{\widehat{E}}(\omega)$, in which according to the rules of mathematical analysis the individual terms of this algebraic sum can be evaluated and compared. Thus the reasons for the neglect of certain terms can be given in a sensible way. Our resulting classical expressions indicate (as will be explicitly shown later) direct correspondence to the quantum expressions for the problem, if the latter are derived by means of the rotating-wave approximation that ignores nonresonant terms and keeps quasiresonant ones (cf. Section 3.2.2). Thus we may call the resulting relations *rotating-wave-approximated* (r.w.a.) relations.

We first study the second derivative d^2E/dt^2 contained in (7.19); we have

$$\frac{d^2E}{dt^2} = \left[-\omega_0^2\widehat{E} - 2i\omega_0\frac{d\widehat{E}}{dt} + \frac{d^2\widehat{E}}{dt^2}\right]\frac{e^{-i\omega_0 t}}{2} + \text{c.c.}. \tag{7.20a}$$

The Fourier transform of the expression in the brackets is

$$\text{FT}[\cdots] = -\omega_0^2\breve{\widehat{E}}(\omega)\left\{1 - 2\frac{\omega}{\omega_0} + \frac{\omega^2}{\omega_0^2}\right\}. \tag{7.20b}$$

Because $\delta\omega \ll \omega_0$, the ratio ω^2/ω_0^2 is smaller than the other terms in the braces. In view of the rotating-wave approximation the third term can be neglected, whereas the second term, which represents the dependence of $\hat{E}(t)$ on time in the lowest order, will be retained. After returning to the time domain, we obtain the r.w.a. expression

$$\left(\frac{d^2E}{dt^2}\right)_{\text{r.w.a.}} = \left[-\omega_0^2\hat{E} - 2i\omega_0\frac{d\hat{E}}{dt}\right]\frac{e^{-i\omega_0 t}}{2} + \text{c.c..} \tag{7.20c}$$

Now we want to apply this procedure to the complete differential equation (7.19). For $E(t)$ the expression (7.19a) is substituted, and for $P(t)$ the expression

$$P(t) = \frac{\hat{P}(t)}{2}e^{-i\omega_0 t} + \text{c.c.}, \tag{7.21}$$

where the amplitude $\hat{P}(t)$ is again assumed to be a relatively slowly varying amplitude. We see that, if (7.19) is satisfied, this is equivalent to satisfying the differential equation

$$\left[-\omega_0^2\hat{E} - i2\omega_0\frac{d\hat{E}}{dt} + \frac{d^2\hat{E}}{dt^2} - i\omega_0 v'\hat{E} + v'\frac{d\hat{E}}{dt} + \omega_0^2\hat{E}\right]$$

$$+\left[-\frac{\omega_0^2}{\varepsilon_0}\hat{P} - i\frac{2\omega_0}{\varepsilon_0}\frac{d\hat{P}}{dt} + \frac{1}{\varepsilon_0}\frac{d^2\hat{P}}{dt^2}\right] = 0, \tag{7.22}$$

where the first bracket (denoted by [LS]) stems from the left-hand side of (7.19), and the second one, (denoted by [RS]) from the right-hand side. We start by considering the expression [LS]. Its Fourier transform is

$$\text{FT[LS]} = -\omega_0^2\underset{\smile}{\hat{E}}(\omega)\left\{1 - 2\frac{\omega}{\omega_0} + \frac{\omega^2}{\omega_0^2} + \frac{iv'}{\omega_0} - \frac{i\omega v'}{\omega_0^2} - 1\right\}. \tag{7.23}$$

A simplification of the sum in braces can be achieved if we know the quantitative relations between $\delta\omega$, v', ω_0. When establishing (7.19) it was stated that $v'/2$ is the decay time of the field strength; this means that the amplitude $\hat{E}(t)$ contains Fourier frequencies up to the order of v', so that $\delta\omega \sim v'$. Thus, in addition to (7.17), it follows that $\omega_0 \gg \delta\omega, v'$. Hence, it is reasonable to neglect the third and the fifth term in comparison with the others. With this approximation, and because the first and the last terms cancel, there results, in the rotating-wave approximation,

$$[\mathrm{LS}]_{\mathrm{r.w.a.}} = -i2\omega_0 \frac{d\hat{E}}{dt} - i\omega_0 v' \hat{E}. \tag{7.24}$$

We now determine $[\mathrm{RS}]_{\mathrm{r.w.a.}}$. To do this, we of course must perform order-of-magnitude evaluations of the same kind as were used for [LS], since the expressions [RS] and [LS] do not act separately but as a sum in the differential equation (7.22). The first term of [RS] contains the polarization amplitude $\hat{P}(t)$; according to (7.12) we have

$$\hat{P}(t) = \varepsilon_0 \chi^{(1)} \left[1 - Td^2 \tau \hbar^{-2} \left|\hat{E}(t)\right|^2\right] \hat{E}(t). \tag{7.25}$$

The frequency bandwidth of $\hat{E}(\omega)$ has been set equal to $\delta\omega$; thus the Fourier transform of $|\hat{E}(t)|^2 \check{\hat{E}}(t)$ has a frequency bandwidth of the order $3\,\delta\omega$, which is of the same order of magnitude as that of $\hat{E}(t)$. Because of this, and also because $\hat{P}(t)$ is roughly proportional to $\hat{E}$, the term $-(\omega_0^2/\varepsilon_0)\hat{P}$ must display the same time behavior as the term $-i\omega_0 v'\hat{E}$ in $[\mathrm{LS}]_{\mathrm{r.w.a.}}$, so it will be kept. The terms in $d\hat{P}/dt$ and $d^2\hat{P}/dt^2$ in [RS] are proportional to $d\hat{E}/dt$ and $d^2\hat{E}/dt^2$, respectively, and can be neglected in comparison with the term $-(\omega_0^2/\varepsilon_0)\hat{P}$. In principle, the reasoning here is similar to that given for the neglect of terms containing $d\hat{E}/dt$ and $d^2\hat{E}/dt^2$ in [LS], though an exact discussion of the convolution integrals occurring in the Fourier transforms of [RS] is rather difficult. Finally, under the approximations described, the differential equation (7.22) goes over into

$$\frac{d\hat{E}}{dt} + \frac{v'}{2}\hat{E} - i\frac{\omega_0}{2\varepsilon_0}\hat{P} = 0, \tag{7.26}$$

which allows us to study in a rather simple way the physical effects described by the basic equation for the real laser resonator. Let us mention here that the equation (7.26) is also reasonable in that it directly corresponds to a fully quantum-theoretical treatment of the problem; this will be shown in Section 7.2.1.

To gain a deeper physical insight, we explicitly introduce $\hat{P}(t)$ of (7.12) into (7.26); this leads to

$$\frac{d\hat{E}}{dt} = \frac{s}{2}\hat{E} \tag{7.27}$$

with the field-dependent quantity

$$s = -v' + \beta - \zeta'|\hat{E}|^2 \tag{7.27a}$$

and the constant parameters

$$\beta = i\omega_0\chi^{(1)} = \omega_0 d^2\gamma_I^e\tau\hbar^{-1}\varepsilon_0^{-1}, \qquad \zeta' = -i\omega_0\chi^{(3)} = \omega_0 d^4 T\tau^2\gamma_I^e\hbar^{-3}\varepsilon_0^{-1}. \tag{7.27b}$$

Thus the relative variation of the field amplitude $(d\hat{E}/dt)\hat{E}^{-1}$ with time, under the influence of loss and amplification, is given by the quantity $s/2$. Since in deriving (7.27) we started from a relatively slow variation of the amplitude, the solutions of (7.27) are only valid under the assumption $s \ll \omega_0$.

Now we consider the behavior of the *electromagnetic energy* U_{em} in the resonator. The time average, with an averaging time of some $2\pi/\omega_0$, is

$$\overline{U}_{\mathrm{em}} = \varepsilon_0 V\overline{E^2} = \varepsilon_0 V\frac{\hat{E}^*\hat{E}}{2}. \tag{7.28}$$

The energy mean value $\overline{U}_{\mathrm{em}}$ bears a slow time dependence, which arises from the slow time dependence of the amplitude $\hat{E}$. From (7.27) and (7.28) one readily obtains

$$\frac{d\overline{U}_{\mathrm{em}}}{dt} = s\overline{U}_{\mathrm{em}}. \tag{7.29}$$

The quantity s thus represents the relative variation of the radiation energy with time. The part $\beta - \zeta'|\hat{E}|^2$ in s results from the generation of radiant energy by the active medium [cf. (7.13)]. Above the threshold where laser oscillation may appear, the inversion γ_I^e, that is, the overpopulation of the upper level, must attain sufficiently high (positive) values that the radiant energy generated is capable of compensating for the losses (denoted by v').

Let us now consider the time behavior of the amplitude $\hat{E}(t)$. To this end we discuss an appropriate solution of (7.27); this relation corresponds to the type of equation given by Van der Pol (Ref. 7.7). We start from a small initial amplitude $\hat{E}(0)$, for which $|\hat{E}(0)|^2 < (\beta - v')/\zeta'$ is assumed to hold. As a solution (cf. Ref. 7.8) we then have

$$\hat{E}(t) = \hat{E}(0)e^{\frac{1}{2}(\beta - v')t}\left[1 + \frac{\zeta'|\hat{E}(0)|^2}{\beta - v'}\left(e^{(\beta - v')t} - 1\right)\right]^{-1/2} \tag{7.30}$$

This equation shows that the phase of $\hat{E}(t)$ agrees with the phase of $\hat{E}(0)$. Below the threshold of laser oscillation we have $\beta - v' < 0$; under this condition $\hat{E}(t)$ goes exponentially to zero for large t. Above threshold, $\beta - v'$ is positive, and for $t \to \infty$ the absolute value of the amplitude rises towards the stationary value

$$\left|\hat{E}(t \to \infty)\right|_{\text{st}} = \left(\frac{\beta - v'}{\zeta'}\right)^{1/2}. \qquad (7.31)$$

Since ζ' is equal to $\omega_0|\chi^{(3)}|$ with $\chi^{(3)}$ the third-order susceptibility, the significance of the nonlinear term in the relation between polarization and field strength becomes obvious for laser operation. Without this term, that is, if $\chi^{(3)} = 0$, $\hat{E}(t)$ would exponentially decrease or increase, according as the process proceeds below or above the threshold, a nonvanishing stationary value would not be attained. In reality, however, because $|\chi^{(3)}| \neq 0$, the stationary value $|\hat{E}(\infty)|_{\text{st}}$ is attained, which guarantees a stable laser oscillation. This corresponds to a stationary field energy U_{st} in the resonator. We have

$$\overline{U}_{\text{st}}(\infty) = \frac{\varepsilon_0 V}{2}\left(\frac{\beta - v'}{\zeta'}\right). \qquad (7.32)$$

The time behavior of the amplitude above threshold is illustrated in Fig. 7.1. It approaches a constant because of the reoccupation connected with the nonvanishing susceptibility $\chi^{(3)}$: Under the (false) assumption that $\chi^{(3)} = 0$ this would not be the case. The important quantity γ_I^e, the equilibrium inversion, which can be adjusted by the experimenter via the pumping rate, is contained in β and ζ'. Since the condition for stable laser oscillation is that $\overline{U}_{\text{st}} > 0$, the actual inversion γ_I^e must exceed the so-called threshold inversion $(\gamma_I^e)_{\text{thr}}$ which is determined by the condition

$$\omega_0 d^2 \tau \hbar^{-1} \varepsilon_0^{-1} (\gamma_I^e)_{\text{thr}} - v' = 0. \qquad (7.33)$$

The original differential equation (7.19) for the field in the laser resonator is a relation for the *rapidly varying* field strength $E(t)$, by which the laser radiation emitted in an interval around the frequency ω_0 is characterized. Therefore, we shall now establish—with the help of the equation (7.27) for the slowly varying r.w.a. field amplitude $\hat{E}(t)$—a relation for the rapidly varying

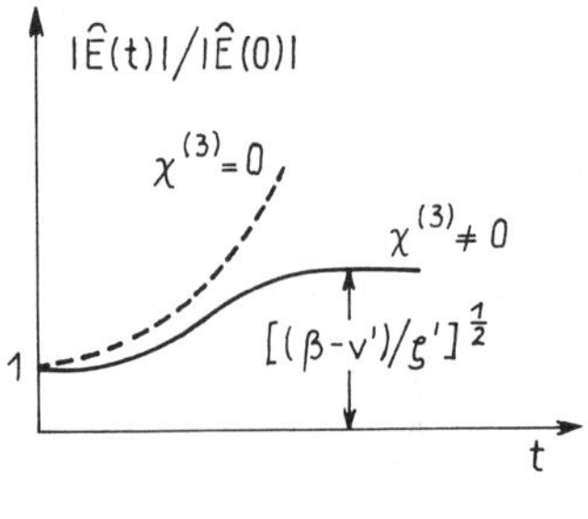

Fig. 7.1. Time dependence of the field amplitude above threshold. Solid curve, $|\chi^{(3)}| \neq 0$; dashed curve, $|\chi^{(3)}| = 0$.

r.w.a. field strength $E(t)$. For the latter the differential equation

$$\frac{d^2E}{dt^2} - (\beta - \upsilon' - \zeta E^2)\frac{dE}{dt} + \omega_0^2 E = 0, \tag{7.34}$$

where $\zeta = 4\zeta' = 4\omega_0 d^4 T\tau^2\gamma_l^e\hbar^{-3}\varepsilon_0^{-1}$, is valid if one can neglect higher harmonics that involve time-dependent functions with the frequency $3\omega_0$. This can be easily proved by substituting the expressions

$$E(t) = \frac{\hat{E}(t)}{2}e^{-i\omega_0 t} + \text{c.c.},$$

$$\left[\frac{dE}{dt}\right]_{\text{r.w.a.}} = -i\omega_0\frac{\hat{E}}{2}e^{-i\omega_0 t} + \text{c.c.},$$

$$\left[\frac{d^2E}{dt^2}\right]_{\text{r.w.a.}} = \left[-\omega_0^2\hat{E} - 2i\omega_0\frac{d\hat{E}}{dt}\right]\frac{e^{-i\omega_0 t}}{2} + \text{c.c.},$$

which have been already explained by means of (7.20a, b, c), into (7.34). This leads to

$$\left[\frac{d\hat{E}}{dt} - \tfrac{1}{2}\left(\beta - \upsilon' - \zeta'|\hat{E}|^2\right)\hat{E}\right]e^{-i\omega_0 t}$$

$$+\frac{\zeta'}{8}\hat{E}^3 e^{-i3\omega_0 t} + \text{c.c.} = 0. \tag{7.35}$$

This relation agrees with (7.27) if higher harmonics are neglected. This neglect is, to a certain extent, physically justified because of special damping mechanisms for oscillations of these higher frequencies. Equation (7.34) gives the stationary solution

$$E(t) = \left(\frac{\beta - \upsilon'}{\zeta}\right)^{1/2} e^{-i\omega_0 t} + \text{c.c.} \tag{7.36}$$

as could be expected in view of (7.31).

This steady-state case represents an exact harmonic oscillation with zero line width. In the real laser, however, there is a nonvanishing line width. To describe it, the additional consideration of *discontinuous* processes, which in reality take part in the generation of laser radiation, is necessary. Such processes are, for example, the absorption and the output of photons, the spontaneous emission of photons, and the pumping processes of atomic systems. To cover this situation, the differential equation (7.34) must be extended with fluctuation forces (cf. Section 3.2.4). These will directly follow from the fully quantum-theoretical description, as will be shown in Section 7.2.

The passing over from the original field equation (7.19) to the rotating-wave-approximated differential equation (7.27) of the field amplitude allowed a simple discussion of the foundations of the physics of lasers. The other basic equations, like that for polarization (7.7) and that for inversion (7.8), can be approximated in an analogous way. We do this not only for the sake of simplification, which of course is an advantage, but also because equations will result that can be directly compared with those provided by a quantum-theoretical treatment of the laser. For the latter reason it is of advantage to carry out the following renormalizations. The field amplitude $\widehat{E}(t)$, the polarization amplitude $\widehat{P}(t)$, and the difference $\Delta\gamma_I(t)$ between the inversion density and the equilibrium value are transformed according to

$$\widehat{E} = 2\sqrt{\frac{\hbar\omega_0}{2\varepsilon_0 V}}\,\mathfrak{a}, \qquad \widehat{P} = -i\frac{2d}{V}\mathfrak{p}, \qquad \Delta\gamma_I + \gamma_I^e = \frac{\Delta N}{V} \tag{7.37}$$

into the dimensionless quantities $\mathfrak{a}(t)$, $\mathfrak{p}(t)$, $\Delta N(t)$. The dependence on the resonator volume shows that we pass over from local quantities to quantities assigned to the resonator as a whole. Thus $\mathfrak{a}(t)$ is associated with the total radiation energy in the resonator by $\overline{U}_{\text{em}}(t) = \hbar\omega_{21}\mathfrak{a}^*(t)\mathfrak{a}(t)$; $\mathfrak{p}$ is the total polarization divided by the transition dipole moment of one atomic system; ΔN is the total inversion in the resonator (i.e. $\Delta N = N_2 - N_1$, where N_2 and N_1 are the occupation numbers in the upper and the lower level, respectively); ΔN^e is the corresponding equilibrium value. From (7.34), (7.7), (7.8) the following set of coupled, rotating-wave-approximated differential equations is obtained:

$$\frac{d}{dt}\mathfrak{a} + \frac{v'}{2}\mathfrak{a} = g\mathfrak{p}, \tag{7.38a}$$

$$\frac{d}{dt}\mathfrak{p} + \frac{1}{\tau}\mathfrak{p} = g\mathfrak{a}\,\Delta N, \tag{7.38b}$$

$$\frac{d}{dt}\Delta N + \frac{1}{T}(\Delta N - \Delta N^e) = g(-2)(\mathfrak{a}^*\mathfrak{p} + \mathfrak{p}^*\mathfrak{a}). \tag{7.38c}$$

Here g is the coupling constant, which, according to

$$g = \sqrt{\frac{d^2\omega_0}{2\varepsilon_0\hbar V}}, \tag{7.39}$$

depends on the atomic quantities d and ω_0 ($\approx \omega_{21}$) and on the volume of the resonator. Like all the terms in the equations (7.38a, b, c), the quantity g has the dimensions of a reciprocal time and expresses the strength of the influence of the driving forces.

An important quantity for describing the real laser process is the mean number Q of photons in the resonator, which is associated with the mean electromagnetic energy U by

$$Q(t) = \frac{\overline{U}(t)}{\hbar\omega_{21}} = \mathfrak{a}^*(t)\,\mathfrak{a}(t). \tag{7.40}$$

From the equations (7.38) we shall derive relations for the time behavior of the photon number Q and the inversion ΔN, which will later on be compared with the basic laser balance equations. From the equations (7.38a, b, c) the relations

$$\frac{d}{dt}\Delta N = \frac{\Delta N^e}{T} - 2g(\mathfrak{a}^*\mathfrak{p} + \mathfrak{a}\mathfrak{p}^*) - \frac{\Delta N}{T}, \tag{7.41}$$

$$\frac{d}{dt}Q = \mathfrak{a}\frac{d\mathfrak{a}^*}{dt} + \mathfrak{a}^*\frac{d\mathfrak{a}}{dt} = g(\mathfrak{a}^*\mathfrak{p} + \mathfrak{a}\mathfrak{p}^*) - v'Q \tag{7.42}$$

immediately follow. Let us now express the term $g(\mathfrak{a}^*\mathfrak{p} + \text{c.c.})$ in terms of Q and ΔN. Because of the large values of τ^{-1} [see (7.11)] and the relatively slow variation of the function $\mathfrak{p}(t)$, $d\mathfrak{p}/dt$ can be neglected in comparison with $\tau^{-1}\mathfrak{p}$. Thus it follows that $\mathfrak{p} = g\tau\mathfrak{a}\,\Delta N$ from (7.38b) and $g(\mathfrak{a}^*\mathfrak{p} + \text{c.c.}) = g^2 2\tau Q\,\Delta N$. Hence, for the generation rate of the inversion and the number of photons, one obtains

$$\frac{d}{dt}\Delta N = \frac{\Delta N^e}{T} - 4g^2\tau Q\,\Delta N - \frac{\Delta N}{T}, \tag{7.43}$$

$$\frac{dQ}{dt} = 2g^2\tau Q\,\Delta N - v'Q. \tag{7.44}$$

Keeping in mind that the average power output of the laser (proportional to the loss parameter v') is

$$L = v'\hbar\omega_{21}Q, \tag{7.44a}$$

it becomes apparent that both the equations (7.43) and (7.44) are of great importance in laser physics. Over a wide range they correctly express the nonstationary behavior (to which, e.g., the so-called relaxation oscillations belong) and the stationary behavior of the laser intensity. For the steady-state case we have

$$Q_{\text{st}} = \frac{1}{T}\left(\frac{\Delta N^e}{2v'} - \frac{1}{4g^2\tau}\right). \tag{7.45}$$

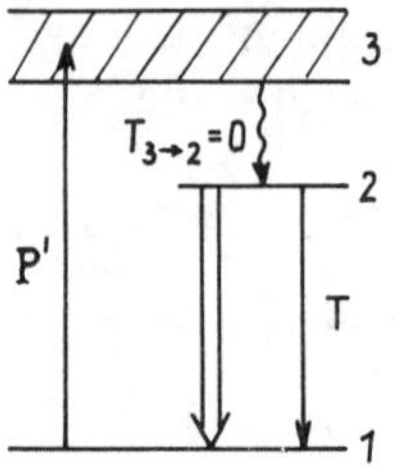

Fig. 7.2. Scheme of a three-level laser: P', pump rate; $T_{3\to 2}$, T, energy relaxation times.

Only if the equilibrium inversion ΔN^e is greater than $v'/2g^2\tau$ does $Q_{\rm st}$ attain a positive value and can a proper laser oscillation occur; $v'/2g^2\tau$ is called the threshold inversion. The way in which ΔN^e and the pumping rate are related will be seen in (7.50).

We now compare the semiclassical equations (7.43) and (7.44) with the *laser rate equations*. These are obtained with the help of transition rates for stimulated and spontaneous emission and absorption for one-photon processes, as they have been derived on a fully quantum-theoretical basis in Section 3.3.1. The establishing of the balance equations is carried out using the scheme given in Fig. 7.2 for a simple three-level laser. We let all atomic systems contribute uniformly, which leads to balance equations for a homogeneously broadened line. The pumping system causes the transition from the ground level to pump level 3. From there a nonradiative transition to the upper laser level takes place in such a short time that the occupation of the pump level can be neglected. Thus we see that at every instant the total number N of the atomic systems can be regarded as being equal to the sum of the number N_2 in the upper laser level and the number N_1 in the ground level. Hence, we have for the occupation numbers and the inversion ΔN

$$N = N_1 + N_2, \qquad \Delta N = N_2 - N_1,$$

$$N_2 = \tfrac{1}{2}(N + \Delta N), \qquad \frac{d}{dt}\Delta N = 2\frac{d}{dt}N_2. \tag{7.46}$$

Here the quantities $N_1, N_2, \Delta N$ have the character of quantum-theoretical expectation values; for example, $N_2(t) = N\langle \hat{B}^\dagger(t)\hat{B}(t)\rangle$, where $\hat{B}^\dagger\hat{B}$ is the occupation-number operator of the upper level of the atomic system [cf. (A.171)]. For the (negative) contribution to the variation of N_2 with time, which is caused by stimulated emission, we obtain

$$\left(\frac{dN_2}{dt}\right)^{\rm st\,em} = -\hbar\omega_{21}B_{21}g_{21}^{\max}\frac{Q}{V}N_2,$$

where B_{21} is the Einstein coefficient and $g_{21}^{\max}$ is the maximum of the line-shape function. Here also Q is a quantum-theoretical expectation value, namely $\langle \hat{a}^\dagger(t)\hat{a}(t)\rangle$, where $\hat{a}$ is the annihilation operator of the field strength of the resonator mode. $N_2^{-1}(dN_2/dt)_{\mathrm{st\,em}}$ may be regarded as the contribution of a single atomic system. It is obtained from the spectral transition rate (3.154) by integration over the spectral distribution

$$\zeta(\omega) = \hbar\omega_{21}\frac{Q}{V}\delta(\omega - \omega_{21})$$

of the mode.

Analogous results can be gained for absorption. Let T be the lifetime of the upper level for spontaneous emission; according to (3.123) we then have

$$\left(\frac{dN_2}{dt}\right)^{\mathrm{sp\,em}} = -\frac{N_2}{T}.$$

Let us now introduce the pump rate P' from level 1 via pump level 3 to level 2. The stimulated emission and absorption not only change N_2 and N_1, but also change the number Q of photons in the mode under consideration. Altogether the following system of laser balance equations results:

$$\frac{d}{dt}\Delta N = \left(2P' - \frac{N}{T}\right) - 2\hbar\omega_{21}B_{21}g_{21}^{\max}V^{-1}Q\,\Delta N - \frac{\Delta N}{T}, \qquad (7.47)$$

$$\frac{dQ}{dt} = \hbar\omega_{21}B_{21}g_{21}^{\max}V^{-1}Q\,\Delta N - v'Q + w\frac{N + \Delta N}{2T}, \qquad (7.48)$$

where $v'Q$ represents the photon-number loss of the resonator per unit time. The last term on the right-hand side of (7.48) indicates which fraction w of all spontaneously emitted photons N_2/T gets into the resonator mode considered.

An explicit expression for w/T can be obtained with the help of (3.118). This equation states that for a single mode the transition rate for stimulated emission, if related to *one* photon, equals the contribution of spontaneous emission. Therefore the factor preceding $Q\,\Delta N$ in the first term on the right-hand side of (7.48) must be equal to the factor preceding $N_2 = (N + \Delta N)/2$ in the last term; this leads to

$$\frac{w}{T} = \hbar\omega_{21}B_{21}g_{21}^{\max}V^{-1}. \qquad (7.48a)$$

In the starting stage of the oscillation the last term is of importance; at a later stage, when there is already a sufficiently high nunber Q of photons in the resonator mode, it can be neglected. Under this condition we compare the equations (7.47) and (7.48) with the equations (7.43) and (7.44).

The first thing to become obvious is that the two equations, though derived in different ways, have exactly the same structure. Moreover, one can also find agreement between the individual terms. In particular, the factors preceding

$Q\,\Delta N$ in (7.44) and in (7.48) are equal; we have

$$2g^2\tau = \hbar\omega_{21}B_{21}g_{21}^{\max}V^{-1}, \tag{7.49}$$

since it is possible, according to (3.146) and (3.95), to set $g_{21}^{\max}$ equal to τ/π. The equations (7.39) and (3.126) confirm that (7.49) is correct, provided that the orientation averaging of the dipole moments described in (3.122) is taken into account.

The unsaturated inversion ΔN^e from (7.41) is related to the pump rate by

$$\frac{\Delta N^e}{TN} = \frac{2P'}{N} - \frac{1}{T}. \tag{7.50}$$

A positive unsaturated inversion can result only if twice the pump rate per particle exceeds T^{-1}.

Finally, we want to stress that the solutions $\mathfrak{a}(t)$, $\mathfrak{p}(t)$, $\Delta N(t)$, $Q(t)$ following from the semiclassical equations (7.38a, b, c), (7.43), and (7.44) are completely predictable functions of time. This is a consequence of the fact that in establishing the basic equations only continuous processes were considered. If discontinuous processes come into play, fluctuation terms must be introduced. These matters will be discussed in the following section.

7.2. INFLUENCE OF FLUCTUATIONS ON LASER PROCESSES

The semiclassical description of continuous laser processes in the preceding section provided relations for classical variables that determine the time behavior of the electric field strength, of the electromagnetic energy in the resonator, and of the inversion of the active medium. There resulted, as solutions of differential equations, temporally completely predicted functions. Relations for completely predicted functions are also provided by the laser rate equations (7.47) and (7.48) for the quantum-theoretical expectation values of the mean number of photons and of the inversion. In reality, discontinuous processes have to be considered in addition to continuous ones. Due to the quantum character of atoms and radiation, there occur unavoidable fluctuations around the average values. These fluctuations are responsible for important and typical properties of laser radiation, such as phase and amplitude fluctuations, line-width effects, and coherence behavior. As will be seen later, these properties are mainly determined by a combination of two phenomena, the *nonlinearities* already displayed in Section 7.1, and the *fluctuations*, which play an important role in laser physics.

In principle, fluctuation phenomena can be adequately described only by a fully quantum-theoretical consideration. The resulting quantum-theoretical equations of motion (for variables in the Heisenberg picture) can be derived in a rather simple way. In Section 7.2.1 they will be interpreted in relation to the system of semiclassical equations (7.38a, b, c) after fluctuation forces with the character of classical random variables have been incorporated appropriately. We shall use these extended semiclassical equations for discussing problems of line width and coherence. This relatively simple procedure yields results that agree over a wide range with those of a fully quantum-theoretical treatment. A quantum-theoretical approach, different from the abovementioned equations of motion, to the explanation of experimental results is the density-operator equation (master equation) for the laser field, which will be treated in Section 7.2.2.

7.2.1. Equations of Motion Containing Fluctuation Forces

This subsection is devoted to the derivation of basic equations of motion for the field strength of the laser light, which accounts for nonlinearities, losses, pumping and relaxation processes, and fluctuations. To this end we have to consider all the mechanisms taking part.

The internal structure of a laser can be illustrated by the model shown in Fig. 7.3 [see, e.g., Haken (Ref. 7.9) and Risken (Refs. 7.10, 7.11)]. The electromagnetic radiation R in a ring resonator (which is assumed to have the same properties as described in Section 7.1) interacts with the pumped active medium. The active medium is assumed to consist of the atomic systems $A_1, A_2, \ldots, A_j, \ldots, A_N$. In the sense of Section 3.2.4.1, each of the systems $R, A_1, A_2, \ldots A_N$ has to be regarded as a dynamical system interacting with a dissipative system. Each of the atomic systems A_j is coupled to a dissipative system D_j, which is capable of affecting the energy and phase relaxation of A_j. In addition, the D_j are assumed to cover the pumping process that gives rise to an enhancement of the number of excited atomic systems beyond the thermal-equilibrium value. The radiation R in the laser resonator is coupled to the dissipative system D_R, which describes the resonator losses.

At first glance the establishment of equations of motion for the relevant quantum-theoretical laser observables seems to be rather difficult because of

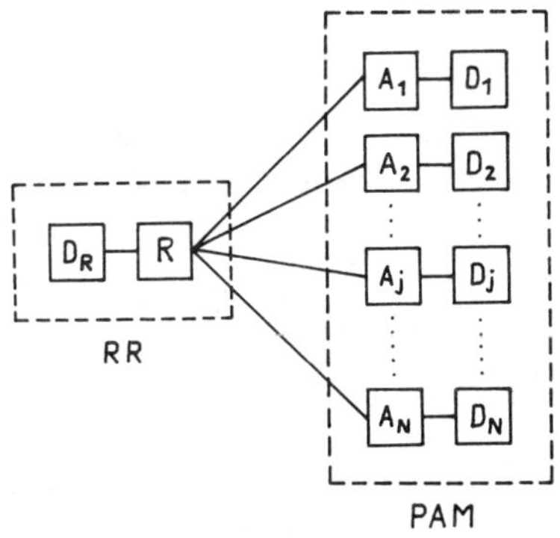

Fig. 7.3. Scheme of the coupling of the field in the ring resonator (RR) with the pumped active medium (PAM).

the complexity of the total system illustrated in Fig. 7.3. But we shall reach our aim without further effort, since we can simply take over results from Part I and Appendix A. For the quantum-theoretical treatment it is necessary to first establish the Hamiltonian $\hat{H}_{\text{tot}}$ of the total system; we have

$$\hat{H}_{\text{tot}} = \hat{H}_R(\hat{a}) + \sum_j \hat{H}_{A_j}(\hat{B}_j) + \hat{H}_{D_R}(\hat{d}_{R,\mu}) + \sum_j \hat{H}_{D_j}(\hat{d}_{j,\mu})$$
$$+ \sum_j \hat{H}^C_{R-A_j}(\hat{a}, \hat{B}_j) + \hat{H}^C_{R-D_R}(\hat{a}, \hat{d}_{R,\mu}) + \sum_j \hat{H}^C_{A_j-D_j}(\hat{B}_j, \hat{d}_{j,\mu}). \quad (7.51)$$

The first four terms on the right-hand side are the Hamiltonians of the free systems. $\hat{H}_R$ refers to the tuned radiation mode in the resonator at the frequency $\omega_0 = \omega_{21}$, where $\hat{a}$ is the annihilation operator. Therefore $\hat{H}_R$ has the form of $\hat{H}_\mu$ in (2.21). $\hat{H}_{A_j}$ represents the energy operator of the jth atomic two-level system as given in (A.173a), where $\hat{B}_j$ is the flip operator that represents the transition from the upper to the lower level. The third and fourth terms stand for the Hamiltonians of the dissipative systems that were discussed in Section 3.2.4.1; the quantities $\hat{d}_{R,\mu}$ and $\hat{d}_{j,\mu}$ are the variables of the μth mode of the reservoirs D_R and D_j, respectively. The fifth term represents the coupling between the radiation field and the atomic systems. In the dipole approximation and with neglect of antiresonant terms we have, according to (3.51),

$$\sum_j \hat{H}^C_{R-A_j}(\hat{a}, \hat{B}_j) = -\left(\frac{\hbar\omega_{21}}{2\varepsilon_0 V}\right)^{1/2} i\hat{a}\sum_j \hat{B}_j d + \text{h.c.}. \quad (7.52)$$

The last two terms in (7.51) belong to the coupling between dynamical and dissipative systems. The sixth takes account of resonator losses, the seventh involves terms [cf. (3.68) and (3.87)] that give rise to real transitions between the atomic levels as well as terms that describe only phase decay [cf. discussion of (3.96)]. It may be added that the operators $\hat{B}_j$ and $\hat{B}_j^\dagger$ allow us to represent the total polarization and the total inversion by

$$\hat{P} = \sum_j \hat{B}_j d + \text{h.c.} \quad \text{and} \quad \widehat{\Delta N} = \sum_j \left(\hat{B}_j^\dagger \hat{B}_j - \hat{B}_j \hat{B}_j^\dagger\right). \quad (7.53)$$

We now go on to establish the equation of motion for the annihilation operator $\hat{a}(t)$ of the field. From the $\hat{a}$-dependent parts in $\hat{H}_{\text{tot}}$, we have, according to (A.70),

$$\frac{d}{dt}\hat{a}(t) = \frac{1}{i\hbar}[\hat{a}(t), \hat{H}_R(t)] + \frac{1}{i\hbar}\left[\hat{a}(t), \sum_j \hat{H}^C_{R-A_j}(t)\right]$$
$$+ \frac{1}{i\hbar}\left[\hat{a}(t), \hat{H}_{D_R}(t)\right] \quad (7.54)$$

as an exact equation of motion for $\hat{a}(t)$. According to (2.24), the first term on the right-hand side gives $-i\omega_0\hat{a}(t)$. Because $[\hat{a}, -i\hat{a} + i\hat{a}^\dagger] = i\hat{I}$, the second term on the right-hand side gives $ig\Sigma_j\hat{B}_j$, where $g = (d^2\omega_{21}/2\varepsilon_0 V\hbar)^{1/2}$. The third term can be inferred from the equation (3.83), which was derived with well-founded approximations for the interaction of a dissipative system with a dynamical one. If the operator $\hat{C}$ in (3.83) is identified with $\hat{a}$ and the damping constant β with $v'/2$, it follows from (3.83) that the third term in (7.54) is equal to $[-(v'/2)\hat{a}(t) + \hat{\Gamma}_{\hat{a}}(t)]$. Thus the equation of motion for the annihilation operator of the radiation field reads

$$\frac{d}{dt}\hat{a}(t) + i\omega_{21}\hat{a}(t) + \frac{v'}{2}\hat{a}(t) = g(-i)\sum_j \hat{B}_j(t) + \hat{\Gamma}_{\hat{a}}(t), \tag{7.55}$$

where $\hat{\Gamma}_{\hat{a}}(t)$ is the fluctuation force assigned to $\hat{a}(t)$. In an analogous way equations of motion for $\hat{P}$ and $\widehat{\Delta N}$ can be established. In doing this, the interaction between a two-level system and a dissipative system comes into play, as it is generally described in (3.88).

In Section 3.2.4.1 we established relations for rapidly varying operators as well as for slowly varying ones; for the latter the main time dependence was eliminated. Here, too, we want to pass from the operator $\hat{a}(t)$ in the Heisenberg picture to the slowly varying amplitude $\hat{\mathfrak{a}}(t)$ by

$$\hat{a}(t) = \hat{\mathfrak{a}}(t)e^{-i\omega_{21}t}. \tag{7.56}$$

By the same procedure the transition from $\hat{P}(t)$ to a slowly varying normalized polarization amplitude $\hat{\mathfrak{p}}(t)$ can be performed. Altogether we obtain the following set of coupled differential equations for the observables of the normalized field amplitude $\mathfrak{a}(t)$ and the normalized polarization amplitude $\hat{\mathfrak{p}}(t)$, and for the total inversion $\widehat{\Delta N}(t)$, where the first equation directly follows from (7.55):

$$\frac{d}{dt}\hat{\mathfrak{a}} + \frac{v'}{2}\hat{\mathfrak{a}} = g\hat{\mathfrak{p}} + \hat{\Gamma}_{\hat{\mathfrak{a}}}, \tag{7.57a}$$

$$\frac{d}{dt}\hat{\mathfrak{p}} + \frac{1}{\tau}\hat{\mathfrak{p}} = g\hat{\mathfrak{a}}\widehat{\Delta N} + \hat{\Gamma}_{\hat{\mathfrak{p}}}, \tag{7.57b}$$

$$\frac{d}{dt}\widehat{\Delta N} + \frac{1}{T}(\widehat{\Delta N} - \Delta N^e\hat{I}) = g(-2)(\hat{\mathfrak{a}}^\dagger\hat{\mathfrak{p}} + \text{h.c.}) + \hat{\Gamma}_{\widehat{\Delta N}}. \tag{7.57c}$$

Here the meaning of the quantities $v'/2, \tau^{-1}, T^{-1}, g, \Delta N^e$ is the same as in (7.38). Note that each of the slowly varying quantities $\hat{\mathfrak{a}}, \hat{\mathfrak{p}}, \widehat{\Delta N}$ has a fluctuation force of its own. The equations are called the Langevin equations of the problem; the operators $\hat{\Gamma}_{\hat{\mathfrak{a}}}, \hat{\Gamma}_{\hat{\mathfrak{p}}}, \hat{\Gamma}_{\widehat{\Delta N}}$ are called the Langevin operators. They depend on the particular reservoir variables. Some general properties of the fluctuation forces are given in Section 3.2.4.1 [e.g. (3.79), (3.80)]. Averaging one of the fluctuation forces $\hat{\Gamma}(t)$ by taking the trace over the reservoir variables leads to zero. The second-order correlation functions in $\hat{\Gamma}$ and/or $\hat{\Gamma}^\dagger$ are either zero or proportional to a delta function of the time difference. This delta-function character mainly originates from the assumption of a large reservoir with an infinite number of degrees of freedom in the limiting case. It expresses mathematically the fact that the correlation times of the fluctuation forces are much less than the time variations of the other quantities involved. The driving forces on the right-hand side reveal that the set of equations includes nonlinearities, as they have already been presented in the semiclassical treatment of Section 7.1.1. To give an example, when substituting the formal solution of $\widehat{\Delta N}$ from (7.57c) into (7.57b), one obtains a nonlinear dependence of the polarization amplitude $\hat{\mathfrak{p}}$ on the field amplitude $\hat{\mathfrak{a}}$.

Disregarding the fluctuation forces and the operator character of the variables, the equations (7.57a, b, c) agree with the semiclassical ones (7.38a, b, c). This outcome is an additional verification of the latter.

In the following we shall establish *semiclassical equations extended by fluctuation forces*. We start by supposing in (7.57a, b, c) the operators $\hat{\mathfrak{a}}, \hat{\mathfrak{p}}, \widehat{\Delta N}$ to be replaced by classical quantities $\mathfrak{a}$, $\mathfrak{p}$, ΔN from (7.37), and the fluctuation operators $\hat{\Gamma}_{\hat{\mathfrak{a}}}, \hat{\Gamma}_{\hat{\mathfrak{p}}}, \hat{\Gamma}_{\widehat{\Delta N}}$ to be replaced by the (classical) fluctuation forces $\Gamma_{\mathfrak{a}}, \Gamma_{\mathfrak{p}}, \Gamma_{\Delta N}$, which are also called Langevin forces. These "*extended semiclassical equations*,"

$$\frac{d}{dt}\mathfrak{a} + \frac{v'}{2}\mathfrak{a} = g\mathfrak{p} + \Gamma_{\mathfrak{a}}, \tag{7.58a}$$

$$\frac{d}{dt}\mathfrak{p} + \frac{1}{\tau}\mathfrak{p} = g\mathfrak{a}\,\Delta N + \Gamma_{\mathfrak{p}}, \tag{7.58b}$$

$$\frac{d}{dt}\Delta N + \frac{1}{T}(\Delta N - \Delta N^e) = g(-2)(\mathfrak{a}^\dagger\mathfrak{p} + \text{c.c.}) + \Gamma_{\Delta N}, \tag{7.58c}$$

yield results that correspond to the fully quantum-theoretical description over a wide range. Differences arise when in the expectation values of operator products the ordering of the factors becomes important. As an example, the product of the amplitude and its conjugate complex or hermitian conjugate may be considered. In the classical description we have $\mathfrak{a}^*\mathfrak{a} = \mathfrak{a}\mathfrak{a}^*$; in the quantum-theoretical description we have $\langle\hat{\mathfrak{a}}^\dagger\hat{\mathfrak{a}}\rangle = \langle\hat{\mathfrak{a}}\hat{\mathfrak{a}}^\dagger\rangle - 1$. The quantities $\langle\hat{\mathfrak{a}}^\dagger\hat{\mathfrak{a}}\rangle$ and $\mathfrak{a}^*\mathfrak{a}$, respectively, correspond to the mean number of photons in the resonator. Therefore the transition from the operators $\hat{\mathfrak{a}}$, $\hat{\mathfrak{a}}^\dagger$ to the commuting classical variables $\mathfrak{a}$, $\mathfrak{a}^*$ is only of minor importance if the number of photons is large enough. From Ref. 7.12 one may estimate that 4000 is an appropriate value for a lower bound on the number of photons in the resonator at threshold, so that the replacement of $\hat{\mathfrak{a}}$ by $\mathfrak{a}$ is accompanied only by a relatively small error.

In the extended equations (7.58a, b, c) the fluctuation forces $\Gamma_{\mathfrak{a}}$, $\Gamma_{\mathfrak{p}}$, $\Gamma_{\Delta N}$ act as classical random variables, the properties of which will now be discussed. $\Gamma_{\mathfrak{a}}$ is connected with the losses of photons in the resonator, $\Gamma_{\mathfrak{p}}$ with the spontaneous emission of the atomic systems. When a single atomic system changes from one level to another (as a result of relaxation or pumping), the total inversion changes rapidly by ± 2; this means that the fluctuation force $\Gamma_{\Delta N}$ describes a process like shot noise. Compared to the fluctuation forces $\Gamma_{\mathfrak{a}}$ and $\Gamma_{\mathfrak{p}}$, the effect of $\Gamma_{\Delta N}$ can be neglected in our discussion. $\Gamma_{\mathfrak{a}}$ and $\Gamma_{\mathfrak{p}}$ exhibit the following general properties, where an overbar indicates the mean value of the ensemble, which under stationary conditions agrees with the time average:

$$\Gamma(t) \text{ is a complex quantity,}$$

$$\overline{\Gamma(t)} = 0,$$

$$\overline{\Gamma(t')\Gamma(t'')} = 0,$$

$$\overline{\Gamma^*(t')\Gamma(t'')} = F\delta(t' - t''). \tag{7.59}$$

Cross correlation functions between $\Gamma_{\mathfrak{a}}$ and $\Gamma_{\mathfrak{p}}$ vanish. The contents of (7.59) characterizes a random variable with the Markov properties. F is called the correlation strength.

First we determine the correlation strength $F_{\mathfrak{a}}$ of the fluctuation force $\Gamma_{\mathfrak{a}}$ from (7.58a). This will be achieved by means of the corresponding quantum equation (7.57a), which in turn, was obtained from (7.54) via (7.55). From the procedure of its derivation we see that the fluctuation force $\Gamma_{\mathfrak{a}}$ results from the third term on the right-hand side of (7.54) in such a way that it is independent of the second term, which is proportional to the coupling constant g. Therefore we may utilize (7.58a) in the following way. We set $g = 0$, so that the solution

$$\mathfrak{a}(t) = \int_0^{\infty} d\tau' \, e^{-(v'/2)\tau'} \Gamma_{\mathfrak{a}}(t - \tau') \tag{7.60}$$

results. This leads to the mean number of photons

$$\overline{\mathfrak{a}^*(t)\mathfrak{a}(t)} = \frac{F_{\mathfrak{a}}}{v'} \tag{7.60a}$$

in the resonator mode, which in thermal equilibrium equals the Bose factor $n_{\text{th}} = [\exp(\hbar\omega_{21}/k_B\mathcal{T}) - 1]^{-1}$. Thus we have

$$F_{\mathfrak{a}} = v' n_{\text{th}}. \tag{7.61}$$

For calculating the correlation strength $F_{\mathfrak{p}}$ of the fluctuation force $\Gamma_{\mathfrak{p}}$, we start from (7.58b) and apply the previously discussed approximation $d\mathfrak{p}/dt \ll$

$\mathfrak{p}/\tau$. This leads to

$$\mathfrak{p} = g\tau\mathfrak{a}\,\Delta N + \tau\Gamma_{\mathfrak{p}},$$

yielding for the normalized slowly varying field amplitude $\mathfrak{a}$ the equation

$$\frac{d}{dt}\mathfrak{a} + \left(\frac{v'}{2} - g^2\tau\Delta N\right)\mathfrak{a} = g\tau\Gamma_{\mathfrak{p}} + \Gamma_{\mathfrak{a}}. \tag{7.62}$$

The strengths of the two fluctuation forces can be determined independently of one another. Thus, for achieving a simple way of determining $F_{\mathfrak{p}}$, we may temporarily disregard the fluctuations of $\mathfrak{a}$ caused by $\Gamma_{\mathfrak{a}}$ and equate $\Gamma_{\mathfrak{a}}$ to zero. Then one obtains

$$\frac{d}{dt}\mathfrak{a}^*\mathfrak{a} = 2g^2\tau\mathfrak{a}^*\mathfrak{a}\,\Delta N - v'\mathfrak{a}^*\mathfrak{a} + g\tau\left(\mathfrak{a}\Gamma_{\mathfrak{p}}^* + \text{c.c.}\right). \tag{7.63}$$

The quantity $\mathfrak{a}^*\mathfrak{a}$ is the number of photons in the resonator mode; obviously it depends on $\Gamma_{\mathfrak{p}}$. We want to find the mean value $\overline{\mathfrak{a}^*\mathfrak{a}}$ by averaging over the random variable $\Gamma_{\mathfrak{p}}$; therefore, using the time derivative of $\mathfrak{a}$, we replace the product $\Gamma_{\mathfrak{p}}^*(t)\mathfrak{a}(t)$ by

$$\Gamma_{\mathfrak{p}}^*(t)\int_{-\infty}^{t} dt'\,\dot{\mathfrak{a}}(t').$$

Inserting $\dot{\mathfrak{a}}$ from (7.62) and averaging over the random variable $\Gamma_{\mathfrak{p}}$, we obtain

$$\overline{\mathfrak{a}\Gamma_{\mathfrak{p}}^*} = g\tau\int_{-\infty}^{t} dt'\,\overline{\Gamma_{\mathfrak{p}}^*(t')\Gamma_{\mathfrak{p}}(t)} = \tfrac{1}{2}g\tau F_{\mathfrak{p}}.$$

Thus (7.63) goes over into

$$\frac{d}{dt}\overline{\mathfrak{a}^*\mathfrak{a}} = 2g^2\tau\overline{\mathfrak{a}^*\mathfrak{a}}\,\Delta N - v'\overline{\mathfrak{a}^*\mathfrak{a}} + g^2\tau^2F_{\mathfrak{p}}. \tag{7.64}$$

We compare this relation with the equation (7.48) for the quantum-theoretical mean value of the number of photons (where also the number of photons n_{th} in thermal equilibrium is disregarded). From this comparison we see that the first two terms on the right-hand side express the contributions of stimulated emission and absorption as well as the resonator losses; the third term must be attributed to the contribution of spontaneous emission. Thus $g^2\tau^2F_{\mathfrak{p}}$ turns out to be equal to w/T times N_2. With (7.48a) and (7.49) we get the correlation strength

$$F_{\mathfrak{p}} = 2\tau^{-1}N_2. \tag{7.65}$$

To approach our aim of establishing a basic relation for the field strength of laser light with nonlinearities and fluctuation taken into account, let us first return to the discussion of the slowly varying field amplitude $\mathfrak{a}(t)$ for which the differential equation (7.62) holds. It must again be emphasized that the random variable $\Gamma_{\mathfrak{a}}$ is responsible for the occurrence of fluctuations connected with resonator losses, whereas the random variable $\Gamma_{\mathfrak{p}}$ describes fluctuations arising from spontaneous emission. The quantitative characterization of the fluctuation forces $\Gamma_{\mathfrak{a}}$ and $\Gamma_{\mathfrak{p}}$ was given in (7.61) and (7.65).

Since, according to (7.37), $\mathfrak{a}$ is proportional to the slowly varying field amplitude $\hat{E}$, the approximations made in establishing (7.27) reveal that the factor $v'/2 - g^2\tau\,\Delta N$ in (7.62) can be replaced by $v'/2 - \beta/2 + (\zeta/2)(\hbar\omega_0/2\varepsilon_0 V)|\mathfrak{a}|^2$. This gives for $\mathfrak{a}$ the differential equation

$$\frac{d}{dt}\mathfrak{a} + \left(\frac{v'}{2} - \frac{\beta}{2} + \frac{\zeta}{2}\frac{\hbar\omega_0}{2\varepsilon_0 V}|\mathfrak{a}|^2\right)\mathfrak{a} = g\tau\Gamma_{\mathfrak{p}} + \Gamma_{\mathfrak{a}} \equiv \Gamma', \tag{7.66}$$

where the parameters β and ζ are given in (7.27b). The total fluctuation force Γ' is composed of two uncorrelated parts; $\Gamma'(t)$ is delta-correlated with the strength

$$F_{\Gamma'} = v'\left(n_{\text{th}} + \frac{2g^2\tau}{v'}\overline{N}_2\right) \tag{7.66a}$$

The factor preceding $\overline{N}_2$ can be expressed, according to the interpretation of (7.45), by the threshold inversion $(\overline{\Delta N})_{\text{thr}}$. On the other hand, in view of (7.46) we have $\overline{N}_2 = \frac{1}{2}(N + \overline{\Delta N})$; thus $\overline{N}_2$ has no constant value, since it depends on the photon number $\mathfrak{a}^*\mathfrak{a}$ via $\overline{\Delta N}$. However, for typical cases, above threshold $\overline{N}_2$ saturates and thus $F_{\Gamma'}$ may be regarded as a constant, to which

$$F_{\Gamma'} = v'\left[n_{\text{th}} + \left(\frac{N_2}{\Delta N}\right)_{\text{thr}}\right] \tag{7.67}$$

is a good approximation.

Let us now again consider the rapidly varying field strength $E(t)$. Proceeding as in deriving (7.34) from (7.27), one obtains from (7.66)

$$\frac{d^2}{dt^2}E + \left(v' - \beta + \zeta E^2\right)\frac{d}{dt}E + \omega_0^2 E = G'. \tag{7.68}$$

The fluctuation force $G'(t)$ originates from the right-hand side of (7.66). Since $\Gamma_\mathfrak{a}$ and $\Gamma_\mathfrak{p}$ are delta-correlated, this also applies to Γ' and thus to G'. For the ensemble mean value of the amplitude-square spectrum $\mathfrak{W}_{G'}(\omega)$ of $G'(t)$, normalized by

$$\int_0^\infty d\omega\, \mathfrak{W}_{G'}(\omega) = \lim_{T\to\infty} \frac{1}{T}\int_{-T/2}^{+T/2} dt\, [G'(t)]^2, \tag{7.69}$$

we get

$$\mathfrak{W}_{G'}(\omega) = \frac{4\hbar}{\pi\varepsilon_0 V}\omega_0^3 v' n_{\text{th}} + \frac{4\hbar}{\pi\varepsilon_0 V}\omega_0^3 v' \left(\frac{N_2}{\Delta N}\right)_{\text{thr}}. \tag{7.70}$$

[Note that because of the normalization by (7.69) we only consider now the frequency range $\omega \geq 0$.]

Finally, on physical grounds we have to subject (7.68) to a correction in order to be able to describe the field strength of the laser light in a reasonably good approximation. The condition that $G'(t)$ is delta-correlated is the mathematical reflection of zero correlation time and of a white spectrum. In reality, an exact white spectrum does not appear; therefore alterations must be made. The first term of $\mathfrak{W}_{G'}$, which results from thermal photon noise, may remain unaltered, since it can be regarded as a sufficiently broadband, quasi-white spectrum. For the second term, which is due to spontaneous emission, the finite line width must be taken into account. This will be achieved by employing the fluorescence line shape $g_{21}(\omega - \omega_{21})$ [cf. (3.153)]. The line width being $\Delta\omega_{\text{fl}}$, the correlation time of the corresponding fluctuation force is about $\Delta\omega_{\text{fl}}^{-1}$. Accordingly, we replace $G'(t)$ by the fluctuation force $G(t)$ with the modified spectral distribution

$$\mathfrak{W}_{G}(\omega) = \frac{4\hbar}{\pi\varepsilon_0 V}\omega_0^3 v' n_{\text{th}} + \frac{4\hbar}{\pi\varepsilon_0 V}\omega_0^3 v' \left(\frac{N_2}{\Delta N}\right)_{\text{thr}} \frac{g_{21}(\omega - \omega_{21})}{g_{21}^{\max}}, \tag{7.71}$$

which is schematically illustrated in Fig. 7.4. We see that it is a superposition of a line with the line shape function g_{21} and the flat thermal noise distribution; in the visible and near infrared regions the thermal noise yields only a relatively small contribution, because there n_{th} is much smaller than $(N_2/\Delta N)_{\text{thr}}$. Thus we end up with the following equation for the field strength belonging to a one-mode laser oscillation:

$$\frac{d^2}{dt^2}E + (v' - \beta + \zeta E^2)\frac{d}{dt}E + \omega_0^2 E = G. \tag{7.72}$$

The quantities v', β, ζ, ω_0 were determined above; the characterization of $G(t)$

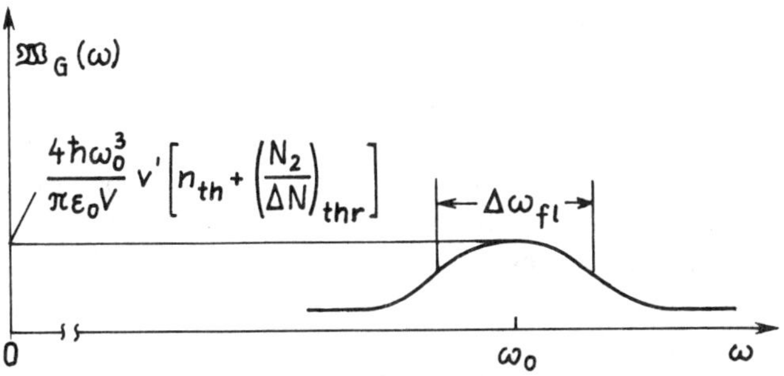

Fig. 7.4. Spectral distribution $\mathfrak{W}_G(\omega)$ of the fluctuation force $G(t)$, with $\Delta\omega_{fl}$ = width of fluorescence line.

by its spectrum was given in (7.71). For the treatment of certain problems it is advantageous (cf. Ref. 7.13) to characterize the total fluctuation force $G(t)$ by slowly varying quantities $G_c(t)$ and $G_s(t)$, which are determined by

$$G(t) = G_c(t)\cos\omega_0 t + G_s(t)\sin\omega_0 t. \tag{7.73}$$

$G_c(t)$ and $G_s(t)$ are uncorrelated, stochastic functions of Gaussian character, which are slowly varying with time. We obtain for their spectrum

$$\mathfrak{W}_{G_c}(\omega) = \mathfrak{W}_{G_s}(\omega) = 2\mathfrak{W}_G(\omega + \omega_0). \tag{7.74}$$

Obviously $\mathfrak{W}_{G_c}$ and $\mathfrak{W}_{G_s}$ are spectral distributions transformed from the range around the high frequency ω_0 to the range near zero frequency. The Gaussianity of the functions $G_c(t)$ and $G_s(t)$ means each of these functions can be represented in the form

$$\sum_l (a_l\cos\omega_l t + b_l\sin\omega_l t), \tag{7.75}$$

where the constant frequency distance $\omega_{l+1} - \omega_l$ of adjacent frequencies is sufficiently small. The quantities a_l, b_l belonging to the frequency ω_l are the statistical variables, which obey the probability distribution

$$p(a_l, b_l)\, da_l\, db_l = (2\pi a^2)^{-1}\exp\left(-\frac{a_l^2 + b_l^2}{2a^2}\right) da_l\, db_l, \tag{7.75a}$$

where $a^2 = \langle a_l^2\rangle = \langle b_l^2\rangle$. The variables a_l and b_l *as* well as the variables of different frequencies are independent of one another.

7.2.2. Density-Operator Equation and Fokker–Planck Equation for the Laser Field

While discussing general methods for the determination of physically relevant quantities in Section 3.1.2, we mentioned that, as a rule, one of the following two procedures is applied. On the one hand, one may start from the equations of motion of the time-dependent operators in the Heisenberg picture; on the other hand, one may calculate the time-dependent density operator of the system, which allows one to evaluate the expectation values wanted. The fundamentals of quantum theory imply that in both cases the results are exactly the same (on the assumption of exact calculation). Depending on the problem to be studied, each of these two methods offers particular advantages. In Section 7.2.1 we proceeded in accordance with the first approach. Under the conditions applying to a laser and on consideration of the characteristics of the interaction between dynamical and dissipative systems, equations of motion [viz. (7.57a, b, c)] were obtained for atomic and field operators, where the unavoidable fluctuations were represented by Langevin operators. In the present subsection we shall derive a density-operator equation and a Fokker–Planck equation for the electric field in the resonator that globally involve all the influences. The resulting equations turn out to have a wide range of validity, covering also the conditions near laser threshold; in particular they may serve for determining the photon distribution of laser light. The derivation of these complicated equations can be handled in a relatively simple way by using important results of Chapters 2 and 3.

For our discussion we start from the laser model already illustrated in Fig. 7.3, where the atomic systems are assumed to be resonant two-level systems. The use of the appropriate Hamiltonian of the total system as given in (7.51) allows us to determine the time behavior of the density operator ρ_{tot} of the total system. Using (A.111), the basic relation in the interaction picture is given by

$$i\hbar \frac{d}{dt} \hat{\rho}_{\text{tot},I}(t) = \left[\hat{H}_{\text{tot},I}(t), \hat{\rho}_{\text{tot},I}(t)\right], \tag{7.76}$$

where, according to the general rules given in Section A.2.1, $\hat{H}_{\text{tot},I}$ originates from the coupling terms of $\hat{H}_{\text{tot}}$, which describe the interactions between the field and the atomic systems, the field and its reservoir, and the atomic systems and their reservoirs. Here we are only interested in the detailed behavior of the dynamical systems, in particular in that of the radiation field, and not in the detailed behavior of the dissipative systems. We therefore pass over to quantities where the influence of all the reservoir variables is averaged out.

In doing this we follow the general pattern discussed in Section 3.2.4. In analogy to the derivation of (3.90) and (3.92), we aim at obtaining quantities that refer only to the vector space of the dynamical systems. Therefore the density operator

$$\hat{\rho}_{\text{dyn}} = \text{Tr}_D\{\hat{\rho}_{\text{tot}}\} \tag{7.77}$$

is introduced, with the trace being taken over all the reservoir variables. Assuming that at time t_0 the dynamical systems are not coupled to the dissipative ones [cf. (3.69a)], the operator $\hat{\rho}_{\text{tot}}(t_0)$ factorizes, at t_0, according to

$$\hat{\rho}_{\text{tot}}(t_0) = \hat{\rho}_{\text{dyn}}(t_0)\hat{\rho}_D(t_0), \tag{7.78}$$

where $\hat{\rho}_D(t_0)$ is that part of $\hat{\rho}_{\text{tot}}(t_0)$, which has to be attributed to the dissipative systems. Now, we want to determine (in a sufficiently good approximation) the time derivative of $\hat{\rho}_{\text{dyn}}$. Using (7.78), the relation (A.116) gives

$$\begin{aligned}\frac{d}{dt}\hat{\rho}_{\text{tot},\,I}(t) &= (i\hbar)^{-1}\left[\hat{H}_{\text{tot},\,I}(t), \hat{\rho}_{\text{dyn}}(t_0)\hat{\rho}_D(t_0)\right]\\ &\quad + (i\hbar)^{-2}\int_{t_0}^{t} dt_1\left[\hat{H}_{\text{tot},\,I}(t), \left[\hat{H}_{\text{tot},\,I}(t_1), \hat{\rho}_{\text{dyn}}(t_0)\hat{\rho}_D(t_0)\right]\right],\end{aligned} \tag{7.79}$$

where, when applying the perturbation theory, the interaction terms up to the second order are kept. The time derivative of $\hat{\rho}_{\text{dyn},\,I}$ is given by

$$\left\{\frac{d}{dt}\hat{\rho}_{\text{dyn},\,I}(t)\right\}_{t_0} = \lim_{t\to t_0}\text{Tr}_D\{\text{right-hand side of (7.79)}\}. \tag{7.80}$$

If in this relation t_0 is replaced by t, a general differential equation results, the so-called *master equation for* $\hat{\rho}_{\text{dyn}}$. As could be expected from the existence of the three coupling terms in (7.51), the relation (7.80) leads to

$$\frac{d}{dt}\hat{\rho}_{\text{dyn},\,I} = \frac{1}{i\hbar}\left[\sum_j \hat{H}^C_{R-A_j,\,I}, \hat{\rho}_{\text{dyn},\,I}\right] + \hat{T}_{R-D_R} + \sum_j \hat{T}_{A_j-D_j} \tag{7.81}$$

The first term on the right-hand side expresses the usual contributions of the interaction of the two dynamical systems, namely those of the field with the ensemble of atomic systems. This term contains the operators $\hat{\rho}_{\text{dyn},\,I}$ and the field and atomic operators. All the reservoirs involved are assumed to be independent of one another. Therefore the contributions $\hat{T}_{R-D_R}$, $\hat{T}_{A_j-D_j}$ originating from them occur as additive terms on the right-hand side. According to statements in Section 3.2.4, their explicit determination can be readily achieved.

This may be exemplified by the coupling of a harmonic oscillator to a dissipative boson system, which was discussed at the beginning of Section 3.2.4. The dynamical system of the harmonic oscillator must be identified with the electromagnetic field of the resonator mode. For determining $\hat{T}_{R-D_R}$ the following assumptions will be used: neglect of antiresonant terms; quasicontinuous spectrum of the reservoir and delta-function-like correlation strength of the fluctuation force; identification of the damping constant β from (3.77) with the resonator loss constant $v'/2$ of (7.57a); introduction of the Bose factor $n_{\text{th}}(\omega_\mu)$ by $\text{Tr}\{\hat{d}^\dagger_{R,\mu}\hat{d}_{R,\mu}\hat{\rho}_{D_R}\} = n_{\text{th},\mu}$ [cf. (3.80)]. Finally, one obtains

$$\hat{T}_{R-D_R} = \frac{v'}{2}\left\{\left[\hat{a}_I\hat{\rho}_{\text{dyn},I}, \hat{a}_I^\dagger\right] + \left[\hat{a}_I, \hat{\rho}_{\text{dyn},I}\hat{a}_I^\dagger\right]\right\}$$
$$+ \frac{v'}{2}n_{\text{th}}\left[\left[\hat{a}_I, \hat{\rho}_{\text{dyn},I}\right], \hat{a}_I^\dagger\right]. \tag{7.82}$$

The contributions $\hat{T}_{A_j-D_j}$ of the interaction of two-level atomic systems with their dissipative systems are obtained in analogy to the discussion of (3.92). The terms $\hat{T}_{A_j-D_j}$ contain the atomic variables $\hat{B}_j$, $\hat{B}_j^\dagger$, the density operator $\hat{\rho}_{\text{dyn}}$, and the energy and phase relaxation times T and τ.

We intend to set up an equation to describe the laser radiation field without being interested in studying the individual atomic systems. Therefore we shall go over from the density operator $\hat{\rho}_{\text{dyn}}$ to the density operator $\hat{\rho}$ that is related to the Hilbert space of the laser field only. Thus we have

$$\hat{\rho} = \text{Tr}_A\{\hat{\rho}_{\text{dyn}}\}, \tag{7.83a}$$

where the trace is taken over all atomic variables. On the other hand, an operator

$$\hat{\rho}_{\text{inv}} = \text{Tr}_A\left\{\frac{1}{N}\sum_j\left(\hat{B}_j^\dagger\hat{B}_j - \hat{B}_j\hat{B}_j^\dagger\right)\hat{\rho}_{\text{dyn}}\right\} \tag{7.83b}$$

in the Hilbert space of the field is introduced; it may be regarded as the mean contribution of one atomic system to the total inversion. Upon making certain approximations [cf. Risken (Ref. 7.11)] whose content is analogous to that used for establishing (7.68), one obtains from (7.81) the following master equation for the density operator $\hat{\rho}_I$ of the field:

$$\frac{d}{dt}\hat{\rho}_I = \frac{g^2\tau N}{2}\left\{\left[\left[\hat{a}_I, \hat{\rho}_I\right], \hat{a}_I^\dagger\right] + \left[\hat{a}_I^\dagger, \hat{\rho}_{\text{inv},I}\hat{a}_I\right] - \left[\hat{a}_I, \hat{\rho}_{\text{inv},I}\hat{a}_I^\dagger\right]\right\}$$
$$+ \frac{v'}{2}n_{\text{th}}\left\{\left[\left[\hat{a}_I, \hat{\rho}_I\right], \hat{a}_I^\dagger\right] + \frac{v'}{2}\left[\hat{a}_I\hat{\rho}_I, \hat{a}_I^\dagger\right]\right\}$$
$$+\text{h.c.}; \tag{7.84}$$

$\hat{\rho}_{\text{inv},\,I}$ is related to $\hat{\rho}_I$ by

$$\hat{\rho}_{\text{inv},\,I} = \frac{T\Delta\sigma\,\hat{\rho}_I}{2}$$

$$-\frac{g^2\tau T}{2}\{[\hat{a}_I^\dagger, \hat{\rho}_I\hat{a}] - [\hat{a}_I, \hat{\rho}_I\hat{a}_I^\dagger] + 2\hat{\rho}_I$$

$$-2\hat{a}_I^\dagger\hat{a}_I\hat{\rho}_{\text{inv},\,I} - \hat{a}_I\hat{\rho}_{\text{inv},\,I}\hat{a}_I^\dagger - \hat{a}_I^\dagger\hat{\rho}_{\text{inv},\,I}\hat{a}_I - \hat{\rho}_{\text{inv},\,I}\}$$

$$+\text{h.c.}. \tag{7.84a}$$

The parameters v', g, τ, T, N have the same meaning as in Section 7.2.1. The parameter $\Delta\sigma$ is the transition rate from the lower to the upper level minus that from the upper to the lower level [cf. (3.94)].

All operators in (7.84) and (7.84a) are operators of the Hilbert space of the electromagnetic field of the resonator mode. Theoretical results that can be related to empirical data may be obtained by writing the operator equations, for example, in the photon-number representation. Multiplying (7.84) by the eigenvectors $\langle n|$ and $|m\rangle$ with fixed photon numbers, the coarse-grained time derivative of $\langle n|\hat{\rho}(t)|m\rangle$ is formed, which (on the right-hand side) is determined by matrix elements $\rho_{n',\,m'}(t)$. By varying n and m a hierarchy of differential equations arises instead of the operator equation (7.84). The diagonal element $\rho_{n,\,n}(t)$ represents the probability of finding n photons in the resonator at time t. The off-diagonal elements can be connected with the expectation values of the field strength and the products of the field strength, respectively.

On the basis of a slightly simplified laser model, master equations for the density matrix in the interaction picture were set up by Scully and Lamb (Ref. 7.14). The time derivative of $\rho_{n,\,n}$ depends on three terms, of which one is proportional to $\rho_{n,\,n}$ and one is proportional to $\rho_{n-1,\,n-1}$. The coefficients of $\rho_{n,\,n}$ and $\rho_{n-1,\,n-1}$ are the constants of the undepleted gain and the saturation, respectively. The third term is proportional to $\rho_{n+1,\,n+1}$; it represents the resonator losses.

So far in this subsection we have treated the properties of the laser field with the aid of the master equation for the density operator $\hat{\rho}$ or for its matrix elements. However, these facts can equivalently be described by continuous functions with the character of c-numbers, since, as we know from Section 2.3.2, $\hat{\rho}$ itself can be connected with continuous functions. Hence, the master equation may be given a more convenient form, which allows a straightforward discussion of physical facts. The continuous functions used are the function $P(\alpha)$ defining the P-representation (2.121) of $\hat{\rho}$, and the function $P_A(\alpha)$, which is equal to $\pi^{-1}\langle\alpha|\hat{\rho}|\alpha\rangle$ [cf. (2.131)]. We recall that, according to (2.128), P is appropriate for representing expectation values of normally ordered products of the field strength, and that, according to (2.133), the function $P_A(\alpha)$ can be used for antinormally ordered products. The introduction of the time-dependent function $P(\alpha, t)$ into (7.84) leads from the master equation for the density operator to a master equation for P. There results a differential equation that contains the partial time derivative of P and the derivatives of P with respect to the complex field amplitudes α and α^* in various orders.

In the following we outline the way to proceed in establishing the master equation for P, since this procedure may also be applied to other problems. The first thing to do in (7.84) is to replace everywhere the density operator $\hat{\rho}_I$ by the characteristic function $\varkappa_I(\lambda)$ defined in (2.129c). Then we pass over to $P_I(\alpha)$. (For simplicity, the index of the interaction picture will be omitted in the following.)

Some typical terms of (7.84) may serve as examples for demonstrating this approach. We multiply (7.84) on the right by the operator $\hat{D}_N(\lambda)$ given in (2.129c) and take the trace. Thus we have to replace the time derivative $d\hat{\rho}/dt$ according to

$$\frac{d}{dt}\hat{\rho} \rightarrow \frac{\partial}{\partial t}\mathrm{Tr}\{\hat{\rho}\hat{D}_N(\lambda)\} = \frac{\partial}{\partial t}\varkappa(\lambda) \tag{7.85}$$

and the expression $[\hat{a}\hat{\rho}, \hat{a}^\dagger]$ according to

$$(\hat{a}\hat{\rho}\hat{a}^\dagger - \hat{a}^\dagger\hat{a}\rho) \rightarrow \mathrm{Tr}\{\hat{a}\hat{\rho}\hat{a}^\dagger\hat{D}_N\} - \mathrm{Tr}\{\hat{a}^\dagger\hat{a}\hat{\rho}\hat{D}_N\}. \tag{7.86}$$

By carrying out an appropriate cyclic permutation under the trace and applying commutation relations between $\hat{D}_N$ and $\hat{a}^\dagger$, $\hat{a}$, the products can be ordered in such a way that $\hat{a}^\dagger$ stands to the left and $\hat{a}$ to the right of $\hat{D}_N$. Then it is possible to replace the trace by a derivative of $\varkappa(\lambda)$ with respect to λ or λ^*. In view of (2.129c) we have

$$\mathrm{Tr}\{\hat{a}\hat{\rho}\hat{a}^\dagger\hat{D}_N\} = \mathrm{Tr}\{\hat{a}^\dagger\hat{D}_N\hat{a}\hat{\rho}\} = -\frac{\partial^2}{\partial\lambda\,\partial\lambda^*}\varkappa(\lambda). \tag{7.87}$$

Using the commutation relation $[\hat{a}^\dagger, \hat{D}_N] = \lambda^*\hat{D}_N$, the derivation of which is analogous to that of the commutation relation (A.29), the other term in (7.86) may be given a suitable form, namely

$$\mathrm{Tr}\{\hat{D}_N\hat{a}^\dagger\hat{a}\hat{\rho}\} = -\lambda^*\mathrm{Tr}\{\hat{D}_N\hat{a}\hat{\rho}\} + \mathrm{Tr}\{\hat{a}^\dagger\hat{D}_N\hat{a}\hat{\rho}\}. \tag{7.88}$$

By virtue of (2.129c) we have $\mathrm{Tr}\{\hat{D}_N\hat{a}\hat{\rho}\} = -\partial\varkappa/\partial\lambda^*$. Therefore we have the replacement

$$(\hat{a}\hat{\rho}\hat{a}^\dagger - \hat{a}^\dagger\hat{a}\hat{\rho}) \rightarrow -\lambda^*\frac{\partial}{\partial\lambda^*}\varkappa(\lambda). \tag{7.89}$$

After introducing the characteristic functions $\varkappa(t)$ and $\varkappa_{\text{inv}}(t)$ into (7.84) and (7.84a) (where $\varkappa_{\text{inv}}$ is constructed from $\hat{\rho}_{\text{inv}}$ in the same way as $\varkappa$ from $\hat{\rho}$), the integral operation

$$\frac{1}{\pi^2}\int d^2\lambda\, e^{-\lambda\alpha^*+\lambda^*\alpha}$$

is carried out on both sides, which means, according to (2.129b), the transition from the function $\varkappa$ to the function P by way of a Fourier transformation. $\partial\varkappa/\partial t$ goes over into $\partial P/\partial t$. The expression on the right-hand side of (7.89) goes over into $(\partial/\partial\alpha)[\alpha P(\alpha)]$, since the Fourier transformation causes the derivative $\partial/\partial\lambda^*$ to become proportional to the Fourier variable α and the factor λ^* to become proportional to the derivative $\partial/\partial\alpha$.

Altogether, for the master equation of P we obtain from (7.84) the differential equation

$$\frac{\partial P}{\partial t}+\left(\frac{\partial}{\partial\alpha}\alpha+\frac{\partial}{\partial\alpha^*}\alpha^*\right)\left(g^2N\tau P_{\text{inv}}-\frac{v'}{2}P\right)$$
$$=v'\left(n_{\text{th}}+\frac{1}{2}\left(\mp\right)\frac{1}{2}\right)\frac{\partial^2}{\partial\alpha\,\partial\alpha^*}P+g^2N\tau\frac{\partial^2}{\partial\alpha\,\partial\alpha^*}\left(P\left(\pm\right)P_{\text{inv}}\right),\quad(7.90)$$

where

$$\left(\frac{\partial}{\partial\alpha}\alpha+\frac{\partial}{\partial\alpha^*}\alpha^*\right)(X)\overset{\text{def}}{=}\frac{\partial}{\partial\alpha}(\alpha X)+\frac{\partial}{\partial\alpha^*}(\alpha^*X).$$

P_{inv} is constructed from $\varkappa_{\text{inv}}$ in the same way as P from $\varkappa$. Equation (7.84a) leads to the following relation between P and P_{inv}:

$$\left\{1+g^2\tau T\left[2(\alpha^2+\alpha^{*2})\left(\mp\right)2\left(\frac{\partial}{\partial\alpha}\alpha+\frac{\partial}{\partial\alpha^*}\alpha^*-1\right)\right]+\frac{\partial^2}{\partial\alpha\,\partial\alpha^*}\right\}P_{\text{inv}}$$
$$=\left\{T\Delta\sigma-g^2\tau T\left[2-\frac{\partial}{\partial\alpha}\alpha-\frac{\partial}{\partial\alpha^*}\alpha^*\left(\pm\right)\frac{\partial^2}{\partial\alpha\,\partial\alpha^*}\right]\right\}P.\quad(7.90a)$$

In determining P from (7.90) and (7.90a), the lower signs in parentheses are ignored.

The master equation for the function P_A needed for the calculation of antinormally ordered products of field operators has the same structure. It can be readily obtained from (7.90) and (7.90a) by replacing P by P_A, P_{inv} by $P_{\text{inv},A}$, and the upper signs by the lower ones in parentheses. Altogether one may say that the one basic equation (7.84) for the density operator is replaced by two equations for the c-number functions P and P_A.

The differential operator acting on P_{inv}, which stands on the left-hand side of (7.90a), contains derivatives with respect to α and α^*. Therefore the elimination of P_{inv} by a differential operator calculus and the insertion of P_{inv} into (7.90) would provide a differential equation for the function P containing derivatives with respect to α and α^* up to infinite order. To get a definite differential equation for P, an expansion up to a certain order must be performed. Following the procedure given by Risken (Ref. 7.11), a normalized time coordinate $\bar{t}$ and a normalized amplitude $\bar{\alpha}$ are introduced instead of t and α, respectively. This enables us to gain a clear insight into the orders of magnitude of the individual terms. The normalized variables are given by

$$\bar{t} = s_t t \quad \text{and} \quad \bar{\alpha} = s_\alpha \alpha \tag{7.91}$$

where the scaling parameters are connected with the parameters in the master equation (7.84) by

$$s_t = g^2\tau\sqrt{\tfrac{1}{2}v'TN} \quad \text{and} \quad s_\alpha = 2\left(\frac{v'T}{2N}\right)^{1/4}. \tag{7.91a}$$

After substituting P_{inv} from (7.90a) into (7.90) and expanding up to terms containing s_α^2, this relation goes over into the following *equation of motion for* P, which is now a function of $\bar{\alpha}$ and $\bar{t}$:

$$\begin{aligned}
&\frac{\partial}{\partial \bar{t}}P + \overline{\frac{\partial}{\partial \alpha}\alpha + \frac{\partial}{\partial \alpha^*}\alpha^*}\left(\tilde{p} - \overline{\frac{\alpha^2 + \alpha^{*2}}{2}}\right)P - \overline{\frac{4\partial^2}{\partial\alpha\,\partial\alpha^*}}P \\
&\quad = s_\alpha^2\left\{\overline{\frac{\partial}{\partial \alpha}\alpha + \frac{\partial}{\partial \alpha^*}\alpha^*}\left[\frac{\tilde{p}}{4\tilde{\sigma}}\,\overline{\frac{\alpha^2 + \alpha^{*2}}{2}} + \frac{1}{4\tilde{\sigma}}\left(2 - \overline{\frac{\partial}{\partial \alpha}\alpha - \frac{\partial}{\partial \alpha^*}\alpha^*}\right)\right.\right. \\
&\qquad \left.(\mp)\,\frac{1}{2}\left(\overline{\frac{\partial}{\partial \alpha}\alpha + \frac{\partial}{\partial \alpha^*}\alpha^*} - 1\right) - \frac{1}{4\tilde{\sigma}}\left(\overline{\frac{\alpha^2 + \alpha^{*2}}{2}}\right)^2\right]P \\
&\qquad \left.(\pm)\,\frac{1}{4}\,\overline{\frac{\partial^2}{\partial\alpha\,\partial\alpha^*}}\left(\tilde{p} - \overline{\frac{\alpha^2 + \alpha^{*2}}{2}}\right)P\right\}.
\end{aligned} \tag{7.92}$$

The bar over derivatives indicates differentiation with respect to $\bar{\alpha}$ and $\bar{\alpha}^*$ and factors containing $\bar{\alpha}$ and $\overline{\alpha^*}$. Near threshold all the barred terms are of the same order. For simplicity, in deriving (7.92) n_{th} was set equal to zero. In addition to the scaling parameter s_α, the pump parameter $\tilde{p}$ and the total equilibrium inversion ΔN^e (related to N) enter the equation of motion for P.

They are connected with the parameters in the master equation (7.84) by

$$\tilde{p} = \left(g^2 \tau \Delta N^e - \frac{v'}{2} \right) \left(\tfrac{1}{2} g^4 v' \tau^2 T N \right)^{-1/2} \tag{7.93a}$$

and

$$\tilde{\sigma} = \frac{\Delta N^e}{N}. \tag{7.93b}$$

If in (7.92) the term P is replaced by P_A and the lower signs in parentheses are used, (7.92) becomes the equation of motion for P_A.

We may call (7.92) the *general Fokker–Planck equation* of the laser field. Setting the right-hand side of (7.92) equal to zero, the relation we are left with is the semiclassical limit of the general Fokker–Planck equation. It may be called *classical Fokker–Planck equation* of the problem, where the second term and the third term on the left-hand side of (7.92) represent the "drift term" and the "diffusion term," respectively. The right-hand side of (7.92) represents —up to order s_α^2—the quantum-theoretical correction of the classical Fokker–Planck equation. From the occurrence of upper and lower signs we see that the expressions for P and P_A in the quantum-theoretical correction are different. Such is found not to be the case with the classical Fokker–Planck equation. In Section 7.3.2 we shall discuss numerical consequences of (7.92), and it will turn out that owing to the smallness of the expansion parameters s_α^2 the classical Fokker–Planck equation may, in general, be regarded as a good approximation to the general Fokker–Planck equation for the laser field.

The classical Fokker–Planck equation of the laser field can be given a clear and easy-to-handle form by replacing the complex field amplitude $\bar{\alpha}$ with $\bar{r}e^{i\bar{\varphi}}$ and thus introducing the modulus $\bar{r}$ and the phase $\bar{\varphi}$ into the equation. Then (7.92) yields (if the quantum-theoretical correction is set equal to zero)

$$\frac{\partial P}{\partial \bar{t}} + \frac{1}{\bar{r}} \frac{\partial}{\partial \bar{r}} \bar{r}^2 (\tilde{p} - \bar{r}^2) P = \frac{1}{\bar{r}} \frac{\partial}{\partial \bar{r}} \left(\bar{r} \frac{\partial P}{\partial \bar{r}} \right) + \frac{1}{\bar{r}^2} \frac{\partial^2 P}{\partial \bar{\varphi}^2}. \tag{7.94}$$

Finally, we compare the statements obtainable from the classical Fokker–Planck equation (as the semiclassical limit of the general Fokker–Planck equation) with those from the semiclassical equation of motion (7.66) for the laser field strength, where fluctuation forces were incorporated.

We start from a general statement. It can be shown (Ref. 7.11) that the classical Fokker–Planck equation, resulting from the equation (7.84) for the density operator in the interaction picture, yields a distribution $P(\bar{\alpha}, \bar{t})$ that reflects the probability distribution for the slowly varying complex field amplitude $\alpha(t)$ from (7.66). This fact is based on the classical description of the field variables, on the delta correlation of the fluctuation force, and on the quantitative conditions of the processes involved.

Further physical insight can be gained from representing the pump parameter, which governs the classical Fokker–Planck equation, by means of terms in the equations of motion for the field amplitude (7.66) and the field (7.72). Equation (7.93a) leads to

$$\tilde{p} = (\beta - v')\left(\frac{\hbar\omega_0}{4\varepsilon_0 V}\zeta\right)^{-1/2}\left(\sqrt{2g^2\tau\frac{N}{2}}\right)^{-1}. \tag{7.95}$$

An expression similar to that in the radical occurred in (7.66a); there the correlation strength of the main part of the fluctuation force was given by $2g^2\tau N_2$. The deviation of this quantity from $2g^2\tau(N/2)$ is, by virtue of (7.46), of the order of s_α^2 and may be neglected in the classical Fokker–Planck equation. According to the explanation of (7.66a) we may take approximately

$$2g^2\tau\frac{N}{2} = v'\left(\frac{N_2}{N}\right)_{\text{thr}}.$$

Taking into account the spectrum (7.71) of the fluctuation force $G(t)$, we obtain the approximation

$$2g^2\tau\frac{N}{2} = \mathfrak{W}_G(0)\frac{\pi\varepsilon_0 V}{4\hbar\omega_0^3}, \tag{7.95a}$$

where now the relatively small term n_{th} is involved that was ignored in (7.92). We finally arrive at

$$\tilde{p} = \frac{4}{\sqrt{\pi}}(\beta - v')\omega_0[\zeta\mathfrak{W}_G(0)]^{-1/2}. \tag{7.95b}$$

Thus the pump parameter depends on the difference between the gain factor β and the loss factor v'. It is negative below threshold, zero at threshold, and positive above threshold. Furthermore it depends on the product of ζ and $\mathfrak{W}_G(0)$. Since ζ is proportional to the third-order susceptibility and $\mathfrak{W}_G(0)$ is proportional to the correlation strength of the fluctuation force, one may say that the pump parameter, among other things, is determined by a *combination of nonlinear and statistical properties*. Note that the line-shape function $g_{21}(\omega)$ and thus the line width of the fluorescence line do not enter into the formula for $\tilde{p}$.

For evaluating the distribution function of the complex field amplitude it is, as a rule, more convenient to start from the classical Fokker–Planck equation, say in the form of (7.94), rather than from the nonlinear differential equation (7.66). This holds in spite of the fact that the classical Fokker–Planck equation, as we have already seen, also rests on a nonlinear laser theory.

7.3. PROPERTIES OF THE LASER OUTPUT RADIATION

The results obtained in Sections 7.1 and 7.2 on continuous and discontinuous laser processes allow us to draw conclusions about the laser output radiation. In Section 7.3.1 a qualitative as well as quantitative treatment of the fluctuations of laser radiation and its line widths is presented. Afterwards we shall discuss in Section 7.3.2 the conclusions on the statistical properties of laser light as represented by correlation functions and photon distributions, and we shall compare them with experimental results. We shall see that the properties of output radiation are determined by a combination of aspects of nonlinearity and fluctuation.

7.3.1. Phase Noise, Amplitude Noise, and Line Width

The laser radiation may be characterized by the field strength $E(t)$, which is determined from the differential equation (7.72). The random variable $G(t)$ of this equation as well as the parameters $v', \beta, \zeta, \omega_0$ were explained in the preceding section and related to parameters of the atomic systems, the resonator, the pump mechanism, and relaxation times.

Before discussing a general solution of (7.72), we seek a special solution under the assumption that $G(t)$ is equal to zero. This gives the completely predicted function

$$E(t) = \hat{E}_0 \cos \omega_0 t, \qquad \text{where} \quad \hat{E}_0 = 2\left(\frac{\beta - v'}{\zeta}\right)^{1/2}, \tag{7.96}$$

as a solution for the stationary field strength under appropriate conditions [cf. (7.34)–(7.36)]. To treat the complete equation (7.72) including the fluctuation force, we start—following Yariv and Caton (Ref. 7.13)—from the ansatz

$$E(t) = \hat{E}_0 \cos \omega_0 t + \hat{E}_0 A(t) \cos \omega_0 t + \hat{E}_0 \varphi(t) \sin \omega_0 t. \tag{7.97}$$

We suppose the functions $A(t)$ and $\varphi(t)$ to be slowly varying, so that d^2A/dt^2 and $d^2\varphi/dt^2$ can be neglected in comparison with $\omega_0\, dA/dt$ and $\omega_0\, d\varphi/dt$ respectively. Since (7.72) is valid also near threshold, we now turn to relations for laser oscillation that hold well above threshold.

To this end let us assume that the terms containing $A(t)$ and $\varphi(t)$ are small compared to the magnitude $\hat{E}_0$ of the fluctuation-free term $\hat{E}_0 \cos \omega_0 t$, so that $|A(t)|, |\varphi(t)| \ll 1$. Upon substituting the ansatz (7.97) into (7.72), using the approximations for $A(t)$ and $\varphi(t)$ given above, and neglecting the terms with higher harmonics, the following equations for $A(t)$ and $\varphi(t)$ result:

$$\frac{d}{dt}A + (\beta - \upsilon')A = \frac{1}{2\omega_0}\frac{G_s}{\hat{E}_0}, \tag{7.98a}$$

$$\frac{d}{dt}\varphi = \frac{1}{2\omega_0}\frac{G_c}{\hat{E}_0}. \tag{7.98b}$$

Here the functions $G_s(t)$ and $G_c(t)$ defined in (7.73) and (7.74) are the slowly varying parts of the rapidly varying total fluctuation force $G(t)$. The condition $|A(t)|, |\varphi(t)| \ll 1$ justifies the replacement of $\arctan[-\varphi/(1 + A)]$ by $-\varphi$, whereby the ansatz (7.97) takes the physically readily interpretable form

$$E(t) = \hat{E}_0[1 + A(t)]\cos[\omega_0 t - \varphi(t)]. \tag{7.99}$$

Obviously the function $E(t)$ represents a cosinelike oscillation which contains, in contrast to (7.96), a (relative) amplitude noise $A(t)$ and a phase noise $\varphi(t)$. Making use of (7.98a, b), the statistical and spectral parameters of $A(t)$ and $\varphi(t)$ can be fairly easily inferred from those of $G_c(t)/\hat{E}_0$ and $G_s(t)/\hat{E}_0$.

If A and φ are regarded, in analogy to a well-known mechanical problem, as displacement coordinates of particles, the relations (7.98a, b) may be interpreted as equations for describing the motion of particles. The first equation corresponds to a motion under the simultaneous influence of a fluctuation force $(G_s/2\omega_0\hat{E}_0)$ and a restoring force whose force constant is $\beta - \upsilon'$. The action of the fluctuation force decreases as $\hat{E}_0$ increases, whereas that of the restoring force increases with increasing difference between the unsaturated gain β and the resonator loss υ'. Qualitatively, the phase equation for the noise displays different behavior. Here the motion results solely from the influence of a fluctuation force $G_c/2\omega_0\hat{E}_0$, which, however, has the same correlation strength as the uncorrelated fluctuation force $G_s/2\omega_0\hat{E}_0$ of the amplitude noise; the restoring force is absent. Consequently mainly the phase noise determines the fluctuations of laser radiation, in particular if $\beta - \upsilon'$ is sufficiently high.

From (7.99) it is seen that the phase noise enters into the formula for the field strength via the function

$$M(t) = \cos[\omega_0 t - \varphi(t)]; \tag{7.100a}$$

thus phase noise can be related to the fluctuations of the so-called instantaneous frequency $\omega(t)$, which is given by

$$\omega(t) \equiv \frac{d}{dt}[\omega_0 t - \varphi(t)] = \omega_0 - \frac{1}{2\omega_0}\frac{G_c(t)}{\hat{E}_0}, \tag{7.100b}$$

where the relation (7.98b) has been used. This means that the spectrum of the deviation $\Delta\omega = \omega(t) - \omega_0$ results in

$$\mathfrak{W}_{\Delta\omega}(\omega) = \frac{1}{\left(2\omega_0 \hat{E}_0\right)^2} \mathfrak{W}_{G_c}(\omega), \tag{7.100c}$$

where the right-hand side can be determined from (7.71) and (7.74).

Our next step is to make quantitative statements on the spectrum of the laser field $E(t)$. This requires that from (7.99) the correlation function

$$\langle E(t)E(t+\tau)\rangle = \hat{E}_0^2\{\langle M(t)M(t+\tau)\rangle + \langle A(t)A(t+\tau)\rangle\langle M(t)M(t+\tau)\rangle\} \tag{7.101}$$

must be established, whose Fourier transform is the spectrum of the laser field strength. In setting up (7.101) use was made of the fact that A and M are independent of one other and the mean value $\langle A\rangle$ is zero.

The first summand $\hat{E}_0^2\langle M(t)M(t+\tau)\rangle$ represents the "pure" phase noise, which, according to the discussion of (7.98a, b) must be regarded as the essential contribution to the second-order correlation function of the field strength $E(t)$. We think it advantageous to sketch the procedure for its calculation, since this approach can be applied analogously to more complicated higher-order correlation functions, for instance, fourth-order correlation functions occurring in resonance fluorescence such as will be discussed in Section 7.3.2.

Because of (7.100a) we have

$$\langle M(t)M(t+\tau)\rangle = \left\langle \operatorname{Re}\{e^{i[\omega_0 t - \varphi(t)]}\}\operatorname{Re}\{e^{i[\omega_0(t+\tau) - \varphi(t+\tau)]}\}\right\rangle. \tag{7.102}$$

From this correlation function the spectrum $\mathfrak{W}_M$ of the function $M(t)$ can be determined by means of Fourier transformation. Using the general definition (7.69), one obtains

$$\mathfrak{W}_M(\omega) = \frac{1}{8\pi}\int_{-\infty}^{+\infty} d\tau\, C(\tau) e^{-i(\omega-\omega_0)\tau}, \tag{7.103}$$

where

$$C(\tau) = \langle e^{-i\varphi(t)} e^{i\varphi(t+\tau)}\rangle \overset{\text{def}}{=} \langle e^{i\Phi}\rangle \tag{7.104}$$

is the correlation function of the phase part $e^{-i\varphi}$. Because of (7.98b) we have

$$\varphi(t) = \int_{-\infty}^{t} dt' \frac{1}{2\omega_0}\frac{G_c(t')}{\hat{E}_0}, \tag{7.105}$$

so that $\Phi = \varphi(t+\tau) - \varphi(t)$ can be written as

$$\Phi = \sum_l \left(a_l \int_t^{t+\tau} dt' \cos \omega_l t' + b_l \int_t^{t+\tau} dt' \sin \omega_l t' \right). \qquad (7.105a)$$

This form takes into account the Gaussianity of $G_c(t')$ [cf. (7.75)]. By applying the corresponding probability distribution (7.75a), the averaging over the function $e^{i\Phi}$ can be readily performed. Since all statistical variables $\{a_l, b_l\}$ are independent of one another, each single factor of $e^{i\Phi}$ can be averaged in terms of one statistical variable. For example, the mean value of the factor $\exp[ia_l F_l(t, \tau)]$ results to be proportional to $\exp[-\langle a_l^2 \rangle F_l^2/2]$. Since the sum of the squares of the two time integrals in (7.105) leads to $4\omega^{-2}\sin^2(\omega\tau/2)$, and also $\langle a_l^2 \rangle = \langle b_l^2 \rangle = \mathfrak{W}_{\Delta\omega}(\omega_l)\, d\omega$, one finally obtains

$$C(\tau) = \exp\left[-\frac{1}{2} \int_0^\infty d\omega\, \tau^2 \mathfrak{W}_{\Delta\omega}(\omega) \operatorname{sinc}^2\left(\frac{\omega\tau}{2}\right) \right]. \qquad (7.106)$$

The function $C(\tau)$ depends quadratically on τ, and so we are allowed to write $C = C(|\tau|)$. The transformation from ω to $\xi = \omega|\tau|/2$ gives

$$C(|\tau|) = \exp\left[-|\tau| \int_0^\infty d\xi\, \mathfrak{W}_{\Delta\omega}\left(\frac{2\xi}{|\tau|}\right) \operatorname{sinc}^2 \xi \right]. \qquad (7.106a)$$

$\mathfrak{W}_{\Delta\omega}(\omega)$ is a low-frequency spectrum consisting of the small quasiconstant thermal contribution and the relatively large contribution originating from spontaneously emitted photons. The latter has its maximum $\mathfrak{W}_{\Delta\omega}(0)$ at $\omega = 0$ and falls to half maximum at $\omega = \frac{1}{2}\Delta\omega_{\text{fl}}$. Therefore in integrating over low frequencies, $\mathfrak{W}_{\Delta\omega}(\omega)$ can in good approximation be regarded as a constant. Thus the exponent on the right-hand side of (7.106a) becomes $-|\tau|\mathfrak{W}_{\Delta\omega}(0)\pi/2$ if $\xi = \frac{1}{4}\Delta\omega_{\text{fl}}|\tau| > \pi$. This leads to

$$C(|\tau|) = \exp\left[-|\tau| \frac{\pi \mathfrak{W}_{G_c}(0)}{8\omega_0^2 \hat{E}_0^2} \right] \quad \text{for} \quad |\tau| > 4\pi\Delta\omega_{\text{fl}}^{-1}. \qquad (7.107)$$

According to the rules of Fourier transformation a correlation function with the form $\exp[-|\tau|\,\Delta\omega']$ for large $|\tau|$ yields approximately a Lorentzian-shaped spectrum around $\omega = 0$ with the line width $\Delta\omega'/2$. In the following we shall assume, as an approximation, that the expression on the right-hand side of (7.107) represents the correlation function for all $|\tau|$. Thus the *spectrum of the phase part* $\hat{E}_0 M(t)$ *of the laser field strength* $E(t)$ is

$$\mathfrak{W}_E(\omega) = \frac{\mathfrak{W}_{G_c}(0)}{16\omega_0^2} \frac{1}{\left(\dfrac{\pi}{8\omega_0^2 \hat{E}_0^2} \mathfrak{W}_{G_c}(0) \right)^2 + (\omega - \omega_0)^2}. \qquad (7.108)$$

This is a Lorentzian line shape around ω_0 with the (fwhm) line width

$$\Delta\omega_E = \frac{\pi}{4}\frac{1}{\omega_0^2 \hat{E}_0^2}\mathfrak{W}_{G_c}(0). \tag{7.109}$$

To discuss this result we make use of the total laser output power including the losses [see (7.44a)], which can be written in the form

$$\mathscr{P}_l = v'\varepsilon_0 V \frac{\hat{E}_0^2}{2} \tag{7.110}$$

in terms of the quantities introduced in Section 7.2. We realize that the maximum of the E-spectrum (at $\omega = \omega_0$) is proportional to $\mathscr{P}_l^2$. On the other hand, the line width $\Delta\omega_E$ is inversely proportional to $\mathscr{P}_l$; thus the total spectral area is proportional to $\mathscr{P}_l$. Note that this result is valid for the pure phase part of the laser field strength. Following this pattern, on the basis of spectral distributions and Gaussianity as given in (7.74) and (7.75), the term with the amplitude noise, which has been ignored so far, can be treated. This part of $E(t)$ has a broad spectral width and a decrease of the spectral area with increasing $\mathscr{P}_l$. Consequently, under normal conditions it is justified to neglect that part which contains the amplitude noise in comparison with the part containing the pure phase noise.

In spite of the incorporated approximations and simplifications, the preceding semiclassical theory with fluctuation forces, which was established as an appropriate modification of the fully quantum-theoretical treatment, provides over a wide range of sufficiently accurate description of laser output radiation. This is, among other things, confirmed by the fact that the expression for $\Delta\omega_E$ in (7.109) represents the accepted line-width formula for the laser,

$$\Delta\omega_E = \frac{\hbar\omega_0}{\mathscr{P}_l}(\Delta\omega_{\mathrm{pr}})^2\left[\left(\frac{N_2}{\Delta N}\right)_{\mathrm{thr}} + n_{\mathrm{th}}\right], \tag{7.111}$$

as given by Schawlow and Townes (Ref. 7.2). The transition from (7.109) to (7.111) is the result of inserting the spectrum $\mathfrak{W}_{G_c}$ from (7.74) and taking into account that the line width $\Delta\omega_{\mathrm{pr}}$ of the passive resonator equals v'. As could be expected from the assumptions given above, the formula (7.111) applies to the limit of homogeneous broadening. Furthermore, $\Delta\omega_E$ can be related to the spectrum $\mathfrak{W}_{\Delta\omega}(\omega)$ of the deviation of the instantaneous frequency from the mean laser frequency ω_0. In view of (7.100c) we have

$$\Delta\omega_E = \pi\mathfrak{W}_{\Delta\omega}(0) \tag{7.112}$$

The laser output power (7.110) can be expressed in terms of the constants v', β, ζ alone:

$$\mathscr{P}_l = \frac{2(v')^2 \varepsilon_0 V}{\zeta}\left(\frac{\beta}{v'} - 1\right). \tag{7.113}$$

The quotient β/v' gives the factor by which pumping exceeds the threshold. Remember that β increases as the unsaturated value of the inversion increases [see (7.27b)], which in its turn increases as the pump rate P' increases [see (7.50)].

All the parameters that enter into the formulas for the spectrum (7.108), the line width (7.111), and the output power (7.113) can be uniquely related by means of (7.27b), (7.50), and (7.34) to the atomic quantities (dipole moment, relaxation times, transition frequency) and the external parameters (resonator losses, resonator volume, pump rate). These relations were derived from the model of the simple three-level laser, as illustrated in Fig. 7.2, assuming a single-mode ring resonator. They can be generalized to a great extent. By an appropriate reinterpretation of the quantities and numerical factors involved they can also be applied to four-level lasers, inhomogeneous broadening, degenerate levels, and other resonator types (standing waves in a cavity) and will lead to the same type of equation for the laser field strength $E(t)$ (Refs. 7.15, 7.16). In the case of degenerate levels the quotient $(N_2/\Delta N)_{\text{thr}}$ in (7.111) must be replaced by $N_2/(N_2 - N_1 g_2/g_1)$, where g_2 and g_1 characterize the degree of degeneracy.

Let us now pass to the discussion of *numerical and measured results* for the abovementioned properties. Manes and Siegman (7.17) have measured the frequency-deviation spectrum $\mathfrak{W}_{\Delta\omega}(\omega)$ of a laser by beating it with a reference laser of much higher stability. The difference-frequency output signal was subjected to a spectral analysis. A typical result is shown in Fig. 7.5.

A well-known and striking property of the laser is its quasi-monochromaticity. For determining the lower bound or limit of the line width of the laser the theoretical treatment based on (7.111) can be used. By an appropriate choice of the parameters V, β, v', ζ in He–Ne lasers and low-pressure CO_2 lasers one

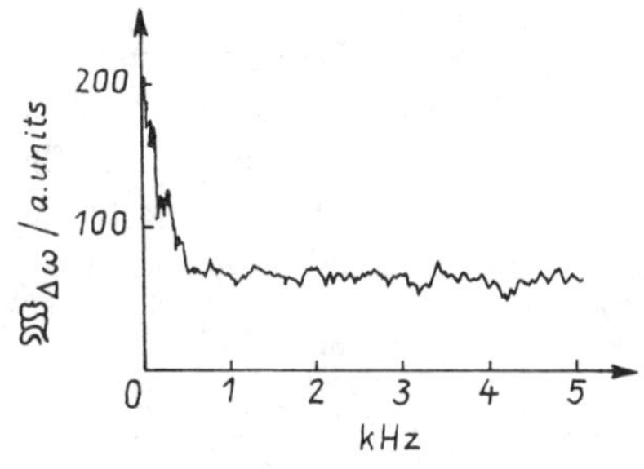

Fig. 7.5. Measured frequency-deviation spectrum [after Manes and Siegman (Ref. 7.17)].

Table 7.1. Internal Parameters and Resulting Theoretical Output Data for Lasers

Laser	He–Ne	Low-pressure CO_2
λ_0 (μm)	0.6328	10.6
Parameters		
Effective resonator volume V (m^3)	5×10^{-7}	6×10^{-5}
Resonator loss v' (s^{-1})	3×10^{6}	6×10^{6}
Saturation parameter ζ ($m^2/s\,V^2$)	9×10^{-1}	3×10^{2}
Output Data		
Output power $\mathscr{P}_l$ (W)	9×10^{-5}	1.3×10^{-4}
Line width $\Delta\omega_E$ (s^{-1})	6.3×10^{-2}	1×10^{-2}
Relative line width $\Delta\omega_E/\omega_0$	2.1×10^{-17}	5.6×10^{-17}

obtains theoretical values of the relative line width $\Delta\omega/\omega_0$ as low as $\sim 10^{-17}$. Table 7.1 gives the values of the line width and the output power obtained under the assumption that $\beta/v' - 1 = 1$ and $(N_2/\Delta N)_{\text{thr}} + n_{\text{th}} = 2$.

The parameters of Table 7.1 concerning the He–Ne laser and CO_2 laser were chosen so that the relative line widths become very small and that there are rather low, but realizable laser output powers.The given line widths have to be understood as a theoretical lower bound calculated by ignoring to a certain extent the experimental problems in their realization. To achieve small line widths in experiments, the first thing to be done is to produce resonance lines as narrow as possible (we shall discuss this problem at the end of this subsection); by means of these narrow lines the resonator length must be stabilized with the help of a powerful servo control device so as to ensure the stability of the resonator frequency, which is determined by the resonator length. Its stabilization is necessary because of mechanical and thermal fluctuations. As a consequence rather stringent requirements have to be imposed on the stabilization of the length: the relative length stability to be reached must nearly be equal to the relative frequency stability required. In the last few years substantial progress has been made. Nowadays, for measuring times less than or approximately equal to 1 s, relative line-width values of 10^{-15} have been measured.

For the ruby laser the orders of magnitude of typical values of phase and energy relaxation times and of resonator losses were given in discussing (7.11) and (7.17), so that they can be simply related to those constants that determine the behavior of the laser field strength $E(t)$. The occupation lifetime in the pump band is about 10^{-7} s, so it is much shorter than the lifetime of ≈ 2 ms in the upper laser level. Both the upper and the lower level are degenerate; the degenerate levels are closely adjacent and nearly equally occupied at room temperature. From (7.46) it is apparent that the positive inversion of the laser operation requires the excitation of more than 50% of all atomic systems, which can only be achieved by supplying a high pump energy.

Let us now consider the relations between the different types of line widths and the experimental implications associated with them. From Section 3.3.2 it follows that the (relative) natural line width of isolated atoms and molecules ranges from 10^{-7} to 10^{-12}. The numerical values in Table 7.1 and the existence of experimentally determined values below 10^{-14} indicate that, as a consequence of the coupling of the invertible active medium with the electric field in a frequency-stable resonator, laser line widths can be achieved that are far below the natural line width and of course far below the fluorescence line width, which exceeds the natural line width to some extent. The existing nonlinearities cause a highly frequency-stabilized oscillation out of the relatively broad spectrum $\mathfrak{W}_G$ corresponding to the fluctuation forces. Typical values of the fluorescence line width which characterizes the spectral width of $\mathfrak{W}_G(\omega)$ are:

1.3×10^{10} s^{-1} for the He–Ne gas laser,
2.4×10^{12} s^{-1} for the ruby laser,
6×10^{13} s^{-1} for the Nd: glass laser (1.06 μm).

Indeed, the fluorescence line width determines the widths of the spectra of the fluctuation forces, as for example $\mathfrak{W}_{G_c}(\omega)$, and hence the directly measurable frequency deviation $\mathfrak{W}_{\Delta\omega}(\omega)$, but the normalized line-shape function g_{21} of the fluorescence line does not enter into the formula for the laser line width or into the formula for the laser output power.

Also, lasers more powerful than those just discussed may have line widths that are much smaller than the natural line width and hence also much smaller than the fluorescence line width. Thus, under appropriate conditions processes within the fluorescence line can be induced.

What is happening in such a process will first be explained in terms of the action of laser radiation on a gas in thermal equilibrium. We start from the model discussed in connection with (3.152). Suppose that the line-shape function $\bar{g}_{21}(\omega)$ of the fluorescence line mainly results from Doppler broadening. The Doppler line width $\Delta\omega_{\mathrm{Do}}$ is assumed to be much greater than the homogeneous line width $\Delta\omega_h$ and thus in turn to be much greater than the width of the laser line. In this sense the acting laser radiation may be regarded as being monochromatic. If the laser frequency ω_L deviates from the central transition frequency ω_{21} in such a way that $|\omega_L - \omega_{21}| \lesssim \Delta\omega_{\mathrm{Do}}$, then the laser radiation can interact with a group of gas particles that possess a certain velocity component in the direction of the laser radiation [see (3.148)]. Suffi-

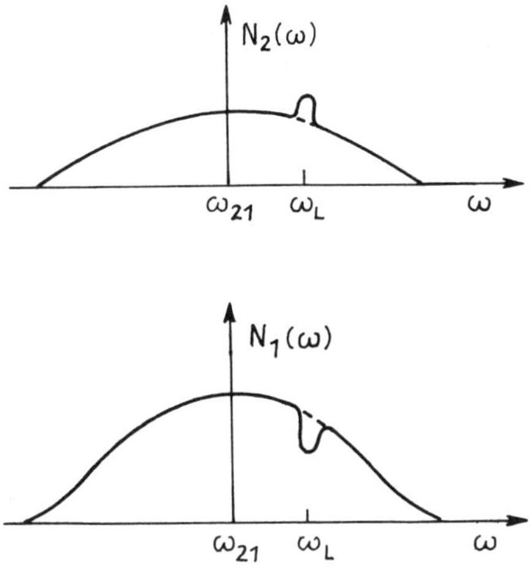

Fig. 7.6. Reoccupation of a Doppler-broadened line due to laser interaction.

ciently intense laser radiation causes a noticeable reoccupation for the particles of this group by absorption.

Figure 7.6 shows the occupation distribution of the lower and upper levels during the interaction of the gas with laser light. Without laser interaction the distribution functions would be symmetric with respect to ω_{21}. The widths of the peak and the dip correspond to the value of the intensity-independent homogeneous line width with an added contribution arising from intensity broadening [cf. Table 3.2]. The number of reoccupied particles depends on the intensity-dependent transition rates between the levels as given in (3.154) and (3.155) as well as on relaxation processes. Below saturation the number of reoccupied particles increases if the laser intensity increases. On measuring the absorption with the help of an additional weak test laser, a decrease of absorption near ω_L results, since it is proportional to $N_1(\omega) - N_2(\omega)$. The appearance of this line, which exhibits a width that is small compared with the fluorescence line, indicates a change in the properties of the gas which is induced by laser radiation.

We now pass from the consideration of a gas in thermal equilibrium to a gas laser. While in the model of Fig. 7.6 the interaction of a "ready-to-absorb" medium with radiation was considered, we now treat the case that the electromagnetic radiation interacts in the laser resonator with an activated and thus "ready-to-emit" medium. For its occupation distributions we know $N_2(\omega) > N_1(\omega)$ as a result of pumping. The interaction with a traveling wave of frequency ω_L causes the inversion to decrease according to the nonlinear mechanism described in Section 7.1.1. Under appropriate conditions (e.g., employing a standing-wave resonator) the laser field in the resonator can be thought of as a superposition of two waves with equal frequencies and intensities traveling in opposite directions. Each of the waves diminishes the inversion at a certain frequency; the two inversion dips lie symmetrically with

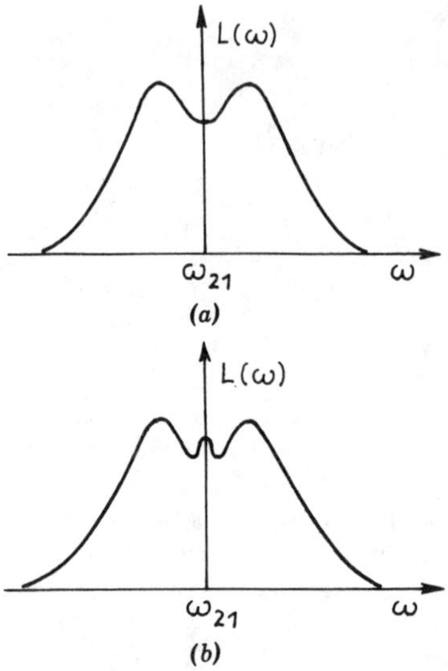

Fig. 7.7. Scheme of (a) Lamb dip and (b) inverse Lamb dip.

respect to the frequency ω_{21} (since the waves travel in opposite directions). In the special case $\omega_L = \omega_{21}$ the two inversion dips coincide, and this sharper decrease in inversion leads to a smaller output power. From this it follows that by tuning the laser frequency ω_L we can achieve the frequency-dependent output power $L(\omega)$ depicted in Fig. 7.7*a* with a dip around ω_{21}, the so-called *Lamb dip* (Ref. 7.15).

These phenomena have proved to be of great importance in applications (Ref. 7.18). One of them is the method of *laser-induced line narrowing*, which is based on the reoccupation of a ready-to-absorb medium (mentioned above); see Fig. 7.6. Another important application is in the technique of the *inverse Lamb dip*. In this, in addition to the active medium in the laser resonator, a cuvette with an absorbing gas is used, which is chosen to exhibit resonance absorption for the laser radiation. This results in an increased output power at the center of the Lamb dip—the so-called inverse Lamb dip, as shown in Fig. 7.7*b*. The width of the inverse Lamb dip agrees with the homogeneous line width of the absorber gas, which can be kept under relatively low pressure, so that a small line width of the inverse Lamb dip can be achieved. The importance of this method is, on the one hand, in nonlinear spectroscopy within the Doppler width, in which homogeneously broadened lines of gas particles can be measured that otherwise are hidden by the Doppler broadening. The values in Table 3.2 show that this method allows the measurement of small relative line widths of the order of 10^{-11}. This means that an increase in resolving power by three orders of magnitude becomes possible. On the other hand, use can be made of the narrow resonance lines resulting from the inverse Lamb dip for frequency stabilization, already mentioned in the discussion on the realization of minimum line widths.

Altogether, it can be stated that these methods, which utilize nonlinear phenomena as well as line-width phenomena based on fluctuations, have great fundamental and practical importance: they can be used for new spectroscopic tasks and they must be regarded as qualitatively improved metrological methods in time and length measuring (see e.g. Ref. 7.19).

7.3.2. Photon Distributions and Correlation Functions

We start with some general remarks on the determination of photon-counting distributions. According to (2.121) the probability of finding n photons inside the laser resonator is

$$\rho_{n,n} = \langle n|\hat{\rho}|n\rangle = \int d^2\alpha \, \frac{|\alpha|^{2n} e^{-|\alpha|^2}}{n!} P(\alpha), \tag{7.114}$$

where $\hat{\rho}$ and $P(\alpha)$ characterize the density operator and the P-representation, respectively, of the laser field. The function $P(\alpha)$ is deducible from the Fokker–Planck equation (7.92). In Section 5.1 we discussed the measurement of photons on the basis of the external photoeffect. Thus from (7.114), by making use of (5.16a) and (5.17), one may obtain the probability $P_n(t, t+T)$ of recording n photoelectrons in the interval $(t, t+T)$, where T is the time of measurement. P_n is a functional of the light intensity $I(t')$, and it contains the quantum efficiency of the detector.

Arecchi, Rodari, and Sona (Ref. 7.12) carried out basic investigations on the statistics of laser radiation; in doing this they also studied the important threshold region. They used a 0.633-μm He–Ne laser supporting a single TEM_{00} mode. In order to achieve high accuracy, they took special care in forcing the dynamical system into the required steady-state conditions. For measuring $P_n(T)$ they used a measuring time T that was shorter than the coherence time τ_c. This condition guarantees that the measurement results may be regarded as being sensitive to the kind of the radiation distribution, as explained in connection with (5.19a, b, c). The physical significance of the results from Ref. 7.12 will be referred to below.

We next turn to the quantitative connection between the general Fokker–Planck equation (7.92) and its semiclassical limit, the "classical" Fokker–Planck equation.

Let us recall that the relation (7.92), because of the substitution of P_{inv} from (7.90a), involves an approximation: it contains quantum corrections up to the order of s_α^2, where s_α is the scaling parameter. The reciprocal of s_α^2 equals approximately the number of photons at threshold; from the experimental results given in Ref. 7.12 one may estimate 4000 as being a typical value under the conditions given. This means that the coupling factor of the quantum corrections to the classical part of the Fokker–Planck equation is very small, so that the classical Fokker–Planck equation is a good approximation to the general Fokker–Planck equation. This fact is also confirmed by comparison of solutions of the Fokker–Planck equation in the stationary case. For example, at the value $\tilde{p} = 3$ of the pump parameter [cf. (7.92), (7.95)] and at the value $s_\alpha^2 = \frac{1}{4000}$ of the squared scaling parameter [cf. (7.91a)], numerical calculation shows that the differences between the functions $P(\bar{I})$, $P_A(\bar{I})$, and $P_{cl}(\bar{I})$ are less than 2%, at least in the range $\bar{I} < 6$, where $\bar{I}$ agrees with $\bar{r}^2$ the normalized intensity from (7.94). Here $P(\bar{I})$ and $P_A(\bar{I})$ are solutions of the untruncated differential equation (7.92), whereas $P_{cl}(\bar{I})$ is the solution of the classical Fokker–Planck equation, where quantum corrections are neglected. Altogether, the usefulness of the classical Fokker–Planck equation in treating typical experimental situations becomes apparent.

In the consideration of solutions of the classical Fokker–Planck equation we start from (7.94). The stationary solution P_{st} does not depend on the phase $\bar{\varphi}$. An analytical solution can be given in the form

$$P_{st}(R;\tilde{p})\cdot I_0 = \begin{cases} \dfrac{2}{\pi}\dfrac{1}{1+\operatorname{erf}\tilde{p}}\exp\left[-\dfrac{1}{\pi}\left(R-\pi^{\frac{1}{2}}\tilde{p}\right)^2\right] & \text{for } R \geq 0, \\ 0 & \text{for } R < 0, \end{cases} \tag{7.115}$$

where I_0 is the mean intensity at threshold ($\tilde{p} = 0$) and R is the reduced intensity I/I_0. This distribution leads to the mean value

$$\langle R\rangle = \pi^{1/2}\tilde{p} + \frac{\exp[-\tilde{p}^2]}{1+\operatorname{erf}\tilde{p}}. \tag{7.115a}$$

Three regions have to be distinguished in the dependence on the value of the pump parameter; for each of them typical distributions are given in Fig. 7.8. Well below threshold ($\tilde{p} \lesssim -2$) we have in good approximation

$$P_{st}(R;\tilde{p}) \propto \exp\left[-\frac{2}{\pi^{1/2}}|\tilde{p}|R\right]. \tag{7.116a}$$

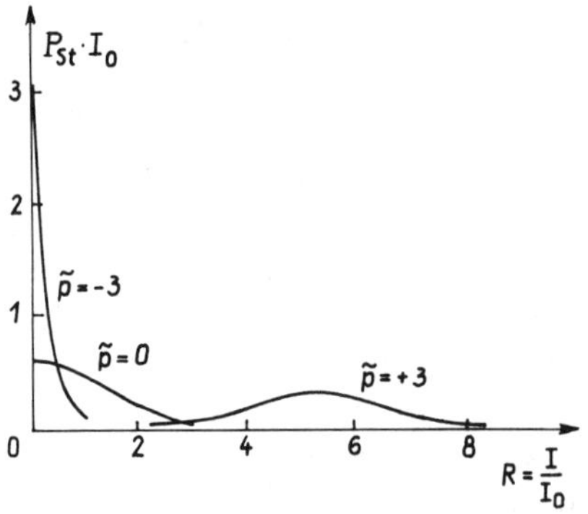

Fig. 7.8. Intensity distribution $P_{st}I_0$ for lasers in stationary operation for the pump parameters $\tilde{p} = -3, 0, +3$ versus $R = I/I_0$ (I_0 denotes mean intensity $\langle I\rangle$ for $\tilde{p} = 0$.)

We see that the square of the modulus of the amplitude decreases exponentially. This means that the amplitude exhibits a normal distribution; thus for $\tilde{p} \lesssim -2$ the laser radiation is of Gaussian character. It should be noted that this result for sufficiently negative $\tilde{p}$ values can easily be explained as well by the Langevin formalism treated in Section 7.2.1: if the gain factor β and the saturation parameter ζ from (7.27b) and (7.34) become sufficiently small, the relation (7.72) for the field strength tends to a linear differential equation with a Gaussian "driving force."

For $\tilde{p} = 0$, at threshold, (7.115) yields

$$P_{st}(R;0) \propto \exp[-\pi^{-1}R^2]. \tag{7.116b}$$

Well above threshold there results

$$P_{st}(R;\tilde{p}) \propto \exp\left[-\frac{\left(\sqrt{R} - \pi^{1/4}\sqrt{\tilde{p}}\right)^2}{\left(\pi^{1/2}/4\right)\tilde{p}^{-1}}\right]. \tag{7.116c}$$

Thus, for large $\tilde{p}$ the absolute value of the amplitude, which is proportional to $\sqrt{R}$, obeys a Gaussian distribution around the center $\pi^{1/4}\sqrt{\tilde{p}}$ with a half-width $\sqrt{\pi^{1/2}\ln 2}\,\tilde{p}^{-1/2}$ (this Gaussian distribution is somewhat truncated for small R, since $P_{st} = 0$ for $R < 0$). The relation (7.116c) reveals that for increasing $\tilde{p}$ the half-width goes to zero, so that P_{st} tends to a delta-function-like distribution. This proves that well above threshold the superposition model, which is illustrated in Fig. 2.2, is a reasonable description of the radiation of a realistic laser.

The results given in the relations (7.116a, b, c) have been confirmed by photon-counting experiments. In accordance with theoretical predictions experimental evidence was given for the transition from the Bose–Einstein distribution for negative $\tilde{p}$ values [see (5.21)] to the Poisson distribution for large positive $\tilde{p}$-values [see (5.20)].

If the photon distribution $P_n(T)$ or the intensity distribution is known, higher moments and higher-order intensity correlation functions can be determined. For instance, from the stationary solution P_{st} of the Fokker–Planck equation the second reduced factorial moment

$$M_2' = \frac{\langle n(n-1)\rangle - \langle n\rangle^2}{\langle n\rangle^2} \tag{7.117}$$

was calculated as a function of the reduced intensity R. The result is shown in Fig. 7.9. Well below threshold ($R < 1$) the value of M_2' tends to unity according to the Gaussian field behavior; well above threshold it tends to zero, since the field tends to a Poisson-distributed one. The experimental results

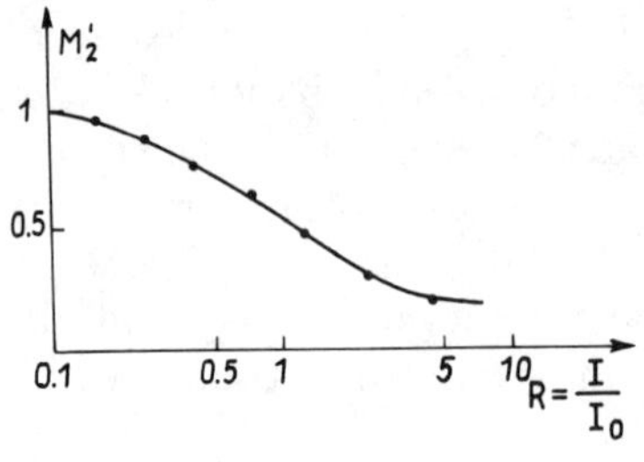

Fig. 7.9. Second reduced factorial moment M_2' versus reduced intensity R (theoretical curve and experimental points). After Arecchi, Rodari, and Sona (Ref. 7.12).

from Ref. 7.12 exhibit very good agreement with the theoretical values mentioned.

We are now going to discuss the *correlation functions of the field strength* $E(t)$ of the laser radiation. From the second-order correlation function (7.101) the spectrum of $E(t)$ can be obtained by Fourier transformation. On taking into account only the phase noise, there resulted the correlation function $C(\tau)$ given in (7.106), which allowed us to determine the spectrum by means of (7.103). The correlation function $C(\tau)$ is a functional of the low-frequency spectrum $\mathfrak{W}_{\Delta\omega}(\omega)$ of the deviation $\Delta\omega$ between the instantaneous frequency and the resonator frequency. $\mathfrak{W}_{\Delta\omega}(\omega)$ contains the line-shape function of the fluorescence line and thus also the line width $\Delta\omega_{\text{fl}}$. In Section 7.3.1 an approximation for $C(\tau)$ was used that yields an exact Lorentzian line shape for the spectrum of $E(t)$. This approximation is called the *phase diffusion model* of the laser field. By means of the complete correlation function (7.106) [if the exact correlation function $C(\tau)$ from (7.106) and the amplitude noise are taken into account], results for the spectrum of $E(t)$ can be obtained that give more information than the phase diffusion model. These general results can also be used for describing the non-Markovian behavior of the laser radiation.

Let us now deal with *fourth-order correlation functions of the laser field*. In the preceding parts of this section more or less complex laser radiation models dependent on the mechanisms involved and the conditions assumed were presented. They provided various expressions for the laser field $E(t)$ according to whether zero or finite line width, pure phase noise, and/or amplitude noise was assumed. This of course leads to different fourth-order correlation functions. We now consider an example of the consequences that may result from employing different models for the interpretation of effects accessible to measurement, namely *resonance fluorescence radiation* (whose experimental and theoretical aspects will be extensively treated in Section 9.2). In the present section we are only interested in how this effect behaves in different models of laser radiation.

Before the problem will be treated quantitatively, some introductory general remarks will be given. The investigation of the intensity correlation of resonance fluorescence radiation became especially important when it provided an improved experimental technique for the detection of the phenomenon of antibunching. Kimble and Mandel (Ref. 7.20) and Carmichael and Walls (Ref. 7.21) predicted that resonance fluorescence radiation of a single atom would exhibit photon antibunching. The experiments carried out (Refs. 7.22, 7.23) used low-density atomic beams. Therefore, because of the fluctuations of the number of emitting atoms, the detection of antibunching from experiments proves to be rather difficult (the connection of these fluctuations with the behavior of a single atom was reported in Ref. 5.17). In the following we shall deal with the resonance fluorescence of a single, fixed atom.

The intensity correlation function of the emitted radiation is determined by the statistical and spectral properties of the laser pump field. The influence of its bandwidth was mentioned in Ref. 7.24. There it was concluded that the intensity correlation function of the emitted radiation factorizes into a product of intensity correlation functions independently of the laser bandwidth. On the other hand Schubert, Süsse, and Vogel (Ref. 7.25) compared a chaotic pump field of finite bandwidth with a coherent laser pump; it was shown that a factorization in the case of chaotic pump radiation cannot occur. Considerable deviations especially appear for small bandwidths of the pump radiation.

The difference between the results in Ref. 7.24 and in Refs. 7.25, 7.26 (where chaotic and laser light have been studied) is as follows: In Ref. 7.24 the starting point is the assumption of factorizability of the averaged product formed by an atomic operator driven by the field and a term depending on the phase of the field; this factorization was justified in Ref. 7.24 by assuming that the atomic variables should be independent of phase fluctuations at subsequent times. However, this can only be correct if the random process causing the phase diffusion of the laser pump has a white power spectrum, that means a vanishing correlation length; under this condition an exact factorization directly follows. But from (7.71) and (7.72) it is known that the spectrum of the stochastic function $G(t)$ as the driving force of the laser field is, among other things, determined by the fluorescence line shape of the active medium. Hence, the fluctuation force has a correlation length of the order of $\Delta\omega_{\mathrm{fl}}^{-1}$, which is given by the fluorescence lifetime of the activated atoms in the laser resonator. For resonance fluorescence the time is of importance during which the intensity correlation function takes on its stationary value; this time is given by the radiative damping constant γ_{21} of the test atom. The correlation time of the laser pump field and the time γ_{21}^{-1} may in principle have the same order of magnitude. At any rate, the factorization of the intensity correlation of the emitted radiation cannot be postulated without further investigations; the influence of the nonwhite phase noise and the amplitude noise of the laser pump field must be estimated and taken into account, respectively.

Investigations of this kind were carried out in Refs. 7.25, 7.26, 7.27. They led to the following relation between the second-order intensity correlation function $\Gamma_{\mathrm{rf}}^{2,2}(t, t+\tau, t+\tau, t)$ of the resonance fluorescence radiation and the second-order correlation function $\Gamma_{\mathrm{pump}}^{2,2}(t-\tau_1, t+\tau', t+\tau'', t-\tau_2)$ of the exciting laser pump field:

$$\Gamma_{\rm rf}^{2,2}(t, t+\tau, t+\tau, t) = \frac{1}{\hbar^4}|d_{21\bullet}e_\bullet|^2 \times \int_0^t d\tau_1 \int_0^t d\tau_2 \int_0^\tau d\tau' \int_0^\tau d\tau'' \, e^{i\omega_{21}(\tau_1-\tau_2+\tau''-\tau')} \times e^{-\gamma_{21}(2\tau+\tau_1+\tau_2-\tau'-\tau'')} \times \Gamma_{\rm pump}^{2,2}(t-\tau_1, t+\tau', t+\tau'', t-\tau_2). \quad (7.118)$$

Here ω_{21} and d_{21} are the transition frequency and the transition dipole moment of the test atom, respectively, while $e_\bullet$ is the polarization unit vector of the laser pump field.

This expression was derived in the lowest order of perturbation theory with respect to the interaction of the laser pump field with the test atom. It holds if the Rabi frequency $d \cdot \hat{E}_0/\hbar$ is smaller than the atomic damping rate γ_{21}. This condition is not crucial for our consideration; since in Ref. 7.24 the factorizability is postulated independently of the strength of the laser pump intensity, we may check this statement also for relatively weak intensity. From (7.99) we have for ideal laser radiation

$$E_{\rm id\,las}^{(+)} = \frac{\hat{E}_0}{2} e^{i\omega_0 t}. \quad (7.119)$$

Thus the corresponding correlation function $\Gamma_{\rm id\,las}^{2,2}$ is

$$\Gamma_{\rm id\,las}^{2,2}(t-\tau_1, t+\tau', t+\tau'', t-\tau_2) = \frac{|\hat{E}_0|^4}{16} e^{i\omega_0(\tau_2-\tau_1+\tau'-\tau'')}. \quad (7.119a)$$

Analogously, we have in the case of pure phase noise

$$E_{\rm ph\,n}^{(+)}(t) = \frac{\hat{E}_0}{2} e^{i[\omega_0 t - \varphi(t)]}. \quad (7.120)$$

This leads to

$$\Gamma_{\rm ph\,n}^{2,2}(t-\tau_1, t+\tau', t+\tau'', t-\tau_2) = \Gamma_{\rm id\,las}^{2,2}\langle e^{i\Phi}\rangle, \quad (7.120a)$$

where Φ can be represented with the help of the fluctuation force $G_c(t)$ from

(7.73) and (7.74) by the relation

$$2\omega_0 \hat{E}_0 \Phi = -\int_t^{t-\tau_1} dt'\, G_c(t') - \int_t^{t+\tau'} dt'\, G_c(t') + \int_t^{t+\tau''} dt'\, G_c(t') + \int_t^{t-\tau_2} dt'\, G_c(t'). \qquad (7.120b)$$

As in establishing (7.105a), the Gaussianity of $G_c(t)$ may be utilized here. Then we may proceed as before from the transition from (7.105a) to (7.106). Finally we obtain

$$\langle e^{i\Phi} \rangle = \exp\Big\{ -\int_0^\infty d\omega\, \mathfrak{W}_{\Delta\omega}(\omega)\omega^{-2} \times [2 + \cos\omega(\tau_1 + \tau') + \cos\omega(\tau_2 + \tau'') - \cos\omega(\tau_1 + \tau'') - \cos\omega(\tau_2 + \tau') - \cos\omega(\tau' - \tau'') - \cos\omega(\tau_1 - \tau_2)] \Big\}. \qquad (7.120c)$$

The *finite* fluorescence line width enters into $\langle e^{i\Phi} \rangle$ via $\mathfrak{W}_{\Delta\omega}(\omega)$. Hence in view of (7.120a) this is also true of the correlation function $\Gamma^{2,2}_{\text{ph n}}$ of the laser radiation.

On evaluating the correlation function $\Gamma^{2,2}_{\text{rf}}$ of the emitted resonance fluorescence radiation by (7.118) with the help of $\Gamma^{2,2}_{\text{ph n}}$, it turns out that the factorization into a product of intensity correlation functions is not correct. As can be expected, deviations occur within a time range of the order of $\Delta\omega_{\text{fl}}^{-1}$. In $\Gamma^{2,2}_{\text{rf}}$ the parameter $\Delta\omega_E/\Delta\omega_{\text{fl}}$ appears, where $\Delta\omega_E$ is the laser line width according to (7.109). Expansion of $\Gamma^{2,2}_{\text{rf}}$ into a power series in $\Delta\omega_E/\Delta\omega_{\text{fl}}$ reveals that the first-order deviation from exact factorizability is proportional to $\Delta\omega_E/\Delta\omega_{\text{fl}}$. From this we see that only in the region well above threshold the factorization may be regarded as a good approximation. In a similar way the influence of the amplitude noise of the laser pump field on $\Gamma^{2,2}_{\text{rf}}$ can be calculated by means of (7.98a): If the laser operates well above threshold, the factorization of the intensity correlation of the fluorescence radiation can be regarded as a good approximation; the more the laser approaches the threshold, the stronger is the nonfactorizability. Near and below threshold the latter becomes predominant.

This discussion exhibited the influence of the quality of various laser-radiation models on the interpretation of higher-order effects (like the intensity correlation of resonance fluorescence).

REFERENCES

7.1. N. G. Basov and A. M. Prokhorov, *Zh. Exp. Theor. Fiz.* **27**, 431 (1954).

7.2. A. L. Schawlow and C. H. Townes, *Phys. Rev.* **112**, 1940 (1958).

7.3. T. H. Maiman, *Nature* **187**, 493 (1960).

7.4. W. G. Wagner and G. Birnbaum, *J. Appl. Phys.* **32**, 1185 (1961).

7.5. M. Schubert and B. Wilhelmi, *Introduction to Nonlinear Optics*, Vol. 1, Teubner, Leipzig, 1971, Appendix 6 (German).

7.6. M. Schubert and B. Wilhelmi, *Introduction to Nonlinear Optics*, Vol. 2, Teubner, Leipzig, 1978, p. 229 (German).

7.7. B. Van der Pol, *Phil. Mag.* **3**, 65 (1927).

7.8. N. N. Bogoljubow and I. A. Mitropolski, *Asymptotische Methoden in der Theorie der Nichtlinearen Schwingungen*, Akademie, Berlin, 1965 (German).

7.9. H. Haken, "Theory of coherence, noise, and photon statistics," in F. T. Arecchi and E. O. Schulz-Dubois (eds.), *Laser Handbook*, Vol. 1, North-Holland, Amsterdam, 1972, p. 116.

7.10. H. Risken, *Fortschr. Phys.* **16**, 261 (1968).

7.11. H. Risken, *Progr. Opt.* **8**, 239 (1970).

7.12. F. T. Arecchi, G. S. Rodari, and A. Sona, *Phys. Lett.* **25A**, 59 (1967).

7.13. A. Yariv and W. M. Caton, *IEEE J. Quant. Electron.* **QE-10**, 509 (1974).

7.14. M. O. Scully and W. E. Lamb, Jr., *Phys. Rev.* **159**, 208 (1967).

7.15. W. E. Lamb, *Phys. Rev.* **134**, 1429 (1964).

7.16. V. Arzt, H. Haken, H. Risken, H. Sauermann, C. Schmid, and W. Weidlich, *Z. Phys.* **197**, 207 (1966).

7.17. K. R. Manes and A. E. Siegman, *Phys. Rev. A* **4**, 373 (1971).

7.18. M. S. Feld and V. S. Letokhov, *Sci. Amer.* **229**, 69 (1973).

7.19. V. S. Letokhov, *Laserspektroskopie* Akademie, Berlin, 1977, Chapter 4 (German).

7.20. H. J. Kimble and L. Mandel, *Phys. Rev. A* **13**, 2123 (1976).

7.21. H. J. Carmichael and D. F. Walls, *J. Phys.* **B9**, L43 (1976).

7.22. H. J. Kimble, M. Dagenais, and L. Mandel, *Phys. Rev. Lett.* **39**, 691 (1977).

7.23. H. J. Kimble, M. Dagenais, and L. Mandel, *Phys. Rev. A* **18**, 201 (1978).

7.24. H. J. Kimble and L. Mandel, *Phys. Rev. A* **15**, 689 (1977).

7.25. M. Schubert, K.-E. Süsse, and W. Vogel, *Opt. Comm.* **30**, 275 (1979).

7.26. M. Schubert, K.-E. Süsee, W. Vogel, and D.-G. Welsch, *Opt. Quant. Electron.* **12**, 65 (1980).

7.27. M. Schubert, K.-E. Süsse, W. Vogel, and D.-G. Welsch, *Opt. Quant. Electron.* **13**, 301 (1981).

8

Nonlinearities in Transient One-Photon Processes

In Chapter 6 we considered nonstationary processes excited by nonresonant interaction between light and matter. Now we deal with *resonant excitation* of matter. The treatment of resonant excitation is here restricted to *one-photon processes*, where however special attention is given to nonlinearities. Such nonlinearities in one-photon interaction processes arise from the coupling between polarization, field strength, and occupation numbers (cf. Chapter 7, where stationary processes have been treated). Similar phenomena in multiphoton processes will be treated in later chapters. In this chapter all processes are treated by semiclassical methods. Some typical quantum effects of transient one-photon processes will be discussed in Chapter 9.

Even in one-photon processes the variety of nonstationary phenomena is far greater than in the case of nonresonant interaction between light and matter. The origin of such complex phenomena is the fact that the atomic response at time t depends not only on the exciting fields at the same time, but also on the fields at earlier times $t' < t$. The excited atomic system can then react back on the field at later times. The atomic response function is characterized by the atomic relaxation times, which describe the decay of nonequilibrium level populations and of polarization. Because of this "memory" of the atomic systems, nonstationary phenomena can be observed even at one point of the sample, and therefore we have to deal with local and nonlocal *transient phenomena*.

We begin our consideration with the derivation of basic equations describing the resonant interaction of two-level systems with light pulses. Then we treat the quasistationary excitation of atoms, the excitation of atoms by very short pulses, and finally the action of pulses comparable in length with the phase decay time of the atomic systems.

8.1. NONSTATIONARY SEMICLASSICAL EQUATIONS

The midfrequency of the exciting pulse is assumed to lie close to one transition frequency of the atomic system. If the incident light contains only spectral components in a narrow frequency range around the center frequency ω_L lying close to the transition frequency ω_{10}, it is often justified to disregard the complex energy-level scheme of the atomic systems and to treat the latter as two-level systems. This possibility will be extensively utilized in the following considerations.

To perform a semiclassical treatment of the interaction processes we start in the first sections from the results of Chapter 4. We presuppose simple geometrical conditions, because this allows a simple presentation of the essential aspects of nonlinearity and nonstationarity. Suppose the irradiating pulses to travel collinearly as plane waves in the z direction, and let the incident light be linearly polarized and described by

$$E(z,t) = \bar{E}(z,t)\cos[\omega t + \varphi(z,t) - kz], \tag{8.1}$$

using here the real amplitude $\bar{E}(z,t)$ and the phase $\varphi(z,t)$, both slowly varying with time and space coordinates. [According to the notation used by us in other places, we should denote the real amplitude function in (8.1) by $|\bar{E}(t,z)|$. Since however no confusion can arise in this chapter, we omit the modulus sign.] The wave vector describes the propagation properties of the material without taking into account the influence of the resonant transition of the atoms under consideration. The atomic systems influenced by radiation and by relaxation processes can be described by means of the equations of motion for their density matrix (cf. Section 3.2). From the density matrix of the atomic systems the polarization of the material is to be calculated and then substituted into the wave equation. Here the approach described in Sections 4.7 and 4.8 will be preferred, starting immediately from the equations of motion for the polarization and for the population inversion in conjunction with the wave equation. In doing so, the procedures presented in Sections 4.7 and 4.8 are equivalent.

For a homogeneously broadened transition we may start from (4.122) in order to give an adequate description of the ensemble of the atomic systems. Here we also have to take into account the dependence of the polarization P and of the density of the population-number inversion γ_I on the space coordinate z. Then we obtain the following system of equations for the amplitudes of the in-phase component of the polarization $\bar{P}_1$, the out-of-phase component $\bar{P}_2$, and the density of the occupation-number inversion, γ_I:

$$\frac{\partial}{\partial t}\bar{P}_1(z,t) = -\left[\Delta + \frac{\partial}{\partial t}\varphi(z,t)\right]\bar{P}_2(z,t) - \frac{1}{\tau_{10}}\bar{P}_1(z,t), \tag{8.2a}$$

$$\frac{\partial}{\partial t}\bar{P}_2(z,t) = \left[\Delta + \frac{\partial}{\partial t}\varphi(z,t)\right]\bar{P}_1(z,t) - \frac{1}{\hbar}d^2\bar{E}(z,t)\gamma_I(z,t) - \frac{1}{\tau_{10}}\bar{P}_2(z,t), \tag{8.2b}$$

$$\frac{\partial}{\partial t}\gamma_I(z,t) = \frac{1}{\hbar}\bar{E}(z,t)\bar{P}_2(z,t) - \frac{1}{T_{10}}[\gamma_I(z,t) + \gamma], \tag{8.2c}$$

$$P(t,z) = \bar{P}_1(z,t)\cos[\omega t + \varphi(z,t) - kz] + \bar{P}_2(z,t)\sin[\omega t + \varphi(z,t) - kz], \tag{8.2d}$$

which, as already explained in Section 4.8, are frequently referred to as the *Bloch equations*. (Note that the differential operators enclosed in brackets do not act on terms outside these brackets.) In these equations $\Delta = \omega - \omega_{10}$ stands for the detuning from resonance, τ_{10} and T_{10} are the transverse and longitudinal relaxation times, and d is an effective value of the real transition moment.

For an inhomogeneously broadened transition we obtain from (4.123) in an analogous way the following system of equations:

$$\frac{\partial}{\partial t}\overline{P}_1'(\omega_{10}', z, t) = -\left[\Delta' + \frac{\partial}{\partial t}\varphi(z, t)\right]\overline{P}_2'(\omega_{10}', z, t) - \frac{1}{\tau_{10}}\overline{P}_1'(\omega_{10}', z, t), \tag{8.3a}$$

$$\frac{\partial}{\partial t}\overline{P}_2'(\omega_{10}', z, t) = \left[\Delta' + \frac{\partial}{\partial t}\varphi(z, t)\right]\overline{P}_1'(\omega_{10}', z, t) - \frac{1}{\hbar}d^2\overline{E}(z, t)\gamma_I'(\omega_{10}', z, t) - \frac{1}{\tau_{10}}\overline{P}_2'(\omega_{10}', z, t), \tag{8.3b}$$

$$\frac{\partial}{\partial t}\gamma_I'(\omega_{10}', z, t) = \frac{1}{\hbar}\overline{E}(z, t)\overline{P}_2'(\omega_{10}', z, t) - \frac{1}{T_{10}}\left[\gamma_I'(\omega_{10}', z, t) - \gamma_I^{e\prime}(\omega_{10}')\right], \tag{8.3c}$$

$$\gamma_I(z, t) = \int d\omega_{10}'\, g_{\text{inh}}(\omega_{10}' - \omega_{10})\gamma_I'(\omega_{10}', z, t), \tag{8.3d}$$

$$\overline{P}_1(z, t) = \int d\omega_{10}'\, g_{\text{inh}}(\omega_{10}' - \omega_{10})\overline{P}_1'(\omega_{10}', z, t), \tag{8.3e}$$

$$\overline{P}_2(z, t) = \int d\omega_{10}'\, g_{\text{inh}}(\omega_{10}' - \omega_{10})\overline{P}_2'(\omega_{10}', z, t), \tag{8.3f}$$

and

$$P(z, t) = \overline{P}_1(z, t)\cos[\omega t + \varphi(z, t) - kz] + \overline{P}_2(z, t)\sin[\omega t + \varphi(z, t) - kz]. \tag{8.3g}$$

Here the detuning for atoms with the transition frequency ω_{10}' is given by $\Delta' = \omega - \omega_{10}'$. The distribution of the resonance frequencies around the mean transition frequency is represented by the line-shape function $g_{\text{inh}}(\omega_{10}' - \omega_{10})$.

We are mainly interested in where, even in one-photon processes, the nonlinearities in the dependence between polarization and field strength may come from. It is evident from (8.2) and (8.3) that the differential equations for the polarization exhibit nonlinearities if the population-number inversion γ_I is dependent on the field strength, that is, if changes in the level occupation due to the effects of the radiation field must be taken into account.

The equations of motion for the ensemble of atomic systems (8.2) or (8.3) have to be solved in conjunction with the one-dimensional wave equation

$$\left(-\frac{\partial^2}{\partial z^2}+\frac{1}{c^2}\frac{\partial^2}{\partial t^2}\right)E(z,t)=-\mu_0\frac{\partial^2}{\partial t^2}P(z,t)-\mu_0\frac{\partial^2}{\partial t^2}P^{\mathrm{NR}}(z,t). \tag{8.4}$$

Let us mention once more that here the quantity $P(z,t)$ represents only the contributions of the resonant transition of the atoms under consideration to the polarization. Contributions of nonresonant transitions of these atoms to the polarization, as well as contributions of a nonresonant material in which these atoms may be embedded, are represented by the term $P^{\mathrm{NR}}(z,t)$.

In the case of field-strength amplitudes and phases varying slowly with space and time, we may pass over from the wave equation for $E(z,t)$ to the following first-order differential equations for $\bar{E}(z,t)$ and $\varphi(z,t)$:

$$\left(\frac{\partial}{\partial z}+\frac{1}{v}\frac{\partial}{\partial t}\right)\bar{E}(z,t)=-\frac{\mu_0\omega^2}{2k}\bar{P}_2(z,t), \tag{8.5a}$$

$$\bar{E}(z,t)\left[\frac{\partial}{\partial z}\varphi(z,t)+\frac{1}{v}\frac{\partial}{\partial t}\varphi(z,t)\right]=-\frac{\mu_0\omega^2}{2k}\bar{P}_1(z,t), \tag{8.5b}$$

where $v=d\omega/dk$ is the group velocity for the nonresonant material, which is related with P^{NR} (cf. Section 1.4.4, where equivalent calculations have been carried out for the slowly varying complex field-strength amplitudes). When dealing with the equations of polarization, field strength, and population inversion, it is often convenient also to pass over from the coordinates t and z to the transformed coordinates

$$\eta=t-\frac{z}{v},\qquad \xi=z, \tag{8.6}$$

since thereby the derivatives

$$\frac{\partial}{\partial t}\quad\text{and}\quad\frac{\partial}{\partial z}+\frac{1}{v}\frac{\partial}{\partial t}$$

occurring in the equations of motion take the very simple form

$$\frac{\partial}{\partial t}=\frac{\partial}{\partial \eta}\quad\text{and}\quad\frac{\partial}{\partial z}+\frac{1}{v}\frac{\partial}{\partial t}=\frac{\partial}{\partial \xi} \tag{8.7}$$

[cf. (1.222) and (1.223)].

If strong pump pulses are acting and if the interaction length within the nonlinear medium is not too large, it is often justified to neglect the reaction of the material on the incident electromagnetic field and to start from given fields in the constitutive equations. Under these assumptions we shall consider quasistationary interactions (pulse duration τ_L much greater than transverse relaxation time τ_{10}) in Section 8.2, the excitation by extremely short pulses ($\tau_L \ll \tau_{10}$) in Section 8.3, and excitation by pulses comparable in duration with τ_{10} in Section 8.4. Then in Sections 8.5 and 8.6 we shall deal with the influence of the interaction process on the incident light.

8.2. QUASISTATIONARY EXCITATION

Quasistationary excitation arises if the strength of pump field is not appreciably changed in time intervals $t \lesssim \tau_{10}$, that is, if $d\overline{E}(t)/dt \ll \overline{E}(t)/\tau_{10}$ and $d\varphi(t)/dt \ll 1/\tau_{10}, \Delta$. We find that under these conditions $d\overline{P}_1(t)/dt \ll \overline{P}_1(t)/\tau_{10}$ and $d\overline{P}_2(t)/dt \ll \overline{P}_2(t)/\tau_{10}$, and thus we obtain from (8.2)

$$\overline{P}_1(t) = \frac{\Delta \cdot \tau_{10}^2 d^2/\hbar}{1 + \Delta^2 \cdot \tau_{10}^2} \overline{E}(t) \gamma_I(t), \tag{8.8a}$$

$$\overline{P}_2(t) = \frac{-\tau_{10} d^2/\hbar}{1 + \Delta^2 \cdot \tau_{10}^2} \overline{E}(t) \gamma_I(t), \tag{8.8b}$$

and

$$\frac{d}{dt}\gamma_I(t) = -\frac{1}{T_{10}}(\gamma_I(t) + \gamma) - \frac{(d^2/\hbar^2)\tau_{10}}{1 + \Delta^2 \cdot \tau_{10}^2} \overline{E}^2(t) \gamma_I(t). \tag{8.8c}$$

This means that we have obtained stationary solutions for the polarization amplitudes and a differential equation for the inversion. Only the square of the amplitude of the field strength, not its phase, enters this equation, and therefore it is both possible and convenient to introduce the photon flux density

$$I(t) = \frac{1}{2}\sqrt{\frac{\varepsilon_0 \varepsilon}{\mu_0}} \cdot \overline{E}^2(t) \frac{1}{\hbar \omega_{10}} \tag{8.9}$$

into (8.8c). This gives

$$\frac{d}{dt}\gamma_I(t) = -\frac{1}{T_{10}}[\gamma_I(t) + \gamma] - 2\sigma_{10}(\Delta) I(t) \gamma_I(t) \tag{8.10a}$$

with the absorption cross section

$$\sigma_{10}(\Delta) = \frac{\tau_{10}\omega_{10}}{\sqrt{\varepsilon}\,\hbar}\sqrt{\frac{\mu_0}{\varepsilon_0}}\,d^2\frac{1}{1+\Delta^2\cdot\tau_{10}^2}. \tag{8.10b}$$

The relation (8.10a) is the rate equation for the two-level system. Since according to the assumptions the photon flux density $I(t)$ is a given function whose change due to the interaction can be neglected, the solution of the rate equation (8.10a) can be immediately written in the form

$$\begin{aligned}\frac{\gamma_I(t)}{\gamma} = &-\frac{1}{T_{10}}\exp\left(-\frac{1}{T_{10}}\int_{-\infty}^{t}dt'\,[1+2\sigma_{10}(\Delta)T_{10}I(t')]\right)\\ &\times\int_{-\infty}^{t}dt''\exp\left(\frac{1}{T_{10}}\int_{-\infty}^{t''}dt'\,[1+2\sigma_{10}(\Delta)T_{10}I(t')]\right)\\ &-\exp\left(-\frac{1}{T_{10}}\int_{-\infty}^{t}dt'\,[1+2\sigma_{10}(\Delta)T_{10}I(t')]\right).\end{aligned} \tag{8.11}$$

If we switch on a constant excitation at $t = 0$, that is, $I(t) = 0$ for $t < 0$ and $I(t) = I_0$ for $t \geq 0$, then (8.11) gives

$$\frac{\gamma_I(t)}{\gamma} = -\frac{1}{1+2\sigma_{10}I_0T_{10}} - \frac{2\sigma_{10}I_0T_{10}}{1+2\sigma_{10}I_0T_{10}}\exp\left(-\frac{(1+2\sigma_{10}I_0T_{10})t}{T_{10}}\right). \tag{8.12}$$

We see that the time-dependent contribution dies away exponentially with the time constant

$$T_{eff} = \frac{\mathrm{T}_{10}}{1+2\sigma_{10}\mathrm{T}_{10}\mathrm{I}_0}. \tag{8.13}$$

For times $t \gg T_{\text{eff}}$, the density of the occupation-number inversion approaches its stationary value

$$\gamma_{I\infty} = -\frac{\gamma}{1+2\sigma_{10}T_{10}I_0}, \tag{8.14}$$

which tends to zero with increasing intensity I_0. For $I_0 = I_S$, where

$$I_S = \frac{1}{2\sigma_{10}T_{10}} \tag{8.15}$$

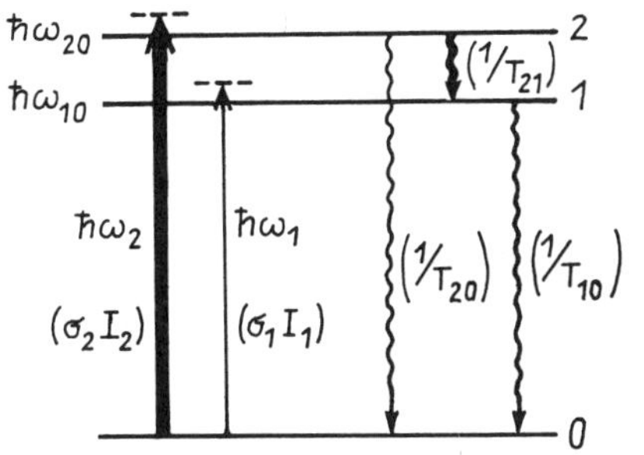

Fig. 8.1. Three-level scheme with light excitation and relaxation transitions.

represents the *saturation photon flux density* or *saturation intensity*, we have $\gamma_{I\infty} = -\gamma/2$. Substituting the solutions for the occupation-number inversion from (8.11) or (8.12) into (8.8a, b), nonlinear relationships between polarization and field strength are obtained. We find that the polarization increases with field strength less than linearly; and at very high fields it even decreases and finally approaches zero.

We finish this section with a remark on more general rate equations. From the density-matrix equations for multilevel systems, with the assumption of quasistationarity, it is possible to obtain appropriate sets of rate equations for the occupation-number densities γ_i of all levels $i = 0, 1, 2, \ldots$ involved in the interaction processes. As an example the resulting rate equations for the three-level system shown in Fig. 8.1 may be presented:

$$\frac{d}{dt}\gamma_2(t) = \sigma_{02}(\omega_2 - \omega_{20})I_2(t)[\gamma_0(t) - \gamma_2(t)]$$

$$-\frac{1}{T_{21}}\gamma_2(t) - \frac{1}{T_{20}}\gamma_2(t), \tag{8.16a}$$

$$\frac{d}{dt}\gamma_1(t) = \sigma_{01}(\omega_1 - \omega_{10})I_1(t)[\gamma_0(t) - \gamma_1(t)]$$

$$+\frac{1}{T_{21}}\gamma_2(t) - \frac{1}{T_{10}}\gamma_1(t), \tag{8.16b}$$

$$\gamma_0(t) + \gamma_1(t) + \gamma_2(t) = \gamma. \tag{8.16c}$$

This example also reveals the conditions under which this multilevel system can be simply described as a two-level system. If the relaxation time T_{21} is very short compared with all other relaxation times as well as compared with the pulse duration and with the time $1/\sigma_{02}I_2(t)$, then from (8.16) we have

$$\frac{d}{dt}\gamma_1(t) = \sigma_{01}(\omega_1 - \omega_{10})I_1(t)[\gamma_0(t) - \gamma_1(t)] + \sigma_{02}(\omega_2 - \omega_{20})I_2(t)\gamma_0(t) - \frac{1}{T_{10}}\gamma_1(t) \tag{8.17a}$$

and

$$\gamma_0(t) + \gamma_1(t) = \gamma. \tag{8.17b}$$

The induced radiation process 0 → 2 together with the fast relaxation 2 → 1 apparently causes a rapid pumping from level 0 to level 1. These two relations (8.17a, b) can be represented in one equation for the occupation-number inversion $\gamma_I(t) = \gamma_1(t) - \gamma_0(t)$:

$$\frac{d}{dt}\gamma_I(t) = -2\sigma_{01}(\omega_1 - \omega_{10})I_1(t)\gamma_I(t) - \sigma_{02}(\omega_2 - \omega_{20})I_2(t)[\gamma_I(t) - \gamma] - \frac{1}{T_{10}}[\gamma_I(t) + \gamma]. \tag{8.18}$$

If we want to discuss the stationary saturation behavior at constant pumping intensity I_2, we set $I_1(t)$ equal to zero and obtain

$$\gamma_{I\infty} = -\gamma\frac{1 - \sigma_{02}T_{10}I_2}{1 + \sigma_{02}T_{10}I_2}. \tag{8.19}$$

In the limiting case of high intensities, this equation implies $\gamma_{I\infty} = \gamma$, which differs from the genuine two-level system, where $\gamma_{I\infty} = 0$ was obtained. We shall return to the relations discussed here when considering the reaction of the atomic systems to light.

For *inhomogeneously broadened transitions* a quasimonochromatic radiation field causes preferential changes of the level occupation within a small region of the spectral transition near the laser frequency. This effect is called *spectral hole burning*, because the quasimonochromatic radiation field burns a "hole" in the equilibrium distribution of molecules within the inhomogeneously broadened line (cf. Section 7.3.1). Under appropriate experimental conditions, that means that if the light field is not too strong and if the interaction time is rather long, the spectral width of the holes is mainly given by the homogeneous line width. Thus, by measuring these holes we obtain spectra with enhanced resolution, which is not limited by the inhomogeneous line width. In the case of gases where the inhomogeneous broadening is mainly given by the Doppler effect, we speak of high-resolution spectroscopy "within the Doppler line" (cf.

e.g. Ref. 10). The investigation of the spectral holes can be performed by using a probe beam with a variable frequency. The absorption of the weak probe beam has then to be measured as a function of the detuning between probe and pump radiation. This procedure, as well as more sophisticated ones, is described in detail for instance in Refs. 10, 13, 22.

8.3. TRANSIENT EXCITATION OF ATOMIC SYSTEMS WITH NEGLIGIBLE RELAXATION

In the preceding Section we discussed stationary and quasistationary excitation. Now extremely nonstationary pumping processes in two-level systems will be treated, supposing $\tau_L \ll \tau_{10}, T_{10}$ for the pulse duration. Under these conditions the relaxation processes during the action of the pulse can be neglected.

8.3.1. Atomic Response for Negligible Relaxation

Let us consider a very simple case in which the incident light does not exhibit phase modulation $[\varphi(t) = 0]$ and the light pulse is of square-wave form $[\bar{E}(t) = \bar{E}$ for $0 \leq t \leq \tau_L$, and $\bar{E}(t) = 0$ for all other times]. Then the relations (8.2) become simpler, so that for $0 \leq t \leq \tau_L$ we have

$$\frac{d}{dt}\bar{P}_1(t) + \Delta \cdot \bar{P}_2(t) = 0, \tag{8.20a}$$

$$\frac{d}{dt}\bar{P}_2(t) - \Delta \cdot \bar{P}_1(t) = -\frac{1}{\hbar}d^2 \bar{E}\gamma_I(t), \tag{8.20b}$$

$$\frac{d}{dt}\gamma_I(t) = \frac{1}{\hbar}\bar{E}\bar{P}_2(t). \tag{8.20c}$$

From these equations there can result extremely nonlinear relations between polarization and field strength. The dependence of the polarization on the field strength may, for instance, be a periodic one. In such cases the expansion of the polarization in powers of the field is not appropriate because of its slow convergence. Differentiating (8.20b) with respect to t with $t > 0$ and inserting the time derivative of $\bar{P}_1$ and γ_I from (8.20a, c) we get the harmonic-oscillator equation

$$\frac{d^2}{dt^2}\bar{P}_2(t) + \Omega_R^2 \bar{P}_2(t) = 0, \tag{8.21a}$$

where

$$\Omega_R = \sqrt{\Delta^2 + \frac{d^2 \bar{E}^2}{\hbar^2}} \tag{8.21b}$$

is called *Rabi frequency* or *Rabi flopping frequency*. The Rabi frequency depends on the *detuning* Δ and on the *field-strength broadening parameter* $d\bar{E}/\hbar$; the latter quantity is also referred to as the Rabi frequency at zero detuning, Ω_{R0}. It becomes apparent from (8.21) that $\bar{P}_2(t)$ changes periodically with time, and according to (8.20) the same is true of the functions $\bar{P}_1(t)$ and $\gamma_I(t)$ (see, e.g., Refs. 3, 8.1, 8.2). With the initial conditions $\bar{P}_1(0) = \bar{P}_2(0) = 0$ and $\gamma_I(0) = -\gamma$ we obtain from (8.20)

$$\bar{P}_1(t) = +\Delta \cdot \frac{d^2}{\hbar}\frac{\bar{E}}{\Omega_R^2}\gamma(\cos\Omega_R t - 1), \tag{8.22a}$$

$$\bar{P}_2(t) = \frac{d^2}{\hbar}\frac{\bar{E}}{\Omega_R}\gamma \sin\Omega_R t, \tag{8.22b}$$

$$\frac{\gamma_I(t)}{\gamma} = -\frac{d^2}{\hbar^2}\frac{\bar{E}^2}{\Omega_R^2}(\cos\Omega_R t - 1) - 1. \tag{8.22c}$$

For strict resonance ($\Delta = \omega - \omega_{10} = 0$) the polarization amplitude $\bar{P}_1(t)$ is identically zero and the Rabi frequency is merely given by the field-strength broadening parameter

$$\Omega_{R0} = \hbar^{-1} d\bar{E}. \tag{8.23}$$

Under this condition, after a time $t = \pi/(2\Omega_{R0}) = \pi\hbar/(2d\bar{E})$ the maximum value of the polarization amplitude $\bar{P}_2(t)$ is reached, and after a time $t = \pi/\Omega_{R0}$ a complete inversion in the two-level system is obtained. Such light pulses of length $\tau_L = \pi/(2\Omega_{R0})$ and $\tau_L = \pi/\Omega_{R0}$ or, more generally, of length $\pi/(2\Omega_R)$ and π/Ω_R, which cause the effects described, are referred to as $\pi/2$ *pulses* and

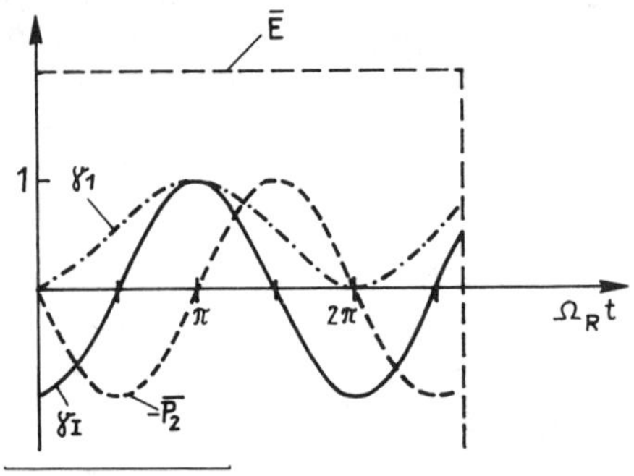

Fig. 8.2. Undamped optical nutation: occupation-number density γ_1 of the excited level, occupation-inversion density γ_I, and polarization amplitude $\bar{P}_2$ for an atomic ensemble being resonantly excited by a light pulse of constant amplitude $\bar{E}$ in the time interval $0 \leq t \leq \tau_L = \frac{21}{8}\pi/\Omega_{R0}$. (All quantities have been normalized.)

π pulses, respectively. Analogously, a *2π pulse* causes the atomic systems to return to their initial state. The periodic changes of occupation-number inversion and polarization (see Fig. 8.2) are termed *Rabi oscillations* or *optical nutations*. The latter term is borrowed from magnetic radio-frequency spectroscopy, where a $\pi/2$ pulse shifts the magnetization vector by 90° from its initial position, a π pulse flips the magnetization vector just in the opposite direction, a 2π pulse transfers this vector into its initial position, and where the term "nutation" is in turn borrowed from gyroscopic theory. It is apparent that the product of field strength and pulse duration—or, with time-dependent field strength, in a more general way the integral

$$\theta(t) = \hbar^{-1} d \int_{t_0}^{t} dt' \, \bar{E}(t'), \tag{8.24}$$

which is termed the *pulse area*—is a measure of the coherent influence of the light pulse on the atomic systems. (Note that we have assumed $t_0 = 0$ so far.)

To briefly discuss the influence of the pulse shape, let us assume highly simplifying conditions—namely, exact resonance ($\Delta = \omega - \omega_{10} = 0$), vanishing phase modulation of the pulses, and as initial conditions $\bar{P}_1(0) = \bar{P}_2(0) = 0$ and $\gamma_I(0) = -\gamma$. Thus the set of equations (8.2) reduces to

$$\frac{d}{dt}\bar{P}_2(t) = -\frac{1}{\hbar} d^2 \bar{E}(t)\gamma_I(t),$$

$$\frac{d}{dt}\gamma_I(t) = \frac{1}{\hbar}\bar{E}(t)\bar{P}_2(t).$$

Its solution is

$$\bar{P}_2(t) = d \cdot \gamma \sin\theta(t) \tag{8.25a}$$

and

$$\gamma_I(t) = -\gamma\cos\theta(t). \tag{8.25b}$$

We can see that the structure of the relations previously obtained for constant field amplitudes is preserved. Note that here the more general definition (8.24) has to be used for the pulse area.

After this short discussion of the effect of the pulse shape, let us for simplicity in the following return to the treatment of pulses with constant field amplitude. Under general initial conditions $\bar{P}_1(t_0)$, $\bar{P}_2(t_0)$, $\gamma_I(t_0)$ the solutions of (8.20) with negligible relaxation are

$$\overline{P}_1(t) = \frac{(\hbar^{-1}d\overline{E})^2 + \Delta^2 \cos \Omega_R \,\delta t}{\Omega_R^2}\overline{P}_1(t_0) - \frac{\Delta}{\Omega_R}(\sin \Omega_R \delta t)\overline{P}_2(t_0)$$

$$+ \frac{d^2 \cdot \overline{E} \cdot \Delta}{\hbar \Omega_R^2}(1 - \cos \Omega_R \,\delta t)\gamma_I(t_0), \tag{8.26a}$$

$$\overline{P}_2(t) = \frac{\Delta}{\Omega_R}(\sin \Omega_R \,\delta t)\overline{P}_1(t_0) + (\cos \Omega_R \,\delta t)\overline{P}_2(t_0)$$

$$- \frac{d^2 \cdot \overline{E}}{\hbar \Omega_R}(\sin \Omega_R \,\delta t)\gamma_I(t_0), \tag{8.26b}$$

$$\gamma_I(t) = \frac{\Delta \cdot \overline{E}}{\hbar \Omega_R^2}(1 - \cos \Omega_R \,\delta t)\overline{P}_1(t_0) + \frac{\overline{E}}{\hbar \Omega_R}(\sin \Omega_R \,\delta t)\overline{P}_2(t_0)$$

$$+ \frac{1}{\Omega_R^2}\left[\Delta^2 + \left(\frac{d\overline{E}}{\hbar}\right)^2 \cos \Omega_R \,\delta t\right]\gamma_I(t_0), \tag{8.26c}$$

where $\delta t = t - t_0$. This relationship between the initial values $\overline{P}_1(t_0)/(d \cdot \gamma)$, $\overline{P}_2(t_0)/(d \cdot \gamma)$, and $\gamma_I(t_0)/\gamma$ and the respective quantities at the time t can likewise be written as a product of the *initial vector* $(\overline{P}_1(t_0)/(d \cdot \gamma), \overline{P}_2(t_0)/(d \cdot \gamma), \gamma_I(t_0)/\gamma)$ with a *time-evolution matrix*. From such a representation it is evident how the solution can be gained if several time intervals follow each other, each of them being characterized by a constant light field. The temporal evolution matrix for the total time interval is then obtained by multiplying the interval matrices, where in individual intervals $\overline{E} = 0$ may hold.

The solutions for general initial conditions, first presented by Rabi (8.2), must for example be applied to materials that previously experienced a transition from thermal equilibrium to another state. This happens for instance in double-pulse experiments. These solutions also exhibit the typical optical nutations. In optical nutation, energy is periodically exchanged with the frequency Ω_R between electromagnetic field and matter. This process is not associated with irreversible absorption. Irreversible absorption does not appear until dissipation, which was here neglected together with the relaxation rates, becomes effective.

It should be mentioned that it is possible, starting from solutions gained for homogeneously broadened transitions, to build up the solutions for inhomogeneously broadened systems. To do this, we have only to carry out in (8.26) the replacements $\bar{P}_1 \to \bar{P}_1'$, $\bar{P}_2 \to \bar{P}_2'$, $\gamma_I \to \gamma_I'$, $\Delta \to \Delta'$, and

$$\Omega_R = \sqrt{\Delta^2 + \frac{d^2\bar{E}^2}{\hbar^2}} \to \Omega_R' = \sqrt{(\Delta')^2 + \frac{d^2\bar{E}^2}{\hbar^2}}$$

and will then obtain the resultant polarization amplitudes $\bar{P}_1$, $\bar{P}_2$ as well as the total occupation-number inversion by integration according to (8.3d, e, f).

Let us now consider the evolution of the polarization and occupation-number inversion after switching off the pumping pulse at the time $t = \tau_L$. Here we again start with homogeneously broadened systems. As long as the relaxation terms in the *Bloch equations* are neglected, (8.20) remains valid with $\bar{E} = 0$. This means that the general solution (8.26) with $\bar{E} = 0$ and $\Omega_R = \Delta$ and with the time $t_0 = \tau_L$ can be used. This solution reads

$$\bar{P}_1(t) = \cos\Delta(t - \tau_L)\,\bar{P}_1(\tau_L) - \sin\Delta(t - \tau_L)\,\bar{P}_2(\tau_L), \tag{8.27a}$$

$$\bar{P}_2(t) = \sin\Delta(t - \tau_L)\,\bar{P}_1(\tau_L) + \cos\Delta(t - \tau_L)\,\bar{P}_2(\tau_L), \tag{8.27b}$$

$$\gamma_I(t) = \gamma_I(\tau_L). \tag{8.27c}$$

We see that the polarization amplitude exhibits undamped oscillations with the detuning frequency $\Delta = \omega - \omega_{10}$, while the occupation-number inversion preserves the value taken on at the end of the pulse. The oscillations of the polarization amplitude with frequency Δ simply mean that now the molecules oscillate at their eigenfrequency $\omega_{10} = \omega - \Delta$. The initial values at the time τ_L can be related to the initial values at the time $t = 0$ by means of the general solution (8.26), choosing $\bar{P}_1(0) = \bar{P}_2(0) = 0$ and $\gamma_I(0) = -\gamma$. There results for $t > \tau_L$, for example, by multiplication of the two time-evolution matrices,

$$\bar{P}_1(t) = -\frac{\Delta \cdot \Omega_{R0}}{\Omega_R^2}\gamma d(1 - \cos\Omega_R\tau_L)\cos\Delta(t - \tau_L)$$
$$-\frac{\Omega_{R0}}{\Omega_R}\gamma d(\sin\Omega_R\tau_L)\sin\Delta(t - \tau_L), \tag{8.28a}$$

$$\bar{P}_2(t) = -\frac{\Delta \cdot \Omega_{R0}}{\Omega_R^2}\gamma d(1 - \cos\Omega_R\tau_L)\sin\Delta(t - \tau_L)$$
$$+\frac{\Omega_{R0}}{\Omega_R}\gamma d(\sin\Omega_R\tau_L)\cos\Delta(t - \tau_L), \tag{8.28b}$$

$$\gamma_I(t) = -\frac{\Delta^2 + \Omega_{R0}^2\cos\Omega_R\tau_L}{\Omega_R^2}\gamma. \tag{8.28c}$$

The set of equations (8.28) gives us the polarization amplitude and the occupation-number inversion for a homogeneously broadened ensemble of atomic systems after excitation by a square pulse between the times 0 and τ_L. The corresponding quantities for inhomogeneously broadened transitions can be obtained according to the procedure already described by replacing in (8.28) the unprimed quantities $\overline{P}_1$, $\overline{P}_2$, γ_I, Δ, and Ω_R by primed ones and calculating the total polarization and occupation-number inversion by integration according to (8.3). In general, complicated time dependences will result, since now the individual molecular groups oscillate at their different resonance frequencies ω'_{10}. In superposition this leads to interference effects (see e.g. Refs. 8.3, 8.4). We shall return to this point after studying the effect of the relaxation terms in the Bloch equations. Here only an extreme limiting case will be considered, which however is of great interest as to physical interpretation. We suppose the homogeneous broadening to be much less than the inhomogeneous ($1/\tau_{10} \ll \Delta\omega_{\text{inh}}$), the $\pi/2$ light pulse with midfrequency $\omega_L = \omega_{10}$ to be extremely short ($\tau_L < 1/\Delta\omega_{\text{inh}}$) and intense ($\Omega_{R0} \gg \Delta\omega_{\text{inh}}$, $\Omega_R \approx \Omega_{R0}$), and the time of observation to be much greater than the pulse duration ($t \gg \tau_L$). Then there follows from (8.28) for the polarization amplitudes

$$\overline{P}'_1(t) = -\overline{P}'_2(0)\sin \Delta' t, \tag{8.29a}$$

$$\overline{P}'_2(t) = \overline{P}'_2(0)\cos \Delta' t, \tag{8.29b}$$

$$\gamma'_I(t) = 0. \tag{8.29c}$$

The integration over the Gaussian distribution $g_{\text{inh}}(\omega'_{10} - \omega_{10})$ of the resonance frequencies then gives

$$\overline{P}_1(t) = 0 \tag{8.30a}$$

and

$$\overline{P}_2(t) = +\gamma d \exp\left[-\left(\frac{\Delta\omega_{\text{inh}}}{4\sqrt{\ln 2}}t\right)^2\right]. \tag{8.30b}$$

This means that the polarization amplitude monotonically decays according to a Gaussian function in time intervals of order $2\pi/\Delta\omega_{\text{inh}}$. Thereby the oscillating contributions of the polarization amplitudes of several molecular groups interfere with each other and cancel one another completely at large times. The drop in polarization following the action of a light pulse is called *free polarization decay* or, in analogy to radio-frequency spectroscopy, *optical free induction decay*. The effect of free polarization decay can be observed by a collective spontaneous emission, the superluminescence, or by the interaction of the polarization with another light wave. We shall return to this later. Under more general conditions with respect to duration and midfrequency of the pulse as well as to the pulse area, but assuming as before $\Omega_{R0} > \Delta\omega_{\text{inh}}$, the procedure described below (8.29) leads for $t > \tau_L$ to

$$\overline{P}_1(t) = -\gamma d \sin \Delta t \exp\left(-\left[\frac{\Delta\omega_{\text{inh}}}{4\sqrt{\ln 2}}(t - \tau_L + \delta)\right]^2\right), \qquad (8.31a)$$

$$\overline{P}_2(t) = \gamma d \cos \Delta t \exp\left(-\left[\frac{\Delta\omega_{\text{inh}}}{4\sqrt{\ln 2}}(t - \tau_L + \delta)\right]^2\right), \qquad (8.31b)$$

and

$$\delta = \frac{1}{\Omega_{R0}} \tan\left(\tfrac{1}{2}\Omega_{R0}\tau_L\right), \qquad (8.31c)$$

where

$$\Delta = \omega_L - \omega_{10}.$$

Thus we obtain for the total polarization

$$P(t) = \gamma d\left[\sin(\omega - \Delta)t\right]\exp\left(-\left[\frac{\Delta\omega_{\text{inh}}}{4\sqrt{\ln 2}}(t - \tau_L + \delta)\right]^2\right).$$

The detuning of the laser from the center frequency of the inhomogeneously broadened line leads to the generation of a component $\overline{P}_1$ in addition to $\overline{P}_2$. The structure of the free induction decay is of the same form as that in (8.30) except for the difference that the maximum of the Gaussian function has been shifted to $t = \tau_L - \delta$, which means that the maximum seems to have appeared at a time before the end of the excitation pulse and thus before the starting point of our calculation. Hence the free polarization behaves for $t > \tau_L$ as if it had its maximum at $t = \tau_L - \delta$ (see Fig. 8.3). In particular, when irradiating

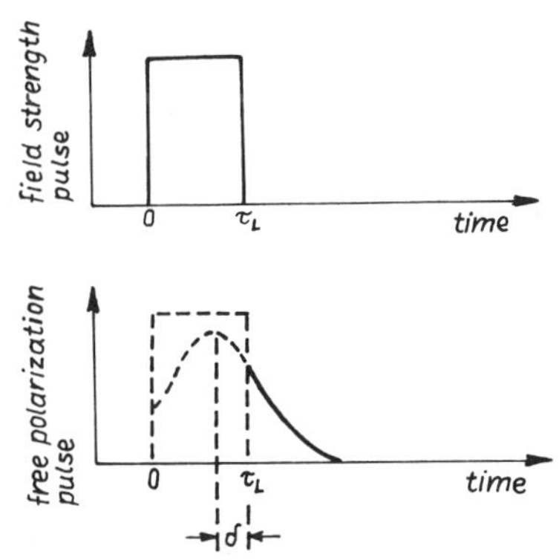

Fig. 8.3. Free polarization decay.

the sample with a $\pi/2$ pulse, we have $\delta = 2\tau_L/\pi$. Such shifts in the free induction decay at finite pulse lengths were discussed for instance in Refs. 8.5–8.8 and 3.

8.3.2. Observation of Oscillations in the Occupation-Number Inversion

The simplest way to observe Rabi oscillations in optical interactions consists in irradiating a sample with very short, intense, and coherent light pulses and measuring the fluorescent light incoherently emitted by this sample after excitation, the power of which is proportional to the number of particles in the excited state. Accordingly, the fluorescence output exhibits the same oscillating dependence on the pulse area as does the occupation-number inversion.

To illustrate this method let us describe a typical experiment performed by Gibbs (8.1) and the results obtained by him. The light of a frequency-stabilized, pulsed ^{202}Hg laser ($\lambda = 794{,}466$ nm) irradiates the sample, which is a ^{87}Rb atomic beam (see Fig. 8.4*a*). By means of a constant magnetic field with $B = 7.45$ T resonance was achieved between the laser radiation and the Zeeman transition 5p $^2P_{1/2}(F = 2,\ M_g = \frac{1}{2},\ M_I = \frac{3}{2}) \rightarrow 5s\ ^2S_{1/2}(F = 1,\ M_g = -\frac{1}{2},\ M_I = \frac{3}{2})$ (see Fig. 8.4*b*). The excited level 1 mainly relaxes by radiative transitions with the transition rate $k_{10} = 1/(84 \text{ ns})$ to the initial level 0, and with the rate $k_{11'} = 1/(42 \text{ ns})$ to the intermediate level 1′. Thus the lifetime of the excited level is $T_1 = (k_{10} + k_{11'})^{-1} \approx 28$ ns, and therefore it is —like the transverse relaxation time—considerably larger than the half-width of the laser pulses with $\tau_L = 7$ ns. The atomic beam has a very low density, and therefore collective coherent emission processes are of minor importance even at strong excitation (cf. Section 9.4). The incoherent fluorescent light from a small solid angle directed less than 90° to the incident light is detected by a photomultiplier followed by a time-correlated photon-counting device. In this way the decay curve of the fluorescence is measured, yielding after the end of the excitation pulse, as expected, a simple exponential decay with a time constant of 28 ns. As a measure for the excitation of level 1, the time-integrated fluorescence intensity has been calculated, where the lower integration limit is chosen as $t_1 = 22$ ns so as to prevent the recording of most of the photons already emitted during the action at the excitation pulse. Now the integral intensity is measured as a function of the pulse area θ. With a transition moment $d = 1.45 \times 10^{-29}$ Asm of the observed transition and $\tau_L = 7$ ns there occur π pulses at a peak intensity of about 0.5 W/cm^2. Figure 8.4*c* shows the Rabi oscillations of the integral fluorescence intensity and hence of the occupation number at the end of the excitation pulse in the excited state. In accordance with the theory described above, the maxima of the emission lie at $\theta = \pi$ and 3π, its minima at 2π and 4π.

However, to allow an exact quantitative interpretation of all the results, the theoretical model needs some refinements. To achieve this, particular consideration must be given to the influence of the third level and the actual form of

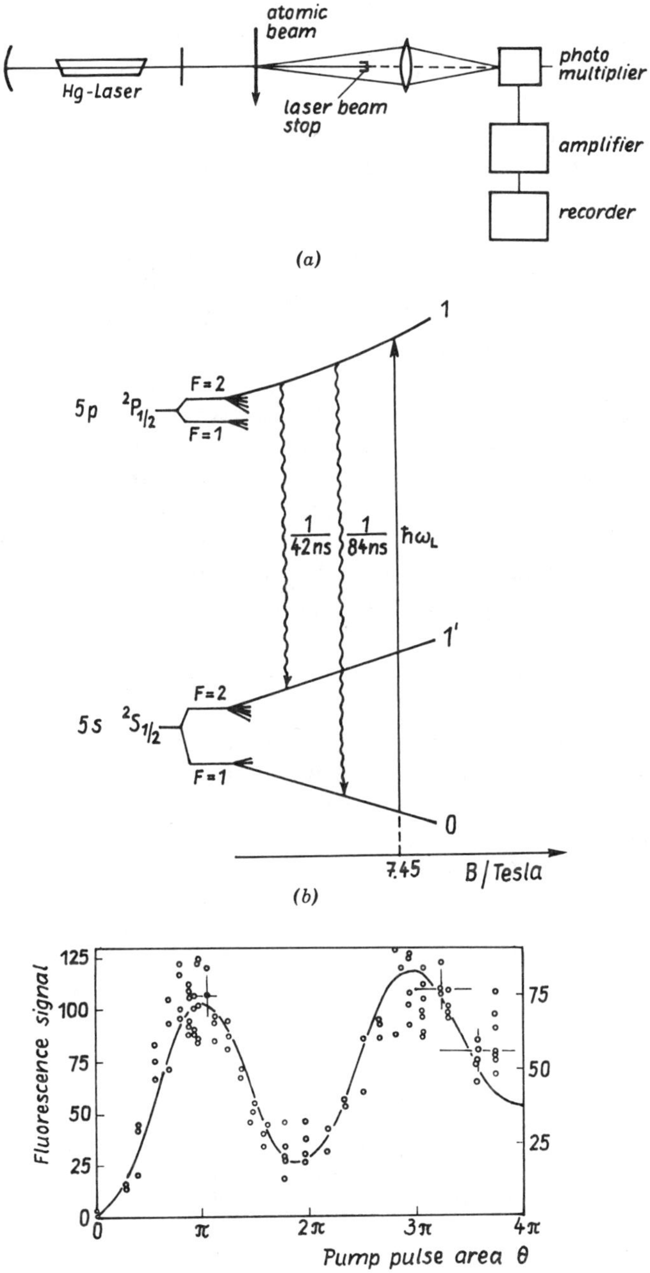

Fig. 8.4. Measurement of the dependence of the incoherent fluorescence signal on the pump pulse area (after Ref. 8.1): (*a*) schematic of the experiment; (*b*) level scheme; (*c*) time-integrated incoherent fluorescence signal versus pump pulse area.

the excitation pulse, whose duration is not completely negligible compared with the relaxation times. On the basis of an appropriate computerized simulation of the experiment, Gibbs, taking into account these conditions, succeeded in achieving excellent agreement with the measuring values. For example, the deviation of integral fluorescence in its minimum at $\theta = 2\pi$ from zero can be mainly explained in terms of the emission processes already occurring during the irradiation by the trailing edge of the excitation pulse.

8.4. TRANSIENT EXCITATION OF ATOMIC SYSTEMS WITH RELAXATION

Let us find out the manner of modifying the solutions of the Bloch equations if we dispense with some of the special assumptions that so far have simplified the calculations very much. In our investigation we shall deal with the effects of relaxation processes during irradiation, which hitherto have been neglected in the original equations. To make the forthcoming discussions as simple as possible, we continue assuming some other restrictive conditions concerning the entrance pulse and the sample.

8.4.1. Atomic Response Affected by Relaxation

For simplicity we assume that a light pulse of constant amplitude $\bar{E}$, phase $\varphi_L = 0$, and frequency $\omega = \omega_{10}$ is switched on at the time $t = 0$ and that the influence of the interaction on the light pulse is negligible. Thus we obtain from (8.2) the set of equations

$$\left(\frac{d}{dt} + \frac{1}{\tau_{10}}\right)\bar{P}_2 = -\frac{d^2}{\hbar}\gamma_I \bar{E}, \tag{8.32a}$$

$$\frac{d}{dt}\gamma_I + \frac{1}{T_{10}}(\gamma_I - \gamma_I^e) = \frac{1}{\hbar}\bar{E}\bar{P}_2. \tag{8.32b}$$

This set has to be solved with the initial conditions $\bar{P}_2 = 0$ and $\gamma_I(0) = \gamma_I^e = -\gamma$. We make the ansatz

$$\bar{P}_2 = p_0 + p_1 e^{\lambda_1 t} + p_2 e^{\lambda_2 t}, \tag{8.33a}$$

$$\gamma_I = \Gamma_0 + \Gamma_1 e^{\lambda_1 t} + \Gamma_2 e^{\lambda_2 t} \tag{8.33b}$$

and obtain for the free parameters

$$\tau_{10}\lambda_{1,2} = -\frac{1}{2}\left(1 + \frac{\tau_{10}}{T_{10}}\right) \pm \sqrt{\frac{1}{4}\left(1 \frac{\tau_{10}}{T_{10}}\right)^2 + (\Omega_{RO}\tau_{10})^2}, \quad (8.33c)$$

$$p_0 = \frac{\Omega_{RO}\tau_{10} d \cdot \gamma}{(T_{10}/\tau_{10})(\Omega_{RO}\tau_{10})^2 + 1}, \quad (8.33d)$$

$$\Gamma_0 = -\frac{\gamma}{(T_{10}/\tau_{10})(\Omega_{RO}\tau_{10})^2 + 1}, \quad (8.33e)$$

$$p_1 = \left(\frac{\tau_{10}\lambda_1 + \tau_{10}/T_{10}}{\Omega_{RO}\tau_{10}}\right)\Gamma_1 d, \quad (8.33f)$$

$$p_2 = \left(\frac{\tau_{10}\lambda_2 + \tau_{10}/T_{10}}{\Omega_{RO}\tau_{10}}\right)\Gamma_2 d, \quad (8.33g)$$

$$\Gamma_1 = \left(\frac{\lambda_2}{\lambda_1 - \lambda_2}\right)\left(\frac{(T_{10}/\tau_{10})(\Omega_{RO}\tau_{10})^2}{1 + (T_{10}/\tau_{10})(\Omega_{RO}\tau_{10})^2}\right)\gamma, \quad (8.33h)$$

$$\Gamma_2 = \left(-1 + \frac{1}{1 + (T_{10}/\tau_{10})(\Omega_{RO}\tau_{10})^2}\right)\gamma - \Gamma_1, \quad (8.33i)$$

where

$$\Omega_{RO} = \frac{\bar{E}d}{\hbar}.$$

Depending on the values of the Rabi frequency Ω_{RO} and of the relaxation parameters, (8.33) represents either monotonically varying or damped oscillating time dependences of the polarization amplitude $\bar{P}_2$ and of the density of the occupation-number inversion, γ_I. For large times we get a stationary solution in any case because of the effect of the relaxation processes. This solution is

$$\bar{P}_2(\infty) = \left(\frac{\Omega_{RO}\tau_{10}}{1 + (T_{10}/\tau_{10})(\Omega_{RO}\tau_{10})^2}\right)d \cdot \gamma, \quad (8.34a)$$

$$\gamma_I(\infty) = -\left(\frac{1}{1 + (T_{10}/\tau_{10})(\Omega_{RO}\tau_{10})^2}\right)\gamma. \quad (8.34b)$$

From (8.34) it can be seen that under strong irradiation ($\Omega_{RO}\tau_{10}\sqrt{T_{10}/\tau_{10}} \gg 1$) saturation of the transition occurs with $\overline{P}_2(\infty) \to 0$, and also $\gamma_I(\infty) \to 0$, which means that we find an equal occupation of the two levels. The transient behavior of $\overline{P}_2$ and γ_I shortly after their being switched on ($t \ll T_{10}$) is of particular interest. In discussing this case, fast phase relaxation ($\tau_{10} \ll T_{10}$) is additionally assumed. From the general solution (8.33) we have

$$\frac{\gamma_I(t)}{\gamma} = -e^{-t/2\tau_{10}}\left\{\cos\left[\sqrt{(\Omega_{RO}\tau_{10})^2 - \tfrac{1}{4}}\,\frac{t}{\tau_{10}}\right] + \frac{\sin\left[\sqrt{(\Omega_{RO}\tau_{10})^2 - \tfrac{1}{4}}\,t/\tau_{10}\right]}{2\sqrt{(\Omega_{RO}\tau_{10})^2 - \tfrac{1}{4}}}\right\} \tag{8.35a}$$

and

$$\frac{\overline{P}_2(t)}{\gamma \cdot d} = \frac{\Omega_{RO}\tau_{10}}{\sqrt{(\Omega_{RO}\tau_{10})^2 - \tfrac{1}{4}}}e^{-t/2\tau_{10}}\sin\left[\sqrt{(\Omega_{RO}\tau_{10})^2 - \tfrac{1}{4}}\,\frac{t}{\tau_{10}}\right]. \tag{8.35b}$$

The occupation-number inversion and polarization amplitude exhibit, according to (8.35), damped oscillations with the oscillation frequency

$$\Omega' = \sqrt{\Omega_{RO}^2 - \frac{1}{4}\left(\frac{1}{\tau_{10}}\right)^2}. \tag{8.36}$$

It is evident that these equations describe an optical nutation, where the amplitudes of the oscillation terms are exponentially damped with the decay parameter $2\tau_{10}$. Note that in contrast to this, after the pulse is switched off, the polarization amplitude is damped with the phase decay time τ_{10}, and the occupation-number inversion is damped with the lifetime T_{10} (see Fig. 8.5).

The general solution for two-level systems with nonresonant irradiation ($\Delta \neq 0$) of a wave of constant amplitude switched on at $t = 0$ was first found by Torrey (8.8); the procedure for the solution is analogous to that in the special case with $\Delta = 0$, which has already been dealt with.

In inhomogeneously broadened systems for each atomic group with given resonance detuning Δ', this general solution must be used according to Ref. 8.8, and the single contributions to polarization and occupation-number inversion must be summed according to (8.3). In such systems, after the pulse is switched off the free polarization proceeds in accordance with more complicated laws, since the phase relations between the individual oscillators are being destroyed not only by phase relaxation but also as a result of the differing resonance frequencies. Let us again discuss the simple limiting case

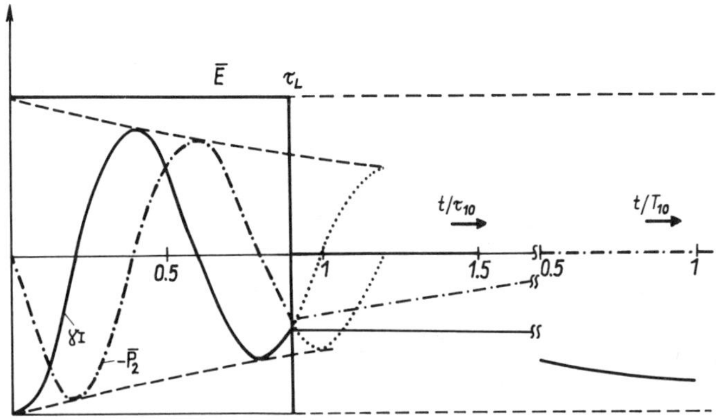

Fig. 8.5. Damped optical nutation and relaxation of polarization and occupation-number inversion: field-strength amplitude $\bar{E}$, polarization amplitude $\bar{P}_2$,and occupation-number inversion γ_I are plotted versus time t; at small and large values of t the phase decay time τ_{10} and the lifetime T_{10} have respectively been used for normalization.

that, without taking into account relaxation, led to (8.30). In this case, after irradiating the sample by an extremely short and intense $\pi/2$ pulse, we have

$$\bar{P}_2(t) = \bar{P}(0)e^{-t/\tau_{10}}\exp\left[-\left(\frac{\Delta\omega_{\text{inh}}}{4\sqrt{\ln 2}}t\right)^2\right]. \tag{8.37}$$

Since we have assumed that $\tau_{10} \gg 1/(\Delta\omega_{\text{inh}})$, the polarization mainly decays according to a Gaussian function, that is, in the same way as in (8.30). Later, when treating the photon echo in Section 8.4, we shall see that the decay of the polarization that is caused by the oscillation frequencies differing from one another and that is described by the Gaussian function can be reversed, whereas this cannot be done with the decay resulting from phase relaxation described by the factor $e^{-t/\tau_{10}}$.

More general cases of the decay of the polarization amplitude in inhomogenously broadened ensembles after excitation by square light pulses of various strength have been calculated in Ref. 8.3 (see Fig. 8.6). We see that the superposition of the polarization components of molecular groups with different transition frequencies may give rise to damped oscillations of the total polarization amplitude where the oscillation frequency increases with time. Such oscillations could already be observed in high-frequency proton resonance spectroscopy, and it should also be possible to detect them in optical experiments (Ref. 8.4).

8.4.2. Observation of Damped Optical Nutation and Free Polarization Decay

Let us now describe representative experiments on optical nutation and free polarization decay, in which the nonlinearity of the interaction process be-

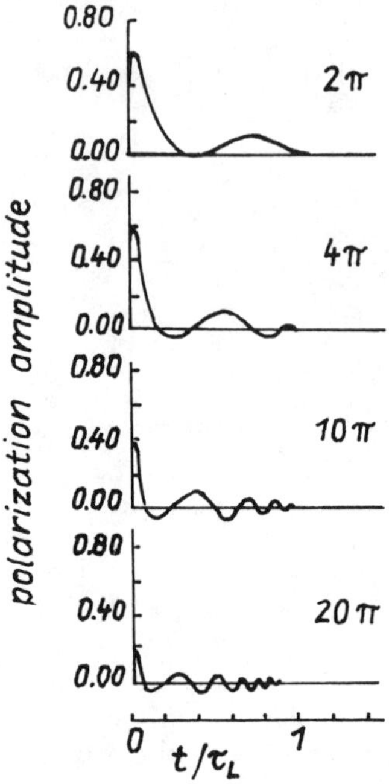

Fig. 8.6. Oscillatory free induction decay (after Ref. 8.3). The time coordinate t begins at the end of the exciting pulse of length τ_L.

comes obvious. Moreover these experiments provide certain atomic parameters such as relaxation times, line widths, and line splittings with high accuracy. The observation of the damped optical nutation just described requires a fast switching on of the interaction between the electromagnetic field and the atomic systems by very fast change of an appropriate parameter of this interaction. The word "fast" here means that the process has to be carried out in a time interval much shorter than the reciprocal Rabi frequency and the phase decay time, and it further means that either the amplitude of the laser (Refs. 8.9, 8.10), its frequency (Refs. 8.11–8.13), its phase (Refs. 8.14, 8.15), or the resonance frequency of the atomic systems (Ref. 8.16) can be switched during a short time interval (see also Refs. 8.17, 8.18). The fast switching of the amplitude can be achieved by irradiating the sample with a pulse having a steep leading edge. Using Pockels and Kerr cells, pulse rise times of about 100 ps are obtainable. Still steeper edges of the pulse can be achieved by shaping the pulses in nonlinear optical interactions (see Section 8.6). The frequency or the phase of continuous-wave (cw) laser radiation can be switched by means of modulators placed within or outside the laser resonator. Modulation outside the resonator offers the advantage of shorter rise times, and modulation within the resonator results in a higher switching efficiency.

Figure 8.7 presents a typical arrangement that allows a variation of these field parameters by means of an electrooptical crystal in the resonator of a single-mode cw dye lasers (Ref. 8.19). As long as the voltage pulses from the pulse generator are acting on the Pockels cell, they give rise to the change δn of

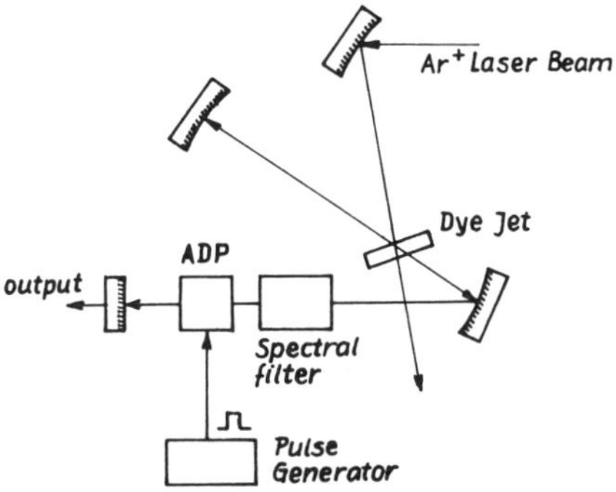

Fig. 8.7. Intracavity frequency switching (phase switching) of a cw dye laser (after Ref. 8.12). (ADP: ammonium dihydrogen phosphate crystal.)

the refractive index n of the electrooptical crystal and, as a consequence, to the change δL of the optical length L of the resonator. This results in a frequency shift $\delta\omega$ such that $\delta\omega/\omega = \delta L/L$. This shift may serve as a means for achieving the resonance between the radiation field and the atomic systems or for canceling it. After the termination of the voltage pulse of length τ_U, the laser frequency returns to its original value.

It has to be noted, however, that meanwhile the phase between the switched light wave and the unswitched one has changed by $\delta\phi = \delta\omega\,\tau_U$. Therefore the same setup can be used for phase switching—where, however, in contrast to frequency switching, the measurements are not being carried out as long as the electric pulse is acting, but only after it has ceased. The advantages of phase switching are the following: first, it needs only relatively low switching voltages, since phase shifts in the order of π are sufficient, and second, all measurements can be carried out at one fixed frequency. The optoelectronic switching of the phase can be done in time intervals of the order of 100 ps. The resonance frequency of the transition itself can be switched through the action of the Stark or Zeeman effect by means of an external electric or magnetic field pulse, and here also switching times of 100 ps are possible. On irradiating the samples by an appropriate picosecond light pulse, the phase of the resonance transition can be switched, even during a time interval of the order of the pulse duration, by nonlinear interaction processes.

To give an example, in Fig. 8.8 a very impressive result on optical nutation is presented (Ref. 8.16), which clearly demonstrates the nonlinearity of the interaction process. Strictly taken, we cannot apply the relations discussed so far to this example, since the optical nutation is observed through the reaction on the light pulse and since the transitions are inhomogeneously broadened. A qualitative description, however, is possible. In this experiment the radiation of a cw CO_2 laser is incident on a gas cuvette containing $^{13}CH_3F$. Using the pulse of an electric field, the interaction between the gas molecules and the radiation is temporarily modified by the Stark effect, which results in a spectral shift larger than the homogeneous line width but smaller than the inhomogeneous one. Before applying the Stark field, because of the Doppler effect, gas particles can take part in the interaction with the extremely monochromatic

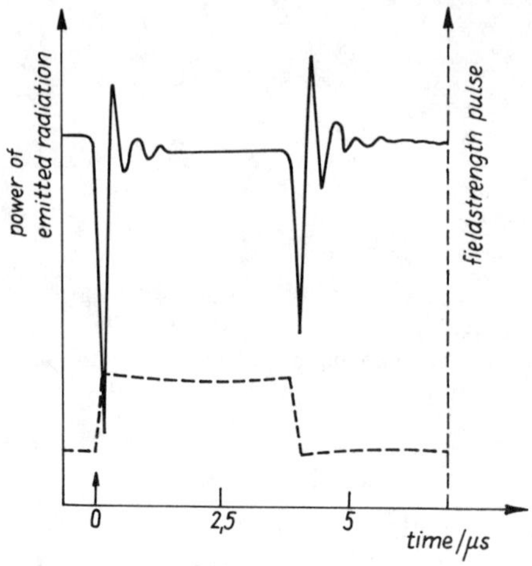

Fig. 8.8. Optical nutation in $^{13}CH_3F$(after Ref. 8.16).

radiation field of frequency ω_L only if they have a specific velocity component in the direction of the incident laser light. From this interaction, absorption saturation results for this group of molecules, since during the excitation the number of particles in the lower level is decreased and that in the upper one is increased. As a result of this saturation the cw absorption is very low. The Stark effect causes this group of particles to shift from resonance and another one, which now has the appropriate velocity component, to shift into resonance. Initially, nearly all the particles of this second group are in the lower level, and their behavior can be described in good approximation by (8.33). The damped oscillations of the polarization and of the occupation-number inversion yield also damped oscillations of the transmitted laser intensity, the appropriate value of the cw absorption being again approached, though after a longer time. From a comparison of the measured and calculated time dependences we can determine the phase decay time and the molecular transition moment.

In addition to optical nutation, the optical free induction decay can be advantageously investigated in this kind of experiment. As already described, immediately after switching off the interaction between the radiation field and a certain group of atomic systems, a macroscopic polarization is still detectable. For the first molecular group, after the Stark field is switched on, the frequency of the free polarization attains the value $\omega_L + \delta\omega_S$. This macroscopic polarization causes the emission of an electromagnetic wave at the same frequency, whereby the emitted power decreases with decreasing polarization. In a completely homogeneously broadened ensemble of atomic systems that are only weakly coupled, the polarization decays with the phase decay time τ_{10}. In the above mentioned experiments on gases with inhomogeneously broadened lines the polarization may decay faster [decay time between τ_{10} and $(\Delta\omega_{\text{inh}})^{-1}$] because of the different resonance frequencies of the molecules

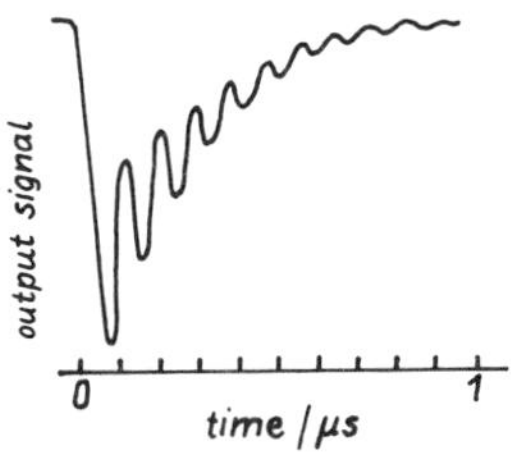

Fig. 8.9. Optical free induction decay (after Ref. 8.16).

taking part in the interaction. In the particular case under study the decay time depends on the spread of the resonance frequencies of the molecules involved in the interaction. This spread, however, hardly depends on the line width of the laser (here assumed to be very narrow), but mainly on its intensity. To guarantee a strong coherent excitation of the molecules, $\Omega_{R0} = d\bar{E}/\hbar \gtrsim 1/\tau_{10}$ must hold, which causes the width of the selected group to be greater than the homogeneous line width. Thus the decay, which in general also depends on the phase decay time τ_{10}, as well as on the reciprocal width of the excitation, is here essentially given by $1/\Omega_{R0}$. Only in the case of very strong excitation are all particles of the Doppler line involved, as was assumed in the derivation of (8.30) and (8.37). In the experiments described above, the fast-decaying polarization of the group with center frequency $\omega_L + \delta\omega_S$ and the field coherently emitted by it interact with the continuous laser wave of frequency ω_L. Thereby a damped output signal is generated, which is modulated at the beat frequency $\delta\omega_S$. This means that the free polarization decay can be measured by means of a heterodyne detection procedure. Figure 8.9 shows a typical experimental result.

Instead of the heterodyne detection procedure for measuring the free polarization decay, a homodyne detection may be used. This employs a phase switching that changes the relative phase between the electric field of an incident cw laser beam and the polarization of the sample. This phase change gives rise to a change of the transmitted laser power, which dies away as a result of phase-destroying processes. An example is provided by Fig. 8.10*b*, which gives results of experiments performed at the transition $({}^3H_4E_2 \leftrightarrow {}^1D_2A_1)$ of the doped crystal 0.1% Pr^{3+}:$LaCl_3$ at $T = 1.6$ K, using a phase-switched cw dye laser ($\lambda = 601.1$ nm) (Ref. 8.15). During a very short time interval, the phase of the laser is subjected to a change Φ. The phase switching is immediately followed by a change $\delta S \propto P^{(+)}E^{(-)} + P^{(-)}E^{(+)}$ in the output signal that depends periodically on the phase difference Φ according to $\delta S \propto \sin^2\Phi/2$ (see Fig. 8.10*a*). For $t > 0$ the signal δS dies off exponentially. In contrast to the investigations discussed so far, this experiment was carried out at rather low laser intensities ($\Omega_R^2 T_{10}\tau_{10} \ll 1$, with Ω_R the Rabi frequency, T_{10} the lifetime, and τ_{10} the phase decay time). This allows the neglect of power broadening: only centers with almost equal resonance frequencies $\omega'_{10} \approx \omega_L$ are excited. Under those conditions and for $\tau_{10} \ll T_{10}$, the decay of polarization is mainly determined by the phase decay time τ_{10}, and hence for

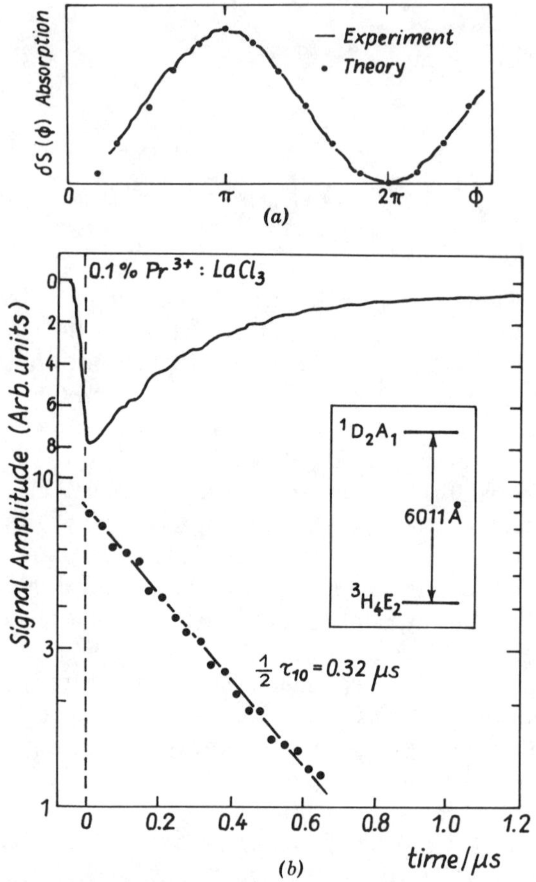

Fig. 8.10. Homodyne detection of free polarization decay (after Ref. 8.15): (*a*) change of the signal δS as function of the phase shift ϕ (sample: I_2 vapor); (*b*) phase-switched decay of the signal in 0.1% Pr^{3+}:$LaCl_3$ at 1.6 K (1, signal; 2, logarithm of signal), giving $\tau_{10} = 0.64$ μs.

$\tau_{10}^{-1} \ll \Delta\omega_{\text{inh}}$ the relation

$$\delta S \propto e^{-2t/\tau_{10}}$$

holds (see ref. 8.15). For the observed transition there results from the decay shown in Fig. 8.10*b* a phase decay time of 0.64 μs, and this corresponds to a homogeneous line width of only 500 kHz.

At low temperatures the crystal Pr^{3+}:LaF_3 exhibits a still much smaller homogeneous line width (Ref. 8.4). The main contribution to the line broadening originates from the magnetic dipolar spin interactions between pairs of ^{19}F and between ^{141}Pr and ^{19}F. This leads to a highly sensitive dependence of the homogeneous line width on external magnetic fields. For example, in the

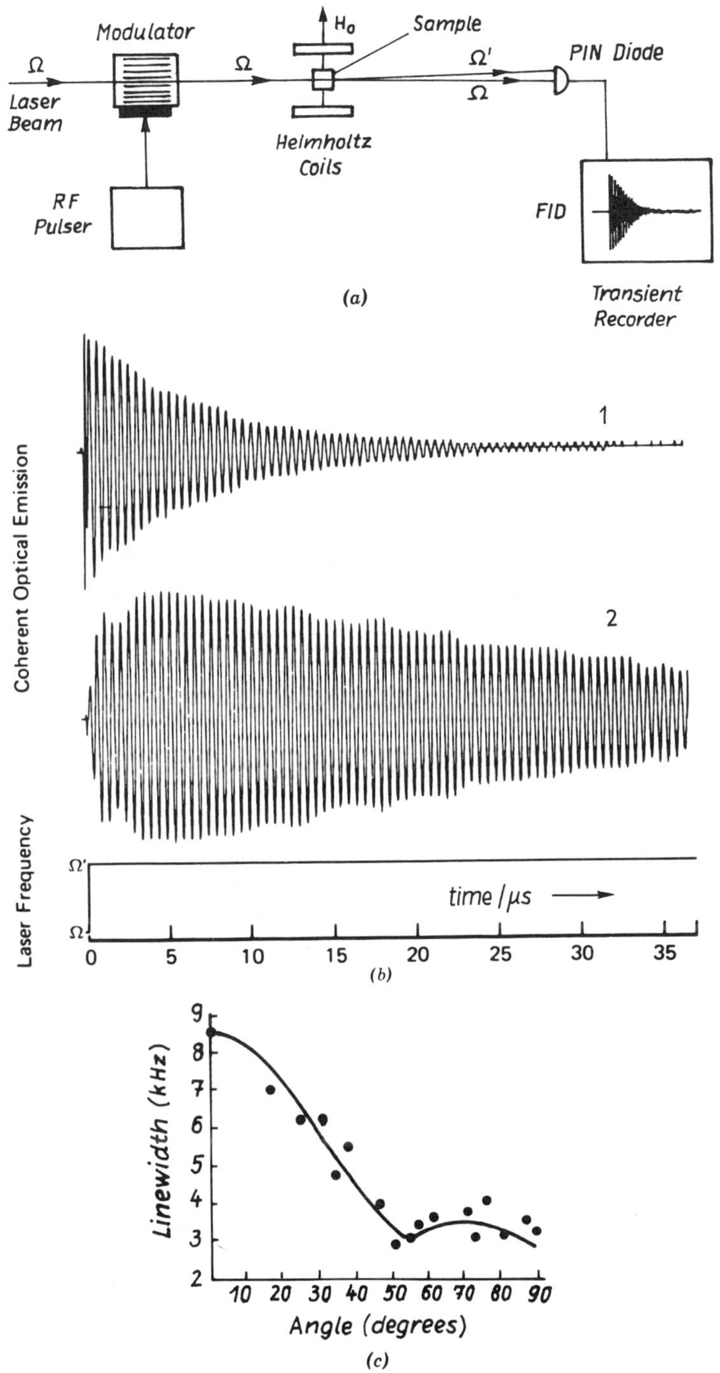

Fig. 8.11. Magic-angle optical free induction decay (after Ref. 8.4): (*a*) experimental setup; (*b*) coherent optical emission from Pr^{3+}:LaF_3 without rf (1) and with rf (2) under the magic-angle condition; (*c*) homogeneous linewidth of Pr^{3+}:LaF_2 versus the angle β (solid circles: experimental points; solid line: theory).

terrestrial magnetic field the homogeneous line width is 44 kHz, and in external magnetic fields which are much stronger than the internal fields it will even drop to 10 kHz. In strong fields the precessional motion is sufficiently fast to cancel the influence of slow chaotic processes caused by fluctuating fields that are directed parallel to the static field. On the other hand the magnetic fluctuations perpendicular to the static field are not affected, and from them a contribution to the residual width results. In Refs. 8.4 and 8.20 the homogeneous line width could be further reduced by applying the *magic-angle line-arrowing method* of NMR to optical spectroscopy. This experiment was performed with an rf and a static magnetic field present (see Fig. 8.11*a*). From Fig. 8.11*b* it is apparent that the coherent optical emission signal is prolonged by an rf field, which means a further reduction of the homogeneous line width. According to

$$\Delta\omega(\beta) = \Delta\omega_0(\cos\beta) \cdot \tfrac{1}{2}(3\cos^2\beta - 1) \tag{8.38}$$

this width depends on the angle β between the effective field and the static field, $\Delta\omega_0$ being the homogeneous width without the rf field. This dependence could be confirmed by experiment (see Fig. 8.11*c*).

8.4.3. Photon Echoes and Stimulated Photon Echoes

As already demonstrated, the polarization of a low-density ensemble of atomic systems decays after the generating field is switched off, as a result of the phase relaxation of the individual particles and (if inhomogeneous broadening is involved) as a result of the different resonance frequencies of the particles. The process of phase relaxation leads to an irreversible decay of the macroscopic free polarization, whereas the decay due to different frequencies is found to be reversible. In the following we discuss appropriate ways for reestablishing this free polarization.

For this, let us assume simple conditions. First, the maximum polarization amplitude is generated at the time $t = 0$ by irradiating the sample with a very short $\pi/2$ pulse with $\omega_L = \omega_{10}$. After the pulse this amplitude will be rapidly destroyed because of free induction decay. For highly *inhomogeneously broadened transitions* and for an excitation with high Rabi frequencies $\Omega_R \gtrsim \Delta\omega_{\text{inh}}$ we obtain, according to (8.30) or (8.37), a decay in time intervals of the order $(\Delta\omega_{\text{inh}})^{-1}$. After a delay time t_D with respect to the first pulse, where t_D is assumed to be longer than $(\Delta\omega_{\text{inh}})^{-1}$ and comparable with τ_{10}, a likewise very short π pulse of the same midfrequency is irradiated on the sample, leaving the polarization amplitude $\overline{P}_1'$ of a certain molecular group unaltered, whereas it reverses the sign of $\overline{P}_2'$. Therefore we obtain from (8.29) (disregarding the phase relaxation)

$$\overline{P}_1'(t_D) = -\gamma d \sin \Delta' t_D, \tag{8.39a}$$

$$\overline{P}_2'(t_D) = -\gamma d \cos \Delta' t_D. \tag{8.39b}$$

After the pulse the polarization continues its evolution according to the relation (8.27), which is known to possess general validity for the field-free case. Now in this relation τ_L has to be replaced by t_D. Substituting (8.39) as initial conditions into (8.27) we obtain

$$\begin{aligned}\bar{P}_1'(t) &= -\gamma d\left[\cos\Delta'(t-t_D)\sin\Delta' t_D - \sin\Delta'(t-t_D)\cos\Delta' t_D\right]\\ &= +\gamma d\sin\Delta'(t-2t_D)\end{aligned}\tag{8.40a}$$

and

$$\begin{aligned}\bar{P}_2'(t) &= +\gamma d\left[-\sin\Delta'(t-t_D)\sin\Delta' t_D - \cos\Delta'(t-t_D)\cos\Delta' t_D\right]\\ &= -\gamma d\cos\Delta'(t-2t_D).\end{aligned}\tag{8.40b}$$

This yields, by integrating over the inhomogeneously broadened line and by taking into account the *phase relaxation*, for $t \geq t_D$,

$$\bar{P}_1(t) = 0 \tag{8.41a}$$

and

$$\bar{P}_2(t) = -\gamma d\exp\left(-\left[\frac{\Delta\omega_{\text{inh}}}{4\sqrt{\ln 2}}(t-2t_D)\right]^2\right)e^{-t/\tau_{10}}. \tag{8.41b}$$

From (8.41) a reestablishment of the phase relations between the molecular groups and hence a reconstruction of the macroscopic polarization that has already decayed becomes apparent. In particular, after twice the retardation time we have

$$\bar{P}_2(2t_D) = -\bar{P}_2(0)e^{-2t_D/\tau_{10}} \tag{8.42}$$

for the polarization amplitude; this means that the initial state has been restored except for the relaxation factor $e^{-t/\tau_{10}}$, and hence again a high free polarization arises. Therefore at the time $t = 2t_D$ there again results a coherent emission of the sample due to a collective radiation effect, whose strength depends on $[\bar{P}_2(2t_D)]^2$. This phenomenon is called *photon echo*. If we increase the delay time t_D, this enables us to determine, according to (8.41), the phase decay time τ_{10} directly from the decrease of the echo signal. In this way it is possible to distinguish the homogeneous line broadening processes from the inhomogeneous ones. The physical reason for the photon echo is that the sign of the polarization component $\bar{P}_2'$ is reversed by the π pulse and that $\bar{P}_1'$ is preserved. This, however, means for the further motion in the $(\bar{P}_1', \bar{P}_2')$ plane that the lead taken by the fastest oscillating molecules was transformed into a lag. After the time $2t_D$ the faster molecules have made up for this lag. This phenomenon can be illustrated by employing the following model. With the first pulse all atoms simultaneously start like runners in a stadium. After several circuits, the synchronism has been completely destroyed, since the fastest runners have taken a lead of about one round trip. In our case the second pulse at time t_D has the effect of a signal that makes the runners reverse their direction of motion, thereby reestablishing sychronism after twice the time t_D.

Under more general assumptions the polarization after the second pulse can be evaluated by calculating the evolution of $\bar{P}_1'$ and $\bar{P}_2'$ during the first pulse, between the first and the second one, and during and after the second pulse according to the Torrey solution (Ref. 8.8). This leads to a multiplication of the corresponding time evolution matrices (cf. e.g. Refs. 3, 8.5, 8.7, 8.21, and the quantum-theoretical treatment in Chapter 9).

Kurnit, Abella and Hartmann were the first to detect (in 1964) photon echoes on irradiating ruby crystals with ruby-laser pulses (Ref. 8.22). In this solid the above mentioned assumption $\tau_L \ll T_{10}, \tau_{10}$ can be satisfied even with nanosecond pulses. Meanwhile various gaseous and solid substances have been investigated with pulses from the microsecond to the picosecond range (see e.g. Refs. 8.4, 8.23, and 8.24). Figure 8.12 presents an experimental scheme and the dependence of the echo signal on the pulse delay t_D. From this dependence the transverse relaxation time can be determined.

When irradiating the sample by more than two pulses, a sequence of echoes can be obtained. As an example of such experiments let us discuss the

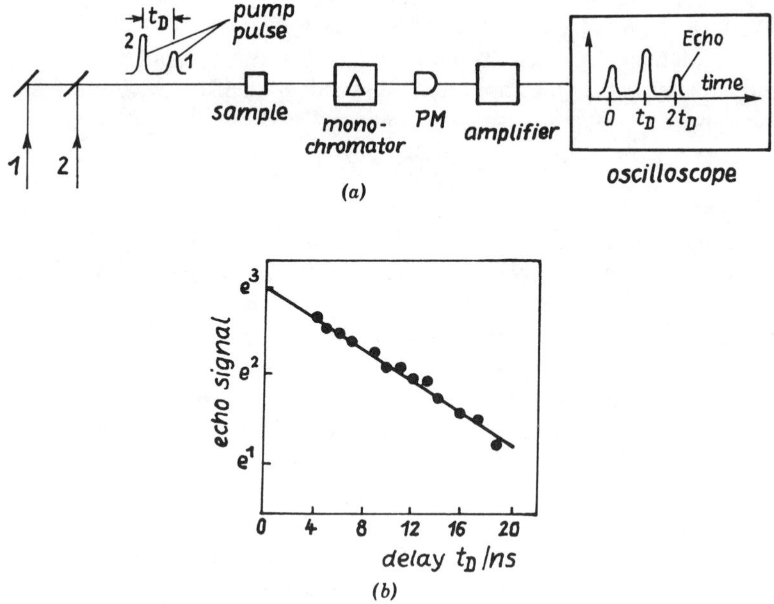

Fig. 8.12. Photon echoes: (*a*) experimental scheme; (*b*) echo signal versus delay time [after Ref. 8.26; sample: *p*-terphenyl crystal doped with pentacen $(1:10^8)$].

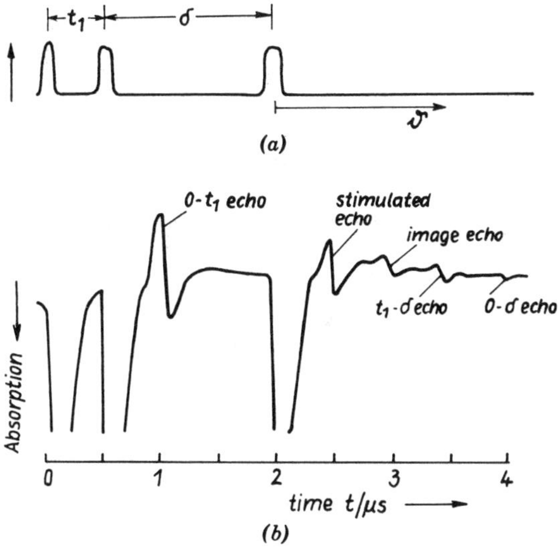

Fig. 8.13. Stimulated and ordinary (two-pulse) photon echoes in $^{13}CH_3F$ gas (b) which result from the sequence of switching pulses (a) (after Ref. 8.4).

interaction of a sample with three extremely short $\pi/2$ pulses of center frequency $\omega_L = \omega_{10}$ (see Fig. 8.13a). Initially the sample is in a state with $\overline{P}_1 = \overline{P}_2 = 0$ and $\gamma_I = -\gamma$, from which it evolves in time under the action of the $\pi/2$ pulses. For calculating the polarization and inversion of a certain molecular group with given resonance frequency, we apply (8.26) in the different time intervals, again assuming the limiting case $\tau_L \to 0$. For $t \geq t_1 + \delta$ and with substitution of $t = \vartheta + \delta + t_1$, the repetitive application of (8.26) is equivalent to the following matrix multiplication:

$$\begin{pmatrix} (1/d)\overline{P}_1'(t) \\ (1/d)\overline{P}_2'(t) \\ \gamma_I(t) \end{pmatrix} = \begin{pmatrix} \cos\Delta'\vartheta & -\sin\Delta'\vartheta & 0 \\ \sin\Delta'\vartheta & \cos\Delta'\vartheta & 0 \\ 0 & 0 & 1 \end{pmatrix} \begin{pmatrix} 1 & 0 & 0 \\ 0 & 0 & -1 \\ 0 & 1 & 0 \end{pmatrix}$$

$$\times \begin{pmatrix} \cos\Delta'\delta & -\sin\Delta'\delta & 0 \\ \sin\Delta'\delta & \cos\Delta'\delta & 0 \\ 0 & 0 & 1 \end{pmatrix} \begin{pmatrix} 1 & 0 & 0 \\ 0 & 0 & -1 \\ 0 & 1 & 0 \end{pmatrix}$$

$$\times \begin{pmatrix} \cos\Delta' t_1 & -\sin\Delta' t_1 & 0 \\ \sin\Delta' t_1 & \cos\Delta' t_1 & 0 \\ 0 & 0 & 1 \end{pmatrix} \begin{pmatrix} 1 & 0 & 0 \\ 0 & 0 & -1 \\ 0 & 1 & 0 \end{pmatrix} \begin{pmatrix} 0 \\ 0 \\ -\gamma \end{pmatrix}, \quad (8.43)$$

from which one gets

$$\frac{\bar{P}_1'(\vartheta)}{\gamma d} = -\tfrac{1}{4}[\sin \Delta'(t_1 + \delta - \vartheta) - \sin \Delta'(\delta + \vartheta - t_1) + \sin \Delta'(\vartheta + t_1 - \delta) + \sin \Delta'(t_1 + \vartheta - \delta) - 2 \sin \Delta(\vartheta - t_1) - 2 \sin \Delta'(\vartheta + t_1)]$$

$$\frac{\bar{P}_2'(\vartheta)}{\gamma d} = -\tfrac{1}{4}[\cos \Delta'(t_1 + \delta - \vartheta) + \cos \Delta'(\delta + \vartheta - t_1) - \cos \Delta'(\vartheta + t_1 - \delta) - \cos \Delta'(t_1 + \delta + \vartheta) + 2 \cos \Delta'(\vartheta - t_1) + 2 \cos \Delta'(\vartheta + t_1)]$$

$$\frac{\gamma_I'(\vartheta)}{\gamma} = -\tfrac{1}{2}[\cos \Delta'(\delta - t_1) - \cos \Delta'(\delta + t_1)] = \text{const}_\vartheta. \tag{8.44}$$

From (8.44) we obtain the polarization amplitude and the population inversion of the ensemble as a whole by integration over the inhomogeneously broadened line, where those contributions vanish that are odd functions of Δ'. The remaining terms have a time dependence in the form of Gaussian functions whose maxima represent the position of the pulse echoes. In particular, the echo at $\vartheta = t_1$ is termed the *stimulated echo*, and that at $\vartheta = \delta - t_1$ the *image echo* (see Fig. 8.13*b*). In addition there are echoes at $t = 2t_1$, $t = 2\delta + t_1$, and $t = 2(\delta + t_1)$, where ordinary two-pulse echoes also occur. The echo at $t = 2t_1$ can be calculated by taking into account only the four matrices that are farthest to the right and by making the replacement $\delta \to t - t_1$ in the matrix containing the pulse distance δ.

At this point attention should be drawn to an interesting difference between the solution after the two $\pi/2$ pulses and that after the usually applied $\pi/2$ and π pulse excitation. While in employing the conventional method only the polarization is excited, the $\pi/2$-pulse method also gives rise to an occupation-number inversion deviating from the equilibrium value; for $t > t_1$, it is given by

$$\gamma_I'(t) = \gamma \cos \Delta' t_1 \tag{8.45}$$

which means that γ_I' remains temporally constant with negligible relaxation. Within the inhomogeneously broadened line, (8.45) expresses a *spectral modulation* with the distance between two maxima on the Δ' scale—that is the modulation period—given by $2\pi/t_1$. This modulation can last for time intervals of the order of the energy relaxation time, when, in the case of fast phase decay, the polarization amplitudes have already died off long before. It is easy to see that the spectral occupation-number modulation suffices to yield an echo with a subsequent pulse. Thus a $\pi/2$ pulse incident on the sample with delay after the decay of the polarization amplitude—that is, for $T_{10} \gg \delta \gg \tau_{10}$—leads to

$$\bar{P}_1' = \gamma d \sin \Delta' \vartheta \cos \Delta' t_1 = \tfrac{1}{2}\gamma d \left[\sin \Delta'(\vartheta + t_1) + \sin \Delta'(\vartheta - t_1)\right], \tag{8.46a}$$

$$\bar{P}_2' = \gamma d \cos \Delta' \vartheta \cos \Delta' t_1 = -\tfrac{1}{2}\gamma d \left[\cos \Delta'(\vartheta + t_1) + \cos \Delta'(\vartheta - t_1)\right], \tag{8.46b}$$

$$\gamma_l' = 0. \tag{8.46c}$$

The integration over the frequencies of the inhomogeneously broadened transition results in a polarization component $\bar{P}_2$ differing from zero, which has its maximum at $\vartheta = t_1$ and induces a stimulated photon echo.

From the modulation of the occupation numbers according to (8.45), further interesting conclusions can be drawn. Since this modulation refers to both the levels involved, with $\gamma_0 + \gamma_1 = \gamma = \text{const}$, the spectral dependence of the occupation numbers of either of the two levels undoubtedly contains full information on the previous coherent interaction. Therefore the third $\pi/2$ pulse produces an echo even if it irradiates the sample at another transition that involves only one of the two levels of the $0 \leftrightarrow 1$ transition (Refs. 8.25, 8.26). In this way, by varying δ, the relaxation of the occupation number can be observed. Thereby not only the relaxation between the levels involved can be detected, but also the *cross relaxation* (i.e., transitions between different Δ' values within one level), since both kinds of relaxation wash out the modulation pattern. In extreme cases (some solids at helium temperature) spectral modulations of the occupation number of the ground level could be detected by means of a stimulated echo even after 30 min.

More general results can again be obtained by applying the Torrey solution to the calculation of the time evolution instead of (8.43). In this procedure the resonance detuning gives rise to additional effects. The propagation properties of the echo pulses will be treated in the next chapter after having dealt with the reaction of the light–matter interaction on the radiation field in nonstationary processes and with the quantum theory of such emission processes.

8.5. SHAPING OF VERY SHORT LIGHT PULSES. SELF-INDUCED TRANSPARENCY

To calculate the effect of the interaction between the field and the sample and of the resultant nonlinear dependence between polarization and field strength on the light pulse, we must solve the wave equation simultaneously with the constitutive equations. This means that we have simultaneously to solve the set of equations (8.2) and (8.5) for homogeneously broadened transitions and (8.3) and (8.5) for inhomogeneously broadened transitions. In this section we shall make the same assumptions as in Section 8.3, namely that the pulses are much shorter than the lifetime T_{10} (longitudinal relaxation time) and the phase decay time τ_{10} (transverse relaxation time) of the observed transition $0 \leftrightarrow 1$. Then, during the action of the pulse, the relaxation terms can be neglected.

For an ensemble of atomic systems with inhomogeneously broadened transitions we start from (8.3) and (8.5), and assume in the following that temporal and spatial changes of the phase of the light wave are of no importance so that (8.5b) can be ignored. Thus we may write

$$\frac{\partial}{\partial t}\bar{P}_1' = -\Delta' \cdot \bar{P}_2', \tag{8.47a}$$

$$\frac{\partial}{\partial t}\bar{P}_2' = \Delta' \cdot \bar{P}_1' - \frac{1}{\hbar}d^2\bar{E}\gamma_I', \tag{8.47b}$$

$$\frac{\partial}{\partial t}\gamma_I' = \frac{1}{\hbar}\bar{E}\bar{P}_2', \tag{8.47c}$$

$$\left(\frac{\partial}{\partial z} + \frac{1}{v}\frac{\partial}{\partial t}\right)\bar{E} = -\frac{\mu_0\omega^2}{2k}\bar{P}_2, \tag{8.47d}$$

and

$$\bar{P}_2(z,t) = \int_{-\infty}^{\infty} d\Delta' g_{\text{inh}}(\Delta' - \Delta)\bar{P}_2'(z,t;\Delta'). \tag{8.47e}$$

For such a system the change of the pulse area during pulse propagation is the first thing to be investigated. This is followed by the treatment of the propagation of distortionless pulses.

8.5.1. Area Theorem for Inhomogeneously Broadened Absorbers

To derive this theorem we substitute (8.47e) into (8.47d), multiply by $d/\hbar$, and integrate this relation over the entire pulse duration. For $\omega_L = \omega_{10}$ (i.e. $\Delta = 0$) this gives

$$\frac{\partial}{\partial z}\theta(z) = -\frac{\mu_0\omega^2}{2k}\frac{d}{\hbar}\int_{-\infty}^{\infty} d\Delta' \int_{-\infty}^{t_M} dt\, g_{\text{inh}}(\Delta')\bar{P}_2'(z,t;\Delta'). \tag{8.48}$$

Into this equation the expression (8.47a) is substituted for $\bar{P}_2'(z,t;\Delta')$, which enables us to carry out a direct integration. We obtain

$$\frac{\partial}{\partial z}\theta(z) = \frac{\mu_0\omega^2}{2k}\frac{d}{\hbar}\int_{-\infty}^{\infty} d\Delta' \frac{1}{\Delta'} g_{\text{inh}}(\Delta')\bar{P}_1'(z,t_M;\Delta'). \tag{8.49}$$

The time t_M is chosen as large as is necessary for the light pulse to have ceased long since. Already in the time interval $t_m < t < t_M$ with $(t_M - t_m) \gg 1/\Delta\omega_{\text{inh}}$

but with $(t_M - t_m) \ll \tau_{10}$, we suppose the second term on the right-hand side of (8.47b) to be negligibly small. Then, in this time interval, $\bar{P}_1'$ and $\bar{P}_2'$ can be determined from their values at $t = t_m$ according to the relations (8.27) for the field-free case. For $\bar{P}_1'$ at the time t_M we find

$$\bar{P}_1'(z, t_M; \Delta') = \bar{P}_1'(z, t_m; \Delta')\cos \Delta'(t_M - t_m) - \bar{P}_2'(z, t_m; \Delta')\sin \Delta'(t_M - t_m). \tag{8.50}$$

This expression is substituted into (8.49). Because of the fast oscillations of the trigonometric functions, only the region in the immediate neighborhood of $\Delta' = 0$ gives a substantial contribution to the integral, where the term with $[\sin \Delta'(t_M - t_m)]/\Delta'$ predominates. With the limiting value

$$\lim_{t_M - t_m \to \infty} \left(\frac{\sin \Delta'(t_M - t_m)}{\Delta'} \right) = \pi\delta(\Delta'),$$

we have

$$\frac{\partial}{\partial z}\theta(z) = -\frac{\pi\mu_0\omega^2 d}{2k\hbar} g_{\text{inh}}(0)\bar{P}_2(z, t_m; 0). \tag{8.51}$$

Now we substitute for $\bar{P}_2(z, t_m; 0)$ the expression previously obtained for strict resonance:

$$\bar{P}_2(z, t_m; 0) = \gamma d \cdot \sin\theta(z, t_m) \tag{8.52}$$

[cf. (8.25a)] and get the *pulse area theorem* of McCall and Hahn (Ref. 8.27) in the form

$$\frac{\partial}{\partial z}\theta(z) = -\tfrac{1}{2}\gamma\sigma_{01}(0)\sin\theta(z), \tag{8.53a}$$

where

$$\sigma_{01}(0) = \frac{\pi\mu_0\omega^2}{k}\frac{d^2}{\hbar} g_{\text{inh}}(0) \tag{8.53b}$$

is the absorption cross section of the inhomogeneously broadened transition at the center frequency. From (8.53) it can be seen that the change of the pulse area obeys a nonlinear differential equation. This nonlinear behavior is directly connected with the oscillations of the polarization and the occupation-number inversion already dealt with in previous sections of this chapter. Only in the case of very weak pulses ($\theta \ll 1$) can the nonlinearities be neglected; then we

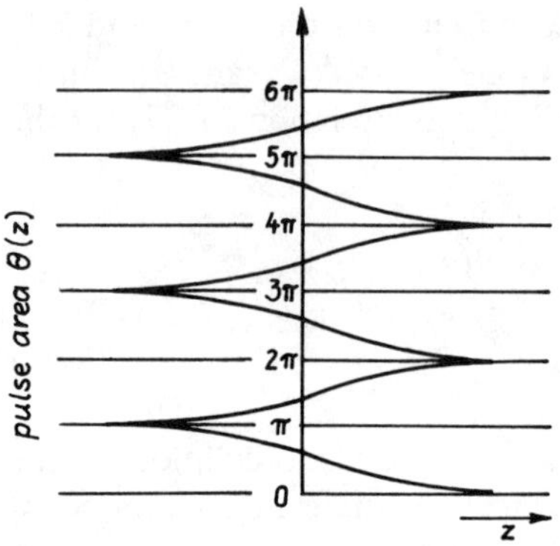

Fig. 8.14. Pulse area $\theta(z)$.

have

$$\frac{\partial}{\partial z}\theta(z) = -\tfrac{1}{2}\gamma\sigma_{01}(0)\theta(z). \tag{8.54}$$

The linearization results from the fact that in weak fields the atomic parameters deviate only slightly from their equilibrium value.

In accordance with our general concern, however, we do not want to continue the discussion of the linearized limit but rather deal with the regions of strong nonlinearity. It is obvious from (8.53) that the pulse area for $\theta = l\pi$ (l integer) does not change under propagation, since for these values $\partial\theta/\partial z = 0$. Figure 8.14 shows several branches of the solution of (8.53). From it we see that during the passage through the sample the pulse area remains stable under perturbations only if its initial value attains $\theta = l \cdot 2\pi$ (l integer). For all other values of θ, the pulse evolves in such a way as to reach an integral multiple of 2π.

8.5.2. Distortionless pulses

We now discuss the propagation of light pulses, which are characterized by certain values of the pulse area at the entrance of the sample. As already mentioned in this chapter, under certain conditions light pulses with a pulse area that is an integral multiple of 2π force the atomic systems they are interacting with to return to their initial states at the end of the interaction. Let us in particular find out whether there are pulses that do not change their shape during propagation. Such pulses are called *distortionless pulses*.

For explaining the procedure of calculation we assume simple conditions. We suppose that all events proceed in time intervals much shorter than the lifetime and the phase relaxation time of the atomic transition. Then we suppose the center frequency ω of the laser pulse to coincide with the transition frequency ω_{10} of the homogeneously broadened transition, and the incoming pulse not to exhibit any phase modulation. In this case the polarization component $\overline{P}_1$ vanishes, and from (8.2) and (8.5) we obtain the simplified equations

$$\frac{\partial}{\partial t}\overline{P}_2(z,t) = -\frac{1}{\hbar}d^2\overline{E}(z,t)\gamma_I(z,t), \qquad (8.55a)$$

$$\frac{\partial}{\partial t}\gamma_I(z,t) = \frac{1}{\hbar}\overline{E}(z,t)\overline{P}_2(z,t), \qquad (8.55b)$$

$$\left(\frac{\partial}{\partial z} + \frac{1}{v}\frac{\partial}{\partial t}\right)\overline{E}(z,t) = -\frac{\mu_0\omega^2}{2k}\overline{P}_2(z,t), \qquad (8.55c)$$

which describe the evolution of the polarization amplitude $\overline{P}_2$, of the inversion of the occupation-number density γ_I of the sample, and of the amplitude $\overline{E}$ of the light pulse. The solutions of the first two equations were already discussed in the preceding sections. With the initial conditions $\overline{P}_2(z,-\infty) = 0$ and $\gamma_I(z,-\infty) = -\gamma$ these equations are satisfied by

$$\overline{P}_2(z,t) = \gamma d \cdot \sin\theta(z,t) \qquad (8.56a)$$

and

$$\gamma_I(z,t) = -\gamma\cos\theta(z,t), \qquad (8.56b)$$

where the pulse area $\theta(z,t)$ must obey the condition

$$\frac{\partial}{\partial t}\theta(z,t) = \frac{1}{\hbar}d \cdot \overline{E}(z,t). \qquad (8.56c)$$

Thus we have

$$\theta(z,t) = \frac{d}{\hbar}\int_{-\infty}^{t} dt'\,\overline{E}(z,t'). \qquad (8.57)$$

This relation is in agreement with (8.24), but here the dependence of the pulse area on the position coordinate z also has to be taken into consideration. Let us now investigate the evolution of the electric field amplitude and the pulse area in space and time. To do this, we replace $\overline{E}(z,t)$ in (8.55c) by $(\partial/\partial t)\theta(z,t)$ from (8.56c), and $\overline{P}_2(z,t)$ by the expression from (8.56a); we obtain

$$\left(\frac{\partial^2}{\partial z\,\partial t} + \frac{1}{v}\frac{\partial^2}{\partial t^2}\right)\theta(z,t) = -\beta\sin\theta(z,t) \qquad (8.58)$$

where

$$\beta = \frac{\mu_0\omega^2 d^2}{2k\hbar}\gamma.$$

As a result of the nonlinear dependence of the polarization on the field strength, we thus obtain a nonlinear partial differential equation for the pulse area $\theta(z, t)$. This relation is called *sine–Gordon equation*.

In accordance with the main topic of this subsection we want to find out whether there exist solutions of these differential equations representing distortionless pulses. A light pulse is called *distortionless* if its amplitude $\bar{E}$ depends on position and time only via the retarded or local time $\vartheta = t - z/v$, v being a constant parameter of the dimension of velocity, which is left to be determined. It represents the propagation velocity of the envelope of the distortionless pulse. With the substitutions

$$\frac{\partial}{\partial t} \to \frac{d}{d\vartheta} \quad \text{and} \quad \frac{\partial}{\partial z} \to -\frac{1}{v}\frac{d}{d\vartheta}$$

we obtain from (8.58) the ordinary differential equation

$$\frac{d^2}{d\vartheta^2}\theta(\vartheta) = \tilde{\omega}^2 \sin\theta(\vartheta), \tag{8.59a}$$

where

$$\tilde{\omega}^2 = \beta v\left(\frac{v}{v} - 1\right)^{-1}. \tag{8.59b}$$

Thus we have obtained a nonlinear differential equation of the mathematical structure of the *pendulum equation*, the solutions of which are well known.

We are mainly interested in single-pulse solutions for which $\bar{E}$ vanishes as $\vartheta \to \pm\infty$, that is, it vanishes at infinitely long times before and after the pulse maximum. According to (8.56c) and (8.59a), together with $\bar{E}$, $(d/d\vartheta)\theta$ and $(d^2/d\vartheta^2)\theta$ also vanish at infinity ($\vartheta \to \pm\infty$). From our treatment of self-induced transparency and from (8.56) we know that the final state of the sample coincides with the initial state if $\theta(\infty) = m \cdot 2\pi$, where m is an integer. By the conditions for the pulse area and its derivatives at infinity, the solution of the pendulum equation is unambiguously determined to be

$$\theta(\vartheta) = 4\arctan(e^{\tilde{\omega}\vartheta}), \tag{8.60}$$

and thus we obtain for the field strength

$$\bar{E}(\vartheta) = \frac{2\hbar\tilde{\omega}}{d}\frac{1}{\cosh\tilde{\omega}\vartheta}. \tag{8.61}$$

It is advantageous to replace the parameter $\tilde{\omega}$ by the maximum value of the field amplitude. Then there results from (8.61)

$$\bar{E}\left(t - \frac{z}{v}\right) = \bar{E}_{\max}\operatorname{sech}\left[\frac{d}{2\hbar}\bar{E}_{\max}\cdot\left(t - \frac{z}{v}\right)\right], \tag{8.62a}$$

where

$$\bar{E}_{\max} = \frac{2\hbar\tilde{\omega}}{d}. \tag{8.62b}$$

From (8.59b) it is evident that also the propagation velocity of the distortionless pulses is connected with its maximum field amplitude; from (8.59b) and (8.62b) we have

$$\frac{v}{v} - 1 = \frac{2\hbar\omega\gamma}{\varepsilon_0\bar{E}^2_{\max}}. \tag{8.63}$$

The interesting properties of these pulses may here be summed up as follows:

1. After the passage of the pulse with amplitude according to (8.62a), the atomic systems have returned to their initial state; that is, $\gamma_I(t \to \infty) = -\gamma$, $\bar{P}_2(t \to \infty) = 0$. As a consequence, there is only a temporary energy exchange between the atomic systems and the electromagnetic field, and hence the sample becomes completely transparent to the pulses with respect to the total energy.
2. The total area of the pulses, $\theta(\vartheta \to \infty) = (d/\hbar)\int_{-\infty}^{\infty}dt'\,\bar{E}(t')$, is 2π. The preceding sections have already shown that with such pulses on the average no energy exchange takes place between light and sample. This phenomenon is called *self-induced transparency*.
3. The passage of the pulses through the sample is distortionless; this means that their field amplitude depends only on the transformed time coordinate $\vartheta = t - z/v$, v being the propagation velocity of the pulse, which differs from the phase velocity ω/k and the linear optical group velocity $v = d\omega/dk$ of the sample. The propagation velocity v depends not only on the properties of the material, but also on the maximum field-strength amplitude $\bar{E}_{\max}$. In absorptive samples with $\gamma_I(t \to -\infty) < 0$, the propagation velocity v of the distortionless pulses is smaller than the linear optical group velocity v.
4. Solutions of nonlinear differential equations that are represented by stationary single pulses with constant propagation velocity are frequently called *solitary waves* (see, e.g., Refs. 8.31 and 8.32).

To conclude this subsection let us add some remarks on more general treatments of self-induced transparency and of the propagation of distortionless pulses. The phenomena of self-induced transparency and shape stability of the sech pulses were first studied by McCall and Hahn (Ref. 8.27). It is not necessary to restrict the treatment of the propagation of distortionless pulses to homogeneously broadened systems as we did. It can be shown that 2π pulses of the sech shape occur in inhomogeneously broadened systems as well. In the preceding subsection we have already stated that all pulses possessing a pulse area of $l \cdot 2\pi$ (l integer) pass through samples with inhomogeneously broadened transitions with no change in pulse area. But unlike the 2π pulses, the pulses with $l \neq 1$ are not distortionless; in particular, they split into 2π pulses after a long travel time (Refs. 8.28, 8.29). The solution for $l = 2$ may serve as an example. It is

$$E(z,t) = \frac{\bar{E}^2_{\max 1} - \bar{E}^2_{\max 2}}{\bar{E}^2_{\max 1} + \bar{E}^2_{\max 2}} \times \frac{\bar{E}_{\max 1}\operatorname{sech}\varphi_1 + \bar{E}_{\max 2}\operatorname{sech}\varphi_2}{1 - \dfrac{2\bar{E}_{\max 1}\bar{E}_{\max 2}}{\bar{E}^2_{\max 1} + \bar{E}^2_{\max 2}}(\tanh\varphi_1\tanh\varphi_2 - \operatorname{sech}\varphi_1\operatorname{sech}\varphi_2)}$$

where

$$\varphi_i = \tilde{\omega}_i\left(t - \frac{z}{v_i}\right) + \delta_i \qquad (i = 1, 2).$$

The substitution of this relation into the basic equations of pulse propagation shows that this pulse induces complete transparency but that it is not distortionless. So for large times and with $\bar{E}_{\max 1} > \bar{E}_{\max 2}$ the shape becomes

$$\bar{E}(z,t) = \bar{E}_{\max 1}\operatorname{sech}(\varphi_1 + \beta_{12}) + \bar{E}_{\max 2}\operatorname{sech}(\varphi_2 - \beta_{12}),$$

where

$$\beta_{12} = \ln\frac{\bar{E}_{\max 1} + \bar{E}_{\max 2}}{\bar{E}_{\max 1} - \bar{E}_{\max 2}}.$$

We see that the 4π pulse splits into two pulses, which propagate at different velocities v_1 and v_2 in accordance with their different amplitudes $\bar{E}_{\max 1}$ and $\bar{E}_{\max 2}$. After passage farther into the sample, two distortionless 2π pulses originate in this process (see Fig. 8.15). In the same way the collision of two 2π pulses can be described; two other 2π pulses are being generated a long time after the collision (see Fig. 8.16).

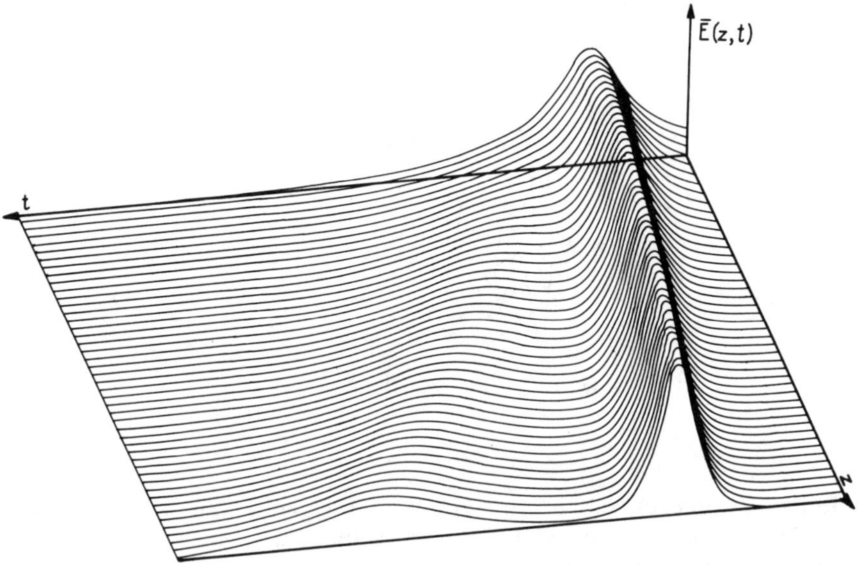

Fig. 8.15. Splitting of a 4π entrance pulse into two 2π pulses (after Ref. 8.29).

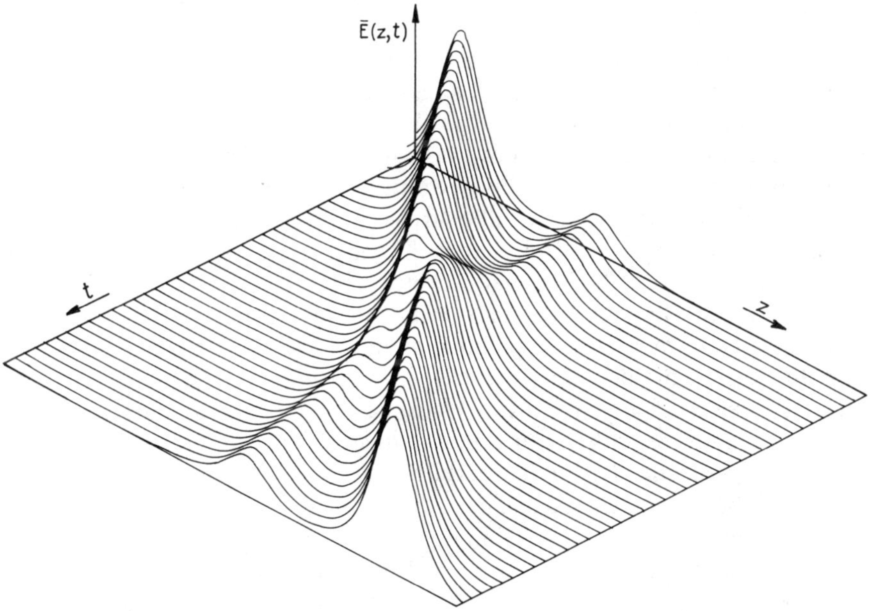

Fig. 8.16. Collision between two 2π pulses (after Ref. 8.29).

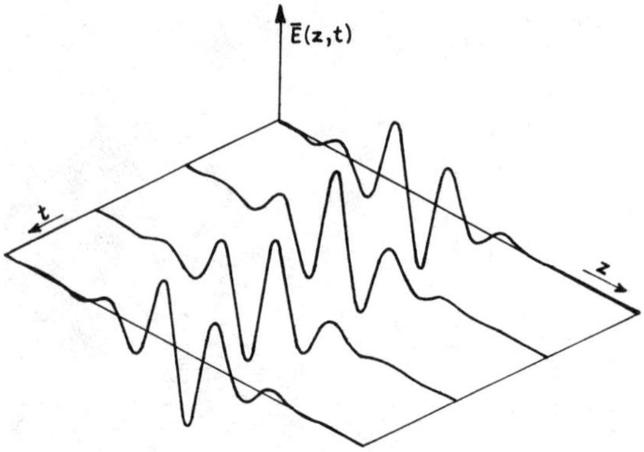

Fig. 8.17. 0π pulse (after Ref. 8.29).

It may be mentioned that not all the pulses split into 2π pulses. For example, in Ref. 8.28 the stability of certain 0π *pulses*—the so-called *fundamental* 0π *pulses*—was investigated; they may be thought of as being composed of a $+2\pi$ pulse and a -2π one. In fact, such a pulse does not exhibit a splitting into partial pulses while passing the medium (see Fig. 8.17).

Solitary waves like the ones considered, which exhibit stability under collisions and which in this sense behave like particles, are called *solitons* (see, e.g., Ref. 8.32).

The preceding discussion concerns the propagation of optical pulses with plane-wave structure in samples that are unbounded perpendicular to the direction of propagation. In Ref. 8.33 the occurrence of surface-wave pulses that reveal soliton-like properties was studied. It was demonstrated that electromagnetic surface-wave pulses with stable envelope and velocity can propagate along dielectric or metallic surfaces if invertible atomic systems with a transition frequency equal to the midfrequency of the surface wave are located on one side of the boundary surface. The basic equations (8.55a) have now to be replaced by more complex ones. Thus in (8.55c) the group velocity v must be replaced by a certain combination of optical parameters of the media on both sides of the boundary surface, since the continuity condition for the tangential field components must be satisfied at the boundary surface. Moreover the dependence of the polarization of the invertible atomic systems on the coordinate perpendicular to the surface must be taken into account. The features of the phenomenon depend on the spatial distribution of the invertible atomic systems. In the case of a thin dopant layer the differential equation for the pulse area has the same mathematical structure as (8.59a). Now, however, the important parameters of the pulse solution depend on geometrical and material parameters of the layer. In the case of a homogeneously doped half space, the basic differential equation deviates qualitatively from (8.59a). Then we have for instance

$$\frac{d^2}{d\vartheta^2}\theta \propto \frac{\sin(\frac{1}{2}\theta + a)\sin\frac{1}{2}\theta}{\frac{1}{2}\theta}$$

with the constant a being dependent on the preparation of the initial state. Under appropriate initial conditions, this means that for appropriate preparation of the sample, soliton-like solutions with novel coherence behavior result.

For simplicity, in our discussions we have not treated *phase modulations* of the pulses. The effect of such modulations—for example that of a chirp—on the pulse propagation was studied in Refs. 8.30 and 3. These investigations reveal that in the case of "sufficiently smooth" single entrance pulses it is often justified to neglect phase modulations during the entire interaction process.

The more recent development of subpicosecond lasers raises the question of how the pulse propagation proceeds under conditions under which the slowly-varying-envelope approximation fails (at present the shortest light pulses—only 8 fs long—consist of about 4 wave periods). This problem was tackled theoretically by Bullough and Ahmed in 1972 (Ref. 8.32). As a distortionless solution to the full wave equation and the Bloch equations (without separation of a carrier frequency) they found a single sech pulse for the field strength $E(z,t)$. Hence the electromagnetic field consists of a single half wave of the specific shape we mentioned. Its passage through the medium is loss-free and distortionless.

In our treatment of self-induced transparency we have so far not discussed the influence of relaxation processes. In reality the energy and phase relaxation cause deviations from the completely loss-free pulse-transmission regime. McCall and Hahn (Ref. 8.27) have taken these effects into account when calculating the pulse propagation (cf. also Refs. 8.35 and 3). The relaxation processes cause a decrease in pulse energy; nevertheless, the pulse area preserves its value of 2π. This fact indicates that there occurs a change in pulse shape, namely a pulse stretching. These authors introduced a critical distance z_c after which a 2π pulse of the initial length τ_{L0} in passing through a homogeneously broadened absorber attains the pulse duration

$$\tau_L(z_c) = \left(\frac{1}{T_{10}} + \frac{2}{\tau_{10}}\right)^{-1}.$$

For z_c they obtained

$$z_c = \frac{3\tau_L(z_c)}{2(k_a)_{\text{eff}}\tau_{L0}},$$

where $(k_a)_{\text{eff}}$ is the effective small-signal absorption coefficient of the material.

Finally, let us mention that in the present decade general procedures for the treatment of nonlinear dynamical problems with given initial conditions have been developed with great success. In this connection the inverse scattering method has proved to be very effective (see, e.g., Ref. 8.32).

8.5.3 Experimental Investigation of Self-Induced Transparency

In their first experiment on self-induced transparency, McCall and Hahn (8.27) studied ruby crystals. They irradiated a ruby crystal at liquid-helium temperature with the radiation of a ruby laser cooled by liquid nitrogen. From this temperature difference coincidence between the frequency of the laser transition $\bar{E}(2E) \leftrightarrow 4A_2(\pm \frac{3}{2})$ and the frequency of the absorption transition $4A_2(\pm \frac{1}{2}) \leftrightarrow \bar{E}(2E)$ could be achieved. The phase relaxation time of the absorption transition amounts to about 50 ns at helium temperature. The pulse duration was chosen to be much less (5–10 ns). At these pulse durations the intensity of 2π pulses is in the order of 1 MW/cm^2. Already in these preliminary experiments self-induced transparency with pulse areas of 2π, pulse retardation, and the splitting of pulses with larger areas could be observed.

These investigations were almost immediately followed by experiments carried out by several groups of workers who, in particular, studied the absorption of gases in the wavelength region of the CO_2 laser (e.g., Ref. 8.34). They established the exact resonance by means of the Stark effect. Very careful measurements, including comparisons with calculations that take into account the real conditions, were also made on Rb vapor excited with Hg lasers. (For experimental details see Section 8.3.2 and Ref. 8.35.) Figure 8.18 gives a

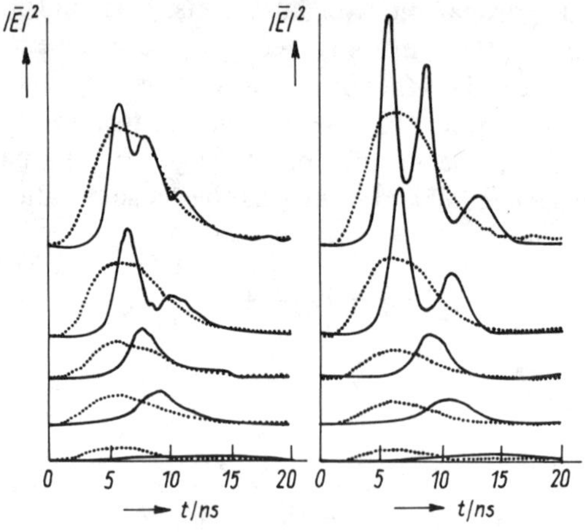

Fig. 8.18. Self-induced transparency in Rb vapor (after Ref. 8.35). Entrance pulses: dotted curves; exit pulses: full curves. (*a*) experimental results; (*b*) calculated results.

comparison between measured and calculated intensities of pulses that have passed through the sample, which has a small-signal transmission of 0.7%. It is obvious that the propagation of 2π pulses is characterized by very small losses because of self-induced transparency. The pulses exhibit a retardation dependent on the pulse area. The agreement between theory and experiment with respect to the splitting of pulses with large pulse areas at the sample entrance is also very impressive. We may conclude that here more complex pulses are present from which, in passing the sample, several 2π pulses originate. The figure further shows that self-induced transparency can be exploited for pulse shaping. In particular, there can be obtained substantial shortenings of the pulse duration with simultaneous increase of the peak intensity.

8.6. SHAPING OF LIGHT PULSES

In the preceding section we dealt with extremely short light pulses ($\tau_L \ll \tau_{10}, T_{10}$). Now we take another direction and discuss pulses much longer than the phase decay time τ_{10}. This means that the field-strength amplitude $\bar{E}(z, t)$ does not change appreciably in time intervals smaller than the phase decay time. Thus we shall restrict our treatment to a quasistationary regime. We start by treating the propagation of light pulses in a sample two-level system. This is followed by a discussion of some applications of such interactions, which may be used for nonlinear filtering of the input signal.

8.6.1. Pulse Shaping by Two-Level Systems

In this subsection we shall first study the propagation of light pulses of midfrequency ω in a sample composed of atomic two-level systems of the resonance frequency ω_{10} where $\omega_{10} \approx \omega$. Let the lifetime T_{10} of the excited two-level systems be appreciably larger than their phase decay time, and let it be comparable to the pulse duration or larger. As we have already mentioned, in condensed samples the phase decay time is in general much shorter than the lifetime of the same transition. Hence the conditions mentioned above are satisfied in a great number of liquids, solutions, and solids. It will be shown that even in such a regime appreciable pulse distortions and in particular pulse shortenings can be achieved, where (in accordance with our assumptions and in contrast to the case treated in Section 8.5) relatively long input pulses can also be used.

Let us assume that the atomic systems under consideration are homogeneously broadened and that the input pulse passes through the sample in the z direction. Under the given assumptions we may start from a set of rate equations for the occupation-number inversion density $\gamma_I = \gamma_1 - \gamma_0$ and the photon flux density $I = \varepsilon_0 cn|\bar{E}|^2/2\hbar\omega$, which, on using the coordinates $\eta = t$

$-z/v$ and $\xi = z$, reads

$$\frac{\partial}{\partial\eta}\gamma_I + \frac{1}{T_{10}}(\gamma_I - \gamma_I^e) = -2\sigma_{10}\gamma_I I, \tag{8.64a}$$

$$\frac{\partial}{\partial\xi}I = \sigma_{10}\gamma_I I, \tag{8.64b}$$

where $\sigma_{10} = \mu_0 c|d_{10}|^2\omega_{10}\tau_{10}/(\hbar n)$ is the one-photon absorption cross section (cf. Section 4.6). From (8.64b) we obtain

$$\gamma_I = \frac{1}{\sigma_{10}}\frac{\partial}{\partial\xi}\ln I$$

for the occupation-number inversion. Substitution of this expression into (8.64a) yields

$$\frac{\partial^2}{\partial\xi\,\partial\eta}\ln I + 2\sigma_{10}\frac{\partial}{\partial\xi}I + \frac{1}{T_{10}}\left\{\frac{\partial}{\partial\xi}\ln I - \sigma_{10}\gamma_I^e\right\} = 0. \tag{8.65}$$

This is a partial differential equation with the boundary condition $I(\xi = 0, \eta) = I_0(\eta)$ and the initial condition $(\partial/\partial\xi)\ln I(\xi, \eta \to -\infty) = \sigma_{10}\gamma_I^e$, $I_0(\eta)$ and γ_I^e being the photon-flux density of the input pulse and the density of the occupation-number inversion in thermal equilibrium. The integration with respect to the coordinate ξ can be performed at once, and we obtain

$$\frac{\partial}{\partial\eta}\ln I + 2\sigma_{10}I + \frac{1}{T_{10}}[\ln I - \sigma_{10}\gamma_I^e\xi] = f(\eta),$$

where $f(\eta)$ is determined from the boundary condition at $\xi = 0$ as

$$f(\eta) = \frac{\partial}{\partial\eta}\ln I_0 + 2\sigma_{10}I_0 + \frac{1}{T_{10}}\ln I_0.$$

Thus we have

$$\frac{\partial}{\partial\eta}\ln\frac{I}{I_0} + 2\sigma_{10}[I - I_0] + \frac{1}{T_{10}}[\ln\frac{I}{I_0} - \sigma_{10}\gamma_I^e\xi] = 0. \tag{8.66}$$

We see that we obtained an ordinary differential equation, which describes the change of the input pulse of the photon-flux density $I_0(\eta)$ into a pulse with the position-dependent photon-flux density $I(\xi, \eta)$ (cf. Ref. 8.36). Except for a free function of ξ, the integral is determined by (8.66). This free function is fixed by the initial condition, which requires that

$$\frac{I(\xi, \eta \to -\infty)}{I_0(\eta \to -\infty)} = e^{\sigma_{10}\gamma_I^e \xi}. \tag{8.67}$$

The ordinary differential equation (8.66) with its initial condition can be solved by simple numerical procedures. Here only two limiting cases will be considered where even this numerical integration can be avoided.

In the stationary limiting case, and even in the case of irradiation by input pulses that are rather long and not too intense ($\tau_L \gg T_{10}$ and $\sigma_{10} I_0 \tau_L \lesssim 1$), the term with the time derivative in (8.66) can be neglected and there results the algebraic relation

$$\ln \frac{I_0(\eta)}{I(\xi, \eta)} + 2\sigma_{10} T_{10}[I_0(\eta) - I(\xi, \eta)] = -\sigma_{10}\gamma_I^e \xi, \tag{8.68}$$

from which it is evident that in this case the output intensity $I(\xi, \eta)$ merely depends on the value of the input intensity at the same time η. Hence, for long pulses the sample represents a *nonlinear filter without memory*. Figure 8.19*a* demonstrates pulse shaping by such a filter, assuming $\gamma_I^e = -\gamma$.

Let us now consider the other limiting case, that of very short input pulses ($\tau_L \ll T_{10}$), where the terms in (8.66) that are multiplied by the factor $1/T_{10}$ can be neglected. The remaining equation can be solved by direct integration, the integral being determined except for an arbitrary function of ξ, which is

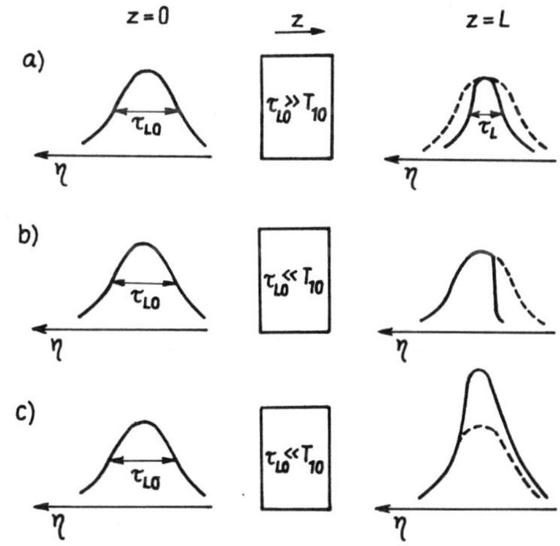

Fig. 8.19. Temporal profile of the pulse at the entrance and the exit of pulse-shaping devices: (*a*) Saturable absorber without memory, (*b*) Saturable absorber with memory, (*c*) Depletable amplifier with memory.

fixed by the initial conditions. This leads to

$$I(\xi,\eta) = I_0(\eta)\frac{\exp\left\{2\sigma_{10}\int_{-\infty}^{\eta} d\eta'\, I_0(\eta')\right\}}{e^{-\sigma_{10}\gamma_I^e \xi} - 1 + \exp\left\{2\sigma_{10}\int_{-\infty}^{\eta} d\eta'\, I_0(\eta')\right\}}. \tag{8.69}$$

According to (8.69) the pulse intensity $I(\xi, \eta)$ at the position ξ depends not only on the value of the input intensity $I_0(\eta)$ at the same instant of time $\eta' = \eta$, but also on its value at all earlier times $\eta' \leq \eta$. Thus for short pulses this sample represents a nonlinear filter with memory.

Providing a Gaussian-shaped input pulse of intensity

$$I_0(\eta) = I_{0m} e^{-(2\sqrt{\ln 2}\,\eta/\tau_{L0})^2} \tag{8.70}$$

and $\gamma_I^e = -\gamma$, we obtain the output pulse

$$I(\xi,\eta) = I_{0m}\frac{\exp\left\{-\left(\dfrac{2\sqrt{\ln 2}\,\eta}{\tau_{L0}}\right)^2 + \dfrac{I_{0m}\sigma_{10}\tau_{L0}\sqrt{\pi}}{\sqrt{4\ln 2}}\left[1+\Phi\left(\dfrac{\sqrt{8\ln 2}\,\eta}{\tau_{L0}}\right)\right]\right\}}{e^{\sigma_{10}\gamma\xi} - 1 + \exp\left\{\dfrac{I_{0m}\sigma_{10}\tau_{L0}\sqrt{\pi}}{\sqrt{4\ln 2}}\left[1+\Phi\left(\dfrac{\sqrt{8\ln 2}\,\eta}{\tau_{L0}}\right)\right]\right\}}, \tag{8.71}$$

where

$$\Phi(x) = \sqrt{\frac{2}{\pi}}\int_0^x dt\, e^{-t^2/2}$$

is the error integral. Figure 8.19*b* demonstrates the pulse shaping due to such a nonlinear filter with memory. In this case the leading edge of the input pulse is heavily absorbed, whereas the trailing edge is hardly attenuated. In contrast, the nonlinear filter without memory described above attenuates the leading and the trailing edges of the pulse in the same way.

8.6.2. Nonlinear Filtering by Saturable Absorption and Gain Depletion

As concerns the practical realization of the two limiting cases mentioned, it may be noted that the excited-state lifetimes of saturable absorbers can take on values that differ by many orders of magnitude. They range from some tens of femtoseconds in dissolved dyes and in solids to some seconds in gases at low pressure. Since the absorption can also be widely varied via the concentration of the active particles and the sample length, nonlinear filters exhibiting very different properties can be realized by employing one-photon processes alone.

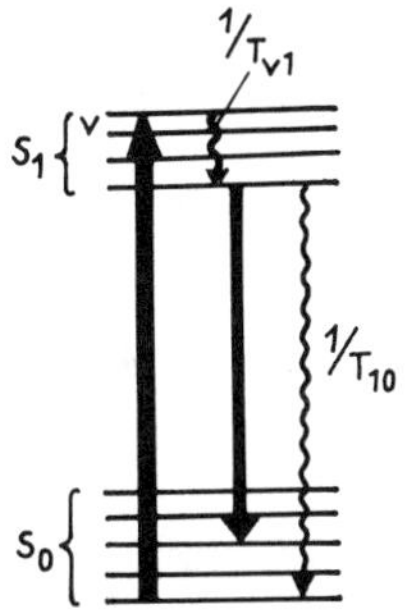

Fig. 8.20. Level scheme of dye molecules with absorber and amplifier transitions.

In this connection the question arises to what extent complex molecules or solids, in whose term scheme various relaxation processes may occur, can be described with respect to their interaction with light pulses that are in resonance with a certain transition by means of the simple model of two-level systems, and under what circumstances a description in terms of multilevel systems is required. As an example the electronic $S_0 \leftrightarrow S_1$ transition in organic dye molecules in condensed matter may be considered (see Fig. 8.20). Such systems are of great use as saturable absorbers in quantum electronics. If the vibrational structure of this electronic transition is resolved and if the radiation field is in resonance with the transition between the two vibrationless electronic states, the interaction can be very well described using an appropriate two-level system.

The case is more complicated if the radiation field is in resonance with a transition between the ground level of the S_0 state and an excited vibrational level of the S_1 state (depending on the constitution of the molecules, such transitions may indeed have large interaction cross sections according to the Franck–Condon principle). The radiationless deactivation of the molecule excited by absorption will most probably proceed via the vibrationless S_1 level as an intermediate level (see Fig. 8.20), since the rate of the vibrational relaxation $1/T_{v1}$, with typical values of 10^{13}–10^{14} s^{-1}, is in general appreciably higher than the electronic relaxation rates $1/T_{v0}$ and $1/T_{10}$, which in typical dye molecules attain values between 10^8 and 10^{11} s^{-1}. Hence because of the relaxation processes, owing to which other levels that do not directly take part in the radiative interaction become occupied, multilevel systems must be used for the description. The population of further levels can be neglected during the light–matter interaction only if the input pulse is much shorter than the vibrational relaxation, since in that case the excited molecules have "no time" during irradiation to pass to other levels. But even with light pulses of 8 fs, which are the shortest generated so far, it is hardly possible to reach this limiting value.

Much more important, however, is the case with $T_{v1} \ll \tau_L \ll T_{10}, T_{v0}$, which is realized in many applications. As a consequence of the fast vibrational relaxation, the upper level of the radiative transition is being depopulated so effectively that its occupation probability can be neglected compared to that of the vibrationless electronic level. Therefore, again the occupation numbers of only two levels have to be taken into account. In this case we obtain

$$I(\xi,\eta) = I_0(\eta)\frac{\exp\left\{\sigma_{10}\int_{-\infty}^{\eta} d\eta'\, I_0(\eta')\right\}}{e^{\sigma_{10}\gamma\xi} - 1 + \exp\left\{\sigma_{10}\int_{-\infty}^{\eta} d\eta'\, I_0(\eta')\right\}}. \qquad (8.72)$$

For the intensity of the output pulse this relation differs from (8.69), which was derived for genuine two-level systems, only by the factor 2 that is missing in front of the time integral of the intensity of the input pulse. This means that the real two-level systems can be saturated more easily. This can be very simply explained from the fact that in the two-level system complete saturation —that is, the disappearance of any further absorption—will be achieved if the occupation numbers in the upper and lower levels become equal, whereas in the three-level system complete saturation will not occur until the ground level is completely depopulated.

In the examples of nonlinear filters with and without memory given in Fig. 8.19*a*, *b*, we have assumed samples for which $\gamma_I^e = -\gamma$, that is, samples where, without irradiation by light, all atomic systems are in the lower level of the resonant transition. In other words, we have dealt with samples that exhibit maximum absorption in their initial states before being affected by the pulses. Filters with highly modified properties can be obtained if this condition is abandoned and indeed positive values of γ_I^e are assumed; in the latter case the sample acts as an amplifier.

Positive values of the occupation-number inversion of the resonant transition can for example be reached by optical pumping via an intermediate level higher than the upper state of the resonant transition. So it is possible to use the levels v in Fig. 8.20 as pumping levels and to irradiate the sample by the signal pulse on the transition $s_1, v = 0 \to s_0$. Let us assume that as a consequence of continuous pumping all atomic systems occupy level 1 as long as no signal pulse on the transition $S_1, v = 0 \to S_0$. Let us assume that as a consequence of continuous pumping all atomic systems occupy level 1 as long as no signal pulse is incident. Further, let the pumping process be so weak that its effect can be neglected during the action of the short signal pulse. Under these assumptions the atomic systems of the sample behave as real two-level systems with respect to their interaction with the signal pulse, and for $\tau_L \ll T_{10}$ the pulse shaping can be calculated by employing (8.69), in which we may set $\gamma_I^e = \gamma$ as the upper limit. This relation describes an amplification of the input pulses where the gain decreases with increasing time, since the occupation-number inversion becomes smaller because of the action of the signal pulse itself. As a result, appropriate conditions provided, the input pulse is highly amplified at its leading edge, whereas the trailing edge does not experience any appreciable amplification (see Fig. 8.19*c*).

Thus we see that, while the application of a sample with saturable absorption results in a suppression of the leading edge of the pulse, a sample with gain depletion causes the trailing edge of the pulse to be suppressed. Therefore it is possible to produce pulses with particularly steep leading or trailing edges in accordance with requirements set by certain applications, using the nonlinear filters described above. Furthermore, by properly combining such filters both the leading and the trailing edge of the pulse can be suppressed, which may lead to an appreciable pulse shortening. Penzkofer (Ref. 8.37) used for example a sequence of such nonlinear filters, where several amplifiers and absorbers were arranged alternately. In this way he succeeded in achieving a shortening of the light pulses by a factor of about 4. It is important that by means of such a combination pulse shortenings in time ranges below the longitudinal relaxation times of absorber and amplifier can be obtained and that limitations towards short times are imposed only by the requirement that the pulses must be much longer than the transverse relaxation times of the transitions. Since in typical dye solutions the transverse relaxation times are of the order of 10^{-14} s, in most applications this requirement is not limiting.

8.6.3. Combined Action of Nonlinear Filters in Passively Modelocked CW-Pumped Dye Lasers

The generation of very short light pulses in passively modelocked dye lasers is based on the combined effect of saturable absorption and gain depletion, which have been described in Section 8.6.2, that is, on nonlinear filtering. The pulses generated in the stationary regime may here become much shorter than the relaxation times of the two materials. This effect can be explained in the following way: once a noise pulse (or a group of noise pulses) has been generated, it will be suppressed in each round trip through the laser resonator (see Fig. 8.21) at its front and rear edges whereas its center experiences net amplification. Therefrom a shortening of the pulse duration and finally the generation of one single very short pulse will result. The continuation of

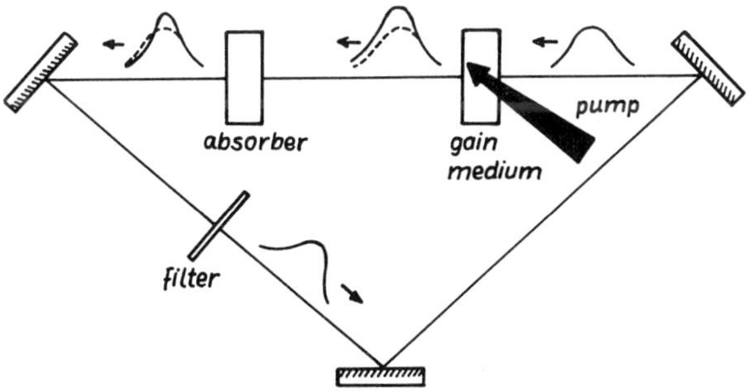

Fig. 8.21. Continuously pumped, passively modelocked dye laser. An ultrashort light pulse circulating through the resonator is schematically shown at several positions in the ring resonator for the stationary single-pulse regime. For clarity the effects of the gain medium, the saturable absorber, and the spectral filter on the pulse shape during one round trip have been exaggerated.

the shortening process is limited by frequency-selective elements present in the resonator, which prolong the pulse during each round trip by suppressing the high frequencies in the spectrum of the slowly varying pulse amplitude. Such frequency filtering may result from a spectral filter, for instance a Fabry–Perot etalon mounted in the resonator, as well as from the gain medium and the absorber medium, whose interaction cross sections are of course dependent on frequency. Finally, owing to the pulse shortening by saturable absorption and gain depletion as well as owing to the pulse lengthening by frequency filtering, a stationary pulse may arise if the parameters have been properly chosen. That is, a pulse may be obtained whose shape is completely reproduced after one round trip in the resonator. Thus these pulses behave similarly to the solitary waves discussed in Section 8.5.2.

Using such arrangements, one can generate extremely short light pulses of duration down to about 10^{-13} s (Refs. 8.38, 8.42) and, by using special modifications that allow the generation of *counterpropagating pulses* (Refs. 8.39–8.42), even down to about 40 fs. From such pulses Fork, Green, and Shank (cf. Section 6.3) by employing an additional pulse compression outside the laser resonator, succeeded in generating the shortest light pulses that had been gained so far. Recently Ippen and Weiner obtained 16 fs (Ref. 8.47) and Shank, Knox, Fork, Downer, and Stolen achieved 8 fs (Ref. 8.48) by using the same method, where the 8-fs pulses consist of only four wave periods of light. For comparison let us mention that the relaxation times of the resonant transitions in the active materials of dye lasers have typical values of some 10^{-9} s and that the relaxation times of the saturable absorbers are of the order of 10^{-10} s. From this it is again obvious that the pulse lengths, with values of some 10^{-14} s, fall far below these relaxation times, whereas on the other hand the pulse length is still long compared with the phase-decay time of typical active media of dye lasers. The basic ideas for the calculation of the self-consistent pulse shapes in the stationary regime were developed by Haus (Ref. 8.43); a discussion of the more recent experimental and theoretical results will be found in Refs. 8.42, 8.44, and 8.45. (The generation of ultrashort light pulses in other lasers is dealt with, e.g., in Ref. 8.42.) Note that the calculation of the evolution of extremely short laser pulses whose durations are comparable with the phase decay time requires one to start from the complete set of Bloch and Maxwell equations (Ref. 8.46).

REFERENCES

8.1. H. M. Gibbs, *Phys. Rev. A* **8**, 446 (1973).

8.2. I. I. Rabi, *Phys. Rev.* **51**, 652 (1937).

8.3. A. Schenzle, N. C. Wong, and R. G. Brewer, *Phys. Rev. A* **21**, 887 (1980).

8.4. R. G. Brewer, R. G. Devoe, S. C. Rand, A. Schenzle, N. C. Wong, S. S. Kano, and A. Wokaun, in A. R. W. McKellar, T. Oka, B. P. Stoicheff (Eds.), *Laser Spectroscopy V*, Springer, Berlin, Heidelberg, 1981, p. 219.

8.5. A. L. Bloom, *Phys. Rev.* **98**, 1105 (1955).

8.6. N. A. Kurnit, I. D. Abella, and S. R. Hartmann, *Phys. Rev.* **141**, 391 (1966).

8.7. U. Kh. Kopvillem, V. R. Nagibarov, V. A. Piroshkov, V. V. Samarzev, and R. G. Usmanov, *JETF Letters* **20**, 139 (1974).

8.8. H. C. Torrey, *Phys. Rev.* **76**, 1059 (1949).

8.9. C. L. Tang and H. Statz, *Appl. Phys. Lett.* **10**, 145 (1968).

8.10. G. B. Hocker and C. L. Tang, *Phys. Rev. Lett.* **21**, 591 (1969).

8.11. J. L. Hall, in S. J. Smith, G. K. Walters, L. H. Volsky (Eds.), *Atomic Physics 3*, Plenum, New York, 1973.

8.12. A. Z. Genack and R. G. Brewer, *Phys. Rev. A* **17**, 1463 (1978).

8.13. R. G. DeVoe and R. G. Brewer, *Phys. Rev. Lett.* **40**, 862 (1978).

8.14. A. Z. Genack, D. A. Weitz, and A. Schenzle, *Bull. Am. Phys. Soc.* **24**, 896 (1979).

8.15. A. Z. Genack, D. A. Weitz, R. M. Macfarlane, R. M. Shelby, and A. Schenzle, *Phys. Rev. Lett.* **45**, 438 (1980).

8.16. R. G. Brewer and R. L. Shoemaker, *Phys. Rev. Lett.* **27**, 631 (1971); *Phys. Rev. A* **6**, 2001 (1972).

8.17. R. G. Brewer, *Phys. Today* **30**, 50 (1977).

8.18. R. G. Brewer, in R. A. Smith (Ed.), *Very High Resolution Spectroscopy*, Academic, London, New York 1976.

8.19. R. G. Brewer and A. Z. Genack, *Phys. Rev. Lett.* **36**, 1959 (1976).

8.20. S. C. Rand, A. Wokaun, R. G. De Voe, and R. G. Brewer, *Phys. Rev. Lett.* **43**, 1868 (1979).

8.21. V. V. Samartsev and A. G. Shagidullin, in *Electromagnetic Superradiance*, Physico-technical Institute of Kasan, Kasan, 1975, p. 10 (Russian).

8.22. N. A. Kurnit, I. D. Abella, and S. R. Hartmann, *Phys. Rev. Lett.* **13**, 567 (1964).

8.23. I. D. Abella, in E. Wolf (Ed.), *Progress in Optics*, Vol. 7, North Holland, Amsterdam, 1969.

8.24. A. N. Orajevskij, *Usp. Fiz. Nauk* **91**, 181 (1973).

8.25. T. Mossberg, A. Kachru, and S. R. Hartmann, *Phys. Rev. Lett.* **42**, 1665 (1979).

8.26. (a) J. B. Morsink and O. A. Wiersma, in H. Walther and K. W. Rothe (Eds.), *Laser Spectroscopy IV*, Springer Series in Optical Sciences, Springer, Berlin (West), Heidelberg, New York, 1979, p. 404. (b) J. B. W. Morsink, T. J. Aartsma, and D. A. Wiersma, *Chem. Phys. Lett.* **49**, 34 (1977).

8.27. S. L. McCall and E. L. Hahn, *Phys. Rev. Lett.* **18**, 986 (1967); *Phys. Rev.* **183**, 457 (1969).

8.28. G. L. Lamb, *Rev. Mod. Phys.* **43**, 99 (1971).

8.29. R. K. Bullough, P. J. Caudrey, J. C. Eilbeck, and J. D. Gibbon, *Optoelectronics* **6**, 121 (1974).

8.30. L. Matulic and J. H. Eberly, *Phys. Rev. A* 6, 822 (1972).

8.31. G. Eilenberger, *Solitons. Mathematical Methods for Physicists*, Springer, Berlin (West), Heidelberg, New York, 1981.

8.32. R. K. Bullough and P. J. Caudrey (Eds.), *Solitons*, Springer, Berlin (West), Heidelberg, New York, 1980.

8.33. H. E. Ponath and M. Schubert, in *Proceedings of the 4th International Conference on Lasers and their Application*, Leipzig, 1981, p. 201; *Optica Acta* **30**, 1139 (1983).

8.34. C. K. N. Patel and R. E. Slusher, *Phys. Rev. Lett.* **19**, 1019 (1967).

8.35. R. E. Slusher and H. M. Gibbs, *Phys. Rev. A* **5**, 1634 (1972); R. E. Slusher, in E. Wolf (Ed.), *Progress in Optics*, Vol. 12, North Holland, Amsterdam, 1974; H. M. Gibbs and R. E. Slusher, *Phys. Rev. A* **6**, 2326 (1972).

8.36. J. Herrmann, J. Wienecke, and B. Wilhelmi, *Optical and Quant. Electr.* **7**, 337 (1975).

8.37. A. Penzkofer, *Opto-Electronics* **6**, 87 (1974).

8.38. D. J. Bradley, in S. L. Shapiro, (Ed.), *Ultrashort Light Pulses*, Springer, Berlin (West), Heidelberg, New York, 1977, p. 17.

8.39. R. L. Fork, B. I. Greene, and C. V. Shank, *Appl. Phys. Lett.* **38**, 197 (1981).

8.40. D. Kühlke, W. Rudolph, and B. Wilhelmi, *Appl. Phys. Lett.* **42**, 325 (1983); *IEEE J. Quant. Electr.* **QE-19**, 526 (1983).

8.41. J. C. Diels, J. J. Fontaine, I. C. McMichael, B. Wilhelmi, W. Dietel, D. Kühlke, and W. Rudolph, in *Proceedings of the Conference "Lasers '82*," New Orleans, 1982; *Kvant. Elektr.* **10**, 2398 (1983).

8.42. J. Herrmann and B. Wilhelmi, *Laser for Ultrashort Light Pulses*, Akademie, Berlin, 1983, Chapter 6 (German); Elsevier, Amsterdam, 1986 (English).

8.43. H. A. Haus, *IEEE J. Quant. Electr.* **QE-13**, 736 (1975).

8.44. J. Herrmann, F. Weidner, and B. Wilhelmi, *Appl. Phys.* **B26**, 187 (1981).

8.45. K. B. Eisenthal, R. M. Hochstrasser, W. Kaiser, and A. Laubereau (Eds.), *Picosecond Phenomena III*, Springer, Berlin (West), Heidelberg, New York, 1982.

8.46. W. Rudolph and B. Wilhelmi, *Appl. Phys.* **B35**, 37 (1984).

8.47. J. G. Fujimoto, A. M. Weiner, and E. P. Ippen, *Appl. Phys. Lett.* **44**, 832 (1984).

8.48. C. V. Shank, W. Knox, R. Fork, M. Downer, and R. Stolen, paper at the Conference on Lasers and Electro-Optics (CLEO), Baltimore 1985.

9

Nonlinearities and Quantum Phenomena in Transient One-Photon Processes

In the preceding chapter we dealt with nonlinear interactions in one-photon absorbers by means of semiclassical methods, with the neglect of spontaneous emission processes. In this chapter, however, we want to focus our attention on the influence of spontaneous emission processes, and this requires a quantum-mechanical treatment. As before, we preferentially discuss transient processes, since many very important nonlinear phenomena are observed under nonstationary conditions.

As early as 1931 Weisskopf (Ref. 9.1) established the theory of *resonance scattering* of optical radiation by atomic systems. For many years this theory was sufficient to satisfactorily describe all the phenomena in this field that had been observed in experiments and to predict others still unknown at that time. It was mainly the development of laser light sources with high electric field strengths that required novel theoretical investigations, in which, without employing perturbation theory, the absorber saturation caused by high fields as well as nonstationary phenomena are taken into account (see, e.g., Refs. 3 and 9.2–9.10). In our treatment we mainly follow the approach used in Refs. 9.4 and 9.9–9.12.

9.1. TRANSIENT FLUORESCENCE OF SINGLE ATOMS

In this section we treat the spontaneous emission of single atomic systems driven by intense laser radiation.

9.1.1. Basic Equations

Let us consider the intense radiation field of the exciting laser pulse as a classical field (for justification see Chapter 7), whose slowly varying amplitude envelope $\bar{E}_L(r_\bullet, t)$ is represented for simplicity by a square pulse whose intensity may take on arbitrary values. In contrast to this, the spectral and time behavior of the fluorescent light is determined with the aid of the spectral energy density and correlation functions calculated on the basis of the quantization of the radiation field. The atom is represented by a two-level system; its interaction with the radiation field is treated on the basis of the dipole approximation (cf. Section 3.2.1) and the rotating-wave approximation (cf. Section 3.2.2). To calculate the fluorescent radiation we determine the spontaneous emission of one atomic system under the influence of the laser pulse. This fluorescent radiation field is represented by the operator of the vector potential in the Heisenberg picture,

$$\hat{A}_{\bullet}(r_{\bullet}, t) = \sum_{\mu} \sqrt{\frac{\hbar}{2\varepsilon_0 V \omega_\mu}}\, e_{\mu\bullet} \hat{a}_\mu^\dagger(t) e^{-ik_{\mu\bullet} r_{\bullet}} + \text{h.c.} \tag{9.1}$$

(cf. Section 2.1), where $k_{\mu\bullet}$, ω_μ, $e_{\mu\bullet}$, and $\hat{a}_\mu^\dagger$, respectively, are the wave vector, the frequency, the unit vector of polarization direction, and the photon creation operator of the mode $\mu = (l, \sigma)$. V is the quantization volume. The creation and annihilation operators $\hat{a}_\mu^\dagger$ and $\hat{a}_\mu$, respectively, satisfy Heisenberg's equations of motion

$$\frac{d}{dt}\hat{a}_\mu^\dagger(t) = \frac{i}{\hbar}\left[\hat{H}(t), \hat{a}_\mu^\dagger(t)\right], \tag{9.2a}$$

$$\frac{d}{dt}\hat{a}_\mu(t) = \frac{i}{\hbar}\left[\hat{H}(t), \hat{a}_\mu(t)\right]. \tag{9.2b}$$

The Hamiltonian operator is built up additively from several terms:

$$\hat{H} = \hat{H}^A + \hat{H}^{AL} + \hat{H}^F + \hat{H}^{AF} \tag{9.3}$$

Here $\hat{H}^A$ denotes the Hamiltonian operator of the atom, which is described as a two-level system with the states $|1\rangle_A$ and $|0\rangle_A$ and which therefore can be represented with the aid of the atomic flip operators $\hat{B}$ and $\hat{B}^\dagger$ as

$$\hat{H}^A = \mathscr{E}_0 \hat{B}\hat{B}^\dagger + \mathscr{E}_1 \hat{B}^\dagger \hat{B}, \tag{9.4}$$

where $\mathscr{E}_1 - \mathscr{E}_0 = \hbar\omega_{10}$ (cf. Section A.2.5.2). At the time t_0, that is, without interaction, the operators $\hat{B}$ and $\hat{B}^\dagger$ are defined by

$$\hat{B}^\dagger(t_0) = |1\rangle_{A\,A}\langle 0| \tag{9.5a}$$

and

$$\hat{B}(t_0) = |0\rangle_{A\,A}\langle 1|, \tag{9.5b}$$

where $|0\rangle_A$ and $|1\rangle_A$ are the eigenstates of $\hat{H}^A$ [cf. (A.160)]. $\hat{H}^{AL}$ describes the interaction between the atom at the position $r_\bullet = 0$ and the incident laser pulse, which, as already mentioned, is treated classically. For simplicity we assume ideal (i.e., nonfluctuating) laser radiation. (In Section 7.3.2, dealing with laser models, we have already discussed the influence of the deviation of real laser radiation from ideal coherent light on the results of resonance fluorescence experiments. The simplifying assumptions used here are justified for lasers working well above threshold.) In the dipole approximation the interaction operator $\hat{H}^{AL}$ can be represented as

$$\hat{H}^{AL}(t) = -\hat{d}_{\bullet} e_{L\bullet} E_L(t), \tag{9.6}$$

where $\hat{d}_{\bullet}$ is the operator of the electric transition moment [cf. Section 3.2 and (A.174)] and $e_{L\bullet}$ the unit vector of the polarization direction of the laser field. Let the field strength of the incident pulse at $r_{\bullet} = 0$ be given by

$$E_L(t) = \begin{cases} \frac{1}{2}\bar{E}_L e^{-i\omega_L t} + \text{c.c.} & \text{for} \quad t_0 \le t \le t_0 + \tau_L \\ 0 & \text{for} \quad t < t_0 \text{ and } t > t_0 + \tau_L. \end{cases} \tag{9.7}$$

$\hat{H}^F$ is the operator of the free radiation field

$$\hat{H}^F(t) = \sum_{\mu} \hbar\omega_{\mu}\left[\hat{a}_{\mu}^{\dagger}(t)\hat{a}_{\mu}(t) + \tfrac{1}{2}\right] \tag{9.8}$$

(cf. Section 2.1). Finally, $\hat{H}^{AF}$ represents the interaction between the atomic system and the quantized radiation field. In dipole approximation and in the Coulomb gauge, which is used throughout in this treatment, it is given by

$$\hat{H}^{AF}(t) = -\sum_{\mu} i\sqrt{\frac{\hbar\omega_{\mu}}{2\varepsilon_0 V}}\, d_{\mu}\hat{a}_{\mu}^{\dagger}(t)\hat{B}(t) + \text{h.c.}, \tag{9.9}$$

where $d_{\mu} = e_{\mu\bullet}\langle 0|\hat{d}_{\bullet}|1\rangle$ has been introduced as an abbreviation. Substituting these terms for the Hamiltonian in (9.2b), we obtain the equation of motion

$$\frac{d}{dt}\hat{a}_{\mu}(t) = -i\omega_{\mu}\hat{a}_{\mu}(t) - \sqrt{\frac{\omega_{\mu}}{2\varepsilon_0 V\hbar}}\, d_{\mu}\hat{B}(t), \tag{9.10}$$

whose formal integration with respect to time leads to

$$\hat{a}_{\mu}(t) = \hat{a}_{\mu}(t_0)e^{-i\omega_{\mu}(t-t_0)} - \sqrt{\frac{\omega_{\mu}}{2\varepsilon_0 V\hbar}}\, d_{\mu}\int_{t_0}^{t} dt'\, \hat{B}(t')e^{-i\omega_{\mu}(t-t')}. \tag{9.11}$$

It is obvious that the first term in (9.10) originates from the *free radiation field* of the mode varying with time. The second term describes the modification resulting from the interaction with the atomic system or, in other words, from the *scattering process*. Substituting (9.11) and the analogous relation for the creation operator $\hat{a}_\mu^\dagger(t)$ of mode μ into (9.1), we obtain an expression for the operator of the vector potential that exhibits a similar composition of two contributions originating from the free field and the scattering process, respectively. From the operator of the vector potential $\hat{A}_\bullet$ the other field operators, for example $\hat{E}_\bullet$ and $\hat{B}_\bullet$ and those operators that can be built up from $\hat{E}_\bullet$ and $\hat{B}_\bullet$, can be determined by means of the relations given in Section 2.11.

If the density operator of the total system is known, the expectation values of measurable quantities of the field can be determined, for example that of the energy density (cf. Chapter 2)

$$\mathscr{E}_V(R_\bullet, t) = \tfrac{1}{2}\mathrm{Tr}\left\{\hat{\rho}\left[\varepsilon_0 \hat{E}_\bullet^2(R_\bullet, t) + \frac{1}{\mu_0}\hat{B}_\bullet^2(R_\bullet, t)\right]\right\}, \tag{9.12}$$

or those of field-strength correlation functions (cf. (5.32)) such as

$$\Gamma^{(2,2)}(t, t+\tau) = \sum_{i,j} \mathrm{Tr}\left\{\hat{\rho}\hat{E}_i^{(+)}(R_\bullet, t)\hat{E}_j^{(+)}(R_\bullet, t+\tau)\hat{E}_j^{(-)}(R_\bullet, t+\tau)\hat{E}_i^{(-)}(R_\bullet, t)\right\}, \tag{9.13}$$

where $\hat{E}^{(+)}$ and $\hat{E}^{(-)}$ are the operators of the positive frequency part and of the negative frequency part of the electric field strength. The energy density and this correlation function are of great importance for the interpretation of measurements, and they are frequently used for characterizing the fluorescence process. In (9.12) and (9.13) $\hat{\rho}$ stands for the density operator of the total system, which here, since the expressions are given in the Heisenberg picture, does not depend on time. The fluorescence correlation function $\Gamma^{(2,2)}(t, t')$ will later be discussed with special attention to the extraordinary coherence properties of the emitted fluorescent light.

In the following we shall first explicitly evaluate the energy density of the field as an example. From (9.12) there follows, using the representation of the field given above, that

$$\mathscr{E}_V(R_\bullet, t) = \frac{\hbar}{4V}\sum_{\mu,\lambda}\sqrt{\omega_\mu\omega_\lambda}\left[\left(\frac{k_{\mu\bullet}}{k_\mu}\times e_{\mu\bullet}\right)\left(\frac{k_{\lambda\bullet}}{k_\lambda}\times e_{\lambda\bullet}\right) + e_{\mu\bullet}e_{\lambda\bullet}\right]$$
$$\times \mathrm{Tr}\left\{\hat{\rho}\left[\hat{a}_\mu^\dagger(t)\hat{a}_\lambda(t)e^{-i(k_{\mu\bullet}-k_{\lambda\bullet})R_\bullet} + \hat{a}_\mu(t)\hat{a}_\lambda^\dagger(t)e^{i(k_{\mu\bullet}-k_{\lambda\bullet})R_\bullet}\right]\right\}. \tag{9.14}$$

Here use has been made of the assumption that the change of the operators

$\hat{a}_\mu^\dagger, \hat{a}_\mu$ during one wave period is very small, which corresponds to the slowly-varying-envelope approximation (SVEA) in the classical description. Into this relation we insert the solution for $\hat{a}_\mu(t)$ from (9.11) and the corresponding expression for $\hat{a}_\mu^\dagger(t)$, assuming the initial condition that at the time $t = t_0$ the radiation field is in the vacuum state. After subtracting from $\mathscr{E}_V(R_\bullet, t)$ the vacuum energy density, which is of no interest for the measuring process, we obtain

$$\mathscr{E}_V(R_\bullet, t) = \frac{1}{8\varepsilon_0 V^2} \sum_{\lambda,\mu} \omega_\mu \omega_\lambda \left[\left(\frac{k_{\mu\bullet}}{k_\mu} \times e_{\mu\bullet} \right) \left(\frac{k_{\lambda\bullet}}{k_\lambda} \times e_{\lambda\bullet} \right) + e_{\mu\bullet} e_{\lambda\bullet} \right]$$

$$\times d_\mu d_\lambda e^{i(\omega_\mu - \omega_\lambda)t} e^{i(k_{\mu\bullet} - k_{\lambda\bullet})R_\bullet}$$

$$\times \int_{t_0}^{t} dt_1 \int_{t_0}^{t_1} dt_2 \left[e^{i(\omega_\lambda t_1 - \omega_\mu t_2)} \left\langle \hat{B}^\dagger(t_2) \hat{B}(t_1) \right\rangle \right.$$

$$\left. + e^{i(\omega_\lambda t_2 - \omega_\mu t_1)} \left\langle \hat{B}^\dagger(t_1) \hat{B}(t_2) \right\rangle \right], \qquad (9.15a)$$

where

$$\left\langle \hat{B}^\dagger(t_2) \hat{B}(t_1) \right\rangle = \mathrm{Tr}\left\{ \hat{\rho} \hat{B}^\dagger(t_2) \hat{B}(t_1) \right\} \qquad (9.15b)$$

is a particular correlation function of the operators $\hat{B}^\dagger$ and $\hat{B}$ of the atomic system we are going to calculate in the following section. We see that the atomic operators in this expression depend on different times. By spatial integration of the energy density in (9.14) we obtain the total energy $\mathscr{E}^F$ of the quantized radiation field as the sum

$$\mathscr{E}^F(t) = \sum_\mu \mathscr{E}_\mu(t)$$

over the energy contributions of all modes, where

$$\mathscr{E}_\mu(t) = \frac{\omega_\mu^2}{4\varepsilon_0 V} |d_\mu|^2 \int_{t_0}^{t} dt_1 \int_{t_0}^{t_1} dt_2 \, e^{i\omega_\mu(t_1 - t_2)} \left\langle \hat{B}^\dagger(t_2) \hat{B}(t_1) \right\rangle + \text{c.c.}. \qquad (9.16a)$$

Moreover, by using the mode density introduced in Chapter 1 let us determine the overall field energy per unit solid angle and unit frequency in the frequency interval $(\omega, \omega + d\omega)$ with the polarization-direction state σ and the direction of propagation in the solid-angle element $d\Omega$ around the direction of $k_\bullet$. It is given as

$$\frac{d^2}{d\omega \, d\Omega} \mathscr{E}(k_\bullet, \sigma, \Omega) = \frac{\omega_{10}^4}{4c^3 \varepsilon_0} |d_\sigma|^2 \int_{t_0}^{t} dt_1 \int_{t_0}^{t_1} dt_2 \, e^{i\omega(t_1 - t_2)} \left\langle \hat{B}^\dagger(t_2) \hat{B}(t_1) \right\rangle + \text{c.c.},$$

$$(9.16b)$$

and for its time derivative we have

$$\frac{d^3}{dt\,d\omega\,d\Omega}\mathscr{E}(k_\bullet,\sigma,\Omega)=\frac{\omega_{10}^4}{4c^3\varepsilon_0}|d_\sigma|^2\int_{t_0}^{t}dt_2\,e^{i\omega(t-t_2)}\left\langle \hat{B}^\dagger(t_2)\hat{B}(t)\right\rangle+\text{c.c.},\tag{9.16c}$$

where ω is the frequency associated with the wave vector $k_\bullet$ and where $d_\sigma = e_{\sigma\bullet\,A}\langle 0|\hat{d}_\bullet|1\rangle_A$, as an analogue to the definition of d_μ, denotes the projection of the transition moment onto the polarization direction of the field. (Note that for nonstationary excitation the time derivative of $d^2\mathscr{E}/d\omega\,d\Omega$ may essentially differ from the photon flux or from the radiation intensity.)

9.1.2. Calculation of Atomic Correlation Functions

For the calculation of the properties of scattered radiation it is necessary to calculate such correlation functions of the atomic systems as $\langle \hat{B}^\dagger(t_2)\hat{B}(t_1)\rangle$, as we have seen in the previous subsection. For carrying out the calculations it is of advantage to relate, by means of the unitary time-development operator $\hat{U}(t,t_0)$ [cf. Section A.14], the two operators bearing the time arguments t_2 and t_1 to their values at the initial time t_0. This yields

$$\left\langle \hat{B}^\dagger(t_2)\hat{B}(t_1)\right\rangle$$
$$=\text{Tr}\left\{\hat{\rho}\hat{U}^{-1}(t_2,t_0)\hat{B}^\dagger(t_0)\hat{U}(t_2,t_0)\hat{U}^{-1}(t_1,t_0)\hat{B}(t_0)\hat{U}(t_1,t_0)\right\}.$$

Using the invariance of the trace under cyclic permutation and employing the property $\hat{U}(t,t_0)=\hat{U}(t,t')\hat{U}(t',t_0)$ it follows that

$$\left\langle \hat{B}^\dagger(t_2)\hat{B}(t_1)\right\rangle$$
$$=\text{Tr}\left\{\hat{U}(t_1,t_2)\hat{U}(t_2,t_0)\hat{\rho}\hat{U}^{-1}(t_2,t_0)\hat{B}^\dagger(t_0)\hat{U}^{-1}(t_1,t_2)\hat{B}(t_0)\right\}.\tag{9.17}$$

The expression

$$\hat{U}(t_2,t_0)\hat{\rho}\hat{U}^{-1}(t_2,t_0)=\hat{\rho}(t_2)\tag{9.18}$$

within the braces in (9.17) represents the density operator in the Schrödinger picture. Thus we may write

$$\left\langle \hat{B}^\dagger(t_2)\hat{B}(t_1)\right\rangle=\text{Tr}\left\{\hat{U}(t_1,t_2)\hat{\rho}(t_2)\hat{B}^\dagger(t_0)\hat{U}^{-1}(t_1,t_2)\hat{B}(t_0)\right\}.\tag{9.19}$$

To evaluate this expression we calculate the matrix elements by using the set of the eigenstates $|i\rangle_A|\cdots\ n_\mu\ \cdots\rangle$ of the undisturbed system, where the $|i\rangle_A$ are the eigenstates of the free atomic system with $i=0,1$ and the $|\cdots\ n_\mu\ \cdots\rangle =$

$|n_1, n_2, \ldots, n_\mu, \ldots\rangle$ are the eigenstates of the free radiation field. With the aid of

$$\hat{B}(t_0)|i\rangle_A = \delta_{1i}|0\rangle_A \quad \text{and} \quad \hat{B}^\dagger(t_0)|i\rangle_A = \delta_{0i}|1\rangle_A$$

[cf. Section A.2.5.2], we get

$$\left\langle \hat{B}^\dagger(t_2)\hat{B}(t_1)\right\rangle = \sum_{j=0,1}\sum_{\alpha,\beta,\gamma} U_{1\alpha,j\beta}(t_1,t_2)\rho_{j\beta,1\gamma}(t_2)U^{-1}{}_{0\gamma,0\alpha}(t_1,t_2), \tag{9.20a}$$

where we have used the abbreviations

$$(n_1,\ldots,n_\mu,\ldots) = \alpha,$$

$$(n'_1,\ldots,n'_\mu,\ldots) = \beta,$$

$$(n''_1,\ldots,n''_\mu,\ldots) = \gamma,$$

and

$$X_{i\alpha,j\beta} = {}_A\langle i|\langle\cdots n_\mu \cdots|\hat{X}|\cdots n'_\mu \cdots\rangle|j\rangle_A, \qquad \text{where} \quad \hat{X} = \hat{\rho}, \hat{U}. \tag{9.20b}$$

We start from the vacuum state of the radiation field with the density operator $|0\rangle_{F\,F}\langle 0|$ and assume that the state of the field in its time evolution will be modified only by the weak field scattered from the single atom, with one photon at most being in the normalization volume at any instant of time. In this case we may perform the calculation of the expectation values by employing the approximate density operator

$$\hat{\rho}(t) = \hat{\sigma}(t)|0\rangle_{F\,F}\langle 0|, \tag{9.21a}$$

where

$$\hat{\sigma}(t) = \mathrm{Tr}_F\{\hat{\rho}(t)\} \tag{9.21b}$$

stands for the reduced density operator of the atomic system under the influence of the classically treated laser field. $\mathrm{Tr}_F\{\cdots\}$ denotes the trace taken with respect to the states of the radiation field. For the following discussion the time evolution of the atomic operator $\sigma(t)$ is needed. If its matrix elements are calculated with the eigenstates of the undisturbed system, it follows from (9.18) by using (9.21a) that

$$\sigma_{ij}(t) = \sum_{k,m} T_{ijkm}(t,t')\sigma_{km}(t') \qquad \text{for} \quad t \geq t', \tag{9.22a}$$

where the time-evolution matrix is given by

$$T_{ijkm}(t, t') = \sum_{\alpha} U^{-1}_{m0,\, j\alpha}(t, t') U_{i\alpha,\, k0}(t, t'). \tag{9.22b}$$

Using these time-evolution matrices we may write the atomic correlation function $\langle \hat{B}^{\dagger}(t_2)\hat{B}(t_1)\rangle$ in (9.20) in the following compact form:

$$\left\langle \hat{B}^{\dagger}(t_2)\hat{B}(t_1)\right\rangle = \sum_{j=0,1} T_{10j0}(t_1, t_2)\sigma_{j1}(t_2).$$

Hence, for calculating the abovementioned quantities of the radiation field, the time-evolution matrices are required. Their determination can be described very briefly because the evolution of the atomic quantities $\sigma_{ij}(t)$ is already known to us. That is to say, for the matrix elements of $\hat{\sigma}(t)$ determined with the eigenstates of the atomic system there hold the equations of motion set up in Chapters 3, 4, and 8 in connection with the semiclassical theory [cf. (4.48), where we used the notation $\hat{\rho}$ instead of $\hat{\sigma}$] and solved in the foregoing chapter under the influence of classically described laser pulses. Let us assume that for the atomic system before switching on the laser pulse we have

$$\sigma_{00}(t_0) = 1, \qquad \sigma_{11}(t_0) = \sigma_{10}(t_0) = \sigma_{01}(t_0) = 0.$$

We further assume that the transition moment ${}_A\langle 1|\hat{d}_{\bullet}|0\rangle_A$ is real and that therefore $d_{10\bullet} = d_{01\bullet}$. We briefly write $d_{10\bullet}e_{L\bullet} = d$, where $e_{L\bullet}$ is the unit vector of the polarization direction of the laser field. Using the notation of this section, the equations for $\sigma_{ij}(t)$ read

$$\frac{d}{dt}\sigma_{11}(t) = -\frac{1}{T_{10}}\sigma_{11}(t) - \frac{d}{i\hbar}\left[\sigma_{01}(t) - \sigma_{10}(t)\right]E_L(t), \tag{9.23a}$$

$$\sigma_{11}(t) + \sigma_{00}(t) = 1, \tag{9.23b}$$

$$\frac{d}{dt}\sigma_{10}(t) = -i\omega_{10}\sigma_{10}(t) - \frac{1}{\tau_{10}}\sigma_{10}(t) - \frac{d}{i\hbar}\left[\sigma_{00}(t) - \sigma_{11}(t)\right]E_L(t), \tag{9.23c}$$

where T_{10} and τ_{10} are the lifetime of the excited level and the phase decay time, respectively. Under the action of a square pulse the solutions of this set of equations are of the same structure as those of the respective macroscopic quantities in Section 8.1. During irradiation ($0 \le t - t_0 \le \tau_L$), in the case of exact resonance ($\omega_L = \omega_{10}$), these solutions are

$$\sigma_{11}(t) = \frac{1}{2}\left\{1 - \frac{1}{T_{10}\tau_{10}\Gamma_1\Gamma_2} - \frac{\left(\dfrac{1}{T_{10}} + \Gamma_1\right)\left(\dfrac{1}{\tau_{10}} + \Gamma_1\right)}{\Gamma_1(\Gamma_1 - \Gamma_2)} e^{\Gamma_1(t-t_0)} - \frac{\left(\dfrac{1}{T_{10}} + \Gamma_2\right)\left(\dfrac{1}{\mathrm{T}_{10}} + \Gamma_2\right)}{\Gamma_2(\Gamma_2 - \Gamma_1)} e^{\Gamma_2(t-t_0)}\right\}, \tag{9.24a}$$

$$\sigma_{10}(t) = -\frac{id\bar{E}_L}{2\hbar}\left\{\frac{\dfrac{1}{T_{10}} + \Gamma_1}{\Gamma_1(\Gamma_2 - \Gamma_1)} e^{\Gamma_1(t-t_0)} + \frac{\dfrac{1}{T_{10}} + \Gamma_2}{\Gamma_2(\Gamma_1 - \Gamma_2)} e^{\Gamma_2(t-t_0)} - \frac{1}{T_{10}\Gamma_1\Gamma_2}\right\} e^{-i\omega_{10}(t-t_0)}, \tag{9.24b}$$

and after irradiation ($t - t_0 > \tau_L$),

$$\sigma_{11}(t) = \frac{1}{2} e^{-(1/T_{10})(t-t_0-\tau_L)}\left\{1 - \frac{1}{T_{10}\tau_{10}\Gamma_1\Gamma_2} - \frac{\left(\dfrac{1}{T_{10}} + \Gamma_1\right)\left(\dfrac{1}{\tau_{10}} + \Gamma_1\right)}{\Gamma_1(\Gamma_1 - \Gamma_2)} e^{\Gamma_1\tau_L} - \frac{\left(\dfrac{1}{T_{10}} + \Gamma_2\right)\left(\dfrac{1}{\tau_{10}} + \Gamma_2\right)}{\Gamma_2(\Gamma_2 - \Gamma_1)} e^{\Gamma_2\tau_L}\right\}, \tag{9.24c}$$

$$\sigma_{10}(t) = -\frac{id\bar{E}_L}{2\hbar} e^{-(1/\tau_{10})(t-t_0-\tau_L)} \times \left\{\frac{\dfrac{1}{T_{10}} + \Gamma_1}{\Gamma_1(\Gamma_2 - \Gamma_1)} e^{\Gamma_1\tau_L} + \frac{\dfrac{1}{T_{10}} + \Gamma_2}{\Gamma_2(\Gamma_1 - \Gamma_2)} e^{\Gamma_2\tau_L} - \frac{1}{T_{10}\Gamma_1\Gamma_2}\right\} e^{-i\omega_{10}(t-t_0)}, \tag{9.24d}$$

where we have with real $\overline{E}_L$

$$\Gamma_{1,2} = -\frac{1}{2}\left[\frac{1}{T_{10}} + \frac{1}{\tau_{10}}\right] \pm \sqrt{\frac{1}{4}\left(\frac{1}{T_{10}} + \frac{1}{\tau_{10}}\right)^2 - \left(\frac{d \cdot \overline{E}_L}{\hbar}\right)^2}. \quad (9.24e)$$

By means of the solution for the matrix elements $\sigma_{ij}(t)$ of the atomic systems, the time-evolution matrices $T_{ijkm}(t, t')$ can be determined by comparing the expressions for $\sigma_{ij}(t)$ with (9.22). They agree in their structure with those matrices introduced in Section 8.1, which describe the evolution of the polarization amplitude and the occupation-number inversion. Here too, we have to distinguish between the different time ranges and therefore we introduce the notation

$$T_{ijkm}(t, t') = T^L_{ijkm}(t, t') \quad \text{for} \quad t_0 \le t, t' \le t_0 + \tau_L \quad (9.25a)$$

(i.e. during the action of the laser pulse) and

$$T_{ijkm}(t, t') = T^0_{ijkm}(t, t') \quad \text{for} \quad t_0 + \tau_L < t, t' \quad (9.25b)$$

(i.e. after the termination of the pulse); then there result for those matrix elements that are needed for our further discussion the following expressions:

$$T^0_{1010}(t, t') = e^{-(1/\tau_{10} + i\omega_{10})(t-t')}, \qquad T^0_{0101} = \left(T^0_{1010}\right)^*, \quad (9.25c)$$

$$T^0_{1111}(t, t') = e^{-(1/T_{10})(t-t')}, \quad (9.25d)$$

$$T^0_{0000}(t, t) = 1 - T^0_{1111}(t, t'), \quad (9.25e)$$

$$T^L_{1010}(t, t') = \frac{1}{2}\left\{e^{-(1/\tau_{10})(t-t')} + \frac{\Gamma_1 + \dfrac{1}{\tau_{10}}}{\Gamma_1 - \Gamma_2} e^{\Gamma_2(t-t')} + \frac{\Gamma_2 + \dfrac{1}{\tau_{10}}}{\Gamma_2 - \Gamma_1} e^{\Gamma_1(t-t')}\right\} e^{-i\omega_{10}(t-t')}, \quad (9.25f)$$

$$T^L_{1000}(t,t') = -\frac{i}{2\hbar} d\bar{E}_L \left\{ \frac{\Gamma_1 + \dfrac{1}{T_{10}}}{\Gamma_1(\Gamma_2 - \Gamma_1)} e^{\Gamma_1(t-t')} + \frac{\Gamma_2 + \dfrac{1}{T_{10}}}{\Gamma_2(\Gamma_1 - \Gamma_2)} e^{\Gamma_2(t-t')} - \frac{\dfrac{1}{T_{10}}}{\Gamma_1\Gamma_2} \right\} e^{-i\omega_{10}(t-t')}, \quad (9.25g)$$

$$T^L_{1100}(t,t') = \frac{1}{2}\left\{ 1 - \frac{\left(\dfrac{1}{T_{10}} + \Gamma_1\right)\left(\dfrac{1}{\tau_{10}} + \Gamma_1\right)}{\Gamma_1(\Gamma_1 - \Gamma_2)} e^{\Gamma_1(t-t')} - \frac{\left(\dfrac{1}{T_{10}} + \Gamma_2\right)\left(\dfrac{1}{\tau_{10}} + \Gamma_2\right)}{\Gamma_2(\Gamma_2 - \Gamma_1)} e^{\Gamma_2(t-t')} - \frac{\dfrac{1}{T_{10}}\dfrac{1}{\tau_{10}}}{\Gamma_1\Gamma_2} \right\}. \quad (9.25h)$$

The time-evolution matrices $T^L_{ijkl}(t,t')$ will now also be given for the following more general case we will make use of in Sections 9.4 and 9.5, but without noting the somewhat lengthy calculations, which have to be carried out in the same way. Let the atom now be at the point $r_\bullet$, where it is excited by the plane laser wave $E_L(r_\bullet, t) = \bar{E}_L e_{L\bullet} \cos(k_{L\bullet} r_\bullet - \omega_L t)$, whose pulse envelope $\bar{E}_L$ is assumed to be constant during the pulse length τ_L. Furthermore we now assume deviations from exact resonance, which means $\Delta = \omega_L - \omega_{10} \neq 0$. Then we have the following matrices

$$T^L_{1010}(t,t',r_\bullet) = \frac{1}{2\Delta\mathscr{N}} e^{i\omega_L(t-t')} \times \left\{ (x_3 - x_2)\left[\Delta + i\left(\frac{1}{\tau_{10}} + x_1\right)\right] \times \left[\frac{1}{\tau_{10}^2} + \frac{1}{\tau_{10}}(x_2 + x_3) + x_2 x_3 - \Delta^2 + i\Delta\left(\frac{2}{\tau_{10}} + x_2 + x_3\right)\right] e^{x_1(t-t')} + \text{cycl.} \right\}, \quad (9.26a)$$

$$T^L_{1001}(t, t', r_\bullet) = \frac{1}{2\Omega_{R0}\Delta\mathcal{N}} e^{i[2k_{L\bullet}r_\bullet - \omega_L(t+t')]}$$

$$\times\left\{(x_3 - x_2)\left[\frac{1}{\tau_{10}^2} - \Delta^2 + \frac{1}{\tau_{10}}(x_2 + x_3) + x_2 x_3 - i\Delta\left(\frac{2}{\tau_{10}} + x_2 + x_3\right)\right]\right.$$

$$\left.\times\left[\left(\frac{1}{\tau_{10}} + x_1\right)^2 + \Delta^2\right] e^{x_1(t-t')} + \text{cycl.}\right\}, \qquad (9.26b)$$

$$T^L_{1000}(t, t', r_\bullet) = e^{i(k_{L\bullet}r_\bullet - \omega_L t)}$$

$$\times\left\{+\frac{\frac{1}{T_{10}}\sigma_I^0\Omega_{R0}}{2\mathscr{R}}\left(\Delta + i\frac{1}{\tau_{10}}\right) - \frac{\Omega_{R0}}{2\mathcal{N}}\left[(x_3 - x_2)\left(\Delta + i\left\{\frac{1}{\tau_{10}} + x_1\right\}\right)\right.\right.$$

$$\left.\left.\times\left(1 + \frac{\frac{1}{T_{10}}\sigma_I^0 x_2 x_3}{\mathscr{R}}\right) e^{x_1(t-t')} + \text{cycl.}\right]\right\}, \qquad (9.26c)$$

$$T^L_{1011}(t, t', r_\bullet) = e^{i(k_{L\bullet}r_\bullet - \omega_L t)}$$

$$\times\left\{-\frac{\frac{1}{T_{10}}\sigma_I^0\Omega_{R0}}{2\mathscr{R}}\left(\Delta + i\frac{1}{\tau_{10}}\right) + \frac{\Omega_{R0}}{2\mathcal{N}}\left[(x_3 - x_2)\left(\Delta + i\left\{\frac{1}{\tau_{10}} + x_1\right\}\right)\right.\right.$$

$$\left.\left.\times\left(1 - \frac{\frac{1}{T_{10}}\sigma_I^0}{\mathscr{R}} x_3 x_2\right) e^{x_1(t-t')} + \text{cycl.}\right]\right\}, \qquad (9.26d)$$

$$T^L_{1111}(t, t', r_\bullet) = \frac{1}{2} + \frac{1}{2}\frac{\frac{1}{T_{10}}\sigma_I^0}{\mathscr{R}}\left(\Delta^2 + \frac{1}{\tau_{10}^2}\right)$$

$$+ \frac{1}{2\mathscr{N}}\left\{(x_3 - x_2)\left[\left(\frac{1}{\tau_{10}} + x_1\right)^2 + \Delta^2\right]\right.$$

$$\left.\times\left(1 - \frac{\frac{1}{T_{10}}\sigma_I^0}{\mathscr{R}}x_3x_2\right)e^{x_1(t-t')} + \text{cycl.}\right\},$$

(9.26e)

$$T^L_{1100}(t, t', r_\bullet) = \frac{1}{2} - \frac{1}{2}\frac{\frac{1}{T_{10}}\sigma_I^0}{\mathscr{R}}\left(\Delta^2 + \frac{1}{\tau_{10}^2}\right)$$

$$+ \frac{1}{2\mathscr{N}}\left\{(x_3 - x_2)\left[\left(\frac{1}{\tau_{10}} + x_1\right)^2 + \Delta^2\right]\right.$$

$$\left.\times\left(1 - \frac{\frac{1}{T_{10}}\sigma_I^0}{\mathscr{R}}x_3x_2\right)e^{x_1(t-t')} + \text{cycl.}\right\}, \quad (9.26f)$$

where the summation of terms with cyclic permutations of their arguments is defined by

$$f(x_1, x_2, x_3) + \text{cycl.} = f(x_1, x_2, x_3) + f(x_2, x_3, x_1) + f(x_3, x_1, x_2). \quad (9.26g)$$

The quantities

$$\mathscr{R} = \frac{1}{T_{10}}\frac{1}{\tau_{10}^2} + \frac{1}{T_{10}}\Delta^2 + \frac{1}{\tau_{10}}\Omega_{R0}^2 \quad (9.26h)$$

and

$$\mathcal{N} = x_1^2(x_3 - x_2) + x_2^2(x_1 - x_3) + x_3^2(x_2 - x_1) = x_1^2(x_3 - x_2) + \text{cycl.} \tag{9.26i}$$

have been used for abbreviation. The x_i $(i = 1, 2, 3)$ are the roots of the cubic characteristic equation of (9.23), and they are given by

$$x_1 = U_+ + U_- - \frac{1}{3}\left(\frac{1}{T_{10}} + \frac{2}{\tau_{10}}\right), \tag{9.26j}$$

$$x_{2,3} = \tfrac{1}{2}(-1 \pm i\sqrt{3})U_+ + \tfrac{1}{2}(-1 \mp i\sqrt{3})U_- - \frac{1}{3}\left(\frac{1}{T_{10}} + \frac{2}{\tau_{10}}\right), \tag{9.26k}$$

where

$$U_\pm = \Bigg(-\tfrac{1}{27}\Gamma^3 + \tfrac{1}{3}\Gamma\left(\tfrac{1}{2}\Omega_{R0}^2 - \Delta^2\right)$$
$$\pm\sqrt{\tfrac{1}{27}\left[\Gamma^4\Delta^2 + \Gamma^2\left(2\Delta^4 - 5\Delta^2\Omega_{R0}^2 - \tfrac{1}{4}\Omega_{R0}^4\right) + \left(\Omega_{R0}^2 + \Delta^2\right)^3\right]}\Bigg)^{1/3} \tag{9.26l}$$

and

$$\Gamma = \frac{1}{T_{10}} - \frac{1}{\tau_{10}}. \tag{9.26m}$$

The quantity $\sigma_I^0 = \sigma_{11}^0 - \sigma_{00}^0$ is given by the value of the diagonal elements before irradiation sets in. If the atomic systems are in thermal equilibrium before irradiation, then $\sigma_I^0 = \sigma^e = \gamma_I^e/\gamma$, where γ_I^e is the equilibrium value of the occupation-number inversion density. Furthermore, $\Omega_{R0} = \overline{E}_L d/\hbar$ denotes the value of the Rabi frequency Ω_R for zero detuning $(\Delta = 0)$.

Using the time-evolution matrices of (9.25) or (9.26), we can now calculate the atomic correlation function $\langle \hat{B}^\dagger(t_2)\hat{B}(t_1)\rangle$. In this section we restrict ourselves to the simple case of exact resonance, where (9.25) holds. Therefore we insert these evolution matrices $T_{10j0}(t_1, t_2)$ into (9.23) and calculate the matrix elements $\sigma_{j1}(t_2)$ from their initial values at $t = t_0$ with the help of the time-evolution matrices $T_{j1km}(t_2, t_0)$. According to the positions of t_1 and t_2 (where $t_1 \geq t_2 \geq t_0$) relative to the end of the irradiation at $t = t_0 + \tau_L$, three different cases must be distinguished, which are shown schematically in Fig. 9.1

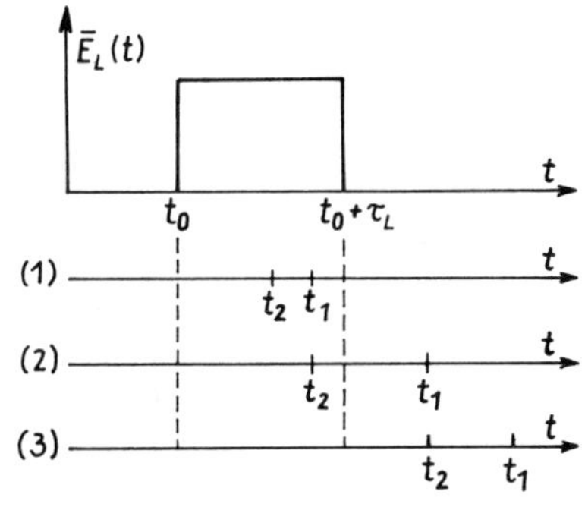

Fig. 9.1. Relative position of the laser pulse and the time points t_1, t_2.

and for which the following results are obtained:

1. $t_1, t_2 \leq t_0 + \tau_L$:

$$\left\langle \hat{B}^\dagger(t_2)\hat{B}(t_1)\right\rangle = T^L_{1010}(t_1, t_2)T^L_{1100}(t_2, t_0) + T^L_{1000}(t_1, t_2)T^L_{0100}(t_2, t_0). \tag{9.27a}$$

2. $t_2 \leq t_0 + \tau_L,\ t_1 \geq t_0 + \tau_L$:

$$\left\langle \hat{B}^\dagger(t_2)\hat{B}(t_1)\right\rangle = T^0_{1010}(t_1, t_0 + \tau_L) \times \left\{ T^L_{1010}(t_0 + \tau_L, t_2)T^L_{1100}(t_2, t_0) + T^L_{1000}(t_0 + \tau_L, t_2)T^L_{0100}(t_2, t_0)\right\}. \tag{9.27b}$$

3. $t_2, t_1 \geq t_0 + \tau_L$:

$$\left\langle \hat{B}^\dagger(t_2)\hat{B}(t_1)\right\rangle = T^0_{1010}(t_1, t_2)T^0_{1111}(t_2, t_0 + \tau_L)T^L_{1100}(t_0 + \tau_L, t_0). \tag{9.27c}$$

9.1.3. Energy and Power of Fluorescence

In Section 9.1.1 relations were given for the spectral energy density $d^2\mathscr{E}/d\omega\, d\Omega$ and its time derivative $d^3\mathscr{E}/d\omega\, d\Omega\, dt$, which are dependent on the atomic correlation functions $\langle \hat{B}^\dagger(t_2)\hat{B}(t_1)\rangle$. Now we are in a position to calculate explicitly the time derivative of the spectral energy density by substituting the expressions for $\langle \hat{B}^\dagger(t_2)\hat{B}(t_1)\rangle$ from (9.27) together with (9.25) into (9.16c). During and after the action of the laser pulse we obtain, respectively,

$$\frac{d^3}{dt\, d\omega\, d\Omega}\mathscr{E}(k_\bullet, \sigma) = \frac{\omega_{10}^4}{4c^3\varepsilon_0}|d_{(k_\bullet)}|^2\tau_{10}D_1(\tilde{t}) \qquad (\tilde{t} \leq \tilde{\tau}_L) \tag{9.28a}$$

and

$$\frac{d^3}{dt\,d\omega\,d\Omega}\mathscr{E}(k_\bullet,\sigma)=\frac{\omega_{10}^4}{4c^3\varepsilon_0}|d_{(k_\bullet)}|^2\tau_{10}$$
$$\times\left\{D_3(\tilde t)+e^{-(\tilde t-\tilde\tau_L)}\left[D_1(\tilde\tau_L)\cos\delta(\tilde t-\tilde\tau_L)\right.\right.$$
$$\left.\left.+D_2(\tilde\tau_L)+D_2(\tilde\tau_L)\sin\delta(\tilde t-\tilde\tau_L)\right]\right\}\qquad(\tilde t>\tilde\tau_L),\quad(9.28\text{b})$$

where $\tilde t=(t-t_0)/\tau_{10}$ and $\tilde\tau_L=\tau_L/\tau_{10}$ are the normalized time of observation and the pulse duration. Furthermore the following set of abbreviations was used for (9.28a, b):

$$D_{1,2}(\tilde t)=\mathscr{D}_{1,2}(\tilde t,\alpha)+\mathscr{D}_{1,2}(\tilde t,-\alpha),\tag{9.28c}$$

$$\mathscr{D}_1(\tilde t,\alpha)=f_1\left[b_2+\left(b_2-b_1+\frac{\gamma_-}{4\alpha^2}\right)f_6+\frac{b_2\delta_+-3b_1\delta_0}{2\alpha}f_5\right]$$
$$+\frac{b_1}{1+\delta^2}(1+\delta f_3-f_4)$$
$$-\frac{4b_1}{4(\delta+\alpha)^2+\delta_-^2}\left[B_1(f_4-f_6)+B_2(f_3+f_5)\right]$$
$$+\frac{b_1}{\delta+2\alpha}\left[-(f_5+f_7)\frac{\delta_-}{4\alpha^2}+(f_6-f_8)\frac{1}{2\alpha}\right]$$
$$-\frac{4}{4(\delta+\alpha)^2+\delta_+^2}$$
$$\times\left[B_3(1-f_8)+B_4f_7+B_5(\cos\delta\tilde t-f_6)+B_6(\sin\delta\tilde t+f_5)\right],\tag{9.28d}$$

$$\mathscr{D}_2(\tilde t,\alpha)=f_2\left[b_2+\left(b_2-b_1+\frac{\gamma_-}{4\alpha^2}\right)f_6+\frac{b_3\delta_+-3b_1\delta_0}{2\alpha}f_5\right]$$
$$+\frac{b_1}{1+\delta^2}(f_3+\delta f_4-\delta)$$
$$-\frac{4b_1}{4(\delta+\alpha)^2+\delta_-^2}\left[B_2(f_4-f_6)-B_1(f_3+f_5)\right]$$
$$+\frac{b_1}{\delta+2\alpha}\left[(f_6-f_8)\frac{\delta_-}{4\alpha^2}+(f_5+f_7)\frac{1}{2\alpha}\right]$$
$$-\frac{4}{4(\delta+\alpha)^2+\gamma_+^2}\left[-B_4(1-f_8)+B_3f_7+B_6(\cos\delta\tilde t-f_6)\right.$$
$$\left.-B_5(\sin\delta\tilde t+f_5)\right];\tag{9.28e}$$

$$D_3(\tilde{t}) = \frac{2b_1}{\delta^2 + \delta_-^2}\left[2 - 2f_6 + \frac{\delta_+}{\alpha}f_5\right]$$
$$\times\left[e^{-(\tilde{t}-\tilde{\tau}_L)}(\delta_-\cos\delta(\tilde{t}-\tilde{\tau}_L) + \delta\sin\delta(\tilde{t}-\tilde{\tau}_L) - \delta_- e^{-\delta_0(\tilde{t}-\tilde{\tau}_L)}\right]; \tag{9.28f}$$

$$\alpha = \frac{1}{2\beta}\sqrt{1 - \beta^2\delta_-^2}, \tag{9.28g}$$

$$\beta = -\frac{i\hbar}{2\tau_{10}d\cdot\bar{E}_L} = -\frac{i}{2\Omega_{R0}\tau_{10}}, \tag{9.28h}$$

$$\delta = (\omega_{10} - \omega)\tau_{10}, \tag{9.28i}$$

$$\delta_0 = \frac{\tau_{10}}{T_{10}}, \qquad \delta_+ = \delta_0 + 1, \qquad \delta_- = \delta_0 - 1, \tag{9.28j}$$

$$b_1 = \frac{1}{4(1 + 4\beta^2\delta_0)}, \tag{9.28k}$$

$$b_2 = \frac{\delta_0^2\beta^2}{(1 + 4\beta^2\delta_0)^2},$$

$$f_1 = \frac{1}{\delta}\sin\delta\tilde{t}, \qquad f_2 = \frac{1}{\delta}[\cos\delta\tilde{t} - 1] \tag{9.28l}$$

$$f_3 = e^{-\tilde{t}}\sin\delta\tilde{t}, \qquad f_4 = e^{-\tilde{t}}\cos\delta\tilde{t} \tag{9.28m}$$

$$f_5 = e^{-\frac{1}{2}\delta_+\tilde{t}}\sin\alpha\tilde{t}, \qquad f_6 = e^{-\frac{1}{2}\delta_+\tilde{t}}\cos\alpha\tilde{t} \tag{9.28n}$$

$$f_7 = e^{-\frac{1}{2}\delta_+\tilde{t}}\sin(\delta + \alpha)\tilde{t}, \qquad f_8 = e^{-\frac{1}{2}\delta_+\tilde{t}}\cos(\delta + \alpha)\tilde{t}, \tag{9.28o}$$

$$B_1 = \delta_0 + \frac{\delta\cdot\delta_+}{2\alpha}, \qquad B_2 = \delta + \alpha - \frac{\delta_0^2 - 1}{4\alpha}, \tag{9.28p}$$

$$B_3 = \left[\frac{1 - 3\delta_0}{2}b_1 + \frac{\delta_+}{2}b_2\right]\frac{\delta}{\alpha} - 2\delta_0 b_1 + \delta_+ b_2, \tag{9.28q}$$

$$B_4 = (b_2 - b_1)\delta + (2b_2 - b_1)\alpha - \frac{\delta_-^2}{4\alpha}b_1, \tag{9.28r}$$

$$B_5 = \left(\frac{\delta_+}{2}b_2 - \delta_0 b_1\right)\frac{\delta}{\alpha} + \delta_+ b_2 - \delta_0 b_1, \tag{9.28s}$$

$$B_6 = \delta b_2 + \alpha b_1 + \frac{1}{\alpha}\left(\frac{\delta_0\delta_+}{2}b_1 - \frac{\delta_+^2}{4}b_2\right). \tag{9.28t}$$

From (9.28a, b) we obtain the spectral energy density $(d^2/d\omega\, d\Omega)\mathscr{E}(k_\bullet, \sigma)$ by integration over the normalized time of observation $\tilde{t} = (t - t_0)/\tau_{10}$. During and after the action of the laser pulse this yields, respectively,

$$\frac{d^2}{d\omega\, d\Omega}\mathscr{E}(k_\bullet, \sigma) = \frac{\omega_{10}^4 \tau_{10}^2}{4c^3\varepsilon_0}|d_{(k_\bullet)\sigma}|^2 \int_0^{\tilde{t}} d\tilde{t}'\, D_1(\tilde{t}') \qquad (\tilde{t} \le \tilde{\tau}_L) \quad (9.29a)$$

and

$$\frac{d^2}{d\omega\, d\Omega}\mathscr{E}(k_\bullet, \sigma) = \frac{\omega_{10}^4 \tau_{10}^2}{4c^3\varepsilon_0}|d_{(k_\bullet)\sigma}|^2 \Bigg\{ \int_0^{\tilde{\tau}_L} d\tilde{t}'\, D_1(\tilde{t}') + D_0(\tilde{t}) + \frac{D_1(\tilde{\tau}_L)}{1 + \delta^2}$$
$$\times \left[1 - e^{-(\tilde{t} - \tilde{\tau}_L)}\{\cos\delta(\tilde{t} - \tilde{\tau}_L) - \delta \sin\delta(\tilde{t} - \tilde{\tau}_L)\}\right]$$
$$+ \frac{D_2(\tilde{\tau}_L)}{1 + \delta^2}\left[\delta - e^{-(\tilde{t} - \tilde{\tau}_L)}\{\sin\delta(\tilde{t} - \tilde{\tau}_L) + \delta\cos\delta(\tilde{t} - \tilde{\tau}_L)\}\right]\Bigg\}, \quad (\tilde{t} > \tilde{\tau}_L) \quad (9.29b)$$

where

$$D_0(\tilde{t}) = b_1\left[2 - 2f_6 - \frac{\delta_+}{\alpha} f_5\right]$$
$$\times \left\{ \frac{2}{\delta_0(1 + \delta^2)} + \frac{2\delta_- e^{-\delta_0(\tilde{t} - \tilde{\tau}_L)}}{\delta_0(\delta^2 + \delta_-^2)} \right.$$
$$\left. - \frac{2e^{-(\tilde{t} - \tilde{\tau}_L)}\left[(\delta^2 + \delta_-)\cos\delta(\tilde{t} - \tilde{\tau}_L) + (2 - \delta_0)\delta \sin\delta(\tilde{t} - \tilde{\tau}_L)\right]}{(\delta^2 + \delta_-)^2 + (2 - \delta_0)^2\delta^2} \right\}. \quad (9.29c)$$

Let us consider two special cases. First, for stationary irradiation or for very long pulses ($\tilde{\tau}_L \gg 1$) and a large time of observation ($\tilde{t} - \tilde{\tau}_L) \gg 1$) we obtain from (9.29) and (9.28)

$$\frac{d^2}{d\omega\, d\Omega}\mathscr{E}(k_\bullet, \sigma) = \frac{\omega_{10}^4 \tau_{10}^2}{4c^3\varepsilon_0}|d_{(k_\bullet)\sigma}|^2$$
$$\times \{\mathscr{G}_1(\tilde{\tau}_L, \alpha) + \mathscr{G}_1(\tilde{\tau}_L, -\alpha)\}, \qquad (9.30a)$$

where

$$\mathscr{G}_1(\tilde{\tau}_L, \alpha) = b_2\left[\frac{1 - \cos\delta\tilde{\tau}_L}{\delta^2} + \frac{\sin\delta\tilde{\tau}_L}{\delta(1+\delta^2)}\right] - \frac{4B_5 \sin\delta\tilde{\tau}_L}{\delta\left[4(\delta+\alpha)^2 + \delta_+^2\right]}$$

$$+ \frac{b_1\tilde{\tau}_L}{1+\delta^2} - \frac{4B_3\tilde{\tau}_L}{4(\delta+\alpha)^2 + \delta_+^2}. \tag{9.30b}$$

Even in this case the dependence of the energy of the fluorescent light on the normalized frequency offset $\delta = (\omega_{10} - \omega)\tau_{10}$ is not given by a Lorentzian profile of natural line width. The main part of the light is emitted within the frequency interval of the incident radiation, which is assumed to be much smaller than the natural line width. This part is proportional to the term $(1 - \cos\delta\tilde{\tau}_L)/\delta^2$ in $\mathscr{G}_1$, which, assuming $\delta \cdot \tilde{\tau}_L \ll 1$, is of the order of $\tilde{\tau}_L^2$, whereas all other terms are only of the order of $\tilde{\tau}_L$. In addition, there are contributions to the fluorescence that give rise to two secondary maxima at $\delta = \pm\alpha$. In the limit $\tilde{\tau}_L \to \infty$ the term $(1 - \cos\delta\tilde{\tau}_L)/\tilde{\tau}_L\delta^2$ becomes proportional to Dirac's delta function, and hence the light is emitted at the sharp frequency of the incident radiation. On the other hand, providing the pulse length τ_L to be short compared with the relaxation time τ_{10}, we obtain

$$\frac{d^2}{d\omega\, d\Omega}\mathscr{E} = \frac{\omega_{10}^4\tau_L^2}{4c^3\varepsilon_0}|d_{(k_\bullet)\sigma}|^2\{\mathscr{G}_2(\tilde{\tau}_L, \alpha) + \mathscr{G}_2(\tilde{\tau}_L, -\alpha)\}, \tag{9.31a}$$

where

$$\mathscr{G}_2(\tilde{\tau}_L, \alpha) = \frac{\tilde{\tau}_L}{1+\delta^2}\left[2b_2 - b_1 + \frac{2\delta^2 b_1}{1+\delta^2} + \frac{4b_1(B_1\delta - B_2)}{4(\delta+\alpha)^2 + \delta_-^2}\right.$$

$$\left. + \frac{4(\delta B_5 - \delta B_3 - B_4 - B_5)(\delta+\alpha)}{4(\delta+\alpha)^2 + \delta_+^2}\right]. \tag{9.31b}$$

Now the energy emitted during the action of the laser pulse can be neglected in comparison with the energy emitted after excitation.

In Fig. 9.2 the spectral energy density $d^2\mathscr{E}/d\omega\, d\Omega$ and its time derivative $d^3\mathscr{E}/d\omega\, d\Omega\, dt$ are plotted against the frequency offset δ for different durations and amplitudes of the incident laser pulses and at different times $t_e = t - \tau_L$. At a given time of observation t_e the spectral behavior of $d^2\mathscr{E}/d\omega\, d\Omega$ is characterized by a primary maximum and two secondary maxima. From comparing the values of the spectral energy density at different times of

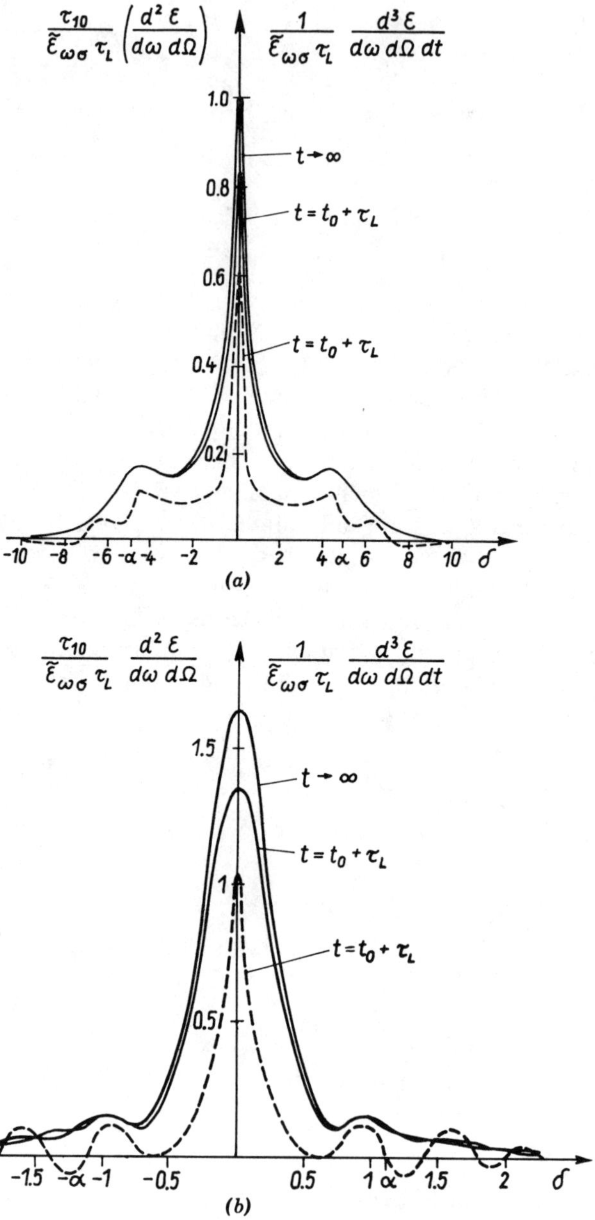

Fig. 9.2. Spectral energy density $d^2\mathscr{E}/d\omega\, d\Omega$ (full line) and its time derivative $d^3\mathscr{E}/d\omega\, d\Omega\, dt$ (dashed line) versus frequency offset $\delta = (\omega_1 - \omega)\tau_{10}$ at $t = t_0 + \tau_L$ and $t \to \infty$. (In the normalization factor the abbreviation $\tilde{\mathscr{E}}_{\omega\sigma} = (\omega_{10}^4 \tau_L^3 / 4c^3 \varepsilon_0 \tau_{10}) |d_{(k_\bullet)\sigma}|^2$ is used.) (a) $\tau_L/\tau_{10} = 10$ and $\beta = \hbar/(2\tau_{10} d\bar{E}_L) = 0.1$; (b) $\tau_L/\tau_{10} = 10$ and $\beta = 0.4$; (c) $\tau_L/\tau_{10} = 2$ and $\beta = 0.4$.

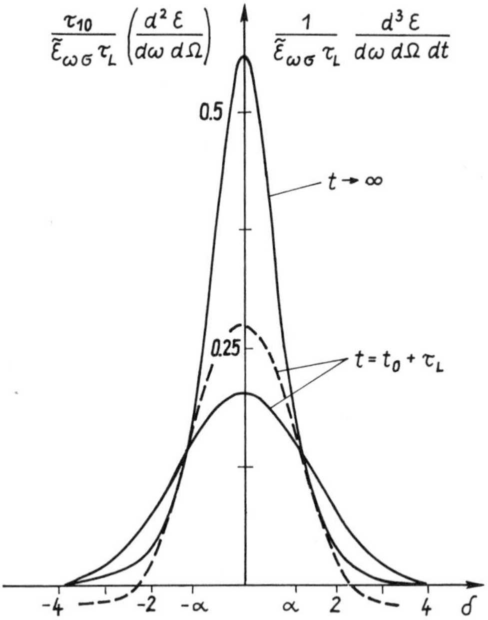

(c)

Fig. 9.2. (*Continued*).

observation t_e, nonstationary spectral effects become obvious. Thus upon excitation the derivative $d^3\mathscr{E}/d\omega\, d\Omega\, dt$ is negative in the outer spectral regions. From this fact a typical change of the spectral content of $d^2\mathscr{E}/d\omega\, d\Omega$ with increasing time t_e results. Immediately upon excitation we find a rather broad and flat frequency distribution of the spectral energy density. Then with increasing time t_e the maximum at $\delta = 0$ rises and the distribution becomes narrower. During this process the energy of the modes far from the center may even decrease.

To describe experiments in which the fluorescent light is observed without spectral resolution, let us now calculate the time derivative $d^2\mathscr{E}/d\Omega\, dt$ of the total radiation energy of fixed polarization emitted into the solid-angle element $d\Omega$ by integration of $d^3\mathscr{E}/d\omega\, d\Omega\, dt$ from (9.16c) with respect to frequency. Using the relations (9.22)–(9.25), we arrive at

$$\frac{d^2}{d\Omega\, dt}\mathscr{E} = \frac{\pi\omega_{10}^4}{c^3\varepsilon_0}|d_{(k_\bullet)\sigma}|^2\sigma_{11}(t) \tag{9.32}$$

for the power of fluorescent light per unit solid angle, where $\sigma_{11}(t)$ is the probability of finding the atom in the atomic eigenstate $|1\rangle_A$.

9.1.4. Intensity Correlation Functions

In Chapter 5 we realized that the interpretation of decisive measurements on the radiation field requires the calculation of correlation functions of field strength and intensity. Let us here investigate the properties of the field-strength correlation function

$$\Gamma^{2,2}_{abcd}(t_1, t_2, t_2, t_1) = \mathrm{Tr}\{\hat{\rho}(0)\hat{E}_a^{(+)}(t_1)\hat{E}_b^{(+)}(t_2)\hat{E}_c^{(-)}(t_2)\hat{E}_d^{(-)}(t_1)\} \quad (9.33)$$

(cf. Section 5.2.2). For simplicity we restrict our consideration to one-component fields and drop the tensor indices. This fourth-order field-strength correlation function is proportional to the second-order intensity correlation function. The intensity correlation functions, which contain important information on the statistics of the radiation field, can for example be determined from time-correlated photon-counting experiments at a fixed place, that is, by making use of one photon detector only (cf. Section 5.2). In the following section experiments on the radiation of fluorescent atoms will be analyzed.

The field-strength correlation function $\Gamma^{2,2}$ can be calculated by using the time-evolution matrices T_{ijkm} of the atomic system in an analogous way to that described for the field energy in Section 9.1.3. Again we assume that the atom is excited by ideally coherent light. (For the influence of fluctuations of the pump radiation on the field-strength correlation functions see Section 7.3.2.) This leads to

$$\Gamma^{2,2}(t, t+\tau, t+\tau, t) = CT_{1100}(t+\tau, t)\sigma_{11}(t), \quad (9.34)$$

where C is a factor independent of the time arguments. From (9.22a) it is evident that

$$T_{1100}(t+\tau, t) = \sigma_{11}^{(c)}(t+\tau), \quad (9.35)$$

where $\sigma_{11}^{(c)}(t+\tau)$ is a special solution of the set of equations (9.23) with the initial condition $\sigma_{11}^{(c)}(t) = \sigma_{10}^{(c)}(t) = \sigma_{01}^{(c)}(t) = 0$ and $\sigma_{00}^{(c)}(t) = 1$, which consequently gives the probability of finding the atomic system in the excited state at the time $t+\tau$, provided that it was in the ground state at the time t. Hence it is possible to represent $\Gamma^{2,2}$ by a product of $\sigma_{11}(t)$ and $\sigma_{11}^{(c)}(t+\tau)$.

Let us now replace $\sigma_{11}(t)$ and $\sigma_{11}^{(c)}(t+\tau)$ by measurable properties of the radiation field. According to (9.32), $\sigma_{11}(t)$ is proportional to the fluorescence power emitted at time t. This power, on the other hand, is proportional to the second-order field-strength correlation function $\Gamma^{1,1}(t, t)$. The quantity $\sigma_{11}^{(c)}(t+\tau)$ is proportional to the average power emitted at the time $t+\tau$ if the interaction between the atom and the laser field was switched on at t. This power in turn is proportional to the correspondingly defined field-strength correlation function $\Gamma^{1,1}_{(c)}(t+\tau, t+\tau)$. Using these relations for $\sigma_{11}(t)$ and

$\sigma_{11}^{(c)}(t+\tau)$, we get the field-strength correlation function

$$\Gamma^{2,2}(t, t+\tau, t+\tau, t) = \Gamma^{1,1}(t, t)\Gamma^{1,1}_{(c)}(t+\tau, t+\tau) \tag{9.36}$$

in factorized form (cf. Ref. 9.6)]. For $t + \tau \leq \tau_L$, that is, during the laser action, we have

$$T_{1100}(t+\tau, t) = T_{1100}(\tau, 0) = \sigma_{11}(\tau),$$

and therefore instead of (9.36) we can write

$$\Gamma^{2,2}(t, t+\tau, t+\tau, t) = \Gamma^{1,1}(t, t)\Gamma^{1,1}_{(c)}(\tau, \tau), \tag{9.37}$$

where $\Gamma^{1,1}_{(c)}(\tau, \tau)$ is proportional to the mean intensity at time τ after switching on the interaction at $\tau = 0$. Furthermore let us calculate the normalized intensity correlation function

$$\gamma^{2,2}(t, t+\tau) = \frac{\Gamma^{2,2}(t, t+\tau, t+\tau, t)}{\Gamma^{1,1}(t, t)\Gamma^{1,1}(t+\tau, t+\tau)} \tag{9.38}$$

(cf. Section 5.2.2). Using (9.37), we have from (9.38)

$$\gamma^{2,2}(t, t+\tau) = \frac{\Gamma^{1,1}_{(c)}(\tau, \tau)}{\Gamma^{1,1}(t+\tau, t+\tau)}. \tag{9.39}$$

By relating the field-strength correlations functions $\Gamma^{1,1}_{(c)}$ and $\Gamma^{1,1}$ to the values of σ_{11} we obtain

$$\gamma^{2,2}(t, t+\tau) = \frac{\sigma_{11}^{(c)}(\tau)}{\sigma_{11}(t+\tau)}. \tag{9.40}$$

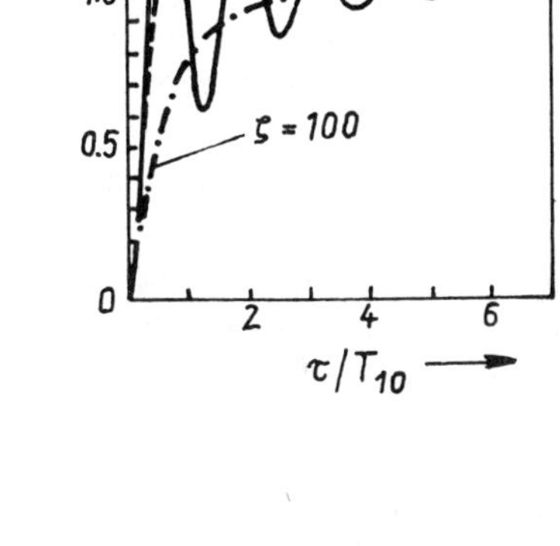

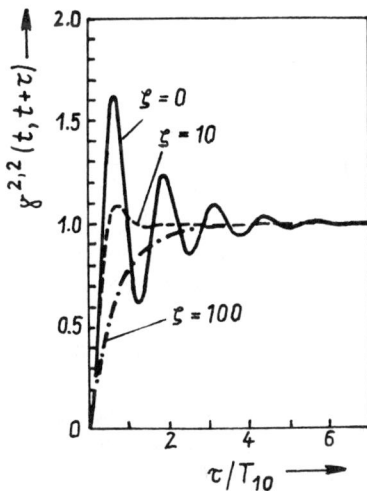

Fig. 9.3. Normalized field-strength correlation function $\gamma^{2,2}(t, t+\tau)$ of fluorescence light for stationary excitation, (after Ref. 9.18). $\Omega_{R0}T_{10} = 5$; $\zeta = 2T_{10}/\tau_{10} - 1 = 0$, 10, and 100.

Thus the normalized correlation function $\gamma^{2,2}(t, t+\tau)$ is given by the ratio of the occupation probabilities of the excited atomic levels at the times τ and $t+\tau$, whose time dependence is already known to us. In the case of stationary excitation we shall average over the time t. In (9.39) and (9.40) this averaging results in the appearance of the time-averaged fluorescence power or of the time-averaged occupation number in the denominator. Figure 9.3 shows the normalized correlation function $\gamma^{2,2}(t, t+\tau)$ against the delay time τ for stationary pump irradiation that builds up the atomic excitation characterized by $\sigma_{11}^{(c)}(\tau)$. Attention should be given to the fact that $\gamma^{2,2}(t, t+\tau)$ vanishes at $\tau = 0$ and tends to unity only after damped oscillations with the Rabi frequency Ω_R [cf. Section 8.1]. The inequality $\gamma^{2,2}(t, t+0) < 1$ indicates the nonclassical character of the radiation source described above; the emitted light exhibits photon antibunching. Immediately after having measured one photon event, the probability of measuring another one per unit time is smaller than the average number of photons per unit time. In the following section we shall deal with the measurement of this property.

9.2. PHOTON ANTIBUNCHING OF FLUORESCENT LIGHT

The detection of light with the property of photon antibunching can be regarded as one of the fundamental proofs of the existence of electromagnetic radiation that has no classical counterpart, so-called "nonclassical light" (see Section 10.6.3). The results gained in the preceding section for the field-strength correlation function $\Gamma^{2,2}$ or for the normalized function $\gamma^{2,2}$ clearly show that the fluorescent light of a single atom that is monochromatically and resonantly excited exhibits photon antibunching. It results from the nonlinear behavior of the atom, owing to which the occupation probability of the excited level vanishes immediately after the measurement of one fluorescence photon, and it will only recover under the influence of the pumping process. According to what was stated in Section 9.1.4, the measurable properties of the radiation field are closely connected with the occupation probability of the excited state.

Now we consider the question, which is of basic physical interest, whether and how in this way a light source can be experimentally established whose light exhibits photon antibunching. The first experimental demonstration of photon antibunching was proposed and performed by Kimble, Dagenais, and Mandel (Refs. 9.13, 9.14, 9.15). A scheme of the setup employed is given in Fig. 9.4. In this experiment the resonance fluorescence of the transition $(3^2S_{1/2}, F = 2, m_F = 2) \leftrightarrow (3^2P_{3/2}, F = 3, m_F = 3)$ of pre-pumped sodium atoms within an atomic beam is investigated. In the interaction volume the pre-pumped sodium atoms are resonantly excited by a cw laser. By calibration of the laser intensity, ratios of the Rabi frequency Ω_R to the Einstein coefficient of spontaneous emission A_{if} between 2 and 2.5 are obtained. Since the relaxation in the atomic beam is determined by the spontaneous emission

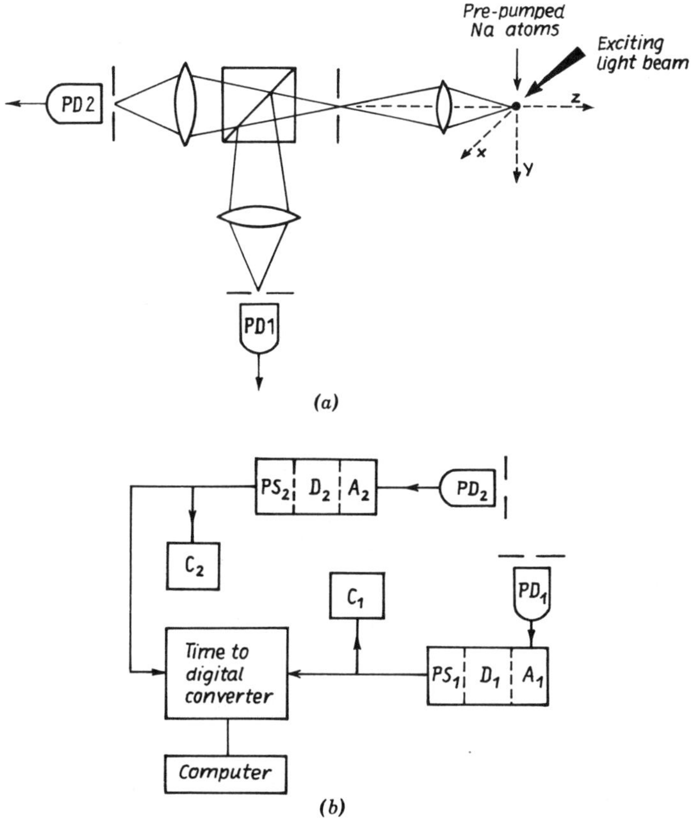

Fig. 9.4. Experimental setup for the observation of photon antibunching (after Ref. 9.13): (*a*) schematic of the fluorescent experiment; (*b*) schematic of the time-interval measuring device. $PD_{1,2}$, photodetectors; $A_{1,2}$, amplifiers; $D_{1,2}$, discriminators; $PS_{1,2}$, pulse-shaping devices; $C_{1,2}$, counters.

alone, the product of the Rabi frequency and the transverse relaxation time is between 4 and 5, and thus damped oscillations of the occupation number of the excited atomic level can arise. By employing a beam splitter the fluorescent light is divided into two beams of equal intensity, ejecting photoelectrons at the cathodes of two photodetectors. After appropriate processing the pulses of these detectors are transmitted to the start and stop entrances of the time-interval measuring device (see Fig. 9.4*b*). The number $n(\tau)$ of events with a time difference between start and stop in the interval $(\tau, \tau + \Delta\tau)$ is stored in the τth channel of a multichannel analyzer. As expected from theory, the number of stop events is seen to increase from its smallest value at $\tau = 0$ (see Fig. 9.5) to a maximum, which is located near $\tau = 25$ ns.

The quantitative analysis of such experiments is rather difficult because the fluctuation of the number of emitting atoms, the finite transit time of the atom

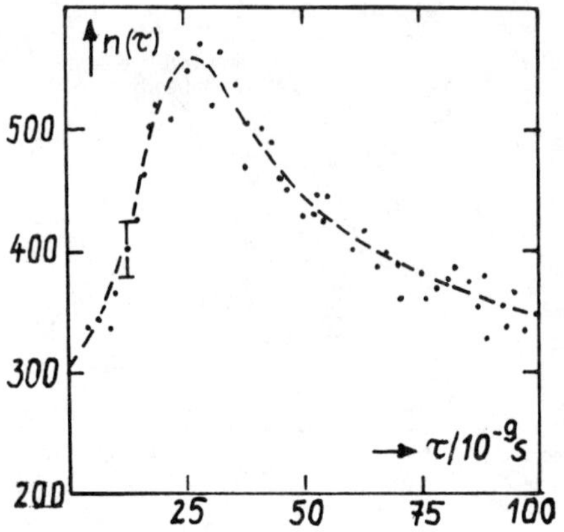

Fig. 9.5. Number of recorded stop pulses, $n(\tau)$, within a time interval of $\Delta\tau = 2 \times 10^{-9}$ s between τ and $\tau + \Delta\tau$ (after Ref. 9.13).

passing the excitation volume, and the mutual coherence between fluorescence and background have to be taken into account. Thus deviations from the behavior calculated for the ideal single-atom light source arise. Since this experiment proved to be of fundamental importance to physics, efforts for interpreting it were undertaken by several authors (Refs. 9.16–9.19). While at present there is complete agreement in the literature that the experimental results can be considered as an indirect proof of the ability of a single atom to emit light with photon antibunching, views diverge in regard to the question raised above: whether the light generated in the experiments reported actually exhibits photon antibunching. Therefore this point will be subjected to a closer examination.

Following Ref. 9.19, we start from the assumption that photon antibunching can be determined as a property of a radiation field at one point in space and in a short time interval, the duration of the latter being determined by the measuring process and assumed to be much smaller than the characteristic time τ_t in which the number of atoms contributing to fluorescence is changed owing to fluctuations.

Let us first assume for a preliminary discussion that there exists stationary excitation and that exactly N atoms are in the excitation volume and contribute to fluorescence. This case is simple in that here the correlation functions do not depend on the time of observation. Then, photon antibunching occurs if the normalized intensity correlation function $\gamma_N^{2,2}(\tau) = \lim_{t\to\infty}\gamma_N^{2,2}(t, t+\tau)$ satisfies the condition

$$\gamma_N^{2,2}(0) < 1. \tag{9.41a}$$

This is equivalent to the *antibunching excess* Δ being negative,

$$\Delta = \Gamma_N^{2,2}(0) - \left[\Gamma_N^{1,1}(0)\right]^2 < 0, \tag{9.41b}$$

where

$$\Gamma_N^{2,2}(\tau) = \lim_{t\to\infty} \Gamma_N^{2,2}(t, t+\tau, t+\tau, t)$$

and

$$\Gamma_N^{1,1}(\tau) = \lim_{t\to\infty} \Gamma_N^{1,1}(t, t+\tau).$$

In this case there exist simple relationships connecting the correlation functions $\Gamma_N^{2,2}$ and $\Gamma_N^{1,1}$ with the respective quantities for the single atom, even if background noise and its interference with the signal are taken into account. They are

$$\Gamma_N^{2,2}(\tau) = N\Gamma_1^{2,2}(\tau) + N(N-1)\left[I_1^2 + |\Gamma^{1,1}(\tau)|^2\right]$$

$$+|\varepsilon|^4 + 2NI_1|\varepsilon|^2 + 2N\,\mathrm{Re}\{\Gamma_1^{1,1}(\tau)\}|\varepsilon|^2 \qquad (9.42a)$$

and

$$\Gamma_N^{1,1}(0) = NI_1 + |\varepsilon|^2, \qquad (9.42b)$$

where $I_1 = \Gamma_1^{1,1}(0)$ and where $|\varepsilon|^2$ represents the intensity of background noise (see Ref. 9.16).

But if there is a statistical fluctuation of the atomic number N, the situation becomes more complicated. The respective correlation functions will generally depend on time, and on the statistical average, measurements at the times t and t' will provide the same results only if $|t - t'| < \tau_t$, where τ_t is the mean transit time of an atom through the scattering volume. Let us now find out whether light, on the statistical average, exhibits photon antibunching. For that to be the case the requirement

$$\overline{\Delta}^t = \overline{\Gamma^{2,2}(t,t,t,t) - \left[\Gamma^{1,1}(t,t)\right]^2}^t < 0 \qquad (9.43)$$

has to be satisfied, where a time averaging has been performed. This time average can be replaced by an ensemble average with respect to N, as the atomic beam represents an ensemble that fluctuates stationarily and in which a definite number of atoms, N, contributes to fluorescence at any instant of time. Hence

$$\overline{\Delta}^t = \overline{\Delta} := \sum_{N=0}^{\infty} p(N)\Delta_N(0) = \sum_{N=0}^{\infty} p(N)\left\{\Gamma_N^{2,2}(0) - \left[\Gamma_N^{1,1}(0)\right]^2\right\}, \quad (9.44)$$

holds where $p(N)$ is the atomic distribution function. The requirement of photon antibunching as imposed in (9.43) is equivalent to

$$\overline{\gamma^{2,2}(\tau = 0)} < 1 \tag{9.45}$$

if the averaged normalized correlation function $\overline{\gamma^{2,2}(\tau)}$ is defined as

$$\overline{\gamma^{2,2}(\tau)} = \overline{\Gamma^{2,2}(t, t+\tau, t+\tau, t)}\Big/\overline{\Gamma^{1,1}(t, t)\Gamma^{1,1}(t+\tau, t+\tau)}. \tag{9.46}$$

The mean number of pulse pairs $n(\tau)$ measured in the experiment can now be calculated as follows. Assuming N atoms to be in the scattering volume, the number $n_N(\tau)$ of pulse pairs is related to the number of starting pulses M_N and the conditional probability $p_N(2, t+\tau|1, t)\,\Delta t$ of the photodetection at detector 2 during the time interval from $(t+\tau)$ to $(t+\tau+\Delta t)$, provided that a photon was detected at detector 1 during the time interval from t to $t+\Delta t$, by

$$n_N(\tau) = M_N p_N(2, t+\tau|1, t)\,\Delta t \tag{9.47}$$

This conditional probability is given by

$$p_N(2, t+\tau|1, t)\,\Delta t = \alpha_2\,\Delta t\,\frac{\Gamma_N^{2,2}(\tau)}{\Gamma_N^{1,1}(0)}, \tag{9.48}$$

where α_2 is the detection efficiency of the second detector. The total number of pulse pairs, $n(\tau)$, becomes

$$n(\tau) = \sum_{N=0}^{\infty} M_N \alpha_2\,\Delta t\,\frac{\Gamma_N^{2,2}(\tau)}{\Gamma_N^{1,1}(0)}. \tag{9.49}$$

For normalization purposes this number is divided by the total number of starting pulses M and by the counting rate $R_2 + r_2$ of the second detector, with R_2 and r_2 resulting from fluorescence and background noise, respectively. Thus we obtain

$$\frac{n(\tau)}{M(R_2 + r_2)\,\Delta t} = \sum_{N=0}^{\infty} \frac{M_N}{M}\alpha_2 \frac{\Gamma_N^{2,2}(\tau)}{\Gamma_N^{1,1}(0)(R_2 + r_2)} \tag{9.50}$$

From (9.50) it is apparent that the number of pulse pairs, $n_N(\tau)$, must not be averaged by use of the atomic distribution function $p(N)$, but by use of the conditional probability M_N/N of finding N atoms in the scattering volume, provided that a starting pulse has been recorded. The conditional probability is given by

$$\frac{M_N}{M} = \frac{\Gamma_N^{1,1}(0)}{\overline{\Gamma^{1,1}(0)}} p(N), \tag{9.51}$$

and hence we obtain from (9.50)

$$\frac{n(\tau)}{M \Delta t (R_2 + r_2)} = \alpha_2 \frac{\overline{\Gamma^{2,2}(\tau)}}{\overline{\Gamma^{1,1}(0)}(R_2 + r_2)}. \tag{9.52}$$

It is now of importance in what way the counting rate $R_2 + r_2$ used for normalization will be determined. In the real experiment detector 2 can only deliver a signal when there was a start signal at detector 1. Hence, for normalization of $n(\tau)$ it is necessary to also use the counting rate that is calculated by using the conditional probability M_N/M rather than the probability $p(N)$. This averaging gives

$$\tilde{R}_2 + \tilde{r}_2 = \alpha_2 \sum_{N=0} \frac{M_N}{M} \Gamma_N^{1,1}(0). \tag{9.53}$$

Using (9.51) this leads to

$$\tilde{R}_2 + \tilde{r}_2 = \alpha_2 \frac{\overline{[\Gamma^{1,1}(0)]^2}}{\overline{\Gamma^{1,1}(0)}} \tag{9.54}$$

Employing this particular counting rate for normalization, we finally arrive at

$$\frac{n(\tau)}{M(\tilde{R}_2 + \tilde{r}_2)\,\Delta t} = \frac{\overline{\Gamma^{2,2}(\tau)}}{\overline{[\Gamma^{1,1}(0)]^2}} = \overline{\gamma^{2,2}(\tau)}. \tag{9.55}$$

That means that the normalized counting rate is equal to the normalized correlation function, which has been discussed above in (9.43)–(9.46) in connection with whether or not light exhibits photon antibunching. Note that in deriving (9.55) use has been made of the assumption that the measured time intervals between the start and stop signals are small compared with the atomic transit time τ_t through the observation volume.

Let us now evaluate $\gamma^{2,2}(\tau)$, assuming $p(N)$ to be the Poisson distribution. Using (9.42) as well as the notation

$$\Gamma_1^{1,1}(\tau) = I_1 \gamma_1^{1,1}(\tau) \tag{9.56a}$$

and

$$\Gamma_1^{2,2}(\tau) = I_1^2 \gamma_1^{2,2}(\tau), \tag{9.56b}$$

we obtain

$$\overline{\gamma^{2,2}(\tau)} = \frac{\overline{N} I_1^2 \gamma_1^{2,2}(\tau) + \overline{N}^2 I_1^2 \left[1 + |\gamma_1^{1,1}(\tau)|^2\right]}{\overline{N}(\overline{N}+1) I_1^2 + 2\overline{N} I_1 |\varepsilon|^2 + |\varepsilon|^4}$$

$$+ \frac{+|\varepsilon|^4 + 2\overline{N} I_1 |\varepsilon|^2 + 2\overline{N} I_1 |\varepsilon|^2 \mathrm{Re}\{\gamma_1^{1,1}(\tau)\}}{\overline{N}(\overline{N}+1) I_1^2 + 2\overline{N} I_1 |\varepsilon|^2 + |\varepsilon|^4} \tag{9.56c}$$

If we restrict ourselves to the zero-noise limit ($|\varepsilon|^2 \to 0$), this leads to

$$\overline{\gamma^{2,2}(\tau)} = \frac{\overline{N} \gamma_1^{2,2}(\tau) + \overline{N}^2 \left[1 + |\gamma_1^{1,1}(\tau)|^2\right]}{\overline{N}(\overline{N}+1)}. \tag{9.57}$$

Provided the sample experience highly monochromatic irradiation and $|\gamma_1^{1,1}(\tau = 0)| = 1$, we arrive at

$$\overline{\gamma^{2,2}(\tau = 0)} = \frac{2\overline{N}}{\overline{N}+1}. \tag{9.58}$$

From this equation it is apparent that photon bunching as well as photon antibunching is possible under the experimental conditions described. But which of these cases will be realized depends on the value of the mean atomic number $\overline{N}$. Thus the light emitted by a resonantly excited atomic beam shows photon antibunching provided that the mean atomic number $\overline{N}$ is smaller than 1. Concerning the background noise, the additional requirement that $|\varepsilon|^2/\overline{N} I_1 \ll 1$ must be satisfied. It should be noted that in the experiments of Kimble, Dagenais, and Mandel the mean atomic number $\overline{N}$ was indeed smaller than 1. We should mention that even such small values as 10^{-2} have been reached in other resonance fluorescence experiments.

In the limiting case $\overline{N} \to 0$ we obtain from (9.57)

$$\lim_{\overline{N} \to 0} \overline{\gamma^{2,2}(\tau)} = \gamma_1^{2,2}(\tau). \tag{9.59}$$

This means that the normalized intensity correlation function of a single atom can be approached. (Concerning the additional influence of spatial coherence in such experiments, see, e.g. Ref. 9.16.)

We have demonstrated that it is possible to establish a genuine light source exhibiting photon antibunching by employing a resonantly pumped atomic beam. Difficulties arise from the fact that the atomic number in the scattering volume shows unavoidable fluctuations. These difficulties can be overcome by employing either a single fluorescent atom in a solid matrix at low temperatures (see Ref. 9.18) or a single ion trapped in a quadrupole field in vacuum, which can be cooled down by resonance fluorescence (see Refs. 9.20, 9.21). But up to now results of such experiments have not been reported.

9.3. THREE-WAVE MIXING AND LIGHT DIFFRACTION BY INDUCED TRANSIENT GRATINGS

The interaction between several light waves in one-photon absorbers or between light and polarization waves may give rise to standing or slowly traveling excitation waves of measurable properties of the sample. These transient grating-like spatial patterns can be measured by employing self-diffraction of the exciting light pulses or diffraction of probe pulses. Figure 9.6 presents the scheme of a typical diffraction experiment. Two strong laser pulses of equal midfrequency ω_l and with different wave vectors $k_{1\bullet}$ and $k_{2\bullet}$ pass a sample in resonance with the transition $0 \leftrightarrow 1$. As a result of this resonant interaction waves of nonlinear polarization are built up in a three-wave mixing process, and the population of levels 0 and 1 undergoes a modulation in space and time by saturation effects. Thus a periodic spatial structure of the occupation-number inversion with the wave vectors $\delta k_{\bullet}^{(\pm)} = \pm(k_{2\bullet} - k_{1\bullet})$ may arise from absorption saturation. As a result of the nonlinear polarization and of the saturation pattern, new light waves are generated at the same midfrequency ω_l but with the wave vectors $2k_{2\bullet} - k_{1\bullet}$ and $2k_{1\bullet} - k_{2\bullet}$. (Radiation with the wave vector $2k_{1\bullet} - k_{2\bullet}$ does not occur if the second pulse enters the sample only after the cessation of the first pulse.) A third weak pulse with wave vector $k_{3\bullet}$, which follows the two strong exciting pulses, may experience "amplifica-

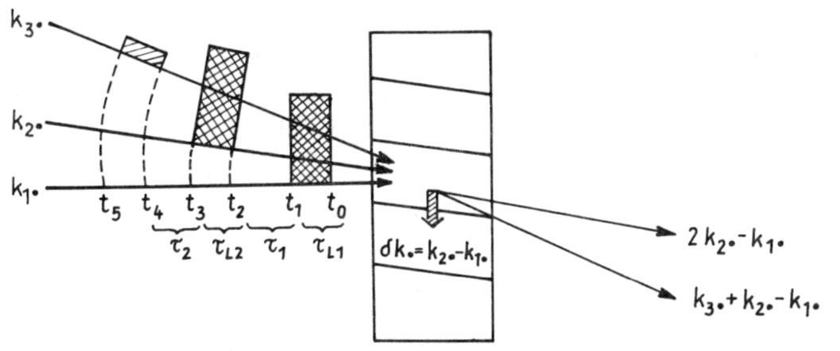

Fig. 9.6. Laser-induced transient grating.

tion" and diffraction, where the "amplification" originates from light of the exciting pulses diffracted into the direction of the probe pulse.

In stationary experiments with monochromatic light beams of equal frequency, the coherent three-wave mixing and the occupation-number grating lead to very similar effects. In contrast to this, under *nonstationary irradiation* quite different phenomena result from the coherently driven polarization and from the occupation-number inversion, as these two physical quantities decay with different relaxation times, namely with the transverse relaxation time τ_{10} and the longitudinal relaxation time T_{10}, respectively. Thus the situation becomes far more complex than it is under stationary irradiation. Therefore on the one hand, excite-and-probe beam measurements in ultrafast spectroscopy have to be carefully evaluated to avoid misinterpretations (see, e.g., Refs. 9.22, 9.23). On the other hand, since both the relaxation parameters τ_{10} and T_{10} influence the experimental result, it should be possible to obtain these parameters from such measurements. So Phillion, Kuizenga, and Siegman (Ref. 9.24) pointed out the possibility of measuring lifetimes as well as molecular reorientation times from the diffraction efficiency of delayed probe pulses. Yajima, Ishida, and Taira (Ref. 9.25) showed that in three-wave mixing experiments, in which two ultrashort light pulses of variable mutual delay τ_1 resonantly excite the sample at the transition $0 \leftrightarrow 1$, the dependence of the energy diffracted into the direction of $k_{\bullet} = 2k_{2\bullet} - k_{1\bullet}$ on the delay yields the phase decay time τ_{10} of this transition. It was emphasized by these authors that the method described of measuring τ_{10} can be employed under less restrictive conditions than the photon echo.

Following the treatment given in Refs. 9.12 and 9.26, we deal with such interaction processes in one-photon absorbers, taking into account phase memory, transient hole burning in inhomogeneously broadened transitions, and specific effects originating from the quantization of the radiation field. This treatment contains the quantum-theoretical description of the photon echo as one special case, which has already been considered semiclassically in Section 8.3. After establishing the basic equations in Section 9.3.1, we calculate in Section 9.3.2 the time behavior of the light emitted into the diffraction direction during and after the action of the second pulse. Section 9.3.3 deals with the diffraction of a third weak probe pulse, and in Section 9.3.4 we finish the section with a discussion of some experiments.

For simplicity let us assume the nonlinear absorber to consist of two-level systems, irradiated by a sequence of square entrance pulses. The sample is assumed to be rather thin, so that the change of the entrance pulses on passage through the sample can be neglected and the diffracted light pulses are too weak to affect the atoms of the sample appreciably.

9.3.1. Energy Density of Fluorescence

For treating three-wave mixing and self-induced diffraction phenomena in saturable two-level absorbers, the results obtained in Section 9.1 on resonance fluorescence of single atoms under strong irradiation can be used. In the approximation already described, we now only need to superpose the electric fields resulting from all the atomic systems of the sample, taking into account the different time delays originating from the spatial position of the individual atomic light sources. Thus we may start from the expression (9.14) for the energy density $\mathscr{E}_V$ of the radiation field where the creation and annihilation operators of the μth mode obey Heisenberg's equation of motion (9.2). Note that now the Hamiltonian operator in these equations is given by

$$\hat{H}(t) = \sum_m \left(\hat{H}_m^A + \hat{H}_m^{AL} + \hat{H}_m^{AF} \right) + \hat{H}^F, \tag{9.60}$$

where the single terms are defined in (9.4)–(9.9) and the summation is to be extended over all atomic systems of the sample. By solving the equations of motion for the annihilation and creation operators and by substituting these solutions into (9.14), we obtain the energy density

$$\begin{aligned}\mathscr{E}_V(R_\bullet, t) = \frac{1}{8\varepsilon_0 V^2} \sum_{\lambda,\mu} \sum_{l,m} \omega_\mu \omega_\lambda \left[\left(\frac{k_{\mu\bullet}}{k_\mu} \times e_{\mu\bullet} \right)\left(\frac{k_{\lambda\bullet}}{k_\lambda} \times e_\lambda \right) + e_{\mu\bullet} e_{\lambda\bullet} \right] \\ \times d_\mu^{(m)} d_\lambda^{(l)*} \int_{t_0}^t dt_1 \int_{t_0}^t dt_2 \left\{ e^{i\omega_\mu(t-t_1)} e^{-ik_{\mu\bullet}(R_\bullet - r_\bullet^{(m)})} + \right. \\ \left. + e^{-i\omega_\lambda(t-t_2)} e^{-ik_{\lambda\bullet}(R_\bullet - r_\bullet^{(l)})} \right\} \\ \times \left\langle \hat{B}_m^\dagger(t_1) \hat{B}_l(t_2) \right\rangle,\end{aligned} \tag{9.61}$$

where the atomic correlation function $\langle B_m^\dagger(t_1) B_l(t_2) \rangle$ contains the operators of the mth and the lth atom at the positions $r_\bullet^{(m)}$ and $r_\bullet^{(l)}$ (cf. Ref. 9.12). Assuming a very large quantization volume V, the summation over the modes can be replaced by an integration over the wave vectors: $\Sigma_\mu \to [V/(2\pi)^3] \int d^3k_\mu$. Provided the fluorescence light is observed in the far field ($r^{(m)} \ll R$), Born's approximation can be applied, and the evaluation of the integrals over $k_\bullet$, $k'_\bullet$, t_1, and t_2 leads to

$$\begin{aligned}\mathscr{E}_V(R_\bullet, t) = \frac{\omega_{10}^4}{32\pi^2 c^4 \varepsilon_0 R^2} \\ \times \sum_{m,l} \left(d_{10\bullet}^{(m)} \times \frac{R_\bullet}{R} \right)\left(d_{01\bullet}^{(l)} \times \frac{R_\bullet}{R} \right) \left\langle \hat{B}_m^\dagger(\tau_m) \hat{B}_l(\tau_l) \right\rangle,\end{aligned} \tag{9.62a}$$

where

$$\tau_m = t - \frac{1}{c}|R_\bullet - r_\bullet^{(m)}| \approx T + \frac{R_\bullet r_\bullet^{(m)}}{Rc} \quad \text{and} \quad T = t - \frac{R}{c}. \tag{9.62b}$$

If the atomic particles change their positions with time, for instance in gases, one has to use $r_{\bullet}^{(m)}$ at time τ_m in (9.62b). But this case will not be discussed here; see for example Ref. 9.10.

Let us now evaluate the atomic correlation functions by using the results of Section 9.1.2. This yields

$$\left\langle \hat{B}_m^{\dagger}(\tau_m)\hat{B}_l(\tau_l)\right\rangle = \begin{cases} \sum_a \sigma_{a1}^{(m)}(\tau_m)T_{10a0}^{(m)}(\tau_m,\tau_m) & \text{for} \quad m=l, \\ \sigma_{01}^{(m)}(\tau_m)\sigma_{10}^{(l)}(\tau_l) & \text{for} \quad m\neq l. \end{cases} \tag{9.63}$$

Substitution of (9.63) into (9.62) leads, on consideration of (9.26), to

$$\mathscr{E}_V(R_{\bullet},t) = \frac{\omega_{10}^4}{32\pi^2c^4\varepsilon_0R^2}\left\{\sum_m \left| d_{10\bullet}^{(m)} \times \frac{R_{\bullet}}{R}\right|^2 \sigma_{11}^{(m)}(\tau_m) + \sum_{m,l}\left(d_{10\bullet}^{(m)} \times \frac{R_{\bullet}}{R}\right)\left(d_{01\bullet}^{(l)} \times \frac{R_{\bullet}}{R}\right)\sigma_{01}^{(m)}(\tau_m)\sigma_{10}^{(l)}(\tau_l)\right\}. \tag{9.64}$$

From (9.64) it is obvious that the energy density of fluorescence, $\mathscr{E}_V(R_{\bullet},t)$, is composed of contributions originating from the incoherent diagonal elements of the density matrix of single atoms and of contributions proportional to the products $\sigma_{01}^{(m)}(\tau_m)\sigma_{10}^{(l)}(\tau_l)$ of the off-diagonal elements of one atom ($l=m$) or of two different atoms ($l\neq m$). The off-diagonal elements of several atoms can be coherently driven by the laser radiation, and it is their contribution to fluorescence that gives rise to interference of the emitted light. Whereas, for large samples, the coherently emitted light is sharply peaked around certain directions that are determined by the wave vectors of the exciting radiation, the incoherent emission is observed in all directions. Note that the coherently emitted light can in approximation be calculated by semiclassical means (see Chapter 8, and especially the treatment of the photon echo in Section 8.4.3).

The decomposition into the coherent and incoherent parts of the energy density leads to

$$\mathscr{E}_V(R_{\bullet},t) = \mathscr{E}_V^{\text{coh}}(R_{\bullet},t) + \mathscr{E}_V^{\text{inc}}(R_{\bullet},t), \tag{9.65a}$$

where

$$\mathscr{E}_V^{\text{coh}}(R_{\bullet},t) = \frac{\omega_{10}^4|d_{10\bullet}\times R_{\bullet}/R|^2}{32\pi^2c^4\varepsilon_0R^2} \times\left|\int_{V_F} d^3r_{\bullet}\,\gamma\int d\omega_{10}'\,g_{\text{inh}}(\omega_{10}')\sigma_{10}(\tau_r)\right|^2, \tag{9.65b}$$

$$\mathscr{E}_V^{\text{inc}}(R_\bullet, t) = \frac{\omega_{10}^4 \left| d_{10\bullet} \times \dfrac{R_\bullet}{R} \right|^2}{32\pi^2 c^4 \varepsilon_0 R^2}$$

$$\times \int_{V_F} d^3 r_\bullet \, \gamma \int d\omega'_{10} \, g_{\text{inh}}(\omega'_{10}) \left[\sigma_{11}(\tau_r) - |\sigma_{10}(\tau_r)|^2 \right], \quad (9.65\text{c})$$

$$\tau_r = t - \frac{|r_\bullet - R_\bullet|}{c} \approx T + \frac{r_\bullet R_\bullet}{cR}, \quad (9.65\text{d})$$

and

$$T = t - \frac{R}{c}.$$

In deriving (9.65b, c) the summation over the molecules in (9.64) has been replaced by the integration over the space co-ordinate $r_\bullet$ within the fluorescent sample of volume V_F and that over the frequency ω'_{10} of the inhomogeneously broadened transition, which is characterized by the center frequency ω_{10} and the line shape function $g_{\text{inh}}(\omega'_{10})$. Thus we have

$$\sum_m \to \int_{V_F} d^3 r_\bullet \, \gamma \int d\omega'_{10} \, g_{\text{inh}}(\omega'_{10}).$$

If γ, the number of fluorescent atomic systems per unit volume, is independent of $r_\bullet$, we may write it in front of the integral. According to the coherent superposition of the field strength originating from different atoms, the energy density $\mathscr{E}_V^{\text{coh}}(R_\bullet, t)$ is proportional to the square of the number density γ of fluorescent molecules, as already noted in the semiclassical description of Chapter 8, whereas the incoherent part $\mathscr{E}_V^{\text{inc}}(R_\bullet, t)$ is only proportional to γ.

In our treatment of the three-wave mixing and diffraction phenomena we are only interested in the coherently emitted light. Therefore let us evaluate $\mathscr{E}_V^{\text{coh}}(R_\bullet, t)$ by employing the time-evolution matrices of (9.26), given that the sequence of pulses shown in Fig. 9.6 interact with the sample. Until the arrival of the first pulse, the sample remains in thermodynamic equilibrium, and we assume that $\sigma_{00}^e = 1$ and $\sigma_{11}^e = \sigma_{01}^e = \sigma_{10}^e = 0$. Then during the first pulse, according to (9.22) and (9.25), we have

$$\sigma_{ij}(r_\bullet, t) = T_{ij00}^{L1}(t, t_0, r_\bullet) \qquad \text{for} \quad t_0 \le t \le t_1, \quad (9.66\text{a})$$

where $T^{Li}(t, t', r_\bullet)$ denotes the time-evolution matrix corresponding to the ith pulse ($i = 1, 2, 3$). Between the first and the second pulse there occur only relaxation processes, and there holds

$$\sigma_{ij}(r_\bullet, t) = T^0_{ijij}(t - t_1) T^{L1}_{ij00}(t_1, t_0, r_\bullet) \qquad \text{for} \quad t_1 \le t \le t_2. \quad (9.66b)$$

As long as the second pulse is acting, $\sigma_{ij}(r_\bullet, t)$ is given by

$$\begin{aligned}
\sigma_{ij}(r_\bullet, t) = {} & T^{L2}_{ij01}(t, t_2, r_\bullet) T^0_{0101}(\tau_1) T^{L1}_{0100}(t_1, t_0, r_\bullet) \\
& + T^{L2}_{ij10}(t, t_2, r_\bullet) T^0_{1010}(\tau_1) T^{L1}_{1000}(t_1, t_0, r_\bullet) \\
& + T^{L2}_{ij00}(t, t_2, r_\bullet) T^0_{0000}(\tau_1) T^{L1}_{0000}(t_1, t_0, r_\bullet) \\
& + T^{L2}_{ij11}(t, t_2, r_\bullet) T^0_{1111}(\tau_1) T^{L1}_{1100}(t_1, t_0, r_\bullet) \\
& \qquad \text{for} \quad t_2 \le t \le t_3, \quad (9.66c)
\end{aligned}$$

where τ_1 and τ_2 are defined in Fig. 9.6. In the time interval between the second and the third pulse we have

$$\begin{aligned}
\sigma_{10}(r_\bullet, t) &= T^0_{1010}(t - t_3) \sigma_{10}(r_\bullet, t_3), \\
\sigma_{11}(r_\bullet, t) &= T^0_{1111}(t - t_3) \sigma_{11}(r_\bullet, t_3) \\
& \qquad \text{for} \quad t_3 \le t \le t_4, \quad (9.66d)
\end{aligned}$$

where $\sigma_{ij}(r_\bullet, t_3)$ from (9.66c) may be inserted. During the action of the probe pulse the off-diagonal element $\sigma_{10}(r, t)$ responsible for the probe-beam diffraction is obtained as

$$\begin{aligned}
\sigma_{10}(r_\bullet, t) = {} & T^{L3}_{1010}(t, t_4, r_\bullet) T^0_{1010}(\tau_2) \sigma_{10}(r_\bullet, t_3) \\
& + T^{L3}_{1001}(t, t_4, r_\bullet) T^0_{0101}(\tau_2) \sigma_{01}(r_\bullet, t_3) \\
& + \left[T^{L3}_{1011}(t, t_4, r_\bullet) - T^{L3}_{1000}(t, t_4, r_\bullet) \right] T^0_{1111}(\tau_2) \sigma_{11}(r_\bullet, t_3) \\
& + T^{L3}_{1000}(t, t_4, r_\bullet). \qquad (9.66e)
\end{aligned}$$

Substituting the solution for the off-diagonal element $\sigma_{10}(r_\bullet, t)$ from (9.66) into (9.65b), we get the energy density of the coherently emitted fluorescent radiation at any time between the beginning of the first pulse at t_0 and the cessation of the probe pulse at t_5. In the following we shall consider some situations of experimental interest in more detail.

9.3.2. Self-Diffraction and Photon Echo

In this subsection we study the coherent emission of light during and after the action of the second pulse. Therefore we need the off-diagonal element $\sigma_{10}(r_\bullet, t)$ in this time interval. During and after the action of the second pulse we respectively obtain from (9.66c) and (9.66d)

$$\sigma_{10}(r_\bullet, \tau_r) = e^{i(2k_{2\bullet}-k_{1\bullet}-k_\bullet)r_\bullet - i\omega_2(T+t_2)}e^{i\omega_1 t_1}e^{i\omega_{10}\tau_1}e^{-\tau_1/\tau_{10}} \times T^{L2}_{1001}(T-t_2)T^{L1}_{0100}(\tau_{L1}) \tag{9.67a}$$

and

$$\sigma_{10}(r_\bullet, \tau_r) = e^{i(2k_{2\bullet}-k_{1\bullet}-k_\bullet)r_\bullet}e^{-i\omega_2(t_3-t_2)}e^{i\omega_1 t_1}e^{i\omega(T-t_3)} \times e^{i\omega_1\tau_1}e^{-(T-t_3)/\tau_{10}}e^{-\tau_1/\tau_{10}}T^{L2}_{1001}(\tau_{L2})T^{L1}_{0100}(\tau_{L1}), \tag{9.67b}$$

where

$$k_\bullet = \frac{\omega'_{10}}{c}\frac{R_\bullet}{R} \quad \text{and} \quad T = t - \frac{R}{c}. \tag{9.67c}$$

Substitution of σ_{10} from (9.67a) or (9.67b) and of the appropriate function $g_{\text{inh}}(\omega'_{10})$ into (9.65b) yields explicit expressions for the energy density of the coherently emitted fluorescent light at $R_\bullet$ and t as an integral over the fluorescent volume and over the inhomogeneously broadened transition. Provided that the number density of fluorescent molecules is independent of space coordinates, the integral over the fluorescent volume is proportional to the factor

$$f(2k_{2\bullet} - k_{1\bullet} - k_\bullet) = \left| \int_{V_F} d^3r_\bullet\, e^{i(2k_{2\bullet}-k_{1\bullet}-k_\bullet)r_\bullet} \right|^2,$$

which thus describes the angular distribution of the emitted light. If both the irradiating pulses travel in nearly the same direction and if the extension of the excited volume V_F is large compared with the wavelength of the radiation, this angular distribution has a sharp maximum in the direction of

$$k_\bullet = k_{2\bullet} + \delta k_\bullet,$$

where

$$\delta k_\bullet = k_{2\bullet} - k_{1\bullet}.$$

Thus the spatial characteristics of the emitted light are determined by both the exciting pulses. This means that the medium stores the knowledge of the spatial structure of the first and the second pulse until the instant of observation, which is during or after the action of the second pulse. Supposing equal wave vectors of the incident pulses ($k_{1\bullet} = k_{2\bullet} = k_{L\bullet}$), the coherent emission is sharply peaked in the forward direction where $k_\bullet = k_{L\bullet}$. Let us now discuss some special cases.

9.3.2.1. Resonant Excitation of Homogeneously Broadened Transitions. The phenomenon of self-diffraction can be most clearly discussed for a sample without any inhomogeneous broadening. Let us assume resonant excitation; this means that the midfrequencies ω_1 and ω_2 of the two exciting pulses equal the midfrequency ω_{10} of the transition. Further we restrict our consideration to the case of exciting pulses whose durations are very small ($\tau_{Li} \ll \tau_{10}, T_{10}$, $i = 1, 2$) and whose amplitudes are rather high ($|d_{10}\bar{E}_L|/\hbar \gg \tau_{10}^{-1}, T_{10}^{-1}$). Given these conditions, by substitution of (9.67a) and (9.67b) into (9.65b) we respectively obtain

$$\mathscr{E}_V^{\text{coh}}(R_\bullet, t) \propto f(2k_{2\bullet} - k_{1\bullet} - k_\bullet)e^{-2\tau_1/\tau_{10}} \times \sin^4 \tfrac{1}{2}\Omega_{R0}^{(2)}(T - t_2)\sin^2\Omega_{R0}^{(1)}\tau_{L1} \tag{9.68a}$$

and

$$\mathscr{E}_V^{\text{coh}}(R_\bullet, t) \propto f(2k_{2\bullet} - k_{1\bullet} - k_\bullet)e^{-2\tau_1/\tau_{10}}e^{-2(T-t_3)/\tau_{10}} \times \sin^4 \tfrac{1}{2}\Omega_{R0}^{(2)}\tau_{L2}\sin^2\Omega_{R0}^{(1)}\tau_{L1} \tag{9.68b}$$

during and after the excitation of the sample by the second pulse, where $\Omega_{R0}^{(i)} = \bar{E}_{Li\bullet}d_{10\bullet}/\hbar$ is the Rabi frequency of the ith pulse.

Note that the energy density of the self-diffracted light varies sinusoidally with the "pulse area" $\Omega_{R0}^{(1)}\tau_{L1}$ of the first pulse and thus vanishes at

$$\Omega_{R0}^{(1)}\tau_{L1} = 1\pi, 2\pi, \ldots .$$

This agrees with the behavior of incoherently emitted light after short-pulse excitation (cf. Section 8.2). The coherently emitted energy decreases exponen-

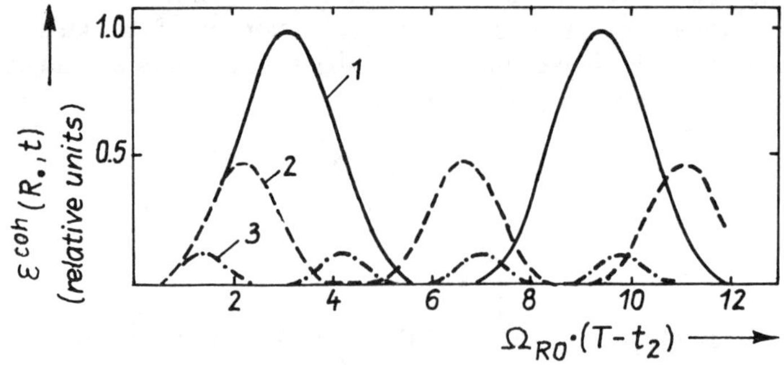

Fig. 9.7. Time dependence of the energy density of the diffracted light during the action of the second pulse for a homogeneously broadened transition (after Ref. 9.12); $\Omega_{R0}^{(1)} = \Omega_{R0}^{(2)} = \Omega_{R0}$, $\Omega_{R0}\tau_{L1} = \pi/2$, $\Delta_1 = \Delta_2 = \Delta$. Curve 1, $\Delta = 0$; curve 2, $\Delta = \Omega_{R0}$; curve 3, $\Delta = 2\Omega_{R0}$.

tially as the time interval between the two exciting pulses increases. The rate parameter of this decrease is given by $\tau_{10}/2$. During the action of the second pulse the energy density varies periodically with the time difference $T - t_2$, the period being given by $2\pi/\Omega_{R0}^{(2)}$ (see Fig. 9.7, curve 1). In summary we may state that even pulses that do not collide within the sample may experience strong interaction leading to diffraction phenomena as long as the time between them is smaller than or comparable with the phase decay time τ_{10}.

9.3.2.2 Nonresonant Excitation of Homogeneously Broadened Transitions. Let us now study self-diffraction for nonresonant excitation, assuming the same conditions to be satisfied as in the previous subsection. Given incident pulses of equal midfrequency $\omega_1 = \omega_2 = \omega_L$ ($\Delta_1 = \Delta_2 = \Delta_L = \omega_L - \omega_{10}$), we obtain from (9.65)

$$\mathscr{E}_V^{\text{coh}}(R_\bullet, t) \propto f(2k_{2\bullet} - k_{1\bullet} - k_\bullet)\frac{\left[\Omega_{R0}^{(1)}\right]^2\left[\Omega_{R0}^{(2)}\right]^4}{\left[\Omega_R^{(1)}\right]^4\left[\Omega_R^{(2)}\right]^4} e^{-2\tau_1/\tau_{10}}\sin^4\tfrac{1}{2}\Omega_R^{(2)}(T - t_2)$$

$$\times\left\{4\Delta_L^2\sin^4\tfrac{1}{2}\Omega_R^{(1)}\tau_{L1} + \left[\Omega_R^{(1)}\right]^2\sin^2\Omega_R^{(1)}\tau_{L1}\right\} \tag{9.69a}$$

and

$$\mathscr{E}_V^{\text{coh}}(R_\bullet, t) \propto f(2k_{2\bullet} - k_{1\bullet} - k_\bullet)\frac{\left[\Omega_{R0}^{(1)}\right]^2\left[\Omega_{R0}^{(2)}\right]^4}{\left[\Omega_R^{(1)}\right]^4\left[\Omega_R^{(2)}\right]^4}$$

$$\times e^{-2\tau_1/\tau_{10}}e^{-2(T-t_3)/\tau_{10}}\sin^4\tfrac{1}{2}\Omega_R^{(2)}\tau_{L2}$$

$$\times\left\{4\Delta_L^2\sin^4\tfrac{1}{2}\Omega_R^{(1)}\tau_{L1} + \left[\Omega_R^{(1)}\right]^2\sin^2\Omega_R^{(1)}\tau_{L1}\right\} \tag{9.69b}$$

during and after irradiation by the second pulse, where

$$\Omega_R^{(i)} = \sqrt{\left[\Omega_{R0}^{(i)}\right]^2 + \Delta_i^2}$$

denotes the Rabi frequency of the ith pulse with detuning Δ_i. As in the case of exact resonance, the coherently emitted energy density varies periodically with the pulse area of the first pulse. Now the diffraction vanishes at $\Omega_R^{(1)}\tau_{L1} = \pi, 2\pi, \ldots$. During the excitation by the second pulse the fluorescence varies periodically with time, the period being $2\pi/\Omega_R^{(2)}$, and it vanishes at

$$\Omega_R^{(2)}(T - t_2) = l\cdot(2\pi),\ l = 1, 2, \ldots .$$

It should be noted that not only does this period decrease with increasing detuning, but so does the maximum energy density (see Fig. 9.7, curves 2 and 3).

9.3.2.3. Inhomogeneously Broadened Transitions. As before, we assume the two incident pulses to be very short ($\tau_{Li} \ll \tau_{10}$) and very strong ($\Omega_{R0}^{(i)} \gg 1/\tau_{10}$) and to be of equal midfrequency, which coincides with the mean transition frequency ω_{10} ($\omega_1 = \omega_2 = \omega_{10}$). For numerical evaluation the inhomogeneous line-shape function $g_{\text{inh}}(\omega'_{10})$ is represented by a Gaussian function as

$$g_{\text{inh}}(\omega'_{10}) = \frac{1}{\sqrt{\pi}\,\Gamma} e^{-(\omega'_{10}-\omega_{10})^2/\Gamma^2},$$

where

$$\Gamma = \frac{1}{2\sqrt{\ln 2}} \Delta\omega_{\text{inh}}$$

is proportional to the inhomogeneous line width $\Delta\omega_{\text{inh}}$. During and after the irradiation by the second laser pulse we respectively have

$$\begin{aligned}\mathscr{E}_V^{\text{coh}}(R_\bullet, t) \propto f(2k_{2\bullet} - k_{1\bullet} - k_\bullet) e^{-2\tau_1/\tau_{10}} \frac{1}{\pi\Gamma^2} \\ \times \Bigg| \int_{-\infty}^{\infty} d\Delta\, e^{-\Delta^2/\Gamma^2} \frac{[\Omega_{R0}^{(1)}][\Omega_{R0}^{(2)}]^2}{[\Omega_R^{(1)}]^2[\Omega_R^{(2)}]^2} [\sin^2 \tfrac{1}{2}\Omega_R^{(2)}(T - t_2)] \\ \times [\Omega_R^{(1)} \cos \Delta\tau_1 \sin \Omega_R^{(1)}\tau_{L1} - 2\Delta \sin \Delta\tau_1 \sin^2 \tfrac{1}{2}\Omega_R^{(1)}\tau_{L1}] \Bigg|^2\end{aligned} \tag{9.70a}$$

and

$$\begin{aligned}\mathscr{E}_V^{\text{coh}}(R_\bullet, t) \propto f(2k_{2\bullet} - k_{1\bullet} - k_\bullet) e^{-(2/\tau_{10})[2\tau_1 + (T - t_3 - \tau_1)]} \frac{1}{\pi\Gamma^2} \\ \times \Bigg| \int_{-\infty}^{\infty} d\Delta\, e^{-\Delta^2/\Gamma^2} \frac{[\Omega_{R0}^{(1)}][\Omega_{R0}^{(2)}]^2}{[\Omega_R^{(1)}]^2[\Omega_R^{(2)}]^2} \sin^2 \tfrac{1}{2}\Omega_R^{(2)}\tau_{L2} \\ \times [\Omega_R^{(1)} \cos \Delta(T - t_3 - \tau_1) \sin \Omega_R^{(1)}\tau_{L1} \\ - 2\Delta \sin \Delta(T - t_3 - \tau_1) \sin^2 \tfrac{1}{2}\Omega_R^{(1)}\tau_{L1}] \Bigg|^2 .\end{aligned} \tag{9.70b}$$

From (9.70) it is apparent that now the time dependence of the fluorescence light is essentially determined by the interference of the coherent radiation emitted by groups of molecules of different frequency detuning with respect to the center frequency of the transition and the laser frequency.

The cases most easily handled are those of extremely short or strong pulses ($\tau_L \ll \Gamma^{-1}$ or $\Omega_{R0} \gg \Gamma$), because then all molecules are forced to take part in the radiation process in almost the same way, irrespective of their detuning from resonance. With $\Omega_{R0}^{(1)} = \Omega_{R0}^{(2)} = \Omega_{R0} \gg \Gamma$—that is, under extremely strong irradiation—we obtain the energy density of fluorescence during the second pulse as

$$\mathscr{E}_V^{\text{coh}}(R_\bullet, t) \propto f(2k_{2\bullet} - k_{1\bullet} - k_\bullet)(\sin^2\Omega_{R0}\tau_{L1}) \times e^{-(2\tau_1/\tau_{10} + \frac{1}{2}\tau_1^2\Gamma^2)}\sin^4\tfrac{1}{2}\Omega_{R0}(T - t_2). \tag{9.71}$$

This relation is of course very similar to that obtained for homogeneously broadened lines in Section 9.3.2.2. On the other hand, for extremely short pulses ($\tau_L \ll 1/\Gamma$) and for $\Omega_{R0}^{(1)} = \Omega_{R0}^{(2)} = \Omega_{R0} \ll \Gamma$ we have

$$\mathscr{E}_V^{\text{coh}}(R_\bullet, t) \propto f(2k_{2\bullet} - k_{1\bullet} - k_\bullet)e^{-2\tau_1/\tau_{10}}(\Omega_{R0})^4(T - t_2)^4 \cdot \left| \tfrac{1}{4}\tau_{L1}\Gamma^2 e^{-\tau_1^2\Gamma^2/8}\left\{ D_{-2}\left(\frac{-i\tau_1\Gamma}{\sqrt{2}}\right) + D_{-2}\left(\frac{i\tau_1\Gamma}{\sqrt{2}}\right)\right\} + \frac{\sqrt{2}\,\tau_{L1}^2\Gamma^3}{8i} e^{-\tau_1^2\Gamma^2/8}\left\{ D_{-3}\left(\frac{i\tau_1\Gamma}{\sqrt{2}}\right) - D_{-3}\left(\frac{-i\tau_1\Gamma}{\sqrt{2}}\right)\right\}\right|^2, \tag{9.72}$$

where $D_j(z)$ with $j = -2, -3$ are the parabolic cylinder functions, which are asymptotically represented by error and exponential functions. From (9.72) we see that the light energy being diffracted during the action of the second pulse tends to zero very quickly as the pulse separation τ_1 increases beyond $1/\Gamma$. In this case most of the fluorescence light will be emitted as a photon echo, which follows some time after the second pulse (see Section 8.4.3 and Fig. 9.9).

Let us now direct our attention to conditions that give rise to *hole-burning* effects. Figure 9.8 presents some numerical results for the light energy density emitted during the irradiation by the second pulse with $\Omega_{R0}^{(2)} = \Omega_{R0}^{(1)} = 0.1\Gamma$. We see that the intensity of the diffracted light decreases with increasing τ_L and τ_1 because of the increasing phase differences between the waves originating from different groups of molecules. Only a special group of molecules of the inhomogeneously broadened transition can now effectively take part in the interaction process. The width of the hole burning is mainly determined by Ω_{R0} and τ_L. The time dependence of the diffracted intensity at times after the cessation of the second pulse is shown in Fig. 9.9, where a typical photon echo is seen to arise. The height and shape of the echo depends on the excitation parameters, above all on $\Omega_{R0}^{(i)}$ and τ_{Li}. From (9.70b) it follows that the shape

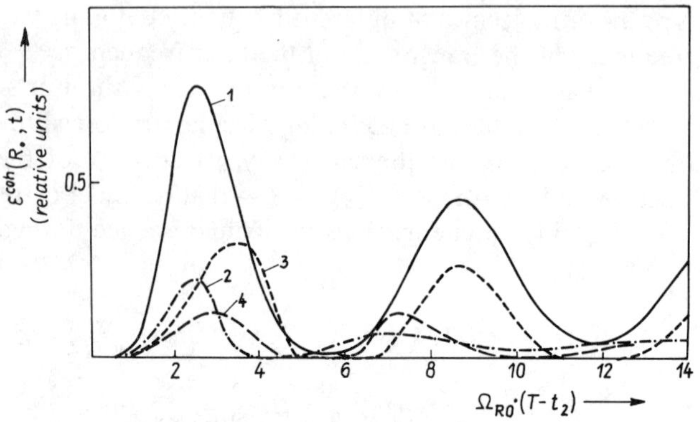

Fig. 9.8. Time dependence of the energy density of the diffracted light during the action of the second pulse for an inhomogeneously broadened transition under hole-burning conditions (after Ref. 9.12); $\Omega_{RO}^{(1)} = \Omega_{RO}^{(2)} = \Omega_{RO}$, $\Omega_{RO} = 0.1\Gamma$, $\tau_1 = 0$, $\Delta_1 = \Delta_2$. Curve 1, $\Omega_{RO}\tau_{L1} = \pi/2$; curve 2, $\Omega_{RO}\tau_{L1} = \pi$; curve 3, $\Omega_{RO}\tau_{L1} = 3\pi/2$; curve 4, $\Omega_{RO}\tau_{L1} = 2\pi$.

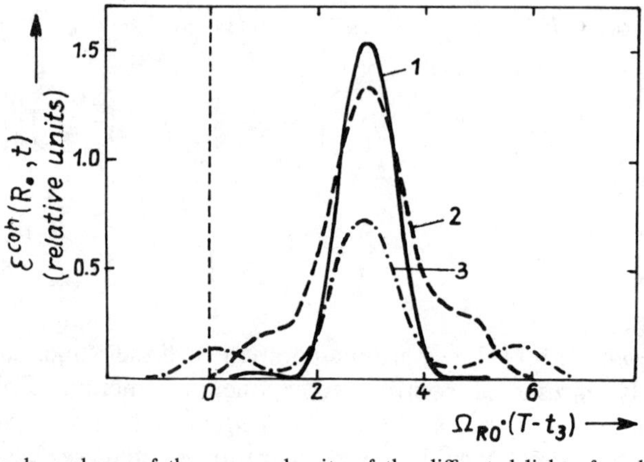

Fig. 9.9. Time dependence of the energy density of the diffracted light after the action of the second incident light pulse for an inhomogeneously broadened transition under hole-burning conditions (after Ref. 9.12); $\Omega_{RO}^{(1)} = \Omega_{RO}^{(2)} = \Omega_{RO}$, $\Omega_{RO} = 0.1\Gamma$, $\Omega_{RO}\tau_{L1} = \pi/2$, $\tau_1 = 2/\Omega_{RO}$, $\Delta_1 = \Delta_2$. Curve 1 $\Omega_{RO}\tau_{L2} = \pi/2$; curve 2 $\Omega_{RO}\tau_{L2} = \pi$; curve 3 $\Omega_{RO}\tau_{L2} = 4$. The dashed vertical line at $\Omega_{RO}(T - t_3) = 0$ marks the trailing edge of the second incident light pulse. [Note that only for $\Omega_{RO}(T - t_3) > 0$ are all curves of physical significance.]

and the amplitude of the echo depend on the time delay τ_1 between the two exciting pulses only through the factor $\exp\{-4\tau_1/\tau_{10}\}$, provided the validity range of (9.70b) is not restricted with respect to the time T. Note that (9.70b) only holds at $T > t_3$. From this limitation no difficulty arises as long as almost all the energy of the echo is emitted at $T > t_3$; this condition is obviously satisfied for instance for curves 1 and 2 of Fig. 9.9.

9.3.2.4. Evaluation of Relaxation Parameters. We have mentioned that the effect of self-diffraction can be exploited for measuring relaxation times. But as we have already learned in this subsection, the dependence of the diffracted intensity on the parameters of the sample and on those of the exciting pulses is rather complex, and full consideration has to be given to which relations apply to the real conditions imposed by the experiment. Let us discuss here the evaluation of the phase decay time τ_{10} for various cases.

In the case of homogeneously broadened transitions the energy of the light diffracted during and after the irradiation by the second pulse decreases with increasing delay time τ_1 as $\exp(-2\tau_1/\tau_{10})$. Thus the phase decay time can be determined by measuring the total energy of the diffracted light as a function of the time delay τ_1 (cf. Ref. 9.25).

In the case of inhomogeneously broadened transitions the situation is relatively simple if no hole-burning effects occur. This happens under irradiation by extremely short ($\tau_L \ll 1/\Gamma$) or extremely strong ($\Omega_{R0} \gg \Gamma$) exciting pulses. Then, for pulse delays $\tau_1 > 1/\Gamma$, the main part of the diffracted light is emitted after the second pulse has ceased, and the diffracted light energy decreases with increasing pulse delay τ_1 as $\exp\{-4\tau_1/\tau_{10}\}$. This just corresponds to the simple case we discussed in our semiclassical treatment of the photon echo in Section 8.4.3. If hole burning occurs, the time dependence becomes more complex, and it must be carefully evaluated by using (9.70). Only at very large delay times τ_1 can the dependence of the diffracted energy on τ_1 be approximated by a single exponential. At such large delay times, however, the light intensity coherently emitted by the sample has already become quite low because of phase-relaxation processes.

9.3.3. Diffraction of Probe Pulses

We are now going to investigate the diffraction of a third pulse, the so-called *probe pulse* or *interrogation pulse*. This diffraction originates from the polarization and the occupation-number pattern generated by the two preceding pulses. The light intensity diffracted during the action of the probe pulse depends on the off-diagonal element $\sigma_{10}(r_\bullet, t)$ of the density matrix given in (9.66e). This equation contains four terms, the first and second of which result from the interaction of the probe pulse with the polarization pattern generated by the exciting pulses. These terms decrease with increasing probe pulse delay τ_2 as $e^{-\tau_2/\tau_{10}}$, and thus they approach zero at $\tau_2 \gg \tau_{10}$. The third term arises from the interaction of the probe pulse with the occupation-number pattern $\sigma_{11}(r_\bullet, t)$, which is caused by the first and the second pulse. This contribution decreases with the probe-pulse delay as $e^{-\tau_2/T_{10}}$, so its decay is characterized by the lifetime T_{10}. The fourth term in (9.66e) describes that part of the coherently emitted light which is caused solely by the interaction of the probe pulse with the sample, and hence it does not contain any information about the exciting pulses.

Let us now discuss the contribution to the diffraction of the third pulse that results from the saturation pattern of $\sigma_{11}(r_\bullet, t)$. At delay times large compared to τ_{10} this is the most important contribution. Starting from (9.66e), we obtain the term responsible for the diffraction by the occupation-number pattern as

$$\sigma_{10}(r_\bullet, t) = e^{-i\omega_3 T} e^{-\tau_2/T_{10}} e^{-\tau_1/\tau_{10}} \left[T^{L3}_{1011}(T - t_4) - T^{L3}_{1000}(T - t_4) \right] e^{i(k_\bullet - k_{3\bullet}) r_\bullet}$$

$$\times \left\{ e^{i(k_{2\bullet} - k_{1\bullet}) r_\bullet} e^{-i\omega_2 t_3} e^{i\omega_1 \tau_1} T^{L2}_{1001}(\tau_{L2}) T^{L1}_{0100}(\tau_{L1}) + \text{c.c.} \right\}. \quad (9.73)$$

It is apparent from (9.73) that the off-diagonal element decreases exponentially as the pulse delays τ_1 and τ_2 increase; the rate parameters are given by τ_{10} and T_{10}, respectively. Thus the information about the coherent interaction between the first and the second pulse, which occurs for mutual pulse delays at most of the order of the phase decay time τ_{10}, is stored for longer times, namely for those of the order of T_{10}. The expression for τ_{10} from (9.73) is then substituted into (9.65b), and the spatial integration over the volume V_F of the fluorescent sample is performed. As expected, the resultant diffraction signal is sharply peaked in the directions of

$$k_\bullet = k_{3\bullet} \pm \delta k_\bullet \quad (9.74)$$

where

$$\delta k_\bullet = k_{2\bullet} - k_{1\bullet}$$

represents the wave vector of the saturation grating. We restrict our consideration to diffraction into the direction of $k_\bullet = k_{3\bullet} + \delta k_\bullet$ and discuss the cases of homogeneously and inhomogeneously broadened transitions.

9.3.3.1. Homogeneously Broadened Transitions. With the assumption of very short and strong excitation by the first and the second pulse ($\tau_L \ll \tau_{10}, T_{10}$, $\Omega_{R0} \gg \tau_{10}^{-1}$), we obtain by substituting (9.73) into (9.65b) the relation

$$\mathscr{E}_V^{\text{coh}}(R_\bullet, t) \propto f(k_{3\bullet} + k_{2\bullet} - k_{1\bullet} - k_\bullet) e^{-2\tau_2/T_{10}} e^{-2\tau_1/\tau_{10}}$$

$$\times \left| \frac{\Omega^{(3)}_{R0}}{\Delta_3 - i/\tau_{10}} \left[1 - e^{-(i\Delta_3 + 1/\tau_{10})(T - t_4)}\right] \right.$$

$$\times \frac{\Omega^{(1)}_{R0}}{[\Omega^{(1)}_R]^2} \left[\Delta_1 (1 - \cos \Omega^{(1)}_R \tau_{L1}) - i\Omega^{(1)}_R \sin \Omega^{(1)}_R \tau_{L1}\right]$$

$$\left. \times \frac{\Omega^{(2)}_{R0}}{[\Omega^{(2)}_R]^2} \left[\Delta_2 (1 - \cos \Omega^{(2)}_R \tau_{L2}) - i\Omega^{(2)}_R \sin \Omega^{(2)}_R \tau_{L2}\right] \right|^2 \quad (9.75)$$

In the case of resonant excitation ($\Delta_i = 0$, $i = 1, 2, 3$) this coherently emitted signal is seen to decrease exponentially with the pulse delays τ_1 and τ_2 as

$e^{-2\tau_1/\tau_{10}}$ and $e^{-2\tau_2/T_{10}}$. Furthermore we note that then the diffraction efficiency varies periodically with the pulse area $\Omega_{R0}^{(i)}\tau_{Li}$ of the first and the second pulse.

9.3.3.2. Inhomogeneously Broadened Transitions. Provided the midfrequencies of all three pulses coincide with the center frequency of the transition, we obtain

$$\mathscr{E}_V^{\text{coh}}(R_\bullet, t) \propto f(k_{3\bullet} + k_{2\bullet} - k_{1\bullet} - k_\bullet)e^{-2\tau_1/\tau_{10}}e^{-2\tau_2/T_{10}}\frac{1}{\pi\Gamma^2}$$
$$\times\Bigg|\int_{-\infty}^{\infty} d\Delta\, e^{-\Delta^2/\Gamma^2}\frac{\Omega_R^{(3)}}{\Delta - i/\tau_{10}}\left[1 - e^{-(i\Delta+\tau_{10}^{-1})(T-t_4)}\right]e^{i\Delta\tau_1}$$
$$\times\frac{\Omega_{R0}^{(1)}}{\left[\Omega_R^{(1)}\right]^2}\left[\Delta\left(1 - \cos\Omega_R^{(1)}\tau_{L1}\right) - i\Omega_R^{(1)}\sin\Omega_R^{(1)}\tau_{L1}\right]\frac{\Omega_{R0}^{(2)}}{\left[\Omega_R^{(2)}\right]^2}$$
$$\times\left[\Delta\left(1 - \cos\Omega_R^{(2)}\tau_{L2}\right) - i\Omega_R^{(2)}\sin\Omega_R^{(2)}\tau_{L2}\right]\Bigg|^2. \tag{9.76}$$

If the conditions $\Omega_{R0}^{(i)} \gg \Gamma$, $\tau_1 > 1/\Gamma$, and $T > 1/\Gamma$ are satisfied, the signal decreases with τ_1 as $e^{-2\tau_1/\tau_{10}}$.

The situation becomes far more complex if the diffraction of a probe pulse of small delay τ_2 is investigated. Then all terms of σ_{10} from (9.66e) contribute to the signal.

9.3.3.3. Evaluation of Relaxation Parameters. As apparent from (9.75) and (9.76), the light intensity diffracted into the direction of $k_\bullet = k_{3\bullet} + k_{2\bullet} - k_{1\bullet}$ depends on the delay time τ_2 as $e^{-2\tau_2/T_{10}}$ for homogeneously and inhomogeneously broadened transitions. Thus the lifetime T_{10} can be determined by the measurement of the total energy diffracted into the direction of $k_\bullet$ at different delay times τ_2. Further, if the first and the second pulse are extremely strong ($\Omega_{R0}^{(i)} \gg \Gamma$, $i = 1, 2$), the phase relaxation time τ_{10} can be directly determined from the dependence of the total energy that is diffracted into the direction of $k_\bullet = k_{3\bullet} + k_{2\bullet} - k_{1\bullet}$ on the pulse delay time τ_1. With T and τ_1 large compared with $1/\Gamma$, the total energy depends on τ_1 as $e^{-2\tau_1/\tau_{10}}$.

In concluding this subsection let us mention that by employing the procedures derived so far we can also calculate the time behavior of $\sigma_{10}(r_\bullet, t)$ and of $\mathscr{E}_V(R_\bullet, t)$ after the cessation of the third pulse. From this, further diffraction and echo phenomena are obtained, which have been discussed qualitatively for the special case of copropagating waves in Section 8.4.3.

9.3.4. Experimental Observations

Experiments demonstrating very clearly self-diffraction and probe-beam diffraction, on the condition that the exciting pulses do not collide with each

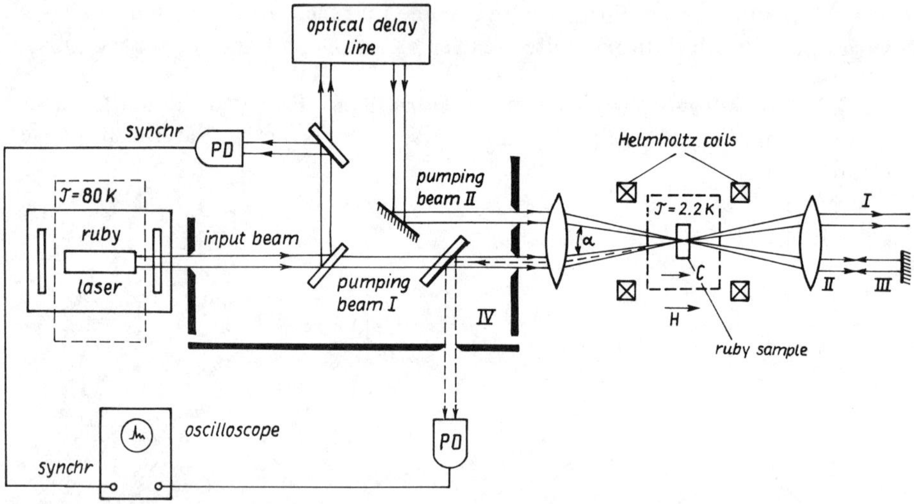

Fig. 9.10. Experimental setup for generating and observing transient gratings in ruby (after Ref. 9.26).

other in the sample and that the phase memory of the absorber is of decisive influence, have been carried out by Styrkov et al. (Ref. 9.26). (Similar phenomena had already been observed in connection with photon-echo experiments where the two pump pulses passed noncollinearily through the sample; see Chapter 8.) Figure 9.10 shows the experimental setup employed for generating and observing transient gratings. This setup resembles those used for stimulated photon-echo experiments (Section 8.4.3), except that the interaction with the third pulse proceeds in a different way. A ruby laser operated in the Q-switched mode and tuned by cooling its active medium is used as the light-pulse generator that delivers pulses of about 1-MW power and 10-ns duration. The laser pulse is split into two exciting pulses, one of which experiences a variable delay. The two pulses are focused into the sample, where the directions of the light beams make an angle α. The third pulse is produced by reflecting the second one at the flat mirror behind the sample. This reflected pulse satisfies the Bragg condition for the transient grating, and hence the diffracted pulse and the diffracted echo satisfy the condition (9.74) with the wave vector $k_{\bullet}$ pointing in the direction of $-k_{1\bullet}$ for all values of the angle α. The signal propagating into the direction of $-k_{1\bullet}$ is observed by means of a fast photodetector and an oscilloscope with small rise time. The overall rise time of the detection device is shorter than 1 ns. The sample, a slice of 1.5-mm thickness, cut perpendicularly to the optical c axis of a low-concentration ruby crystal, is irradiated by a sequence of light pulses. To achieve a long phase decay time, the sample is cooled down to 2.2 K. At this temperature the phase decay time of the transition ${}^4A_2 \leftrightarrow {}^2E(\bar{E})$ is of the order of 10^{-7} s, whereas $1/\Delta\omega_{\text{inh}} \sim 10^{-10}$ s and $T_{10} \approx 3 \times 10^{-3}$ s. Resonance between the laser transi-

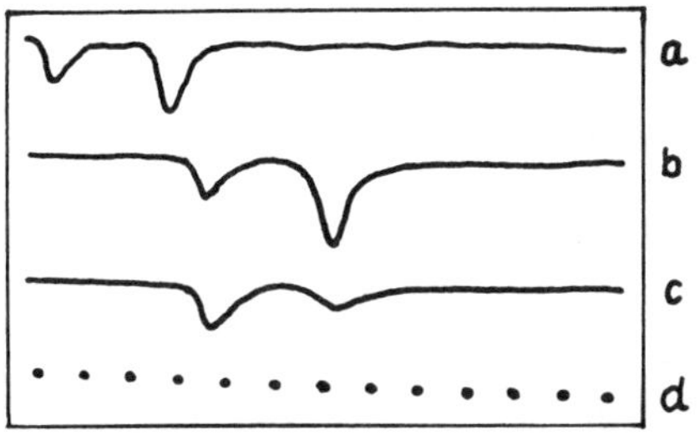

Fig. 9.11. Oscilloscope traces of photodetector signals originating from (a) the exciting light pulses I and II, (b) the self-diffracted signal and the stimulated photon-echo signal (the angle β between $c_{\bullet}$ and $H_{\bullet}$ equals $+12°$), (c) the same signals as in (b) where $\beta = -15°$, (d) time calibration (time between adjacent points 20 ns.) (After Ref. 9.26.)

tion ${}^4A_2(M_S = \pm \frac{1}{2}) \leftrightarrow {}^2E(\bar{E})$ and the sample transition ${}^4A_2(M_S = \pm \frac{3}{2})$ $\leftrightarrow {}^2E(\bar{E})$ is achieved by cooling the laser rod to 80 K and the sample to 2.2 K. All light beams are polarized perpendicularly to the plane of incidence.

Figure 9.11 presents typical experimental results. The upper trace of the oscilloscope shows the two exciting pulses, whose duration is of about 10 ns; the delay between them is 50 ns; the two lower traces show the signal emitted into the direction IV for two different orientations of the weak magnetic field with respect to the c axis of the ruby crystal. The signal consists of the diffracted probe pulse and the stimulated photon echo. Note that the signal will disappear either if one of the exciting pulses is absent or if the resonance condition is violated, for example, by heating the laser rod.

Let us add that the diffracted wave and the echo wave possess *phase-conjugated wave fronts* with respect to the input beam. This is evident from the fact that the signal wave IV is a parallel light beam, though it has passed through the lens L. (We shall return to the phenomenon of phase conjugation in Chapter 14.)

In Ref. 9.25 the time resolution of such experiments has strongly been increased by employing the subpicosecond light pulses of passively mode-locked dye lasers. Figure 9.12 gives the experimental scheme for observing the self-diffraction phenomenon. At wavelengths around 610 nm the cw-pumped and passively modelocked dye laser generates pulses with a peak power of 100 W, a duration of 0.15–0.3 ps, and a pulse repetition rate of the order of 100 MHz. The pulse train is mechanically chopped and split into the two beams 1 and 2 by means of a beam splitter. One of the two beams undergoes a variable optical delay. Then both the pulses are focused by one lens of $f = 15$ cm into the 1-mm-thick sample from different directions, which are characterized by the wave vectors $k_{1\bullet}$ and $k_{2\bullet}$. These wave vectors make a small angle of 2°. At a rather large distance from the sample the diffracted light beams traveling into the directions of $k_{3\bullet} = 2k_{1\bullet} - k_{2\bullet}$ and $k_{4\bullet} = 2k_{2\bullet} - k_{1\bullet}$ are measured through an aperture. Each of the two detection systems consists of a monochromator, a photomultiplier, and a lock-in amplifier, the latter being operated synchronously with the chopper wheel. The final correlation traces are recorded on a pen recorder against the delay between the two pulse trains.

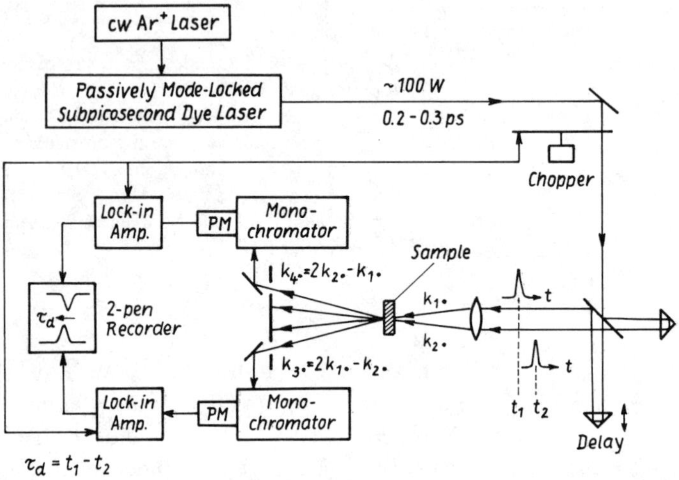

Fig. 9.12. Experimental scheme for observing self-diffraction of subpicosecond light pulses (after Ref. 9.25).

Figure 9.13 shows a representative result gained for the dye DODCI solved in ethanol. The measured correlation traces exhibit a nearly symmetric dependence of the signals diffracted into the $k_{3\bullet}$ and $k_{4\bullet}$ directions on the pulse delay. This symmetry implies that the signal shape is almost pulsewidth-limited, and hence the signal does not provide evidence on the influence of relaxation effects. The pulse duration calculated from this signal, which is a third-order correlation function of the pulse, accords with the result of other correlation measurements, for instance, with results from second-harmonic generation (cf.

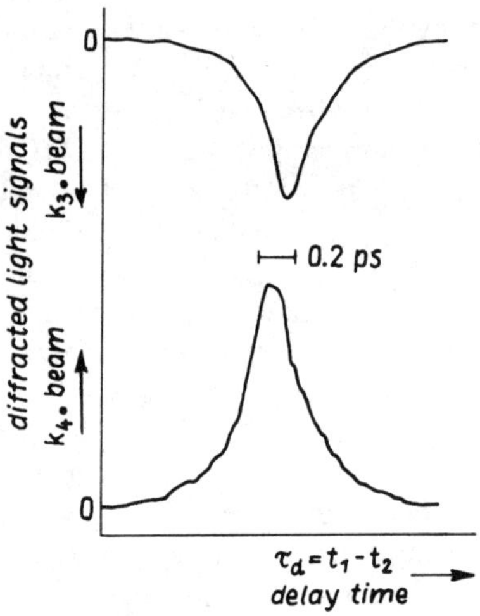

Fig. 9.13. Measured correlation traces of two diffracted signals (after Ref. 9.25).

Section 11.3). As concerns the relaxation parameters, only an upper limit on the phase decay time τ_{10} can be estimated from the signal width. This limit is 0.5 ps for a pulse width of 0.25 ps. It can be decreased by about one order of magnitude if one employs the shortest light pulses now available. But by evaluating the shift of the signal maximum with respect to $t_D = t_1 - t_2 = 0$ rather than the width of the signal, it is possible to estimate very short relaxation parameters without applying extremely short light pulses. Such shifts can be measured fairly precisely, since the light pulses diffracted into the direction of $k_{3\bullet}$ and $k_{4\bullet}$ experience shifts of equal magnitude but with opposite signs. Comparing the measured shift with the calculated one, it is possible to estimate $\tau_{10} \lesssim 0.05$ ps for the dye DODCI. (Note that in such calculations, which follow the procedure explained in Section 9.3.2, the real pulse shape has to be taken into account.) Thus this experiment gives evidence of extremely short phase decay times in dye solutions. This is in agreement with results obtained by employing more indirect methods of cw spectroscopy, as for instance the resonant Rayleigh-type mixing spectroscopy (cf. Ref. 9.27).

The two experiments described above demonstrate the interaction of rather short and strong pulses in saturable absorbers. Now we consider pulses that are long compared with the phase decay time ($\tau_L \gg \tau_{10}$) and not extremely strong ($\sigma_{10}^{(1)} I \lesssim 1/T_{10}$ with $\sigma_{10}^{(1)}$ the one-photon cross section). Supposing these pulse properties, we are allowed to use the stationary solution for the off-diagonal element σ_{10} of the density matrix. Now information on the interacting coherent light pulses can only be stored if the two exciting light pulses overlap in the sample. The interference of the coherent light waves yields a wave pattern that generates a saturation pattern of the occupation-number densities of the atomic levels. This pattern, which is caused by nonlinear interaction between light and matter, then causes self-diffraction of the two exciting pulses as well as the diffraction of a third interrogating pulse.

Thus such diffraction phenomena may considerably affect experiments in excite-and-probe pulse spectroscopy performed in order to measure the lifetimes of molecular transitions. On condition that the excitation and probe pulses are incident on the sample from different directions and are overlapping in space and time, they may cause a grating in the molecular occupation numbers. The diffraction of the two incident pulses gives rise to new light waves, and furthermore light is diffracted from one pulse into the other. From this second effect an "amplification" of the weaker pulse may arise. Thus for delay times allowing an overlap of the pulses, the dependence of the probe-beam transmission on the pulse delay t_D is modified by a *coherent coupling*. If the incident light pulses are not transform-limited but show a strong phase modulation, the maximum delay time up to which the pulses experience coherent coupling is no longer determined by the pulse overlap alone but also by this modulation. Figure 9.14 gives representative results for the probe-beam transmission of a sample composed of saturable dye molecules in the case of transform-limited and phase-modulated pulses (Ref. 9.28). "Coherence peaks"

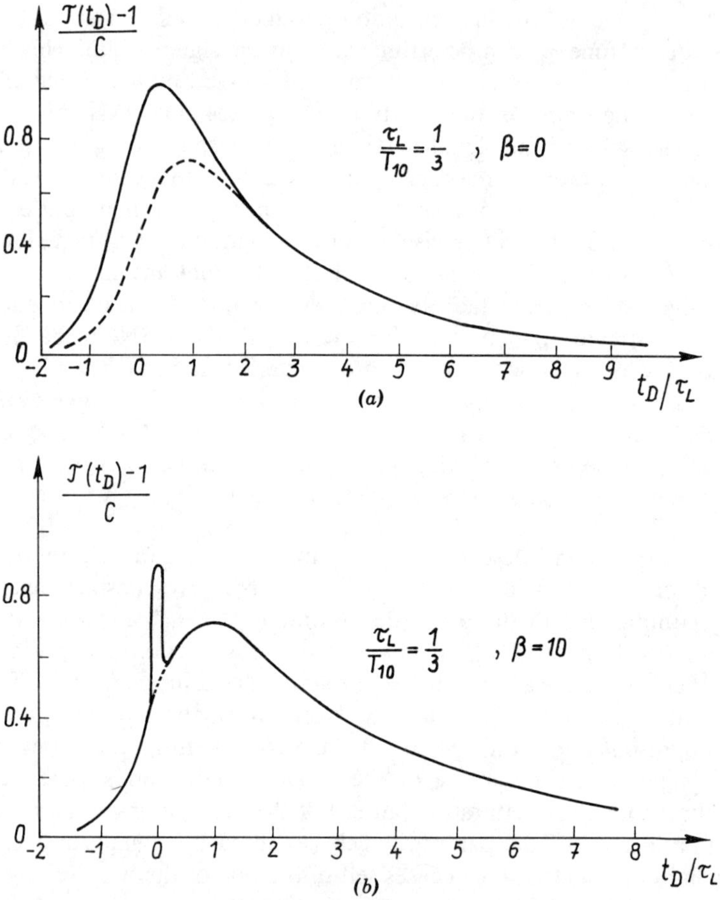

Fig. 9.14. Energy transmission for Gaussian-shaped light pulses of maximum intensity I_0 and duration τ_L against the delay t_D between the exciting pulse and the probe pulse (after Ref. 9.28) in the case of (a) transform-limited light pulses, (b) phase-modulated light pulses. The phase modulation $\varphi(\eta)$ of the entrance signal as a function of the local time η is given as $\varphi(\eta) = 2\sqrt{\ln 2}\,\beta(\eta/\tau_L)^2$, where β is a scaling factor and $C = (\sigma_{10}^{(1)} I_0 \tau_L)(\gamma \sigma_{10}^{(1)} L)$ is a normalization parameter ($\sigma_{10}^{(1)}$ = absorber cross section, T_{10} = lifetime, γ = number of absorbing molecules per unit volume, L = sample length).

such as are shown in Fig. 9.14*b* have been measured by several authors (e.g. Refs. 9.22, 9.23, 9.29).

The diffraction of a third weak probe pulse following the two exciting pulses has been used by Phillion, Kuizenga, and Siegman (9.24) in order to determine the decay of the induced transient grating. The grating decays away in time as the molecules that are photoexcited by the two strong pulses relax or as the excitation spreads in space by diffusion or energy transfer. The influence of diffusion and energy transfer can be changed by varying the grating spacing via the angle at which the exciting beams intersect in the sample. Thus the local and the nonlocal decay processes can be determined separately (see, e.g., Ref. 9.30). Furthermore, by employing light pulses of definite polarization direction it is possible to measure the influence of the orientational motion of the excited molecules (e.g., Ref. 9.31).

9.4. SUPERFLUORESCENCE

In this chapter we have so far neglected the retroaction of the electromagnetic field of the emitted light on the ensemble of oscillating atomic dipoles. In the following we discuss the assumptions that justify this neglect and investigate the effects of such retroaction on the emission of the ensemble of radiating atoms. We are particularly interested in knowing under what conditions after incoherent pulse excitation the specific coherent radiation phenomenon of superfluorescence may occur in atomic ensembles. Dicke was the first to direct his attention to the buildup of such a cooperative radiation phenomenon (Ref. 9.32). He investigated an ensemble of atoms in a small volume whose maximum linear extensions were much less than the wavelength. It was shown by him that a macroscopic polarization and an emission signal whose power is proportional to the square of the particle number may originate not only from the coherent excitation of atoms (cf. Chapter 8), but also from incoherent excitation. In the latter case the atoms of the ensemble start to radiate while uncoupled, and a macroscopic polarization evolves only in the course of time.

So far it has not been possible to establish experimental conditions that precisely correspond to the assumptions made in Ref. 9.32. However more elaborate theories (e.g., Refs. 9.33–9.41) and experiments (e.g., Refs. 9.42–9.48) have shown that under appropriate, but not so restrictive conditions such cooperative emission phenomena may occur in samples whose linear dimensions are much greater than λ. To give an adequate theoretical description of such ensembles is however extremely difficult, since the quantum-theoretical equations of motion must be solved for the spatially extended system of atoms and field. To date, this complex problem has not been solved satisfactorily. There are however various approximate descriptions available by means of which it is possible to explain the basic experimental phenomena (see, e.g., Refs. 9.39, 9.49–9.51).

In the present discussion we restrict our attention to some rough estimates, following Refs. 9.35 and 9.49. To establish geometrical conditions that are both realizable in experiment and as simple as possible, we assume a pencil-shaped sample of length L and cross section A with a Fresnel number $\mathsf{F} = A/\lambda L = 1$. (In case $\mathsf{F} > 1$, the problem becomes more difficult because of the occurrence of many transverse modes; in case $\mathsf{F} < 1$, because of heavy diffraction losses of the mode.) We assume the sample to consist of atoms that at the time $t = 0$ are excited by a short pulse of length τ_{ex} to the upper level 1 of the radiation transition under study.

First we evaluate the time after which one single atom on the average spontaneously emits a photon within such a small solid angle around the longitudinal axis of the sample as to enable this photon to pass the entire length L of the sample and, in consequence, to interact with a large number of excited atoms. Provided that the atomic lifetime, denoted as T_{10}, is only determined by the spontaneous emission of the single atom, this time is

$$\tau_{R1} = T_{10}\left(\frac{4\pi L^2}{A}\right). \tag{9.77}$$

If there are altogether N atoms in the sample, N being given by the number density γ and the sample volume AL, we have

$$\tau_R = fT_{10}\left(\frac{4\pi L^2}{A}\right)\frac{1}{N} = fT_{10}\frac{4\pi L}{\gamma A^2} \tag{9.78}$$

for the average time after which one arbitrary atom of the entire ensemble emits a photon into the appropriate solid angle. Here f is a numerical factor of the order of unity that accounts for the different positions of the atoms of the ensemble; with homogeneous particle distribution more detailed calculations yield $f = \frac{2}{3}$ (Ref. 9.40). Assuming $\mathsf{F} = A/(L\lambda) = 1$, we obtain

$$\tau_R = \frac{8\pi}{3\gamma L\lambda^2}T_{10}. \tag{9.79}$$

This time τ_R is needed to build up, by way of spontaneous emission, a coupling between atomic systems and to emit radiation by way of cooperative emission. This cooperative effect can come into play only if τ_R is smaller than the single-atom relaxation times τ_{10} and T_{10}, after which, respectively, the phase of the single oscillating dipoles is destroyed and the excitation of the single atoms decays altogether. Furthermore, in inhomogeneously broadened samples τ_R must be smaller than the macroscopic phase decay time $1/\Delta\omega_{\text{inh}}$, where $\Delta\omega_{\text{inh}}$ is the width of the inhomogeneous line. For cooperative radiation phenomena to occur, it is hence indispensable that

$$\tau_R < T_{10}, \tau_{10}, \frac{1}{\Delta\omega_{\text{inh}}}. \tag{9.80}$$

For the case of naturally broadened atomic systems, where $\tau_{10} = 2T_{10}$, the behavior of the ensemble can be described by the single parameter $C = T_{10}/\tau_R$. For $C < 1$ the atoms mainly radiate independently of each other; for $C > 1$ cooperative behavior is predominant. According to (9.79) the time τ_R can be varied widely through its dependence on the number density γ and the sample length L, and thus it is possible to establish the case $C < 1$, to which our consideration was restricted in the preceding sections of this chapter.

For pronounced and easily interpretable superfluorescence to occur, other conditions besides (9.80) must be satisfied. In particular the sample is required to be not too long in order to guarantee that the escape time

$$\tau_E = \frac{L}{c} \tag{9.81}$$

is smaller than τ_R. Instead of τ_E, the so-called *cooperation length*

$$L_C = \sqrt{\frac{8\pi c T_{10}}{3\gamma\lambda^2}}, \tag{9.82a}$$

which results from (9.79) with $\tau_R = \tau_E$, is used, and it is required that $L < L_C$. The quantity

$$\tau_C = \frac{L_C}{c} \tag{9.82b}$$

is called the *cooperative time*, and it gives the value of τ_R for a sample of length L_C. For $L < L_C$ the cooperative emission approximately yields one single pulse, whereas at $L > L_C$ several pulses may occur in succession.

To give a rough description of the time evolution of superfluorescence after excitation, another time, the so-called *delay time* τ_D, may be evaluated. While the maximum of the incoherent single-atom emission coincides with the position of the δ pulse excitation, in cooperative emission the radiation pulse builds up rather slowly, and its maximum is delayed by τ_D relative to the excitation (analogous statements apply to the more complex cases, where a pulse sequence is emitted instead of a single pulse).

To perform a rough evaluation of τ_D in a simple way, we examine a sample with $L \ll L_C$ and with completely inverted, homogeneously broadened atomic two-level systems. We assume that initially a field-strength pulse with a pulse area θ_0' originates from spontaneous emission. While the buildup of the initial pulse can be only described fully quantum-theoretically, the following processes can be treated semiclassically. In doing this, we are allowed to start from the relaxation-free equation of motion (8.58) for the pulse area θ, since the processes are assumed to proceed in time intervals that are much shorter than τ_{10}. In the local coordinates $\eta = t - z/v$, $\xi = z$ this equation reads

$$\frac{\partial^2\theta}{\partial\eta\,\partial\xi} = -\frac{\sin\theta}{\tau_R L}. \tag{9.83}$$

In Section 8.5 we were only interested in the distortionless solutions of (9.83), which led us from the partial differential equation to an ordinary one, but now we want to consider the evolution of the pulse in space and time.

Since we again aim at employing a very simple treatment, we shall only investigate the time evolution of the output pulse at $\xi = L$, roughly approximating the spatial variation of the signal within the sample. For this purpose we integrate (9.83) with respect to ξ, neglect in the result $(\partial\theta/\partial\eta)_{\xi=0}$ compared with $(\partial\theta/\partial\eta)_{\xi=L}$, and perform the estimation

$$\int_0^L d\xi \sin\theta(\xi, \eta) \approx \frac{L}{2}\sin\theta(L, \eta). \tag{9.84}$$

This is of course justified only for very short samples and weak output pulses. We now obtain for $\theta(\eta) \equiv \theta(L, \eta)$ the ordinary differential equation

$$\frac{d}{d\eta}\theta(\eta) = -\frac{\sin\theta(\eta)}{2\tau_R}. \tag{9.85}$$

It is easy to show that (9.85) is equivalent to the pendulum equation for the quantity $\alpha = 2\theta$. We differentiate with respect to η, replace the first derivative of θ on the right-hand side of the differentiated equation according to (9.85), and replace 2θ by α [which is connected with the field-strength envelope $\bar{E}$ by way of $(\partial/\partial\eta)\alpha = (2d/\hbar)\bar{E}$]. Thus we obtain

$$\frac{d^2}{d\eta^2}\alpha(\eta) = \frac{\sin\alpha(\eta)}{(2\tau_R)^2}. \tag{9.86}$$

According to the results of Section 8.5.2 this equation is solved by the field-strength pulse

$$\bar{E}(\eta) = \frac{\hbar}{d\tau_R}\operatorname{sech}\left(\frac{\eta - \tau_D}{2\tau_R}\right), \tag{9.87}$$

where τ_D is the delay of the pulse maximum. For $\tau_D \gg \tau_R$ and with $\theta(0) \ll 1$ we obtain from the equations (9.85) to (9.87), by integrating $\bar{E}(\eta)$ from $-\infty$ to 0,

$$\tau_D \approx 2\tau_R \ln\frac{2}{\theta(0)}. \tag{9.88}$$

It is apparent that the delay time is connected with the initial value of the pulse area $\theta(0)$, which, as mentioned above, cannot be determined within the frame of our semiclassical consideration. For small samples quantum-theoretical calculations and estimates (see, e.g., Refs. 9.53–9.55, 9.35) reveal that $\theta(0)$ is in approximation inversely proportional to the square root of the number N of active particles, and in Ref. 9.35

$$\theta(0) \approx \frac{2}{\sqrt{N}}$$

is given as an approximate expression. Thus we finally have

$$\tau_D \approx 2\tau_R \ln\sqrt{N}\,. \tag{9.89}$$

We see that for the generation of a pronounced superfluorescence pulse the following time hierarchy results for the escape time τ_E, the cooperation time τ_C, the buildup time τ_R for cooperation by radiation, the delay time τ_D, the relaxation times T_{10}, τ_{10}, and the reciprocal value of the inhomogeneous line width $\Delta\omega_{\text{inh}}$:

$$\tau_E < \tau_C < \tau_R < \tau_D < T_{10},\, \tau_{10},\, \frac{1}{\Delta\omega_{\text{inh}}}\,. \tag{9.90}$$

A superfluorescence pulse was first unambiguously observed by Skribanowitz et al. (9.42) (see Fig. 9.15). By means of a short laser pulse at $\lambda = 2.5\ \mu\text{m}$ an inversion was generated in hydrogen fluoride gas on the rotational transition $(v = 1,\ J = 3) \rightarrow (v = 1,\ J = 2)$. While in noncooperative spontaneous emission one may expect the maximum of the fluorescence radiation to appear at the end of the excitation pulse, followed by an exponential decay (see Fig. 9.15*b*), in the experiment strong pulses of about 200-ns duration were observed that followed the excitation pulse after a time of about 1 μs. The pulses propagated in both directions along the sample axis with small divergence.Note that in superradiance, when the macroscopic polarization results from one coherent light pulse with given direction of propagation, the radiation will be emitted in this one direction only. The experiments showed that the superfluorescence signal increases with the square of the particle number. In contrast

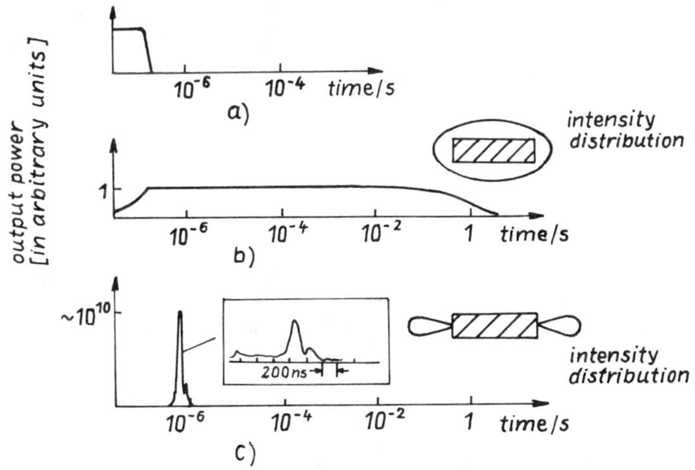

Fig. 9.15. Superfluorescence experiment in hydrogen fluoride gas (after Ref. 9.42): (*a*) excitation pulse, (*b*) output expected from noncooperative spontaneous emission, (*c*) observed output.

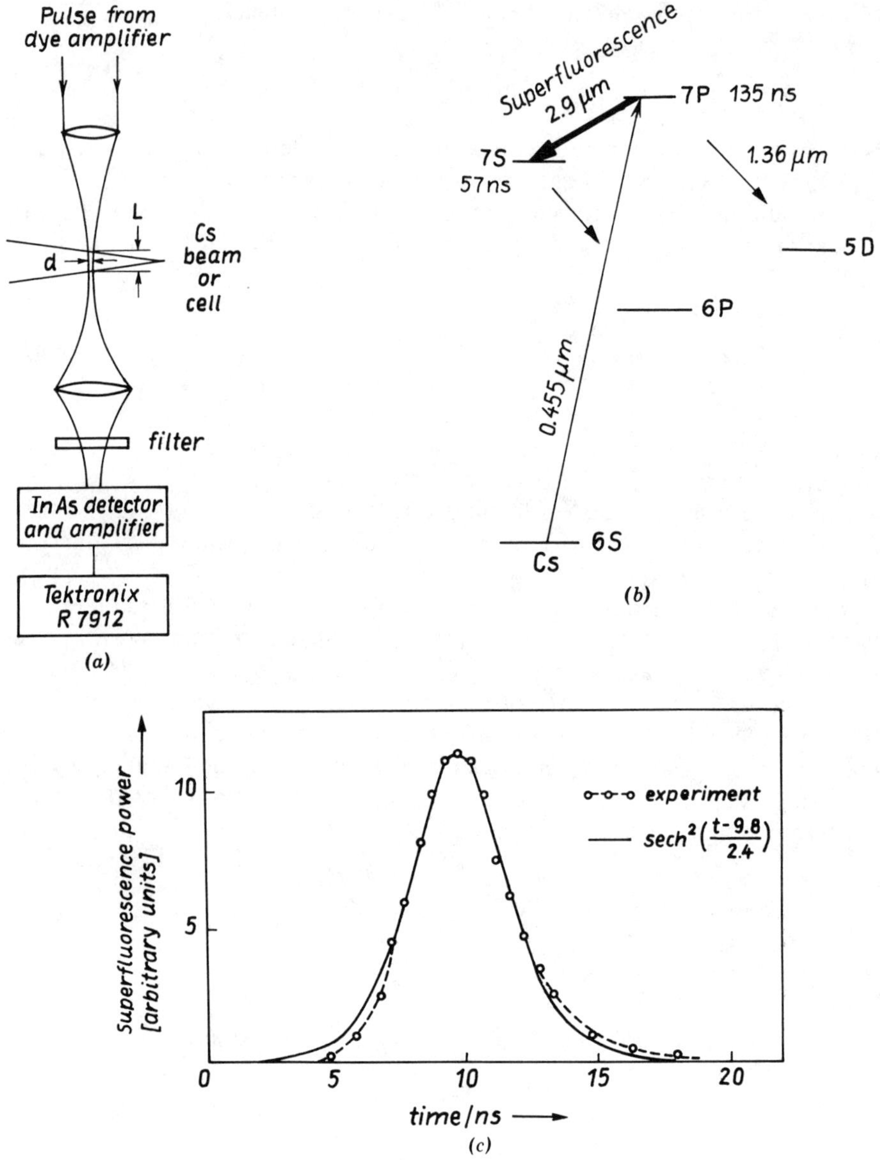

Fig. 9.16. Superfluorescence experiment in a cesium beam (after Ref. 9.43): (*a*) experimental scheme, (*b*) level diagram with excitation transition and superfluorescence transition, (*c*) superfluorescence power versus time (the measured delay τ_D is about 10 ns).

to our simplified model the experiment showed that in general no symmetrical single pulses are emitted, but a succession of pulses may appear, which is referred to as pulse ringing.

In Ref. 9.43 experiments are reported on the $(7\,^2P_{3/2}) \to (7\,^2S_{1/2})$ transition of cesium (see Fig. 9.16), with special care being taken that the conditions given above be satisfied ($\tau_E \approx 0.067$ ns, $\tau_C \approx 0.18$ ns, $\tau_R \approx 0.5$ ns, $\tau_D \approx 10$ ns, $T_{10} \approx 70$ ns, $\tau_{10} \approx 80$ ns, $1/\Delta\omega_{\text{inh}} \approx 32$ ns). Under these conditions symmetrical single pulses with sech^2 temporal profile (see Fig. 9.16*c*) are obtained in good approximation. This fact, however, must not be regarded as a proof that our simple model could provide a satisfactory quantitative description of the time evolution. Actually, under the given conditions and assuming plane waves, computer simulations in general reveal that pulse ringing occurs. It is supposed that in the actual experiment the influence exerted by relaxation, transverse mode structure, and diffraction on the evolution of cooperation prevents the occurrence of pronounced pulse ringing (cf. Refs. 9.49, 9.51). Only more detailed quantum-theoretical investigations of the entire space–time-dependent process can be expected to yield full quantitative agreement with experiment.

REFERENCES

9.1 V. Weisskopf, *Ann. d. Phys.* **9**, 23 (1931).

9.2 S. G. Rautian and I. I. Sobelman, *Zh. Eksp. i Teor. Fiz.* **41**, 456 (1961).

9.3 M. C. Newstein, *Phys. Rev.* **167**, 89 (1967).

9.4 B. R. Mollow, *Phys. Rev.* **188**, 1969 (1969).

9.5 C. R. Stroud, *Phys. Rev. A* **3**, 1044 (1971).

9.6 H. J. Kimble and L. Mandel, *Phys. Rev. A* **13**, 2123 (1976).

9.7 H. J. Kimble and L. Mandel, *Phys. Rev. A* **15**, 698 (1977).

9.8 H. J. Carmichael and D. F. Walls, *J. Phys. B* **9**, L43 (1976).

9.9 J. Herrmann, K. E. Süsse, and D. Welsch, *Ann. d. Phys.* **80**, 37 (1973).

9.10 H. Paerschke, K. E. Süsse, and D. Welsch, *Ann. d. Phys.* **33**, 215, 228 (1976).

9.11 M. Schubert, K. E. Süsse, and W. Vogel, *Opt. Comm.* **30**, 275 (1979).

9.12 H. Paerschke, K. E. Süsse, and B. Wilhelmi, *Opt. and Quant. Electron.* **15**, 41 (1983).

9.13 H. J. Kimble, M. Dagenais, and L. Mandel, *Phys. Rev. Lett.* **39**, 691 (1977).

9.14 H. J. Kimble, M. Dagenais, and L. Mandel, *Phys. Rev. A* **18**, 201 (1978).

9.15 M. Dagenais and L. Mandel, *Phys. Rev. A* **18**, 2217 (1978).

9.16 E. Jakeman, E. R. Pike, P. N. Pusey, and J. M. Vaughan, *J. Phys. A* **10**, L257 (1977).

9.17 H. J. Carmichael, P. Drummond, and D. F. Walls, *J. Phys. A* **11**, L121 (1978).

9.18 K. E. Süsse, W. Vogel, D. G. Welsch, and B. Wilhelmi, *Opt. Comm.* **28**, 389 (1979).

9.19 M. Schubert, K. E. Süsse, W. Vogel, D. G. Welsch, and B. Wilhelmi, *Kvant. Electron.* **9**, 495 (1982); Forschungsergebnisse der Universität Jena, Preprint N/79/36, 1979.

9.20 W. Neuhauser, M. Hohenstatt, P. Toschek, and H. Dehmelt, *Phys. Rev. Lett.* **41**, 233 (1978).

9.21 D. J. Wineland and W. M. Itano, *Phys. Lett.* **82A**, 75 (1981).

9.22 C. V. Shank and D. H. Auston, *Phys. Rev. Lett.* **34**, 479 (1976).

9.23 C. V. Shank, E. P. Ippen, and R. Bersohn, *Science* **193**, 50 (1976).

9.24 D. W. Phillion, D. J. Kuizenga, A. E. Siegman, *Appl. Phys. Lett.* **27**, 85 (1975).

9.25 T. Yajima, Y. Ishida, and Y. Taira, "Investigation of subpicosecond dephasing processes by transient spatial parametric effect in resonant media," in R. M. Hochstrasser, W. Kaiser, and C. V. Shank (Eds.), *Picosecond Phenomena II*, Springer, Berlin (West), Heidelberg, New York, 1980.

9.26 E. I. Styrkov, N. L. Nevelskaya, V. S. Kobkov, and N. G. Yarmukhamotov, *Phys. Stat. Sol.* (*B*) **98**, 473 (1980).

9.27 T. Yajima, H. Souma, and Y. Ishida, *Phys. Rev. A* **17**, 309, 324 (1978).

9.28 B. Wilhelmi and J. Herrmann, *Kvant. Elektron.* **7**, 1876 (1980).

9.29 M. A. Vasil'eva, J. Vichakas, V. Gulbinas, V. I. Malyshev, A. V. Maslov, V. Kabelka, and V. Syrus, *IEEE J. Quant. Electron.* **QE-19**, 724 (1983).

9.30 V. L. Vinezkij, N. V. Kukhtarev, S. G. Odulov, and M. S. Soskin, *Usp. Fiz. Nauk* **129**, 113 (1979).

9.31 A. von Jena and H. E. Lessing, *Opt. Quant. Electron.* **11**, 419 (1979).

9.32 R. H. Dicke, *Phys. Rev.* **93**, 99 (1954).

9.33 R. Bonifacio and P. Schwendimann, *Lett. Nuovo Cimento* **3**, 509 (1970).

9.34 R. Bonifacio, P. Schwendimann, and F. Haake, *Phys. Rev. A* **4**, 382 (1971).

9.35 R. Bonifacio and L. A. Lugiato, *Phys. Rev. A* **11**, 1507 (1975); **12**, 587 (1975).

9.36 D. Polder, M. F. H. Schuurmans, and Q. H. F. Vrehen, *Phys. Rev. A* **19**, 1192 (1979).

9.37 R. Glauber and F. Haake, *Phys. Lett.* **68A**, 29 (1978).

9.38 M. F. H. Schuurmans, D. Polder, and Q. H. F. Vrehen, *J. Opt. Soc. Am.* **68**, 699 (1978).

9.39 F. Haake, J. W. Hans, H. King, G. Schröder, and R. Glauber, *Phys. Rev. A* **23**, 1322 (1981).

9.40 J. C. MacGillivray and M. S. Feld, "Superradiance," in M. S. Feld and V. S. Letokhov (Eds.), *Coherent Nonlinear Optics*, Springer, Berlin (West), Heidelberg, New York, 1980.

9.41 R. Bonifacio, J. D. Farina, and L. M. Narducci, *Opt. Comm.* **31**, 377 (1979).

9.42 N. Skribanowitz, I. P. Herman, J. C. MacGillivray, and M. S. Feld, *Phys. Rev. Lett.* 30, 309 (1973).

9.43 H. M. Gibbs, Q. H. F. Vrehen, and H. M. J. Hikspoors, *Phys. Rev. Lett.* **39**, 547 (1977).

9.44 A. Flusberg, T. Mossberg, and S. R. Hartmann, *Phys. Lett.* **58A**, 373 (1976).

9.45 M. Gross, C. Fabre, P. Pillet, and S. Haroche, *Phys. Rev. Lett.* **36**, 1035 (1976).

9.46 M. Gross, J. M. Raimond, and S. Haroche, *Phys. Rev. Lett.* **40**, 1711 (1978).

9.47 J. Okada, K. Ikeda, and M. Matsuoka, *Opt. Comm.* **26**, 189 (1978); **27**, 321 (1978).

9.48 J. C. MacGillivray and M. S. Feld, *Contemp. Phys.* **22**, 299 (1981).

9.49 Q. H. F. Vrehen and H. M. Gibbs, "Superfluorescence Experiments," in R. Bonifacio (Ed.), *Dissipative Systems in Quantum Optics*, Springer, Berlin (West), Heidelberg, New York, 1982, p. 111.

9.50 R. Bonifacio and L. A. Lugiato, in R. Bonifacio (Ed.), *Dissipative Systems in Quantum Optics*, Springer, Berlin (West), Heidelberg, New York, 1982, p. 1.

9.51 E. A. Watson, H. M. Gibbs, F. P. Mattar, M. Cormier, Y. Claude, S. L. McCall, and M. S. Feld, *Phys. Rev. A*, 27, 1427 (1983).

9.52 F. T. Arecchi and E. Courtens, *Phys. Rev. A* **2**, 1730 (1970).

9.53 N. E. Rehler and J. H. Eberly, *Phys. Rev. A* **3**, 1735 (1971).

9.54 J. H. Eberly, *Am. J. Phys.* **40**, 1374 (1972).

9.55 J. C. MacGillivray and M. S. Feld, *Phys. Rev. A* **14**, 1169 (1976).

10

Multiphoton Absorption and Emission

Whereas the last two chapters dealt with nonlinearities in one-photon interaction processes, this chapter is devoted to multiphoton emission and absorption. First we describe some basic phenomena. Then we calculate transition probabilities and the absorption and amplification of laser waves in multiphoton interaction processes. This is followed by the discussion of some specific processes: multiphoton fluorescence, multiphoton ionization, and two-photon lasing processes.

10.1. BASIC PHENOMENA

A long time before lasers became available, processes had been known in which an atomic system absorbed several photons and became excited, the energy of the excited atomic level being approximately equal to the sum of the photon energies. These processes turned out to be a sequence of one-photon processes by which the atom experiences stepwise one-photon absorptions from one level to the next (see Fig. 10.1).

In 1961, immediately after the invention of the laser, Kaiser and Garrett (Ref. 10.1) were successful in observing two-photon absorption of ruby laser photons in the system $Eu^{2+}:CaF_2$, which does not possess any intermediate state at or near resonance with the one-photon energy $\hbar\omega'$. This means that instead of Fig. 10.1 the level scheme of Fig. 4.6a, with $\hbar\omega'$ and $\hbar\omega''$ equal, applies to the interaction process. Kaiser and Garrett observed a blue fluorescence, the intensity of which was found to be proportional to the square of the laser intensity. Direct measurements of the attenuation of one of the incident waves by two-photon absorption were first described in Ref. 10.2. Such genuine multiphoton processes, in which one active system is excited at the expense of the *simultaneous* annihilation of several photons in one or in various modes of the radiation field, had already been predicted in 1931 by Göppert-Mayer (Ref. 10.3), who treated the atomic systems and the radiation field, as well as their interaction, by quantum-theoretical means. From this theory it is apparent that also the reverse process is possible in which the atom flips from the upper to the lower level and several photons are simultaneously

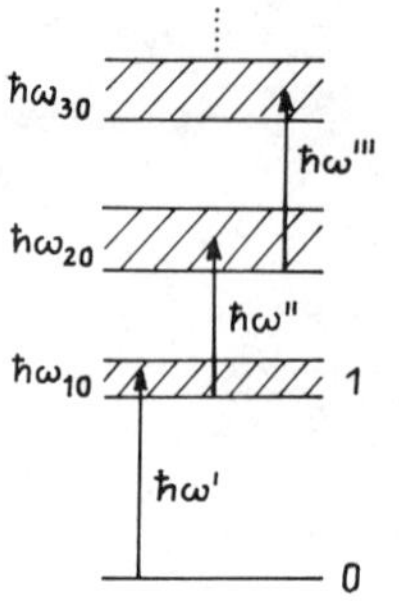

Fig. 10.1. Excitation of an atomic system by multistep absorption.

created. In the two decades that have passed since the experiment of Kaiser and Garrett, several interesting phenomena of multiphoton absorption and emission have been observed.

In particular it has become possible to detect the interaction process using various methods. This is schematically illustrated in Fig. 10.2 for the case of two-photon absorption. The sample is excited by the radiation of two lasers with the frequencies ω' and ω'', which satisfy the resonance condition $\omega_{10} = \omega' + \omega''$. As is known from the Kaiser–Garrett experiment, in the special case $\omega_{10} = 2\omega'$, one laser is sufficient. The most direct detection method is to measure with photodetectors the intensity of one of the two laser beams before and behind the sample and, upon amplifying the electric signals, to evaluate the transmission ratio $I(L)/I(0)$. A general difficulty arising in this approach is that the two-photon absorption produces very small intensity changes that have to be detected in the presence of the high total intensity. To increase the detection sensitivity, the radiation of frequency ω' is often modulated and the modulated part of the signal of frequency ω'', which originates from the two-photon interaction, is detected by applying phase-sensitive noise discrimination. Also pulsed lasers can be used to advantage, as the efficiency of multiphoton processes strongly increases with the intensity of the input signals. But then the arrival and the propagation of the pulses with frequencies ω' and ω'' must of course be sufficiently accurately synchronized in the sample.

With transmission techniques the highest detection efficiency can be obtained by employing methods that combine the advantages of pulse excitation and modulation techniques. To achieve this, the infinite pulse trains of two synchronously pumped dye lasers are superposed in the sample, the pulse repetition frequency being on the order of 100 MHz. Now the intensity of the pulse train with light of frequency ω' is modulated, and the modulated light of frequency ω'' is measured by a phase-sensitive detector. Such a modulation technique was first applied to measure the Raman gain (see Ref. 10.4 and Chapter 12). Precise stabilization of the parameters and proper choice of the modulation frequency in the megahertz range allowed the detection of relative signal changes as small as 10^{-8}.

In addition to transmission measurements, other techniques—for example, those where certain effects arising from the energy absorbed in the sample are

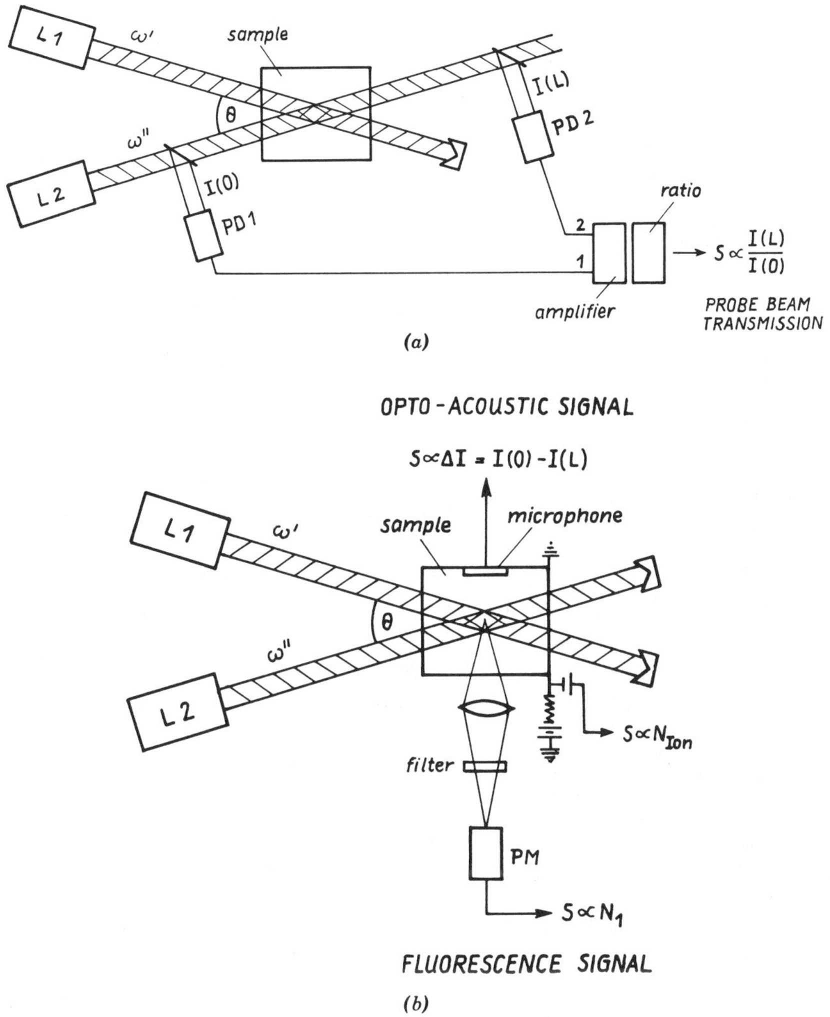

Fig. 10.2. Scheme of multiphoton absorption devices: (*a*) measurement of probe beam transmission, (*b*) measurement of fluorescence signal and optoacoustic signal.

detected—can successfully be employed. A very sensitive method is the measurement of fluorescent radiation that originates from the transition of the excited atomic systems to the ground state or to another state between the excited and ground levels (see Fig. 10.3*a*). A spectral filter is used for separating the fluorescence signal, whose wavelength is often shorter than that of the exciting light, from the scattered laser light. The resulting signal is proportional to the number of fluorescent particles, and thus to the number N_1 of atomic systems in the excited level 1. To increase the sensitivity, modulation

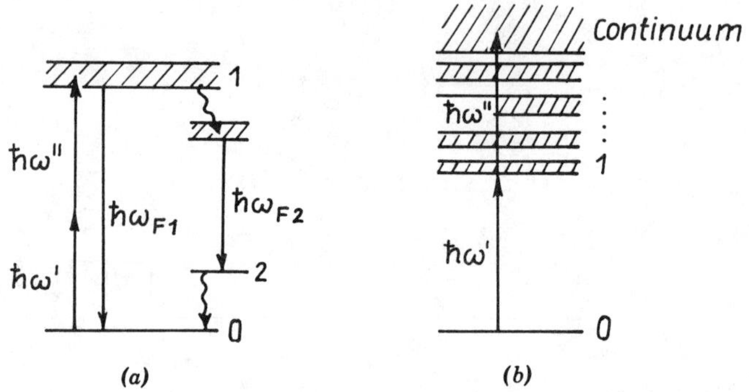

Fig. 10.3. Level scheme of two-photon excitation: (*a*) two-photon fluorescence, (*b*) two-photon ionization.

techniques can be employed also in this case. Thus, one of the pump beams can be modulated and the fluorescence signal at this frequency can be detected with a phase-sensitive detector. In continuous laser excitation successful use can be made of photon counting, and for continuous pulse-train excitation time-correlated photon counting is of great advantage.

A high sensitivity as high as in two-photon fluorescence can be achieved by employing *two-photon ionization*, in which the atomic system is excited by the multiphoton process into the ionization continuum (see Fig. 10.3*b*). The resulting free charges can now be electrically detected, especially in gases, where the counting of the charge carriers is possible (see Fig. 10.2*b*). The detection signal is proportional to the number N_{ion} of ionized particles (see, e.g., Ref. 10).

If the two-photon process does not yield any free charge carriers and if the greatest part of the excitation power is not converted into fluorescent radiation, there remains the possibility of measuring the heat released by relaxation processes in the sample after irradiation. In modulated cw excitation and in pulse excitation the time-varying heat sources cause time-varying expansions and compressions in the material, which result in the generation of pressure waves. These pressure waves can be received by a microphone, where the signal is proportional to the power absorbed in the sample (Fig. 10.2b; see, e.g., Refs. 10 and 21).

The great variety of multiphoton absorption and emission processes have given rise to numerous interesting applications in the fields of spectroscopy, quantum electronics, measuring techniques, material excitation, and material processing.

Special mention should be made of the spectroscopic applications of two-photon and multiphoton absorption, because they yield results one can hardly achieve, if at all, with other spectroscopic techniques. There are several reasons for this fact. First, high energy levels can be reached without having to work in the vacuum-ultraviolet spectral region, which is experimentally rather inaccessible. So, for instance, in a fairly simple manner, by second-harmonic generation (SHG) of dye-laser radiation, tunable radiation in the spectral region up to about 220 nm can be generated, which enables us to make transitions with three-photon absorption that correspond to wavelengths as short as 73 nm in one-photon spectroscopy. Second, the excitation of atoms can be affected by transitions which in one-photon processes are forbidden by selection rules. Note that in multiphoton processes different selection rules hold. This point will be pursued further in Section 10.2. Third, the free choice of polarization directions of the various radiation sources yields a greater number of degrees of freedom in polarization-direction spectroscopy, which allows the determination of the symmetry of the levels involved. Fourth, by varying the directions of the incident laser beams a wave-vector spectroscopy becomes possible, which has interesting applications in solid-state physics (Section 10.4). In particular, the directions of incidence can be chosen so that in interactions occurring in gases the resultant impulse transferred to the single atomic system is zero. In this way resonances of extremely narrow bandwidth can be achieved in high-resolution gas spectroscopy. This method is referred to as *sub-Doppler two-photon spectroscopy* (see, e.g., Refs. 13 and 22).

In addition to spectroscopy, multiphoton absorption and emission have found applications in some other fields. Because of the weak attenuation of the exciting light beams, it is for instance possible to obtain highly homogeneous excitations of samples or homogeneous charge-carrier distributions, which are of importance for various experimental purposes. Thus the homogeneous two-photon excitation of electron-hole pairs can for instance be applied in semiconductor lasers (see, e.g., Ref. 10.5). Furthermore, homogeneously ionized plasmas can be produced (e.g., Ref. 10.6).

Multiphoton ionization can be applied to induce *avalanche processes*, in which a few charge carriers generated by quantum processes induce further ionization processes, after being accelerated in the radiation field or in external static fields. Such processes are of great importance in plasma generation by lasers (see, e.g., Ref. 10.7). On the other hand, multiphoton ionization in solids

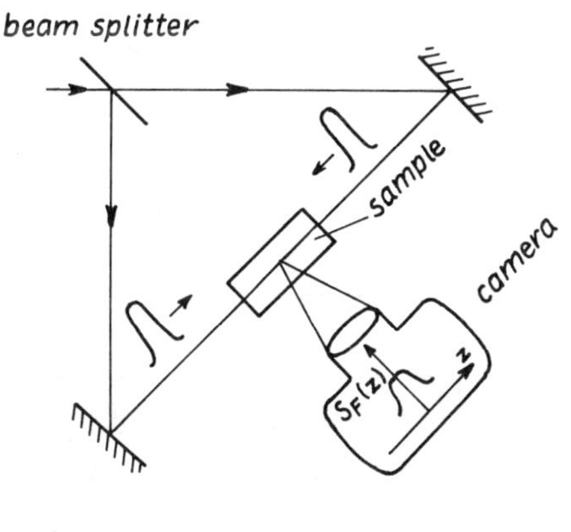

Fig. 10.4. Triangular setup for measuring pulse correlation functions by TPF. $S_F(z)$ is the fluorescence signal as a function of the space coordinate z.

followed by an avalanche process may be the source of undesired radiation damage (see, e.g., Ref. 10.8).

Multiphoton fluorescence is used with great success for determining correlation functions of light pulses and hence for determining their temporal profile (see Section 10.5). Figure 10.4 shows the basic experimental setup. Two identical counterpropagating pulses generated by beam splitting of the laser pulse are superposed in a cuvette, where by way of two-photon absorption they cause the solute molecules to fluoresce. The density of the excited molecules is higher in the region where the pulses overlap than it is in regions where the two pulses pass through in succession. We measure the spatial distribution of the fluorescent radiation, whose power is proportional to the number of excited molecules in the region of emission. Thus the width of the region of pulse superposition can be measured, and the pulse duration can be inferred from it by utilizing the light velocity in the material under study.

Moreover, the stimulated multiphoton emission from excited systems can be employed for the generation of coherent light at particular wavelengths ω'', which are determined by the resonance condition $\omega'' = \omega_{10} - \omega'$ if the sample is irradiated with light of frequency ω' (see Section 10.6).

10.2. TRANSITION PROBABILITY OF MULTIPHOTON ABSORPTION

Having described in the preceding section some basic phenomena, we can now proceed to a more detailed study of the transition probabilities in multiphoton absorption. We start with the discussion of semiclassical treatments and continue with some consideration of the fully quantum-theoretical approach. Finally, in Section 10.2.3 we discuss some aspects of the use of solids as multiphoton absorbers.

10.2.1. Semiclassical Treatment

In this subsection we rely on the semiclassical calculations of Sections 4.5 and 4.6, where we calculated the nonlinear optical susceptibilities [see (4.63)] and, in a direct manner, also the transition probability for two-photon absorption [see (4.79)]. Making use of the expressions derived there, we shall discuss the selection rules for two-photon absorption.

As in one-photon spectroscopy, the selection rules in multiphoton processes are as well determined by the matrix elements of the interaction operator $\hat{H}^C$, which in the dipole approximation are proportional to the matrix elements $d_{\alpha\beta}$ of the transition moment operator. In one-photon processes between levels 0 and 1 only the matrix element d_{01} enters the interaction probability or the corresponding interaction cross section of the transition $0 \to 1$ [see (4.78)]. In contrast to this, the matrix element d_{01} does not appear in the expression for the transition probability of two-photon processes between levels 0 and 1 or in the corresponding nonlinear susceptibilities or interaction cross sections. Instead of d_{01} the matrix elements $d_{0\beta}$ and $d_{\beta 1}$ between the initial state $|0\rangle$ and all the other states $|\beta\rangle$ (the so-called *intermediate states*), as well as between the final state $|1\rangle$ and the $|\beta\rangle$, appear in these expressions. This can be seen from the polarizability tensor

$$\alpha_{01jl}(\omega', \omega'') = \frac{1}{\hbar}\sum_{\beta}\left\{\frac{d_{0\beta j}d_{\beta 1l}}{\omega_{\beta 0} - \omega'} + \frac{d_{0\beta l}d_{\beta 1j}}{\omega_{\beta 0} - \omega''}\right\} \qquad (10.1)$$

of (4.61a), on which the transition probability and the interaction cross section depend in a simple manner. Accordingly, the two-photon transition is allowed only if both the initial and the final state are associated with the same intermediate states $|\beta\rangle$ by way of allowed one-photon processes. Following Placzek, one may refer to this fact as the *third-common-level rule*. Thus, starting from the molecular ground level, certain excited states may be populated by means of two-photon absorption that cannot be reached by a one-photon process (see, e.g., Refs. 10.9, 10.10).

As a simple example several selection rules for alkali-like one-electron systems will be discussed. Figure 10.5 shows a level scheme with one-, two-, and three-photon transitions. In the case of one-photon transitions the selection rule $\Delta L = \pm 1$ holds for the orbital angular momentum. Correspondingly, transitions do not occur within the S, P, D, or F "ladders", but they may occur between an S level and a P level or between a P level and a D level. These selection rules for one-photon transitions, which make predictions as to the vanishing of certain matrix elements $d_{\alpha\beta}$ of the transition moment, also imply selection rules for multiphoton transitions. This is what we attempt to explain for transitions starting from the ground state of the atom. The only atomic states $|\beta\rangle$ that are of importance as intermediate states are those that one-photon processes can reach from the ground state, where $L = 0$. Thus only levels of the P ladder appear as intermediate levels. These intermediate states are connected only with final states in the S ladder or in the D ladder by nonvanishing matrix elements. Therefore, starting from the ground state, only states of the S and D ladders can be reached by two-photon processes.

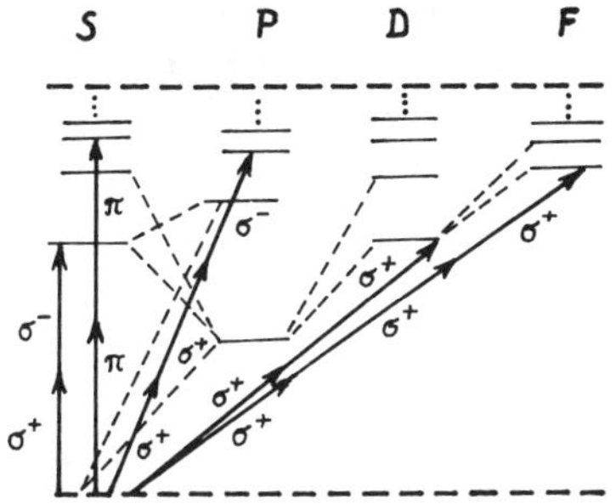

Fig. 10.5. Level scheme of an alkali atom with some one-photon and multiphoton transitions. Full lines: multiphoton transitions; dashed lines: one-photon transitions. π, σ^+, and σ^- indicate that the radiation is linearly, right-hand circularly, and left-hand circularly polarized, respectively.

In a more general way it can be concluded that the selection rule $\Delta L = 0, \pm 2$ holds. Applying the angular-momentum conservation law of quantum theory to the total system of the atoms and the radiation field, these selection rules can be interpreted as follows: for two-photon transitions $\Delta L = 0$ means that the angular momenta of the two absorbed (emitted) photons compensate each other, while $\Delta L = \pm 2$ implies the absorption (emission) of two photons with equal angular momenta. Thus, from the example of two-photon absorption in one-electron systems it is obvious that certain final states as well as the corresponding intermediate states can be selected, not only by choice of the frequencies but also by choice of the polarization states of the two incident light beams. In this way information on the symmetry of unknown states can be gained.

By analogous considerations other selection rules for two-photon processes can be obtained, where profitable use can be made of group theory. Since the transition probabilities for two-photon absorption and for two-photon emission depend equally on the matrix elements of a general polarizability tensor, as does the transition probability for Raman processes, the results on selection rules and symmetry relations derived for the Raman effect (see, e.g., Refs. 10.11, 10.12) can be applied to two-photon absorption and emission.

10.2.2. Quantum-Theoretical Treatment

In Section 3.2.2, in treating the rotating-wave approximation, we have already presented a fully quantum-theoretical calculation of the transition probability for two-photon absorption for the special case that initially the radiation field is in an energy eigenstate. We are now going to discuss the transition probability of an atom under the more general condition that the radiation field, when switched on at time $t = 0$, is characterized by the density operator $\hat{\rho}_F(0)$. We do not intend to discuss the dependence of the transition probability on the atomic parameters in detail, since we have already determined it semiclassically in the preceding subsection, and the fully quantum-theoretical treatment does not offer essential new aspects.

In this subsection we start from the time-dependent perturbation theory and proceed as in Sections 5.1 and 5.2, where the treatment of photoelectric counting by n one-photon processes occurring in n atoms was described in detail. It will be seen that under appropriate conditions both the transition probabilities depend on the field properties in the same way. In the order $2n$ of the perturbation theory we obtain for the probability $P_{0\to 1}^{(n)}(t)$ that before time t the atom was raised from the ground state $|0\rangle_A$ to the excited state $|1\rangle_A$ as a result of n-photon absorption

$$P_{0\to 1}^{(n)}(t) = \int_0^t dt_1' \int_0^t dt_1 \cdots \int_0^t dt_n \int_0^t dt_n'$$

$$\times \varphi_A^*(t_1', \ldots, t_n')\varphi_A(t_1, \ldots, t_n)\Gamma^{n,n}(t_1', \ldots, t_n), \quad (10.2a)$$

where

$$\Gamma^{(n,n)}(t_1',\ldots,t_n) = \mathrm{Tr}\{\hat{\rho}_F(0)\hat{E}^{(+)}(t_1')\cdots\hat{E}^{(+)}(t_n')\hat{E}^{(-)}(t_1)\cdots\hat{E}^{(-)}(t_n)\} \tag{10.2b}$$

is the *field-strength correlation function* of order $2n$. The field-strength operators $\hat{E}^{(\pm)}(t)$ represent the positive- and negative-frequency components of $\hat{E}(t)$ in the Heisenberg picture and refer to the position of the atom. The function φ_A contains the same characteristic parameters of the atom which the matrix elements of the polarization tensor in (10.1) depend on, namely the transition moments $d_{\alpha\beta}$ and the transition frequencies $\omega_{\alpha\beta}$ [cf. (5.7) and (5.10)]. As in the semiclassical treatment, the dipole and rotating-wave approximations were used for deriving (10.2); the vector character of the field was ignored in order to simplify and elucidate the principal nonlinear relationships. (More general cases are for instance treated in Refs. 10.9, 10.10, 10.13, 10.14.)

We recognize that (10.2) is of the same structure as a simple specialization of (5.22) that gives the joint probability $P(t_1,\ldots,t_n)$ for the case that in a photon counter made up of n atoms the first atom undergoes a counted one-photon absorption in the interval $(0, t_1)$, the second one in the interval $(0, t_2)$, and so forth. The specialization consists in equating all the times $(t_1 = t_2 = \cdots = t_n = t)$, that is, in asking for the probability $P^{(n)}(t)$ of the transition of n atoms with each atom absorbing one photon in the time interval $(0, t)$, and in supposing that the atoms are so close to each other that their space coordinates appearing in the field-strength operators can be equated. Then in both cases the transition probability of the atomic system depends only on the local field-strength correlation function $\Gamma^{(n,n)}(t_1',\ldots,t_n', t_1,\ldots,t_n)$. Thus the transition probability of the n-photon process depends on the coherence state of the field just like the counting probability of n photons in n one-photon processes.

Assuming the fields to be stationary and the conditions of time coarse-graining to be satisfied (cf. Section A.2.3), the time-independent multiphoton transition rate $\Delta P_{0\to 1}^{(n)}/\Delta t$ can be obtained from (10.2). For this case and for quasimonochromatic fields near resonance, this means that the line width $\Delta\omega_R$ of the incident radiation field is small compared to the atomic line width $\Delta\omega_{10}$ and that $2\hbar\omega \approx \hbar\omega_{10}$, we obtain

$$\frac{\Delta P_{0\to 1}^{(n)}}{\Delta t} \propto \left\langle (\hat{a}^\dagger)^n (\hat{a})^n \right\rangle, \tag{10.3}$$

where $\hat{a}^\dagger$ and $\hat{a}$ are the creation and annihilation operators of the mode under consideration. The expectation value $\langle(\hat{a}^\dagger)^n(\hat{a})^n\rangle$ depends strongly on the *statistical properties of the radiation* of this mode. So we obtain

$$\left\langle (\hat{a}^\dagger)^n (\hat{a})^n \right\rangle = n!\langle \hat{a}^\dagger \hat{a}\rangle^n \tag{10.4a}$$

for chaotic light,

$$\left\langle (\hat{a}^\dagger)^n (\hat{a})^n \right\rangle = \langle \hat{a}^\dagger \hat{a} \rangle^n \tag{10.4b}$$

for coherent light, and

$$\left\langle (\hat{a}^\dagger)^n (\hat{a})^n \right\rangle = \left(1 - \frac{1}{\langle \hat{a}^\dagger \hat{a} \rangle}\right) \cdots \left(1 - \frac{n-1}{\langle \hat{a}^\dagger \hat{a} \rangle}\right) \langle \hat{a}^\dagger \hat{a} \rangle^n \tag{10.4c}$$

for light in the photon-number state $|m\rangle$, where $\langle \hat{a}^\dagger \hat{a} \rangle = m$ and $n < m$.

Clearly the transition rate for monochromatic thermal light is $n!$ times higher than that for fully coherent light, that is, for ideal laser light. The physical reason of this shortcoming of coherent light in multiphoton absorption is as follows. In its intensity, chaotic radiation exhibits more pronounced deviations from the average value than does laser light. Because of the nonlinearity of the process, the positive deviations, which appear as intensity peaks, yield very large contributions to the response, and hence the transition rate is enhanced by the factor $n!$ compared with that for coherent light of the same mean intensity. The expectation value of $\langle (\hat{a}^\dagger)^n (\hat{a})^n \rangle$ for light in a photon-number state is even smaller than that for coherent light of the same mean intensity. This fact can be related to the phenomenon of photon antibunching: the probability of simultaneously detecting two or more photons with radiation in a photon-number state is smaller than that of detecting the same number of photons with coherent radiation (cf. Section 9.2). Only at rather high mean intensities ($m = \langle \hat{a}^\dagger \hat{a} \rangle \gg n$) does the transition probability approach that of coherent light.

Having discussed the special case of quasimonochromatic laser radiation, we now present in Fig. 10.6 results due to Chrostowski and Karcewski (Refs. 10.15, 10.16) concerning the influence of the finite bandwidth $\Delta\omega_R$ of the incident light. These authors calculated the ratio

$$r_n(t) = \frac{\left(\dfrac{1}{t} P_{0 \to 1}^{(n)}(t)\right)_{\text{chaotic}}}{\left(\dfrac{1}{t} P_{0 \to 1}^{(n)}(t)\right)_{\text{coherent}}} \tag{10.5}$$

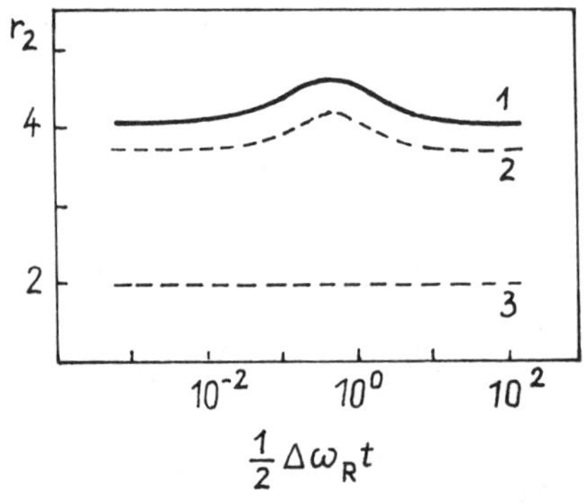

Fig. 10.6. Influence of the bandwidth $\Delta\omega_R$ of the incident radiation on the ratio r_2 of the two-photon transition probabilities for chaotic light to that for coherent light (after Refs. 10.15, 10.16): curve 1, $\Delta\omega_R \gg \Delta\omega_{10}$; curve 2, $\Delta\omega_R = \Delta\omega_{10}$; curve 3, $\Delta\omega_R \ll \Delta\omega_{10}$.

as a function of time for several values of the ratio of the radiation bandwidth $\Delta\omega_R$ to the transition bandwidth $\Delta\omega_{10}$. The incident coherent and chaotic fields are supposed to be characterized by the same first-order correlation function

$$\Gamma^{1,1}(t', t) = \langle \hat{a}^\dagger \hat{a} \rangle e^{+i\omega_0(t'-t) - \frac{1}{2}\Delta\omega_R|t'-t|}. \tag{10.6}$$

The ratio r_n is seen to reach its maximum at $\Delta\omega_R t \sim 1$. For $t \to \infty$ and $\Delta\omega_R \ll \Delta\omega_{10}$ the result is $r_n = n!$, which is in agreement with the calculations given above. The ratio r_n becomes larger as $\Delta\omega_R$ increases; with $\Delta\omega_R \gg \Delta\omega_{10}$ and for $t \to \infty$, for instance, the value $r_2 = 4$ is obtained.

10.2.3. Transition Probability in Solids

In Section 10.1 we have mentioned that in particular in solids multiphoton spectroscopy yields interesting results. In analogy to multiphoton absorption by individual atoms and molecules, also in solids information can be gained about energy levels and their distribution, about transition moments, about the occupation of particular states, and about interaction parameters. We shall therefore present in the following a calculation of the transition probability in solids. We restrict the treatment to insulators and semiconductors.

For simplicity we consider the solid in the one-electron approximation and investigate *direct optical transitions* only, that is, transitions with no phonons taking part. We start from the model of energy bands as it is given in Fig. 10.7. We assume the valence band (V) to be completely filled, and the conduction band (C) to be completely empty. Higher conduction bands denoted by γ represent virtual intermediate states. The wave vectors of one-electron states in the particular bands are denoted by $k_{V\bullet}, k_{C\bullet}, k_{\gamma\bullet}$, and the energies at $k_{V\bullet}, k_{C\bullet}, k_{\gamma\bullet}$ by $\mathscr{E}_{Vk\bullet}, \mathscr{E}_{Ck\bullet}, \mathscr{E}_{\gamma k\bullet}$. The band structure in the $k_\bullet$ space is assumed to be such that all band edges are at $k = 0$. To calculate the transition probability we employ the semiclassical theory, since no investigation of the influence of the coherence properties of the field, which has already been studied in the preceding subsection, is intended. In contrast to the

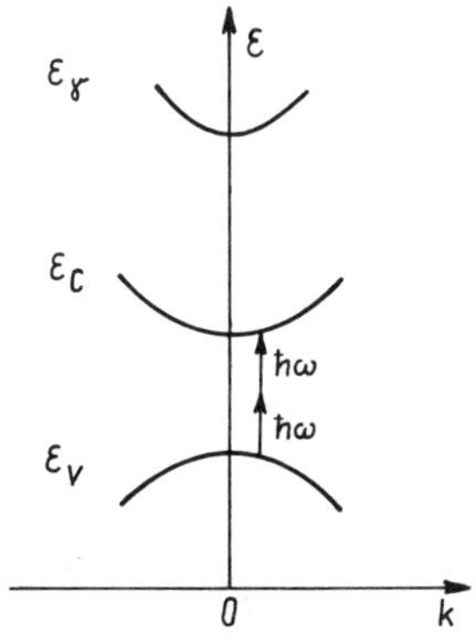

Fig. 10.7. Two-photon transition between the valence band (V) and the conduction band (C) of a solid. The energy $\mathscr{E}$ is given as a function of k, the component of the wave vector in a certain direction.

calculations carried out in Sections 3.2.2 and 4.5, we do not start from the dipole approximation but use the interaction operator $\hat{H}^C = -\hat{A}_\bullet(\hat{r}_\bullet, t)\hat{p}_\bullet$, where $\hat{p}_\bullet$ is the momentum operator of the electron. In spite of these differences in the interaction operator, the calculation will be carried out in a similar manner to the above.

Let the solid sample be irradiated by two quasimonochromatic laser light sources with coherent light of frequencies ω_1 and ω_2, wave vectors $k_{1\bullet}$ and $k_{2\bullet}$, and photon flux densities I_1 and I_2. For the transition rate from a certain state of the valence band with the wave vector $k_{V\bullet}$ to a certain state of the conduction band with the wave vector $k_{C\bullet}$, we obtain in the fourth order of the density matrix perturbation calculation

$$\frac{\Delta P_{i\to f}^{(4)}}{\Delta t} = \zeta''(I_1, I_2)|\Lambda|^2\delta(\mathscr{E}_{Ck_\bullet} - \mathscr{E}_{Vk_\bullet} - \hbar\omega_1 - \hbar\omega_2), \tag{10.7a}$$

where

$$\Lambda = \left\{\sum_\gamma \int d^3k_{\gamma\bullet} \frac{\left(\langle\mathscr{E}_{Ck_\bullet}|e^{ik_{2\bullet}\hat{r}_\bullet}\hat{p}_\bullet|\mathscr{E}_{\gamma k_\bullet}\rangle e_\bullet^{(2)}\right)\left(\langle\mathscr{E}_{\gamma k_\bullet} e^{ik_{1\bullet}\hat{r}_\bullet}\hat{p}_\bullet|\mathscr{E}_{Vk_\bullet}\rangle e_\bullet^{(1)}\right)}{\mathscr{E}_{\gamma k_\bullet} - \mathscr{E}_{Vk_\bullet} - \hbar\omega_1}\right.$$

$$\left. + \sum_\gamma \int d^3k_{\gamma\bullet} \frac{\left(\langle\mathscr{E}_{Ck_\bullet}|e^{ik_{1\bullet}\hat{r}_\bullet}\hat{p}_\bullet|\mathscr{E}_{\gamma k_\bullet}\rangle e_\bullet^{(1)}\right)\left(\langle\mathscr{E}_{\gamma k_\bullet}|e^{ik_{2\bullet}\hat{r}_\bullet}\hat{p}_\bullet|\mathscr{E}_{Vk_\bullet}\rangle e_\bullet^{(2)}\right)}{\mathscr{E}_{\gamma k_\bullet} - \mathscr{E}_{Vk_\bullet} - \hbar\omega_2}\right\} \tag{10.7b}$$

and

$$\zeta''(I_1, I_2) \propto (\hbar\omega_1 I_1)(\hbar\omega_2 I_2). \tag{10.7c}$$

In most cases we are not interested in the transition rate $\Delta P_{i\to f}^{(4)}/\Delta t$ of two-photon absorption between the initial state $|i\rangle$ of the electron and one distinct final state $|f\rangle$, but in the total rate $\Delta P_i^{(4)}/\Delta t$ of two-photon absorption due to an electron that is initially in the state $|i\rangle$ and gets excited to an arbitrary final state. This transition rate is obtained by summing $\Delta P_{i\to f}^{(4)}/\Delta t$ over all possible final states:

$$\frac{dP_i^{(4)}}{dt} \propto \int d^3k_{C\bullet}|\Lambda(k_{C\bullet}, k_{V\bullet})|^2\delta(\mathscr{E}_{Ck_\bullet} - \mathscr{E}_{Vk_\bullet} - \hbar\omega_1 - \hbar\omega_2).$$

In deriving this relation we have replaced the summation over the final states by an integration over the wave vector of the conduction band. If we are only interested in the total transition probability that, independently of the initial and final states of the electronic transition involved, two photons are absorbed by the entire solid sample, we additionally have to sum $dP_i^{(4)}/dt$ over all the initially occupied one-electron states. Thus we obtain the total transition rate for two-photon absorption of the solid as

$$\frac{dP^{(4)}}{dt} = \zeta'(I_1, I_2) \int d^3k_{V\bullet} \int d^3k_{C\bullet} |\Lambda(k_{C\bullet}, k_{V\bullet})|^2$$

$$\times \delta(\mathscr{E}_{Ck_\bullet} - \mathscr{E}_{Vk_\bullet} - \hbar\omega_1 - \hbar\omega_2), \tag{10.8a}$$

where

$$\zeta'(I_1, I_2) \propto (\hbar\omega_1 I_1)(\hbar\omega_2 I_2). \tag{10.8b}$$

We now take into account the specific properties of solids, in particular, the periodicity of crystal lattices. In coordinate representation the state functions to be used are

$$\langle r_\bullet | \mathscr{E}_{\gamma k_\bullet} \rangle = \frac{1}{\sqrt{V}} u(\gamma, k_{\gamma\bullet}; r_\bullet) e^{ik_{\gamma\bullet} r_\bullet}, \tag{10.9}$$

where V is the crystal volume, u a lattice-periodic function of $r_\bullet$, and the exponential expression is the *Bloch factor*; analogous representations hold for the states $|\mathscr{E}_{Vk_\bullet}\rangle$ and $|\mathscr{E}_{Ck_\bullet}\rangle$. This gives for the matrix elements that appear in the coupling parameter

$$\langle \mathscr{E}_{\alpha k_\bullet} | e^{ik_{L\bullet}\hat{r}_\bullet} \hat{p}_\bullet | \mathscr{E}_{\beta k_\bullet} \rangle$$

$$= \frac{1}{V_u} \left\{ \int_{V_u} d^3r_\bullet \, u^*(\alpha, k_{\alpha\bullet}, r_\bullet) e^{ik_{L\bullet} r_\bullet} (\hbar k_{L\bullet} - i\hbar\nabla_\bullet) u(\beta, k_{\beta\bullet}, r_\bullet) \right\}$$

$$\times \delta(k_{\beta\bullet} + k_{L\bullet} - k_{\alpha\bullet}), \tag{10.10}$$

where V_u is the volume of the unit cell of the crystal (see, e.g., Ref. 10.17), and $k_{L\bullet}$ stands for $k_{1\bullet}$ or $k_{2\bullet}$. The appearance of the delta function means that with a given wave vector $k_{V\bullet}$ of the initial state only intermediate states at $k_{\gamma\bullet} = k_{V\bullet} + k_{L\bullet}$ give contributions to Λ. From an intermediate state with this wave vector only final states can be achieved that lie at $k_{C\bullet} = k_{\gamma\bullet} + k'_{L\bullet}$, with $k'_{L\bullet}$ also standing for $k_{1\bullet}$ or $k_{2\bullet}$. In the present case, where we assume the absorption of only one photon of energy $\hbar\omega_1$ and another of energy $\hbar\omega_2$ to be possible because of the resonance conditions $\hbar\omega_1 + \hbar\omega_2 = \hbar\omega_{10}$, there results the momentum conservation law

$$k_{C\bullet} = k_{V\bullet} + k_{1\bullet} + k_{2\bullet}. \tag{10.11}$$

From this it is apparent that only those states are connected by two-photon absorption transitions whose wave number vectors satisfy the conservation law (10.11).

In contrast to one-photon absorption, we are here given the possibility of applying *wave-vector-selective spectroscopy*, or *momentum-selective spectroscopy*, since the momentum transferred to the solid during an absorption process can

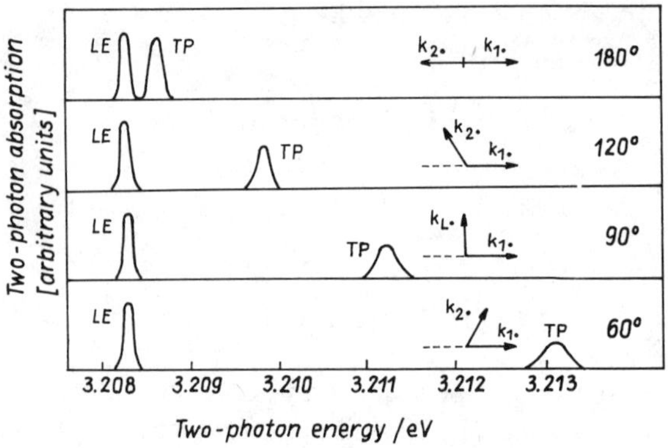

Fig. 10.8. Two-photon absorption in CuCl (after Ref. 10.18).

be varied between $\hbar(k_1 + k_2)$ and $\hbar(k_1 - k_2)$ by varying the angle between the laser beams with given photon energies. However, as the wave numbers for electromagnetic radiation in the visible range are small, this method allows only a small fraction of the Brillouin zone to be covered. While with $\lambda_L \approx 0.5$ μm the wave number of the photons is about 10^7 m^{-1}, with a lattice constant of some 10^{-10} m the extension of the Brillouin zone is of the order of 10^{10} m^{-1}.

But even in this narrow range of the Brillouin zone which can be covered, there occur very interesting interactions of photons with the elementary excitations of the solid, such as phonons and excitons. As an example Fig. 10.8 presents results of two-photon measurements on the first exciton of CuCl performed at a temperature of 1.4 K. While within the measuring accuracy the position of the longitudinal exciton transition (LE) does not depend on the angle between the laser beams, the transverse excitonlike polariton transition (TP) exhibits with increasing angle a shift and a broadening of the resonance line (Ref. 10.18). Results obtained from such measurements make it possible to construct the dispersion curve of the polaritons and to draw conclusions on the dielectric properties of the material.

After these brief remarks on wave-vector-selective spectroscopy, we now make use of the *dipole approximation* in a more detailed discussion of the two-photon transition probability. In this case the integrations in (10.8) over $k_{C\bullet}$ and $k_{V\bullet}$ reduce to an integration over $k_\bullet = k_{C\bullet} = k_{V\bullet}$. In addition, we start from the following simplifying assumption for the $k_\bullet$ dependence of the band energies:

$$\mathscr{E}_{Vk\bullet} = \mathscr{E}_{V0} - \frac{\hbar^2 k_V^2}{2\tilde{m}_V}, \tag{10.12a}$$

$$\mathscr{E}_{Ck\bullet} = \mathscr{E}_{C0} + \frac{\hbar^2 k_C^2}{2\tilde{m}_C}; \tag{10.12b}$$

that is, parabolic bands and isotropic effective masses $\tilde{m}_V$ and $\tilde{m}_C$ of the valence and the conduction band are assumed. Here $\mathscr{E}_G = \mathscr{E}_{C0} - \mathscr{E}_{V0}$ denotes the band gap. Because of the isotropic parabolic wave-vector dependence of the band energy, the integration over $k_{C\bullet}$ in (10.8) can be easily performed. Upon substitution of the reduced mass

$$\tilde{m} = \left(\frac{1}{\tilde{m}_V} + \frac{1}{\tilde{m}_C} \right)^{-1} \tag{10.13}$$

and using the general relation

$$\delta(f(k)) = \sum_n \frac{\delta(k - k_n)}{\left(\dfrac{d}{dk} f(k) \right)_{k=k_n}},$$

where the k_n are the zeros of $f(k)$, the delta function in (10.8) goes over into

$$\begin{aligned} &\delta\left(\frac{\hbar^2 k^2}{2\tilde{m}} - [\hbar\omega_1 + \hbar\omega_2 - \mathscr{E}_G] \right) \\ &= \frac{1}{2\sqrt{(\hbar\omega_1 + \hbar\omega_2) - \mathscr{E}_G}} \left\{ \delta\left(\frac{\hbar k}{\sqrt{2\tilde{m}}} - \sqrt{(\hbar\omega_1 + \hbar\omega_2) - \mathscr{E}_G} \right) \right. \\ &\qquad \left. + \delta\left(\frac{\hbar k}{\sqrt{2\tilde{m}}} + \sqrt{(\hbar\omega_1 + \hbar\omega_2) - \mathscr{E}_G} \right) \right\}. \end{aligned}$$

Thus the integral in (10.8) becomes proportional to

$$\sqrt{(\hbar\omega_1 + \hbar\omega_2) - \mathscr{E}_G}\, |\Lambda(k_\bullet)|^2,$$

and we obtain

$$\frac{dP^{(4)}}{dt} = \zeta(I_1, I_2) \sqrt{(\hbar\omega_1 + \hbar\omega_2) - \mathscr{E}_G}\, |\Lambda(k_\bullet)|^2 \tag{10.14}$$

where $\zeta(I_1, I_2)$ is proportional to the product of the photon flux densities I_1 and I_2 of the two laser beams and, apart from universal constants, is dependent on material parameters such as the refractive index and effective mass, which for the most important semiconductors are known. As in the case with molecules, greater difficulties however have to be overcome in the exact determination of $|\Lambda(k_\bullet)|^2$, because here more subtle data on the solid, in particular those on all the intermediate states, enter. As with atoms and molecules, attempts are also made with solids to summarize the influence of all intermediate states by means of an effective intermediate state or an effective intermediate band. For typical semiconductors and dielectrics such evaluations provide two-photon absorption coefficients on the order of $k_a^{(2)} = (10^{-27}\ \mathrm{cm}^{-1}) \cdot I$, where I is in $\mathrm{cm}^{-2}\ \mathrm{s}^{-1}$, that is, absorption coefficients of about 1 cm^{-1} at intensities of $4 \times 10^8\ \mathrm{W/cm}^2$ with $\lambda_L = 0.5\ \mu\mathrm{m}$. Measurements on semiconductors and dielectrics are in agreement with these estimates.

The rough model employed so far can of course offer only a preliminary insight into the problems of interest. The fact that we have to take into account the actual structure of the solid (anisotropy of effective masses, interaction between electrons and holes, interaction between electrons and phonons) actually implies a change in the selection rules and the wave-vector relations. Thus the precise calculation of the transition probabilities turns out to be extremely difficult (see, e.g., Refs. 10.9, 10.10, 10.19, 10.20). Therefore the treatment up to now in most cases has been restricted to the calculation of the functional dependence of the transition rates on the parameters of the incident wave and on parameters of the solid, followed by the determination of numerical values of the rates and parameters by comparison with experimental results.

10.3. ATTENUATION OF THE ELECTROMAGNETIC FIELD

We have so far calculated the transition probability of atomic systems for two-photon and multiphoton processes in given radiation fields. Now we want to study in detail how the electromagnetic field is changed as a result of two-photon processes. We start with the quantum-theoretical calculation of the change in photon numbers in one mode of the radiation field.

10.3.1 The Decrease of the Number of Photons in One Radiation Mode

In principle, the alteration of the electromagnetic field can be calculated by solving the quantum-mechanical equations of motion of the interaction process under consideration. This can be done by starting from the equation of motion for the density operator $\hat{\rho}$ of the total system that consists of the radiation field and the atomic systems. Then an equation is obtained for the density operator $\hat{\rho}_F$ of the radiation field alone by taking the trace over all the atomic variables (cf. Section 7.2.2). In doing so, one may advantageously apply the rotating-wave approximation and the concept of effective nonlinear interaction operators (cf. Section 3.2.2), where the interaction Hamiltonian of the n-photon process is given as

$$\hat{H}^C = -\sum_j \eta^{(n)} (\hat{B}_{10})_j \left[\hat{E}^{(-)}(r_{j\bullet})\right]^n + \text{h.c.} \tag{10.15}$$

Here $(\hat{B}_{10})_j$ is the flip operator of the transition $0 \leftrightarrow 1$ of the jth atom at the position $r_{j\bullet}$, and $\eta^{(n)}$ describes the nonlinear coupling. Furthermore the irreversibility approximation may be used, which means that the thermal equilibrium of the atomic ensemble is supposed to be not disturbed by the radiation field. Under these assumptions, and provided that only one mode of the radiation field is involved, the equation of motion

$$\frac{d}{dt}\hat{\rho}_F = -\frac{\zeta^{(n)}}{2}\{N_0[(\hat{a}^\dagger)^n\hat{a}^n\hat{\rho}_F - 2\hat{a}^n\hat{\rho}_F(\hat{a}^\dagger)^n + \hat{\rho}_F(\hat{a}^\dagger)^n\hat{a}^n] + N_1[\hat{a}^n(\hat{a}^\dagger)^n\hat{\rho}_F - 2(\hat{a}^\dagger)^n\hat{\rho}_F\hat{a}^n + \hat{\rho}_F\hat{a}^n(\hat{a}^\dagger)^n]\} \tag{10.16}$$

is obtained, where $\zeta^{(n)}$ is proportional to the square of the modulus of the nonlinear coupling parameter $\eta^{(n)}$ and where N_0, N_1 are the occupation numbers of the levels $i = 0$ and 1. Evidently the first part of the right-hand side of (10.16), which is proportional to the number of atomic systems in the lower state, describes the alteration of the field by n-photon absorption, whereas the second part represents the influence of n-photon emission processes. In spite of the simplifying assumptions used in deriving (10.16), this relation is rather difficult to treat, and exact solutions of it are not yet known.

Starting from (10.16), a set of differential equations can be derived for the probabilities $p_m(t)$ of finding m photons in the mode of the radiation field under consideration. With respect to the physical interpretation this set of differential equations is of great advantage, but the mathematical difficulties in solving them of course remain. Following Simaan and Loudon (10.21), we derive such equations of motion for the $p_m(t)$ in another way, namely by applying elementary physical considerations to the interaction process. We explain this procedure for the case of two-photon absorption where $n = 2$. The change of p_m with time is expressed by four terms (see Fig. 10.9), which represent (i) the decrease of the photon number from m to $m - 2$, (ii) the decrease of the photon number from $m + 2$ to m, (iii) the increase of the photon number from m to $m + 2$, and (iv) the increase of the photon number from $m - 2$ to m. The first and the second term are associated with the transition of the atom from state 0 to state 1, and the third and the fourth term are associated with the transition $1 \to 0$. Taking into consideration these four

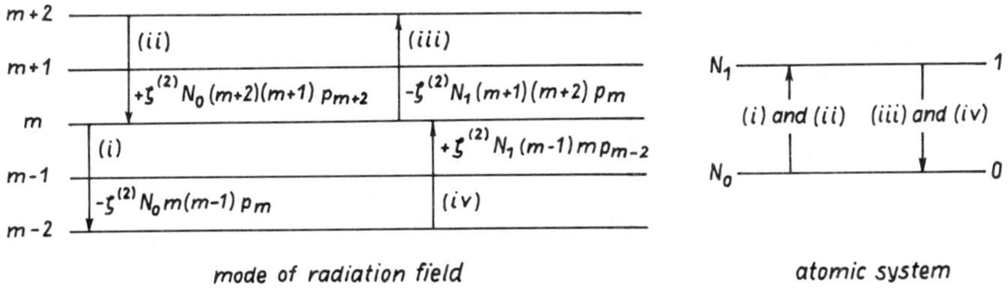

Fig. 10.9. Two-photon absorption and emission processes.

terms, assuming again N_0 atoms in the lower level and N_1 atoms in the upper one, and making use of (10.3) and (10.4) for the elementary atomic transition in the m-photon field, we obtain the relation

$$\frac{d}{dt}p_m = -\zeta^{(2)}N_0 m(m-1)p_m - \zeta^{(2)}N_1(m+1)(m+2)p_m$$
$$+\zeta^{(2)}N_1(m-1)mp_{m-2} + \zeta^{(2)}N_0(m+2)(m+1)p_{m+2}, \quad (10.17)$$

where $m = 0, 1, 2, \ldots$. From (10.17) the following set of equations for the kth moment $\langle m^k \rangle$ of the photon-number distribution can be derived:

$$\frac{d}{dt}\langle m \rangle = \sum_m m \frac{d}{dt}p_m$$
$$= 2\zeta^{(2)}(N_1 - N_0)\langle m^2 \rangle + 2\zeta^{(2)}(3N_1 + N_0)\langle m \rangle + 4\zeta^{(2)}N_1, \quad (10.18a)$$

$$\frac{d}{dt}\langle m^2 \rangle = \sum_m m^2 \frac{d}{dt}p_m$$
$$= 4\zeta^{(2)}(N_1 - N_0)\langle m^3 \rangle + 8\zeta^{(2)}(2N_1 + N_0)\langle m^2 \rangle$$
$$+4\zeta^{(2)}(5N_1 - N_0)\langle m \rangle + 8\zeta^{(2)}N_1, \quad (10.18b)$$

and so forth. It is obvious from (10.17) and (10.18) that the change of p_m depends on p_{m+2} and that the alteration of $\langle m^k \rangle$ is dependent on the higher moment $\langle m^{k+1} \rangle$. Thus all the probabilities p_m of finding a certain number of photons in the field mode, and also the photon-number moments, appear coupled, and it is from this that the difficulties in solving either (10.17) or (10.18) result.

A preliminary insight into the tendencies of the temporal evolution of the probabilities and the moments can be gained by using the *short-time approximation*. Provided that $N\zeta^{(2)}t \ll 1$ holds, where $N = N_1 + N_0$ is the total number of atomic systems, it is possible to expand the solutions in powers of t and to neglect terms nonlinear in t. In this way the desired parameters that characterize the radiation field are obtained from (10.17) or (10.18) by substituting the initial values $p_m(0)$ and $\langle m_0^k \rangle$ into the right-hand sides of these equations. Thus, from (10.18b), we obtain for instance

$$\langle m^2 \rangle = \langle m_0^2 \rangle + 4\zeta^{(2)}t\{N_1[\langle m_0^3 \rangle + 4\langle m_0^2 \rangle + 5\langle m_0 \rangle + 2]$$
$$-N_0[\langle m_0^3 \rangle - 2\langle m_0^2 \rangle + \langle m_0 \rangle]\}. \quad (10.19)$$

The number of atoms in the excited level can often be neglected in thermal

Table 10.1. Influence of Two-Photon Absorption on Light Beams with Different Initial States in the Short-Time Approximation ($\tilde{t} = \zeta^{(2)} N t \ll 1$)

	initial state			final state		
	$\langle m_0^2 \rangle$	$\langle m_0^3 \rangle$	$\gamma_0^{2,2}$	$\langle m \rangle$	$\langle m^2 \rangle$	$\gamma^{2,2} = \frac{\langle m^2 \rangle - \langle m \rangle}{\langle m \rangle^2}$
number	$\langle m_0 \rangle^2$	$\langle m_0 \rangle^3$	$1 - \frac{1}{\langle m_0 \rangle}$	$\langle m_0 \rangle - 2\tilde{t}[\langle m_0 \rangle^2 - \langle m_0 \rangle]$	$\langle m_0 \rangle^2 - 4\tilde{t}[\langle m_0 \rangle^3 - 2\langle m_0 \rangle^2 + \langle m_0 \rangle]$	$\left(1 - \frac{1}{\langle m_0 \rangle}\right)(1 + 2\tilde{t})$
coherent	$\langle m_0 \rangle^2 + \langle m_0 \rangle$	$\langle m_0 \rangle^3 + 3\langle m_0 \rangle^2 + \langle m_0 \rangle$	1	$\langle m_0 \rangle - 2\tilde{t}\langle m_0 \rangle^2$	$\langle m_0 \rangle^2 + \langle m_0 \rangle - 4\tilde{t}[\langle m_0 \rangle^3 + \langle m_0 \rangle^2]$	$1 - \tilde{t}$
pulsed coherent	$\langle m_0 \rangle^2 \frac{T_0}{\tau_L} + \langle m_0 \rangle$	$\langle m_0 \rangle^3 \left(\frac{T_0}{\tau_L}\right)^2 + 3\langle m_0 \rangle^2 \frac{T_0}{\tau_L} + \langle m_0 \rangle$	$\frac{T_0}{\tau_L}$	$\langle m_0 \rangle - 2\tilde{t}\left(\frac{T_0}{\tau_L}\right)\langle m_0 \rangle^2$	$\left(\frac{T_0}{\tau_L}\right)\langle m_0 \rangle^2 + \langle m_0 \rangle - 4\tilde{t}\left[\left(\frac{T_0}{\tau_L}\right)^2 \langle m_0 \rangle^3 + \left(\frac{T_0}{\tau_L}\right)\langle m_0 \rangle^2\right]$	$\frac{T_0}{\tau_L}(1 - 2\tilde{t})$
chaotic	$2\langle m_0 \rangle^2 + \langle m_0 \rangle$	$6\langle m_0 \rangle^3 + 6\langle m_0 \rangle^2 + \langle m_0 \rangle$	2	$\langle m_0 \rangle - 4\tilde{t}\langle m_0 \rangle^2$	$2\langle m_0 \rangle^2 + \langle m_0 \rangle - 8\tilde{t}[3\langle m_0 \rangle^3 + \langle m_0 \rangle^2]$	$2 - 4\tilde{t}[2\langle m_0 \rangle + 1]$

Source: Ref. 10.23.

equilibrium. Then in the irreversibility approximation and with $N_1 = 0$ and $N_0 = N$, only two of the four terms considered in (10.17) remain. Provided that this neglect is justified, we calculate from (10.19) the evolution of the normalized correlation function $\gamma^{2,2}$ with identical space–time arguments of the four field operators as

$$\gamma^{2,2} = \gamma_0^{2,2} - \frac{2\tilde{t}}{\langle m_0\rangle^2}\left\{2\langle m_0^3\rangle - \langle m_0^2\rangle + \langle m_0\rangle - 2\frac{\langle m_0^2\rangle^2}{\langle m_0\rangle}\right\}, \quad (10.20a)$$

where

$$\tilde{t} = \zeta^{(2)}Nt \quad (10.20b)$$

is the normalized time and characterizes the interaction strength.

This relation reveals the strong dependence of the temporal evolution of the correlation function on the coherence of the radiation initially present in the interaction volume. Table 10.1 gives some results for characteristic initial probability distributions. For comparison, the results for a special nonstationary radiation field have been included in Table 10.1. This is a field with square pulses of length τ_L, time T_0 between adjacent pulses, and a mean photon number $\langle m_0\rangle T_0/\tau_L$ during the pulse length, where $\langle m_0\rangle$ is the time average of the mean photon number. Even from these approximate solutions the trend of the change in the coherence properties becomes apparent. In particular, when starting from a coherent state, a decrease of the normalized correlation function below unity can be observed. As we have already learned in Chapter 5 and Chapter 9, this means that the radiation exhibits photon antibunching. In this case the antibunched radiation is achieved by the action of the two-photon absorption on ideal laser light. From the experimental standpoint the difficulty in generating antibunched radiation in this manner is that the observation of antibunching requires very small mean photon numbers where the two-photon transition probability is small.

To study this behavior in detail, one has to go beyond the region of the short-time approximation. Provided all the atoms are in the ground state ($N_0 = N$, $N_1 = 0$), then the remaining set of rate equations (10.17) for the $p_m(t)$ can be solved exactly by employing the generating function method. This was shown by Agarwal (Ref. 10.22), who used the special generating function

$$F(\lambda, \tilde{t}) = \sum_{m=0}^{\infty} \lambda^m p_m(\tilde{t}). \quad (10.21)$$

Multiplying the mth equation of the set (10.17) by λ^m and summing all the resultant equations from $m = 0$ to ∞, we obtain one differential equation for the generating function:

$$\frac{\partial}{\partial\tilde{t}}F = [1 - \lambda^2]\frac{\partial^2}{\partial\lambda^2}F. \quad (10.22)$$

From the solution of (10.22), which can be easily obtained by separation of the variables, the factorial moments

$$\langle m(m-1)\cdots(m-k+1)\rangle = \left(\frac{\partial^k}{\partial\lambda^k}F\right)_{\lambda=1} \tag{10.23a}$$

as well as the probability

$$p_m(\tilde{t}) = \frac{1}{m!}\left(\frac{\partial^m}{\partial\lambda^m}F\right)_{\lambda=0} \tag{10.23b}$$

of finding m photons in the radiation mode can be calculated. As an example, Fig. 10.10 shows some probabilities $p_m(\tilde{t})$ versus the normalized time $\tilde{t}$ for radiation fields initially in definite photon-number eigenstates. The results

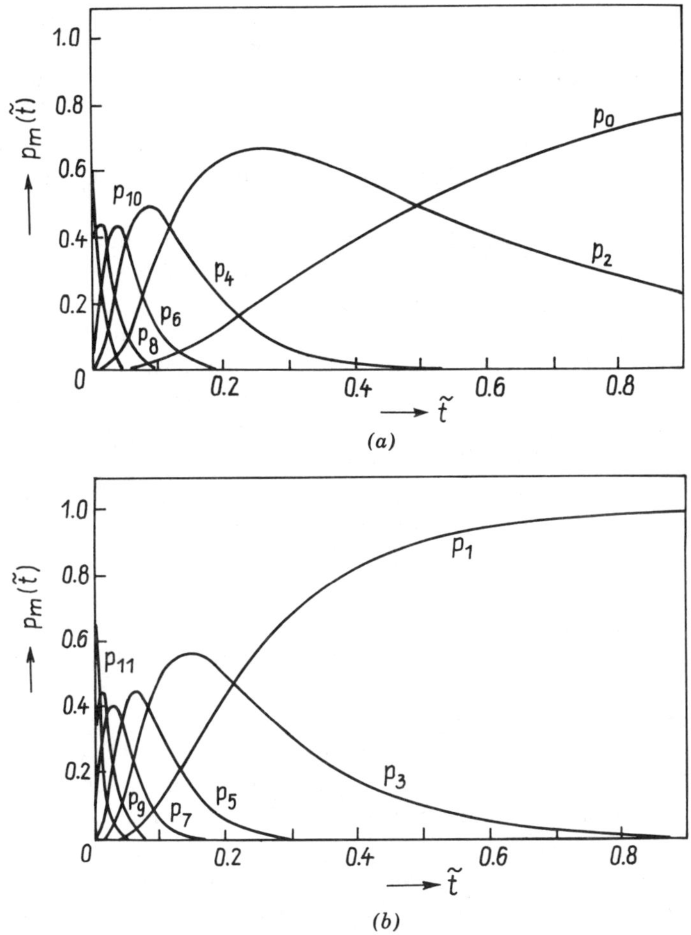

Fig. 10.10. Evolution of the probability distributions $p_m(\tilde{t})$ for radiation initially in the photon-number states (a) $|m = 10\rangle$ and (b) $|m = 11\rangle$ (after Ref. 10.21).

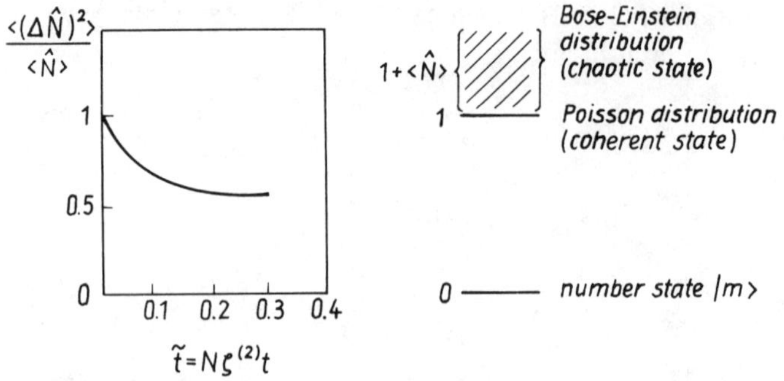

Fig. 10.11. Evolution of the reduced variance $\sigma = \langle(\Delta\hat{N})^2\rangle/\langle\hat{N}\rangle$ of a radiation field initially in the coherent state with $\langle\hat{N}\rangle_0 = 10$ under the action of two-photon absorption (after Ref. 10.23).

differ significantly according to whether $\langle\hat{N}\rangle_0 = m_0$ is even or odd. Thus the quantum character of two-photon absorption becomes obvious at low quantum numbers. Note that, for instance, after a long interaction time the mean number of photons tends to zero and to one for m_0 even and odd, respectively. The physical reason for this is simply that a one-photon field cannot be attenuated by two-photon absorption. For the same reason the degree of second-order coherence tends to zero or infinity, depending on the initial condition.

The change in the coherence properties of a light field initially in a coherent state has been calculated in Ref. 10.23. Figure 10.11 gives the resulting time-dependent reduced variance $\langle(\Delta\hat{N})^2\rangle/\langle\hat{N}\rangle$ for an initial state with $\langle\hat{N}\rangle_0 = 10$. Note that the reduced variance decreases considerably with increasing time $\tilde{t} = N\zeta^{(2)}t$ in the interval $0 \leq \tilde{t} \lesssim 0.2$. In addition, for comparison, the reduced variances of some fundamental probability distributions are presented in Fig. 10.11, namely those of the Bose–Einstein distribution describing chaotic thermal radiation fields with $\langle(\Delta\hat{N})^2\rangle/\langle\hat{N}\rangle = 1 + \langle\hat{N}\rangle$, of the Poisson distribution describing the coherent field with $\langle(\Delta\hat{N})^2\rangle/\langle\hat{N}\rangle = 1$, and of photon-number states with $\langle(\Delta\hat{N})^2\rangle/\langle\hat{N}\rangle = 0$. The calculated curve lies between the value of the coherent state and that of the photon-number state. Exact calculations lead to the limiting value $\langle(\Delta\hat{N})^2\rangle/\langle\hat{N}\rangle = 0.5(1 + \exp[-2\langle\hat{N}\rangle_0])$ as $\tilde{t} \to \infty$. Values of the reduced variance between zero and unity indicate radiation fields exhibiting the effect of photon antibunching. This confirms the result already obtained in the short-time approximation.

In summary, we may say that two-photon absorption leads from states having no photon correlation to states with photon anticorrelation. Because of the fundamental importance of this quantum-mechanical phenomenon, the problem of the photon antibunching obtainable with two-photon absorbers has been widely discussed since 1970 (see, e.g., Refs. 10.24–10.30).

The general case of n-photon absorption has also been treated (Ref. 10.30). Starting from the rate equations for the $p_m(t)$ that correspond to (10.17), differential equations for the higher photon-number moments have to be set up and solved. Thus the normalized second-order intensity correlation function $\gamma^{2,2}$ with equal space–time arguments is obtained as

$$\gamma^{2,2} = 1 - \frac{n-1}{2n-1}\frac{1}{\langle \hat{N} \rangle}.$$

This expression is less than 1 for $n \geq 2$, and the deviation from the value 1, which is attained by the coherent state, increases with increasing order n of the multiphoton process.

In spite of the great number of theoretical studies on the generation of radiation with photon antibunching by employing two-photon or multiphoton absorption, there have not yet been made any realistic proposals for experimental realization. As already mentioned, the physical reason for this difficulty is the low efficiency of multiphoton effects at low intensities. If the statistical properties of the resulting radiation are measured using the common photon-counting techniques, the accuracy of the ratio $\sigma' = \langle(\Delta\hat{N})^2\rangle/\langle\hat{N}\rangle^2$ will be prescribed by the experimental conditions, where, for instance, σ' takes on the values $1/\langle\hat{N}\rangle$ and 0 for coherent states and photon-number states respectively. Thus the maximum possible change of σ' to be expected as a result of two-photon absorption is of the order of $1/\langle\hat{N}\rangle$, and as a consequence, changes smaller than $1/\langle\hat{N}\rangle$ have to be measured, which is expected to be rather difficult at mean photon numbers above 100 in the coherence volume. Moreover, with partially coherent driving fields in place of the ideal coherent radiation assumed so far, the antibunching effect achievable by two-photon absorption is decreased (Ref. 10.31).

10.3.2. Wave Attenuation

In the preceding subsection we considered a one-mode field, employing quantum theory, and calculated the mode excitation as a function of time only. For treating wave propagation in multiphoton absorbers the change in the wave parameters and in the statistical properties of the radiation field as functions of the space and time coordinates must be considered. It was shown by Shen (10.32) that, in approximation, the stationary propagating problem can be treated in analogy to the one-mode cavity problem by replacing the time coordinate t with the space coordinate z divided by the velocity of light. Thus the results obtained so far in this section can be applied to stationary space-dependent processes.

Let us follow here another line by describing the field within the framework of classical electrodynamics. In the case of multiphoton absorption with its normally ordered interaction Hamiltonian such a description, which is much simpler than the quantized one, can be utilized at high intensities. This means that the expectation values of normally ordered operators are replaced by the respective classical quantities. The classical treatment proves to be very useful in many applications, especially when complex space–time-dependent processes are treated. With respect to application we shall discuss multiphoton absorbers as nonlinear filters that change the parameters and especially the coherence properties of given incident signals. In our discussion we start with the calculation of the attenuation of coherent waves.

10.3.2.1. Attenuation of Coherent Waves. In this sub-subsection we study the attenuation of monochromatic coherent waves within the framework of classical electrodynamics. We start from the relation for the spatial derivative of the time-averaged modulus $S_1(z)$ of the Poynting vector of a signal wave of frequency ω_1 and field-strength amplitude $E_1(z)$ propagating in the z direction:

$$\frac{d}{dz} S_1(z) = \frac{i\omega_1}{4} \widehat{E}_1^*(z) \widehat{P}_1(z) + \text{c.c.}, \tag{10.24}$$

where $\widehat{P}_1(z)$ is the amplitude of the nonlinear optical polarization at the frequency ω_1 (cf. Section 1.4.2). Let us first consider multiphoton absorbers without resonant intermediate levels. In the following evaluation of this relation, attention must be paid to whether the nonlinear polarization at ω_1 is solely produced by the electromagnetic field at this frequency or whether it is influenced by fields at another frequency (or at several other frequencies). For simplicity we assume equal polarization directions of all waves and isotropic materials.

A most simple case is that in which the signal at ω_1 itself is very weak and the nonlinearity is only caused by a single strong wave at the frequency ω_2. Then, in the case of two-photon absorption, (10.24) takes the form

$$\frac{d}{dz} S_1(z) = -\kappa_a^{(2)} S_1(z), \tag{10.25a}$$

where

$$\kappa_a^{(2)} = -\frac{2\mu_0^2 c^2 \omega_1}{(n^L)^2} \operatorname{Im}\{\chi^{(3)}(\omega_1; \omega_1, \omega_2, -\omega_2)\} S_2(z) \tag{10.25b}$$

is the absorption coefficient for two-photon absorption (cf. Section 4.5), which is proportional to the intensity of the strong wave, and n^L is the linear optical refractive index. According to the structure of this relation we may say that the strong wave at ω_2 induces the absorption for the weak wave at ω_1, or that the optical "constants" for the signal at ω_1 are modified by the strong pump wave. Particularly simple relations result if the intensity of the pump wave can be assumed to be independent of z. It is in this case that (10.25) agrees in its structure with the relation for one-photon absorption processes and we obtain an exponential spatial attenuation of the wave according to

$$S_1(z) = S_1(0)e^{-\ell_a^{(2)}z}. \tag{10.26}$$

As we see from this relation, the attenuation of the wave at ω_1 can be controlled by means of another wave. This fact is of importance in experimental application.

In contrast to the case described, for the attenuation of only one single wave of frequency ω_1 we obtain from (10.24)

$$\frac{d}{dz}S_1(z) = -b[S_1(z)]^2, \tag{10.27a}$$

where

$$b = -\frac{2\mu_0^2c^2\omega_1}{(n^L)^2}\operatorname{Im}\{\chi^{(3)}(\omega_1;\omega_1,\omega_1,-\omega_1)\}. \tag{10.27b}$$

For atomic systems that do not interact with one another we obtain at resonance ($\omega_1 = \omega_{10}$)

$$b = \frac{\tilde{\sigma}_{10}^{(2)}(\omega_{10})\gamma}{\hbar\omega_{10}},$$

where $\tilde{\sigma}_{10}^{(2)}$ is the two-photon absorption cross section [cf. (4.82)] and γ is the number density of the particles. The solution of (10.27) is

$$S_1(z) = S_1(0)\frac{1}{1 + bzS_1(0)}. \tag{10.28}$$

We see that for large values of $bzS_1(0)$ the absorption process results in an output signal $S_1(z) = 1/bz$ that is independent of the input signal. This already shows that signal fluctuations can be reduced by means of multiphoton absorption. In the following sub-subsection this possibility will be discussed in more detail.

Let us mention here that it has been possible to experimentally verify the two different relationships for two-photon absorption (10.26) and (10.28). In experimental investigations of this kind special attention has to be given to the need to see that the attenuation due to the two-photon transition $0 \to 1$ is not superposed on an attenuation due to an absorption transition $1 \to x$ from the excited state to higher levels (see Fig. 10.12). Processes like these were for example studied in Refs. 10.33 and 10.34.

On the other hand, such an absorption from the excited state can also be favorably utilized for modifying signals, because the efficiency of one-photon processes even at low occupation-number densities in the initial level 1 may

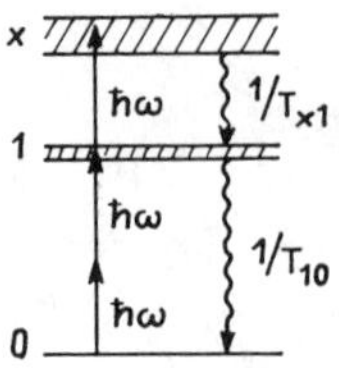

Fig. 10.12. Level scheme of a special two-step absorber.

cause significant attenuation of the signal if each excited atom repeatedly takes part in one-photon absorption processes. This will be studied in the following because of its importance as an effective nonlinear filtering process. Obviously, it is necessary to depart from the irreversible approximation and to start from the rate equations for the occupation-number densities of the levels 0, 1, and x involved in the process and from the rate equation for the photon flux density. This set is

$$\frac{\partial}{\partial\eta}\gamma_0 = -\tfrac{1}{2}\tilde{\sigma}_{01}^{(2)}I^2(\gamma_0-\gamma_1)+\frac{1}{T_{10}}\gamma_1, \tag{10.29a}$$

$$\frac{\partial}{\partial\eta}\gamma_1 = \tfrac{1}{2}\tilde{\sigma}_{01}^{(2)}I^2(\gamma_0-\gamma_1)-\frac{1}{T_{10}}\gamma_1-\sigma_{1x}^{(1)}I(\gamma_1-\gamma_x)+\frac{1}{T_{x1}}\gamma_x, \tag{10.29b}$$

$$\frac{\partial}{\partial\eta}\gamma_x = \sigma_{1x}^{(1)}I(\gamma_1-\gamma_x)-\frac{1}{T_{x1}}\gamma_x, \tag{10.29c}$$

$$\frac{\partial}{\partial\xi}I = -\tilde{\sigma}_{01}^{(2)}(\gamma_0-\gamma_1)I^2-\sigma_{1x}^{(1)}(\gamma_1-\gamma_x)I, \tag{10.29d}$$

where $\gamma_0+\gamma_1+\gamma_x=\gamma$ is the total occupation-number density, $\tilde{\sigma}_{01}^{(2)}$ and $\sigma_{1x}^{(1)}$ are the cross sections for two-photon absorption and one-photon absorption of the transitions denoted by the subscripts, T_{10} and T_{x1} are lifetimes, $\eta = t - z/v$ is the *retarded time*, and $\xi = z$ is a coordinate in the moving frame.

Now the parameters of the two-photon absorption step and the sample length L are supposed to satisfy the condition $\tilde{\sigma}_{01}^{(2)}\gamma_0 IL \ll 1$. Provided this condition is obeyed, the light intensity is only slightly changed by the two-photon absorption itself. The one-photon transition is supposed to affect the incident light strongly without being saturated. This means the two conditions $\sigma_{1x}^{(1)}\gamma_1 L \sim 1$ and $\sigma_{1x}^{(1)}IT_{x1} \ll 1$ have to be satisfied, which is possible if $\sigma_{1x}^{(1)}$ is fairly large and T_{x1} is very small. Experimentally, the condition $\sigma_{1x}^{(1)}IT_{x1} \ll 1$ has been demonstrated to be fulfilled for some dissolved organic molecules up to photon flux densities of $I \sim 10^{28}\ \mathrm{cm}^{-2}\,\mathrm{s}^{-1}$. We note that such conditions cannot only be satisfied by dissolved organic molecules; there exist for instance solids suitable for these purposes with large cross sections $\sigma_{1x}^{(1)}$, very short lifetimes T_{x1}, and times T_{10} varying over a broad range. Under the conditions mentioned we obtain from (10.29) the following much simpler set of rate equations:

$$\frac{\partial}{\partial \eta}\gamma_1(\xi,\eta) = \tfrac{1}{2}\tilde{\sigma}_{01}^{(2)}\gamma I^2(\xi,\eta) - \frac{1}{T_{10}}\gamma_1(\xi,\eta), \tag{10.30a}$$

$$\frac{\partial}{\partial \xi}I(\xi,\eta) = -\sigma_{1x}^{(1)}\gamma_1(\xi,\eta)I(\xi,\eta), \tag{10.30b}$$

where we have made use of the stationary solution of (10.29c), of the relation $\gamma_0 + \gamma_1 \approx \gamma$, and of the inequality $\tilde{\sigma}_{01}^{(2)}I \ll \sigma_{1x}^{(1)}$.

We multiply (10.30b) by $I(\xi,\eta)$ and eliminate $I(\xi,\eta)$ by means of (10.30a). Thus we get, with neglect of $(1/T_{10})\gamma_1$ compared to $(\partial/\partial\eta)\gamma_1$, the differential equation

$$\frac{\partial^2}{\partial \eta\,\partial \xi}\gamma_1(\xi,\eta) = -2\sigma_{1x}^{(1)}\gamma_1(\xi,\eta)\frac{\partial}{\partial \eta}\gamma_1(\xi,\eta). \tag{10.31}$$

Integration with respect to η gives

$$\frac{\partial}{\partial \xi}\gamma_1(\xi,\eta) = -\sigma_{1x}^{(1)}[\gamma_1(\xi,\eta)]^2 + f(\xi). \tag{10.32}$$

From the initial condition $\gamma_1(\eta \to -\infty) = 0$ and $(\partial/\partial\eta)\gamma_1(\eta \to -\infty) = 0$ we have $f(\xi) = 0$. Taking into account the boundary condition

$$\gamma_1(0,\eta) = \tfrac{1}{2}\tilde{\sigma}_{01}^{(2)}\gamma\int_{-\infty}^{\eta} d\eta'\, I^2(0,\eta'), \tag{10.33}$$

the solution of (10.32) is obtained as

$$\gamma_1(\xi,\eta) = \frac{1}{\sigma_{1x}^{(1)}\xi + \left[\tfrac{1}{2}\tilde{\sigma}_{01}^{(2)}\gamma\int_{-\infty}^{\eta} d\eta'\, I^2(0,\eta')\right]^{-1}}. \tag{10.34}$$

This occupation-number density of the first excited level is substituted into (10.30b) and then the resulting differential equation is integrated with respect to ξ, which yields the result

$$I(\xi,\eta) = \frac{I(0,\eta)}{1 + \tfrac{1}{2}\tilde{\sigma}_{01}^{(2)}\sigma_{1x}^{(1)}\gamma\xi\int_{-\infty}^{\eta} d\eta'\, I^2(0,\eta')}. \tag{10.35}$$

We see from (10.35) that the transmission decreases with increasing time, and thus an incident light pulse is suppressed at its trailing edge. Hence pulses with steep rear edges can be obtained.

If the laser pulse passes in succession through a slowly relaxing one-photon absorber as described in Section 8.6.1 and the two-step absorber described above, it can be shortened substantially (see Fig. 10.13) because the leading edge and the trailing edge are heavily suppressed, whereas the pulse maximum is only slightly attenuated under appropriate conditions (Ref. 10.35). Note that both absorbers used represent nonlinear filters with memory.

From the examples discussed in this subsection it is obvious that multiphoton absorption with and without intermediate resonant levels offers the opportunity, even if the conditions for the rate equation approximation are satisfied, to construct nonlinear filters with properties as required for pulse shaping and signal processing in various applications.

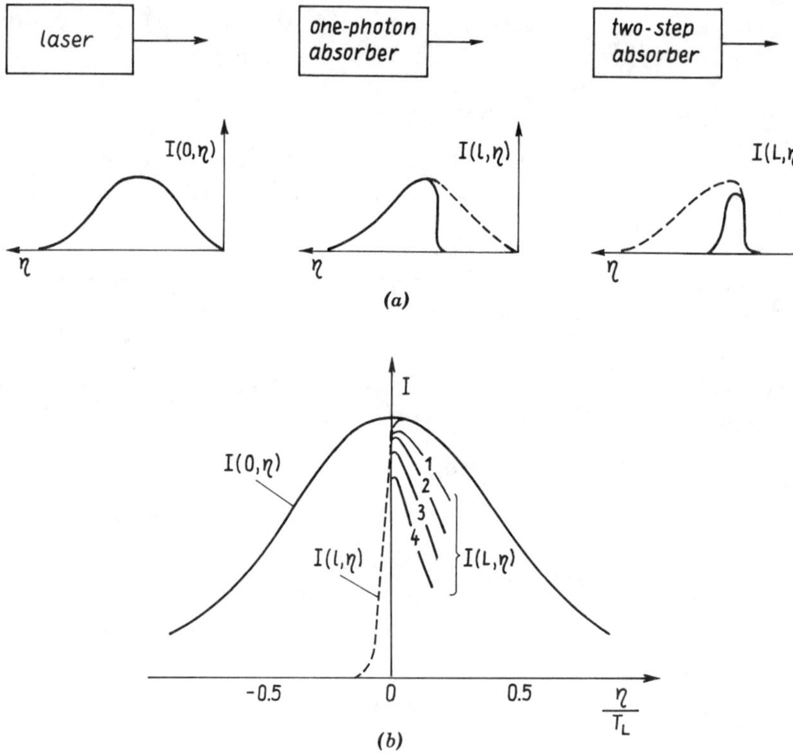

Fig. 10.13. Pulse shaping by a slow saturable absorber and a special two-step absorber (after Ref. 10.35): (*a*) arrangement of the laser and the absorbers, (*b*) photon flux densities $I(0, \eta)$, $I(l, \eta)$, and $I(L, \eta)$ at the entrance of the slow saturable absorber, at the entrance of the two-step absorber, and at the exit of the entire device, respectively. The parameters of the one-photon absorbers are $\sigma_{10}^{(1)}\gamma l = 30$, $\sigma_{10}^{(1)} I_0 \tau_{L0} = 30$, and the parameters of the two-step absorber are $\frac{1}{2}\tilde{\sigma}_{01}^{(2)}\sigma_{1x}^{(1)} L I_{L0}^2 \tau_L = 1, 2, 4, 8$ for curves 1 to 4, respectively.

The variety of observable phenomena, and also that of possible filtering processes, increases appreciably if transient multiphoton processes are included in which the characteristic quantities like field amplitudes and occupation-number densities change appreciably in times that are smaller than or comparable to the phase decay time of the transitions involved. Then coherent phenomena can be observed in multiphoton absorption analogous to those in one-photon absorption, in particular self-induced transparency and the photon echo. Note however that only in materials with very large two-photon absorption cross sections is it possible to achieve strong transient excitations with intensities nondestructive for the material. Favorite materials for this purpose, having high nonlinear susceptibilities, are the semiconductors. Under simplifying conditions the transient multiphoton phenomena can be described in full analogy to those of one-photon interactions treated in Section 8.6. In the case of two-photon absorption, if the resonance condition $2\omega_L = \omega_{10}$ is satisfied and if the field-strength amplitude $\overline{E}(\xi, \eta)$ is real, we obtain, starting from (4.87), for the occupation-number inversion γ_I and the imaginary part $\overline{Q}_2$ of the polarizability amplitude, which is defined in full analogy to that of the stimulated Raman effect in Section 4.7, the following set of equations:

$$\left(\frac{\partial}{\partial \eta} + \frac{1}{\tau_{10}}\right)\overline{Q}_2(\xi, \eta) = -\frac{1}{4\hbar}\overline{|\alpha_{01}|^2}^{0}\gamma_I(\xi, \eta)\overline{E}^2(\xi, \eta), \tag{10.36a}$$

$$\frac{\partial}{\partial \eta}\gamma_I(\xi, \eta) + \frac{1}{T_{10}}[\gamma_I(\xi, \eta) - \gamma_I^e] = \frac{1}{2\hbar}\overline{Q}_2(\xi, \eta)\overline{E}^2(\xi, \eta), \tag{10.36b}$$

$$\frac{\partial}{\partial \xi}\overline{E}(\xi, \eta) = -\frac{\mu_0\omega_{10}^2}{4k_L}\overline{Q}_2(\xi, \eta)\overline{E}(\xi, \eta). \tag{10.36c}$$

The analogy with the corresponding equations for one-photon processes becomes even more apparent if we assume the variations of $\overline{E}$ to be small. Then we may regard the field amplitude $\overline{E}(\eta)$ in (10.36a, b) as a given function of time, and hence these relations are of the same structure as (8.2) rewritten for exact resonance ($\Delta = 0$) and without phase modulation ($d\varphi/dt = 0$), provided we use the formal substitutions $\overline{P}_2 \to (\text{const} \cdot Q_2)$ and $\overline{E} \to (\text{const} \cdot \overline{E}^2)$. Thus it is possible to use the results gained in Section 8.3 immediately for such nonstationary processes as optical nutation.

10.3.2.2 Attenuation of Fluctuating Waves. The general consideration of multiphoton absorption has already allowed us to infer that during the passage of radiation through a multiphoton absorber its intensity fluctuations decrease and a stabilization of the intensity results. Therefore, assuming the same conditions concerning the radiation field and the average attenuation in the material, the resultant effect is the more pronounced the higher is the order of the nonlinear process.

Based on the abovementioned classical treatment a simple example will now be studied which is of importance for practical purposes. The fluctuation of the input signal is Gaussian, and therefore its intensity $I(0)$ satisfies the probability distribution

$$P[I(0)]\,dI(0) = \frac{1}{\langle I(0)\rangle} e^{-I(0)/\langle I(0)\rangle}\,dI(0). \tag{10.37}$$

More exactly, $I(0)$ represents the instantaneous photon flux density $I(0, \eta)$ of the entrance signal; the measurement has to be made in times much shorter than the correlation time τ_c of the signal. The assumption made about the statistical behavior is satisfied by all thermal radiation sources; also the radiation of lasers emitting on a great number of modes with uncorrelated phases satisfies these statistics to a good approximation. The correlation time of the signals of such lasers is given by the reciprocal of the width of the frequency band that the laser is oscillating in. In the following investigations we assume that the conditions for applying the irreversibility approximation are satisfied; this means that the occupation numbers of the atomic levels preserve the equilibrium values attained before the sample has been irradiated by the signal. Since in general $\mathscr{E}_1 - \mathscr{E}_0 \gg k_B \mathscr{T}$, for all times the relations $\gamma_1 \approx 0$ and $\gamma_0 \approx 1$ can be inferred. The use of the irreversibility approximation in the investigation of multiphoton absorption is justified in many experimental situations, since the absorption is so weak that within the lifetime of the excited state the individual atomic system is excited with only a probability much smaller than 1. Quantitatively we require the condition

$$\tilde{\sigma}_{10}^{(2)}[I(0)]^2 T_{10} \ll 1. \tag{10.38}$$

With $\tilde{\sigma}_{10}^{(2)} \sim 10^{-48}\ \mathrm{cm}^4\ \mathrm{s}$ and $T_{10} \sim 10^{-9}$ s, which are typical values for strong electronic transitions (cf. Section 4.6), this inequality is satisfied up to photon flux densities of about $10^{28}\ \mathrm{cm}^{-2}\ \mathrm{s}^{-1}$. Under these conditions the absorber has no memory of the previous signals, and thus the output signal $I(L, \eta)$ at the position L only depends on $I(0, \eta)$. To take into account the finite transit time through the absorber we have here used the local time η. In such two-photon absorbers an arbitrary instantaneous input intensity $I(0, \eta)$ yields the output signal

$$I(L, \eta) = \frac{I(0, \eta)}{1 + \tilde{\sigma}_{10}^{(2)} \gamma L I(0, \eta)}, \tag{10.39}$$

which corresponds to the result (10.28) for the stationary case; in three-photon absorbers it yields the output signal

$$I(L, \eta) = \frac{I(0, \eta)}{\sqrt{1 + 2\tilde{\sigma}_{10}^{(3)} \gamma L I^2(0, \eta)}}. \tag{10.40}$$

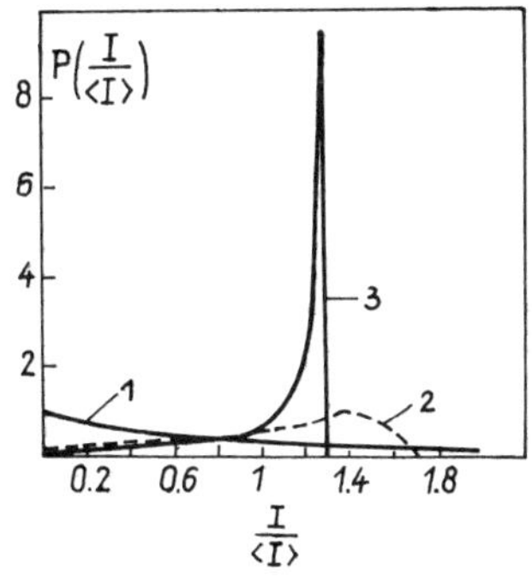

Fig. 10.14. Influence of multiphoton absorbers on the intensity distribution $P(I)$ of fluctuating signals (after Ref. 10.37). Curve 1: Gaussian input signal; curve 2 (3): output signal of two-photon absorber (three-photon absorber) with average transmission $\mathscr{T}_{\text{eff}} = \langle I(L, \eta)\rangle / \langle I(0, \eta)\rangle = 0.27$.

In general, for n-photon absorbers we have

$$I(L, \eta) = \frac{I(0, \eta)}{\left[1 + (n-1)\gamma\tilde{\sigma}_{10}^{(n)} L \{I(0, \eta)\}^{n-1}\right]^{1/(n-1)}} \tag{10.41}$$

In this way we obtain a simple memory-free relation between input and output signals allowing a direct transformation of the probability distribution (10.37). From (10.39) and (10.40) respectively we determine the input signal $I(0, \eta)$ as a function of $I(L, \eta)$ and substitute it into (10.37), thereby obtaining the probability distribution of the output signals. Figure 10.14 shows this resultant probability distribution for a two-photon absorber (curve 2) and a three-photon absorber (curve 3), assuming the same effective transmission $\mathscr{T}_{eff} = \langle I(L, \eta)\rangle / \langle I(0, \eta)\rangle = 0.27$ (see Refs. 10.23, 10.36, 10.37). For comparison the probability distribution of the input signal (curve 1) is given. We see that the intensity-stabilizing effect of the multiphoton absorption is very strong. So the radiation at the output of the three-photon absorber, for example, exhibits only small fluctuations around the mean intensity, and in its statistical behavior it resembles the radiation of a laser above threshold (cf. Section 7.2).

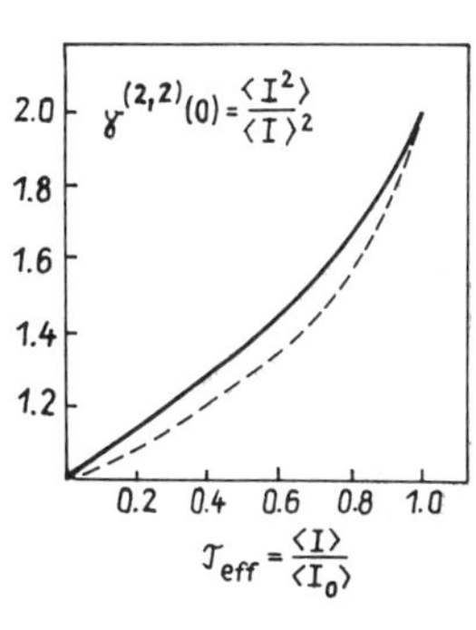

Fig. 10.15. Normalized intensity correlation function $\gamma^{2,2}(0) = \langle I^2\rangle / \langle I\rangle^2$ at the exit of an n-photon absorber ($n = 2$, solid line, $n = 3$, broken line) for Gaussian light at the entrance (after Ref. 10.37).

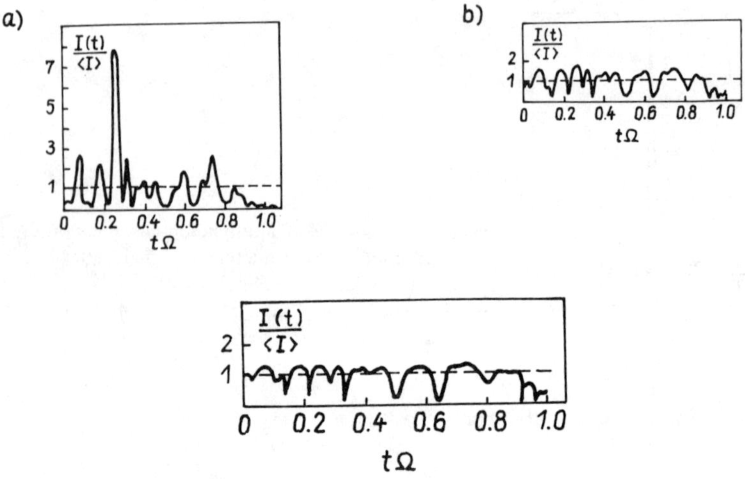

(c)

Fig. 10.16. Influence of n-photon absorbers on a fluctuating signal (after Refs. 10.36 and 10.37): (a) input signal I_0, (b) output signal at the exit of a two-photon absorber ($\langle I \rangle / \langle I_0 \rangle = 0.27$), ($c$) output signal at the exit of a three-photon absorber ($\langle I \rangle / \langle I_0 \rangle = 0.27$).

As already mentioned in Chapters 5 and 9 and Section 10.3.1, the normalized second-order intensity correlation function $\gamma^{2,2}$ at identical space–time points may be regarded as a measure of the intensity fluctuations, where the value 2 corresponds to Gaussian behavior and the value 1 corresponds to fully intensity-stabilized light. (Note that in the classical description the value of 1 is a lower limit, in contrast to the results of quantum theory, where the inequality $\gamma^{2,2} < 1$ indicates photon antibunching.) The normalized second-order intensity correlation function $\gamma^{2,2}(0) = \langle I^2 \rangle / \langle I \rangle^2$ of the signal that has passed through a two-photon absorber or a three-photon absorber is shown in Fig. 10.15 as a function of the effective transmission $\mathscr{T}_{eff} = \langle I \rangle / \langle I(0) \rangle$. It is apparent that the same stabilizing effect can be obtained with the three-photon absorber at an appreciably higher transmission than with the two-photon absorber. The intensity-stabilizing effect becomes still more obvious if we study the influence of multiphoton absorbers on one realization of the stochastically fluctuating input signal. Figure 10.16a shows such an input, and Fig. 10.16b and c present this signal after its passage through a two-photon absorber and a three-photon absorber, respectively. Note the strong reduction of the high fluctuation peaks of the signal in this process.

So far our interpretation has shown that under appropriate conditions multiphoton absorbers act as nonlinear, memory-free filters. Their practical application is limited by the fact that the required influence on the signal can only be achieved at very high input intensities satisfying the condition

$$(n-1)\gamma \tilde{\sigma}_{01}^{(n)} L I^{n-1} \sim 1. \tag{10.42}$$

Limitations are imposed on an arbitrary enlargement of L in that linear losses, which were disregarded in our discussion, also increase with L. Therefore in the first place it is of advantage to choose materials with multiphoton absorption cross sections as high as possible. The two-photon absorption cross section takes particularly high values, as already stated in Section 4.6, if near the one-photon energy there exist intermediate levels. Analogous statements are valid for multiphoton absorbers with $n > 2$.

The effect of such intermediate levels on the interaction process becomes even stronger if these levels will be reached by resonant processes and will become populated during the interaction. In general, such multiphoton absorbers with resonant intermediate levels act as nonlinear filters with memory. The memory effect becomes negligible only if the lifetime of the resonant intermediate level is short in comparison with the signal correlation time τ_c. Generally it is very difficult to calculate the influence of nonlinear filters with memory on fluctuating signals (see Ref. 10.38).

As a comparatively simple example of nonlinear, noninstantaneous filtering we give an approximate calculation of the modification of the statistical properties of a Gaussian signal with center frequency ω_L in a three-photon absorber with an intermediate level at $2\hbar\omega_L$ (see Fig. 10.12), which in reality is a two-step absorber. Such an absorber can be described by the set of rate equations (10.29) and, provided all the conditions given below (10.29) are met, by the simplified system (10.30). In our discussion we shall follow Ref. 10.39. Substituting the formal solution of (10.30a) into (10.30b), we obtain

$$\frac{\partial}{\partial\xi}I(\xi,\eta) = -\tfrac{1}{2}\tilde{\sigma}_{01}^{(2)}\sigma_{1x}^{(1)}\gamma\int_{-\infty}^{\eta} d\eta'\, I^2(\xi,\eta')I(\xi,\eta)e^{(\eta'-\eta)/T_{10}}. \quad (10.43)$$

In order to describe the modification of the statistical properties of light within the absorber, the changes in the classical intensity correlation functions $G^{(p)}(\xi,\eta_1;\ldots;\xi,\eta_p) = \langle I(\xi,\eta_1)\cdots I(\xi,\eta_p)\rangle$ of all orders p, which are proportional to the field-strength correlation functions $\Gamma^{p,p}(\eta_1,\ldots,\eta_p,\eta_p,\ldots\eta_1)$ [cf. (5.26)], have to be calculated as a function of the coordinate ξ. Equations for these correlation functions are obtained from (10.43) by multiplying it by the photon flux densities of certain time arguments and taking ensemble averages. They are

$$\frac{\partial}{\partial\xi}\langle I(\xi,\eta)\rangle = -\tfrac{1}{2}\tilde{\sigma}_{01}^{(2)}\sigma_{1x}^{(1)}\gamma\int_{-\infty}^{\eta} d\eta'\,\langle I^2(\xi,\eta')I(\xi,\eta)\rangle e^{(\eta'-\eta)/T_{10}},$$

$$\begin{aligned}\frac{\partial}{\partial\xi}&\langle I(\xi,\eta)I(\xi,\eta+\tau)\rangle\\ &= -\tfrac{1}{2}\tilde{\sigma}_{01}^{(2)}\sigma_{1x}^{(1)}\gamma\left\{\int_{-\infty}^{\eta} d\eta'\,\langle I^2(\xi,\eta')I(\xi,\eta)I(\xi,\eta+\tau)\rangle e^{(\eta'-\eta)/T_{10}}\right.\\ &\qquad\left.+\int_{-\infty}^{\eta+\tau} d\eta'\,\langle I^2(\xi,\eta')I(\xi,\eta)I(\xi,\eta+\tau)\rangle e^{(\eta'-\eta-\tau)/T_{10}}\right\},\end{aligned}$$

(10.44)

and so forth. Obviously the equations for the intensity correlation function of different orders are coupled with one another. Therefore the system can be solved only approximately. Aiming at this solution, we expand the correlation functions in terms of ξ and obtain from (10.44) in the first order of ξ

$$G^{(1)}(\xi,\eta) \equiv \langle I(\xi,\eta)\rangle = G_0^{(1)}(\eta) - \tfrac{1}{2}\tilde{\sigma}_{01}^{(2)}\sigma_{1x}^{(1)}\gamma\xi \times \int_{-\infty}^{\eta} d\eta'\, G_0^{(3)}(\eta',\eta',\eta)\, e^{(\eta'-\eta)/T_{10}}, \tag{10.45a}$$

$$\begin{aligned} G^{(2)}(\xi,\eta;\xi,\eta+\tau) &\equiv \langle I(\xi,\eta)I(\xi,\eta+\tau)\rangle \\ &= G_0^{(2)}(\eta,\eta+\tau) \\ &\quad - \tfrac{1}{2}\tilde{\sigma}_{01}^{(2)}\sigma_{1x}^{(1)}\xi\left\{\int_{-\infty}^{\eta} d\eta'\, G_0^{(4)}(\eta',\eta',\eta,\eta+\tau)\, e^{(\eta'-\eta)/T_{10}}\right. \\ &\quad \left. + \int_{-\infty}^{(\eta+\tau)} d\eta'\, G_0^{(4)}(\eta',\eta',\eta,\eta+\tau)\, e^{(\eta'-\eta-\tau)/T_{10}}\right\}, \end{aligned} \tag{10.45b}$$

and so forth, where $G_0^{(p)}$ denotes the correlation function at the position $\xi = 0$.

An insight into the process of intensity stabilization can be gained by calculating for $\eta = 0$ and $\tau = 0$ the degree of second-order coherence, $\gamma_\xi^{2,2}$, as a function of ξ, which is the same as the normalized second-order correlation function; it is given by

$$\gamma_\xi^{2,2} = \frac{G^{(2)}(\xi,0;\xi,0)}{\left[G^{(1)}(\xi,0)\right]^2} = \frac{\langle I^2(\xi)\rangle}{\langle I(\xi)\rangle^2}. \tag{10.46}$$

This function describes the efficiency of nonlinear optical processes such as for instance the nonlinear absorption of the filtered light in another two-photon experiment.

We now assume the light at the entrance of the absorber to be a stationary Gaussian signal with Lorentzian band shape. Hence the entrance signal is entirely characterized by the field-strength correlation function

$$\Gamma_0^{1,1}(\eta,0) \propto \langle I_0\rangle e^{i\omega_0\eta - |\eta|/\tau_c}, \tag{10.47}$$

where τ_c is the correlation time of the signal, which is related to the half-width $\Delta\omega$ of the Lorentzian line by $\Delta\omega = 2/\tau_c$. Then the entrance correlation functions of interest are calculated to be

$$G_0^{(2)}(\eta, 0) = \langle I_0 \rangle^2 \{ e^{-2|\eta|/\tau_c} + 1 \}, \tag{10.48a}$$

$$G_0^{(3)}(\eta, \eta, 0) = \langle I_0 \rangle^3 \{ 4e^{-2|\eta|/\tau_c} + 2 \}, \tag{10.48b}$$

$$G_0^{(4)}(\eta, \eta, 0, 0) = \langle I_0 \rangle^4 \{ 4e^{-4|\eta|/\tau_c} + 16e^{-2|\eta|/\tau_c} + 4 \}. \tag{10.48c}$$

By substituting these functions $G_0^{(p)}$ into (10.45) and (10.46) we obtain

$$\mathscr{T}_{\text{eff}}(L) \equiv \frac{\langle I \rangle}{\langle I_0 \rangle} = 1 - \tilde{\sigma}_{01}^{(2)} \sigma_{1x}^{(1)} \gamma L \langle I_0 \rangle^2 T_{10} \left\{ 1 + \frac{2}{1 + 2T_{10}/\tau_c} \right\}, \tag{10.49a}$$

$$\frac{\langle I^2 \rangle}{\langle I_0 \rangle^2} = 2 - 4\tilde{\sigma}_{01}^{(2)} \sigma_{1x}^{(1)} \gamma L \langle I_0 \rangle^2 T_{10} \left\{ 1 + \frac{4}{1 + 2T_{10}/\tau_c} + \frac{1}{1 + 4T_{10}/\tau_c} \right\} \tag{10.49b}$$

and

$$\gamma_\xi^{2,2} = 2 - 4\tilde{\sigma}_{01}^{(2)} \sigma_{1x}^{(1)} \gamma L \langle I_0 \rangle^2 T_{10} \left\{ \frac{2}{1 + 2T_{10}/\tau_c} + \frac{1}{1 + 4T_{10}/\tau_c} \right\} \tag{10.49c}$$

at the exit of the sample with $\xi = L$. The special case of a nonlinear filter without memory is obtained from (10.49) in the limit $T_{10}/\tau_c \to 0$. Note that these solutions are only valid if $x = \tilde{\sigma}_{01}^{(2)} \sigma_{1x}^{(1)} \gamma L \langle I_0 \rangle^2 T_{10}$ is small compared to one.

In order to extend the range of validity, the higher orders in x have to be calculated in an iteration procedure. But typical trends of the coherence properties can be discussed by taking into account only the first order. So we calculate the parameter

$$\imath = \left[\frac{\gamma_\xi^{2,2} - \gamma_0^{2,2}}{\mathscr{T}_{\text{eff}}(\xi) - 1} \right]_{\xi \to 0},$$

which is of special interest because, without changing the mean intensity very strongly, we want to exert a great influence on the second-order correlation function describing intensity stabilization. In Fig. 10.17 the parameter $\imath$ is

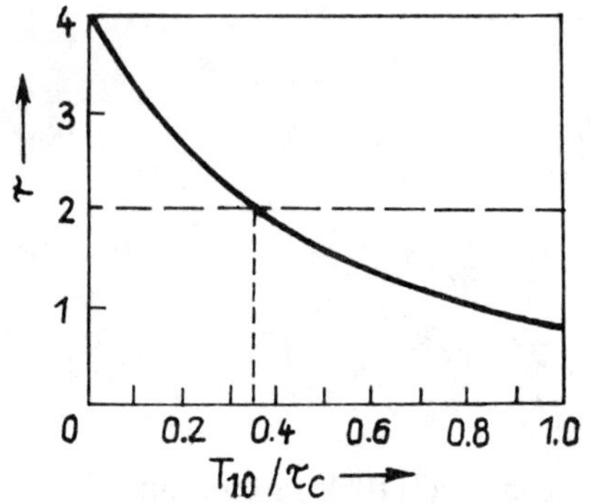

Fig. 10.17. Coherence parameter $\imath = [(\gamma_\xi^{2,2} - \gamma_0^{2,2})/(\mathcal{T}_{\text{eff}}(\xi) - 1)]_{\xi \to 0}$ of a three-photon absorber with intermediate resonant level 1 (full line) and of a two-photon absorber without intermediate resonant levels (horizontal dashed line) (after Ref. 10.39).

plotted against the ratio of the lifetime T_{10} to the correlation time τ_c. Obviously the two-step absorber with $T_{10}/\tau_c \to 0$ is the most efficient filter. Note that in the case of $T_{10}/\tau_c \to 0$ the normalized correlation function can be calculated without approximations. The parameter $\imath$ decreases monotonically with increasing T_{10}/τ_c and becomes zero at infinity. This means that the statistical properties of the incident light do not change in the interaction process as $T_{10}/\tau_c \to \infty$.

10.4. MULTIPHOTON IONIZATION

In the introduction to this chapter mention was made of multiphoton ionization as a special case of multiphoton absorption. The specific phenomena of multiphoton ionization originate from the structure of the final state of the atomic system, where there are unbound electrons, as well as from the resulting electrical detectability of the absorption process.

As an example we calculate the transition probability of *two-photon ionization* of atomic systems under irradiation with a monochromatic radiation field of frequency ω and the photon flux density $I(\omega)$. To do this, we start from the transition probability of two-photon absorption from a certain initial state $|i\rangle_A$ of the atomic system to a certain final state $|f\rangle_A$. According to (3.53) this is

$$\frac{\Delta P_{i \to f}^{(4)}}{\Delta t} = [\hbar\omega I(\omega)]^2 |\tilde{\Lambda}_{if}^{(2)}|^2 \delta(\mathscr{E}_{fi}), \tag{10.50}$$

where $\mathscr{E}_{fi} = \mathscr{E}_{Af} - \mathscr{E}_{Ai} - 2\hbar\omega$ and where $\tilde{\Lambda}_{if}^{(2)}$ is proportional to the matrix element α_{if} of (4.61). [Note the analogy to (10.7a).] If the transitions do not only lead to a single final state, but to a continuum with regard to the states of the atomic system with the state density $\sigma(\mathscr{E}_{Af})$, from (10.50) the total transition probability per unit time is obtained as

$$\frac{\Delta P^{(4)}}{\Delta t} = [\hbar\omega I(\omega)]^2 |\tilde{\Lambda}^{(2)}|^2 \sigma(\mathscr{E}_{Af}) \tag{10.51}$$

(cf. Section 3.3.2.2). In deriving (10.51) we have assumed the coupling parameters $\tilde{\Lambda}_{if}^{(2)}$ to be equal to each other for all possible final states $|f\rangle$.

For the specific example of two-photon ionization, the state density $\sigma(\mathscr{E}_{Af})$ needed in (10.51) must be calculated at the energy $\mathscr{E}_{Af}$ determined by

$$\mathscr{E}_{Af} \equiv \mathscr{E}_{\text{ion}} + \frac{\hbar^2 k^2}{2m} = \mathscr{E}_{Ai} + 2\hbar\omega, \tag{10.52}$$

where $\mathscr{E}_{\text{ion}}$ is the *ionization energy* of the atom, k is the length of the wave vector of the electron, and m is its mass. The eigenfunctions of the free electrons released by ionization, which we further ascribe to the atomic system according to our subdivision of the total system into the field and the atomic system, are assumed to be normalized in the volume V so that

$$\tilde{\sigma}(k_\bullet)\, d^3k_\bullet = V\left(\frac{1}{2\pi}\right)^3 d^3k_\bullet \tag{10.53}$$

holds for the state density $\tilde{\sigma}(k_\bullet)$ as a function of the wave vector $k_\bullet$ (see, e.g., Ref. 10.40), if the spin of the electrons is fixed. Hence, using $\tilde{\sigma}(k_\bullet)\, d^3k_\bullet = \sigma(\mathscr{E}_{Af})\, d\mathscr{E}_{Af}$, $d^3k_\bullet = k^2\, dk\, d\Omega = k(m/\hbar^2)\, d\mathscr{E}_{Af}\, d\Omega$, and (10.53), $\sigma(\mathscr{E}_{Af})$ becomes

$$\sigma(\mathscr{E}_{Af}) = \frac{V}{(2\pi)^3}\frac{mk}{\hbar^2}\Delta\Omega, \tag{10.54}$$

where $\Delta\Omega$ is the solid-angle element around the wave vector $k_\bullet$. Thus we obtain for the probability of transitions into a group of final states with the electron flying into the solid-angle element $\Delta\Omega$

$$\frac{\Delta P_\Omega^{(4)}}{\Delta t}\Delta\Omega = \left(\frac{1}{2\pi}\right)^3 V\frac{m}{\hbar^2}k|\tilde{\Lambda}^{(2)}|^2[\hbar\omega I(\omega)]^2\,\Delta\Omega, \tag{10.55}$$

where $\Delta P_\Omega/\Delta t$ is a transition rate per unit solid angle. The total ionization rate can be obtained by integrating (10.55) over the entire solid angle.

The essential difference in the calculation of the matrix elements $\tilde{\Lambda}^{(2)}$ of (10.55) from those discussed above [cf. (4.79)] is that the final states $|f\rangle$ belong to the continuum. For hydrogenlike one-electron systems the final states can be expanded in eigensolutions of the Coulomb problem for positive energies (see, e.g., Ref. 10.41). After this, only the matrix elements between the discrete and continuous states of the hydrogen atom remain to be calculated, and this can be done without major difficulties (see, e.g., Ref. 10.42).

Note that both in performing these calculations and in carrying out measurements one must pay attention to whether and what intermediate resonances have to be taken into account (see, e.g., Ref. 10.43). Far from all intermediate resonances and for small atomic systems, typical values for $\tilde{\sigma}^{(2)}$ on the order of 10^{-50} cm^4 s are obtained, and the calculated values roughly agree with those measured. If the one-photon energy $\hbar\omega$ is near an intermediate level, $\tilde{\sigma}^{(2)}$ will rise by several orders of magnitude. As an example, Fig. 10.18*a* gives the two-photon interaction cross section calculated for hydrogen atoms as a function of the energy of the incident photons. Below this curve the energy levels of the hydrogen atom with the principal quantum numbers $2, 3, \ldots$ are plotted; these may act as resonant intermediate levels. The sharp increase of $\tilde{\sigma}^{(2)}$ in the vicinity of these one-photon resonances is obvious.

The problem of n-photon absorption with $n > 2$ can be treated analogously. Instead of (10.55) one obtains

$$\frac{\Delta P_{\Omega}^{(2n)}(\Omega, t)}{\Delta t}\Delta\Omega = V\left(\frac{1}{2\pi}\right)^3 \frac{m}{\hbar^2}k|\tilde{\Lambda}^{(n)}|^2[\hbar\omega I(\omega)]^n\,\Delta\Omega, \qquad (10.56)$$

where $\tilde{\Lambda}^{(n)}$ is the corresponding interaction matrix element of the n-photon process. Note that in calculating and measuring the multiphoton ionization rates or the corresponding multiphoton interaction cross sections the possibilities for the occurrence of intermediate resonances become far more frequent than with the two-photon process since now those intermediate levels can be reached not only by one-photon transitions but also by m-photon transitions with $m < n$. Figure 10.18*b* is an example of the interaction cross section in the case of twelve-photon absorption. In the investigated energy interval of the incident photons between 1.14 and 1.22 eV there are intermediate resonances, where the levels with the principal quantum numbers between 4 and 8 can be reached by an eleven-photon absorption and the level with the principal quantum number 3 by a ten-photon absorption. For high-order multiphoton processes as well, the results of the easily performed calculations on hydrogen can be regarded as representative of other atoms and small molecules with respect to the order of magnitude of the multiphoton cross section far from intermediate resonances and near resonance.

It is worth noting that these high-order effects can in fact be observed and can even play an important role in some experiments. We can see this clearly by the following rough estimation. We evaluate the photon flux of a Nd-laser pulse, which is needed to generate at least one electron in the focus volume under representative experimental conditions. Thus we assume a number density of atoms equal to 10^{20} cm^{-3}, a 10-μm-diameter beam waist near the focus, an interaction length of 100 μm, and a pulse length of 10 ns. With the interaction cross section $\tilde{\sigma}^{(12)} \sim 10^{-374}$ cm^{24} s^{11} (see Fig. 10.18*b*) we obtain

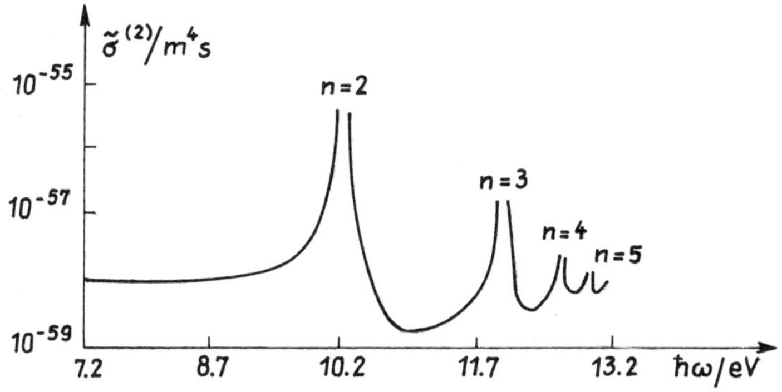

(a)

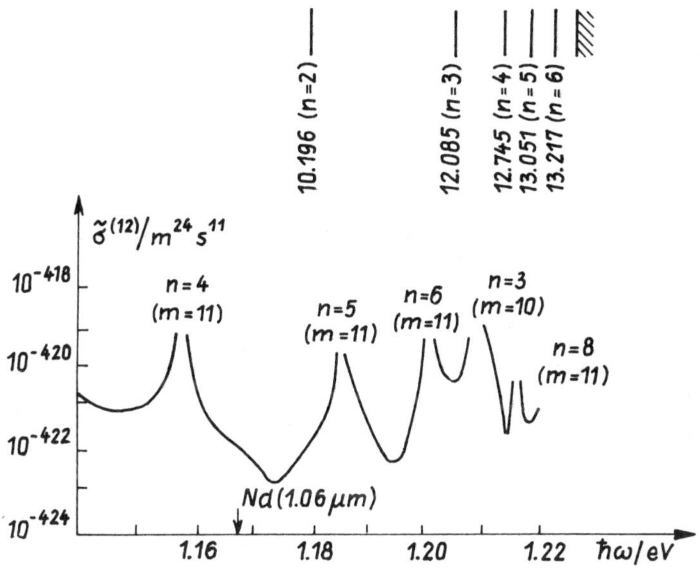

(b)

Fig. 10.18. Multiphoton ionization cross sections of hydrogen (after Ref. 10.42): (*a*) two-photon ionization (for comparison the lines of the one-photon spectrum are given); (*b*) twelve-photon ionization. The principal quantum numbers n of the resonant intermediate levels and the number *m* of photons required to reach these levels are given.

the required photon flux density of about 10^{31} cm^{-2} s^{-1}, and hence a laser power of the order of 10^8 W and a pulse energy of about 1 J. From this it is apparent that it is possible to accomplish multiphoton ionization processes of very high order by focusing laser pulses with commonly available parameters. These statements apply not only to interaction-free atoms or molecules, but also to liquids and solids where we have to take into account the correspondingly higher densities. Of course, at high particle density, the influence of ion–electron recombination on the experimental results may become significant and must then be taken into account.

Multiphoton processes of very high order have for instance been investigated in Refs. 10.44 and 10.45. It has been observed that in addition to singly ionized atoms two-, three-, and fourfold ionized rare-gas atoms can be generated by multiphoton absorption. To explain the experimental findings several excitation channels have to be compared. Processes with the absorption of $n, n+1, n+2, \ldots$ photons of energy $\hbar\omega_L$ may compete among one another, where the integer n and the energy $\mathscr{E}_{\text{ion}, j}$ required for the j-fold ionization satisfy the conditions $n \geq \mathscr{E}_{\text{ion}, j}/\hbar\omega > n - 1$.

In the experiments, single picosecond pulses with $\tau_L \approx 50$ ps from a Nd:glass laser are amplified and focused in the sample, where the power per unit area, J, is up to 10^{14} W cm^{-2}. The ions generated by multiphoton absorption are detected and analyzed with respect to mass and charge by means of a time-of-flight mass spectrometer. The number of singly ionized krypton atoms is proportional to J^n with $n = 13 > \mathscr{E}_{\text{ion},1}/\hbar\omega_L = (13.99 \text{ eV})/(1.165 \text{ eV}) > 12$ as long as saturation effects are avoided. By analogy, with $\mathscr{E}_{\text{ion},2} = 38.35$ eV, the number of twofold ionized krypton atoms should be proportional to J^{33}. The measured exponent, however, is somewhat smaller because there exist several ionization channels. Thus Kr^{2+} ions can be generated either by extracting two electrons from neutral Kr atoms in one process or by releasing one electron from Kr^+. In general saturation effects have to be taken into account in order to explain all experimental findings. (Note that the ratio of the number of Kr^{2+} ions to that of Kr^+ ions has been observed to be as high as 10% at the maximum intensity.) By comparison between calculation and experiment the contribution of different channels to the ionization yield can be determined. The authors of Refs. 10.44 and 10.45 concluded that under the given experimental conditions the direct n-photon j-fold ionization with $n > E_{\text{ion}, j}/\hbar\omega_L$ is the main process. For Kr^{3+} this process leads to a relationship where the ionization yield is approximately proportional to J^{65}. The transition rate for direct j-fold ionization has been found to be surprisingly high in comparison with that for the excitation of j electrons in the discrete spectrum.

It is of importance for the interaction of laser pulses with matter that the small number of electrons resulting from multiphoton ionization are capable of triggering and inducing an *avalanche process*, particularly in condensed matter. This process arises from the acceleration of the electron in the radiation field. A high velocity can be reached only by those electrons that reverse their direction of motion by collisions at the right moment, that is, at the time when the field changes its direction. Bloembergen and coworkers (Ref. 10.8, 10.46) calculated the probability for an initial electron to gain the kinetic energy required to induce a new ionization process. It turns out that in condensed matter this avalanche process can actually occur with rather high probability. This avalanche effect is thought to be the essential cause of laser damage in transparent optical materials. In this effect, however, not only the electrons released by multiphoton ionization may act as initial electrons, but also those that, mainly owing to defects, were already almost free before the incidence of the laser light.

10.5. MEASUREMENT OF INTENSITY CORRELATION FUNCTIONS BY TWO-PHOTON FLUORESCENCE

The temporal profile of ultrashort light pulses cannot be recorded using the common photoelectric detection methods. Photodetectors can provide time resolutions of about 50 ps. They are surpassed by photoelectronic streak cameras, which are capable of measuring time intervals as short as 1 ps. To obtain subpicosecond time resolutions, one has to resort, for the most part, to nonlinear optical correlation methods by means of which the autocorrelation function of the (intensity) signal or the cross-correlation function with another light pulse can be gained. To achieve this, any nonlinear optical effect may in principle be used, for example, second-harmonic generation, up-conversion, or resonant multiphoton processes.

If we intend to measure the temporal profile of single nonrepeatable or only slowly repeatable pulses, we have to record the entire correlation function with only one single pulse. A method that can advantageously be used for this purpose and besides yields high time resolution and sensitivity is the recording of two-photon fluorescence in the arrangement presented in Fig. 10.4. The molecules are excited by the simultaneous absorption of two photons of frequency ω_L, and subsequently they emit light by fluorescence. The absorption process can be regarded as inertialess, provided that no resonant intermediate levels occur (cf. Section 10.3) and that the reciprocal of the homogeneous line width is much smaller than the pulse duration. Under these assumptions the transition rate of two-photon absorption at time t is proportional to the square of the light intensity $I(z, t)$ at the location z of the molecule, that is to say, to the fourth power of the magnitude of the field-strength amplitude $\hat{E}(z, t)$. If samples are chosen whose lifetime in the excited state is much greater than the pulse duration, the radiative and nonradiative transitions can be neglected during irradiation. Thus, for the occupation-number density $\gamma_1(z)$, immediately upon the action of the light pulse, we obtain

$$\gamma_1(z) \propto \int_{\text{pulse}} dt \left|\hat{E}(z, t)\right|^4, \tag{10.57}$$

where the integration has to be extended over the time range of the pulse action. In the arrangement presented in Fig. 10.4, the field-strength amplitude $\hat{E}(z, t)$ is given by the superposition of the fields of the counterpropagating pulses 1 and 2 with the slowly varying temporal amplitudes $\hat{E}_1(z, t) =$

$\bar{E}_1(z,t)e^{+ik_Lz}$ and $\hat{E}_2(z,t) = \bar{E}_2(z,t)e^{-ik_Lz}$. The two counterpropagating pulses result from the beam splitting of an input pulse with wave amplitude $\bar{E}(t)$.

Let us now assume that the attenuation of the light pulses in the sample arising from two-photon absorption or other losses can be neglected. Then the pulse amplitudes $\bar{E}_1$ and $\bar{E}_2$ will only depend on the local times $t - z/v$ and $t + z/v$, not explicitly on the space coordinate z. Thus we obtain

$$\gamma_1(z) \propto \int_{-\infty}^{\infty} dt \left| -\bar{E}\left(t - \frac{z}{v}\right)e^{+ik_Lz} + \bar{E}\left(t + \frac{z}{v}\right)e^{-ik_Lz} \right|^4, \qquad (10.58)$$

where the extension of the integration interval to infinity is of no importance, provided that there the field strength decreases sufficiently fast (the negative sign before one of the two amplitudes results from the phase shift during the reflection of the positively directed wave by the beam splitter. According to (10.58) the occupation-number density contains terms exhibiting very rapid spatial oscillations in the z direction that have the structure of a standing wave with the spatial frequency $2k_L$, as well as terms whose slow dependence on z results solely from the temporal profile of the pulses and their mutual retardation $\tau = 2z/v$. The fluorescence signal is photographically recorded in its z dependence, which originates from the spatial structure of γ_1. If the photographic resolution interval is much greater than the wavelength λ_L, which for example is in general satisfied in silver halide photography and also with optical multichannel analyzers based on vidicons or photodetector matrices, then the high-spatial-frequency terms in γ_1 will be averaged out. (These contributions to γ_1 exert an appreciable influence on the recorded signal only if the genuine two-photon fluorescence is overlaid by competitive processes; see Refs. 10.47, 10.48.)

Under the conditions mentioned, we have for the time integral F of the fluorescence signal, which is recorded as a function of the space coordinate z,

$$F\left(\tau = \frac{2z}{v}\right) \propto \int_{-\infty}^{\infty} dt \left\{ I^2\left(t - \frac{\tau}{2}\right) + I^2\left(t + \frac{\tau}{2}\right) + 4I\left(t - \frac{\tau}{2}\right)I\left(t + \frac{\tau}{2}\right) \right\}. \qquad (10.59)$$

From this, using the *intensity autocorrelation function*

$$\tilde{G}^{(2)}(\tau) = \int_{-\infty}^{\infty} dt\, I(t) I(t+\tau)$$

—which, unlike the correlation functions used in Section 10.3, enables us to describe signals of finite duration—we obtain

$$F\left(\tau = \frac{2z}{v}\right) = 2F_0\left\{1 + \frac{2\tilde{G}^{(2)}(\tau)}{\tilde{G}^{(2)}(0)}\right\}, \qquad (10.60)$$

where F_0 is that fluorescence signal which is measured with one of the two light paths interrupted. With pulses of finite duration, F_0 is simply related by

$$F(\infty) = 2F_0$$

to the value of the measured function F for large delays.

As is apparent from (10.60), the z and the τ dependence of the fluorescence signal are determined by the intensity autocorrelation function $\tilde{G}^{(2)}(\tau)$ (see Fig. 10.19). The measured signal, however, does not directly represent this function; there occurs a background resulting from the emission of fluorescent light at z even if the pulses do not meet at this space point but pass it successively. The height, width, and shape of the spatial distribution of the light emission are related to the pulse parameters. In principle, however, there is no unambiguous way of determining these pulse parameters from the autocorrelation measurements (see Ref. 10.49). As a consequence, other experimental results have to be taken into account to remove the ambiguity (see, e.g., Ref. 10.47). In Fig. 10.19 the autocorrelation functions for some representative signals are given.

From these examples we realize that the complete characterization of light pulses cannot only be based on the measurement of the length of the autocorrelation trace, but that additional quantities have to be employed for it. From Fig. 10.19 it becomes clear that the contrast defined as

$$K = F(0)/F(\infty)$$

is an important measure for the evaluation of signals. From (10.60) we thus obtain $K = 3$ for unmodulated pulses of finite duration, whereas a noise signal of infinite duration gives $K = 1.5$. Additional information about the pulse signal can be obtained from the product of pulse duration and bandwidth, $\tau_L \cdot (\Delta\omega/2\pi)$, for which we know that

$$\tau_L \cdot \frac{\Delta\omega}{2\pi} \geq C_B, \tag{10.61}$$

where C_B is a constant characteristic of the particular pulse shape. It is for

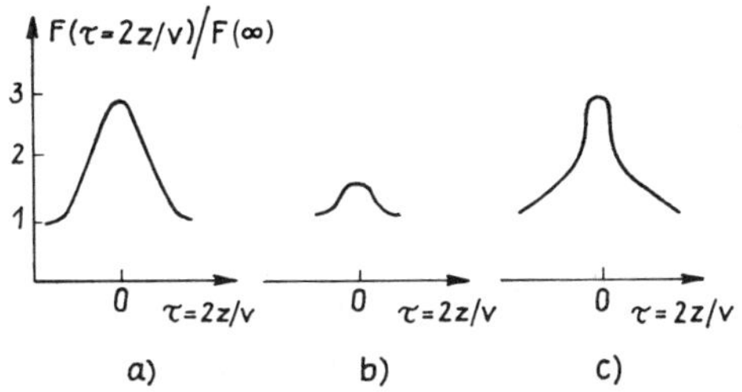

Fig. 10.19. Typical TPF autocorrelation traces for (a) single pulses, (b) narrowband noise of long duration, and (c) "rugged" pulses (noise bursts within an envelope).

example 0.441 for Gaussian pulses and 0.315 for sech^2 pulses. Only for *bandwidth-limited pulses*, that is, only for pulses without phase modulation, does the pulse-bandwidth product take the value C_B. Hence, these pulses possess the smallest possible duration for given spectral widths. If more detailed knowledge of the pulse shape is required for example, on the asymmetry, the measurement of autocorrelation functions of higher orders in addition to $\tilde{G}^{(2)}$ is indispensable. They can be determined, among other methods, by employing n-photon fluorescence with $n > 2$ (see, e.g., Refs. 10.50, 10.51).

10.6. TWO-PHOTON EMISSION AND TWO-PHOTON LASING PROCESSES

In this section we start with a brief description of some basic phenomena originating from two-photon emission and of their potential application. Then we go on to discuss the threshold, the self-sustaining condition, and the dynamical behavior of two-photon lasers, and finally we give a short account of the coherence properties of light emitted in two-photon lasing processes.

10.6.1. Basic Phenomena

The discussion in Section 10.2 of the Hamiltonian (3.58) and of the transition probabilities of multiphoton processes revealed that, in addition to multiphoton absorption processes, multiphoton emission processes may also occur (see Fig. 10.9). Quantum amplifiers based on two-photon transitions from the upper level of the active atomic systems, where the population has been produced by an appropriate pumping process, were discussed very early (Refs. 10.52, 10.53). Almost at the same time, spontaneous (10.54, 10.55) and enhanced spontaneous (10.56–10.58) two-photon emission were experimentally established.

In spite of these early advances, only slow progress was made in amplifier and laser design. This fact is due to the experimental difficulties connected with the generation of a sufficiently high two-photon emission. The development of multiquantum amplifiers and oscillators based on stimulated Raman scattering proceeded much more rapidly and with greater success (cf. Section 12.2). Nevertheless, there has been continuing strong interest in two-photon emission processes and two-photon lasers arising from different aspects of application and basic research.

Research in the field of laser-induced nuclear fusion has been the most important motive in the continuous search for more effective radiation sources with appropriate parameters. The fact that, in spite of its small atomic-interaction cross section, two-photon emission is taken into consideration in these efforts is for two main reasons.

The first is that the pulse shape can be varied within wider limits than it can with one-photon lasers. That means that pulses can be produced having a tailored shape (see, e.g., Ref. 10.60), which is of importance for achieving the optimum heating regime of the targets. In applications like these, the smallness of the coupling constants must be compensated for by employing sufficiently large input fields. Thus it was estimated that, using two-photon lasers, it should be possible to generate intense light pulses with a duration of some nanoseconds and an extremely short rise time of only a few picoseconds (Refs. 10.52, 10.59). Such tailored laser pulses are of course of great interest in other fields, for instance, in spectroscopy and measuring techniques. The main differences between the operation principles of two-photon and of one-photon lasers will be treated in Section 10.6.2.

The second reason for discussing the application of two-photon lasing processes in laser-induced nuclear fusion is the following. By combining two-photon emission and the stimulated generation of anti-Stokes Raman radiation, pulses can be produced exhibiting an extremely large wavelength shift within their duration. This is of importance for the penetration of the radiation into the plasma, whose density is strongly increasing with time (Ref. 10.61).

In Fig. 10.20 the energy levels of a two-level system with the corresponding radiation transitions are shown. We assume the atomic systems to be initially in the upper level. The inverted active material is irradiated by a pulse of frequency ω_1. Under appropriate conditions the stimulated two-photon emission processes will initially predominate, and with sufficiently high two-photon gain this will result in the buildup of two strong coherent waves at the frequencies ω_1 and $\omega_2 = \omega_{10} - \omega_1$, or in the degenerate case, which is of particular interest here, it will cause a strong wave to be built up at the frequency $\omega_1 = \omega_{10}/2$. Now the intense pulse at ω_1 is capable of stimulating the Raman process, thereby generating an anti-Stokes pulse at the frequency of the third harmonic, $\omega_3 = 3\omega_1 = \omega_1 + \omega_{10}$. With sufficiently high Raman gain this wave can continue building up, so that, in turn, it produces an anti-Stokes radiation at the frequency $\omega_5 = 5\omega_1 = \omega_3 + \omega_{10}$, and so forth. If this goes on, hope arises that it will be possible to generate coherent pulses by means of

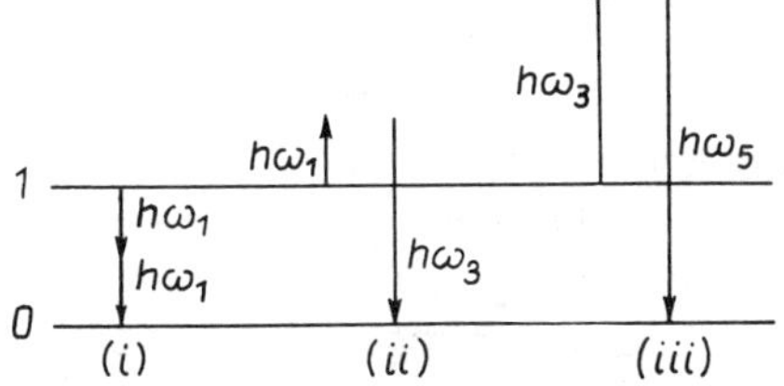

Fig. 10.20. Two-quantum transitions employed in succession to obtain light pulses whose center frequency increases with time (after Ref. 10.61): (i), Two-photon emission; (ii), (iii), anti-Stokes Raman emission.

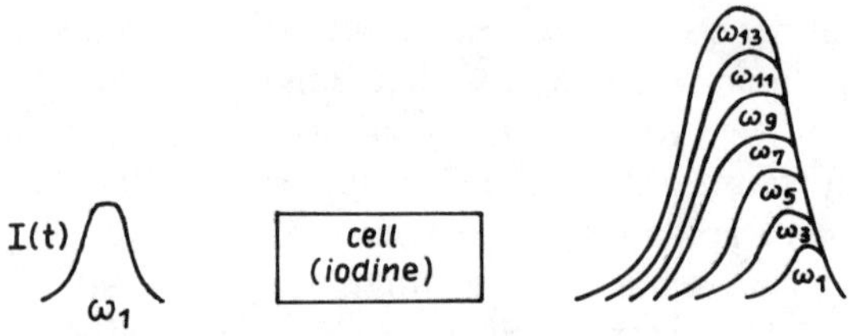

Fig. 10.21. Schematic of the generation of light pulses with large frequency sweep from ω_1 to $\omega_{13} = 13\omega_1$ (after Ref. 10.61).

laser light whose wavelengths, as long as the pulses continue, will shift from the visible to the VUV or soft X-ray region (see Fig. 10.21). An effective interaction of the pulses of frequencies $\omega_1, 3\omega_1 \cdots (2m+1)\omega_1$ with one another as well as with the pulse of atomic excitation can only be achieved if the respective wave-vector conservation laws are satisfied with sufficient accuracy in the forward direction (cf. Chapter 12). This can be accomplished in gases (Ref. 10.61).

It is of advantage to base the description of these interaction processes on the equations given in Section 4.8 for the macroscopic parameters of the ensemble of atomic systems. In Section 4.8 the example of stimulated Raman scattering was discussed. Similar relationships can also be established for two-photon emission, where the polarizability amplitude describes an exciting wave at the sum frequency, that is, at $\omega_1 + \omega_2 = \omega_{10}$, or in the degenerate case at $2\omega_1 = \omega_{10}$. These equations can be evaluated in order to calculate the nonstationary two-photon emission and Raman processes, which may be coupled to each other.

Such evaluations (Ref. 10.61) reveal that atomic iodine in the $^3P_{1/2}$ state, which can for example be obtained from CF_3I by photodissociation, is suitable for use as a two-photon amplifying material. If laser radiation at 2.63 μm is employed, an energy amplification of about 100 can be achieved in an amplifier of about 2 m in length and at inversion densities of about 2×10^{16} cm^{-3}. From excitation by these amplified pulses the anti-Stokes Raman process may originate.

From the viewpoint of basic research the two-photon processes deserve particular interest in that the radiation emitted by them may exhibit extraordinary coherence behavior differing from that of a one-photon laser. A discussion of these properties will be given in Section 10.6.3.

Let us close this subsection with some remarks concerning the measurement of two-photon emission. In the nondegenerate case (i.e., if two photons of unequal energy are simultaneously emitted), stimulated two-photon emission has been observed from potassium vapor irradiated by ruby-laser light and by Stokes-shifted light originating from stimulated Raman scattering in organic liquids (Ref. 10.62). Two types of coherent two-photon emission have been observed for the $2s\,^2S \leftrightarrow 3d\,^2D$ transition of atomic lithium (Ref. 10.63). Here nondegenerate two-photon generation of coherent light at frequency $\omega_2 = \omega_{2s \leftrightarrow 3d} - \omega_1$ stimulated by injected light pulses of frequency $\omega_1 \neq \omega_2$ has been found, and moreover degenerate single-pass amplification of injected light of frequency $\omega_3 = \omega_{2s \leftrightarrow 3d}/2$ has been observed. In this experiment, lithium vapor contained in an Al_2O_3 heat pipe was irradiated by two light pulses of 4-ns duration from two different nitrogen-laser-pumped dye lasers in which coumarin 1 and DCM served as active media. The maximum power of the pulses at ω_2 was about 30 W. The light emitted at ω_2 and ω_3 exhibited no phase correlations with the pump light.

10.6.2. Self-Sustained Light Generation in Two-Photon Lasers

Not only two-photon amplifiers, but also two-photon lasers are of great interest as radiation sources exhibiting unique features. Here we discuss first of all the self-sustaining condition, which displays remarkable differences from that for one-photon lasers. Consider the equation describing the photon balance in the resonator. In analogy to the treatment of one-photon lasers, we obtain for lasers with degenerate two-photon emission the number N_{ph} of photons in the resonator:

$$\frac{d}{dt}N_{ph} = 2P_{10}^{(4)}N_1 - 2P_{01}^{(4)}N_0 - \frac{N_{ph}}{T_{res}}, \tag{10.62a}$$

where the first term describes the increase in the number of photons owing to two-photon emission; the second and the third term represent, respectively, the effect of two-photon absorption and the decrease in the number of photons owing to resonator losses; and T_{res} gives the mean lifetime of the photons in the unpumped resonator (cf. Section 1.1.3). The transition rates $P_{10}^{(4)}$ and $P_{01}^{(4)}$ of two-photon emission and absorption are connected, according to Section 10.3, with a two-photon interaction cross section $\tilde{\sigma}_{01}^{(2)}$ by way of

$$P_{10}^{(4)} = \tilde{\sigma}_{10}^{(2)}\frac{c^2}{\left(n^{(L)}\right)^2V^2}\left(N_{ph}+1\right)\left(N_{ph}+2\right) \tag{10.62b}$$

and

$$P_{01}^{(4)} = \tilde{\sigma}_{01}^{(2)}\frac{c^2}{\left(n^{(L)}\right)^2V^2}N_{ph}\left(N_{ph}-1\right), \tag{10.62c}$$

respectively. For simplicity we suppose the resonator of volume V to be homogeneously filled with the active material. The phase velocity of light in the resonator is denoted by $c/n^{(L)}$. Using the abbreviation $\varkappa = 2\tilde{\sigma}_{10}^{(2)}c^2/[(n^{(L)})^2V^2]$ and $\tilde{\sigma}_{10}^{(2)} = \tilde{\sigma}_{01}^{(2)}$, we obtain from (10.62)

$$\frac{d}{dt}N_{\text{ph}} = \varkappa(N_1 - N_0)N_{\text{ph}}^2 - \left[\frac{1}{T_{\text{res}}} - 3\varkappa N_1 - \varkappa N_0\right]N_{\text{ph}} + 2\varkappa N_1. \quad (10.63)$$

With representative parameters ($\tilde{\sigma}_{10}^{(2)} \sim 10^{-52}$ cm^4 s, $n^{(L)} \sim 1$, $V \sim 10^2$ cm^3, $\gamma = \gamma_0 + \gamma_1 \sim 10^{20}$ cm^{-3}, $T_{\text{res}} \sim 10^{-8}$ s) we can neglect $3\varkappa N_1$ and $\varkappa N_0$ compared with $1/T_{\text{res}}$ in (10.63).

We next consider the threshold condition of two-photon lasers. As with one-photon lasers, we have to require

$$\frac{d}{dt}N_{\text{ph}} > 0$$

above threshold. From (10.63) we see that this derivative has the very small positive value $2\varkappa N_1$ at $N_{\text{ph}} = 0$; then, with increasing N_{ph}, it becomes negative, and only for

$$N_{\text{ph}} \geq (N_{\text{ph}})_S, \quad (10.64\text{a})$$

where

$$(N_{\text{ph}})_S = \frac{1}{2\varkappa(N_1 - N_0)T_{\text{res}}} + \sqrt{\frac{1}{[2\varkappa(N_1 - N_0)T_{\text{res}}]^2} - \frac{2N_1}{N_1 - N_0}}$$

$$\approx \frac{1}{\varkappa(N_1 - N_0)T_{\text{res}}}, \quad (10.64\text{b})$$

does it attain positive values throughout. Hence, in contrast to one-photon lasers, it is not sufficient to operate at high occupation-number inversions; in addition, there must initially exist a number of photons much larger than the quantum noise. Only if both these requirements are met will the laser be capable of oscillating in a self-substained regime. For this reason the additional requirement (10.64) is termed the *self-sustaining condition* (Ref. 10.52). Note that this condition merely requires an initial photon pulse to be injected into the cavity, and the laser will operate without external help thereafter. To illustrate the self-sustaining regime, we calculate the dynamical behavior of such lasers provided there are initially $(N_{\text{ph}})_0$ photons in the cavity. Neglecting the last term, which is very small compared with the others at $(N_{\text{ph}})_0 > (N_{\text{ph}})_S$, and assuming the occupation-number inversion to remain constant during the process under consideration, we obtain from (10.63)

$$\frac{d}{d(t/T_{res})}\left[\frac{N_{ph}}{(N_{ph})_S}\right] = \left[\frac{N_{ph}}{(N_{ph})_S}\right]^2 - \left[\frac{N_{ph}}{(N_{ph})_S}\right]. \tag{10.65}$$

Integration yields

$$\left[\frac{N_{ph}}{(N_{ph})_S}\right]_t = \left[\frac{(N_{ph})_0}{(N_{ph})_S}\right] \times\left\{\left[\frac{(N_{ph})_0}{(N_{ph})_S}\right] - \left(\left[\frac{(N_{ph})_0}{(N_{ph})_S}\right] - 1\right)e^{t/T_{res}}\right\}^{-1}. \tag{10.66}$$

For $(N_{ph})_0/(N_{ph})_S < 1$ the number of photons initially present in the cavity decays exponentially. For $(N_{ph})_0/(N_{ph})_S = 1$ the number of photons remains constant, and only for $(N_{ph})_0/(N_{ph})_S > 1$ does it increase and even approach infinite values at the time

$$\frac{t_{div}}{T_{res}} = \ln\left\{\left[\frac{(N_{ph})_0}{(N_{ph})_S}\right] \Big/ \left[\frac{(N_{ph})_0}{(N_{ph})_S} - 1\right]\right\}. \tag{10.67}$$

This divergence does of course not exist in reality. The photon number is increasing rapidly only as long as the requirement of $N_1 - N_0$ being constant is satisfied. At high photon numbers the inversion depletion has to be taken into account, which yields a slowdown in the photon-number growth and finally leads to a constant photon number if the laser is continuously pumped. If, on the other hand, the laser is not pumped during its action, the photon number will rise to its maximum value and will then decrease below the value $(N_{ph})_S$.

Figure 10.22 shows the resulting photon pulse obtained through numerical solution to the complete set of rate equations (Ref. 10.60). As long as the self-sustaining requirement is not satisfied ($t < 0$), the photon number is seen to increase rather slowly. For $t > 0$ it then rises rapidly in agreement with (10.66), and $N_{ph} > (N_{ph})_S$ until inversion depletion becomes significant.

10.6.3. Coherence Properties of Radiation Generated by Two-Photon Emission. Nonclassical Light

Many theoretical investigations (see, e.g., Refs. 10.64–10.74) deal with the coherence properties of radiation originating from two-photon emission processes, because such light is known to exhibit extraordinary properties; which, if certain conditions are satisfied, essentially differ from those of

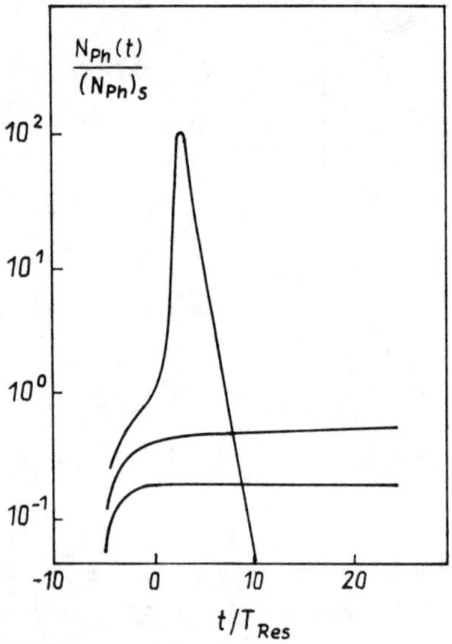

Fig. 10.22. Normalized photon number $N_{\mathrm{ph}}/(N_{\mathrm{ph}})_s$ versus normalized time t/T_{res} (after Ref. 10.60). The increase in the photon number before $t=0$ originates from an additional photon source with $((d/dt)N_{\mathrm{ph}})_{\mathrm{add}} T_{\mathrm{res}} \sqrt{2\tilde{\sigma}_{10}^{(2)} c^2 T_{\mathrm{res}}/V^2} = 0.025$. The initial value of the occupation-number inversion is given by $(N_1 - N_0)/\sqrt{\sigma_{10}^{(2)} c^2 T_{\mathrm{res}}/V^2} = 15$.

conventional thermal light as well as from those of ordinary one-photon-laser light.

Following the same pattern as in the case of the one-photon laser [cf. (7.92)], under simplifying conditions the Fokker–Planck equation for the one-mode two-photon laser has been established (Ref. 10.73). In the particular case of very high power of the two-photon radiation, essential statistical properties turn out to be similar to those of one-photon lasers: If stationarity is assumed, then the photon distribution is approximately a Poisson distribution, and the structure of the line-width formula agrees with the expression derived in the case of the phase-diffusion model introduced in Subsection 7.3.2, except that the line width exceeds that in the one-photon case by the factor two.

The investigations performed by Yuen (Refs. 10.64–10.66) have shown that one must consider whether or not certain states of the radiation field which had been unknown until that time—namely, the so-called *two-photon coherent states* with extraordinary properties—appear in light originating from two-photon emission. Following Yuen, let us study this problem by utilizing a quantum-theoretical treatment of a comparatively simple *two-photon laser model*. As shown in Section 10.6.2, in contrast to ordinary lasers, two-photon lasers cannot start from only the photons emitted by the two-photon emission process itself. Initially, a sufficiently strong field at the laser frequency ω_L must be present. In our model this initial field of the single-mode laser is supplied by one-photon emission proceeding within the laser cavity. Thus we have two kinds of lasing media in the cavity. The laser-active atoms of species 1 deliver the initial radiation by one-photon lasing processes, where the interaction between one individual atom and the radiation mode is described by the interaction Hamiltonian

$$\left(\hat{H}_1^C\right)_i = -\hbar\chi_1'\hat{B}_{lu}^{(1)}\hat{a}^\dagger + \text{h.c.}, \tag{10.68}$$

where $\hat{a}^\dagger$ is the photon creation operator of the laser mode, $\hat{B}_{lu}^{(1)}$ is the flip operator that flips the upper atomic state $|u\rangle_1$ into the lower atomic state $|l\rangle_1$, and $\hbar\chi_1'$ is the corresponding matrix element of the one-photon transition (cf. Section 3.2.2.1). In addition, the atoms of species 2 emit photons into the laser mode through two-photon emission processes. The interaction between one atom of species 2 with the radiation field is described by the operator

$$\left(\hat{H}_2^C\right)_i = -\hbar\chi_2'\hat{B}_{lu}^{(2)}\left(\hat{a}^\dagger\right)^2 + \text{h.c.}, \tag{10.69}$$

where $\hat{B}_{lu}^{(2)}$ is the flip operator of the atom of species 2 and $\hbar\chi_2'$ denotes the matrix element of the two-photon emission process (cf. Section 3.2.2.2).

If losses are neglected in our simple consideration, the operator

$$\hat{H} = \hat{H}_1^A + \hat{H}_2^A + \hat{H}^F + \hat{H}_1^C + \hat{H}_2^C \tag{10.70}$$

can be regarded as the basic Hamiltonian of our two-photon laser model. Where $\hat{H}_1^A$, $\hat{H}_2^A$, and $\hat{H}^F$ denote the Hamiltonians of the free atomic ensembles 1 and 2 and of the free radiation field, respectively. $\hat{H}_1^C$ and $\hat{H}_2^C$ respectively describe the interaction of the ensembles 1 and 2 with the radiation field; they are composed of the operators $(\hat{H}_1^C)_i$ and $(\hat{H}_2^C)_i$. For simplicity we assume that the density operator of the total system factorizes into the atomic parts $\hat{\rho}_1$ and $\hat{\rho}_2$ and the field part $\hat{\rho}_F$, with only the latter being of interest to us. This factorization is justified in representative experimental situations (cf. Section 7.2). Then we obtain the equation of motion for $\hat{\rho}_F$ as

$$i\hbar\frac{d}{dt}\hat{\rho}_F = \left[\hat{H}', \hat{\rho}_F\right]. \tag{10.71}$$

The Hamiltonian $\hat{H}'$ is obtained from $\hat{H}$ of (10.70) by taking the trace Tr_{atom} over all atomic variables. This gives

$$\begin{aligned}\hat{H}' = \hbar\omega\hat{a}^\dagger\hat{a} &- \left[\hbar\chi_1'\text{Tr}_{\text{atom}}\left\{\hat{\rho}_1 B_{lu}^{(1)}\right\}\hat{a}^\dagger + \text{h.c.}\right]\\ &- \left[\hbar\chi_2'\text{Tr}_{\text{atom}}\left\{\hat{\rho}_2 B_{lu}^{(2)}\right\}\left(\hat{a}^\dagger\right)^2 + \text{h.c.}\right].\end{aligned} \tag{10.72}$$

The temporal evolution of the radiation field can now be found by solving (10.71). This can be done through the calculation of the time-development operator $\hat{U}(t, t_0)$ associated with the Hamiltonian $\hat{H}'$ by general relationships of quantum mechanics (see Section A.1.4). To achieve two-photon lasing, we must start with an occupation-number inversion in the two atomic ensembles at time t_0. The result is that at the beginning the one-photon lasing process evolves. High above the threshold of the one-photon laser, the resulting radiation field can be described approximately by an ordinary coherent state $|\alpha\rangle$ as long as the contribution of the two-photon emission is negligible. Suppose that the radiation field is in the state $|\alpha_i\rangle$ at time t_1. In this case, as was first shown by Yuen (Ref. 10.66), the specific time-development operator $\hat{U}(t, t_1)$ describes transitions from the state $|\alpha_i\rangle$ into the state $|y_f\rangle$ for $t > t_1$, where $|y_f\rangle$ belongs to a more general class of coherent states $|y\rangle$, which are referred to as *two-photon coherent states*.

Let us now discuss the properties of this class of final states $|y\rangle$, because they represent the extraordinary coherence properties mentioned above. Generally, the states $|y\rangle$ are defined as being right-hand eigenstates of the operator

$$\hat{y} = \mu\hat{a} + \nu\hat{a}^{\dagger} \tag{10.73a}$$

according to

$$\hat{y}|y\rangle = y|y\rangle, \tag{10.73b}$$

where

$$|\mu|^2 - |\nu|^2 = 1, \tag{10.73c}$$

and where $\mu = |\mu|\exp(i\varphi_\mu)$, $\nu = |\nu|\exp(i\varphi_\nu)$, and $y = |y|\exp(i\varphi_y)$ are complex numbers. With respect to the coherence properties it proves useful to represent the two-photon coherent states $|y\rangle$ with the aid of the (ordinary) coherent states $|\alpha\rangle$. This representation is

$$|y\rangle = \frac{1}{\pi}\int d^2\alpha\,|\alpha\rangle\frac{1}{\sqrt{|\mu|}}\exp\left[-\tfrac{1}{2}|\alpha|^2 - \tfrac{1}{2}|y|^2 - \frac{\nu}{2\mu}(\alpha^*)^2 + \frac{\nu^*}{2\mu}y^2 + \frac{1}{\mu}\alpha^* y\right]. \tag{10.74}$$

For the special case $\mu \to 1$, $\nu \to 0$ the states $|y\rangle$ coincide with ordinary coherent states $|\alpha\rangle$ and only for $|\nu| > 0$ do important differences become apparent in the counting statistics and the minimum of the measurement uncertainty. With respect to the quadrature components $\hat{a}_1 = (\hat{a}^\dagger + \hat{a})/2$ and $\hat{a}_2 = i(\hat{a}^\dagger - \hat{a})/2$ of the annihilation and creation operators, the mean squared deviations of fields in the ordinary coherent state $|\alpha\rangle$ are

$$\left\langle \alpha\left|\left(\widehat{\Delta a_1}\right)^2\right|\alpha\right\rangle = \left\langle \alpha\left|\left(\widehat{\Delta a_2}\right)^2\right|\alpha\right\rangle = \tfrac{1}{4} \tag{10.75}$$

(cf. Section 2.3). In contrast to this, we obtain for the two-photon coherent states $|y\rangle$ the variances

$$\left\langle y\left|\left(\widehat{\Delta a_1}\right)^2\right|y\right\rangle = \tfrac{1}{4}|\mu - \nu|^2 \tag{10.76a}$$

and

$$\left\langle y\,|\left(\widehat{\Delta a_2}\right)^2|\,y\right\rangle = \tfrac{1}{4}|\mu + \nu|^2. \tag{10.76b}$$

If the relation $\mu = g\nu$ holds with g any real number, we obtain from (10.76)

$$\left\langle y\,|\left(\widehat{\Delta a_1}\right)^2|\,y\right\rangle\left\langle\, y|\left(\widehat{\Delta a_2}\right)^2|\,y\right\rangle = \tfrac{1}{16}. \tag{10.77}$$

This means that these $|\,y\rangle$ are minimum-uncertainty states with respect to $\hat{a}_1$ and $\hat{a}_2$, as are the ordinary coherent states $|\,\alpha\rangle$ (cf. Section A.1.2). The most important deviation of the coherence properties of particular states $|\,y\rangle$ from those of the states $|\,\alpha\rangle$ is that $\langle y\,|(\widehat{\Delta a_1})^2|\,y\rangle$ can be less than $\frac{1}{4}$ whereas the other variance $\langle y\,|(\widehat{\Delta a_2})^2|\,y\rangle$ exceeds the value $\frac{1}{4}$. This property of unequal variances (uncertainties) of the quadrature components is referred to as *squeezing*.

The expectation values $\langle y\,|(\widehat{\Delta a_{1/2}})^2|\,y\rangle$ of (10.76) are not directly related to measurable parameters, and therefore the measurable spatial distribution and the measurable higher-order correlation functions of fields in two-photon coherent states have been studied in Ref. 10.68 by calculating the expectation value and the probability distribution of the electric field strength with the help of (10.74) and (10.76). The expectation value

$$\langle y\,|\hat{E}(z)|\,y\rangle = 2\sqrt{\frac{\hbar\omega}{2\varepsilon_0 V}}\,\langle y\,|\hat{a}|\,y\rangle\cos\!\left(kz + \frac{\pi}{2} - \arg\langle y\,|\hat{a}|\,y\rangle\right), \tag{10.78}$$

is a sine-wave function of the space coordinate z whose amplitude $|\hat{E}|$ and phase φ depend on the complex numbers μ, ν, and y and on the real number $\sqrt{\hbar\omega/2\varepsilon_0 V}$. The probability distribution is Gaussian at every space point z, with the distribution parameters depending on z. Within certain z intervals the mean-square deviation $\langle\, y|(\widehat{\Delta E}(z))^2|y\rangle$ is seen to attain values below $\langle 0\,|(\widehat{\Delta E}(z))^2|0\rangle$. These results suggest the introduction of a *general definition of squeezed states*, regardless of whether the radiation originates from two-photon emission or from other processes. Let the field-strength operator of a single mode be expressed as

$$\hat{E}(\Phi) = i\sqrt{\frac{\hbar\omega}{2\varepsilon_0 V}}\,\hat{a}e^{i\Phi} + \text{h.c.}, \tag{10.78a}$$

where Φ is the phase $kz - \omega t - \varphi_0$ with φ_0 the initial phase value. The density operator $\hat{\rho}_{\text{SQU}}$ characterizes a squeezed state, if there are phases Φ, for which

$$\operatorname{Tr}\left\{\hat{\rho}_{\text{SQU}}\left[\widehat{\Delta E}(\Phi)\right]^2\right\} - \left\langle 0\left|\left[\widehat{\Delta E}(\Phi)\right]^2\right|0\right\rangle < 0. \tag{10.78b}$$

This means that in a squeezed state the field-strength fluctuations are, within a certain space-time interval, less than those of the photon vacuum whose fluctuations equal those of the coherent state. Because of this fundamental property and because of the possibility of a considerable noise reduction in important applications, squeezed states recently received a great deal of attention. Expected applications concern, e.g., qualitatively new aspects in optical communication theory and the detection of gravitational waves; there are, in principle, several promising methods of the Nonlinear Optics for the production of squeezed states (see e.g., Ref. 10.74). Further remarks will be given in Sections 11.4.2 and 14.3.

After these general remarks we return to properties of two-photon coherent states. From $\langle y\,|[\widehat{\Delta E}(\Phi)]^2|\,y\rangle$ one can infer that a state $|\,y\rangle$ satisfies the squeezing condition (10.78b) for Φ values, where $|\nu| + |\mu|\cos(\Phi + \varphi_\nu - \varphi_\mu)$ is negative.

An additional insight into the statistical properties of two-photon states can be gained by calculating the normalized intensity correlation function $\gamma^{2,2}(\tau)$ at $\tau = 0$. The result is

$$\gamma^{2,2}(0) = 1 + \frac{|y|^2\left\{8|\nu|^4 + 6|\nu|^2 - \left(8|\nu|^3|\mu| + 2|\nu|\,|\mu|\right)\cos\left(\varphi_y - \varphi_\mu - \varphi_\nu\right)\right\}}{\langle y\,|\hat{a}^\dagger \hat{a}|\,y\rangle}. \tag{10.79}$$

For small $|\nu|$ and appropriate values of the phases φ_y, φ_μ, and φ_ν we have $\gamma^{2,2}(0) < 1$, which indicates photon antibunching. The minimum value of $\gamma^{2,2}(0)$, however, cannot attain the value zero, which corresponds to photon-number eigenstates (see Ref. 10.69). On the other hand, in certain regions of the parameters y, μ, and ν, values of $\gamma^{2,2}(0)$ greater than 1 can be obtained, and even greater than 2, which represents enhanced photon bunching (cf. Section 5.2 and 9.2).

The Yuen model of the two-photon laser is based on some simplifications—namely, on the neglect of laser losses, which in Chapter 7 have been shown to play an important role in one-photon lasers, and on the averaging with respect to atomic variables, as a result of which the atomic operators $\hat{B}_{lu}^{(1)}$ and $\hat{B}_{lu}^{(2)}$ appear only implicitly in c-numbers. This means among other things that fluctuations originating from the quantization of the atomic systems are neglected. Because of these simplifications one has to consider the results from the the Yuen model as limiting expressions.

A more general approach to the description of two-photon emission processes has been given by Schubert and Vogel (Ref. 10.71). Starting with an arbitrary initial state, they solved the exact equations of motion of the field operators and the atomic operators in the short-time approximation. Representative quantities, as for instance the time-dependent mean photon number, the second-order intensity correlation functions, and the field-strength variance, show different behavior from that in coherent states. In particular, this quantum-statistical analysis has shown that squeezed states [in the sense of (10.78b)] can be produced by dominant two-photon absorption processes, but not under dominant two-photon emission conditions (see Ref. 10.74). Reid and Walls (Ref. 10.72) confirmed these results for two-photon optical lasers.

Let us finish this chapter with some summarizing remarks on *classical and nonclassical light*. In preceding chapters and in the present one we occasionally treated photon states that exhibit, in comparison with other states (like thermal light and ideal laser light), an extraordinary behavior; sub-Poissonian light (cf. Section 2.3.2), antibunched light [cf. (5.49) and Section 9.2], and squeezed light [cf. (10.78b)]. Thermal (chaotic) light is super-Poissonian, bunched, nonsqueezed. Its Glauber-Sudarshan P representation can be interpreted as a proper (positive definite) probability density (cf. Fig. 2.3); thus this light has a classical counterpart and it is called classical light. Light of coherent states (laser light well above threshold) is Poissonian, neither bunched nor antibunched, nonsqueezed. Although it has a deltafunction-like P representation, it can be regarded as classical light. If light exhibits one or more of the attributes sub-Poissonian, antibunched, squeezed, it has no positive definite, mathematically well-behaved P representation and is called nonclassical light. Let us consider two examples. A photon number state $|n\rangle$ exhibits antibunching [cf. (5.49)] and its P representation is the nth derivative of the delta function [cf. discussion following (2.130)]. If $P_{\text{squ}}(\alpha)$ is the P representation of a squeeezed state, the squeezing condition (10.78b) reads

$$\int d^2\alpha P_{\text{squ}}(\alpha)\left[(e^{i\Phi}\alpha + \text{c.c.}) - \left(e^{i\Phi}\int d^2\alpha' P_{\text{squ}}(\alpha')\alpha' + \text{c.c.}\right)\right]^2 < 0.$$

Obviously this requirement leads to nonpositive definite functions $P_{\text{squ}}(\alpha)$.

REFERENCES

10.1. W. Kaiser and C. G. B. Garrett, *Phys. Rev. Lett.* **7**, 229 (1961).

10.2. J. J. Hopfield, J. M. Worlock, and K. Park, *Phys. Rev. Lett.* **11**, 414 (1963).

10.3. M. Goeppert-Mayer, *Ann. Phys.* **9**, 273 (1931).

10.4. J. P. Heritage, "Surface Picosecond Raman Gain Spectra of a Molecular Monolayer and Ultrathin Films," in R. M. Hochstrasser, W. Kaiser, and C. V. Shank (Eds.), *Picosecond Phenomena*, Springer, Berlin (West), Heidelberg, New York, 1980.

10.5. N. G. Basov, A. Z. Grasyk, I. G. Zubarev, and V. A. Katulin, *JETP Lett.* **1**, 86 (1965).

10.6. C. K. N. Patel, P. A. Fleury, R. E. Slusher, and H. L. Frisch, *Phys. Rev. Lett.* **16**, 971 (1966).

10.7. H. Hora, *Laser Plasmas and Nuclear Energy*, Plenum, New York, London, 1975.

10.8. N. Bloembergen, *IEEE J. Quant. Electron.* **QE-10**, 375 (1974); *Appl. Opt.* **12**, 661 (1979).

10.9. D. Fröhlich, "Two-Photon Spectroscopy in Solids," in O. Madelung (Ed.), *Festkörperprobleme* (Advances in solid state physics), Vol. 10, Akademie, Berlin, 1970, p. 227 (German).

10.10. J. M. Worlock, "Two-Photon Spectroscopy," in F. T. Arecchi and E. O. Schulz-Dubois (Eds.), *Laser Handbook*, Vol. 2, North Holland, Amsterdam, 1972, p. 1323.

10.11. G. Herzberg, *Infrared and Raman Spectra of Polyatomic Molecules*, Van Nostrand, New York, 1945.

10.12. R. London, *Adv. in Phys.* **13**, 423 (1964).

10.13. A. M. Bonch-Bruevich and V. A. Khodovoi, *Usp. Fiz. Nauk* **85**, 3 (1965) (Russian); English transl. *Soviet Phys.—Usp.* **8**, 3 (1965).

10.14. W. L. Peticolas, *Ann. Rev. Phys. Chem.* **18**, 233 (1967).

10.15. J. Chrostowski and B. Karcewski, *Phys. Lett.* **63A**, 239 (1977).

10.16. B. Karczewski, "Two-Photon Absorption and Light Statistics," in S. Kielich, F. Kaczmarek, and T. Bancewicz (Eds.), *Quantum Electronics and Nonlinear Optics* (Proceedings of EKON 78), University Press, Poznan, 1978, p. 267.

10.17. A. S. Davydov, *Theory of Solids*, Nauka, Moscow, 1976 (Russian).

10.18. D. Fröhlich, *Phys. in u. Zeit* **6**, 47 (1975).

10.19. L. V. Keldysh, *Zh. Eksp. Teor. Fiz.* **47**, 1945 (1965); English transl., *Sov. Phys.—JETP* **20**, 1307 (1965).

10.20. M. S. Bespalov, L. A. Kulevskij, V. P. Makarov, A. M. Prokhorov and A. A. Tikhonov, *Zh. Eksp. Teor. Fiz.* **55**, 144 (1969); English transl., *Sov. Phys.—JETP* **28**, 77 (1969).

10.21. H. D. Simaan and R. Loudon, *J. Phys. A* **8**, 539 (1975).

10.22. G. S. Agarwal, *Phys. Rev. A* **1**, 1445 (1970).

10.23. M. Schubert and B. Wilhelmi, "On Coherence Problems in Nonlinear Optical Effects under Special Consideration of Ultrashort-Time Phenomena," in B. Havelka and J. Blabla (Eds.), *Recent Advances in Optical Physics*, Proceedings of the Tenth Congress of the International Commission for Optics, Prague, 1975, p. 225.

10.24. N. Chandra and H. Prakash, *Phys. Rev. A* **1**, 1696 (1970).

10.25. N. Tornau and A. Bach, *Opt. Comm.* **11**, 46 (1974).

10.26. H. D. Simaan and R. Loudon, *J. Phys. A* **11**, 435 (1978).

10.27. I. M. Every, *J. Phys. A* **8**, L69 (1975).

10.28. A. Bandilla and H.-H. Ritze, *Ann. d. Phys.* **33**, 207 (1976).

10.29. A. Bandilla and H. H. Ritze, *Opt. Comm.* **19**, 169 (1976).

10.30. H. Paul, U. Mohr, and W. Brunner, *Opt. Comm.* **17**, 145 (1976).

10.31. S. Chaturvedi, P. Drummond, and D. F. Walls, *J. Phys. A* **10**, L187 (1977).

10.32. Y. R. Shen, *Phys. Rev.* **155**, 921 (1967).

10.33. J. Kleinschmidt, S. Rentsch, W. Tottleben, and B. Wilhelmi, *Chem. Phys. Lett.* **24**, 133 (1974).

10.34. B. Wilhelmi, E. Heumann, and W. Triebel, *Kvant. Elektron.* **3**, 732 (1976).

10.35. J. Herrmann, J. Wienecke, and B. Wilhelmi, *Opt. Quant. Electron.* **7**, 337 (1975).

10.36. H. P. Weber, *IEEE J. Quant. Electron.* **QE-7**, 189 (1971).

10.37. B. Wilhelmi, *Wiss. Z. Univ. Jena Math.-Nat.* **25**, 429 (1976).

10.38. L. A. Zadeh, *Proc. IRE* **50**, 856 (1952).

10.39. M. Schubert and B. Wilhelmi, *Exp. Tech. Phys.* **27**, 201 (1979).

10.40. A. Messiah, *Quantum Mechanics*, Vol. II, North Holland, Amsterdam, 1966.

10.41. S. Flügge, *Practical Quantum Mechanics*, Springer, New York, Heidelberg, Berlin (West), 1974.

10.42. H. B. Bebb and A. Gold, *Phys. Rev.* **143**, 1 (1966).

10.43. C. Lecompte, G. Mainfray, C. Manus, and F. Sanchez, *Phys. Rev. Lett.* **32**, 265 (1965).

10.44. P. Agostini, F. Fabre, G. Mainfray, G. Petite, and N. K. Rahmann, *Phys. Rev. Lett.* **42**, 1127 (1979).

10.45. A. L'Huillier, L. A. Lompré, G. Mainfray, and C. Manus, *J. Phys. B* **16**, 1363 (1983).

10.46. W. L. Smith, J. H. Bechtel, and N. Bloembergen, *Phys. Rev. B*, **15**, 4039 (1977).

10.47. D. J. Bradley, "Methods of Generation," in S. L. Shapiro (Ed.), *Ultrashort Light Pulses*, Springer, Berlin (West), Heidelberg, New York, 1977, p. 19.

10.48. J. Herrmann, M. Palme, and K. E. Süsse, *Opt. Quant. Electron.* **10**, 195 (1978).

10.49. T. I. Kusnetsova, *Zh. Eksp. Teor. Fiz.* **28**, 1303 (1969).

10.50. H. P. Weber and R. Dändliker, *Phys. Lett.* **28A**, 77 (1968).

10.51. Z. Bauman, *IEEE J. Quant. Electron.* **QE-13**, 875–880 (1977).

10.52. P. P. Sorokin and N. Braslau, *IBM J. Res. Dev.* **8**, 177 (1964).

10.53. A. M. Prokhorov, *Science* **149**, 828 (1965).

10.54. M. Lepeles, R. Novick, and N. Tolk, *Phys. Rev. Lett.* **15**, 690 (1965).

10.55. R. C. Elton, L. J. Palumbo, and H. R. Green, *Phys. Rev. Lett.* **20**, 783 (1968).

10.56. S. Yatsiv, M. Rokni, and S. Basak *Phys. Rev. Lett.* **20**, 1282 (1968).

10.57. P. Braünlich and P. Lambropoulos, *Phys. Rev. Lett.* **25**, 135 (1970).

10.58. A. P. Veduta, M. D. Galanin, B. P. Kirsanov, and Z. A. Chizhikova, *Zh. Eksp. Teor. Fiz. Lett.* **11**, 182 (1970).

10.59. R. L. Garwin, *IBM. J. Res. Dev.* **8**, 338 (1964).

10.60. M. Schubert, and G. Wiederhold, *Exp. Tech. Phys.* **27**, 217 (1979).

10.61. R. L. Carman, *Phys. Rev. A* **12**, 1048 (1975).

10.62. S. Barak, M. Rokni, and S. Yatsiv, *IEEE J. Quant. Electr.* **QE5**, 448 (1969).

10.63. B. Nikolaus, D.Z. Zhang, and P. E. Toschek, *Phys. Rev. Lett.* **47**, 171 (1981).

10.64. H. P. Yuen, *Phys. Lett.* **51A**, 1–2 (1975).

10.65. H. P. Yuen, *Appl. Phys. Lett.* **26**, 505 (1975).

10.66. H. P. Yuen, *Phys. Rev. A* **13**, 2226–2243 (1976).

10.67. A. M. Perelomov, *Usp. Fiz. Nauk* **123**, 23 (1977).

10.68. M. Schubert and W. Vogel, *Phys. Lett.* **68A**, 321 (1978).

10.69. O. Hirota and S. Ikehara, *Phys. Lett.* **57A**, 317 (1976).

10.70. K. J. McNeil and D. F. Walls, *Phys. Lett.* **51A**, 233 (1975).

10.71. M. Schubert and W. Vogel, *Opt. Comm.* **36**, 164 (1981).

10.72. M. D. Reid and D. F. Walls, *Phys. Rev. A* **28**, 332 (1983).

10.73. Yu. M. Golubev, *Opt. i Spektr.* **46**, 1 (1979).

10.74. M. Schubert, in *Proc. of 6th International School on Quantum Optics*, Polish Academy of Sciences, Ustron, 1985.

11

Generation of Harmonics and Sum and Difference Frequencies. Parametric Amplification and Oscillation

In the preceding chapter we dealt with the absorption and emission of light quanta by an atomic system with the sum of the converted photon energy equal or approximately equal to the atomic transition energy. This means that resonant or quasiresonant processes were being considered. In contrast to this, the present chapter is concerned with the interaction of three or—in the case of second-harmonic generation—two light waves whose frequencies can, in typical cases, be assumed to be far from the atomic resonance frequencies. This, in particular, accounts for the fact that the classical theory of strong, coherent fields is capable of providing an acceptable explanation of empirical findings in a wide range. It is the second-order susceptibility function $\varkappa^{(2)}_{\bullet\bullet\bullet}$ that plays an essential role as material property. This function was introduced in the relations (1.109) and (1.110) as a phenomenological quantity, and it was connected with the parameters of the atomic system in (4.43). However, for describing the coherence behavior of radiation and its generation processes, quantum-theoretical means are needed.

The generation of second harmonics was the first effect to be detected in nonlinear interactions between optical waves. We may say that this discovery (Franken, Hill, Peters, and Weinreich, Ref. 11.1) was the starting point of the rapid development of nonlinear optics in the sixties. This effect is of great importance in the generation of intense coherent radiation in the short-wave-length region. Similar statements may be made for the generation of sum frequencies and difference frequencies. By utilizing the former in experiments that mix the light from different sources of the visible region, we can achieve intense short-wave radiation or (under certain conditions) intense ultraviolet radiation; by means of the latter long-wave (infrared) radiation can be obtained. By mixing experiments it also becomes possible to transform infrared radiation into the visible spectral region, thereby facilitating signal detection.

In parametric amplification and oscillation, the material is irradiated by a light wave, the pump wave, which causes the *system parameters of the material* to be changed in such a way that the generation or the amplification of two light waves with different frequencies becomes possible. By employing special experimental methods—for example, by locating the crystal irradiated by the pump wave in a resonator—it is possible to pass from the amplification effect occurring in one of the two light waves to a stable oscillation. The parametric oscillator is of particular importance in practice because it can be used for obtaining intense, narrowband light sources that are tunable over a wide frequency range.

Both in frequency conversion and in parametric processes sufficiently large conversion rates between the waves of different frequencies can be achieved in many cases, so that in this way practical needs for the generation of strong light waves can be satisfied. In discussing this property and other features of the interaction of light waves, we shall use the solutions of the amplitude equations given in Section 11.1, which are systems of differential equations. To perform order-of-magnitude estimations of the susceptibilities entering the amplitude equations, we shall establish, using the semiclassical representation given in its general form in Chapter 4, a relation with the relevant atomic parameters of the transition frequencies and dipole moments. Section 11.2 presents a quantum-theoretical treatment of those processes that are needed for an adequate description of mixing experiments and parametric processes, which cannot be covered by classical description. Section 11.3 continues with the discussion of application aspects and of quantitative experimental results including parameters of the media. Coherence aspects of frequency conversion and parametric amplification are dealt with in Section 11.4.

11.1. AMPLITUDE EQUATIONS FOR TWO AND THREE INTERACTING LIGHT WAVES

In the present section we describe the radiation fields and their interaction with the medium by classical susceptibility functions introduced phenomenologically. The waves involved are assumed to be sufficiently intense and quasi-monochromatic so that the spatially slowly varying wave amplitudes satisfy coupled amplitude equations [cf. (1.200a)]; noise terms may be neglected. While second-harmonic generation is connected with the interaction of two light waves (see Section 11.1.1), the generation of sum and difference frequencies as well as that of parametric processes rests on the interaction of three light waves (see Section 11.1.2).

11.1.1. Interaction of Two Light Waves

In this subsection we discuss second-harmonic generation (SHG). Suppose the material under consideration to be loss-free at the fundamental frequency ω and the harmonic 2ω in the sense of Section 1.3.2.6, and the nonlinear medium to occupy the half space $z \geq 0$. We may assume here that the waves are infinitely extended in the x, y directions and that the slowly varying wave amplitudes are only dependent on the z coordinate, which is perpendicular to the boundary plane. Those amplitudes of second-order polarization and of the field that are essential to the generation of second harmonics are, according to (1.112), given by

$$\overline{P}_i^{(2)}(\omega, z) = \varepsilon_0 \sum_{j,k} \chi_{ijk}^{(2)}(\omega; 2\omega, -\omega)\overline{E}_j(2\omega, z)\overline{E}_k^*(\omega, z), \qquad (11.1)$$

$$\overline{P}_i^{(2)}(2\omega, z) = \varepsilon_0 \sum_{j,k} \chi_{ijk}^{(2)}(2\omega; \omega, \omega)\overline{E}_j(\omega, z)\overline{E}_k(\omega, z). \qquad (11.2)$$

Under conditions given in Section 1.3.2.7 the Kleinman symmetry relations [cf. (1.173)] reveal that the susceptibility functions of the frequencies ω and 2ω are connected by

$$2\chi_{ijk}^{(2)}(2\omega; \omega, \omega) = \chi_{ijk}^{(2)}(\omega; 2\omega, -\omega). \qquad (11.3)$$

Under the assumption of the time dependence of a quasimonochromatic wave and a slow spatial alteration in the z direction, the general differential equation (1.200a) for the slowly varying amplitude was derived. Denoting the fundamental wave by $\mu = 1$ (frequency ω, wave vector $k_{1\bullet}$, unit vector $e_{1\bullet}$ of the polarization direction) and the second harmonic by $\mu = 2$ (2ω, $k_{2\bullet}$, $e_{2\bullet}$), we obtain the following coupled system of equation:

$$\frac{d\overline{E}(\omega, z)}{dz} = \frac{i\omega^2 e_{1\bullet}}{2c^2 k_1 \cos\alpha_1 \cos(\alpha_1 - \beta_1)} \chi_{\bullet\bullet\bullet}^{(2)}(\omega; 2\omega, -\omega)$$

$$\times e_{2\bullet} e_{1\bullet} \overline{E}(2\omega, z)\overline{E}^*(\omega, z) e^{-i\Delta k z}, \qquad (11.4a)$$

$$\frac{d\overline{E}(2\omega, z)}{dz} = \frac{i(2\omega)^2 e_{2\bullet}}{2c^2 k_2 \cos\alpha_2 \cos(\alpha_2 - \beta_2)} \chi_{\bullet\bullet\bullet}^{(2)}(2\omega; \omega, \omega)$$

$$\times e_{1\bullet} e_{1\bullet} \overline{E}^2(\omega, z) e^{+i\Delta k z}, \qquad (11.4b)$$

where $\Delta k = k_{2,z} - 2k_{1,z}$ denotes the momentum mismatch, which leads to the phase mismatch $\Delta k z$. The angle α_j designates the angle between the vectors $D_{j\bullet}$ and $E_{j\bullet}$ as well as between the vectors $k_{j\bullet}$ and the Poynting vector $S_{j\bullet}$, whereas β_j is the angle between $k_{j\bullet}$ and the z axis. Multiplying the complex conjugate of (11.4a) by $-\overline{E}(\omega, z)$ and (11.4b) by $\overline{E}^*(2\omega, z)/2$, and adding the resulting equations, we have

$$k_1 \cos\alpha_1 \cos(\alpha_1 - \beta_1)\overline{E}(\omega, z)\frac{d\overline{E}^*(\omega, z)}{dz}$$

$$+ \tfrac{1}{2} k_2 \cos\alpha_2 \cos(\alpha_2 - \beta_2)\overline{E}^*(2\omega, z)\frac{d\overline{E}(2\omega, z)}{dz} = 0. \qquad (11.5)$$

Since $\cos\alpha \cdot \bar{E}\,d\bar{E}^*/dz$ is proportional to the spatial derivative of the Poynting vector $S_\bullet$, we have, by integrating over z,

$$\frac{|S_{1\bullet}(z)|\cos(\alpha_1 - \beta_1)}{\omega} + \frac{2|S_{2\bullet}(z)|\cos(\alpha_2 - \beta_2)}{2\omega} = \frac{\mathscr{P}}{\omega}, \tag{11.6}$$

with $\mathscr{P}$ the total power flux per unit area normal to the z direction. Since the medium is assumed to be loss-free, $\mathscr{P}$ must be regarded as a constant. When the number of higher-energy photons increases by 1, the number of lower-energy photons decreases by 2. Thus the relation (11.6) corresponds to the Manley–Rowe relations (Section 1.3.2.8).

Of course, the explicit z dependence of $\bar{E}(2\omega, z)$ under given initial conditions is of interest. However, its determination is accompanied by several problems. In anisotropic materials the amplitude equations depend in a complex way on the directions of propagation and polarization of the individual waves. This can be seen for the general case and for SHG from (1.200a) and (11.4), respectively. Since, in general, we have to start from a nonvanishing momentum mismatch, this means from $\Delta k \neq 0$, we mostly fail to get complete analytical solutions of the set of differential equations.

The structure of the amplitude equations becomes simpler for an isotropic medium with waves of only one direction of polarization. In the following the general amplitude equations will be applied to anisotropic materials; however, they will be altered in such a way as to make their structure resemble that of the simpler equations of an isotropic medium. To this end, we make the following assumptions: Suppose the wave vectors of all the waves involved lie in one direction, which is chosen to be the z axis of the laboratory coordinate system. This can be achieved by normal incidence of the waves onto the appropriately cut surface of the crystal. Moreover, we confine ourselves to optically uniaxial crystals and set the y axis of the laboratory coordinate system in the principal section, that is, in the plane formed by the axis of incidence and the optical axis. So the x axis is perpendicular to the principal axis, and the x component of the wave with the fundamental frequency ω propagates as an ordinary wave with the wave number $k^{[x]}(\omega) = k_o(\omega)$; the y component propagates as an extraordinary wave with the wave number $k^{[y]}(\omega) = k_{eo}(\omega)$. In the following the wave number of light with polarization direction j is denoted by $k^{[j]}$.

Finally we shall utilize the assumption, which in most cases is well satisfied, that the deviation of the ellipsoid of the linear refractive index from sphericity is a small one. Therefore, in most cases the component of the electric field strength in the direction of propagation can be ignored. This means among other things that the Poynting vector lies in the direction of propagation of the wave. Separate reference will be made to phenomena that originate from the differences between the direction of the Poynting vector and that of propagation. They may have some effect in the case of long interaction lengths of the waves.

In contrast to what was stated above, it is indispensable to take into account the dependence of the phase velocity on the direction of polarization, since the phase difference of the individual waves affects the generation process most critically.

Under the abovementioned assumptions the amplitude equations for the SHG take the form

$$\frac{d\overline{E}_i(\omega, z)}{dz} = \sum_{j,k} \frac{i\omega^2}{2c^2 k^{[i]}(\omega)} \chi^{(2)}_{ijk}(\omega; 2\omega, -\omega)$$

$$\times \overline{E}_j(2\omega, z)\overline{E}^*_k(\omega, z) e^{-iz[k^{[i]}(\omega) - k^{[j]}(2\omega) + k^{[k]}(\omega)]}, \qquad (11.7a)$$

$$\frac{d\overline{E}_i(2\omega, z)}{dz} = \sum_{j,k} \frac{i2\omega^2}{c^2 k^{[i]}(2\omega)} \chi^{(2)}_{ijk}(2\omega; \omega, \omega)\overline{E}_j(\omega, z)\overline{E}_k(\omega, z) e^{-iz\Delta k}. \qquad (11.7b)$$

Given the amplitudes and susceptibilities, the exponential factor $e^{iz\Delta k}$ in (11.7b) determines the phase of $d\overline{E}_i(2\omega, z)/dz$ at a fixed position z; it depends on

$$\Delta k = k^{[i]}(2\omega) - k^{[j]}(\omega) - k^{[k]}(\omega). \qquad (11.7c)$$

The wave-number component $k^{[i]}(\omega)$ can equal only the wave number of the ordinary wave $[k_o(\omega)]$ or that of the extraordinary wave $[k_{eo}(\omega)]$. Inspection of (11.2) shows that $[k^{[j]}(\omega) + k^{[k]}(\omega)]z$ is the phase value of the polarization wave with frequency 2ω at z. Thus $\Delta k z$ can be regarded as the difference between the phase values of the field-strength wave and the polarization wave of the second harmonic. Hence, at a certain position z, this quantity determines the phase difference and therefore also the absolute value and the sign of the power amplification of the harmonic wave. If Δk vanishes, the situation is called *exact phase match*; otherwise, *phase mismatch*.

Let us now use the set of equations (11.7a, b) for describing the growth of the electric wave with frequency 2ω on its passage through the nonlinear medium. Of course this is connected with an attenuation of the fundamental wave. For the present we neglect this depletion and assume $\overline{E}(\omega, z) = \overline{E}(\omega, 0)$. We may proceed in this way if a low conversion rate and a relatively strong fundamental wave can be assumed. If this so-called *small-signal approximation* can be employed, then the relation (11.7b) can be immediately integrated, and we have

$$\overline{E}_i(2\omega, z) = -\sum_{j,k} \frac{i2\omega^2}{c^2 k^{[i]}(2\omega)} \chi^{(2)}_{ijk}(2\omega; \omega, \omega)$$

$$\times \overline{E}_j(\omega, 0)\overline{E}_k(\omega, 0)\frac{e^{-iz\Delta k} - 1}{i\,\Delta k}, \tag{11.8}$$

where $\overline{E}_i(2\omega, 0)$ has been neglected in comparison with $\overline{E}_i(2\omega, z)$.

The wave-number difference Δk can be expressed in terms of the *linear* refractive indices $n^{[\cdots]}$ for the corresponding frequencies and polarization directions in the following way:

$$\Delta k = \frac{\omega}{c}\left[2n^{[i]}(2\omega) - n^{[j]}(\omega) - n^{[k]}(\omega)\right]. \tag{11.9}$$

When all waves propagate with uniform polarization direction—either as ordinary or as extraordinary waves—then (11.9) goes over into one of the two relations

$$\Delta k_{\mathrm{o}} = \frac{2\omega}{c}\left[n_{\mathrm{o}}(2\omega) - n_{\mathrm{o}}(\omega)\right], \qquad \Delta k_{\mathrm{eo}} = \frac{2\omega}{c}\left[n_{\mathrm{eo}}(2\omega) - n_{\mathrm{eo}}(\omega)\right], \tag{11.9a}$$

where n_{o}, n_{eo} are the linear refractive indices of the ordinary and the extraordinary wave, respectively.

In the following we shall not assume the special case of only one polarization direction, but we suppose that only one of the various combinations (j, k) of the amplitude components $\overline{E}_{\bullet}(\omega, z)$ contributes to the generation of $\overline{E}_i(2\omega, z)$. Thus the squared amplitude at z becomes

$$|\overline{E}_i(2\omega, z)|^2 = \left|\frac{2\omega^2}{c^2 k^{[i]}(2\omega)} \chi^{(2)}_{ijk}(2\omega; \omega, \omega)\right|^2$$

$$\times |\overline{E}_j(\omega, 0)|^2 |\overline{E}_k(\omega, 0)|^2 \frac{\sin^2[z\,\Delta k/2]}{(\Delta k/2)^2}. \tag{11.10}$$

This quantity is associated with the generated power flux by

$$|S_i(2\omega, z)| = \sqrt{\varepsilon^{(L)}(2\omega)\varepsilon_0/4\mu_0}\,|\overline{E}_i(2\omega, z)|^2.$$

Inspection of (11.10) shows that the generated power flux depends periodically on z. Within the interval $0 \le z \le L$ the quantity $|S_i(2\omega, z)|$ increases, where L

is given by

$$L = \frac{\pi}{\Delta k}; \tag{11.11}$$

L is called phase-coherence length. At the position L the modulus of the Poynting vector attains its maximum value. It can be seen that it decreases with increasing $(\Delta k)^2$. For higher z, in the interval $L < z < 2L$, the phase between polarization and field strength of the harmonic is such that energy is extracted from this wave and pumped back into the fundamental wave. At $z = 2L$ the intensity of the harmonic again becomes zero; more precisely, it attains the value $\bar{E}_i(2\omega, 0)$, which was neglected in (11.10). For $z > 2L$ the amplification process of the harmonic restarts.

To let the maximum value of the generated power flux attain the largest possible value, one mainly utilizes birefringence in certain nonlinear crystals. If the path lengths in nonlinear crystals are long, the generation rate will be very sensitive to small changes of Δk; details on the methods to be employed and the potential results are given in Section 11.3.2.

Now we leave the small-signal approximation and take into account the z dependence of the fundamental wave during the interaction with the second harmonic. From (11.6), with the assumption of a loss-free passage through the material, we can obtain a relation between the z-dependent Poynting vectors of the two waves. To achieve an explicit determination of the z dependence of the amplitudes, the set of the equations (11.4a, b) or (11.7a, b) must be solved simultaneously. General solutions were given in Ref. 11.2, but it was difficult to derive them. Important features can be seen under simplifying conditions. Following Bloembergen (Ref. 11.3), we assume exact phase matching $\Delta k = 0$ and $\alpha_1 = \alpha_2 = \beta_1 = \beta_2 = 0$. We choose $\bar{E}(\omega, z)$ to be real; then $\bar{E}(2\omega, z)$ becomes purely imaginary. By using symmetry relations, (11.4a, b) go over into

$$\frac{d\bar{E}(\omega, z)}{dz} = iQ\bar{E}(\omega, z)\bar{E}(2\omega, z), \tag{11.12a}$$

$$\frac{d\bar{E}(2\omega, z)}{dz} = iQ\bar{E}^2(\omega, z), \tag{11.12b}$$

where $Q = \omega^2 e_{1\bullet}\chi^{(2)}_{\bullet\bullet\bullet}(\omega; 2\omega, -\omega)e_{2\bullet}e_{1\bullet}/2c^2k_1$. Under the initial condition $E(2\omega, 0) = 0$, which means negligible second-harmonic amplitude at $z = 0$, we have the solution functions

$$\bar{E}(\omega, z) = \bar{E}(\omega, 0)\operatorname{sech}\{Q\bar{E}(\omega, 0)z\},$$

$$\bar{E}(2\omega, z) = i\bar{E}(\omega, 0)\tanh\{Q\bar{E}(\omega, 0)z\}. \tag{11.13}$$

In Fig. 11.1 the ratios $|\bar{E}(\omega, z)|/|\bar{E}(\omega, 0)|$ and $|\bar{E}(2\omega, z)|/|\bar{E}(\omega, 0)|$ are plotted as functions of the reduced length $\tilde{z} = z/L_{\text{int}}$. The parameter $L_{\text{int}} =$

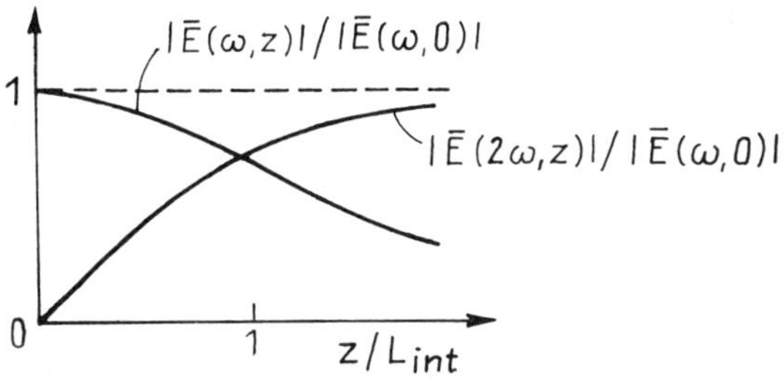

Fig. 11.1. Normalized amplitudes of the fundamental and the second-harmonic wave as functions of the propagation distance z in units of the interaction length L_{int}.

$[QE(\omega,0)]^{-1}$ may be called the characteristic interaction length: within it, 58% of the initial power of the fundamental wave is converted into the second harmonic; for $2L_{\text{int}}$ this proportion rises to 93%. The interaction length varies inversely as the square root of the initial power of the fundamental wave. Note that exact phase matching, as assumed for deriving (11.13), leads not only to the balance of energy $\omega_{\text{SH}} = 2\omega$, but also to the balance of momentum, since $k_{\bullet}(2\omega) = 2k_{\bullet}(\omega)$.

11.1.2. Interaction of Three Light Waves

In this subsection we discuss interaction effects between three quasimonochromatic light waves arising from nonlinear second-order polarization in a crystal. Let the frequencies of the three waves be related by $\omega_1 = \omega_2 + \omega_3$. In the basic amplitude equations there are analogies to those for the interaction of two waves, which were discussed in connection with second-harmonic generation. Making the same assumptions as were described in detail when introducing the equations (11.7a, b), for the case under study the following set of three amplitude equations is obtained:

$$\frac{d\bar{E}_i(\omega_1, z)}{dz} = \sum_{j,k} \frac{i\omega_1^2}{2c^2k^{[i]}(\omega_1)} \chi_{ijk}^{(2)}(\omega_1; \omega_2, \omega_3)\bar{E}_j(\omega_2, z)\bar{E}_k(\omega_3, z)e^{-iz\Delta k_1}, \tag{11.14a}$$

where $\Delta k_1 = k^{[i]}(\omega_1) - k^{[j]}(\omega_2) - k^{[k]}(\omega_3)$;

$$\frac{d\bar{E}_i(\omega_2, z)}{dz} = \sum_{j,k} \frac{i\omega_2^2}{2c^2k^{[i]}(\omega_2)} \chi_{ijk}^{(2)}(\omega_2; \omega_1, -\omega_3)$$

$$\times \bar{E}_j(\omega_1, z)\bar{E}_k^*(\omega_3, z)e^{-iz\Delta k_2}, \tag{11.14b}$$

where $\Delta k_2 = k^{[i]}(\omega_2) - k^{[j]}(\omega_1) + k^{[k]}(\omega_3)$;

$$\frac{d\bar{E}_i(\omega_3, z)}{dz} = \sum_{j,k} \frac{i\omega_3^2}{2c^2 k^{[i]}(\omega_3)} \chi_{ijk}^{(2)}(\omega_3; -\omega_2, \omega_1)$$

$$\times \bar{E}_j^*(\omega_2, z) \bar{E}_k(\omega_1, z) e^{-iz\Delta k_3}, \tag{11.14c}$$

where $\Delta k_3 = k^{[i]}(\omega_3) + k^{[j]}(\omega_2) - k^{[k]}(\omega_1)$. The symbols used here stand for the same quantities as in the system (11.7). As in the treatment of second harmonics, it is assumed that the medium is loss-free in the sense of Section 1.3.2.6.

11.1.2.1. Sum and Difference Frequencies. Let us first solve the set of equations (11.14) by assuming that the medium is irradiated by two strong light sources with frequencies ω_2 and ω_3 and that within the medium a weak wave of frequency ω_1 is amplified by nonlinear effects. Since by assumption the conversion rate is small, we can use the small-signal approximation with $\bar{E}_j(\omega_2, z) = \bar{E}_j(\omega_2, 0)$ and $\bar{E}_k(\omega_3, z) = \bar{E}_k(\omega_3, 0)$. Suppose again that only one combination (j, k) of the amplitude components $\bar{E}_j(\omega_2, z)$, $\bar{E}_k(\omega_3, z)$ contributes to the amplification of $\bar{E}_i(\omega_1, z)$. If these conditions are satisfied, (11.14a) can be readily integrated and one has

$$\bar{E}_i(\omega_1, z) - \bar{E}_i(\omega_1, 0) = -i \frac{\omega_1^2}{2c^2 k^{[i]}(\omega_1)} \chi_{ijk}^{(2)}(\omega_1; \omega_2, \omega_3)$$

$$\times \bar{E}_j(\omega_2, 0) \bar{E}_k(\omega_3, 0) \frac{e^{-iz\Delta k_1} - 1}{i\,\Delta k_1}. \tag{11.15a}$$

as an equation for the amplification of a wave with the sum frequency ω_1 and the initial amplitude $\bar{E}_i(\omega_1, 0)$. Analogously, strong irradiation of waves with frequencies ω_1 and ω_3 results in the amplification of a wave with the difference frequency ω_2 and the initial amplitude $\bar{E}_i(\omega_2, 0)$ according to

$$\bar{E}_i(\omega_2, z) - \bar{E}_i(\omega_2, 0) = -i \frac{\omega_2^2}{2c^2 k^{[i]}(\omega_2)} \chi_{ijk}^{(2)}(\omega_2; \omega_1, -\omega_3)$$

$$\times \bar{E}_j(\omega_1, 0) \bar{E}_k^*(\omega_3, 0) \frac{e^{-iz\Delta k_2} - 1}{i\,\Delta k_2}. \tag{11.15b}$$

If in (11.5a) the initial amplitude $\bar{E}_i(\omega_1, 0)$ and in (11.15b) the initial amplitude $\bar{E}_i(\omega_2, 0)$ are equal to zero, two relations result, which describe the generation process of a wave with the sum frequency or difference frequency. In this case the structure of the relations (11.15a, b) agrees with that of the amplitude equation (11.8) for the second harmonic, except that the right-hand side of the latter gives the products of two field amplitudes with *different* frequencies. In particular, this structural analogy becomes apparent in the dependence of the amplitudes of the generated waves on z if there exists nonvanishing momentum mismatch Δk_1 or Δk_2. Therefore, what was stated for second-harmonic generation also holds in the present case: for achieving a high conversion rate, which now depends on the power flux of two waves of different frequencies, Δk_1 or Δk_2 must be chosen to be as low as possible; the methods to be employed are similar to those used in second-harmonic generation and will be discussed in Section 11.3.2.

11.1.2.2. Parametric Amplification. Another important special case of interaction between three light waves is parametric amplification. It is the amplification or generation of *two* light waves of frequencies ω_2 and ω_3 in a medium which is irradiated by *one* intense pump wave of frequency $\omega_1 = \omega_2 + \omega_3$. Each of the three waves is assumed to be polarized in one fixed direction of the coordinate system, which for $\omega_1, \omega_2, \omega_3$ is denoted by i, j, k. Furthermore, in the process a relatively small depletion of the pump wave is assumed and hence the space dependence of its amplitude is neglected. Without loss of generality $\bar{E}_i(\omega_1, 0)$ can be chosen real. The susceptibilities $\chi^{(2)}_{ijk}(\omega_2; \omega_1, -\omega_3)$ and $\chi^{(2)}_{ijk}(\omega_3; -\omega_2, \omega_1)$ from (11.14a, b) can be rewritten by means of symmetry relations (cf. Section 1.3.2.7) in terms of $\chi^{(2)}_{ijk}(\omega_1; \omega_2, \omega_3)$, so that from these two amplitude equations we obtain

$$\frac{d\bar{E}_j(\omega_2, z)}{dz} = \frac{i\omega_2^2}{2c^2 k^{[j]}(\omega_2)} \chi^{(2)}_{ijk}(\omega_1; \omega_2, \omega_3) \bar{E}_i(\omega_1, 0) \bar{E}_k^*(\omega_3, z) e^{iz\Delta k}, \tag{11.16a}$$

$$\frac{d\bar{E}_k(\omega_3, z)}{dz} = \frac{i\omega_3^2}{2c^2 k^{[k]}(\omega_3)} \chi^{(2)}_{ijk}(\omega_1; \omega_2, \omega_3) \bar{E}_i(\omega_1, 0) \bar{E}_j^*(\omega_2, z) e^{iz\Delta k}, \tag{11.16b}$$

where $\Delta k = k^{[i]}(\omega_1) - k^{[j]}(\omega_2) - k^{[k]}(\omega_3)$. We see that the amplitude-amplification rates of $\bar{E}_j(\omega_2, z)$ and $\bar{E}_k(\omega_3, z)$ are proportional to $\chi^{(2)}_{ijk} \bar{E}_i(\omega_1, 0)$. Therefore an amplitude-amplification coefficient

$$\eta = \sqrt{\frac{\omega_2^2 \omega_3^2}{k^{[j]}(\omega_2) k^{[k]}(\omega_3)}} \frac{1}{2c^2} \chi^{(2)}_{ijk}(\omega_1; \omega_2, \omega_3) \bar{E}_i(\omega_1, 0) \tag{11.17}$$

is introduced. Thus, from the set (11.16) the following set of equations can be obtained:

$$\frac{d\bar{E}_j(\omega_2, z)}{dz} + iG\eta \bar{E}_k^*(\omega_3, z)e^{-iz\Delta k} = 0,$$

$$\frac{d\bar{E}_k^*(\omega_3, z)}{dz} - i\frac{1}{G}\eta \bar{E}_j(\omega_2, z)e^{iz\Delta k} = 0, \tag{11.18}$$

where the constant G is given by $G = [\omega_2^2 k^{[k]}(\omega_3)/\omega_3^2 k^{[j]}(\omega_2)]^{1/2}$; both G and η are real quantities.

Now we solve the set of equations on the assumption of negligible phase mismatch (i.e., $\Delta k = 0$), whereby (11.18) becomes a system of linear differential equations with constant coefficients. By means of the ansatz e^{Qz} for $\bar{E}_j(\omega_2, z)$ and $\bar{E}_k(\omega_3, z)$ the general solution can be derived using the well-known method for differential equations of this kind. Since the condition $Q = \pm\eta$ follows from the characteristic polynomial, the solution contains the z-dependent terms $e^{\eta z}$ and $e^{-\eta z}$. The initial conditions are determined by the given amplitudes $\bar{E}_j(\omega_2, 0)$ and $\bar{E}_k(\omega_3, 0)$. Assuming a negligibly small initial amplitude of one of the frequencies ω_2, ω_3—say $\bar{E}_k(\omega_3, 0) = 0$—the solution is

$$\bar{E}_j(\omega_2, z) = \bar{E}_j(\omega_2, 0)\cosh \eta z, \tag{11.19a}$$

$$\bar{E}_k^*(\omega_3, z) = -i\bar{E}_j^*(\omega_2, 0)\frac{1}{G}\sinh \eta z. \tag{11.19b}$$

Even if $\Delta k \neq 0$, the set (11.18) of equations can be solved. Then, instead of the ansatz e^{Qz} the ansatz $e^{Qz - iz\Delta k}$ (for the wave with frequency ω_2) and the ansatz $e^{Qz + iz\Delta k}$ (for the wave with frequency ω_3) has to be introduced. For Q we obtain the relation $Q = \pm\eta'$, where η' is given by

$$\eta' = \sqrt{\eta^2 - \left(\frac{\Delta k}{2}\right)^2}. \tag{11.20}$$

In form the solutions for $\Delta k \neq 0$ are similar to those for $\Delta k = 0$: in (11.19a, b) the quantity η must be replaced by η' and phase factors $e^{\pm iz\Delta k}$, according to the trial solutions $e^{Qz \pm iz\Delta k}$, must be added.

Let us begin the physical interpretation of the solutions obtained with the case $\Delta k = 0$. The solution functions reveal that the amplitudes of the two waves increase exponentially if z is much greater than the coefficient η. This, of course, only holds insofar as by this process the pump wave is not yet appreciably attenuated; recall that this was one of our basic assumptions in establishing (11.16a, b). In the case of $\Delta k \neq 0$, amplification can only happen if the amplification coefficient η' does not vanish and if it is real; that means we have to require $2\eta > \Delta k$. For a certain orientation of the crystal (and thus for a fixed value Δk) and a certain nonlinear interaction coefficient $\chi^{(2)}_{ijk}$ of the crystal, this requirement must be regarded as a condition on the strength of the pump wave.

In summary, the physical situation characterized by the solutions (11.19a) and (11.19b) can be interpreted as follows: If a sufficiently strong pump wave is fed in, a wave (frequency ω_2) with nonvanishing initial amplitude can be amplified. Simultaneously in this process a wave of frequency $\omega_3 = \omega_1 - \omega_2$ may be generated whose initial amplitude is zero. The wave of frequency ω_2 is called the *signal*; the wave of frequency ω_3 is called the *idler*.

Furthermore, there exists quite another experimentally realizable process that is characterized by the irradiation of a *single* wave of frequency ω_1, whereby *two* waves (of frequencies ω_2 and ω_3) are generated in the nonlinear medium. Obviously, this result cannot be achieved by solutions of the classical relations (11.19a, b). If in addition to $\overline{E}_k(\omega_3, 0)$ also $\overline{E}_j(\omega_2, 0)$ were set equal to zero, no wave phenomena at the frequencies ω_2 and ω_3 would result. An adequate description of this phenomenon, which is connected with the decay of one higher-energy photon into two lower-energy photons, can only be given on a quantum-theoretical basis. We shall refer to it in Section 11.2.2.

11.2. QUANTUM FUNDAMENTALS OF THE PROCESSES

In the preceding section we employed purely classical means for describing the amplification and generation of harmonics and sum and difference frequencies as well as parametric amplification; we used the phenomenologically introduced, classical susceptibility functions and studied the interaction of classical electromagnetic waves. The subject of this section is, on the one hand, the treatment of the interrelations between susceptibilities and the relevant quantities of atomic systems wherefrom, for example, estimations of the order of magnitude of susceptibilities will result. On the other hand it includes the quantum-theoretical description of the interaction between the waves involved, which in particular for parametric amplification will yield new information how waves may originate from fluctuations.

In Section 11.2.1 we shall consider the frequency conversion. This is followed by the discussion of parametric amplification in Section 11.2.2.

11.2.1. Frequency Conversion

We are allowed to study the susceptibilities for the second-order effects to be treated under the following simple conditions: Let the radiation field interact with a system of weakly coupled, uniform atomic systems so that the description of the total effect can be performed by summing the contributions of the individual systems; this can be done by using the number density γ.

Suppose that the atomic systems are in the ground state. The frequencies ω, ω_1, ω_2 of the incident waves and the frequencies 2ω, $\omega_1 + \omega_2$ of the amplified or generated waves are supposed to be much smaller than the atomic transition frequencies. Section 4.4 shows how in a semiclassical approach the time-dependent perturbation theory can be used for determining the susceptibility functions of different orders. Under our conditions (in particular $\rho^e_{00} = 1$) the general relation (4.44) for the second-order susceptibility in the frequency domain reads

$$\underset{\smile}{\varkappa}^{(2)}_{kim}(\omega_1, \omega_2) = \frac{1}{2\varepsilon_0 \hbar^2} \gamma \mathsf{P}_2 \sum_{\beta, \beta'} \left\{ \frac{d_{0\beta k} d_{\beta\beta' i} d_{\beta' 0 m}}{(\omega_{\beta' 0} + \omega_2)(\omega_{\beta 0} + \omega_1 + \omega_2)} + \frac{d_{0\beta i} d_{\beta\beta' k} d_{\beta' 0 m}}{(\omega_{\beta' 0} + \omega_2)(\omega_{\beta 0} - \omega_1)} + \frac{d_{0\beta i} d_{\beta\beta' m} d_{\beta' 0 k}}{(\omega_{\beta 0} - \omega_1)(\omega_{\beta' 0} - \omega_1 - \omega_2)} \right\}, \tag{11.21}$$

where P_2 is a permutation operator that gives rise to an exchange between the tensor components i, m, and the assigned frequencies ω_1, ω_2; γ is the number density; $d_{\beta\beta' k}$ is the kth vector component of the atomic dipole moment, and $\omega_{\beta\beta'}$ the corresponding transition frequency between the states $|\beta\rangle$ and $|\beta'\rangle$.

Let us now make an order-of-magnitude estimation, for which it suffices to consider only one of the tensor components. This may be $\underset{\smile}{\varkappa}^{(2)}_{xxx}$, which we denote by $\underset{\smile}{\varkappa}^{(2)}$. We have

$$\underset{\smile}{\varkappa}^{(2)}(\omega_1, \omega_2) = \frac{\gamma}{\varepsilon_0 \hbar^2} \sum_{\beta, \beta'} d_{0\beta} d_{\beta\beta'} d_{\beta' 0} \left\{ \frac{1}{(\omega_{\beta' 0} + \omega_2)(\omega_{\beta 0} + \omega_1 + \omega_2)} + \frac{1}{(\omega_{\beta' 0} + \omega_2)(\omega_{\beta 0} - \omega_1)} + \frac{1}{(\omega_{\beta 0} - \omega_1)(\omega_{\beta' 0} - \omega_1 - \omega_2)} \right\}. \tag{11.22}$$

The estimation is based on the assumption that $\omega_1, \omega_2, \omega_1 + \omega_2$ are considerably smaller than ω_{10}. For making the estimation we keep those terms in (11.22) that give the largest contributions; the corresponding three-level scheme is illustrated in Fig. 4.4. Taking into account the atomic ground level (0) and the first two excited terms (1, 2), one obtains from (11.22)

$$\underset{\smile}{\varkappa}^{(2)}(\omega_1, \omega_2) = \frac{6\gamma}{\varepsilon_0 \hbar^2} d_{01} d_{12} d_{20} \frac{1}{\omega_{10} \omega_{20}}. \tag{11.23}$$

With the values for the atomic parameters given in Table 3.1 we obtain $\underset{\smile}{\varkappa}^{(2)}(\omega_1, \omega_2) \sim 10^{-11}$ m/V. Let us remark that, when taking into account the Lorentzian field correction, the value given for the second-order susceptibility must be multiplied, according to Section 1.2.3, by the factor $[(\varepsilon^{(L)}_{NR} + 2)/3]^2$. In passing over from sum-frequency generation to second-harmonic generation we have to note that the susceptibilities $\chi^{(2)}(2\omega; \omega, \omega)$ related to the discrete frequency spectrum are associated with the corresponding Fourier transform in the frequency domain by $\chi^{(2)}(2\omega; \omega, \omega) = \frac{1}{2}\underset{\smile}{\varkappa}^{(2)}(\omega, \omega)$.

Let us now consider the case where the sum frequency $\omega_1 + \omega_2$ or 2ω approaches the atomic transition frequency ω_{10}, strong absorption however being still disregarded. Then we have

$$\underset{\smile}{\varkappa}^{(2)}(\omega_1, \omega_2) \approx \frac{\gamma}{\varepsilon_0 \hbar^2} d_{01} d_{12} d_{20} \frac{1}{\omega_{20}[\omega_{10} - (\omega_1 + \omega_2)]}, \tag{11.24}$$

and it becomes apparent that $\underset{\smile}{\varkappa}^{(2)}$, because of the factor $\omega_{10} - (\omega_1 + \omega_2)$, may attain values that are markedly higher than 10^{-11} m/V. If the sum frequency approaches the point of resonance even more closely, the damping has to be incorporated into the susceptibility function according to Section 4.5. In our case this leads to

$$\underset{\smile}{\varkappa}^{(2)}(\omega_1, \omega_2) = \frac{\gamma}{\varepsilon_0 \hbar^2} d_{01} d_{12} d_{20} \times \frac{1}{(\omega_{20} - \omega_1)[\omega_{10} - (\omega_1 + \omega_2) - i/\tau_{10}]}. \tag{11.25}$$

Then, in addition to the real part of the susceptibility, also its imaginary part has to be taken into quantitative consideration.

Susceptibility estimations for the generation of higher-order sum frequencies can be found in an analogous manner. As an example we treat third-order sum frequencies, confining ourselves to the off-resonance case. From the expression (4.46) for the loss-free third-order susceptibility it can be inferred that

$$\underset{\smile}{\varkappa}^{(3)}(\omega_1, \omega_2, \omega_3) \sim \frac{\gamma Z}{\varepsilon_0 \hbar^3} |d_{\alpha\beta}|^4 \frac{1}{|\omega_{\alpha\beta}|^3}. \tag{11.26}$$

The dimensionless factor Z depends on the particular atomic model and the frequencies ω_1, ω_2, and ω_3; in typical cases it is of the order of 10. With values of the dipole moments typical of electronic transitions $\underset{\smile}{\varkappa}^{(3)}(\omega_1, \omega_2, \omega_3)$ turns out to be of the order of 10^{-22} m^2/V^2. The factor $|\underset{\smile}{\varkappa}^{(2)}/\underset{\smile}{\varkappa}^{(3)}|$ given in the third-order calculations of Section 4.4 is of the order of $\hbar\omega_{10}/|d_{10}| \sim 10^{11}$ V/m and thus is of the same order as $|\underset{\smile}{\varkappa}^{(1)}/\underset{\smile}{\varkappa}^{(2)}|$.

In the quantum-theoretical description of the generation or amplification of sum-frequency waves we may start from the interaction operator

$$\hat{H}_S^C = \sum_m \tilde{\eta}_m^{(S)} \hat{E}_\mu^{(+)}(r_{m\bullet}) \hat{E}_{\mu_1}^{(-)}(r_{m\bullet}) \hat{E}_{\mu_2}^{(-)}(r_{m\bullet}) + \text{h.c.}, \tag{11.27}$$

by means of which the equations of motion for the operators of the field-strength amplitudes can be determined. In the same sense as it does in multiphoton absorption [cf. (3.58) and (10.15)], $\hat{H}_S^C$ represents an effective operator composed of the basic dipole-interaction operators. The operators $\hat{E}^{(-)}_{\mu_1}(r_{m\bullet})$ and $\hat{E}^{(-)}_{\mu_2}(r_{m\bullet})$ represent the field strength containing the annihilation operators for the modes μ_1 and μ_2 (with the frequencies ω_{μ_1} and ω_{μ_2} and the polarization unit vectors $e_{\mu_1\bullet}$ and $e_{\mu_2\bullet}$) at the point $r_{m\bullet}$ of the mth atomic system; $\hat{E}^{(+)}_{\mu}(r_{m\bullet})$ characterizes the field strength operator containing the creation operators for the mode μ (frequency $\omega_{\mu_1} + \omega_{\mu_2}$, polarization unit vector $e_{\mu\bullet}$). The sum involves all the contributions of the individual atomic systems. The term $\tilde{\eta}^{(S)}_m$ represents the interaction parameter of the mth atomic system for the generation of the sum frequency (index S), which is proportional to the abovementioned susceptibility function

$$\sum_{p,q,r} \chi^{(2)}_{pqr}(\omega_{\mu_1}, \omega_{\mu_2}) e_{\mu p} e_{\mu_1 q} e_{\mu_2 r}.$$

In the case of uniform, identically oriented, and homogeneously distributed single systems or for homogeneous solids it is possible to let the interaction constant precede the sum and then (if bulk compact media are studied) to integrate over the volume V. We obtain the interaction operator

$$\hat{H}^C_S = \eta^S \left\{ \frac{1}{V} \int_V d^3 r_\bullet \exp\left[-i\left(k_{\mu\bullet} - k_{\mu_1\bullet} - k_{\mu_2\bullet}\right) r_\bullet \right] \right\} \hat{a}^\dagger_\mu \hat{a}_{\mu_1} \hat{a}_{\mu_2} + \text{h.c.}. \quad (11.28)$$

The interaction parameter η^S is proportional to $\tilde{\eta}^{(S)}$ and, owing to the transition from the field-strength operators to the creation operator $\hat{a}^\dagger_\mu$ and the annihilation operators $\hat{a}_{\mu_1}$ and $\hat{a}_{\mu_2}$, also proportional to $(\omega_\mu \omega_{\mu_1} \omega_{\mu_2}/V^3)^{1/2}$. From the structure of the interaction operator it is already evident that the processes can proceed with high efficiency only if the magnitude of $\Delta k_\bullet \equiv k_{\mu\bullet} - k_{\mu_1\bullet} - k_{\mu_2\bullet}$ is zero or so small that its volume integral over the basic region will not be much less than V.

The interaction operator $\hat{H}^C_S$ contains only operators that act on the electromagnetic field but not on the atomic systems. Thus in generating and amplifying harmonics and mixed frequencies (if the frequencies of the incident and generated waves are not in resonance with the atomic transitions), the atomic systems remain in their initial states. On the time average there is no energy exchange between the electromagnetic field and the atomic systems; this means that the process is a *loss-free* one.

With the use of the operator $\hat{H}^C_S$ as given in (11.28), the methods described in Section 3.2 permit the calculation of transition probabilities and of expectation values of operators, such as $\hat{a}_\mu$ or $\hat{a}^\dagger_\mu \hat{a}_\mu$. Since the processes under study can be considered to be loss-free, the operator $\hat{H}^C_S$ can also be used, for example, in the form of (11.27), as an effective interaction operator, thus enabling us to derive equations of motion for the field operators.

The sum-frequency mode may serve as an example. The equation of motion for $\hat{E}^{(-)}_\mu(t)$ in the Heisenberg picture (cf. Section A.1.4) is given by

$$i\hbar \frac{d\hat{E}^{(-)}_\mu}{dt} = \left[\hat{E}^{(-)}_\mu, \hat{H}^F + \hat{H}^C_S\right] = \hbar\omega_\mu \hat{E}^{(-)}_\mu + \left[\hat{E}^{(-)}_\mu, \hat{H}^C_S\right], \quad (11.29)$$

where $\hat{H}^F$ is the Hamiltonian of the free system. The commutator can be expressed by

$$\left[\hat{E}_\mu^{(-)}, \hat{H}_S^C\right] = C_{\mu\mu_1\mu_2}\hat{E}_{\mu_1}^{(-)}\hat{E}_{\mu_2}^{(-)}, \qquad (11.29a)$$

where $C_{\mu\mu_1\mu_2}$ is proportional to $\tilde{\eta}^S$ and thus also proportional to the second-order susceptibility. In view of the statements on coherent states in Section 2.3.1.2, in the case of intense radiation exhibiting a high degree of coherence the field operators $\hat{E}_{\mu_1}(t)$, $\hat{E}_{\mu_2}(t)$ can be replaced by the c-numbers $E_{\mu_1}(t)$, $E_{\mu_2}(t)$. Then from (11.29) the solution

$$\hat{E}_\mu^{(-)}(t) = \left[\hat{E}_\mu^{(-)}(0) - \frac{i}{\hbar}C_{\mu\mu_1\mu_2}\hat{I}E_{\mu_1}^{(-)}(0)E_{\mu_2}^{(-)}(0)t\right]\exp\left[-i(\omega_{\mu_1} + \omega_{\mu_2})t\right] \qquad (11.30)$$

can be obtained.

We discuss the expectation value of the intensity operator $\langle\hat{E}_\mu^{(+)}(t)\hat{E}_\mu^{(-)}(t)\rangle$ of the sum frequency mode by assuming that initially the sum frequency is not excited; then $\langle\hat{E}_\mu^{(+)}(0)\hat{E}_\mu^{(-)}(0)\rangle = 0$. Under this condition, from (11.30) the resulting expectation value of the intensity of the sum frequency radiation can be derived as

$$\left\langle\hat{E}_\mu^{(+)}(t)\hat{E}_\mu^{(-)}(t)\right\rangle = \frac{t^2}{\hbar^2}|C_{\mu\mu_1\mu_2}|^2E_{\mu_1}^{(+)}(0)E_{\mu_1}^{(-)}(0)E_{\mu_2}^{(+)}(0)E_{\mu_2}^{(-)}(0). \qquad (11.31)$$

In accordance with the classical treatment [cf. (11.10)], the intensity of the sum-frequency wave is proportional to the product of the intensities in the two modes of the pump radiation. Provided no fully coherent pump radiation can be assumed, then the correlation function $\langle E_{\mu_1}^{(+)}(0)E_{\mu_1}^{(-)}(0)E_{\mu_2}^{(+)}(0)E_{\mu_2}^{(-)}(0)\rangle$ has to be substituted for the fourfold field-strength product on the right-hand side of (11.31). From the example of sum-frequency generation it becomes obvious that the frequency conversion is essentially determined by the coherence properties of the pump radiation. These effects on the radiation transformation will be treated in the next subsection, where parametric amplification serves as an example.

11.2.2. Parametric Processes

In parametric amplification the place of the interaction operator $\hat{H}_S^C$ in (11.27) is taken by the operator

$$\hat{H}_{PA}^C = -\int_V d^3r_\bullet\,\chi_{PA}\hat{E}^{(+)}(\omega_S)\hat{E}^{(+)}(\omega_I)\hat{E}^{(-)}(\omega_p) + \text{h.c.}, \qquad (11.32)$$

where the frequencies of the signal, idler, and pump waves are denoted by the indices S, I, and P (the terms signal, idler, and pump are used in the sense of Section 11.1.2). The effective susceptibility χ_{PA} is made up of second-order susceptibility components and polarization unit-vector components, namely

$$\chi_{PA} = \sum_{\alpha, \beta, \gamma} \chi^{(2)*}_{\alpha, \beta, \gamma}(\omega_P; \omega_S, \omega_I) e_\alpha(\omega_P) e_\beta(\omega_S) e_\gamma(\omega_I). \quad (11.33)$$

If the field-strength amplitudes are formulated in terms of creation and annihilation operators, we obtain $\hat{H}^C_{PA}$ in the form

$$\hat{H}^C_{PA} = -\hbar\eta_{PA}\left\{\frac{1}{V}\int_V d^3r_\bullet\, e^{i(k_{P\bullet} - k_{S\bullet} - k_{I\bullet})r_\bullet}\right\}\hat{a}_P\hat{a}^\dagger_S\hat{a}^\dagger_I + \text{h.c.}. \quad (11.34)$$

This form corresponds to the relation (11.28) for frequency conversion. The interaction constant η_{PA} is proportional to χ_{PA}, $V^{-3/2}$, and the three factors $(\omega_{S,I,P}/\varepsilon_{S,I,P})^{1/2}$. The susceptibility components incorporated in χ_{PA} can again be related to the quantities of the medium (transition energies, transition moments) as was shown at the beginning of Section 11.2.1; therefore no separate semiclassical treatment for parametric amplification is needed. The parameter η_{PA} is normalized so that it has the dimension of a frequency. As with frequency conversion, here again the aspect of phase matching is important. In exact phase match the expression in braces takes on the value of unity; in mismatch, which for the present will be disregarded, its value is less than unity. The first term on the right-hand side of (11.34) represents the annihilation of a pump photon with simultaneous creation of one signal photon and one idler photon. The second, hermitian-conjugate term represents the reverse process. Thus, also in this respect, agreement with the structure of the corresponding relation (11.28) for frequency conversion becomes apparent.

The equations of motion for the time-dependent photon creation operators of the signal and idler waves in the Heisenberg picture can be determined according to general rules from (11.34); this leads to

$$\frac{d}{dt}\hat{a}^\dagger_S = i\omega_S\hat{a}^\dagger_S - i\eta^*_{PA}\hat{a}^\dagger_P\hat{a}_I, \quad (11.35a)$$

$$\frac{d}{dt}\hat{a}^\dagger_I = i\omega_I\hat{a}^\dagger_I - i\eta_{PA}\hat{a}^\dagger_P\hat{a}_S. \quad (11.35b)$$

It can be seen that the temporal variation of the signal wave depends on the amplitudes of the pump and the idler wave; that of the idler wave depends on the amplitudes of the pump and the signal wave.

As pump wave, in general the radiation of a laser well above threshold is used. This fact allows us to describe the pump wave in good approximation as a function with defined time dependence (cf. Section 7.3). Moreover, in many cases the attenuation of the pump wave may be considered negligible. Under these conditions the operator describing the pump wave can be regarded as a time-dependent function with c-number character; this results in a substantial formal simplification. The following substitution is used:

$$\eta_{PA}\hat{a}_P \to \nu(t)e^{-i\omega_P t}, \qquad \text{where} \quad \nu(t) = \nu_0 \exp\left[i\varphi_{\eta_{PA}} - i\varphi_P(t)\right]. \quad (11.36)$$

For simplicity, the interaction parameter η_{PA} has been connected with the pump-wave operator, from which the main time dependence has been eliminated. The temporally defined function $\nu(t)$ contains the phase of η_{PA} and that of the pump wave as well as the constant term ν_0, which is equal to the product of the magnitude of η_{PA} times the magnitude of the pump-wave amplitude.

For the following we assume the pump wave to have a time-independent fixed phase (more general conditions will be discussed in Section 11.4). Then $\nu(t)$ becomes a constant complex number ν. If this condition is satisfied, exact analytical solutions of the differential equations (11.35a, b) can be obtained. Separating the main time dependence according to

$$\hat{a}_{S,I}(t)e^{i\omega_{S,I}t} \stackrel{\text{def}}{=} \underset{\sim}{\hat{a}}_{S,I}(t) \qquad (11.37)$$

from the operators in the Heisenberg picture employed in (11.35a, b), the differential equations

$$\frac{d}{dt}\underset{\sim}{\hat{a}}_S = i\nu\underset{\sim}{\hat{a}}_I^\dagger, \qquad (11.38a)$$

$$\frac{d}{dt}\underset{\sim}{\hat{a}}_I = i\nu\underset{\sim}{\hat{a}}_S^\dagger \qquad (11.38b)$$

result for the slowly time-varying operators $\underset{\sim}{\hat{a}}_{S,I}(t)$ [they vary slowly in comparison with time intervals of the order $(\omega_{S,I})^{-1}$]. From these coupled differential equations there immediately follow the two decoupled differential equations for the signal and idler photons:

$$\frac{d^2}{dt^2}\underset{\sim}{\hat{a}}_S = |\nu|^2\underset{\sim}{\hat{a}}_S, \qquad (11.39a)$$

$$\frac{d^2}{dt^2}\underset{\sim}{\hat{a}}_I = |\nu|^2\underset{\sim}{\hat{a}}_I. \qquad (11.39b)$$

Let us now calculate the variation of the number of signal and idler photons per unit time. Since

$$\frac{d}{dt}\left\{\underset{\sim}{\hat{a}}_{S,I}^\dagger\underset{\sim}{\hat{a}}_{S,I}\right\} = \frac{d}{dt}\left\{\hat{a}_{S,I}^\dagger\hat{a}_{S,I}\right\},$$

(11.38) yields

$$\frac{d}{dt}\left\{\hat{a}_S^\dagger\hat{a}_S\right\} = \frac{d}{dt}\left\{\hat{a}_I^\dagger\hat{a}_I\right\}. \qquad (11.40)$$

Regarding the situation existing at the beginning of the process at $t = 0$, we make the following plausible assumption: let the signal and the idler wave be decoupled and the total density operator be composed of two factors, one assigned to the signal mode and the other one to the idler mode. Under this assumption there follows from (11.40) the equality of the time derivatives of the signal and idler photon-number expectation values and the time invariance of the difference of these expectation values. These statements represent a special case of the Manley–Rowe relations, which comprise general conservation laws for the mean photon numbers in nonlinear optical processes (see Section 1.3.2.8).

Now we are going to determine the temporal dependence of the signal amplitude. The solution of the differential equations (11.38) and (11.39) yields

$$\underset{\sim}{\hat{a}}_S(t) = \hat{a}_S(0)\cosh(|\nu|t) - i\hat{a}_I^\dagger(0)\frac{|\nu|}{\nu}\sinh(|\nu|t), \tag{11.41}$$

where the initial conditions are given by $\hat{a}_S(0)$ and $\hat{a}_I(0)$. From the expression (11.41) the particle-number operator can be obtained as a function of time. In a time interval $(0, t)$ for which $e^{2|\nu|t}$ is much greater than 1—that is, for $t \gg |\nu|^{-1}$—one obtains

$$\begin{aligned} \hat{N}_S(t) &= \hat{a}_S^\dagger(t)\hat{a}_S(t) \\ &= \tfrac{1}{4}e^{2|\nu|t}\left\{1 + \hat{N}_S(0) + \hat{N}_I(0) + \left[\frac{|\nu|}{i\nu}\hat{a}_S^\dagger(0)\hat{a}_I^\dagger(0) + \text{h.c.}\right]\right\}. \end{aligned} \tag{11.42}$$

From this the expectation value of the signal photons

$$\langle \hat{N}_S(t)\rangle = \frac{e^{2|\nu|t}}{4}\mathfrak{A} \tag{11.43}$$

follows, where

$$\mathfrak{A} = 1 + \langle \hat{N}_S(0)\rangle + \langle \hat{N}_I(0)\rangle + \left[\frac{|\nu|}{i\nu}\langle \hat{a}_S^\dagger(0)\rangle\langle \hat{a}_I^\dagger(0)\rangle + \text{c.c.}\right].$$

This relation shows that the expectation value of the number of signal photons, $\langle \hat{N}_S(t)\rangle$, increases (exponentially) with time; this means *parametric amplification* occurs. The (temporal) amplification coefficient $2|\nu|$ is proportional to the interaction parameter in the interaction operator as well as to the absolute value of the pump amplitude. The range of validity of (11.43) is bounded at high values of t insomuch as the number of the created signal photons must not result in a noticeable depletion of the pump mode. At a given time t the signal photon number is proportional to $\mathfrak{A}$.

We now discuss the fourth term in $\mathfrak{A}$, containing products of signal and idler parts. This summand vanishes in the following cases: if initially no signal or idler photons are present; if initially signal and idler photons do exist, but one or both of these waves have a statistically undetermined phase; if the signal and idler waves are uncorrelated (Ref. 11.4). In case both the signal and the idler wave can be described by coherent states (cf. Section 2.3.1.2), the fourth term of $\mathfrak{A}$ is given by

$$\sqrt{\langle \hat{N}_S(0)\rangle}\sqrt{\langle \hat{N}_I(0)\rangle}\,[e^{-i(\pi/2+\varphi_\nu+\varphi_S+\varphi_I)} + \text{c.c.}], \tag{11.44}$$

where φ_ν is the phase of ν, which is determined by the phase of the interaction constant η_{PA} and by the phase of the pump wave; φ_S and φ_I are the phases of the signal and idler waves, respectively. The maximum value of the expression in (11.44) is attained if the phases satisfy the condition $\pi/2 + \varphi_\nu + \varphi_S + \varphi_I = 0$. Taking into account the minimum value and the maximum value of the fourth term, the following interval for $\mathfrak{A}$ is obtained:

$$1 + \langle \hat{N}_S(0)\rangle + \langle \hat{N}_I(0)\rangle \le \mathfrak{A} \le 1 + \left[\sqrt{\langle \hat{N}_S(0)\rangle} + \sqrt{\langle \hat{N}_I(0)\rangle}\right]^2. \tag{11.45}$$

The exact value of $\mathfrak{A}$ in this interval depends on the statistical properties of signal and idler, in accordance with the considerations made above, and on the relative phase with respect to the pump wave.

We emphasize that irrespective of the initial states of signal and idler, we have $\mathfrak{A} \ge 1$ in any case, which means that parametric generation is always possible. This even holds for $\langle \hat{N}_S(0)\rangle = \langle \hat{N}_I(0)\rangle = 0$, that is for an initially vanishing number of signal and idler photons. In this case the generation and amplification processes of the signal wave are started by the spontaneous decay of pump photons, which is represented by the term 1 in $\mathfrak{A}$.

Thus it has been shown that by using quantum-theoretical means the important empirical findings on the generation of waves can be clarified without additional assumptions. In a classical treatment however (see end of Section 11.1.2), the generation of a wave originating from "noise" would necessitate the incorporation of a fluctuation force, which must be regarded, from the classical point of view, as a more or less artificial addition.

So far, we have assumed exact phase matching ($\Delta k_\bullet = 0$) and moreover a fixed phase of the pump wave ($\varphi_p = \text{const}$). Under actual conditions deviations from these assumptions occur. The right-hand side of (11.34) indicates that in the case of momentum mismatch not only must η_{PA} and the amplitude of the pump wave be inserted into the factor ν from (11.36), but also the factor in braces. This factor causes a decrease of the temporal amplification coefficient $2|\nu|$ if $\Delta k_\bullet = k_{P\bullet} - k_{S\bullet} - k_{I\bullet}$ does not vanish. The assumption of a fixed phase of the pump wave implies special coherence properties, namely those of ideal laser light. The deviations resulting from the coherence properties of real laser radiation (phase fluctuations) will be described in Section 11.4.

11.3. MATERIAL PARAMETERS AND APPLICATIONS

In the preceding sections of this chapter we treated the essential physical principles of frequency conversion and parametric amplification on the basis of classical and quantum-theoretical relations. The subject of the present section is the application of these processes and the discussion of typical quantitative results. This includes the classical model of the anharmonic oscillator (Section 11.3.1). Its use allows to give straightforward explanations of higher-order contributions to the polarization. Phase matching and focusing (Section 11.3.2) have proved to be important methods for achieving sufficiently high efficiency. The way in which the mode structure influences efficiency is dealt with in Section 11.3.3. The process of up-conversion and the results obtained from it and from third-harmonic and higher-harmonic generation are discussed in Section 11.3.4; uses of parametric oscillation and fluorescence are presented in Section 11.3.5.

11.3.1. The Model of the Anharmonic Oscillator and Its Application

The classical amplitude equations used in Section 11.1 for describing wave interactions that lead to the generation and amplification of frequency-converted waves contain the susceptibility functions that specify the relation between polarization and field strength. Since for treating some problems of second-harmonic generation the classical treatment proves to be transparent and useful, we shall employ such a treatment for the susceptibilities involved as well. To this end we extend the model of *linear polarization* introduced by P. Drude and H. A. Lorentz. This relatively unsophisticated model, which is based on classical concepts, has achieved great importance in representing optoelectric parameters of materials because many of the quantum-theoretically derived relations (cf. Section 11.3 and Chapter 4) exhibit the same functional dependence of the relevant physical quantities as they do in the classical theory. Thus, though some basic assumptions made by Drude and Lorentz, such as the existence of electrons elastically bound to equilibrium positions in atoms, are not satisfied in nature, equivalence between the quantum-theoretical and classical theories of linear polarization can be achieved to a certain extent by suitable interpretation of the respective quantities and constants. The same is true of the extension to nonlinear polarization.

In the description of linear polarization in an electric field we start from the assumption that in the material under study charge carriers (charge q_e, mass m) are bound by an elastic force. The deviation from equilibrium in x direction induces a dipole moment; the interaction between the individual dipole moments is regarded as being negligible. Next, we consider one charge carrier, which is assumed to be bound to the position $x = 0$ by means of the linear force $F_l(x) = -kx$. In an electric field $E(t)$ in x direction this charge carrier obeys the equation of motion

$$m\ddot{x} = F_l(x) + q_e E(t) - \Gamma'\dot{x}, \qquad (11.46)$$

where Γ' is a damping constant to be explained later. The deviation causes a dipole moment $p = q_e x$, assuming that at $x = 0$ there is an effective (immovable) charge $-q_e$. Since we assume the individual charge carriers to be independent, the polarization $P = \gamma p = \gamma q_e x$ can be introduced, where γ is the number density. Thus from (11.46) the differential equation

$$\ddot{P} + \Gamma\dot{P} + \omega_0^2 P = \frac{\gamma q_e}{m} E(t) \tag{11.47}$$

immediately follows for the polarization $P(t)$. Here Γ stands for Γ'/m, and ω_0^2 is equal to k/m.

We now extend the relation (11.47) to nonlinear conditions, which include relatively large deviations from equilibrium. To this end, we replace the linear restoring force F_l by the nonlinear restoring force $F(x) = -kx + k'x^2$; terms of higher than second order in x may be neglected. This leads from (11.46) immediately to the differential equation

$$\ddot{P}(t) + \Gamma\dot{P}(t) + \omega_0^2 P(t) = \frac{\gamma q_e^2}{m} E(t) + \frac{k'}{\gamma q_e m} P^2(t), \tag{11.48}$$

from which the nonlinear polarization $P(t)$ is to be determined.

Before drawing conclusions from this differential equation, the physical meaning of the quantities composing it has to be elucidated. The charge carriers affected by the electric field $E(t)$ are assumed to be electrons; hence we set $q_e = -e$, and m equal to the electron mass. The function $x(t)$, introduced in (11.46) for the "deviation of the electron," in reality stands for the center-of-charge deviation of a smeared charge cloud of one electron in one atom, in an ensemble of atoms, or in a solid. In Section 1.3.2.5 it was mentioned that second-order effects such as SHG do not occur in systems with inversion symmetry. This implies that models whose structure resembles that of an individual inversion-symmetrical hydrogen or alkali atom cannot be employed. Instead, a model is needed in which the center of charge of the electron under study has an acentric equilibrium position (on the x axis) caused by molecular or crystal forces.

Suppose that the electron considered is one of the bonding electrons between the molecular or lattice elements. For small deviations from equilibrium the restoring force is proportional to x, that is, the potential energy is proportional to x^2. The resonance frequency $\omega_0 = \sqrt{k/m}$ appearing in the Drude–Lorentz theory has to be identified with the quantum-theoretical transition frequency of the electrons. For the corresponding transition frequencies in the ultraviolet region one has to set $\omega_0 \sim 10^{16}\ \mathrm{s}^{-1}$. This means that the force constant k is of the order 10^2 N/m. This constant may be connected with the force eE_{atom} which represents the acting force of the atomic core or nucleus on the electron under study. Applying the virial theorem concerning the quantum-theoretical relation between the mean total energy and the mean potential energy, one obtains the result

$$m\omega_0^2 R \sim e|E_{\text{atom}}|, \tag{11.49a}$$

where R is the mean distance of the electron from the center of the atom. The so-called atomic field strength E_{atom} is of the order of 10^{11} V/m.

The x dependence of the potential energy of the center of charge of the smeared charge cloud is determined by the electrostatic and quantum-mechanical forces of all the participating charge carriers as well as by the symmetry of the molecule or lattice. It is therefore difficult to make generally valid statements. However, it can be assumed that one property common to all potential functions involved is that the potential energy of the electron approaches the electronic transition energy or dissociation energy as the deviation x from equilibrium tends to the equilibrium distance R' of the lattice elements; thus

$$k \sim |k'|R'. \tag{11.49b}$$

With $R' \sim 10^{-10}$ m we therefore have $|k'| \sim 10^{12}$ N/m^2. This value enables us to give order-of-magnitude estimates of important frequency-conversion effects for typical groups of materials.

In Fig. 11.2 the potential function $U_{\text{pot}}(x)$ of an electron in an unsymmetrical force field is schematically plotted; the dashed line represents the harmonic oscillator part, the solid line the anharmonic one. A fictitious, immovable, positive charge—made up of all the other charge carriers of the molecule or crystal—is assumed to be located at $x = 0$. It creates a dipole moment together with the electron, if a nonvanishing displacement x occurs. We assume that in the molecule or the crystal as a whole the charge carriers are so arranged that the polarization P without an external field is zero. The damping constant Γ is connected with the homogeneous line width $\Delta\omega$ and with the transverse relaxation time τ according to

$$\Gamma = \Delta\omega = \frac{2}{\tau} \tag{11.50}$$

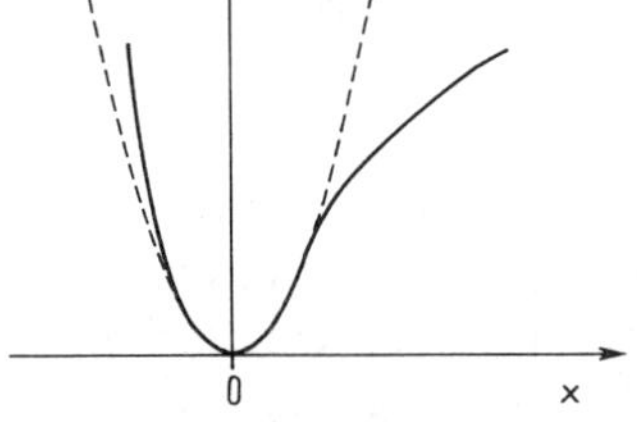

Fig. 11.2. Potential energy $U_{\text{pot}}(x)$ of an electron in an unsymmetric field (schematic).

(cf. Section 4.6). Here τ describes the damping of the nondiagonal terms of the density matrix, and if one assumes radiation damping only can be taken to be of the order of 10^{-8} s. Under the influence of other relaxation mechanisms, however (see Table 3.2), times may result that are smaller by several orders; nevertheless one still has $\Gamma \ll \omega_0$.

To obtain an iterative solution of the differential equation (11.48) we pass over to an integral relation and assume that the functions $E(t)$ and $P(t)$ are Fourier-transformable. In (11.48) the time variable t is replaced by $t - \tau$; both sides are multiplied by the function

$$Q(t) = \begin{cases} 0 & \text{for} \quad \tau < 0, \\ \dfrac{\exp[-\Gamma\tau/2]}{\sqrt{\omega_0^2 - \Gamma^2/4}} \sin\left(\tau\sqrt{\omega_0^2 - \Gamma^2/4}\right) & \text{for} \quad \tau \geq 0 \end{cases} \tag{11.51}$$

and then integrated over the whole range of τ. The function $Q(t)$ is the response function of the damped harmonic oscillator in the time domain. These operations lead to

$$P(t) = \int d\tau\, Q(\tau) \frac{\gamma e^2}{m} E(t - \tau) + \int d\tau\, Q(\tau) \frac{-k'}{\gamma e m} P^2(t - \tau), \tag{11.52}$$

which may be regarded as an implicit solution for $P(t)$. Only the first term on the right-hand side is P-independent, the second term contains $P(t - \tau)$. Representing $P(t - \tau)$ according to

$$\begin{aligned} P(t - \tau) = &\int d\tau'\, Q(\tau') \frac{\gamma e^2}{m} E(t - \tau - \tau') \\ &+ \int d\tau'\, Q(\tau') \frac{-k'}{\gamma e m} P^2(t - \tau - \tau') \end{aligned} \tag{11.53}$$

and substituting this expression into (11.52), we arrive at

$$\begin{aligned} P(t) = &\frac{\gamma e^2}{m} \int d\tau\, Q(\tau) E(t - \tau) \\ &- \frac{k'\gamma e^3}{m^3} \int d\tau\, Q(\tau) \left[\int d\tau'\, Q(\tau') E(t - \tau - \tau')\right]^2 + \mathscr{R}. \end{aligned} \tag{11.54}$$

Here the first two terms are independent of P and contain the field strength in the first and second orders. The third is a remainder term, which, on proceeding further iteratively, is found to contain the field strength in an order higher than second. We compare (11.54) with the general expression for $P(t)$ [cf.

(1.73)], summing up to the second order (inclusive). Then we have

$$P(t) = \varepsilon_0 \int d\tau\, \varkappa^{(1)}(\tau_1) E(t - \tau_1)$$
$$+ \varepsilon_0 \int d\tau_1 \int d\tau_2\, \varkappa^{(2)}(\tau_1, \tau_2) E(t - \tau_1) E(t - \tau_2). \quad (11.55)$$

From the contributions of the first-order polarization $P^{(1)}(t)$ and of the second-order polarization $P^{(2)}(t)$, the corresponding susceptibilities result:

$$\varkappa^{(1)}(\tau_1) = \frac{\gamma e^2}{\varepsilon_0 m} Q(\tau_1), \quad (11.56a)$$

$$\varkappa^{(2)}(\tau_1, \tau_2) = \begin{cases} -\dfrac{k'\gamma e^3}{\varepsilon_0 m^3} \int d\tau' Q(\tau') Q(\tau_1 - \tau') Q(\tau_2 - \tau') & \text{for} \quad \tau_1, \tau_2 \geq 0, \\ 0 & \text{otherwise.} \end{cases} \quad (11.56b)$$

It is obvious from (11.56b) that the second-order susceptibility is formed by an integrand that contains a triple product of first-order susceptibilities with different arguments. Note that, $\varkappa^{(2)}$ contains the force constant k' of the quadratic part of the restoring force.

The calculation of higher-order contributions to the polarization by proceeding in the iteration presented above turns out to be rather troublesome and only weakly transparent. But the method employed can be easily explained with a block diagram (Fig. 11.3). One or two input quantities enter a certain block, and one output quantity (→) results from the operation carried out in the particular block. Two different kinds of blocks occur. There are those that are denoted by $\boxed{\mathscr{A}\varkappa^{(1)}}$. They represent a linear filter with the response function $\mathscr{A}\varkappa^{(1)}(\tau)$ in the time domain or, what is equivalent, with the response function $\mathscr{A}\breve{\varkappa}^{(1)}(\omega)$ in the frequency domain. The factor $\mathscr{A}$ preceding $\varkappa^{(1)}$

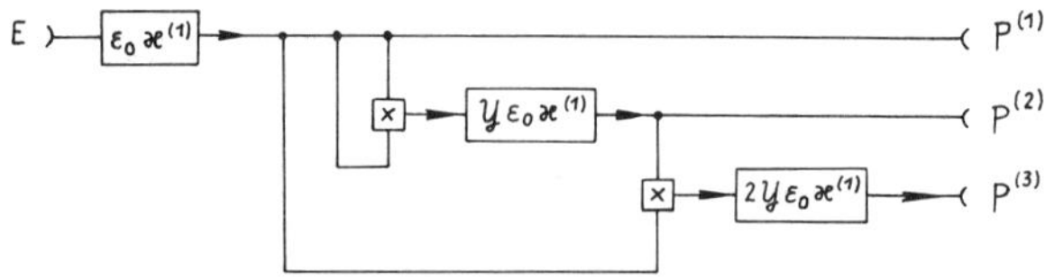

Fig. 11.3. Block diagram for the calculation of polarization contributions up to the third order. $\mathcal{Y} = -k'/\gamma^2 e^3$.

stands for a constant amplification factor. In the blocks denoted by $\boxed{\times}$ the two input quantities are multiplied by one another.

Thus, from the simple concept of the motion of an electron in the anharmonic potential we are led to the block diagram in Fig. 11.3, and from there to the straightforward determination of higher-order contributions of polarization. This holds both in the time domain and in the frequency domain. Given a discrete frequency ω with the amplitude of the field strength $\hat{E}(\omega)$ as input amplitude, the output amplitude $\mathscr{A}\varkappa^{(1)}(\omega)\hat{E}(\omega)$ will result after the linear filter $\mathscr{A}\varkappa^{(1)}$. If at the input of a block $\boxed{\times}$ there are two oscillations with frequencies ω_1 and ω_2 and input amplitudes $\hat{E}(\omega_1)$ and $\hat{E}(\omega_2)$, then an oscillation of frequency $\omega_1 + \omega_2$ with the amplitude $\frac{1}{2}\hat{E}(\omega_1)\hat{E}(\omega_2)$ and a frequency $\omega_1 - \omega_2$ with the amplitude $\frac{1}{2}\hat{E}(\omega_1)\hat{E}^*(\omega_2)$ will result as output quantities.

The diagram of Fig. 11.3 can be used for simply obtaining order-of-magnitude estimates in the generation of higher harmonics. To this end we assume that a field-strength oscillation of frequency ω with the amplitude $\hat{E}(\omega)$ acts on a material characterized by the linear susceptibility function $\varkappa^{(1)}$. From the diagram it follows directly that

$$\hat{P}^{(1)}(\omega) = \hat{E}(\omega)\varepsilon_0\breve{\varkappa}^{(1)}(\omega),$$

$$\hat{P}^{(2)}(2\omega) = \left[\hat{E}(\omega)\varepsilon_0\breve{\varkappa}^{(1)}(\omega)\right]\left[\hat{E}(\omega)\varepsilon_0\breve{\varkappa}^{(1)}(\omega)\right]\tfrac{1}{2}\mathscr{Y}\varepsilon_0\breve{\varkappa}^{(1)}(2\omega),$$

$$\hat{P}^{(3)}(3\omega) = \hat{P}^{(1)}(\omega)\hat{P}^{(2)}(2\omega)\tfrac{1}{2}\cdot 2\mathscr{Y}\varepsilon_0\varkappa^{(1)}(3\omega). \tag{11.57}$$

This yields for the susceptibilities $\chi^{(j)}$ in the relations between polarization and field strength for monochromatic waves (cf. Section 1.3.2) with $\mathscr{Y} = -k'/\gamma^2 e^3$:

$$\chi^{(1)}(\omega;\omega) = \breve{\varkappa}^{(1)}(-\omega), \tag{11.58a}$$

$$\chi^{(2)}(2\omega;\omega,\omega) = \frac{-\varepsilon_0^3}{2}\frac{k'}{\gamma^2 e^3}\breve{\varkappa}^{(1)}(-2\omega)\left[\breve{\varkappa}^{(1)}(-\omega)\right]^2, \tag{11.58b}$$

$$\chi^{(3)}(3\omega;\omega,\omega,\omega) = \frac{\varepsilon_0^5}{2}\frac{k'^2}{\gamma^4 e^6}\breve{\varkappa}^{(1)}(-3\omega)\breve{\varkappa}^{(1)}(-2\omega)\left[\breve{\varkappa}^{(1)}(-\omega)\right]^3. \tag{11.58c}$$

We see that the second-order susceptibility, as could be expected from (11.56b), is dependent on a triple product of first-order susceptibilities; furthermore $\chi^{(2)}$ is proportional to k'. For the ratios of polarization contributions of successive orders we obtain

$$\frac{1}{|\hat{E}(\omega)|}\frac{|\hat{P}^{(2)}(2\omega)|}{|\hat{P}^{(1)}(\omega)|} = \frac{1}{2}\frac{|k'|}{\gamma^2 e^3}\varepsilon_0^2|\underset{\smile}{\varkappa}^{(1)}(-\omega)|\cdot|\underset{\smile}{\varkappa}^{(1)}(-2\omega)|$$

$$= \frac{|\chi^{(2)}(2\omega;\,\omega,\omega)|}{|\chi^{(1)}(\omega;\,\omega)|}, \tag{11.59}$$

$$\frac{1}{|\hat{E}(-\omega)|}\frac{|\hat{P}^{(3)}(-3\omega)|}{|\hat{P}^{(2)}(2\omega)|} = \frac{|k'|}{\gamma^2 e^3}\varepsilon_0^2|\underset{\smile}{\varkappa}^{(1)}(-\omega)|\cdot|\underset{\smile}{\varkappa}^{(1)}(-3\omega)|$$

$$= \frac{|\chi^{(3)}(3\omega;\,\omega,\omega,\omega)|}{|\chi^{(2)}(2\omega;\,\omega,\omega)|}. \tag{11.60}$$

According to the conditions usually satisfied in the generation of higher harmonics we suppose that ω, 2ω, and 3ω are well below the atomic resonance frequency ω_0; this may be expressed by the condition $\omega, 2\omega, 3\omega \leq \frac{3}{4}\omega_0$. Then the susceptibility values $|\underset{\smile}{\varkappa}^{(1)}(-\omega)|$, $|\underset{\smile}{\varkappa}^{(1)}(-2\omega)|$, $|\underset{\smile}{\varkappa}^{(1)}(-3\omega)|$ are of the order $\gamma e^2/\varepsilon_0 m\omega_0^2$. Then, by virtue of (11.49a) and (11.49b), the right-hand sides of the equations (11.59) and (11.60) can in their orders of magnitude be set equal to $|E_{\text{atom}}|^{-1}$. Altogether we obtain

$$\frac{|\hat{P}^{(2)}(2\omega)|}{|\hat{P}^{(1)}(\omega)|} \sim \frac{|\hat{E}(\omega)|}{|E_{\text{atom}}|}, \qquad \frac{|\hat{P}^{(3)}(3\omega)|}{|P^{(2)}(2\omega)|} \sim \frac{|\hat{E}(\omega)|}{|E_{\text{atom}}|}. \tag{11.61}$$

Thus it can be seen that the polarization amplitudes of successive orders behave as the quotient of the external field given by the experimenter and the internal field by which the unaffected atomic system is characterized. This statement corresponds to that on the comparison of susceptibilities derived by the quantum-theoretical perturbation theory in Section 4.4. Under the above-discussed condition that the frequency of the fundamental wave and the sum frequencies are small with respect to the atomic transition frequency, one obtains (for typical values of the number density, the atomic transition frequency, and the nonlinear force constant k') the estimate $|\chi^{(2)}| \sim 10^{-11}$ m/V for the second-order susceptibility from (11.58). A comparison with the semiclassical estimate following (11.23) reveals that $|\chi^{(2)}|$ is of the same order. This indicates that the classical model sketched in the present subsection, which can advantageously be used owing to its great simplicity, is capable of providing an order-of-magnitude estimation of the relevant nonlinear properties.

In concluding this subsection some remarks on third-order and higher-order susceptibilities should be made. From (11.60) the value $\chi^{(3)} \sim 10^{-22}\ \mathrm{m^2/V^2}$ follows; it corresponds with that obtained from quantum-theoretical perturbation theory [cf. (11.26)]. In a more precise classical model additional terms should be inserted on the right-hand side of (11.58c). The model presented by us is based on an ansatz for the restoring force that includes only terms up to the second order. Higher-order force constants do not enter $\chi^{(1)}$ and $\chi^{(2)}$. But in addition to the given term in (11.58c), which is proportional to $(k')^2$, a third-order force constant would enter this expression. Generally verifiable statements on third-order or higher-order force constants can scarcely be made on a classical basis. Therefore it is convenient to start from the semiclassical expression for $\chi^{(3)}$ given in (4.46).

11.3.2. Phase Matching and Focusing

In the classical treatment of second-harmonic generation we realized in (11.10) that the generated power is critically dependent on the difference Δk between the wave numbers of the field strengths and of the polarization [see (11.8)]. The power decreases considerably if a relatively large momentum mismatch Δk occurs. To avoid this, special measures must be taken. An analogous situation arises with the generation of sum and difference frequencies as well as with parametric amplification. This is shown by the relations (11.15b) and (11.16). In the following, when discussing phase matching in second-harmonic generation, we thereby lay a basis for its treatment in connection with frequency conversion in general and also with parametric amplification.

Provided we succeed in increasing the phase coherence length L [see (11.11)]—that is, in decreasing Δk—the rate of conversion from the fundamental wave to the harmonic will increase. What results is called phase matching, because over a long distance there is no considerable change in the (relative) phases of field strength and polarization waves at the frequency 2ω. As can be seen from (11.9), Δk depends on the refractive indices at the frequencies ω and 2ω in various directions of polarization. They are to be investigated below.

In Fig. 11.4 the spectral profile of the refractive index for a fixed polarization direction as a function of ω is schematically presented for a substance with two absorption regions. In an absorption-free region (i.e., a region of normal dispersion) the refractive index increases as the frequency increases.

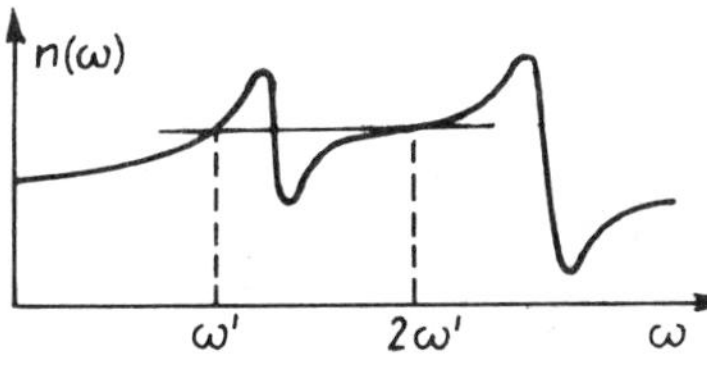

Fig. 11.4. Linear refractive index for a substance with two absorption regions (schematic).

Thus, if fundamental and harmonic waves with equal propagation and polarization directions lie in a region without an absorption band, there holds $n(2\omega) > n(\omega)$ for their linear refractive indices. According to (11.9) a complete phase matching $\Delta k = 0$ cannot be achieved in this case. But the wave number difference Δk can be made to vanish if the frequencies ω and 2ω are separated by an absorption region of appropriate strength (see the frequencies ω' and $2\omega'$ in Fig. 11.4). However, the exploitation of this method poses certain difficulties for the generation of second harmonics, because it is based on crystal systems of certain symmetries in which absorption bands of the desired intensity cannot be used without satisfying certain conditions. With third-harmonic generation, which also occurs in liquids, there is a completely different situation with respect to phase matching, which we shall discuss in Section 11.3.4.

A highly effective method of phase matching in second-harmonic generation relies on the utilization of double refraction in anisotropic crystals; optically uniaxial crystals are frequently used. As is well known, in a uniaxial crystal two waves of equal frequency may propagate in every direction—with the exception of that along the optical axis—with different phase velocities: the ordinary and the extraordinary wave. These waves are polarized perpendicularly to one another. The phase velocity w_{eo} and the linear refractive index n_{eo} of the extraordinary beam exhibit directional dependence and agree with the values of the ordinary beam only in the direction of the optical axis. Crystals with $w_o > w_{eo}$ are called *positively birefringent*; those with $w_o < w_{eo}$ are called *negatively birefringent*.

In Fig. 11.5 the refractive index of a negatively birefringent crystal has been plotted for the frequencies ω and 2ω and for all directions of propagation in one plane. It is a schematic representation in which the deviation from sphericity and the dispersion have been greatly exaggerated. Qualitatively Fig. 11.5 represents, for example, the situation in KDP under the incidence of ruby laser light. It is obvious that there exists a direction inclined to the optical axis

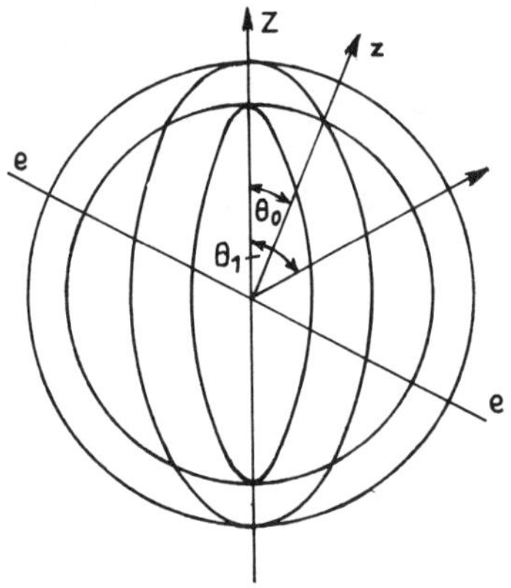

Fig. 11.5. Schematical representation of the refractive indices of a negatively birefringent crystal for all directions of propagation in one plane (frequencies ω and 2ω): θ_0, θ_1, phase-matching angles; *ee*, intersection line.

at the angle θ_0 with $n_{eo}(2\omega) - n_o(\omega)$ equal to zero. In the direction θ_0, complete phase matching will result if the fundamental wave propagates as an ordinary beam, and the harmonic propagates as an extraordinary beam. Moreover, there is a direction θ_1 in which $2n_{eo}(2\omega) - n_{eo}(\omega) - n_o(\omega)$ vanishes. Thus in this case complete phase matching will result, with the fundamental wave propagating both as an ordinary wave and as an extraordinary one in the crystal.

For generating second harmonics, plates (rods) are cut from the crystal whose surface normals coincide with the phase-matching directions. As an example, in Fig. 11.5 the intersection of such a plane with the plane of the drawing is represented by *ee*. Upon this phase-matching direction, which makes the angle θ_0 with the optical axis, light of frequency ω polarized in the x direction (hence, normal to the principal section) is incident as an ordinary wave in the crystal. Then the generated radiation of frequency 2ω propagates as the extraordinary wave polarized in the y direction. We see that in this case, in the relations (11.7) and (11.8) only the susceptibility component $\chi^{(2)}_{yxx}(2\omega; \omega, \omega)$ appears.

Having described the susceptibility components in the laboratory coordinate system (x, y, z), which is determined by the directions of incidence and polarization of the radiation, let us now pass over to the description in a system (X, Y, Z) fixed in the crystal. So far we have mentioned only the orientation of the Z axis; it coincides with the optical axis of the crystal and was put in the (y, z) plane. Using a suitable transformation of this kind, the susceptibility components for arbitrary orientations can be expressed in terms of relatively few components, which can be readily listed in tables.

The transformation equation between the components of the laboratory system and the crystal system reads

$$\chi^{(2)}_{ijk} = \sum_{I,J,K} s_{iI} s_{jJ} s_{kK} \chi^{(2)}_{IJK}, \tag{11.62}$$

where the transformation matrix

$$s_{iI} = \begin{pmatrix} \cos\vartheta & \sin\vartheta & 0 \\ -\sin\vartheta\cos\theta & \cos\vartheta\cos\theta & \sin\theta \\ \sin\vartheta\sin\theta & -\cos\vartheta\sin\theta & \cos\theta \end{pmatrix} \tag{11.63}$$

is expressed by the Eulerian angles according to Fig. 11.6. To give an example we express the component $\chi^{(2)}_{yxx}(2\omega; \omega, \omega)$ for a medium of the point group $\bar{4}2m$ by the coordinate system (X, Y, Z). We have

$$\chi^{(2)}_{yxx}(2\omega; \omega, \omega) = \sin\vartheta\cos\vartheta\sin\theta\left[\chi^{(2)}_{ZXY}(2\omega; \omega, \omega) + \chi^{(2)}_{ZYX}(2\omega; \omega, \omega)\right]. \tag{11.64}$$

This means that the nonlinear second-order polarization has its maximum at the crystal orientation $\vartheta_0 = 45°$. In accordance with this relation experiments

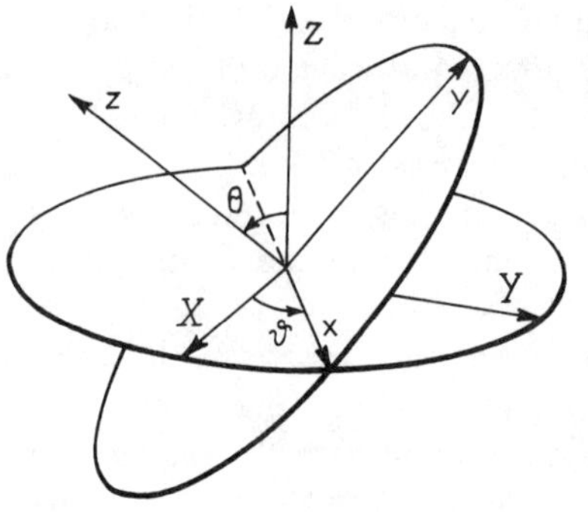

Fig. 11.6. Laboratory (x, y, z) and crystal (X, Y, Z) system. θ and ϑ are Eulerian angles.

reveal that, besides the pronounced dependence of the intensity of the harmonic on the angle θ resulting from phase matching, there also exists a weaker dependence of it on the angle ϑ (Ref. 11.5), which is illustrated in Fig. 11.7. Thus it can be inferred that for generating harmonics with high efficiency it is necessary to insure appropriate orientation of all crystal axes relative to the directions of incidence and polarization, and not only of the optical axis.

We now elaborate on our statement that the number of the susceptibility components $\chi^{(2)}_{IJK}$ related to the material system (X, Y, Z) can be reduced. Among the 27 components of the third-rank tensor $\chi^{(2)}_{XYZ}$ there exist the same symmetry relations as among those of the third-rank tensor characterizing the piezoelectric effect. None of the substances with an inversion center is piezoelectric, and thus only crystals without inversion centers are of importance for second-order nonlinear optical effects. Because of the symmetry relations (1.133), there exist certain relations between the tensor components. To simplify the notation for the second-order susceptibility components we may use the abbreviation

$$\chi^{(2)}_{IJK}(2\omega;\, \omega, \omega) = d_{Il}(2\omega;\, \omega, \omega), \tag{11.65}$$

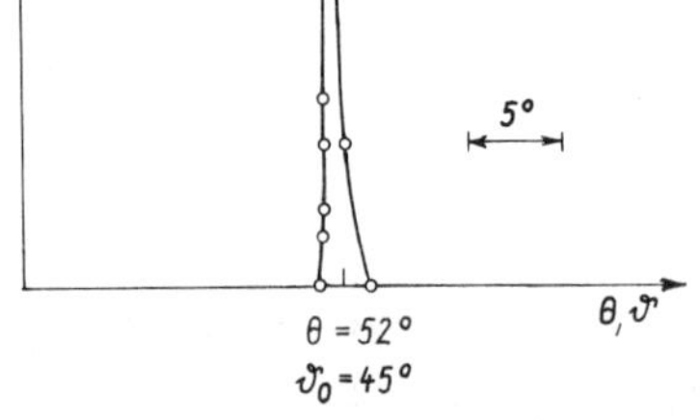

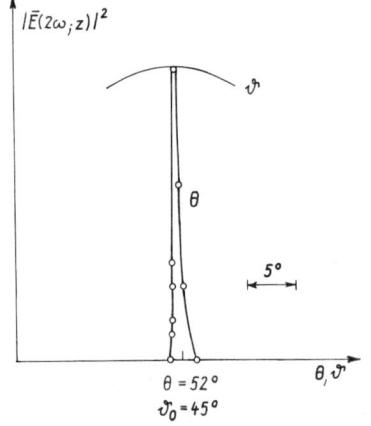

Fig. 11.7. Intensity of the second harmonic as a function of the of crystal orientation (angles θ and ϑ) of KDP (after Ref. 11.5).

which was introduced in (1.113) as well as in (1.114a), and which is frequently employed in nonlinear optics. Note that, in comparing (11.65) with (1.113), now the Cartesian coordinates X, Y, Z (corresponding to $I = 1, 2, 3$) are related to the material. The abbreviated notation already contains the symmetry relations (1.133).

As an example we consider the point group $\bar{4}2m$, characterizing among others the crystals KDP (KH_2PO_4) and ADP ($NH_4H_2PO_4$), which are especially important in nonlinear optics. The point group $\bar{4}2m$ has three diad symmetry axes perpendicular to one another. They may be chosen as axes of a Cartesian system XYZ; the components $\chi^{(2)}_{XYZ}$ can be related to the components d_{Il}. Because of the symmetry relations of the group, only two independent components, namely d_{14} ($= d_{25}$) and d_{36}, appear, so that the relation for the nonlinear polarization amplitude from (11.2) goes over into

$$\begin{aligned}
\bar{P}_X(2\omega, z) &= 2\varepsilon_0 d_{14}(2\omega; \omega, \omega)\bar{E}_Y(\omega, z)\bar{E}_Z(\omega, z),\\
\bar{P}_Y(2\omega, z) &= 2\varepsilon_0 d_{14}(2\omega; \omega, \omega)\bar{E}_Z(\omega, z)\bar{E}_X(\omega, z),\\
\bar{P}_Z(2\omega, z) &= 2\varepsilon_0 d_{36}(2\omega; \omega, \omega)\bar{E}_X(\omega, z)\bar{E}_Y(\omega, z).
\end{aligned} \tag{11.66}$$

The factors 2 stem from the summation over the contributions with $\bar{E}_I\bar{E}_J$ and $\bar{E}_J\bar{E}_I$. The components of the second-order susceptibilities of various substances are frequently represented relative to the component d_{36} of KDP. The corresponding quotients are given in the Table 11.1. The absolute value of

Table 11.1. Susceptibilities for Second-Harmonic Generation

Substance	Crystal Class[a]	Wavelength, μm	d_{mn}	d_{mn}/d_{36} (KDP)
KDP	$\bar{4}2m$	0.69	d_{36}	1
		0.69	d_{14}	0.95 ± 0.06
		1.06	d_{36}	1
		1.06	d_{14}	1.01
ADP	$\bar{4}2m$	1.06	d_{36}	0.93 ± 0.06
		1.06	d_{14}	0.89 ± 0.04
$LiNbO_3$	$3m$	1.06	d_{22}	6.3 ± 0.6
		1.06	d_{31}	11.9 ± 1.7
CdS	$6mm$	1.06	d_{31}	35 ± 2
		1.06	d_{15}	32 ± 2
		1.06	d_{33}	63 ± 4
GaAs	$\bar{4}3m$	1.06	d_{14}	560 ± 140
		10.6	d_{14}	294 ± 100
Te	32	10.6	d_{11}	4230 ± 670

[a] International notation.

d_{36}(KDP) is $(9.5 \pm 0.6) \times 10^{-13}$ m/V for He–Ne-laser light at 0.6328 μm. From the numerical values we can see that the model of the motion of electrons in an anharmonic potential $kx^2/2 + k'x^3/3$ presented in Section 11.3.1 allows us to give a plausible explanation of the orders of the measured values.

In the present subsection we have so far discussed the difficulties arising from momentum mismatch (11.7c) and the methods to overcome them in the case of second-harmonic generation. These problems also appear in other areas of nonlinear optics. This is for example true of the interaction of three light waves, as can be seen from the relations (11.14) holding for sum- and difference-frequency generation. These equations reveal that the factor $e^{iz\Delta k}$—with Δk a nonvanishing wave number mismatch and z the propagation coordinate—prevents the complete conversion of the wave. This applies also to certain situations in Raman scattering, for instance in the amplification of the anti-Stokes wave, where the wave number difference is given by

$$\Delta k_\bullet = 2k_\bullet(\omega_{\text{las}}) - k_\bullet(\omega_{\text{anti-Stokes}}) - k_\bullet(\omega_{\text{Stokes}}).$$

This case will be discussed in detail in the next chapter [cf. (12.68)].

We have to remark that the relations discussed for Δk or $\Delta k_\bullet$, if expressed in terms of (field-independent) refractive indices of linear optics [cf., e.g., (11.9)], are only valid insofar, as the refractive indices are not strongly affected by nonlinear contributions. Otherwise the phase-matching conditions become field-dependent. This may happen in nonlinear optical effects at higher intensities (see Sections 5.8 and 5.9 in Ref. 7).

Up to now we have considered phase matching for beams of infinite extent, but in order to meet real conditions it is necessary to discuss the influence of a finite cross section of the beam. In such a case, even with exact phase matching $\Delta k_\bullet = 0$, there is only a limited region in which the waves overlap and thus are capable of mutual interaction. This *walkoff* of the waves results from double refraction, which causes the ordinary and extraordinary light beams to propagate in slightly different directions; as a result, the overlapping of the beams in the crystal is no longer sufficient after they have traveled a certain distance. This effect decreases if the angle θ_0 is close to 90°, for the two beams are parallel to one another if they are normal to the optical axis. The condition $\theta_0 = 90°$ can for example be satisfied for argon-laser light ($\lambda = 515$ nm) in KDP at a temperature of 259 K.

From (11.10) it can be easily seen that under simple conditions (i.e., phase matching, unaltered diameter of the laser beam throughout the interaction interval) the efficiency of SHG increases with decreasing area A, since the power of the second harmonic is inversely proportional to A for a given power of the fundamental wave. This statement, however, will only hold if the area A is not too small, since otherwise the beam will diverge due to diffraction. In practice, optimum focusing is needed. If the beam is focused too weakly, the concentration of the beams becomes too weak for the second harmonic to be generated efficiently. On the other hand, excessively strong focusing results in diminishing efficiency because of strong beam divergence and walkoff of the fundamental wave and the second harmonic.

The optimization of the crystal length in which phase matching is possible and of the focal-spot radius is a rather complex problem. In Ref. 11.6 physical conditions required for simplifying the problem and several methods needed for quantitative calculation are presented. A special result is that (for a uniaxial crystal at a fixed angle to the (x, y) plane) the knowledge of the crystal parameters will permit the unique solution of the optimization problem.

The conversion efficiency in SHG can be increased if a resonator is used for the generation of the second harmonic. In this resonant case the power gain is considerably higher ($\sim 10^2$) than in the nonresonant case; see e.g. Refs. 11.7, 11.8. In this connection the technique of internal frequency doubling may be mentioned, where the laser resonator contains the laser medium and a nonlinear crystal. The laser operates at the frequency ω. One of the mirrors is constructed to be transparent to the frequency 2ω, and there the power emerges that is converted from ω to 2ω in the nonlinear crystal; the power depends on loss factors and on the mode structure (see, e.g., Refs. 11.9, 11.10). By means of internal frequency doubling in a Nd : YAG laser with a Ba–Na–Nb crystal it was possible to generate monochromatic TEM_{00} beams at 0.53 μm of more than 1 W in the CW regime. We add that in Ref. 11.11 a connection is reported between noncollinear focused SHG and the registration of autocorrelation functions.

We also mention here that by using the results from (11.13) it has been experimentally established that the fundamental wave can be converted almost completely. This can be done with crystal lengths of $\gtrsim 10^{-2}$ m, at $\chi^{(2)} \sim 10^{-11}$ m/V, $\lambda = 1\ \mu$m, and with powers per unit area of the fundamental wave up to 10^{10} W/m^2.

11.3.3. Mode-Structure Effects

So far we have considered the amplification or generation of the second harmonic under the tacit assumption that the fundamental wave behaves as ideal radiation, that is, a wave with a single frequency and a Gaussian intensity profile in space. We now study effects that are due to deviations from these ideal assumptions.

Let us first discuss SHG by a *modelocked laser*. We suppose that its electric field E_{ml} is given in the form

$$E_{ml}(t) = \sum_{j=1}^{N} E_j e^{i(\omega + j\Delta\omega)t}, \tag{11.67}$$

where E_j is assumed to be the real amplitude of the jth longitudinal mode, so that all modes have the same phase. We may assume that all the N modes involved have the same amplitude (and intensity J_0 is set E_j^2). The output of the modelocked laser is a train of uniform pulses with $2\pi/\Delta\omega$ spacings and a pulse width of $2\pi/N\Delta\omega$. For what follows we assume exact phase match for all modes and neglect depletion of the pump wave.

According to (11.8) the field strength of the second harmonic is given by

$$(E_{2\omega})_{ml} = \eta \sum_{j=1}^{N} \sum_{j'=1}^{N} E_j E_{j'} e^{i[2\omega + (j+j')\Delta\omega]t}, \tag{11.68}$$

where η is the effective interaction constant. The intensity $J_{2\omega}$ of the second harmonic is proportional to the square of $E_{2\omega}$. Its time average is

$$\overline{(J_{2\omega})_{ml}}^{t} = \eta^2 J_0^2 \sum_{j=1} \sum_{j'=1} \sum_{m=1} \sum_{m'=1} \delta(j + j' - m - m')$$

$$= \eta^2 J_0^2 N \frac{2N^2 + 1}{3}. \tag{11.69}$$

If we had used the same power for generating the second harmonic at a single frequency ω, the average intensity of the second harmonic would have been

$$\overline{(J_{2\omega})_{\mathrm{sf}}}^{t} = \eta^2 J_0^2 N^2. \tag{11.70}$$

A comparison thus yields

$$\overline{(J_{2\omega})_{ml}}^{t} = \overline{(J_{2\omega})_{\mathrm{sf}}}^{t} \left(\tfrac{2}{3} N + \frac{1}{3N} \right). \tag{11.71}$$

It is obvious that the field-strength function (11.67), which characterizes a pulse train and which exhibits a modelocked spectrum of bandwidth $N\Delta\omega$, exerts a very strong effect on the generation of the second harmonic. For a Nd:YAG laser $N = 10^2$ may be assumed as an effective number of modes with quasiconstant intensity; thus by mode locking a gain of about 60 can be achieved experimentally.

We now proceed to discuss another limiting case of the properties of the fundamental wave used for SHG. We assume a laser operating in the TEM_{00} mode whose field strength E_{rp} is again a sum of mode contributions with equal frequency distances and with each mode having the intensity J_0, but now the phases φ_j are randomly distributed. This leads to

$$E_{\mathrm{rp}} = \sum_{j=1}^{N} E_j e^{i[(\omega + j\Delta\omega)t + \varphi_j]}. \tag{11.72}$$

These properties are approximately exhibited by many types of gas lasers and solid-state lasers. For the generation of the second harmonic exact phase matching is again assumed. Then calculating the intensity of the second harmonic in analogy to the procedure employed above, we see that there appear an average phase-independent term $\overline{(J_{2\omega})_{rp}}^t$ and additive fluctuations around this value arising from noncanceling random terms. Relating $\overline{(J_{2\omega})_{rp}}^t$ to $\overline{(J_{2\omega})_{sf}}^t$, from (11.70) we now obtain

$$\overline{(J_{2\omega})_{rp}}^t = \overline{(J_{2\omega})_{sf}}^t \left(2 - \frac{1}{N}\right). \tag{11.73}$$

Ignoring the abovementioned intensity fluctuations, we obtain a gain of 2 if N takes large values.

It can thus be stated that the radiation sources discussed, which deviate from single-mode light, yield a higher efficiency for SHG than in the single-mode case. This can be accounted for by the fact that the radiation sources characterized by (11.67) and (11.72) display more pronounced deviations from the average value than does ideal single-mode light. Since the positive deviations (the intensity peaks) yield larger contributions to the nonlinear response than the negative ones, the generation rate is enhanced. An analogous statement was made for multiphoton absorption in the discussion of (10.4). Although in (10.4) the behavior of quantum states was considered, whereas here classically described radiation sources are concerned, in both cases the nonlinearity results in the same reasoning as to the enhancement of the response.

11.3.4. Up-Conversion and Third-Harmonic and Higher-Harmonic Generation

In (11.15a) it can be seen how one can amplify the field strength of a wave with the sum frequency $\omega_1 = \omega_2 + \omega_3$, and in (11.15b) the same for a wave with the difference frequency $\omega_2 = \omega_1 - \omega_3$. Quantitative estimates of the susceptibilities can be found in Section 11.2.1.

Usually, for the practical application of these processes to the generation of waves with new frequencies, the orders of magnitude of the powers of the generating waves are chosen to be equal; in the case of sum-frequency generation these are the powers at the frequencies ω_2 and ω_3. If a set of coherent radiation sources with fixed and tunable frequencies (e.g. lasers, parametric oscillators) is available, the formation of the sum frequency can result in coherent radiation sources covering a wide frequency range. It may be mentioned that in commercial applications there are systems that provide coherent radiation in the whole wavelength range between 250 nm and 3 μm by exploiting sum-frequency generation.

An important special application of sum-frequency generation is that where (deviating from the abovementioned conditions) the power of the wave at frequency ω_2 is assumed to be much greater than that of the wave at frequency

ω_3. Then in the set of equations (11.14) the derivative $d\bar{E}(\omega_2, z)/dz$ can be set equal to zero. With the assumption of exact phase matching and a nonvanishing initial amplitude at frequency ω_3 one obtains the z-dependent wave amplitudes

$$\bar{E}(\omega_1, z) = \left(\frac{\omega_1^2/k_1}{\omega_3^2/k_3}\right)^{1/2} \bar{E}(\omega_3, 0)\sin\frac{z\pi}{2L'}, \qquad (11.74a)$$

$$\bar{E}(\omega_3, z) = \bar{E}(\omega_3, 0)\cos\frac{z\pi}{2L'}, \qquad (11.74b)$$

with the length L' being proportional to $(\omega_1^2\omega_3^2/k_1k_2)^{-1/2}[\bar{E}(\omega_2, 0)]^{-1}$ (the polarization indices are ignored in these relations). At $z = L'$ the power of the wave with frequency ω_3 is completely transferred to the sum-frequency wave. Thus there is frequency conversion of radiation with high efficiency if a mixing crystal of appropriate length is used. Such a (frequency) up-converter can be used successfully for detecting infrared radiation. Because of the high efficiency of the photon detectors in the visible (and near infrared) region, the minimum number of photons per second that can be detected is much smaller than that of the photons in the infrared region. Therefore it is advantageous to mix in the mixing crystal an intense laser pump wave at frequency ω_2 with the infrared radiation to be detected and to measure the resultant up-converted radiation with a high-efficiency photomultiplier. For example, it is possible to mix infrared radiation at about 10 μm in proustite, a crystal transparent from the visible through to the infrared region, with the radiation of a ruby laser; the radiation obtained can be measured with extremely high detectivity in the visible region. It should be noted that according to (11.74a) the amplitude of the up-converted radiation exhibits a maximum at the crystal length $L = L'$ and is proportional to the *amplitude* of the radiation to be measured, so that the coherence properties, too, are transferred. Since L' is inversely proportional to $\bar{E}(\omega_2, 0)$, this length can be significantly affected by the intensity of the laser pump.

Let us now discuss phenomena connected with nonvanishing momentum mismatch. The wave number mismatch Δk is associated with dispersion. This can be inferred explicitly from the relations (11.9) and (11.9a) for SHG, and it holds analogously for sum-frequency generation. The exploitation of this relationship was proposed for *nonlinear spectroscopy* and the design of nonlinear monochromators in Ref. 11.2. Through a system of mirrors and lenses the light of a laser (frequency ω_L) together with the light to be investigated (with the frequency components $\omega_1, \omega_2, \ldots$) is sent into the nonlinear crystal. There waves with the sum frequencies $\omega_L + \omega_1, \omega_L + \omega_2, \ldots$ are generated. Optimum generation of these sum frequencies is achieved in those directions in which the phase-matching conditions are satisfied best in each case. Thus spreading of the angles of the emerging sum-frequency radiation results, according to the frequency components in the radiation to be investigated. This effect can be exploited in nonlinear spectroscopy.

The generation of the *third harmonic* is described with the help of the susceptibility function $\chi^{(3)}_{ijkl}(3\omega;\ \omega, \omega, \omega)$. Its connection with parameters of the atomic systems and a numerical estimate were given in (11.26). Terhune, Maker, and Savage (Ref. 11.13) gave a point-group representation of the fourth-rank tensor. They were among the first scientists to succeed in observing the third-harmonic generation. It was observed in (isotropic) liquids, metal vapors, and cubic crystals. By dissolving a dye of appropriate concentration in a liquid, the absorption and thus the dispersion can be regulated so that $n(3\omega) - n(\omega) = 0$, and thus $\Delta k = k(3\omega) - 3k(\omega)$ becomes zero. An analogous method has been successfully applied to third-harmonic generation in metal vapors in which narrow, but strong absorption lines are situated between ω and 3ω. Here exact phase matching is achieved by adding to the vapor an absorption-free buffer gas, for instance, a noble gas of suitable pressure. Another advantage of this method is the possibility of extending higher-harmonic generation to the far ultraviolet region.

In liquids and crystals the maximum photon energy achievable in frequency conversion is determined by the first electronic absorption band because of its large spectral range. On the other hand, metal vapors (for instance alkali vapors) have very narrow absorption bands, which are separated by loss-free regions extending up to the VUV region. Hence with a properly chosen fundamental frequency ω, it is possible to generate higher harmonics at the frequencies $3\omega, 5\omega, \ldots$ in loss-free regions. By varying the pressure of the buffer gas, phase matching and thus maximization of the conversion efficiency of a certain higher harmonic can be achieved. In this way coherent radiation with a wavelength as short as 38 nm has been obtained (Refs. 11.14, 11.15). In calcite there are two directions in which momentum matching may happen. At 47° from the trigonal axis the relation $3k_o(\omega) = k_o(3\omega)$ holds for the ordinary beam (index o). At the direction of 57° a relation appears that contains both momentum terms of the ordinary and of the extraordinary beam (index eo) namely $2k_o(\omega) + k_{eo}(\omega) = k_o(3\omega)$. The intensity observed at 3ω is found to be proportional to the cube of the intensity of the fundamental wave.

Wave-mixing processes, such as sum- and difference-frequency mixing, have been applied to the generation of phase-conjugate waves, too; this will be discussed in Section 14.2.2.

11.3.5. Parametric Amplification, Oscillation, and Fluorescence

In Section 11.1.2.2 we derived fundamental relations for the parametric amplification of the field-strength amplitudes of signal and idler wave under the influence of an intense laser pump wave. The parametric amplification and generation of the signal and idler photons resulting from the spontaneous decay of pump photons was described in (11.43).

From (11.19a) and (11.19b) it can be seen that the amplification of the amplitudes of signal and idler waves increases with increasing path length z at fixed amplitude of the pump wave. The path length can effectively be increased by allowing the beams to pass through the mixing crystal repeatedly. When upon a first amplification and with proper phase the signal and the idler repeat their passage through the crystal, amplification again occurs. Thus we can see that by adding an appropriate feedback to the amplifying system, it can be made to operate as a *parametric oscillator*. To achieve this, the mixing crystal is placed in a resonator whose mirrors are highly transparent to the pump wave (reflection coefficient of the resonator plates $r_P \approx 0$) while they are scarcely transparent to the frequencies of the signal wave (frequency ω_S) and the idler wave (frequency ω_I), so that $r_S, r_I \approx 1$. The threshold condition for the stationary generation of a signal wave is satisfied if on a round trip through the resonator the losses of the wave are compensated for by its amplification. These losses can be thought of as concentrated at the two mirrors and taken into account by the effective reflective coefficients $\bar{r}_1$, $\bar{r}_2$, where $\bar{r}_1\bar{r}_2 = \bar{r}^2$ is real. For a round trip of length $2l$ the threshold condition is then

$$e^{2\eta' l}\bar{r}^2 = 1, \tag{11.75a}$$

where l is the length of the crystal (resonator) and η' the amplification coefficient introduced in (11.20). Assuming weak losses for the signal and the idler wave, this means $\bar{r}^2 \approx 1$, the relation (11.75a) leads to the threshold condition

$$2\eta' l \geq 1 - \bar{r}^2 \tag{11.75b}$$

by taking the logarithm.

In a given arrangement of the crystal and the resonator the relation (11.75b) is a condition for the strength of the incident pump wave, since according to (11.17) and (11.20) only for a sufficiently strong pump wave can a sufficiently large amplification coefficient η' be obtained. When the pump amplitude is raised, the oscillator starts oscillating at the first frequencies for which the condition (11.75b) becomes satisfied. Provided that the dependence of the reflection coefficients on the frequency is negligible, the oscillations start at those frequencies that exhibit phase matching for a fixed specific crystal orientation, as can be seen from (11.20).

Giordmaine and Miller (Ref. 11.16) were the first to succeed in detecting parametric oscillation. As a pump wave the second harmonic of a Q-switched $Nd{:}CaWO_4$ laser was used, which had been generated in a lithium niobate crystal (parameters of the pump wave: wavelength 529 nm, power about 7 kW, pulse duration some nanoseconds). The mixing crystal was also lithium niobate; it was used in a plane-parallel resonator. The required threshold power was high, so that pulsed pump sources were needed.

Parametric oscillation was later achieved also with cw-lasers (Ref. 11.17). To this end, instead of lithium niobate the crystal barium sodium niobate was used (because of its higher nonlinearity coefficient), and a confocal resonator with a high Q value was employed.

The parametric oscillator is of great practical importance, since it can be used for establishing intense narrowband and tunable radiation sources. The frequency ratio ω_S/ω_I for which $\Delta k_\bullet = 0$ (and for which the oscillation threshold is achieved) can be easily altered by rotating the nonlinear birefringent mixing crystal, by varying its temperature, or by applying an electric d.c. field to it. Thus, by thermally varying the refractive index of lithium niobate, a tuning range from 0.7 to 2 μm could be achieved for a parametric oscillator (Ref. 11.16). Angular change and temperature change permitted in lithium niobate a tuning of the signal wave from about 0.5 to 3 μm. Having chosen a suitable material, in a cw regime with an input power of $\sim$ 10 W, several tenths of the input power could be recovered as signal power.

The radiation called *parametric fluorescence* can be observed on irradiating a thin crystal slice (with negligible reflection at its surface) with pump light. We are interested in the numbers of signal and idler photons emitted per unit time, unit frequency, and unit solid angle as a result of the excitation by the pump wave (Ref. 11.18).

The explanation of parametric fluorescence requires the extension of the Hamiltonian of (11.34). Instead of one signal mode and one idler mode, we now have to take into consideration, according to the experimental conditions, a variety of signal modes (denoted by S_l) and idler modes (denoted by $I_{l'}$), where the mode indices l and l' vary. Consequently the basic interaction operator is now

$$\hat{H}^C_{PF} = -\hbar \sum_{l,l'} \eta_{PA}(S_l, I_{l'})$$

$$\cdot \left\{ \frac{1}{V} \int_V d^3 r_\bullet \exp[i(k_{P\bullet} - k_{Sl\bullet} - k_{Il'\bullet}) r_\bullet] \right\} \hat{a}_P \hat{a}^\dagger_{S_l} \hat{a}^\dagger_{I_{l'}} + \text{h.c.}. \quad (11.76)$$

The pump wave is represented by one single mode. For the signal and idler photons the initial state must be chosen to be the vacuum state. The calculations provide the following results, which are in agreement with experiment. The signal radiation that is associated with the (approximate) fulfilment of the phase-matching condition displays a relatively narrow line in the frequency spectrum. Its midfrequency depends on the direction of observation. If the deviation of the angle from the phase-matching condition exceeds a certain limiting value, directed emission will no longer occur. As a rule, radiation processes that are associated with a phase mismatch provide noticeably weaker radiation, uniformly distributed over all angles.

The methods presented in the preceding sections of this chapter can be used analogously for the description of parametric processes of higher orders, for instance of those with four interacting waves (cf. Ref. 11.19 and Section 14.2.2).

11.4. COHERENCE PROPERTIES

From multiphoton absorption and emission, where the alteration of the radiation field is associated with the transition between real levels of the atomic systems, we have already seen that there exists a mutual relationship between the nonlinear optical effect and the coherence properties of light (Section 10.6). This is also true of parametric amplification and frequency conversion, which may be considered under the assumption that, although annihilation and creation of photons take place, no change of the population of the atomic states results.

To describe coherence problems of parametric amplification and frequency conversion in the lowest order, we may start from an interaction operator of the form (11.34). We write

$$\hat{H}^C = -\hbar\chi'\left\{\frac{1}{V}\int_V d^3r_\bullet\, e^{i(k_{1\bullet}-k_{2\bullet}-k_{3\bullet})r_\bullet}\right\}\hat{a}_1\hat{a}_2^\dagger\hat{a}_3^\dagger + \text{h.c.} \tag{11.77}$$

where χ' is substituted for η_{PA} in (11.34); its dependence on the second-order susceptibility, the frequencies, and the volume is the same. If the momentum-matching condition is satisfied, as will be assumed in the subsequent treatment, the expression in braces is equal to unity. Energy conversion is reflected by the relation $\omega_1 = \omega_2 + \omega_3$.

Using this expression, all the effects of the preceding sections can be covered by the following specializations. In the case of *ordinary parametric amplification* modes 2 and 3 are identified with the signal and the idler mode and mode 1 is identified with the pump mode. In the case of *degenerate parametric amplification* the signal mode equals the idler mode. The second term on the right-hand side of (11.77), containing the product $\hat{a}_1^\dagger\hat{a}_2\hat{a}_3$, represents *frequency conversion* in that a wave with the sum frequency ω_1 is formed. In case the two modes 2 and 3 are identical, *second-harmonic generation* is represented.

Besides the interaction operator (11.77), which only describes the dynamical system, in a general treatment the influence of dissipative systems has to be considered. In analogy to the description of the lasing process [cf. (7.51)], the formation of the total-interaction part of the Hamiltonian $\hat{H}$ requires the addition of a term

$$\hat{H}^R = \sum_{l=1}^{3}\sum_{\mu}\xi_{l\mu}\hat{a}_l^\dagger\hat{R}_{l\mu} + \text{h.c.} \tag{11.78}$$

to $\hat{H}^C$. This term represents the interaction of the radiation modes 1, 2, 3 with the particular dissipative system. Let each of the three modes be coupled to an independent reservoir that may be assumed to be a boson system with a broadband quasicontinuous spectrum. In (11.78) $\hat{R}_{l\mu}$ is the annihilation operator of the μth reservoir mode that is coupled to the radiation mode l; $\zeta_{l\mu}$ is the corresponding interaction parameter. It may be mentioned that the form of the operator $\hat{H}^R$ guarantees that the absorption/emission of one photon of mode l is accompanied by the emission/absorption of one reservoir quantum, the photon energy being approximately equal to the energy of the reservoir quantum because of the quasiresonance condition.

From the total Hamiltonian the equations of motion for the photon annihilation operators (in the Heisenberg picture) can be derived:

$$\frac{d\hat{a}_1(t)}{dt} = -i\omega_1\hat{a}_1(t) + i\chi'\hat{a}_2(t)\hat{a}_3(t) - \frac{\gamma_1}{2}\hat{a}_1(t) + \hat{\Gamma}_1(t),$$

$$\frac{d\hat{a}_2(t)}{dt} = -i\omega_2\hat{a}_2(t) + i\chi'\hat{a}_1(t)\hat{a}_3^\dagger(t) - \frac{\gamma_2}{2}\hat{a}_2(t) + \hat{\Gamma}_2(t),$$

$$\frac{d\hat{a}_3(t)}{dt} = -i\omega_3\hat{a}_3(t) + i\chi'\hat{a}_1(t)\hat{a}_2^\dagger(t) - \frac{\gamma_3}{2}\hat{a}_3(t) + \hat{\Gamma}_3(t). \quad (11.79)$$

While the second terms on the right-hand sides describe the interaction within the dynamical system, the two last terms represent the interaction with the dissipative system; the situation and the structure of the relations are analogous to those given in (7.55) for the lasing process. The coefficients $\gamma_1, \gamma_2, \gamma_3$ are (positive) damping constants. $\hat{\Gamma}_l(t)$ is the fluctuation force assigned to mode l. The nonvanishing correlators of the fluctuation forces have a delta-function-like time dependence; the strength of their correlators may be taken as being independent of the field operators in a certain approximation (Ref. 11.20).

Let us now sketch some general considerations for solving the basic relations. The problems can be formulated as equations of motion for the annihilation and creation operators or as equations for the P-representation and its Fourier transform, respectively (a similar procedure was applied, for instance, to the explanation of laser operation in Sections 7.2.1 and 7.2.2). However, in general the solutions of these basic equations cannot be obtained in a compact form; only under certain assumptions or special physical conditions may one succeed in doing so. As a rule one has to resort to numerical calculations and/or approximations. Peřina and Peřinova (Ref. 11.21) showed that recursion procedures leading to solutions of the characteristic functions for any time can be applied.

In a relatively simple way important information can be gained from employing the so-called *short-time approximation*. For this purpose the characteristic functions are expanded in powers of the interaction time t, where it is assumed that the interaction is switched on at $t = 0$. The coefficients of the powers of t have a physical meaning and depend on the initial coherence conditions, the intensity of the incident waves, and the interaction strength. The time dependence of the density operator was evaluated for various conditions in Ref. 11.22.

Up to now we have discussed only the cavity problem. But, real experimental situations are mostly related to propagation problems. Shen (Ref. 10.32) pointed out that quasistationary propagation problems can be treated in the same way as cavity problems if the time coordinate t is replaced by z/u, where u is the (group) velocity of light in the medium under study and z is the propagation length. When separating the fast main time dependence $\exp(-i\omega_l t)$ from the operators $\hat{a}_l(t)$ and thus passing over to the slowly varying operators $\hat{c}_l$, the replacement of t by z yields differential equations for $\hat{c}_l(z)$ that correspond to the classical amplitude equations as given in Section 1.1. In analogy to the abovementioned short-time approximation the *short-path approximation* can be applied to the equations for the operators $\hat{c}_l(z)$. In Ref. 11.23 the convergence behavior of the z expansion for special situations was discussed.

In the following subsections some examples concerning the interrelationship between coherence and frequency conversion or parametric amplification will be discussed. In Section 11.4.1 we are going to consider the coherence behavior under the assumption of an intense pump field whose depletion can be neglected. In Section 11.4.2 we drop this simplifying assumption and consider general problems.

11.4.1. Coherence Behavior with Unaltered Pump Field

In the present subsection the depletion of the pump wave will be neglected. First we recall the case of ordinary parametric amplification, discussed in Section 11.2.2, in which the pump laser could be described by a function with prescribed time dependence, that is, with a fixed magnitude of its amplitude and a predicted phase function $\phi_P(t)$ [cf. (11.36)]. From (11.43) it became apparent how the coherence properties entered the expression for the number of the signal photons generated at time t. The time dependence of the signal photon annihilation operator is explicitly given in (11.41). From $\underset{\sim}{\hat{a}}_S(t)$ and $\hat{a}_S(t)$, respectively, the normalized second-order correlation function $\gamma^{2,2}$ of the signal wave can be evaluated for large t values also. Assuming a coherent initial signal state and a vacuum state for the idler wave, the function tends to 1 at large initial numbers of signal photons and 2 at small initial numbers. The meaning of these values becomes clear from the discussion of (5.47) and (5.48). In the first case the resulting light becomes coherent; in the latter it becomes incoherent. It should be mentioned that a detailed discussion of the case $\langle \hat{N}_S(0) \rangle = \langle \hat{N}_I(0) \rangle = 0$ is given in Ref. 11.24, and that in Ref. 11.25 a method is presented that yields the density matrix by means of the unitary time-development operator for arbitrary initial conditions of the signal and idler states.

In Section 5.2.2 attention was drawn to the fact that antibunched radiation may exist. The following consideration will demonstrate that also for *degenerate parametric amplification* (in which the operators $\hat{a}_S$ of the signal and $\hat{a}_I$ of the idler are identical) the possibility of antibunching can be predicted. The statistical behavior of degenerate parametric amplification is essentially different from that of ordinary parametric amplification. Stoler (Ref. 11.26) established the equation for the signal operator $\hat{a}_S(t)$ under the assumption of an initial coherent state for the signal and an ideal laser pump wave with the temporally constant phase φ_P; in doing this, the quantity ν from (11.36) enters the equation as a real constant proportional to the pump amplitude and the interaction constant. From the solution of the equation of motion for $\hat{a}_S(t)$ the reduced variance can be calculated, which results in the form

$$\frac{\langle [\Delta \hat{N}_S(t)]^2 \rangle}{\langle \hat{N}_S(t) \rangle} = A(\nu t) + B(\nu t)\sin(2\varphi_S - \varphi_P), \tag{11.80}$$

where $A(\nu t)$ and $B(\nu t)$ are nonnegative functions. At $t = 0$ the reduced variance is unity. Antibunching occurs if the variance falls below unity. Only if $\sin(2\varphi_S - \varphi_P) < 0$ is this result possible.

The time behavior of the reduced variance is illustrated in Fig. 11.8; we have chosen the optimum phase relation $2\varphi_S - \varphi_P = -\pi/2$ and an initial number of signal photons that exceeds 10. The curve has its minimum value at $\nu t \approx 0.35$ and again rises to unity at $\nu t \approx \ln\sqrt{2\langle \hat{N}_S(0)\rangle^{1/3}}$. The initial coherent state, however, is no longer attained; this becomes apparent from the values of the higher factorial moments. For larger νt the reduced variance remains greater than unity.

Let us imagine the following experimental realization. An intense cw laser generates second-harmonic light. Both the fundamental wave and the second-harmonic wave are guided into the nonlinear crystal. There the second-harmonic wave acts as pump wave, and the fundamental wave acts as signal. In using (11.80) for our description we have to substitute $\nu' z$ for νt, where z is the interaction length and ν' an intensity-dependent material parameter. For barium niobate, which may be regarded as a suitable material, we have $\nu' \approx 9 \times 10^{-4}(I_P/\mathrm{W})^{1/2}$. Thus, with a pump intensity I_P of 10^8 W m^{-2} the optimum condition $\nu' z \approx 0.35$ for antibunching would be reached at $z \approx 4$ cm.

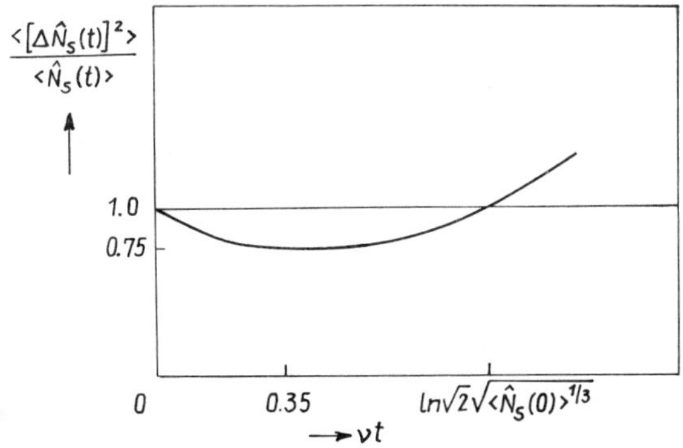

Fig. 11.8. Reduced variance of the signal photons versus time in the case of degenerate parametric amplification.

The advantage of the antibunching realization suggested above is that the mean photon number is high compared with other effects under discussion (e.g. two-photon absorption or resonance fluorescence). However, the observability of antibunching crucially depends on the stability of the relation between the phases φ_S and φ_P; in real radiation sources, in principle, phase fluctuations occur.

The pump source we have considered so far possessed a fixed magnitude and phase of the complex amplitude, as is the case with ideal laser light. We now turn to a real pump laser whose phase exhibits a statistical dependence on time. Details of this case are discussed in Section 7.3.1. The time derivative $f \equiv d\varphi_P/dt$ of the pump phase will be interpreted as a random variable (Ref. 11.27).

We first discuss the parametric amplification. As can be seen from (11.36) and (11.41), the signal photon operator $\hat{a}_S(t)$ contains the laser pump phase $\varphi_P(t)$; a corresponding statement can be made for the quantum-theoretical expectation values that are calculated by means of $\hat{a}_S(t)$. The operators and expectation values have to be averaged with respect to the random variable f. If there occur only relatively slow phase variations (r.m.s. deviations of f much less than $|\nu|$), the gain rate $2|\nu|$, as given for an ideal laser in (11.43), must be replaced by the expression $2|\nu|[1 - (f^2)_{\text{av}}/2|\nu|^2]$. This means that a nonzero mean-square deviation $(f^2)_{\text{av}}$ diminishes the gain rate of parametric amplification.

Let us now pass to the case of frequency up-conversion with a real laser pump wave. The conditions to be satisfied for its generation were discussed in Section 11.3.4. Since $\omega_1 = \omega_2 + \omega_3$, we have to regard the wave with index 1 as the signal wave and the wave with index 2 as the pump wave; the wave with index 3 can be termed the idler wave. Here, too, slow phase variations $\varphi_P(t)$ of the pump wave due to statistical fluctuations are assumed, so that $(f^2)_{\text{av}} = (d\varphi_P/dt)^2_{\text{av}} \ll |\nu|^2$. In Fig. 11.9 the time behavior of $\langle \hat{N}_S(t)\rangle_{\text{av}}$ is schematically illustrated following Ref. 11.27. The solid line represents the case of fully coherent pump radiation; obviously a complete energy exchange between the signal and the idler takes place periodically. The dashed line represents the case of slowly varying phase; asymptotically the energy exchange becomes negligible and the average number of signal photons becomes equal to that of idler photons.

11.4.2. General Treatment

The purpose of the present subsection is to consider the coherence problems under more general conditions, in particular by taking into account the depletion of the pump wave.

Second-harmonic generation is a special case of sum-frequency formation; the energy-conservation relation $\omega_1 = \omega_2 + \omega_3$ changes into $\omega_1 = 2\omega_2$, where the indices 1 and 2 characterize the second harmonic (index S) and the

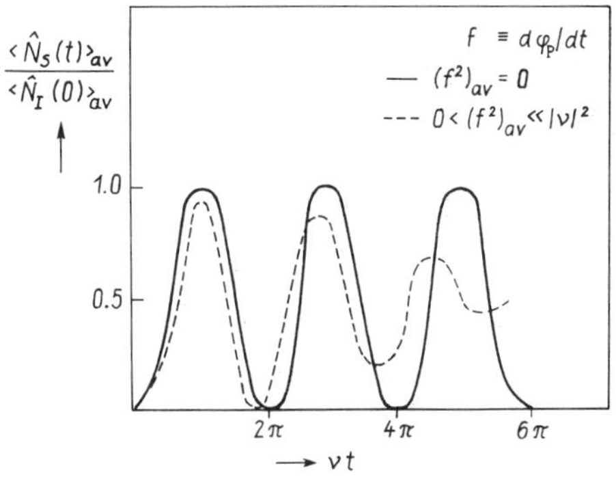

Fig. 11.9. Influence of phase fluctuations on the time evolution of the averaged signal-photon number in the case of frequency up-conversion (after Ref. 11.27).

fundamental wave (index P), respectively. Using the abovementioned spatially slowly varying operators $\hat{c}$ instead of $\hat{a}$, one obtains the following equations

$$\frac{d}{dz}\hat{c}_S(z) = i\varkappa\hat{c}_P^2(z), \tag{11.81a}$$

$$\frac{d}{dz}\hat{c}_P(z) = 2i\varkappa^*\hat{c}_P^\dagger(z)\hat{c}_S(z), \tag{11.81b}$$

which correspond to the classical amplitude equations. The quantity $\varkappa$ is proportional to the interaction parameter χ' in (11.79).

In contrast to the previous treatment, we now take into account the alteration of the laser pump (fundamental) wave and seek simultaneous solutions of the two differential equations. Let us assume that at $z = 0$ only photons of the fundamental wave exist, whereas $\langle\hat{c}_S(0)\rangle$ and $\langle\hat{c}_S^+(0)\hat{c}_S(0)\rangle$ are zero. With the solutions for $\hat{c}_S(z)$ and $\hat{c}_P(z)$ the field correlation functions

$$\Gamma_{P,S}^{l,l}(0; z) = \left(\frac{\hbar\omega_{P,S}}{2\varepsilon_0 V}\right)^l \left\langle \left[\hat{c}_{P,S}^\dagger(z)\right]^l \left[\hat{c}_{P,S}(z)\right]^l \right\rangle \tag{11.82}$$

can be calculated. The argument 0 in $\Gamma^{l,l}$ indicates that the space and time coordinates of all $2l$ operators are the same. But it turns out that through the z dependence of the operators $\hat{c}^\dagger(z)$ and $\hat{c}(z)$ the correlation functions $\Gamma^{l,l}(0; z)$ become z-dependent. From Section 5.2.2 it follows that the bunching excess

$$\Delta^2\Gamma_{P,S}(z) \equiv \Gamma_{P,S}^{2,2}(0; z) - \left[\Gamma_{P,S}^{1,1}(0; z)\right]^2 \tag{11.83}$$

allows us to draw conclusions as to the occurrence of bunching or antibunching. $\Delta^2\Gamma(z) > 0$ indicates bunching, whereas $\Delta^2\Gamma(z) < 0$ indicates antibunching. In accordance with the short-path approximation sketched at the beginning of this section, series for $\hat{c}_P(z)$ and $\hat{c}_S(z)$ in powers of z are formed by means of which the correlation functions can be obtained as expansions in z (Ref. 11.28). For the fundamental wave one has

$$\Delta^2\Gamma_P(z) = -\left(\frac{\hbar\omega_P}{2\varepsilon_0 V}\right)^2 2|\varkappa|^2\langle \hat{N}_P(0)\rangle^2 z^2 + \cdots \tag{11.84a}$$

if the incident light is coherent. If there is incident chaotic light one obtains

$$\Delta^2\Gamma_P(z) = \left(\frac{\hbar\omega_P}{2\varepsilon_0 V}\right)^2 \langle \hat{N}_P(0)\rangle^2\left[1 - 4|\varkappa|^2\{4\langle \hat{N}_P(0)\rangle + 1\}z^2 + \cdots\right]. \tag{11.84b}$$

The sign on the right-hand side of (11.84a) implies an antibunching behavior for positive z values. The right-hand side of (11.84b) of course reveals a bunching effect at $z = 0$. However, with increasing z this bunching effect diminishes. The corresponding relations for the second-harmonic waves are given by

$$\Delta^2\Gamma_S(z) = -\left(\frac{\hbar\omega_S}{2\varepsilon_0 V}\right)^2 \tfrac{8}{3}|\varkappa|^6\langle \hat{N}_P(0)\rangle^4 z^6 + \cdots \tag{11.85a}$$

for an incident coherent wave, and by

$$\Delta^2\Gamma_S(z) = \left(\frac{\hbar\omega_S}{2\varepsilon_0 V}\right)^2 20|\varkappa|^4\langle \hat{N}_P(0)\rangle^4 z^4 \times\left[1 - \tfrac{4}{15}|\varkappa|^2\{54\langle \hat{N}_P(0)\rangle + 17\}z^2 + \cdots\right] \tag{11.85b}$$

for an incident chaotic wave. Recalling that the number of second-harmonic photons is assumed to be zero at $z = 0$, the relations (11.85a) and (11.85b) express qualitatively the same behavior as was found for the fundamental wave, namely, a tendency towards antibunching and towards diminishing of bunching with increasing z. In a way similar to that employed here for describing second-harmonic generation, other cases, such as sum-frequency generation with two different subharmonics, have also been treated.

As an example of the research on spatial coherence properties of second-harmonic generation, the theoretical and experimental investigations in Ref. 11.29 may be mentioned. By measuring Michelson's visibility [cf. (5.40)], the modulus of the normalized mutual coherence function $|\gamma^{1,1}(x_1, x_2)|$ was determined. For KDP a strong influence of the orientation of the crystal was

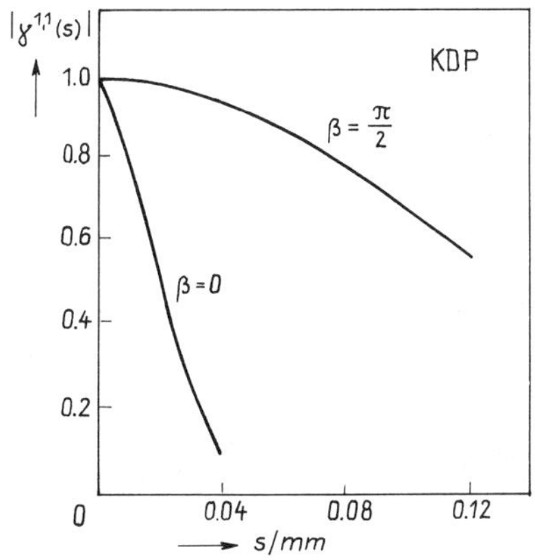

Fig. 11.10. Measured normalized correlation function $|\gamma^{1,1}(x_1, x_2)|_{KDP}$ as a function of the path difference s for the angles $\beta = 0, \pi/2$ (after Ref. 11.29).

found. Figure 11.10 gives $|\gamma^{1,1}(x_1, x_2)|$ as a function of the path difference s for different angles β with respect to the axis Y.

Finally we want to remark that squeezing [cf. (10.78b)] is predicted in the process of second-harmonic generation [cf. Refs. 11.30 and 10.74].

REFERENCES

11.1. P. A. Franken, A. E. Hill, C. W. Peters, and G. Weinreich, *Phys. Rev. Lett.* **7**, 118 (1961).

11.2. J. A. Armstrong, N. Bloembergen, J. Ducuing, and P. S. Pershan, *Phys. Rev.* **127**, 1918 (1962).

11.3. N. Bloembergen, *Non-linear Optics*, W. A. Benjamin, Inc., New York, Amsterdam, 1965, Section 4.2.

11.4. M. Schubert, in *Proceedings of the 3rd International Conference on Lasers and Their Applications*, Dresden, 1977, p. 19.

11.5. P. D. Maker, R. W. Terhune, M. Nisenhoff, and C. M. Savage, *Phys. Rev. Lett.* **8**, 21 (1962).

11.6. G. D. Boyd and D. A. Kleinman, *J. Appl. Phys.* **39**, 3597 (1968).

11.7. A. Ashkin, G. D. Boyd, and J. M. Dziedzic, *IEEE J. Quant. Electron.* **QE2**, 109 (1966).

11.8. M. Brieger, H. Büsener, A. Hese, F. v. Moers, and A. Renn, *Opt. Comm.* **38**, 423 (1981).

11.9. R. G. Smith, *IEEE J. Quant. Electron.* **QE6**, 215 (1970).

11.10. R. G. Smith, *IEEE J. Quant. Electron.* **QE7**, 150 (1971).

11.11. R. Fischer and C. Rempel, in *Proceedings of the International Conference "Lasers '82,"* New Orleans, 1982, p. 427.

11.12. A. G. Akhmanov, S. A. Akhmanov, R. V. Khokhlov, A. I. Kovrigin, A. S. Piskarskas, and A. P. Sukhorukov, *IEEE J. Quant. Electron.* **QE4**, 828 (1968).

11.13. R. W. Terhune, P. Maker, and C. M. Savage, in P. Grivet and N. Bloembergen (Eds.), *Proceedings of the 3rd Conference on Quantum Electronics, Paris 1963*, Columbia University Press, New York, 1964.

11.14. D. C. Hanna, M. A. Yuratich, and D. Cotter, *Nonlinear Optics of Free Atoms and Molecules*, Springer, Berlin (West), Heidelberg, New York, 1979.

11.15. H. Pummer, T. Srinivasan, H. Egger, K. Boyer, T. S. Luk, and C. K. Rhodes, *Opt. Lett.* **7**, 93 (1982).

11.16. J. A. Giordmaine and R. C. Miller, *Phys. Rev. Lett.* **14**, 975 (1965).

11.17. R. G. Smith, J. E. Geusic, H. J. Levinstein, J. J. Rubin, S. Singh, and L. G. van Uitert, *Appl. Phys. Lett.* **12**, 308 (1968).

11.18. D. A. Kleinman, *Phys. Rev.* **174**, 1027 (1968).

11.19. A. Penzkofer and W. Kaiser, *Opt. and Quant. Electron.* **9**, 315 (1977).

11.20. H. E. Ponath and M. Schubert, *Ann. d. Phys.* **33**, 1413 (1976).

11.21. V. Peřinova and J. Peřina, *Czech. J. Phys.* **B28**, 306 (1978).

11.22. G. P. Agrawal and C. L. Mehta, *J. Phys.* **A7**, 607 (1974).

11.23. M. Kozierowski, R. Tanas, and S. Kielich, in *Proceedings of the 8th Conference on Quantum Electronics and Nonlinear Optics* (Poznań, Poland), Vol. B, 1978, p. 126.

11.24. P. Chmela, *Acta Phys. Pol.* **A52**, 835 (1977).

11.25. V. T. Trung and F. J. Schütte, *Ann. d. Phys.* **34**, 262 (1977).

11.26. D. Stoler, *Phys. Rev. A* **33**, 1397 (1974).

11.27. B. Crosignani, P. DiPorto, U. Ganiel, S. Solimeno, and A. Yariv, *IEEE J. Quant. Electron.* **QE-8**, 731 (1972).

11.28. M. Kozierowski and R. Tanas, *Opt. Comm.* **21**, 229 (1977).

11.29. S. A. Akhmanov, Yu. D. Goljaev, V. G. Tunkin, and A. S. Chirkin, *Kvant. Elektron.* **2**, 1171 (1975).

11.30. C. K. Hong and L. Mandel, *Phys. Rev. Lett.* **54**, 323 (1985).

12

Stimulated Raman Scattering

Experimenting with Q-switched lasers, Woodbury and Ng (Ref. 12.1) discovered in 1962 that in addition to the normal spectral content of the laser radiation (as described in Section 7.3), under certain conditions frequency-shifted radiation may occur. It can be observed if the laser cavity contains certain materials; the frequency shifts are identical with molecular vibrational frequencies of the material or with its integral multiples. This phenomenon is connected with inelastic scattering of light by molecules, discovered by Raman in 1928 in his experiments on scattering by liquids. The latter effect is known as the *spontaneous Raman effect* and occurs at low excitation intensity.

Starting from the investigations by Woodbury and Ng, a lot of systematic studies were performed in which certain types of materials—both microphysically ordered and disordered systems—were exposed to *intense* exciting (laser) radiation. The resulting scattered radiation exhibited properties essentially different from those of the spontaneous Raman effect. This scattering is called the *stimulated* or *induced Raman effect*.

Essential differences between the stimulated and the spontaneous Raman effect become apparent with respect to interference: in the stimulated Raman effect interference of the radiation occurs; in the spontaneous Raman effect it does not. Another difference concerns the relation between the intensities of scattered and exciting radiation. While the scattering intensity of the spontaneous Raman effect is proportional to the excitation intensity, that of the stimulated Raman effect deviates essentially from this proportionality, so that the latter obviously must be regarded as a nonlinear optical phenomenon.

Stimulated Raman scattering by polarizable molecules was not only experimentally verified prior to such scattering by other groups of substances, but it may also be regarded as a guideline for the systematic description of the Raman effect. This vibrational Raman scattering by molecules will be dealt with in Section 12.1. In spite of the abovementioned differences of their behavior, the spontaneous and the stimulated Raman effect can both be explained on the basis of a uniform model of the interaction of radiation with matter. When explaining the generation of second and higher harmonics (Section 11.3.1) and related effects it was necessary to treat the behavior of outer-shell electrons and of bonding electrons that were affected by potentials of immovable atomic cores or atomic nuclei; these potentials were assumed to be constant with respect to time. In the present case we have to consider the interaction of radiation with outer-shell electrons by taking into account the nuclear motion (the oscillatory motion of molecular and lattice elements). A straightforward extension of the successful classical polarizability theory of Placzek used for explaining the spontaneous Raman effect enables us to describe also the nonlinear stimulated Raman effect under a wide range of conditions. The quantum-theoretical consideration is particularly needed for explaining the interference capability of scattered light originating from fluctuations. Furthermore we shall discuss amplification of waves in a Raman-active medium and such important applications as active Raman scattering.

Section 12.2 is devoted to Raman scattering of light by phonons and polaritons. This enables us to make qualitative and quantitative statements on interaction mechanisms between (genuine) elementary particles (photons) and quasiparticles (phonons, polaritons). It provides insight into optically excited wave propagation processes in crystals, and simultaneously it allows us to determine optoelectronic crystal parameters in regions that are not accessible by other procedures. As for specific spectroscopic methods the utilization of material properties is of importance for tunable radiation sources.

In Section 12.3 the scattering of electromagnetic waves by acoustic waves in condensed matter will be treated. By means of this stimulated Brillouin scattering, some nonlinear optical processes can be clarified that are connected with the propagation of hypersonic waves of high intensity.

The investigation of radiation-excited spin-flip processes of free carriers in semiconductors, to be described in Section 12.4, yields a deeper insight into general properties of semiconductors as well as knowledge about applications of practical interest as spin-flip lasers.

12.1. STIMULATED RAMAN SCATTERING BY POLARIZABLE MOLECULES

Placzek's classical polarizability theory is capable of explaining from an uniform basic concept, the interaction mechanisms for the spontaneous Raman effect and for the stimulated Raman effect. We deal with these aspects of scattering by molecules in Section 12.1.1. Section 12.1.2 presents a quantum-theoretical treatment that permits a quantitative explanation of electrooptical constants, which are introduced phenomenologically in Section 12.1.1. It also covers the unavoidable fluctuation phenomena. Subsection 12.1.3 is devoted to the amplification and generation of waves in a Raman-active medium, and in Subsection 12.1.4 specific Raman scattering processes and applications (in particular spectroscopic ones such as active Raman spectroscopy) are described.

12.1.1. Classical Model for the Interaction of Radiation with Molecules

In establishing the basic relationship between the polarization P and the field strength E for the explanation of Raman scattering, we may start from a simple molecular structure, namely a vibrating molecule consisting of two

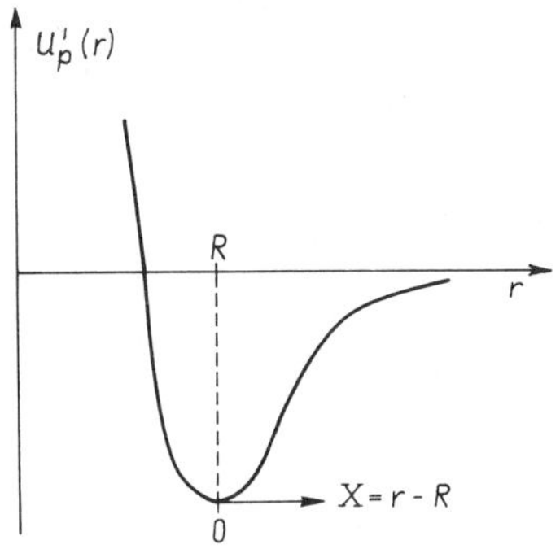

Fig. 12.1. Scheme of the potential energy for molecular vibration.

equal atoms (for more complex molecules analogously structured equations can be obtained by introducing normal coordinates for the vibrational modes).

We consider the unperturbed molecule. According to general principles of molecular physics, let us assume that the two atoms are capable of altering their positions relative to one another and that the potential energy as a function of the internuclear distance r is $U_p'(r)$. The form of the function $U_p'(r)$ is known from the fundamentals of molecular physics, and in a stable electronic ground state it corresponds to the curve given in Fig. 12.1. If we denote the deviation $r - R$ from the equilibrium position R by X, the harmonic approximation is given by

$$U_p'(X) = \frac{k_M X^2}{2}, \tag{12.1}$$

where k_M represents the linear force constant.

First the atoms are imagined to be fixed at a certain distance r. If an electric field $E_\bullet$ is applied, a dipole moment $p_\bullet$ is induced in the molecule: the center of charge of the electron shell shifts under the influence of an external force. This property can be described by the relation

$$p_\bullet = \alpha_{\bullet\bullet} E_\bullet, \tag{12.2}$$

where $\alpha_{\bullet\bullet}$ is the second-rank polarizability tensor of the molecule.

The polarizability is a function of r and X; the dependence of the distortion of the electron shell on the external field $E_\bullet$ itself changes with the absolute value and sign of X. For example, if a field $E_\bullet$ of magnitude $|E_\bullet|$ is supposed to act in the direction of the internuclear axis, we can easily understand that the shift of the center of the electron shell for $r = R + |X|$ differs from that for $r = R - |X|$. Thus we write $\alpha_{\bullet\bullet} = \alpha_{\bullet\bullet}(X)$. For the following we may assume that the function $\alpha_{\bullet\bullet}(X)$ can be described by the linear expression

$$\alpha_{\bullet\bullet}(X) = \alpha_{(0)\bullet\bullet} + \alpha_{(1)\bullet\bullet} X, \tag{12.3}$$

where $\alpha_{(0)\bullet\bullet}$ is the polarizability at the equilibrium position and $\alpha_{(1)\bullet\bullet}$ is its derivative $(d\alpha_{\bullet\bullet}(X)/dX)_{X=0}$. The possibility to make the linear ansatz (12.3) depends on the fact that only deviations that are not too large need to be taken into account.

Passing over from temporally fixed to time-varying deviations $X(t)$, it can be assumed for molecules that in a temporally determined electric field $E_\bullet(t)$ the assigned dipole moment $p_\bullet(t)$ builds up in very short times, during which no noticeable change of X occurs. This behavior corresponds to the Born–Oppenheimer approximation and has its origin in the fact that the electrons, because of their relatively small mass, adjust rapidly to the state resulting from the molecular potential and the external field. Therefore we obtain

$$p_\bullet(t) = \alpha_{\bullet\bullet}[X(t)]E_\bullet(t). \tag{12.4}$$

It is to be noted that the *resonance* Raman effect will be excluded in the following considerations: the frequencies of the external field are assumed to be well below the electronic transition frequencies but well above the vibrational frequencies of the molecule.

Making these assumptions, we obtain at the deviation X the energy required for building up the induced dipole moment:

$$U_p''(X) = -\int_0^E dE'_\bullet\, p_\bullet[X;E'] = -\tfrac{1}{2}[\alpha_{\bullet\bullet}(X)E_\bullet]E_\bullet. \tag{12.5}$$

If X_1 and X_2 are the coordinates of the two atomic nuclei on the X axis and m_A is the mass of one atomic nucleus, the system under study has the following kinetic energy $\mathscr{E}_{\text{kin}}$ and potential energy $\mathscr{E}_{\text{pot}}$:

$$\mathscr{E}_{\text{kin}} = \frac{m_A}{2}\left(\frac{dX_1}{dt}\right)^2 + \frac{m_A}{2}\left(\frac{dX_2}{dt}\right)^2, \qquad \mathscr{E}_{\text{pot}} = U_p'(X) + U_p''(X). \tag{12.6}$$

Substituting $X_1 - X_2 - R = X$, we have the following equation of motion:

$$\frac{d^2X}{dt^2} + \Gamma_M\frac{dX}{dt} + \omega_M^2 X = \frac{1}{2\mathscr{M}}[\alpha_{(1)\bullet\bullet}E_\bullet]E_\bullet \tag{12.7}$$

for $X(t)$. In formal agreement with (11.46), a damping term has been inserted on the left-hand side, so that this expression describes a damped harmonic oscillator. The frequency depends on the force constant k_M and the reduced mass $\mathscr{M} = m_A/2$ according to $\omega_M^2 = k_M/\mathscr{M}$. The general solution of this differential equation can be written in the form

$$X(t) = \tilde{X}e^{-\frac{1}{2}\Gamma_M t}\cos\left[t\sqrt{\omega_M^2 - \tfrac{1}{4}\Gamma_M^2} + \varphi\right]$$

$$+\frac{1}{2\mathscr{M}}\int_0^\infty d\tau\, Q_M(\tau)[\alpha_{(1)\bullet\bullet}E_\bullet(t-\tau)]E_\bullet(t-\tau). \tag{12.8}$$

Here $Q_M(\tau)$ behaves analogously to $Q(\tau)$ in the first term on the right-hand side of (11.52), where ω_0 has to be replaced by ω_M and Γ by Γ_M. The quantities $\tilde{X}$ and φ are real constants of the homogeneous part of the solution of (12.7) and can be freely chosen. Using the relations (12.3) and (12.4), we obtain

$$p_\bullet(t) = \alpha_{(0)\bullet\bullet}E_\bullet(t)$$
$$+\tilde{X}e^{-\frac{1}{2}\Gamma_M t}\cos\left[t\sqrt{\omega_M^2 - \tfrac{1}{4}\Gamma_M^2} + \varphi\right]\alpha_{(1)\bullet\bullet}E_\bullet(t)$$
$$+\frac{1}{2\mathscr{M}}\alpha_{(1)\bullet\bullet}E_\bullet(t)\int_0^\infty d\tau\, Q_M(\tau)\left[\alpha_{(1)\bullet\bullet}E_\bullet(t-\tau)\right]E_\bullet(t-\tau) \quad (12.9)$$

for the time-dependent dipole moment. The dipole moment contains two linear terms and a third-order term in the field strength. Each term accounts for a specific effect, as will be illustrated below. To explain the effects described by the linear terms it suffices to assume one single frequency in the spectrum of the field strength (e.g. the frequency ω_L of the laser radiation). The associated amplitude is designated by $\hat{E}(\omega_L)$; since the phase factor is of no importance in discussing the linear terms, we assume $\hat{E}(\omega_L)$ to be real.

Rayleigh scattering of molecules is described by the first term on the right-hand side of (12.9). Obviously the induced dipole moment has the same frequency spectrum as the incident radiation. *Spontaneous Raman scattering* is represented by the second term, which will be denoted by $[p_\bullet(t)]_{\mathrm{spRS}}$. We now consider the numerical values of the parameters. We may choose $k_M = 5 \times 10^2$ N/m (the force constant of a C–C single bond) as a typical value. If the reduced mass is taken to be $\mathscr{M} = 10^{-26}$ kg, a vibrational frequency of $\omega_M \approx 2 \times 10^{+14}$ s^{-1} results; this corresponds to a wavelength of about 10 μm. The estimation of the damping constant Γ_M is somewhat more difficult than it was when considering the damping of the pure electronic effect in Section 11.3.1, because now the frequency bandwidth of the *vibrational* co-ordinate $X(t)$ is concerned. Equation (12.9) reveals that in the case of a narrow excitation (laser) line the width $\Delta\omega_S$ of the Stokes line roughly corresponds with the bandwidth Γ_M of $X(t)$. Thus, the following estimate of an upper bound for Γ_M can be given. In investigating liquids, the width of a Stokes line is found to be $\Delta\omega_S \lesssim 10^{12}$ s^{-1}. In the gaseous state $\Delta\omega_S$ decreases by 2 to 3 orders of magnitude. Therefore in gases and liquids $\Delta\omega_S \lesssim 10^{12}$ s^{-1}. Since we may set in numerical examples $\omega_L = 2.5 \times 10^{15}$ s^{-1} for the laser frequency ω_L, the resulting frequencies and damping constants are connected in the following order-of-magnitude relation:

$$\omega_L \gg \omega_M \gg \Gamma_M. \quad (12.10)$$

Because of the estimate $\Gamma_M \lesssim 10^{12}$ s^{-1}, over a time interval of at least 10^{-12} s the second summand in (12.9) can be represented by

$$[p_\bullet(t)]_{\text{spRS}} = \tilde{X}\alpha_{(1)\bullet\bullet}\hat{E}_\bullet(\omega_L)\cos(\omega_M t + \varphi)\cos\omega_L t$$

$$= \tfrac{1}{2}\tilde{X}\alpha_{(1)\bullet\bullet}\hat{E}_\bullet(\omega_L)\cos[(\omega_L - \omega_M)t + \varphi]$$

$$+ \tfrac{1}{2}\tilde{X}\alpha_{(1)\bullet\bullet}\hat{E}_\bullet(\omega_L)\cos[(\omega_L + \omega_M)t + \varphi]. \quad (12.11)$$

Thus it becomes apparent that the expression for $[p_\bullet(t)]_{\text{spRS}}$ contains an oscillation with a frequency smaller than ω_L, namely the Stokes frequency $\omega_S = \omega_L - \omega_M$, and an oscillation with a frequency larger than ω_L, namely the anti-Stokes frequency $\omega_A = \omega_L + \omega_M$. Note that the function $[p_\bullet(t)]_{\text{spRS}}$ expresses the dipole moment of a single molecule. The fact that the phase for each molecule of an ensemble can be freely chosen means that the scattering *intensities* originating from the single molecules add up to the total intensity. This is a phenomenon characteristic of the spontaneous Raman effect.

Now we perform an estimation of the value $\alpha_{(1)\bullet\bullet}$ of the derivative of the polarizability. According to the fundamentals of electrodynamics, the overall power $\langle P_S \rangle$ emitted by the dipole moment of a molecule into the entire solid angle at the frequency $\omega_L - \omega_M$ is proportional to the quantities $\tilde{X}^2$, $|\alpha_{(1)\bullet\bullet}E_\bullet(\omega_L)|^2$, and $(\omega_L - \omega_M)^4$; the proportionality factor is determined by universal constants.

For estimating $\omega_L - \omega_M$ and thus also $(\omega_L - \omega_M)^4$ the numerical values given above can be used.

The value of the square $\tilde{X}^2$ of the vibrational amplitude originates from the following reasoning, which makes use of certain quantum-theoretical concepts and results (to be discussed in detail in Section 12.1.2). The starting point of our discussion of Stokes scattering is that the radiation of frequency ω_L interacts with a molecule in its vibrational ground state. Molecules possess the zero-point energy $\hbar\omega_M/2$, which is equally composed of the expectation values of the kinetic and the potential energy. From this results a mean squared deviation from the equilibrium $\hbar/2\omega_M\mathcal{M}$, leading, from our representative numerical values, to the effective amplitude $\tilde{X} \approx 7 \times 10^{-12}$ m.

Among other conditions the direction of the vector $\hat{E}_\bullet$ relative to the direction of the molecular axis determines the value of $|\alpha_{(1)\bullet\bullet}\hat{E}_\bullet(\omega_L)|^2$. Of interest to us is the value averaged over all molecular directions, which we set equal to $\alpha_1^2|E_\bullet(\omega_L)|^2$, where α_1 is an effective value of the polarizability derivative. $|E_\bullet(\omega_L)|^2$ is proportional to the intensity $\langle J_L \rangle$ of the incident radiation.

Thus we finally obtain

$$\langle P_S \rangle = C_L\alpha_1^2\langle J_L \rangle \quad (12.12)$$

for the scattering intensity of the Stokes line. The proportionality factor C_L can be calculated from universal constants and $\tilde{X}$ and $\omega_L - \omega_M$; its order of magnitude is 10^{26} $(\mathrm{V/A\,s})^2$. The measurement of the scattering intensity due to a macrophysical quantity of substance allows us to determine the molecular scattering cross section $\langle P_S \rangle / \langle J_L \rangle = C_L \alpha_1^2$ of a single molecule. From measurements on liquids, for strong Stokes lines we get $C_L \alpha_1^2 \sim 10^{-32}$ m^2. This gives the numerical value $\alpha_1 \sim 10^{-29}$ $\mathrm{A\,s\,m/V}$. In the following subsection quantum-theoretical values will be given.

The connection between the scattered intensity due to a macrophysical quantity of matter and that due to a single molecule results from the following reasoning. For a large number ($N \gg 1$) of uncoupled molecules the total scattered intensity

$$\langle P_{S,\mathrm{tot}} \rangle = \sum_j \langle P_{S,j} \rangle = \sum_j C_L' \left[\alpha_{(1)\bullet\bullet}(j) \hat{E}_\bullet(\omega_L) \right]^2 \tag{12.13}$$

is obtained from the addition of the individual scattered intensities $\langle P_{S,j} \rangle$, because, under the conditions mentioned, each molecule vibrates with its own phase φ_j, which is independent of the phases of all the other molecules. C_L' is a proportionality factor analogous to C_L. The quantity $\alpha_{(1)\bullet\bullet}(j)$ characterizes the derivative of the polarizability tensor of the jth molecule. Suppose that the medium is macrophysically isotropic, so that the molecules are homogeneously distributed over all orientation directions. If we assume that the incident radiation of frequency ω_L is not so intense as to cause this equipartition to be remarkably disturbed, we eventually obtain

$$\langle P_{S,\mathrm{tot}} \rangle = N C_L' \overline{\left[\alpha_{(1)\bullet\bullet} \hat{E}_\bullet(\omega_L) \right] \left[\alpha_{(1)\bullet\bullet} \hat{E}_\bullet(\omega_L) \right]}^{\mathrm{or}}, \tag{12.14}$$

where the averaging on the right-hand side is over all orientations of an arbitrary molecule.

The tensor components $\alpha_{(1)\bullet\bullet}$ on the right-hand side, which are known to be related to the laboratory system (x, y, z), can be transformed to a reference system (X, Y, Z) fixed in the molecule. This procedure is analogous to that employed in introducing a reference system fixed in the crystal for SHG [cf. (11.62)]. This transformation is represented by

$$\alpha_{(1)pr} = \sum_{P,R} S_{pP} S_{rR} \alpha_{(1)PR}. \tag{12.15}$$

For the diatomic molecule under study the internuclear axis is chosen to be the X axis. Assuming that the tensor $\alpha_{(1)\bullet\bullet}$, in the reference system inherent to the molecule, has only one element differing from zero, namely $\alpha_{(1)XX} \equiv \alpha_1$, then the orientation averaging provides

$$\overline{\alpha_{(1)xx}\alpha_{(1)xx}}^{\mathrm{or}} = \tfrac{1}{5}\alpha_1^2, \qquad \overline{\alpha_{(1)xy}\alpha_{(1)xy}}^{\mathrm{or}} = \tfrac{1}{15}\alpha_1^2. \tag{12.16}$$

These results will be made use of in the following, when susceptibility components are calculated.

The *stimulated Raman effect* is described by the third term on the right-hand side of (12.9), that is, by the third-order term with respect to the field strength, which, if arising from the jth molecule, will be denoted by $p_{\bullet}^{(3)}(j)$. All the parameters contained in this term have already been subjected to an order-of-magnitude estimation in discussing the spontaneous Raman effect. If $p_{\bullet}^{(3)}(j)$ is written in the form of one vibration or a sum of vibrations, then, according to (12.9), the phases of $p_{\bullet}^{(3)}(j)$ are obviously determined uniquely by $E_{\bullet}(t)$. Considering a large number ($N \gg 1$) of molecules that are under the action of the same field strength $E_{\bullet}(t)$, all the phase values of the dipole moments are equal. Hence, for such an ensemble of molecules the third-order polarization associated with the stimulated Raman effect is

$$\begin{aligned} P_{\bullet}^{(3)}(t) &= \frac{1}{\Delta V} \sum_j p_{\bullet}^{(3)}(j) \\ &= \frac{1}{\Delta V} \sum_j \frac{1}{2\mathscr{M}} \alpha_{(1)\bullet\bullet}(j) E_{\bullet}(t) \\ &\quad \times \int_0^{\infty} d\tau\, Q_M(\tau) \left[\alpha_{(1)\bullet\bullet}(j) E_{\bullet}(t-\tau)\right] E_{\bullet}(t-\tau). \end{aligned} \tag{12.17}$$

The summation extends over the molecules in a volume element ΔV, which is chosen on the one hand to contain a large number of molecules, and on the other hand to be so small that the field strength of the incident radiation can be considered constant over it.

The expression (12.17) holds exactly for uncoupled molecules only. In many cases, a coupling of the molecules can be readily taken into account by using effective values for α_1 and Γ_M rather than the values for the isolated molecule.

In the following we want to show that under certain assumptions for the frequency spectrum of the field strength $E_{\bullet}(t)$ of the incident radiation, polarization contributions from (12.17) can be derived that give rise to the phenomena of the stimulated Raman effect. A one-dimensional model suffices to derive the basic features; therefore we henceforth disregard the spatial transformation properties of polarization, field strength, and polarizability. By inserting the number density γ, from (12.17) we obtain the relatively simple expression

$$P^{(3)}(t) = \frac{\gamma \alpha_1^2}{2\mathscr{M}} E(t) \int_0^{\infty} d\tau\, Q_M(\tau) E^2(t-\tau). \tag{12.18}$$

For the sake of clarity, we again use a block diagram (see Fig. 12.2, where the notation in the operator blocks is the same as in Fig. 11.3) for discussing how third-order polarization arises. It is obvious that large polarization amplitudes can be achieved only if the input of the linear filter $\mathscr{Y}_M Q_M$ contains a

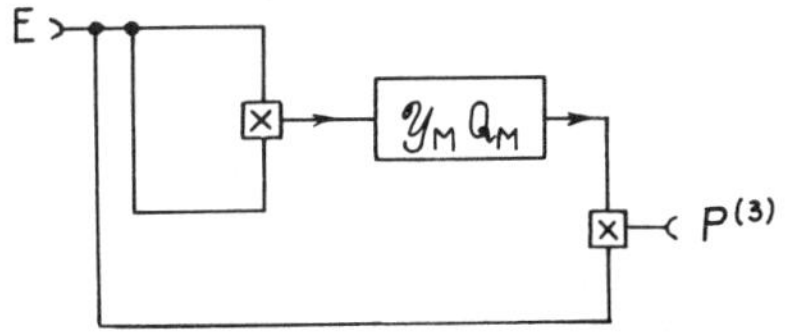

Fig. 12.2. Block diagram for the stimulated Raman effect; $\mathcal{Y}_M = \gamma\alpha_1^2/2\mathcal{M}$.

frequency component in the spectrum lying near the resonance frequency ω_M (i.e. near a molecular vibrational frequency). If only one single (laser) frequency ω_L is contained in the input field strength $E(t)$, this cannot be accomplished. In $E(t)$ at least a second frequency ω_S must be present for which $\omega_L - \omega_S \approx \omega_M$. Provided $E(t)$ contains two oscillations with the complex amplitudes $\hat{E}(\omega_L)$ and $\hat{E}(\omega_S)$, the linear filter is preceded by a component with frequency $\omega_L - \omega_S$ and amplitude $\hat{E}(\omega_L)\hat{E}^*(\omega_S)/2$. Since the other frequency components are far from the molecular vibrational frequency ω_M, they give only negligibly small contributions after passing the linear filter. The amplitude of the difference-frequency component is given by

$$\hat{A}_M(\omega_L - \omega_S) = \frac{\gamma\alpha_1^2}{2\mathcal{M}} \underset{\smile}{Q}_M(\omega_L - \omega_S)\frac{\hat{E}(\omega_L)\hat{E}^*(\omega_S)}{2}. \tag{12.19}$$

This component leads by sum and difference mixing with the frequencies ω_L, ω_S of the input field strength, to third-order contributions to the polarization with the following frequencies:

$$\omega_L + (\omega_L - \omega_S) \approx \omega_L + \omega_M, \qquad \omega_S - (\omega_L - \omega_S) \approx \omega_S - \omega_M,$$

$$\omega_S + (\omega_L - \omega_S) = \omega_L, \qquad \omega_L - (\omega_L - \omega_S) = \omega_S.$$

Now the power amplification of an electromagnetic wave on its passage through a layer of matter will be determined. A nonvanishing power amplification occurs only for components of the polarization whose frequencies agree with those of the field strength. Therefore we restrict our attention to polarization components with the frequencies $\omega_S + (\omega_L - \omega_S)$ and $\omega_L - (\omega_L - \omega_S)$. The polarization amplitude of the former component is

$$\begin{aligned}\hat{P}^{(3)}(\omega_L) &= \tfrac{1}{2}\hat{E}(\omega_S)\hat{A}_M(\omega_L - \omega_S)\\ &= \varepsilon_0\chi^{(3)}(\omega_L; \omega_L, \omega_S, -\omega_S)\hat{E}(\omega_L)\hat{E}(\omega_S)\hat{E}^*(\omega_S),\end{aligned} \tag{12.20}$$

the amplitude of the latter is

$$\begin{aligned}\hat{P}^{(3)}(\omega_S) &= \tfrac{1}{2}\hat{E}(\omega_L)\hat{A}_M^*(\omega_L - \omega_S)\\ &= \varepsilon_0\chi^{(3)}(\omega_S; \omega_S, \omega_L, -\omega_L)\hat{E}(\omega_S)\hat{E}(\omega_L)\hat{E}^*(\omega_L).\end{aligned} \tag{12.21}$$

The susceptibilities involved are given by

$$\chi^{(3)}(\omega_S;\, \omega_S, \omega_L, -\omega_L) = \chi^{(3)*}(\omega_L;\, \omega_L, \omega_S, -\omega_S) = \frac{\gamma\alpha_1^2}{4\mathscr{M}\varepsilon_0} \underset{\smile}{Q}_M(\omega_L - \omega_S). \tag{12.22}$$

Because

$$\underset{\smile}{Q}_M(\omega) = \frac{1}{-\omega^2 - i\Gamma_M\omega + \omega_M^2},$$

we find near resonance ($\omega_L - \omega_S \approx \omega_M$), the relation

$$\chi^{(3)}(\omega_S;\, \omega_S, \omega_L, -\omega_L) = -\chi^{(3)}(\omega_L;\, \omega_L, \omega_S, -\omega_S) = -i|\chi_R|, \tag{12.23}$$

where

$$|\chi_R| = \frac{\gamma\alpha_1^2}{4\varepsilon_0\mathscr{M}\Gamma_M\omega_M}. \tag{12.23a}$$

The imaginary part $|\chi_R|$ of $\chi^{(3)}(\omega_S;\, \omega_S, \omega_L, -\omega_L)$ determines the amplification for the stimulated Raman scattering. For the scattered Stokes wave (frequency ω_S) the power amplification per unit length is given by

$$\frac{d}{dz}\langle S(\omega_S)\rangle = \varepsilon_0\frac{\omega_S}{2}|\chi_R|\hat{E}(\omega_S)\hat{E}(\omega_L)\hat{E}^*(\omega_S)\hat{E}(\omega_L)\hat{E}^*(\omega_L), \tag{12.24a}$$

and for the laser wave it is given by

$$\frac{d}{dz}\langle S(\omega_L)\rangle = -\frac{\varepsilon_0\omega_L}{2}|\chi_R|\hat{E}^*(\omega_L)\hat{E}(\omega_S)\hat{E}^*(\omega_S). \tag{12.24b}$$

In both cases the power amplification is proportional to $|\chi_R|$; it is positive for the scattered wave, and negative for the laser wave. Using the abovementioned values of α_1 (10^{-29} A s m/V), γ (10^{28} m^{-3}), $\mathscr{M}$ (10^{-26} kg), Γ_M (10^{12} s^{-1}), and ω_M (2×10^{14} s^{-1}), one obtains $|\chi_R| \sim 10^{-20}$ m^2/V^2.

From (12.23a) it is apparent that the constant Γ_M characterizing the damping is of great importance for the stimulated Raman effect. The susceptibility $|\chi_R|$ and thus the power amplification are inversely proportional to Γ_M. Note that in contrast to this, the damping constants of the susceptibilty functions accounting for the common case of frequency conversion and parametric amplification may be ignored.

In establishing the basic equation (12.17) for the stimulated Raman effect we started from the harmonic approximation for the relation between the molecular potential energy and the deviation from equilibrium [i.e., (12.1)]. We now provide evidence that this approximation, which is very common in linear spectroscopy, is also valid for the nonlinear effect of stimulated Raman scattering. The amplitude of the induced deviation from equilibrium—that is, the second term on the right-hand side of (12.8)—is given by

$$|\tilde{X}| = \frac{\alpha_1}{4\mathscr{M}\Gamma_M\omega_M}\frac{\hat{E}(\omega_L)\hat{E}(\omega_S)}{2}. \tag{12.25}$$

Thus it increases as the field strength amplitudes and Γ_M^{-1} increase. Even in the case of large field strength amplitudes (of the order of 10^8 V/m) and small values of Γ_M (of the order of 10^9 s^{-1}), $|\tilde{X}|$ attains values that hardly exceed one-tenth of the equilibrium distance of the atoms. This justifies the employment of the harmonic approximation for the stimulated Raman effect in a very wide range.

Having discussed the basic facts by means of a one-dimensional model, we may simply transfer the results achieved above to the realistic three-dimensional conditions. The susceptibility describing the stimulated Raman effect as a third-order effect is a fourth-rank tensor. Because of (12.17) the expression $\chi^{(3)}(\omega_S; \omega_S, \omega_L, -\omega_L)$ in (12.23) can be replaced by a tensor with the components

$$\chi^{(3)}_{pqrs}(\omega_S; \omega_S, \omega_L, -\omega_L) = \frac{-i}{4\mathscr{M}\omega_M\varepsilon_0\Gamma_M}\frac{1}{\Delta V}\sum_j \alpha_{(1)pr}(j)\alpha_{(1)qs}(j), \tag{12.26}$$

where $\alpha_{(1)pr}(j)$ is a tensor component of the polarization derivative for the jth molecule; the summation is taken over all molecules in the volume ΔV. If the medium is macrophysically isotropic, the averaging of the tensor products over the volume can be readily performed. It can be replaced by orientation averaging over one molecule, so that

$$\chi^{(3)}_{pqrs}(\omega_S; \omega_S, \omega_L, -\omega_L) = \frac{-i\gamma}{4\mathscr{M}\omega_M\varepsilon_0\Gamma_M}\overline{\alpha_{(1)pr}\alpha_{(1)qs}}^{\,\mathrm{or}} \tag{12.27}$$

is obtained. The results for $\overline{\alpha_{(1)pr}\alpha_{(1)qs}}^{\,\mathrm{or}}$ are the same as those obtained for the spontaneous Raman effect given in (12.16).

In the preceding considerations *two* frequencies (ω_L, $\omega_S \approx \omega_L - \omega_M$) were assumed in the spectrum of the field strength of the incident radiation. It may be mentioned here that, if more than two different frequencies

$$\omega_b \approx \omega_L + b\omega_M, \qquad \text{where} \quad b = 0, \pm 1, \pm 2, \ldots, \tag{12.28}$$

are present in the spectrum of the incident radiation, results on the stimulated

Raman effect can also be obtained for the higher harmonics of the molecular vibrational frequency, where $b < 0$ characterizes Stokes and $b > 0$ anti-Stokes lines.

We have seen that polarizable molecules capable of vibration can be successfully described with respect to the spontaneous and stimulated Raman effects by employing the extended classically based Placzek polarizability theory. In doing this, we made use of quantum-theoretical concepts and results at two important points, namely, when introducing temperature-independent zero-point vibrations and when making quantitative estimations of the polarizability derivative with respect to the vibrational coordinate. The fully quantum-theoretical treatment of the following subsection provides—in addition to other results—unambiguous insight and numerical values concerning these two points.

12.1.2. Quantum Description of the Vibrational Raman Effect

In this section we present a fully quantum-theoretical treatment of the spontaneous and the stimulated Raman effect, which was given for the first time by Dirac (Ref. 12.2). Our approach is analogous to that employed for spontaneous and stimulated one-photon emission (cf. Section 3.3.1); now we determine the probabilities of spontaneous and stimulated emission of Stokes quanta with energy $\hbar\omega_S$ under the irradiation of matter by (laser) light of photon energy $\hbar\omega_L$. As in Section 12.1.1 the relation $\omega_L - \omega_S \approx \omega_M$ is assumed to be satisfied, ω_M being the frequency of the molecular vibration by which the light is scattered.

Suppose that if a Stokes quantum is emitted, the molecule under study flips from the ground state $|\mathscr{E}_{A,0}\rangle$ via an intermediate state $|\mathscr{E}_{A,l}\rangle$ to an excited vibrational state $|\mathscr{E}_{A,1}\rangle$ of the electronic ground state (as illustrated in Fig. 12.3, where $|\mathscr{E}_{A,l}\rangle$ is a vibrational state of the first excited electronic level). In this process the emission of the Stokes quantum is associated with the absorption of a laser quantum. Thus it is a second-order process, and we are allowed to base the formulation of the present approach on results of Section 3.2.2.2. We may assume that only two modes of the radiation field take part in this process and that at the beginning of the process n_S photons of the Stokes mode and n_L photons of the laser mode are in the radiation field. Thus the

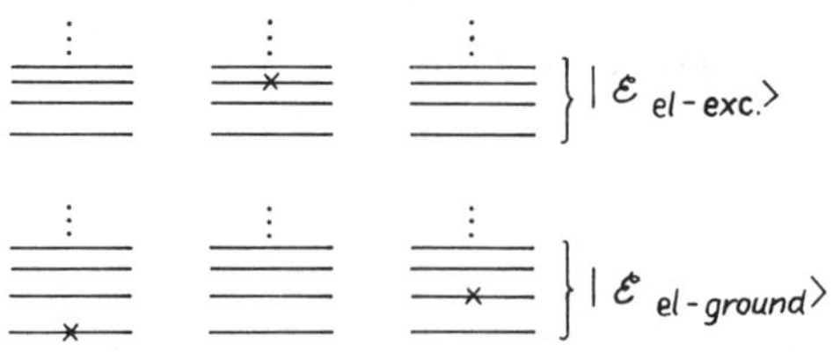

Fig. 12.3. Initial, intermediate, and final molecular state in the emission of a Stokes photon.

initial state $|\mathscr{E}_i\rangle$ and the final state $|\mathscr{E}_f\rangle$ are to be represented as follows:

$$|\mathscr{E}_i\rangle = |\mathscr{E}_{A,0}\rangle|n_S, n_L\rangle,$$

$$|\mathscr{E}_f\rangle = |\mathscr{E}_{A,1}\rangle|n_S + 1, n_L - 1\rangle. \tag{12.29}$$

We subdivide the process into two one-photon processes. As intermediate states only states of the type

$$|\mathscr{E}_j\rangle = |\mathscr{E}_{A,l}\rangle|n_S + 1, n_L\rangle \qquad \text{(case I)}$$

$$|\mathscr{E}_j\rangle = |\mathscr{E}_{A,l}\rangle|n_S, n_L - 1\rangle \qquad \text{(case II)} \tag{12.30}$$

are of interest, because other states cannot contribute to the transition probability. In case I, a Stokes quantum is created in the first step and a laser quantum is annihilated in the second step; in case II the sequence is reversed. For calculating the transition probability we use (3.53). According to which of the two cases is considered, we have to represent the energy difference in it by

$$\mathscr{E}_i - \mathscr{E}_j = \begin{cases} \mathscr{E}_{A,0} - \mathscr{E}_{A,l} - \hbar\omega_S & \text{(case I)} \\ \mathscr{E}_{A,0} - \mathscr{E}_{A,l} + \hbar\omega_L & \text{(case II)}. \end{cases} \tag{12.31}$$

Because $\mathscr{E}_{A,0} - \mathscr{E}_{A,l} < 0$ and $(-\hbar\omega_S) < 0$, the energy difference $\mathscr{E}_i - \mathscr{E}_j$ in case I is always unequal to zero. In case II, with appropriate energy of the laser photon (i.e., if $\hbar\omega_L \approx \mathscr{E}_{A,l} - \mathscr{E}_{A,0}$), the energy difference in the denominators in the formulas (3.52) or (3.53) for the transition probability may vanish, so that additional considerations are necessary; the condition $\hbar\omega_L \approx \mathscr{E}_{A,l} - \mathscr{E}_{A,0}$ characterizes the *resonance* Raman effect. It will be excluded from the following discussion. Our intention is to consider the *normal* Raman effect, where the vibrational energy $\hbar\omega_M = \mathscr{E}_{A,1} - \mathscr{E}_{A,0}$ is assumed to be much less than both the laser photon energy and the difference $(\mathscr{E}_{A,\text{exc}} - \mathscr{E}_{A,0}) - \hbar\omega_L$. Figure 12.4 illustrates the transition diagrams for the Stokes and anti-Stokes processes.

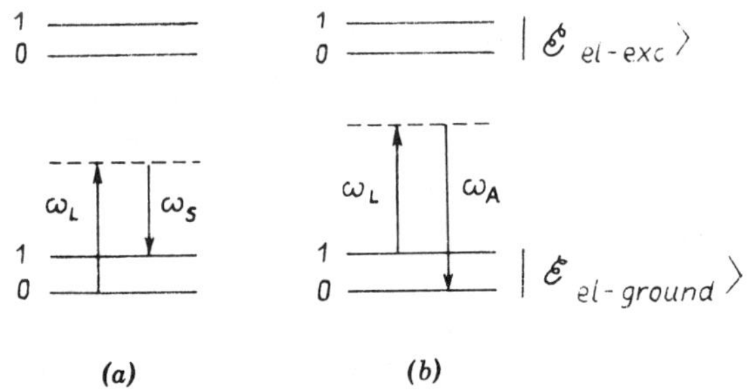

Fig. 12.4. Transition scheme for Raman scattering: (*a*) Stokes process, (*b*) anti-Stokes process.

We choose $-\hat{d}_{\bullet}\hat{E}_{\bullet}$ as the operator for the interaction between one molecule and the radiation, and let the energy resonance between the final state $|\mathscr{E}_f\rangle$ and one of the possible intermediate states $|\mathscr{E}_j\rangle$ be excluded [cf. derivation of (3.53)], whereas because $\mathscr{E}_{A,1} - \mathscr{E}_{A,0} \approx \hbar\omega_L - \hbar\omega_S$, resonance exists between the initial and the final state. Making use of second-order perturbation theory, we obtain, in analogy to (3.57) and (3.58), for the transition probability of the normal Raman effect (index R)

$$p_{i\to f}^{(R)}(t) = \frac{t^2}{\hbar^2}\frac{1}{4\varepsilon_0^2 V^2}\hbar\omega_L n_L \hbar\omega_S(n_S+1)|\eta^{(R)}|^2$$

$$\times \operatorname{sinc}^2\left[\frac{\mathscr{E}_i - \mathscr{E}_f}{2\hbar}t\right], \tag{12.32}$$

where

$$\eta^{(R)} = \sum_l \frac{(d_{l0})_S(d_{1l})_L}{-(\mathscr{E}_{A,l} - \mathscr{E}_{A,0}) - \hbar\omega_S} + \frac{(d_{l0})_L(d_{1l})_S}{\hbar\omega_L - (\mathscr{E}_{A,l} - \mathscr{E}_{A,0})}. \tag{12.32a}$$

Here $(d_{l0})_S$ is the scalar product of the vector $\langle\mathscr{E}_{A,l}|d_{\bullet}|\mathscr{E}_{A,0}\rangle$ with the polarization unit vector of the Stokes mode, with analogous definitions for the quantities with the index L referring to the laser mode. Arising from the factor $n_S + 1$, the probability $p_{i\to f}^{(R)}$ contains two terms. Concerning the photon numbers, one of them is proportional to $n_S n_L$, and the other one is proportional to n_L. The former gives a contribution only if Stokes photons are present at the beginning, that is, if $n_S > 0$; this contribution represents the stimulated Raman effect (index SRS). The latter is independent of the number of Stokes photons; it only depends on the number of laser photons. Thus, this term represents the spontaneous Raman effect (index spR).

The argument of the sinc function indicates that a noticeable amount of the transition probability can only occur for small values of

$$(\mathscr{E}_{A,1} - \mathscr{E}_{A,0}) - (\hbar\omega_L - \hbar\omega_S).$$

Under this assumption the *total transition probability* is now calculated in the same way as after (A.124). A summation is carried out over all the final states that satisfy the condition $\mathscr{E}_f \approx \mathscr{E}_i$ for one and the same initial state $|\mathscr{E}_i\rangle$.

Let us first treat the *stimulated effect*. We introduce the spatial energy density $\sigma_S(\omega_S, \Omega_S)$ per unit frequency and per unit solid angle of the Stokes photons for a certain polarization direction. This gives the spatial energy density of the Stokes photons in the intervals $\Delta\omega_S$ and $\Delta\Omega_S$ in the form

$$\sum_{\Delta\omega_S,\,\Delta\Omega_S} \frac{\hbar\omega_S n_S}{V} = \int_{\Delta\Omega_S} d\Omega_S \int_{\Delta\omega_S} d\omega_S\, \sigma_S(\omega_S, \Omega_S) \tag{12.33}$$

assuming a small volume V. From this the total transition probability

$$P^{(\mathrm{SRS})}(t) = \sum_{\mathscr{E}_f \approx \mathscr{E}_i} p_{i \to f}^{(\mathrm{SRS})}(t)$$

$$= \frac{t^2}{\hbar^2} \frac{\hbar \omega_L n_L}{4 \varepsilon_0^2 V} \int_{\Delta \Omega_S} d\Omega_S \int_{\Delta \omega_S} d\omega_S \, \sigma_S(\omega_S, \Omega_S) |\eta^{(R)}|^2 \operatorname{sinc}^2 \left[\frac{[\mathscr{E}_f - \mathscr{E}_i] t}{2\hbar} \right] \tag{12.34}$$

results. Because $\mathscr{E}_f - \mathscr{E}_i = \mathscr{E}_{A,1} + \hbar\omega_S - \mathscr{E}_{A,0}$, the variable ω_S is also included in the argument of the sinc function. The square of the sinc function is rapidly varying in ω_S; only in an $\mathscr{E}_f$ energy interval $4\pi\hbar/t$ around $\mathscr{E}_i$ does it differ substantially from zero. In contrast, σ_S and $|\eta^{(R)}|^2$ are slowly varying factors and may be written in front of the integral over ω_S. After integrating over ω_S, where the limits of the integral may be chosen to be $-\infty$ and $+\infty$ without noticeable error, we obtain

$$P^{(\mathrm{SRS})}(t) = t \frac{2\pi}{\hbar^2} \frac{\hbar \omega_L n_L}{4 \varepsilon_0^2 V} \int_{\Delta \Omega_S} d\Omega_S \, \sigma_S(\omega_S, \Omega_S) |\eta^{(R)}|^2. \tag{12.35}$$

Note that the term $|\eta^{(R)}|^2$ is a function of Ω_S because of the scalar products $(d_{l0})_S$ and $(d_{1l})_S$, and therefore it must not be written in front of the integral.

In calculating the contribution of the *spontaneous Raman effect*, the summation over the final states must now be carried out by means of the mode density in a way analogous to that of the treatment of the spontaneous one-photon emission. Assuming that

$$\bar{\sigma} = \frac{\hbar \omega_S^3}{8\pi^3 c^3} \tag{12.36}$$

is the spatial energy density per unit frequency and per unit solid angle, we then have

$$\sum_{\Delta\omega_S, \Delta\Omega_S} \frac{\hbar \omega_S}{V} = \int_{\Delta \Omega_S} d\Omega_S \int_{\Delta \omega_S} d\omega_S \, \bar{\sigma}. \tag{12.37}$$

If the integral over ω_S is evaluated in the same way as in (12.34), the corresponding total transition probability is

$$P^{(\mathrm{spR})}(t) = \sum_{\mathscr{E}_f \approx \mathscr{E}_i} p_{i \to f}^{(\mathrm{spR})}(t)$$

$$= t \frac{2\pi}{\hbar^2} \frac{\hbar \omega_L n_L}{4 \varepsilon_0^2 V} \int_{\Delta \Omega_S} d\Omega_S \frac{\hbar \omega_S^3}{8\pi^3 c^3} |\eta^{(R)}|^2. \tag{12.38}$$

Thus, for both the stimulated and the spontaneous Raman effect the total transition probability has been determined by summation over all $\mathscr{E}_f$ for which $\mathscr{E}_f \approx \mathscr{E}_i$.

So far we have represented the incident radiation by one mode (ω_L) with n_L photons. In general, this does not correspond to reality. Therefore we introduce a spatial energy density $\sigma_L(\omega_L, \Omega_L)$ per unit frequency and per unit solid angle, which allows us to consider the incident photons in a frequency band $\Delta\omega_L$ and a solid-angle element $\Delta\Omega_L$. The transition probability given in (12.35) and (12.38) must then be summed over ω_L and Ω_L as well. Altogether one obtains the rate

$$\frac{\Delta P^{(R)}(t)}{\Delta t} = \frac{2\pi}{\hbar^2 4\varepsilon_0^2} \int_{\Delta\omega_L} d\omega_L \int_{\Delta\Omega_L} d\Omega_L\, \sigma_L(\omega_L, \Omega_L) \times \int_{\Delta\Omega_S} d\Omega_S \left\{\sigma_S(\omega_S, \Omega_S) + \frac{\hbar\omega_S^3}{8\pi^3 c^3}\right\} |\eta^{(R)}|^2. \tag{12.39}$$

This is the probability per unit time that the molecule passes over from the state $|\mathscr{E}_{A,0}\rangle$ into the state $|\mathscr{E}_{A,1}\rangle$ with the absorption of a laser photon in the frequency interval $\Delta\omega_L$ and the solid angle element $\Delta\Omega_L$, and with the emission of a Stokes photon of frequency ω_S into the solid-angle region $\Delta\Omega_S$, provided that at the position of the molecule the radiation densities are $\sigma_L(\omega_L, \Omega_L)$ and $\sigma_S(\omega_S, \Omega_S)$. The *mean scattered power* is $(\hbar\omega_S)(\Delta P^{(R)}/\Delta t)$. In case the radiation density σ_S of the Stokes photons is equal to zero, one obtains

$$\frac{\Delta P^{(\mathrm{spR})}}{\Delta t} = \frac{2\pi}{\hbar^2 4\varepsilon_0^2} \int_{\Delta\omega_L} d\omega_L \int_{\Delta\Omega_L} d\Omega_L \int_{\Delta\Omega_S} d\Omega_S\, \sigma_L(\omega_L, \Omega_L) \frac{\hbar\omega_S^3}{8\pi^3 c^3} |\eta^{(R)}|^2, \tag{12.40}$$

which is called the Kramers–Heisenberg formula for spontaneous Raman scattering.

Equation (12.39) may be called the rate of the stimulated and spontaneous emission of Stokes quanta. We now dispense with favoring particular solid angles and polarization directions of the radiation, and therefore we replace $\sigma_S(\omega_S, \omega_S)$ by $\sigma_S(\omega_S)/8\pi$ and $\sigma_L(\omega_L, \Omega_L)$ by $\sigma_L(\omega_L)/8\pi$. Then (12.39) can (for sufficiently small $\Delta\omega_L$) be written in the form

$$\frac{\Delta P^{(R)}}{\Delta t} = \Delta\omega_L\, \sigma_L(\omega_L) \{ B'_{01}\sigma_S(\omega_S) + A'_{01} \}, \tag{12.41}$$

where $\Delta\omega_L\, \sigma_L(\omega_L)$ is the energy of the incident radiation per unit volume. A comparison of (12.41) with the formula for the stimulated and spontaneous one-photon emission derived from (3.125) and (3.123), namely

$$\frac{\Delta P^{(\text{one-ph})}}{\Delta t} = B_{ul}\zeta(\omega_{ul}) + A_{ul}, \tag{12.42}$$

shows the same structure. The terms A'_{01} and B'_{01} depend not only on the molecular transition frequencies and transition moments, but also on the frequencies ω_S and ω_L. The quotient A'_{01}/B'_{01} depends only on universal constants and on ω_S, and it has the same value as the quotient A_{ul}/B_{ul} of the Einstein coefficients provided that the Stokes frequency is identified with the atomic transition frequency ω_{ul}. This reveals the general significance of the result (12.41).

In the factor in braces in (12.39), the second term represents the contribution of the spontaneous Raman effect. In this quantum-theoretical treatment it arises naturally, whereas in Section 12.1.1 the temperature-independent zero-point vibration had to be introduced artificially and is inconsistent with the classical treatment.

In the classical treatment the derivative $\alpha_{(1)\bullet\bullet}$ of the polarizability with respect to the vibration coordinate played an essential role. It was introduced phenomenologically in Section 12.1.1, where also an order-of-magnitude estimation of its value from empirical findings was performed. Information on this important quantity may be gained by a quantum-theoretical basis if we, so to speak, translate the classical Placzek theory into the quantum theory, which will be done in the following.

Since we now want to direct our attention only to the basic features, we start from simple conditions. The problem is treated one-dimensionally. We consider a molecule with only one normal vibration; the deviations from the equilibrium position are assumed to be small. Therefore the Hamiltonian operator $\hat{H}_M$ of the unperturbed molecule (in harmonic approximation) can be set equal to

$$\hat{H}_M = \frac{\hat{p}^2}{2\mathscr{M}} + \frac{\mathscr{M}\omega_M^2\hat{x}^2}{2}. \tag{12.43}$$

Here $\mathscr{M}$ is the reduced mass, and $\hat{x}$ and $\hat{p}$ are the canonically conjugate operators assigned to the position (deviation from the equilibrium) and the momentum. In establishing the interaction operator for the Raman scattering we take into account (12.3) and (12.5) and obtain

$$\hat{H}^C = -\frac{1}{2}\left(\frac{\partial \alpha}{\partial x}\right)\hat{x}\hat{E}^2, \tag{12.44}$$

into which $\partial\alpha/\partial x$ (as a c-number) as well as the distortion $\hat{x}$ and the field strength $\hat{E}$ (as operators) enter. According to (A.136b) $\hat{x}$ is related to the creation and annihilation operators of the molecular vibration according to

$$\hat{x} = \sqrt{\frac{\hbar}{2\mathscr{M}\omega_M}}\,(\hat{a}^\dagger + \hat{a}). \tag{12.45}$$

$\hat{p}$ is treated analogously.

Suppose that the radiation field again is made up of a radiation mode of frequency ω_L and a radiation mode (Stokes mode) of frequency ω_S. Therefore the total field strength is

$$\hat{E} = i\sqrt{\frac{\hbar}{2\varepsilon_0 V}}\,\left\{\omega_S^{1/2}\left(\hat{a}_S - \hat{a}_S^\dagger\right) + \omega_L^{1/2}\left(\hat{a}_L - \hat{a}_L^\dagger\right)\right\}. \tag{12.46}$$

In analogy to (12.29) we can write the initial and final states in the form of

$$\begin{aligned} |\mathscr{E}_i\rangle &= |v' = 0\rangle |n_S, n_L\rangle, \\ |\mathscr{E}_f\rangle &= |v' = 1\rangle |n_S + 1, n_L - 1\rangle, \end{aligned} \tag{12.47}$$

where v' is the vibrational quantum number. From this the transition probability, according to (A.120), we have

$$\begin{aligned} p_{i\to f}^{(R)}(t) &= \frac{t^2}{\hbar^2}\left|\langle\mathscr{E}_f\,|\hat{x}\hat{E}^2|\,\mathscr{E}_i\rangle\right|^2 \\ &\quad\times\frac{1}{4}\left(\frac{\partial\alpha}{\partial x}\right)^2 \operatorname{sinc}^2\left[\frac{(\mathscr{E}_i - \mathscr{E}_f)t}{2\hbar}\right]. \end{aligned} \tag{12.48}$$

The operator $\hat{x}\hat{E}^2$ consists of 24 terms with different operator products. For the given initial and final states only the term $\hat{a}^\dagger\hat{a}_S^\dagger\hat{a}_L$ can yield a nonvanishing contribution to the matrix element, it is proportional to $\sqrt{n_L}\sqrt{n_S + 1}$. Therefore we have

$$\begin{aligned} p_{i\to f}^{(R)} &= \frac{t^2}{\hbar^2}\,\frac{1}{4\varepsilon_0^2 V^2}\hbar\omega_L n_L \hbar\omega_S(n_S + 1) \\ &\quad\times\frac{1}{8}\left(\frac{\partial\alpha}{\partial x}\right)^2 \frac{\hbar}{\mathscr{M}\omega_M}\operatorname{sinc}^2\left[\frac{(\mathscr{E}_i - \mathscr{E}_f)t}{2\hbar}\right]. \end{aligned} \tag{12.49}$$

This relation is of exactly the same structure as (12.32), and therefore it is possible to give an atomistic interpretation of the phenomenological quantity $\partial\alpha/\partial x$:

$$\left(\frac{\partial\alpha}{\partial x}\right)^2 = \frac{8\mathscr{M}\omega_M}{\hbar}|\eta^{(R)}|^2. \tag{12.50}$$

Typical values for $\mathscr{M}$ and ω_M were given above; for the determination of $|\eta^{(R)}|^2$ a model of the molecular system is needed. This can be accomplished by means of a four-level system (where two vibrational levels are in the electronic ground state and two in the first excited state); see Ref. 12.3. The transition moments in $\eta^{(R)}$ can be determined on the basis of the so-called adiabatic approximation, in which the total eigenfunction of the molecule in the position representation is a product of an electronic eigenfunction with fixed nuclear distance and an eigenfunction assigned to the motion of the nuclei. In the off-resonance case $|\eta^{(R)}|^2$ can be estimated to be of the order of 10^{-81} $(\mathrm{A\,s\,m})^4(\mathrm{W\,s})^{-2}$. Altogether one obtains $\partial\alpha/\partial x \sim 10^{-29}$ $\mathrm{A\,s\,m\,V^{-1}}$. This quantum-theoretical estimate corresponds to values obtained from determinations of the cross section of intense Stokes lines when performing measurements on the spontaneous Raman effect [cf. the interpretation of the parameters in (12.11)].

The discussion of general methods in Part I indicated (Section 3.2.2.2) that for treating multiphoton processes it is often helpful to establish effective interaction operators that in the first order of perturbation theory yield results on transition probabilities, transition rates, and so on. From the considerations presented above it can immediately be seen that, by virtue of (12.48), the (hermitian) interaction operator tailored to the Raman effect can be set equal to

$$\hat{H}^{(R)} = \mathfrak{R}\hat{a}^{\dagger}\hat{a}_S^{\dagger}\hat{a}_L + \mathfrak{R}^*\hat{a}\hat{a}_S\hat{a}_L^{\dagger}. \tag{12.51}$$

From (12.48) and (12.50) it can be deduced how to choose $\mathfrak{R}$. The interaction operator $\hat{H}^{(R)}$ is a three-boson operator containing creation and annihilation operators of the molecular vibration and the Stokes and laser modes of the electromagnetic field. The first term yields the transition rate for the emission of a Stokes quantum with simultaneous annihilation of a laser photon and vibrational excitation. Its contribution to the total transition probability is proportional to $n_L(n_S + 1)$. The second term gives the transition rate for the emission of a laser photon with the molecule making a transition to the vibrational ground level and a Stokes photon being annihilated simultaneously. The latter process reveals a certain feature of anti-Stokes scattering, namely the generation of photons of higher frequencies at the expense of photons of lower frequencies. Its contribution to the total transition probability is proportional to $(n_L + 1)n_S$. In both processes the terms proportional to the products $n_L n_S$ are assigned to stimulated processes, whereas the terms proportional to either n_L or n_S alone represent the corresponding spontaneous processes. It will be shown in Section 12.2.2 that three-boson operators of the form (12.51) are also capable of representing the scattering of light by other excitation states, namely the scattering by quasiparticles such as phonons or polaritons.

12.1.3. Behavior of Stokes and Anti-Stokes Waves in a Medium

Now we turn our attention to relationships and process parameters that can directly be related to empirical findings. Our presentation is based on the results of Sections 12.1.1 and 12.1.2. It is however necessary to generalize the transition probabilities and rates derived there for "sharp" energy levels and photon energies by introducing line-shape functions, as we have already done in an analogous way when dealing with one-photon processes (Section 3.3.2). We start with the treatment of the stimulated Raman scattering and ignore the contributions of the spontaneous Raman effect in the transition probabilities. Under these conditions we can use the semiclassical theory with classical fields as well as with susceptibility functions that are related, by means of the third-order perturbation theory (see end of Section 4.5), to atomic quantities such as transition moments, transition energies, and relaxation times. It provides results that are equivalent, over a wide range of conditions, to those of the fully quantum-theoretical description.

The following experimental condition is required for the semiclassical description: at least two coherent waves are present in the medium, the laser wave and the Stokes wave. We shall also mention the process where the medium is only irradiated by the laser wave and a scattered stimulated Stokes wave is generated "from fluctuations." However, in principle this process must be described on a fully quantum-theoretical basis (see Section 12.2.3).

12.1.3.1. Amplification and Generation of the Stokes Wave. Empirical findings reveal that if a Raman-active medium is irradiated with a high-power laser beam, stimulated first-order or higher-order Stokes and anti-Stokes waves result [cf. (12.28)]. If the laser power is low enough, then only the first-order Stokes wave with the frequency $\omega_{b=-1} = \omega_L - \omega_M$ appears. Under this condition it suffices to deal only with the interaction between laser wave and Stokes wave, which in the following will be accomplished by means of simple classical description.

We start from the assumption that at $z = 0$ a laser wave and a (weak) Stokes wave enter the medium. The interaction between Stokes wave and laser wave, both of which are assumed to be polarized in the x direction, is represented by the relations

$$\hat{P}_x^{(3)}(\omega_S) = \varepsilon_0 \chi_{xxxx}^{(3)}(\omega_S; \omega_S, \omega_L, -\omega_L)\left|\hat{E}_x(\omega_L)\right|^2 \hat{E}_x(\omega_S), \quad (12.52a)$$

$$\hat{P}_x^{(3)}(\omega_L) = \varepsilon_0 \chi_{xxxx}^{(3)}(\omega_L; \omega_L, \omega_S, -\omega_S)\left|\hat{E}_x(\omega_S)\right|^2 \hat{E}_x(\omega_L), \quad (12.52b)$$

where $\chi_{xxxx}^{(3)}$ are the corresponding third-order susceptibility functions. From the general relation (1.200a) we obtain for the special process under study the equations

$$\frac{d\bar{E}_x(\omega_S, z)}{dz} = +\frac{i\omega_S^2}{2c^2k(\omega_S)}\chi^{(3)}_{xxxx}(\omega_S; \omega_S, \omega_L, -\omega_L)|\bar{E}_x(\omega_L, z)|^2\bar{E}_x(\omega_S, z), \tag{12.53a}$$

$$\frac{d\bar{E}_x(\omega_L, z)}{dz} = +\frac{i\omega_L^2}{2c^2k(\omega_L)}\chi^{(3)}_{xxxx}(\omega_L; \omega_L, \omega_S, -\omega_S)|\bar{E}_x(\omega_S, z)|^2\hat{E}_x(\omega_L, z) \tag{12.53b}$$

for the slowly varying wave amplitudes.

We have to emphasize that in describing the amplification of the Stokes wave a factor of the form $e^{i\Delta k z}$ does not occur, whereas it appears in the general equation (1.200a) and plays an important practical role in the frequency-conversion and parametric-amplification processes discussed in the preceding chapter. Thus special measures to increase the coherence length are not necessary for SRS as they were for SHG.

Let us study the near-resonance case where the frequency deviation $\Delta\omega = |(\omega_L - \omega_S) - \omega_M|$ is much less than the line width of the spontaneous Raman transition and (12.23) holds; thus the susceptibilities in (12.53) are purely imaginary. Introducing $i|\chi_R|$, upon integration (12.53a) and (12.53b) yield

$$\bar{E}_x(\omega_S, z) = \bar{E}_x(\omega_S, 0)\exp\{\frac{\omega_S^2}{2c^2k(\omega_S)}|\chi_\text{R}|\int_0^z dz'\,|\bar{E}_x(\omega_L, z')|^2\}, \tag{12.54a}$$

$$\bar{E}_x(\omega_L, z) = \bar{E}_x(\omega_L, 0)\exp\{-\frac{\omega_L^2}{2c^2k(\omega_L)}|\chi_R|\int_0^z dz'\,|\bar{E}_x(\omega_S, z')|^2\}. \tag{12.54b}$$

When considering the amplitude of the Stokes wave described in (12.54a), we recognize that it is amplified *independently* of the phase. In a more general way this result can be expressed as follows: In the interaction between two waves (with the frequencies ω_1, ω_2) in a Raman-active medium, where $\omega_1 - \omega_2 \approx \omega_M$, the low-frequency wave is amplified at the cost of the high-frequency wave (this is always true if an occupation-number inversion of the vibrational levels involved can be excluded). The process proceeds independently of the phases φ_L, φ_S of the waves involved. This independence indicates that the molecule is capable of taking up the momentum difference $\hbar k(\omega_L) - \hbar k(\omega_S)$ between laser and Stokes photon.

From the set of equations (12.53) one may draw conclusions as to the energy balance. Since the phase of the amplitudes $\bar{E}_x(\omega_L, z)$ and $\bar{E}_x(\omega_S, z)$ does not depend on z, the set of equations preserves the same structure if we substitute for the amplitudes their absolute values. If we then multiply the first

(Stokes) equation by $n^{(L)}(\omega_S)|\bar{E}(\omega_S, z)|/\omega_S$ as well as the second equation by $n^{(L)}(\omega_L)|\bar{E}(\omega_L, z)|/\omega_L$, the resulting set of equations reveals that the quantity

$$\frac{n^{(L)}(\omega_S)}{\omega_S}|\bar{E}_x(\omega_S, z)|^2 + \frac{n^{(L)}(\omega_L)}{\omega_L}|\bar{E}_x(\omega_L, z)|^2$$

does not depend on z. This fact can be interpreted as the conservation law of the total photon number

$$\gamma_{\text{ph}}(\omega_L, z) + \gamma_{\text{ph}}(\omega_S, z) = \text{const}, \tag{12.55}$$

because the number density $\gamma_{\text{ph}}(\omega, z)$ of photons with frequency ω incident on a plane at right angles to the direction of propagation is proportional to the intensity $\varepsilon_0 c|\bar{E}_x(\omega, z)|^2 n^{(L)}(\omega)/2$. In stimulated Raman scattering from each of the annihilated laser quanta one Stokes quantum is created, but not all the energy is transferred to the Stokes quanta. The energy difference $\hbar(\omega_L - \omega_S)$ per quantum excites one molecule. Finally, relaxation processes convert this excitation energy into heat.

The spatial behavior of the amplitudes of the Stokes and laser waves described in (12.54a) and (12.54b) results from the general solution of the system of differential equations (12.53). In the following we restrict ourselves to low conversion rates, for which the laser amplitude can be considered constant. Thus we have

$$\bar{E}_x(\omega_S, z) = \bar{E}_x(\omega_S, 0)\, e^{\frac{1}{2}g_S z}, \tag{12.56}$$

where the coefficient g_S of the power amplification is given by

$$g_S = 2\mu_0 \frac{\omega_S}{\varepsilon^{(L)}(\omega_S)} |\chi_R| S(\omega_L, 0). \tag{12.56a}$$

Here $|\chi_R|$ is the Raman susceptibility near resonance from (12.23) and $S(\omega_L, 0)$ is the Poynting vector modulus of the laser wave. Using typical values of the parameters as given in Sections 12.1.1 and 12.1.2, the power-amplification coefficient of the Stokes wave is of the order of $g_S \sim (10^{-11}\ \text{m/W}) S(\omega_L, 0)$.

The Stokes amplitude and thus the power of the Stokes wave increase exponentially with z. If z is fixed, the power of the Stokes wave increases exponentially with the laser power $S(\omega_L, 0)$. With large values of the propagation distance and/or large values of the laser power, the assumption of a z-independent conversion rate is violated because of the high conversion occurring in such cases. Then the saturation region is reached as shown in Fig. 12.5; $\log S(\omega_S, z)$ becomes constant. The limit of the Stokes intensity is reached when all the laser photons are converted into Stokes photons. It should be mentioned that this consideration ignores other nonlinear effects, such as self-focusing.

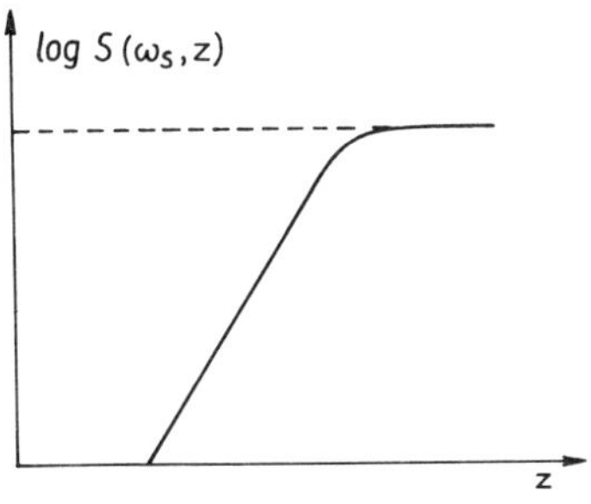

Fig. 12.5. Stokes intensity as function of the length z of the sample (schematic).

Extending the consideration to arbitrary angles between the propagation directions of the laser and Stokes wave, calculations show that under excitation by an infinitely extended plane laser wave a *continuous angular distribution* of the stimulated Stokes radiation results (as with the spontaneous Raman effect). For a finite cross section of the laser beam the interaction length of the two waves, which determines the amplification of the Stokes wave, depends on the angle between the two propagation directions $k_\bullet(\omega_S)$ and $k_\bullet(\omega_L)$. Hence, the angular distribution of the stimulated Stokes radiation depends strongly on the geometry of the experimental setup.

Now the susceptibilities so far introduced phenomenologically will be related to parameters of the atomic systems of the Raman-active medium. This is done by making use of the semiclassical presentation given at the end of Section 4.5 and of the results of Sections 12.1.1 and 12.1.2. It is however necessary to generalize the transition probabilities and rates derived there for "sharp" energy levels and photon energies by introducing line-shape functions (as we did for one-photon and multiphoton processes in preceding chapters). We may use the Raman susceptibility given in (4.64). On the assumption that initially all particles are in the ground state, that the vibration frequency of the transition is approximately equal to $\omega_L - \omega_S$, and that the process takes place in the off-resonance region of one-photon transitions, the Stokes susceptibility in the frequency domain becomes

$$\underset{\smile}{\varkappa}^{(3)}_{kpqr}(\omega_S, \omega_L, -\omega_L) = \frac{\gamma}{\varepsilon_0 6\hbar} \frac{\omega_{10} - \omega_L + \omega_S + i\tau_{10}^{-1}}{(\omega_{10} - \omega_L + \omega_S)^2 + \tau_{10}^{-2}}$$

$$\times \alpha_{01kq}(-\omega_S, \omega_L)\alpha^*_{01pr}(-\omega_S, \omega_L) \qquad (12.57)$$

containing the *Raman tensor*

$$\alpha_{01kq}(-\omega_S, -\omega_L) = \frac{1}{\hbar}\sum_\beta \left\{ \frac{d_{0\beta,k}d_{\beta 1,q}}{\omega_{\beta 0} + \omega_S} + \frac{d_{0\beta,q}d_{\beta 1,k}}{\omega_{\beta 0} - \omega_L} \right\}. \qquad (12.57a)$$

It should be mentioned that the connection with the susceptibilities for discrete frequencies is given by

$$\chi^{(3)}_{kpqr}(\omega_S; \omega_S, \omega_L, -\omega_L) = \tfrac{3}{4}\underset{\smile}{\varkappa}^{(3)}_{kpqr}(-\omega_S, -\omega_L, +\omega_L). \tag{12.58}$$

Between the interaction constant $\eta^{(R)}$ [cf. (12.32a)] characteristic of the Raman effect [which according to (12.50) also determines the polarizability derivative] and the Raman tensor α_{01kq}, the following relationship can be established:

$$\eta^{(R)} = \sum_{k,q} \alpha^{*}_{01kq}(-\omega_S, \omega_L)(e_S)_k(e_L)_q, \tag{12.59}$$

where the unit vectors $e_{L\bullet}$ and $e_{S\bullet}$ specify the polarization direction of the laser and Stokes waves. In Fig. 4.8 the real and imaginary parts of $\underset{\smile}{\varkappa}^{(3)}$ were plotted. In discussing (4.65) estimations of the resonant and nonresonant terms were presented. From (12.57) it can be seen that $\mathrm{Im}\{\underset{\smile}{\varkappa}^{(3)}(\omega_S, \omega_L, -\omega_L)\}$ is greater than zero. Thus, by using a quantity determined by atomic parameters, the previously mentioned fact is confirmed that a low-frequency signal wave (ω_S) is amplified in the field of a higher-frequency pump wave $(\omega_L \approx \omega_{10} + \omega_S)$. Neglecting the attenuation of the pump wave and considering fields in an anisotropic medium that are polarized in the same direction, we obtain the amplification coefficient

$$g_S = \frac{\mu_0^2 c^2 \omega_L \gamma}{2(n^{(L)})^2 \hbar} \left\{ \frac{\pi \tau_{10}^{-1}}{(\omega_{10} - \omega_L + \omega_S)^2 + \tau_{10}^{-2}} \right\} |\alpha_{01}|^2 S(\omega_L). \tag{12.60}$$

In other physical situations the line-shape functions with Lorentz profile, appearing here in braces, can be replaced with more general line-shape functions, as was explained in solving analogous problems in one-photon and multiphoton processes.

So far we have dealt with the behavior of two coherent waves in a Raman-active medium; in the following we want to sketch what happens when the Raman-active medium is irradiated by only one coherent wave (the laser wave) and a coherent Stokes wave is generated via the *spontaneous* Raman effect.

The fully quantum-theoretical foundations can be inferred from (12.39), where the second term in the braces describes the generation of Stokes photons even if their initial density is zero. In this way a nonzero Stokes-photon density is built up, and by way of the first term in the braces on the right-hand side of (12.39), a stimulated Raman effect occurs. The foundations of the connection between the quantum-theoretical and the classical description are given in (12.49) and (12.50).

The classical description of the Raman effect is based on two terms in the formula (12.9) for the dipole moment of the molecules. One of them, the second, is characterized by a linear dependence of the dipole moment on the field strength and by the fact that a molecule vibrates with a fixed phase, independent of that of other molecules. This leads to the spontaneous Raman effect, in which radiation emerging from different molecules is incoherent even though the laser light is coherent. The scattering intensity $S(\omega_S)$ is linear in the excitation intensity $S(\omega_L)$. The other relevant term in (12.9) is the third; it leads to the mechanism of the stimulated Raman effect. A Stokes wave at the entrance to the sample at $z = 0$, which may be weak, can be amplified by interacting with a laser wave according to (12.54a). The weak Stokes wave at $z = 0$ can be supposed either to have been provided by a second suitable light source or to have been generated by the spontaneous Raman effect near $z = 0$. In the latter case we see that by irradiating the Raman-active medium with the laser wave alone, a Stokes wave essential to the stimulated process is generated. This was only a preliminary explanation. More correct quantitative information on this generation of stimulated scattered radiation will be given later on: From the discussion following (12.95) we shall learn that the behavior of isolated molecules can be regarded as a limiting case of the behavior of long-wave optical phonons, whose stimulated scattering interaction with an electromagnetic wave will be treated in detail in Section 12.2.3.1.

If a Raman-active medium is irradiated by laser light of small intensity $S(\omega_L)$, one observes a *linear* increase of the scattering intensity as a function of $S(\omega_L)$. For higher laser intensities, an *exponential* dependence of the scattering intensity on $S(\omega_L)$ follows, which characterizes the region of the stimulated Raman effect under the conditions leading to (12.56). At still higher laser intensities a saturation region appears. In experiments laser intensities between 10^{11} and 10^{15} W/m^2 are used.

If the Stokes wave in the Raman-active medium sustains linear losses (represented by the absorption coefficient $\mathscr{k}_{\text{abs}}$), the effective amplification $g_{\text{eff},S}$ of the Stokes wave is no longer given by the quantity g_S from (12.60), but by

$$g_{\text{eff},S} = g_S - \mathscr{k}_{\text{abs}}. \tag{12.61}$$

Only if $g_{\text{eff},S}$ is positive is an effective amplification of the Stokes wave possible.

In the stimulated Raman effect the light scattered from a volume element that is excited by a spatially coherent laser wave is coherent. Observations of the resulting interferences have been carried out with double-slit arrangements.

12.1.3.2. Amplification and Generation of the Anti-Stokes Wave. This subsubsection is concerned with the interaction between three waves: the laser wave, the Stokes wave, and the anti-Stokes wave. The anti-Stokes wave is assumed to be amplified. We shall proceed as in Section 12.1.3.1 in discussing stimulated Raman scattering of the Stokes wave, and assume three waves with nonzero amplitudes at $z = 0$.

The nonlinear third-order polarization of frequency $\omega_{b=+1} = \omega_A$ can be obtained according to

$$\hat{P}_x^{(3)}(\omega_A) = \hat{P}_x^{(3)}(\omega_A; \omega_A, \omega_L, -\omega_L) + \hat{P}_x^{(3)}(\omega_A; -\omega_S, \omega_L, \omega_L) \quad (12.62)$$

in two ways. The polarization part $\hat{P}_x^{(3)}(\omega_A; \omega_A, \omega_L, -\omega_L)$ reveals that the higher-frequency anti-Stokes wave is attenuated in favor of the low-frequency laser wave, and thus it displays the same basic features as the amplification of the Stokes wave dealt with in the preceding subsection. We intend to study the process by assuming that the amplitude of the anti-Stokes wave is much less than the amplitudes of the laser and the Stokes wave. Then the polarization part $\hat{P}_x^{(3)}(\omega_A; \omega_A, \omega_L, -\omega_L)$ can be neglected. For the susceptibility assigned to the second polarization part in (12.62) we have near resonance

$$\chi_{xxxx}^{(3)}(\omega_A; -\omega_S, \omega_L, \omega_L) = +i|\chi_R|, \quad (12.63)$$

where $|\chi_R|$ is the quantity given in (12.11). The amplitude equation for $\bar{E}_x(\omega_A, z)$ is

$$\frac{d\bar{E}_x(\omega_A, z)}{dz} = -\frac{\omega_A^2}{2c^2k(\omega_A)}|\chi_R|\bar{E}_x^2(\omega_L, z)\bar{E}_x^*(\omega_S, z)e^{-iz\Delta k}, \quad (12.64)$$

where $\Delta k = 2k(\omega_L) - k(\omega_A) - k(\omega_S)$.

In contrast to the treatment of the Stokes wave in Section 12.1.3.1 here the relations between the momenta (and therefore those between the propagation vectors as well) obviously play an important role. To display this fact we may separate (12.64) into an equation for the magnitude of the wave amplitude and another one for the phase φ_A of the anti-Stokes wave. The first equation is

$$\frac{d|\bar{E}_x(\omega_A, z)|}{dz} = -\frac{\omega_A^2}{2c^2k(\omega_A)}|\chi_R||\bar{E}_x(\omega_L, z)|^2|\bar{E}_x(\omega_S, z)|$$

$$\times\cos\{z\,\Delta k - (2\varphi_L - \varphi_A - \varphi_S)\}. \quad (12.65)$$

The amplification of the anti-Stokes wave requires that

$$\frac{d}{dz}|\bar{E}_x(\omega_A, z)| > 0$$

This holds if

$$\frac{\pi}{2} < z\,\Delta k - (2\varphi_L - \varphi_A - \varphi_S) < \tfrac{3}{2}\pi. \quad (12.66)$$

It can be seen that at the entrance of the sample—that is, at $z = 0$—an anti-Stokes wave experiences its maximum amplification if $\varphi_A = \pi + 2\varphi_L - \varphi_S$. The amplification of the wave continues along a length

$$l_c = \frac{1}{4\Delta k}. \tag{12.67}$$

This is followed by an attenuation. Examining the amplitude equation of the Stokes wave corresponding to (12.64), it becomes apparent that the amplification of the anti-Stokes wave is accompanied by an attenuation of the Stokes wave.

In order to make the interaction length for (positive) amplification of the anti-Stokes wave as large as possible and thus to achieve high anti-Stokes intensities, the wave number mismatch Δk has to be made as small as possible. Methods for this purpose were presented in the discussion of frequency conversion in Section 11.3.2. They can be employed in analogous cases. In our ansatz the wave vectors all lie in the z direction. Because of the behavior of the dispersion curve, (cf. Fig. 11.4), in forward scattering the value $\Delta k = 0$ cannot be reached. If we do not restrict ourselves to exact forward-scattered waves but extend our observation to arbitrary angles between the wave vectors, then the requirement for maximum amplification yields the more general relation

$$2k_{\bullet}(\omega_L) - k_{\bullet}(\omega_A) - k_{\bullet}(\omega_S) = 0. \tag{12.68}$$

In general, at given absolute values of the vectors $k_{\bullet}(\omega)$ this relation can be satisfied provided the vectors make certain angles with one another. This explains the existence of an "emission cone" of the anti-Stokes radiation in the case of the presence of laser and Stokes waves of appropriate directions of propagation. The amplification of the anti-Stokes radiation propagating in the direction denoted by $k_{\bullet}(\omega_A)$ causes attenuation of the Stokes waves propagating in the direction $k_{\bullet}(\omega_S)$.

If the experimental setup is chosen so that in the neighborhood of the $k_{\bullet}(\omega_S)$ direction there exists a continuous angular distribution of the stimulated Stokes radiation (cf. Section 12.1.3.1), then the Stokes intensity is lower than at neighboring angles. Figure 12.6 presents a diagram of the propagation vectors of the laser, Stokes, and anti-Stokes waves for exact phase matching, the angles θ_A and θ_S between scattered and laser waves being greatly exag-

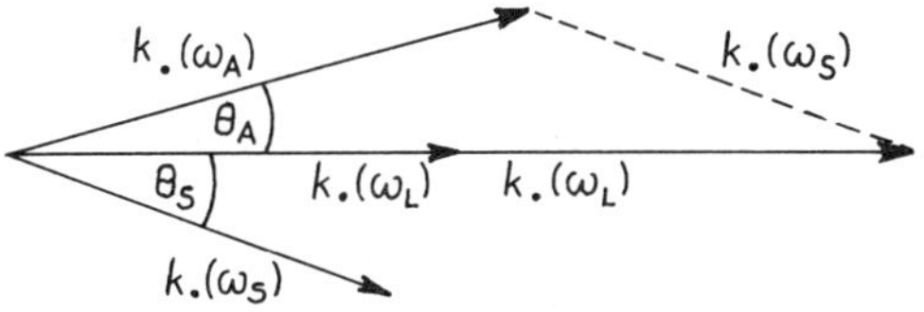

Fig. 12.6. Propagation vectors of the laser wave, Stokes wave, and anti-Stokes wave in the phase-matching case $\Delta k_{\bullet} = 0$.

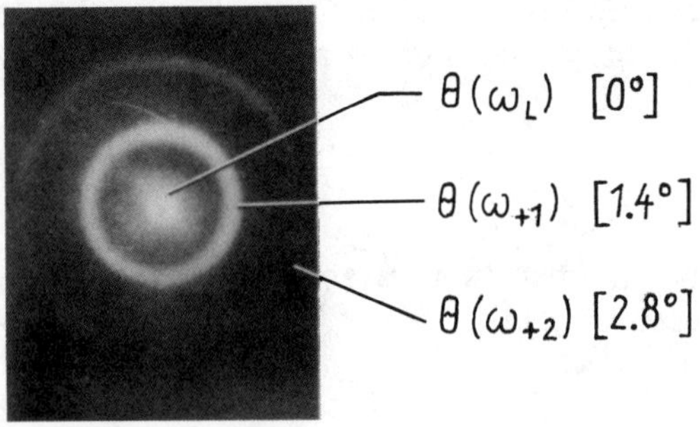

Fig. 12.7. Emission cones of anti-Stokes radiation for calcite (after Ref. 12.4).

gerated; their actual sizes amount to several degrees. These distinguished emission and absorption angles can be found in liquids and gases, that is, in disordered systems. However, this phenomenon may also occur in crystals: following Chiao and Stoicheff (Ref. 12.4), Fig. 12.7 gives the emission cones of the stimulated anti-Stokes radiation for the frequencies $\omega_{b=+1}$ and $\omega_{b=+2}$ for calcite. These experimental results can be obtained only if the stimulated Raman scattering is hardly influenced by other nonlinear effects.

Equation (12.62) reveals that it was stated that there are two different polarization parts at the frequency of the anti-Stokes wave. While the polarization part $\hat{P}_x^{(3)}(\omega_A; \omega_A, \omega_L, -\omega_L)$ describes a two-wave interaction process, $\hat{P}_x^{(3)}(\omega_A; -\omega_S, \omega_L, \omega_L)$ characterizes a three-wave interaction process whose amplitude equation is given by (12.64). The nature of the latter process differs from that of the two-wave processes, where the radiation was always accompanied by an *excitation* of the molecule involved. In the three-wave process, however, the energy of the initial state of the molecule equals that of its final state. We want to calculate the susceptibility corresponding to the more general case of nondiscrete frequencies, and therefore, according to

$$\chi^{(3)}(\omega_A; -\omega_S, \omega_L, \omega_L) = \tfrac{3}{4}\underset{\smile}{\varkappa}^{(3)}(+\omega_S, -\omega_L, -\omega_L),$$

we pass over to the susceptibility in the frequency domains $\underset{\smile}{\varkappa}^{(3)}(-\omega_S, \omega_L, \omega_L)$. From (4.59) we have

$$\underset{\smile}{\varkappa}^{(3)}(-\omega_S, \omega_L, \omega_L) = \frac{\gamma}{3\varepsilon_0 \hbar} \frac{(\omega_{10} - \omega_L + \omega_S) - i\tau_{10}^{-1}}{(\omega_{10} - \omega_L + \omega_S)^2 + \tau_{10}^{-2}} |\alpha_{01}(-\omega_S, \omega_L)|^2 \tag{12.69}$$

for the resonance term. This quantity is to be compared with the susceptibility of the Stokes wave. With negligible frequency dependence of the polarizability matrix elements α_{01} in the range from ω_S to ω_L one has

$$\chi^{(3)}(\omega_A; -\omega_S, \omega_L, \omega_L) = \chi^{(3)*}(\omega_S; \omega_S, \omega_L, -\omega_L). \tag{12.70}$$

It remains to determine the spatial behavior of the anti-Stokes wave in the three-wave process. With the assumption of exact resonance $\omega_L - \omega_S = \omega_{10}$, (12.64) holds; on dropping this assumption it is necessary to substitute $\chi^{(3)}(\omega_A; -\omega_S, \omega_L, \omega_L)$ for $i|\chi_R|$ in (12.64). Then the spatial derivative of the anti-Stokes amplitude will be proportional to $+i\chi^{(3)}\bar{E}^*(\omega_S, z)\bar{E}^2(\omega_L, z)$. Thus optimum matching (i.e., maximum amplification) requires not only that the wave-vector condition (12.68) be satisfied, but also the phase condition

$$\varphi_A = -\frac{\pi}{2} + \varphi_{\chi^{(3)}} - \varphi_S + 2\varphi_L. \tag{12.71}$$

Provided that (12.68) and (12.71) are fulfilled, and assuming that the variation of the (strong) amplitudes of the laser and Stokes waves can be neglected, the spatial dependence of the anti-Stokes amplitude is given by

$$|\bar{E}(\omega_A, z)| - |\bar{E}(\omega_A, 0)| = \frac{\omega_A^2}{2c^2 k(\omega_A)} |\chi^{(3)}(\omega_A; -\omega_S, \omega_L, \omega_L)| \times |\bar{E}(\omega_L, 0)|^2 |\bar{E}(\omega_S, 0)| z. \tag{12.72}$$

It can be seen that the power gain of the anti-Stokes wave is proportional to the square of $|\chi^{(3)}(\omega_A; -\omega_S, \omega_L, \omega_L)|$.

12.1.4. Specific Raman Scattering Processes and Applications

The stimulated Raman effect has been applied to the solution of many problems, above all to spectroscopic investigations and to the construction of new coherent radiation sources. Some of these important aspects will be discussed in the following.

12.1.4.1. Ordinary (Off-Resonance) Vibrational Raman Effect. After having given a general description of the behavior of Stokes and anti-Stokes waves in the ordinary vibrational Raman effect—the conditions of which were formulated in connection with (12.31)—we shall now consider some aspects of its application.

According to what was stated above, we may use the measurements of the Stokes amplification [see (12.60)] to determine the imaginary part, and those of the anti-Stokes amplification [see (12.72)] to determine the absolute value of the Raman susceptibility as a function of $\omega_L - \omega_S$. By virtue of the Kramers–Kronig relation (see, e.g., Ref. 38, Vol. 1, Appendix 3), from these two quantities the complex susceptibility can be calculated. This permits a comparison of the results of the two measurement methods and a discussion of the dispersion of the matrix elements of the polarizability derivative. The numerical estimation of the susceptibility follows the explanation given below (12.50); $\alpha_{10} \approx 3 \times 10^{-41}$ (A s m)^2 V^{-1} can be taken as a typical value.

Another important quantity is the transverse relaxation time τ_{10}; in condensed matter it is of the order of 10^{-12} s, and here it is set equal to 2×10^{-12} s. Hence with the number density 10^{28} m^{-3} we obtain

$$\operatorname{Im}\{\varepsilon_0 \chi^{(3)}(\omega_S = \omega_L - \omega_{10}; \omega_S, \omega_L, -\omega_L)\} \approx 5 \times 10^{-32} \text{ A s m/V}^3,$$

and with the ruby laser frequency $\omega_L = 2.7 \times 10^{15}$ s^{-1} we have

$$g_s/S(\omega_L) \approx 1.5 \times 10^{-11} \text{ m/W}$$

for the power amplification coefficient. That means that for laser intensities of 10^{12} W/m^2 there occur power amplification coefficients of ~ 10 m^{-1}.

The selection rules for the Raman effect differ qualitatively from those for one-photon processes. This can be seen from the Raman susceptibilities (12.57) and (12.59), which contain double products of the form $d_{0\beta}\, d_{\beta 1}$. This reveals that the initial and final molecular states may have prescribed parities (or the same parity). Therefore even in inversion-symmetric systems the Raman effect can be observed; in particular it provides information on transitions where (in the first order of perturbation theory) emission and absorption are forbidden (since the dipole-moment transition element vanishes for equal-parity states). Under appropriate experimental conditions the spontaneous Raman effect shows all the molecular eigenvibrations in which the derivative of the polarizability with respect to the corresponding normal coordinate does not vanish for symmetry reasons (these are the "Raman active" eigenvibrations).

In general, in the stimulated spectrum only lines of one or a few eigenvibrations can be observed (when starting from fluctuations at the Stokes frequencies). Mostly these are totally symmetrical eigenvibrations (generating intense narrow lines in the case of the spontaneous effect). As concerns the transitions associated with larger changes in the vibrational quantum numbers, the situation is as follows. For the spontaneous Raman effect additional transitions can be observed that are connected with changes $|\Delta n_v| > 1$, where Δn_v is the difference between the quantum numbers of the vibrational states involved. Because of the anharmonicity of the vibrational potential (see Fig. 12.1), the vibrational levels are not equidistant. These lines possess only small $\chi^{(3)}$ values, and therefore they do not appear in the stimulated effect. In contrast to the spontaneous effect, the stimulated effect shows the *exact* harmonics $\omega_L + b\omega_M$ with $b = \pm 1, \pm 2, \ldots$ of the fundamental vibration. Furthermore, the intensity even of higher-order Stokes lines can reach the order of the intensity of the exciting laser, whereas in the spontaneous effect the intensity of these lines decreases drastically with increasing $|\Delta n_v|$. The intensity of the spontaneous anti-Stokes radiation is proportional to the particle number in the excited state; it therefore vanishes at lower temperatures. This does not happen with the stimulated effect, where strong anti-Stokes waves arise from three-wave interaction in systems at thermodynamic equilibrium. It is confirmed by experiments that the stimulated anti-Stokes radiation can be caused by molecules in the vibrational ground state. This corresponds with what was stated in Section 12.1.3.2 about the relative strength of the contribution of the three-wave process to the anti-Stokes scattering.

The above mentioned empirically verifiable generation of higher-order Stokes lines with $b = -2, -3, -4, \ldots$ can be accounted for in terms of the two-wave process (see Section 12.1.3.1) that transfers power from the higher-frequency wave to the lower-frequency one. When the bth-order Stokes wave ($b < 0$) has been sufficiently increased as a result of amplification, this wave itself acts as a "laser wave" and builds up its first-order Stokes line, which possesses the frequency $(\omega_L + b\omega_M) - \omega_M = \omega_L + (b-1)\omega_M$. If a molecule gives rise to several Raman-active vibrations, it is the Stokes wave with the largest amplification coefficient that is built up fastest; that means, in general, the wave whose Raman susceptibility is the highest. According to (12.24a) and (12.60) this is the case for large values α_1^2 and $|\alpha_{01}|^2$, respectively, and for small values Γ_m and τ_{10}^{-1}. In the spontaneous Raman effect such lines give rise to intense and narrow lines. The Stokes waves with the strongest amplification are capable of decreasing the laser intensity greatly before the intensities of the other lines have reached the experimental detection limit, so that in the spectrum only lines belonging to very strong vibrational transitions appear.

Deviating from the conditions of the ordinary vibrational Raman effect, in the *resonance vibrational Raman effect* (Ref. 12.5) the laser frequency ω_L approaches the one-photon transition frequency of an excited electronic level (cf. Fig. 12.3). When calculating the corresponding susceptibilities, the relaxation processes of the resonance intermediate states must not be neglected. Estimations reveal that the susceptibilities of the resonance Raman effect may be increased by some orders of magnitude over those of the ordinary Raman effect.

12.1.4.2. Inverse Raman Effect. The results on the interaction of a laser wave with a Stokes wave given in Section 12.1.3.1 permit a straightforward explanation of the inverse Raman effect. From the set of equations (12.54a, b) and (12.55) it can be directly seen that observation of the stimulated Raman effect is feasible by amplifying the lower-frequency Stokes wave as well as by attenuating the higher-frequency laser wave. The latter method is what is called the *inverse Raman effect*.

Let us illustrate this by discussing the following experiment (Ref. 12.6). A laser wave (frequency ω_L) and additional radiation of continuous spectral distribution with a broad frequency band pass through a Raman-active medium. The frequency band covers the frequencies $\omega_L + \omega_{M,j}$, where the $\omega_{M,j}$ are meant to represent the frequencies of various Raman-active molecular vibrations. At the frequency $\omega = \omega_L + \omega_{M,j}$ an absorption line can be observed; the absorption strength depends on the intensity of the laser radiation.

An interpretation can be given with the help of the set of equations (12.54). Since the phase relations are of no importance in the process, it is not necessary to use two coherent waves. It suffices to consider one coherent wave (the laser wave) and one incoherent wave (the radiation with the spectral continuum). The laser wave used in the experiment here acts as the lower-frequency wave, whereas the continuum radiation acts as the higher-frequency one; it is attenuated in the inverse Raman effect. This process gives rise to absorption from the continuous spectrum, which allows the detection of the Raman lines.

A particular advantage of the inverse Raman effect is that it yields a large number of lines (in principle, all the lines appearing in the spontaneous Raman effect). Experimental realization requires that a substance be simultaneously irradiated by an intense laser pulse and a radiation pulse with a short-wavelength continuous spectrum. This can be brought about (Ref. 12.6) in the following way: the frequency of the radiation of a ruby laser is doubled in a KDP crystal, the frequency-doubled radiation excites the fluorescence of a dye, and this fluorescent radiation together with the radiation from the ruby laser (ω_L) is fed into the medium to be investigated (a schematic representation of the experiment and the spectra is given in Fig. 12.8). Thereby the required temporal and spatial superposition of the two waves is guaranteed. In this way, using for example pulse lengths of the order of 10^{-8} s, it is possible to record the benzene and toluene spectra of the CH– and C–C compounds with sufficiently high spectral resolving power. On the transition to shorter pulses

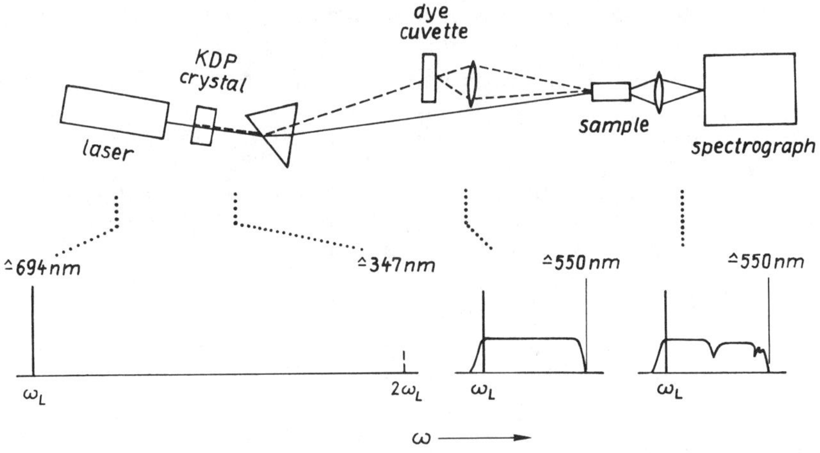

Fig. 12.8. Schematic representation of an experimental device and spectra in inverse Raman scattering (after Ref. 12.6).

(picosecond pulses) the resolving power can be increased. An increase in sensitivity, which is particularly important for the detection of molecules in mixtures, can be achieved by employing the method of intracavity absorption, which is for example described in Ref. 12.7.

12.1.4.3. Active Raman Scattering. The term *active spectroscopy* describes spectral investigation of a medium whose ordinary state (thermodynamic equilibrium) has been changed by *activation* via external influences. In the case of active Raman scattering, a Raman-active vibration of molecules is excited by two intense laser waves (with frequencies ω_1, ω_2, wave vectors $k_{1\bullet}$, $k_{2\bullet}$, wave amplitudes $\bar{E}_{1\bullet}$, $\bar{E}_{2\bullet}$, and intensities I_1, I_2) whose frequency difference is nearly equal to the molecular vibrational frequency ω_M [cf. (12.19)]. The vibrations excited in the medium by the laser waves exhibit fixed phase differences for molecules in different regions. The wave number of such a spatially coherent wave is $k_{v\bullet} = k_{1\bullet} - k_{2\bullet}$, its amplitude is proportional to $|\bar{E}_{1\bullet}\bar{E}_{2\bullet}|$. If the activated medium is irradiated by a probe beam (ω_L, $k_{L\bullet}$), Raman scattering by the spatially coherent molecular vibrations gives rise to Stokes and anti-Stokes radiation with the frequencies $\omega_S = \omega_L - \omega_M$ and $\omega_A = \omega_L + \omega_M$. This *active Raman scattering* combines the advantages of spontaneous and stimulated Raman scattering (Refs. 12.8, 12.9).

Because of momentum conservation the amplitudes of the corresponding scattered waves reach their maximum intensities in the directions of the vectors

$$k_{S\bullet} = k_{L\bullet} - (k_{1\bullet} - k_{2\bullet}), \tag{12.73a}$$

$$k_{A\bullet} = k_{L\bullet} + (k_{1\bullet} - k_{2\bullet}). \tag{12.73b}$$

The conditions of propagation for Stokes scattering are demonstrated in Fig. 12.9. Under the conditions given in (12.73), in the amplitude equations for the scattered waves the high-spatial-frequency terms vanish in the directions $k_{S\bullet}$ and $k_{A\bullet}$, and the amplitude of the scattered wave is proportional to the length L passed by the beam in the medium. Thus the scattering intensity is found to be proportional to L^2. The scattering intensities $I_{S,A}$ are equal for the directions $k_{S,A\bullet}$ and proportional to the product $I_1 I_2 I_L$.

The phenomenon just discussed is to be regarded as the coherent part of the active Raman scattering. In addition to the coherent part there is an incoherent part originating from the incoherent vibrations of the molecules under a

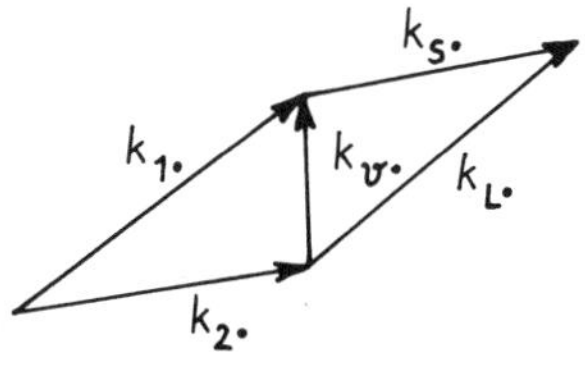

Fig. 12.9. Propagation vectors in the Stokes case.

predetermined occupation of the ground state and the excited vibrational level. The two parts differ from each other not only in their directional dependence (the scattered radiation of the coherent part is sharply directed in the directions of $k_{S\bullet}$ and $k_{A\bullet}$, and that of the incoherent part is continuously distributed), but also in their time behavior. For illustration we consider the anti-Stokes radiation. When the two strong laser beams with frequencies ω_1 and ω_2 are suddenly switched off, the scattered radiation directed along $k_{A\bullet}$ decays with the phase decay time of the molecular vibrations; this happens because [cf. (3.92)] the coherent active Raman radiation is due to the *polarization* of the medium at frequency ω_M. The continuously distributed Raman radiation, however, decays with the longitudinal (energy) relaxation time [cf. (3.89)]. This is the time needed for the *reoccupation* of the vibrational levels from the excitation caused by the strong laser waves (frequencies ω_1, ω_2) to the equilibrium state.

Active Raman spectroscopy can be realized in several variants. Some investigations have been undertaken with lasers operating at fixed frequencies. The field of application can be extended if tunable lasers are employed for exciting the molecular vibrations, for example in such a way as to fix ω_1 while leaving ω_2 tunable. In order to simplify the experimental arrangement, instead of a separate probe beam of frequency ω_L, one of the waves exciting the molecular vibrations (e.g., that with the fixed frequency ω_1) can be used as the probe beam.

We now discuss the last-mentioned method. It has gained particular importance as *coherent anti-Stokes Raman spectroscopy* (CARS; see, e.g., Ref. 12.10). CARS was first observed by Maker and Terhune (Ref. 12.11). It is one of several well-known third-order processes, which are the strongest nonlinear optical processes in centrosymmetric media. For investigating a gaseous sample the arrangement to be chosen for CARS may be the following. Two collinear pump beams of frequencies ω_1 and ω_2 (where $\omega_1 - \omega_2 \approx \omega_M$) pass through the sample, where a new wave is generated with the anti-Stokes frequency $\omega_A = \omega_1 + (\omega_1 - \omega_2) = 2\omega_1 - \omega_2$ propagating in the forward direction collinear with the pump beams. [As can be expected from our general discussion on active Raman scattering, a coherent Stokes wave is generated in a similar way (see, e.g., Ref. 12.12); because of the difficulties arising in their technical application we do not treat coherent Stokes Raman scattering, but restrict ourselves to CARS.] In addition to the interaction resulting in the generation of an anti-Stokes wave with frequency $\omega_A = 2\omega_1 - \omega_2$, there is a coupling of these waves according to the mechanism of the stimulated Raman scattering described in Section 12.1.3.

The quantitative description proceeds as in Section 12.1.3.2, where a general discussion of the behavior of the anti-Stokes wave was provided. The second term on the right-hand side of (12.62) is to be interpreted as the polarization of the CARS wave, where ω_L must be identified with ω_1, and ω_S with ω_2. Under the assumption of exact phase matching (12.73), the equation (12.64) for the electric field amplitude of the anti-Stokes wave is calculated to be

$$\frac{d\bar{E}_x(\omega_A, z)}{dz} = \frac{\omega_A^2}{2c^2k(\omega_A)}\chi^{(3)}_{xxxx}(\omega_A; -\omega_2, \omega_1, \omega_1)\bar{E}_x^2(\omega_1, z)\bar{E}_x^*(\omega_2, z), \tag{12.74}$$

where the general third-order susceptibility function rather than the near-resonance approximation has been inserted. The first term on the right-hand side of (12.62) describes the polarization part of the stimulated Raman effect. It is proportional to the amplitude product $|\bar{E}(\omega_1, z)|^2\bar{E}(\omega_A, z)$ and also to the susceptibility function $\chi^{(3)}_{xxxx}(\omega_A; \omega_A, \omega_1, -\omega_1)$. Since the components of $\chi^{(3)}_{xxxx}(\omega_A; \omega_A, \omega_1, -\omega_1)$ and $\chi^{(3)}_{xxxx}(\omega_A; -\omega_2, \omega_1, \omega_1)$ are of comparable magnitude and since the condition

$$|\bar{E}(\omega_A, z)| \ll |\bar{E}(\omega_1, z)|, |\bar{E}(\omega_2, z)| \tag{12.75}$$

may be assumed, in a CARS experiment the polarization amplitude of the stimulated Raman scattering part is smaller than the CARS polarization amplitude.

From Section 12.1.3.2 we learnt about the way in which the CARS susceptibility and the atomic quantities are interconnected; to this end, in the semiclassical susceptibility functions $\chi^{(3)}$ and $\breve{\varkappa}^{(3)}$ obtained by means of quantum perturbation theory the arguments ω_L and ω_S in (12.69) have to be replaced by ω_1 and ω_2. To obtain the total effective susceptibility it is necessary to add the nonresonant part $\chi^{(3)}_{\mathrm{nr}\bullet\bullet\bullet\bullet}$.

In a pure gas the tensor components $\chi^{(3)}_{\mathrm{nr}\bullet\bullet\bullet\bullet}$ are smaller by a factor of about 10^2 to 10^3 than those of the resonant part $\chi^{(3)}_{\mathrm{r}\bullet\bullet\bullet\bullet}$ of the susceptibility tensor. This corresponds to the values given in discussing (4.65). The nonresonant part interferes with the resonant part of the susceptibility and leads, in principle, to a distortion of the line shape; in practice in pure gases it only slightly affects the line profile and can be considered constant in the region of one line. However, in mixtures the nonresonant part $\chi^{(3)}_{\mathrm{nr}\bullet\bullet\bullet\bullet}$ contains contributions from the probed molecules as well as from the diluent molecules. If the number density of the probed molecules is much less than that of the diluent molecules, $\chi^{(3)}_{\mathrm{nr}\bullet\bullet\bullet\bullet}$ is mainly determined by the contribution of the latter, and in this case its components may exceed those of $\chi^{(3)}_{r\bullet\bullet\bullet\bullet}$. Most of the CARS experiments are performed off resonance, so that the frequencies of the optical beams lie well below the absorption bands of the molecules. The term *resonance-enhanced CARS spectroscopy* is used for the kind where one or more of the frequencies ω_1, ω_2, ω_A are close to a one-photon transition frequency.

We now pass over to the pulsed CARS generation, basing our discussion on relatively simple conditions (frequencies off resonance with respect to electronic levels, disregard of pump depletion, neglect of the terms due to stimulated Raman scattering). Furthermore we assume the field envelopes of the three waves with the main frequencies $\omega_1, \omega_2, \omega_A$ to be slowly varying functions of the propagation coordinate z and time t, where in analogy to Section 6.1 a local time $\eta = t - z/v_A$ is introduced. On the basis of the CARS polarization part, the wave equation leads to a temporal–spatial differential equation for the field-strength wave amplitudes. As could be expected from the structure of the stationary amplitude equation (12.74), there results the relation

$$I_A(z, \eta) \propto \left|\chi^{(3)\mathrm{CARS}}\right|^2 z^2 I_1^2(\eta) I_2(\eta) \operatorname{sinc}^2\left(\Delta k \frac{z}{2}\right) \tag{12.76}$$

between the intensities of the CARS wave and the two pump waves, if certain conditions on the interaction length and the dispersive behavior of the medium are fulfilled [cf. Section 6.2]. The quantity Δk represents the wave number mismatch $2k_1 - k_2 - k_A$ in case the condition (12.73b) is not exactly satisfied.

In summary, the following conclusions may be drawn. The CARS pulse propagates along with the pump waves. Its temporal profile at a fixed position z is given by the product $I_1^2(\eta) I_2(\eta)$. Thus, if the pump pulses are temporally Gaussian-shaped with the duration τ_L, then the CARS pulse also is Gaussian-shaped but with a shorter duration, namely $\tau_L/\sqrt{3}$. Due to $z^2 \operatorname{sinc}^2(z\Delta k/2)$, the peak amplitude of the CARS wave exhibits a sinusoidal variation with respect to the propagation distance z. The maximum intensity increases with the coherence length $\pi/\Delta k$. Under certain conditions the dispersion in gases is so weak that the CARS signal intensity increases as z^2 over a long distance z (one meter or more). Since I_A is proportional to $|\chi^{(3)\mathrm{CARS}}|^2$, the CARS signal intensity depends on the squared particle number density of the medium. For spectroscopic investigations the output intensity can be monitored as a function of $\omega_1 - \omega_2$ by keeping ω_1 fixed and varying ω_2 so that $\omega_1 - \omega_2$ is swept across the resonance bands of interest. In this way it becomes possible to determine the dispersive behavior of $\chi^{(3)\mathrm{CARS}}$, to identify the Raman resonances of molecules, and to measure the densities of the species involved.

Let us finish our considerations of CARS by presenting some examples of its application in spectroscopy. In practice, high-power pulsed lasers are used and one of the following two techniques of recording: scanning spectroscopy with lasers of narrow but finite bandwidths, and point-to-point measuring and multiplex spectroscopy with a broadband $\tilde{\omega}_2$ source and a narrowband $\tilde{\omega}_1$ laser. Figure 12.10 shows a scanning CARS spectrum of atmospheric CO_2 as given in Ref. 12.10; here I_A and I_A^{ref} are the sample and reference anti-Stokes intensities, respectively. For each data point the results of ten pulses were averaged. The experimental curve indicates a high signal-to-noise ratio and background elimination, in that the spectrum even reveals two faint oxygen (O) lines.

Short-lived molecular states have also been investigated by means of CARS (see, e.g., Ref. 12.13). These investigations require high time resolution and high sensitivity because of the low concentration of the species involved. The

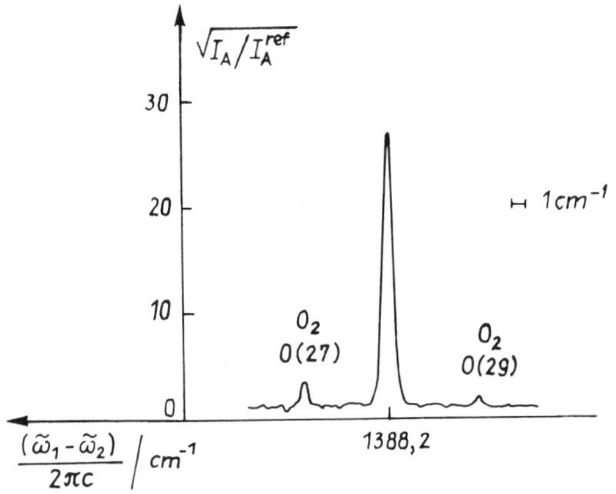

Fig. 12.10. Atmospheric CO_2 spectra obtained with a CARS spectrometer (after Ref. 12.10).

first requirement can easily be met by using laser pulses of appropriate duration, whose value largely determines the time resolution. The high sensitivity that is needed requires that one makes use of resonance enhancement. A detailed consideration (Ref. 12.14) reveals that in contrast to the off-resonance case, under resonance conditions CARS has a very high sensitivity. The investigation of fluorescent short-lived states (e.g. molecules as pyrene, chrysene, and rhodamine 6G and B in the S_1 state) proves that resonance CARS is an appropriate method for time-resolved spectroscopy. The use of a multiplex setup offers the advantage of recording the CARS spectrum spread over about 200 cm^{-1} with only one laser pulse. The pulse duration is of the order of some nanoseconds. The population of the excited states can be achieved either by means of a separate laser or by the CARS pump beams themselves. The delay time between molecular excitation and the CARS investigation is set by a trigger and is of the order of some nanoseconds.

In contrast to the phase-matched case usually treated in CARS spectroscopy, collinear generation of anti-Stokes radiation without phase matching has been dealt with experimentally and theoretically in Ref. 12.15; in benzene collinear anti-Stokes radiation was observed in addition to phase-matched anti-Stokes radiation.

In addition to the just-discussed anti-Stokes scattering by coherent molecular vibrations, CARS can be employed for studying crystals. For example, diamond (Ref. 12.16) has been investigated by measuring the resonant and nonresonant parts of the intensity of the anti-Stokes wave resulting from scattering by coherent phonon vibrations. Scattering by phonons will be dealt with in the Section 12.2.3.

Let us now direct our attention to a method of coherent Raman spectroscopy that makes use of defined input and output polarizations, the *Raman-induced Kerr effect* (RIKE). When a Raman-active medium is simultaneously irradiated by two exciting radiation sources, namely by an elliptically polarized wave (called the laser wave, with frequency ω_L) and a linearly polarized Stokes wave (frequency ω_S) or anti-Stokes wave (frequency ω_A), the nonlinear third-order interaction gives rise to a new Stokes wave or anti-Stokes wave that is polarized perpendicularly to the wave originally incident on the medium. This phenomenon exhibits resonance if $|\omega_L - \omega_{S,A}|$ is near the frequency ω_M of the Raman-active molecular vibration.

The empirically accessible line in the resonance region near ω_M and the signal in the spectral region of nonresonance are determined on the one hand by the polarization properties of the elliptically polarized laser wave, and on the other hand by the resonant and nonresonant terms in the nonlinear susceptibility. A circularly polarized laser wave leads to a background-free Raman spectrum if one of the exciting radiation sources is being tuned or if broadband irradiation is used. The first studies of this "pure" RIKE were performed with organic liquids (Ref. 12.17). Compared with other methods of active Raman scattering, the advantage of RIKE is that the phase-matching condition in isotropic media is satisfied in a broad-band spectrum so that under fixed excitation geometry, spectral tuning can be achieved.

RIKE with an elliptically polarized laser wave proves to be an appropriate polarization-spectroscopic method for determining optical material parameters. This was shown for benzene in Ref. 12.18, where a narrow-band pulse excitation was performed with point-by-point scanning of the spectrum. In Ref. 12.19 the resonant and nonresonant parts of the third-order susceptibility were experimentally determined for benzene and bromobenzene. For pulse excitation, in addition to ruby-laser radiation, the radiation of a Stokes continuum was used; thus all the information on a spectrum in a given polarization state of the laser wave with only one laser pulse.

The calculation of the line shape was performed in Ref. 12.19 under general experimental conditions—in particular, including the instrumental depolarization effects, the background radiation, and the dependence on polarization of the signal. Let the laser and Stokes waves propagate in the z direction in an isotropic medium. For the positive-frequency part $E_{L\bullet}^{(+)}$ of the elliptically polarized plane wave and the positive-frequency part $E_{S\bullet}^{(+)}$ of the Stokes wave we set

$$E_{L\bullet}^{(+)} = \tfrac{1}{2} e_{L\bullet} \bar{E}_L e^{i(\omega_L t - k_L z)},$$

$$E_{S\bullet}^{(+)} = \tfrac{1}{2}\left[e_{x\bullet} \bar{E}_S^{(1)} + e_{y\bullet} \bar{E}_S^{(2)} \right] e^{i(\omega_S t - k_S z)}. \tag{12.77}$$

The unit vectors $e_{x\bullet}$, $e_{y\bullet}$, $e_{z\bullet}$ span a space-fixed rectangular coordinate system. The ellipticity of the laser wave is determined by the complex components f_x, f_y of the unit vector $e_{L\bullet}$, according to

$$e_{L\bullet} = f_x e_{x\bullet} + f_y e_{y\bullet}. \tag{12.77a}$$

The wave amplitude $\bar{E}_S^{(1)}$ belongs to the intense incident Stokes wave, whereas $\bar{E}_S^{(2)}$ belongs to the signal wave with perpendicular polarization that is being built up. In small-signal amplification $\bar{E}_L$ and $\bar{E}_S^{(1)}$ can be considered as being constant, whereas $\bar{E}_S^{(2)}$ is slowly varying with z.

In accordance with results on stimulated Raman scattering given above [cf. (12.53a)] the amplitude equation is calculated to be

$$\frac{d\bar{E}_S^{(2)}}{dz} = +iC_S|\bar{E}_L|^2\big[\bar{\chi}^{(3)}(\omega_S;\,\omega_S,\omega_L,-\omega_L)\bar{E}_S^{(1)} + \bar{\bar{\chi}}^{(3)}(\omega_S;\,\omega_S,\omega_L,-\omega_L)\bar{E}_S^{(2)}(z)\big], \tag{12.78}$$

where C_S is proportional to ω_S/n_S. Because of the fixed polarization conditions the quantities $\bar{\chi}^{(3)}$ and $\bar{\bar{\chi}}^{(3)}$ can be calculated from the components of the susceptibility tensor $\chi^{(3)}_{\bullet\bullet\bullet\bullet}$ to be

$$\bar{\chi}^{(3)} = \chi^{(3)}_{yxxy}f_xf_y^* + \chi^{(3)}_{yxyx}f_x^*f_y,$$

$$\bar{\bar{\chi}}^{(3)} = \chi^{(3)}_{yyxx}f_xf_x^* + \chi^{(3)}_{yyyy}f_yf_y^*. \tag{12.79}$$

The integration of (12.78) is carried out by taking into account the physical fact that there a nonvanishing signal amplitude already appears at $z = 0$. Thus we have

$$\bar{E}_S^{(2)}(z) = \frac{\bar{\chi}^{(3)}}{\bar{\bar{\chi}}^{(3)}}(e^{-Gz} - 1)\bar{E}_S^{(1)} + e^{-Gz}\bar{E}_S^{(2)}(0), \tag{12.80}$$

where

$$G = -iC_S\bar{\bar{\chi}}^{(3)}|\bar{E}_L|^2.$$

For $|Gz| \ll 1$ and $|\bar{E}_S^{(2)}(0)| \ll |\bar{E}_S^{(1)}|$ there results

$$\bar{E}_S^{(2)}(z) = +iC_S\bar{\chi}^{(3)}z|\bar{E}_L|^2\bar{E}_S^{(1)} + \bar{E}_S^{(2)}(0). \tag{12.81}$$

In (12.80) and (12.81) the first term on the right-hand side represents the true RIKE contribution to the signal amplitude $\bar{E}_S^{(2)}$ caused by the strong amplitudes $\bar{E}_L$ and $\bar{E}_S^{(1)}$ of the incident waves. The second term describes the contribution of the initial amplitude present in the signal polarization. We recognize that the contribution of a small initial amplitude $\bar{E}_S^{(2)}(0)$ must be added to the RIKE signal and thus gives rise to interference terms in the intensity $I_S^{(2)} \propto |\bar{E}_S^2|^2$. Considering that the third-order susceptibility $\chi^{(3)}_{\bullet\bullet\bullet\bullet}$ is composed of a nonresonant and a resonant contribution,

$$\chi^{(3)}_{\bullet\bullet\bullet\bullet} = \chi^{(3)}_{\mathrm{nr}\bullet\bullet\bullet\bullet} + \chi^{(3)}_{\mathrm{r}\bullet\bullet\bullet\bullet}, \tag{12.82}$$

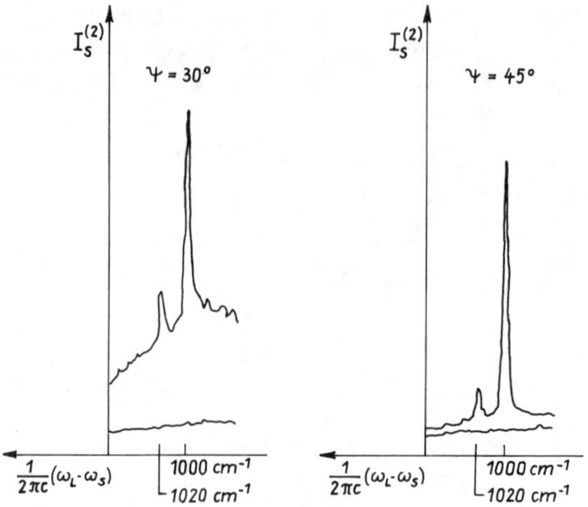

Fig. 12.11. Molecular resonance of bromobenzene at $\omega_M/2\pi c = 1000$, 1020 cm^{-1} (after Ref. 12.19). Upper curves: scattering intensity $I_S^{(2)}$ in arbitrary units; lower curves: zero level of the optical multichannel analyzer. ψ is the angle characterizing the polarization state.

one has the signal intensity

$$I_S^{(2)} \propto I_L^2 I_S^{(1)} z^2 \left| \frac{B}{2}\left(\chi^{(3)}_{r,\,yxyx} - \chi^{(3)}_{r,\,yxxy}\right) + \frac{A}{2}\left(\chi^{(3)}_{r,\,yxyx} + \chi^{(3)}_{r,\,yxxy} + \tfrac{2}{3}\chi^{(3)}_{nr,\,xxxx}\right) + ib \right|^2, \quad (12.83)$$

where $A = f_x f_y^* + f_x^* f_y$ and $B = f_x^* f_y - f_x f_y^*$. The quantity $b = \bar{E}_S^{(2)}(0) c n_S / 2\pi\omega_S \bar{E}_S^{(1)} |\bar{E}_L|^2 z$ is a measure of the background radiation. The quantity $|b|^2$ can be determined experimentally by measuring the signal intensity in the nonresonance region.

In the experimental setup in Ref. 12.19 a Q-switched ruby laser (5 MW, 40 ns) was used. One part of its radiation was employed for exciting a broadband dye laser in the Stokes region (745 ± 10 nm). The detection of the signals was performed with an optical multichannel analyzer. Figure 12.11 shows the signal intensities of bromobenzene as a function of $(1/2\pi c)(\omega_L - \omega_S)$ in a region around 1010 cm^{-1} for two different polarization states of the incident laser wave. In the experiment they were produced from the originally linearly polarized wave ($\psi = 0°$) by rotating a $\lambda/4$ platelet by the angle ψ. A strong line could be noticed at 1000 cm^{-1} and a weak line at 1020 cm^{-1}. The case $\psi = 45°$ corresponds to circularly polarized light, where under ideal conditions according to (12.83) a spectrum can be expected that does not contain any nonresonance contributions. (In practice deviations from ideal instrumental performance can be numerically corrected by using the formulas given in Ref. 12.19.)

The RIKE method has been applied not only to liquids but also to crystals, namely to the coupling with polaritons. This will be dealt with in Section 12.2.3.

12.1.4.4. Comparison of Various Methods. Having discussed in the preceding sub-subsections the ordinary Raman effect, the inverse Raman effect, and the active Raman scattering, we now discuss how these methods differ from one another regarding the insights offered by them and the conclusions that can be drawn from their employment. One advantage of stimulated Raman scattering over spontaneous Raman scattering—even if the latter results from a strong laser as exciting light source—is that in general it provides appreciably higher scattering intensities and that it permits the measurement of lines in much shorter times. However, the existence of the threshold condition (12.61) and the concurrence between the lines may be regarded as a disadvantage; in the spectrum of the stimulated scattering only the strongest lines (or the strongest line) appear. In contrast to this, in spontaneous and inverse Raman scattering even the weaker lines can be recorded. But recording by spontaneous scattering has also disadvantages, among them the long recording time required and the relatively low sensitivity. Using the inverse Raman effect, we can avoid these disadvantages.

In active Raman scattering the advantages of spontaneous Raman scattering (nearly the complete number of lines appearing) are combined with those of stimulated Raman scattering (high scattering intensity). Even if the condition (12.73) is not fully satisfied and the coherence length L is only of the order of 1 mm, the sensitivity can be drastically increased (by a factor 10^4 or more) over that of spontaneous Raman spectroscopy. Another advantage of active Raman spectroscopy is that it can be used for investigating substances with considerable broadband fluorescence (in particular in molecular-biological compounds). In observing the anti-Stokes signal (ω_A, k_A), the isotropic fluorescence in the Stokes region is practically suppressed.

In the preceding sub-subsection it was shown that the nonresonant background can also be suppressed by making use of polarization techniques. Thus, with respect to background elimination, RIKE may be regarded as an alternative to CARS. Besides, RIKE has the advantage that the phase-matching condition can be satisfied without special experimental measures.

Altogether, the most advantageous methods just described require correspondingly large experimental efforts.

12.2. STIMULATED RAMAN SCATTERING BY PHONONS AND POLARITONS

Whereas the preceding section provided insights into the scattering of photons by polarizable molecules, the present consideration will deal with the scattering of photons by quasiparticles, such as phonons and polaritons. Our treatment is again based on the concept that an unperturbed atomic system—in the present case the vibrations of the lattice elements of a crystal—interacts with an external electromagnetic field. The relation between lattice vibrations and electron motion is similar to that between the latter and the molecular vibrations described in Section 12.1: in the adiabatic approximation the effective potential for the motion of the atomic cores is equal to the electronic energy in the ground state, calculated at fixed nuclear positions. The treatment of the unperturbed atomic system goes beyond the assumptions used in Section 12.1 in that the internal electric crystal fields associated with the vibrations of the lattice elements are taken into account.

Fundamental insights into the optical properties, in particular into the nonlinear optical ones, can be gained from fairly simple crystal models, so we shall consider one-dimensional waves in a linear chain, and then three-dimensional waves with fixed polarization and propagation directions in optically isotropic crystals with two ions per unit cell. This enables us to display the interaction of long-wave optical lattice vibrations (the optical phonons) and of polarization waves (namely the so-called phonon polaritons, which can be regarded as a mixture of mechanical lattice vibrations and electromagnetic oscillations) with the external radiation field. Section 12.2.1 deals with the unperturbed atomic systems; the interaction with the radiation field is dealt with in Section 12.2.2. Applications will be discussed in Section 12.2.3.

12.2.1. Phonons and Polaritons

We first treat *long-wave optical phonons* and use as a starting-point the basic relationships and the results of the classical treatment of lattice vibrations for simple crystal models (see, e.g., Ref. 12.20).

We discuss the longitudinal vibrations of a simple diatomic chain using the model given in Fig. 12.12*a*. We assume equal distances $l/2$ and equal force constants C between adjacent atoms or ions, and negligible force constants between nonadjacent crystal elements. Let the two masses be M_1, M_2 with $M_1 > M_2$; thus the reduced mass is $M = M_1M_2/(M_1 + M_2)$. The dispersion relation for the vibrational frequency versus the wave number k for this chain is given in Fig. 12.12*b*. The upper and the lower function present the optical and the acoustical branch, respectively. Typical numerical values are $M \sim 10^{-25}$ kg, optical frequencies of some 10^{13} s^{-1}, and maximum k values $k_{max} \sim 10^{10}$ m^{-1}. For the optical branch in the neighborhood of $k = 0$ we have

$$\omega^2(k) = \frac{2C}{M} - \beta k^2 + O(k^4), \tag{12.84}$$

where $\beta = Cl^2/2(M_1 + M_2)$ and $O(k^4)$ is an order symbol. At $k = 0$ the two atoms exhibit a vibration with a phase shift of π. Even for the more general case of a chain with unequal distances and unequal force constants [see Fig. 12.12*c*] the dispersion relation displays a qualitatively similar function $\omega(k)$

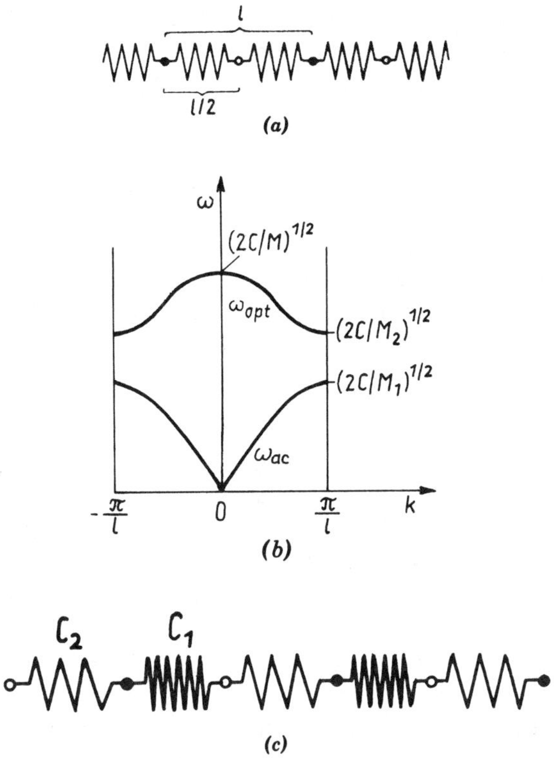

Fig. 12.12. Two-atom chain: (*a*) model of a special two-atom chain, (*b*) dispersion relation of the chain presented in part (*a*), (*c*) model of a general two-atom chain.

with numerical values of the same orders of magnitude. For $|k| \ll k_{\max}$ one obtains on the one hand a vibration of the entire elementary cell where the distances between the atoms remain approximately unchanged; this vibration must be attributed to the *acoustic* branch with $\omega(k \to 0) = 0$. On the other hand, in the same limit one obtains a relative countermovement of the lattice elements, attributable to the *optical* branch with $\omega(k \to 0) > 0$.

In the three-dimensional case with s atoms (ions) in the elementary cell there exist $3(s-1)$ optical branches with nonvanishing frequencies ω for $|k_\bullet| \to 0$. In diatomic crystals ($s = 2$) at small $|k_\bullet|$ values similar atoms, forming a sublattice, move with approximately equal amplitudes and phases, whereas different atoms in an elementary cell are in a countermovement. Thus, at small $|k_\bullet|$ values only the coordinates of the relative motion are of relevance. The three optical branches for $|k_\bullet| \ll k_{\max}$ can be subdivided into a longitudinal branch (LO: motion in the direction of $k_\bullet$) and two transverse branches (TO: motion perpendicular to the direction of $k_\bullet$). For a given direction of polarization and of particle motion the dispersion relation of the optical branches qualitatively resembles the upper function in Fig. 12.12*b*.

The next thing we have to do is to establish the equations of motion for the diatomic crystal, where the macrophysical electric field in the crystal interior must be included. The continuum theory (see Refs. 12.21, 12.22) is based (for the optical branches) on the Hamiltonian density

$$\mathfrak{H}^F = \tfrac{1}{2}\mathfrak{p}_\bullet^2 + \tfrac{1}{2}\omega_{\mathrm{TO}}^2\mathfrak{q}_\bullet^2 - \sigma\mathfrak{q}_\bullet E_\bullet - \tfrac{1}{2}\varepsilon_0\chi_{\mathrm{el}}E_\bullet^2, \tag{12.85}$$

where the third and fourth terms on the right-hand side arise from the IR activity and electronic susceptibility, respectively, as will be shown in the following. Here the vector $\mathfrak{q}_\bullet$ designates the (reduced) relative shift of the elements of the two sublattices. We have

$$\mathfrak{q}_\bullet = (r_{1\bullet} - r_{2\bullet})(M\gamma)^{-1/2}, \tag{12.86}$$

where $M\gamma$ is the reduced mass density and $r_{1\bullet}$, $r_{2\bullet}$ are the position vectors of the elements of the two sublattices. $\mathfrak{p}_\bullet = d\mathfrak{q}_\bullet/dt$ is the momentum conjugate to $\mathfrak{q}_\bullet$. The quantity $E_\bullet$ is the macrophysical electric field in the crystal interior and consists of the field resulting from the deformation of the crystal lattice and, possibly, an externally applied field. Note that in the expression for the Hamiltonian density $\mathfrak{H}^F$ the influence of the term βk^2 from (12.84) can be neglected because of the assumption $|k_\bullet| \ll k_{\mathrm{max}}$. Terms with more than two factors $\mathfrak{q}_\bullet$, $E_\bullet$ are omitted, since small position-vector differences $\mathfrak{q}_\bullet$ and a weak field $E_\bullet$ are assumed. From the Hamiltonian density $\mathfrak{H}^F$ the equation of motion for $\mathfrak{q}_\bullet$ can be obtained according to the general procedure (Section A.1.6): at a given space point,

$$\frac{d^2}{dt^2}\mathfrak{q}_\bullet + \omega_{\mathrm{TO}}^2\mathfrak{q}_\bullet = \sigma E_\bullet. \tag{12.87}$$

The application of the gradient operator to the Hamiltonian density with respect to $E_\bullet$ yields the components of the polarization according to

$$P_\bullet = -\nabla_{E_\bullet}\mathfrak{H}^F = \sigma\mathfrak{q}_\bullet + \varepsilon_0\chi_{\mathrm{el}}E_\bullet. \tag{12.88}$$

Since $\mathfrak{q}_\bullet$ represents the mutual displacement of the two sublattices, the first summand on the right-hand side represents the polarization part due to the displacement of the two sublattices; the parameter χ_{el} describes the susceptibility arising from the distortion of the electrons in the atoms (ions) by the field E. We have

$$\varepsilon_0\chi_{\mathrm{el}} = \varepsilon_0(\varepsilon_{\mathrm{el}} - 1), \tag{12.89}$$

where $\varepsilon_{\mathrm{el}}$ is the dielectric constant at the electronic frequencies. The nature of the parameter σ can be elucidated by supposing that an external electric field acts on the system. With $d^2 \mathfrak{q}_\bullet / dt^2 = 0$ in the static case, the relation

$$\varepsilon_0(\varepsilon_{st} - 1) = \frac{\sigma^2}{\omega_{TO}^2} + \varepsilon_0 \chi_{el} \tag{12.90}$$

is obtained for the dielectric constant ε_{st}. This leads to

$$\sigma = \omega_{TO} \varepsilon_0^{1/2} (\varepsilon_{st} - \varepsilon_{el})^{1/2}. \tag{12.91}$$

The constant σ represents the IR activity of the material.

We now return to a discussion of the situation in the absence of an external field. Since the transverse vibration is not connected with a depolarization field, ω_{TO} is the frequency of the TO vibration at $|k_\bullet| \to 0$. For the long-wave longitudinal vibration we obtain $E_\bullet = -\varepsilon_0^{-1} P_\bullet$ for the polarization field, because $D_\bullet = 0$. This yields

$$\omega_{LO} = \omega_{TO} \left(\frac{\varepsilon_{st}}{\varepsilon_{el}} \right)^{1/2} \tag{12.92}$$

for the frequency of the LO vibration at $|k_\bullet| \to 0$. As an example the numerical values for NaCl may be given: $\omega_{TO} = 3.1 \times 10^{13}\ s^{-1}$, $\varepsilon_{st} = 5.0$, $\varepsilon_{el} = 2.3$, $\omega_{LO} = 4.6 \times 10^{13}\ s^{-1}$.

The equations of motion of the coordinates $\mathfrak{q}_{TO\bullet}$ and $\mathfrak{q}_{LO\bullet}$ for long-wave optical lattice vibrations are

$$\frac{d^2}{dt^2} \mathfrak{q}_{TO\bullet} + \omega_{TO}^2 \mathfrak{q}_{TO\bullet} = 0, \tag{12.93}$$

$$\frac{d^2}{dt^2} \mathfrak{q}_{LO\bullet} + \omega_{LO}^2 \mathfrak{q}_{LO\bullet} = 0. \tag{12.94}$$

We assume the solutions of these equations to be of the form

$$\mathfrak{q}_\bullet(r_\bullet, t) = \bar{\mathfrak{q}}_\bullet^{(-)} e^{i[k_\bullet r_\bullet - \omega(k_\bullet)t]} + \text{c.c.}. \tag{12.95}$$

The insertion of the factor $e^{ik_\bullet r_\bullet}$ expresses the wave character explicitly. We suppose that the wave amplitude $\bar{\mathfrak{q}}_\bullet^{(-)}$ is constant. In TO vibrations $\bar{\mathfrak{q}}_\bullet^{(-)}$ is perpendicular to $k_\bullet$; in LO vibrations $\bar{\mathfrak{q}}_\bullet^{(-)}$ is parallel to $k_\bullet$. Owing to (12.85), the Hamiltonian density depends on $\mathfrak{q}_\bullet$ *and* $E_\bullet$. Since these two quantities are *linearly* interrelated by (12.87), $\mathfrak{H}^F$ can be expressed as a quantity that depends only on $\mathfrak{q}_\bullet$ in second order.

The constancy of the wave amplitude is connected with the conservation of the total energy. This suggests an interpretation as quasiparticles: these waves can be considered as *long-wave optical phonons* of energy $\hbar\omega(k_\bullet)$ and of (quasi-) momentum $\hbar k_\bullet$. The quantization can be performed according to the rules of the quantum field theory (Section A.1.6). By expanding $\mathfrak{q}_\bullet(r_\bullet, t)$ in traveling waves, the classical wave amplitude $\mathfrak{q}^{(-)}$ goes over—apart from a constant factor—into the *phonon annihilation operator* $\hat{a}_{\text{ph}}$. This quantization is formally similar to that of the photons, where the wave amplitude $\overline{A}^{(-)}$ became proportional to the photon annihilation operator $\hat{a}$. The boson quantization relations for the photons [cf. (2.20)] correspond to those for the phonons.

What still remains to be done is to compare the long-wave optical lattice vibrations with the molecular vibrations dealt with in the preceding subsection. Because in the long-wave optical lattice vibrations the function $\omega(k)$ of the form given in (12.84) could be replaced by the tangent of the function $\omega(k)$ for $k = 0$, spatial interconnections do not explicitly enter the equations of motion for $\mathfrak{q}_\bullet$. Vibrations of isolated, spatially distributed molecules may exhibit an arbitrary relationship between the phases of the vibrational coordinates of the individual molecules, which can be enforced by external conditions. As an example we recall the active Raman scattering (Section 12.1.4.3). Correspondingly also a coherent motion according to (12.95) is possible. From this it can be inferred that under appropriate conditions the vibrations of the molecules in a liquid, if their translational movement is not too fast, can be treated as the motion of long-wave optical phonons. This situation may be illustrated in the following way. The relations for the long-wave optical lattice vibrations will not change if we interpret the model of Fig. 12.12*c* as a molecular crystal by taking the force constant C_2 to be much less than C_1; this limiting case yields the model of isolated molecules.

Let us now pass to the *phonon polaritons*. When dealing with the classical relations for phonons, we used (12.87) and (12.88) without taking into account that there is an additional relation, namely the wave equation

$$\nabla_\bullet \times (\nabla_\bullet \times E_\bullet) + \frac{1}{c^2}\frac{\partial^2}{\partial t^2}E_\bullet = -\mu_0 \frac{\partial^2}{\partial t^2}P_\bullet, \tag{12.96}$$

between the field $E_\bullet$ and the polarization $P_\bullet$. The simultaneous treatment of the three equations (12.87), (12.88), (12.96) means physically that the interaction between lattice vibrations and electromagnetic vibrations is taken into consideration. As a rule, in the coupling of waves highly pronounced effects occur in those regions where the frequencies and wave numbers approximately agree. If in Fig. 12.12*b* the straight line corresponding to the dispersion function $\omega(k)$ for photons were plotted, at the given values of C, M, β the intersection point of the straight line with the optical branch would lie at a wave-number value of the order of $10^5\ \text{m}^{-1}$. This value is much less than $k_{\max} \sim 10^{10}\ \text{m}^{-1}$ in any case. This means that the relations (12.87) and (12.88) may be used also in the case of coupling with electromagnetic vibrations. In the following we concentrate on TO vibrations with both $\mathfrak{q}_\bullet$ and $E_\bullet$ perpendicular to $k_\bullet$. Thus the vectors $\mathfrak{q}_\bullet$, $E_\bullet$, and $P_\bullet$ all point in the same direction; therefore the vector symbol can be omitted.

For the electric field we make the ansatz

$$E = \tfrac{1}{2}\bar{E}^{(-)}e^{i(kz-\omega t)} + \text{c.c.}, \tag{12.97}$$

where z is the coordinate in the direction of propagation and $k = k_z$. Let the complex wave amplitude $\bar{E}^{(-)}$ be constant. Substitution into (12.87) yields

$$\mathfrak{q} = \tfrac{1}{2}\bar{\mathfrak{q}}^{(-)}e^{i(kz-\omega t)} + \text{c.c.}, \qquad \text{where} \quad \bar{\mathfrak{q}}^{(-)} = \frac{\sigma}{\omega_{\text{TO}}^2 - \omega^2}\bar{E}^{(-)}. \tag{12.98}$$

Substitution of (12.97) into (12.88) leads to

$$P = \tfrac{1}{2}\bar{P}^{(-)}e^{i(kz-\omega t)} + \text{c.c.}, \qquad \text{where} \quad \bar{P}^{(-)} = \sigma\bar{\mathfrak{q}}^{(-)} + \varepsilon_0\chi_{\text{el}}\bar{E}^{(-)}. \tag{12.99}$$

Furthermore, from (12.96) and (12.97) we may infer the relation

$$\left(k^2 - \frac{\omega^2}{c^2}\right)\bar{E}^{(-)} = \mu_0\omega^2\bar{P}^{(-)}. \tag{12.100}$$

After eliminating the quantities $\bar{\mathfrak{q}}^{(-)}$ and $\bar{P}^{(-)}$ from the preceding three equations we finally obtain

$$\left(k^2 - \frac{\varepsilon_{\text{el}}}{c^2}\omega^2 - \frac{\varepsilon_{\text{st}} - \varepsilon_{\text{el}}}{c^2}\omega_{\text{TO}}^2\frac{\omega^2}{\omega_{\text{TO}}^2 - \omega^2}\right)\bar{E}^{(-)} = 0. \tag{12.101}$$

If the dependence $\omega(k)$ is such that the expression in parentheses vanishes, (12.101) is obviously satisfied for arbitrary nonvanishing $\bar{E}^{(-)}$. This fact suggests, according to (12.98) and (12.85), that eigensolutions exist. Physically, polarization waves originate that are mixtures of lattice waves and electromagnetic waves. The Hamiltonian density of these polarization waves may be given a form that depends either on $\bar{\mathfrak{q}}^{(-)}$ or on $\bar{E}^{(-)}$ in second order.

These polarization waves can be quantized using arguments analogous to those following (12.95) that were employed for the quantization of phonons. But now the phonon annihilation operators $\hat{a}_{\text{ph}}$ are replaced by polariton annihilation operators $\hat{a}_P(k_P)$ dependent on the wave numbers k_P. The operators $\hat{a}_P(k_P)$, $\hat{a}_P^\dagger(k_P)$ obey the same quantization relations as do the phonon operators $\hat{a}_{\text{ph}}$, $\hat{a}_{\text{ph}}^\dagger$ and the respective photon operators. The resulting quasiparticles are called *phonon polaritons* to distinguish them from other polaritons, such as the exciton polaritons, which we shall not deal with. One polariton has the energy $\hbar\omega_P(k_P)$ and the (quasi-) momentum $\hbar k_P$. The Hamilton operator of the total system of polaritons is

$$\hat{H}^F = \sum_{k_P}\hbar\omega(k_P)\left[\hat{a}_P^\dagger(k_P)\hat{a}_P(k_P) + \tfrac{1}{2}\hat{I}\right]. \tag{12.102}$$

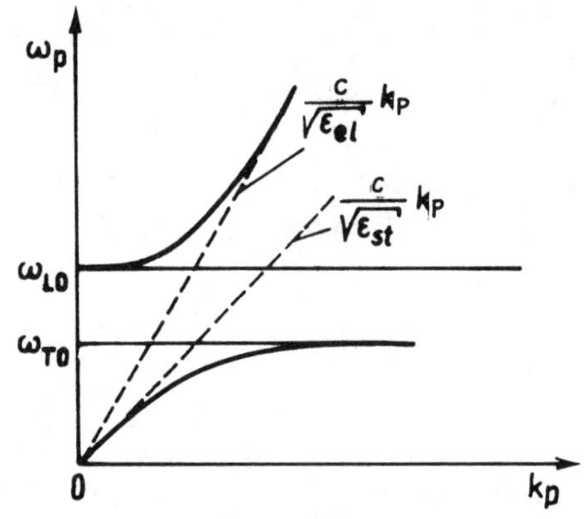

Fig. 12.13. Dispersion relation of phonon polaritons (schematic).

The polariton annihilation operator $\hat{a}_P(k_P)$ corresponds to the (classical) field-strength amplitude $\overline{E}_P^{(-)}(k_P)$ of the polarization wave. Expansion in plane waves in a periodicity volume yields the operator of the electric field strength (in the Schrödinger picture)

$$\hat{E}_P(k_P) = i\sqrt{\frac{\hbar\omega_P(k_P)}{2\varepsilon_0 V}}\sqrt{\frac{v_P(k_P)\tilde{v}_P(k_P)}{c^2}}\,\hat{a}_P(k_P)e^{ik_P z} + \text{h.c.}. \quad (12.103)$$

The first radical corresponds to a photon factor as given in (2.51). The second is a dimensionless factor characterizing the particular behavior of the polaritons with the phase velocity $v_P(k_P)$ and the group velocity $\tilde{v}_P(k_P)$, which can be derived from the dispersion relation $\omega_P(k_P)$ for polaritons. We get the function $\omega_P(k_P)$ when the expression in parentheses in (12.101) is set equal to zero. There results

$$\begin{aligned}\omega_P^2 = \frac{1}{2\varepsilon_{\text{el}}}\{&[\omega_{\text{TO}}^2\varepsilon_{\text{st}} + c^2k_P^2]\\ &\pm[(\omega_{\text{TO}}^2\varepsilon_{\text{st}} + c^2k_P^2)^2 - 4\omega_{\text{TO}}^2k_P^2c^2]^{1/2}\}.\end{aligned} \quad (12.104)$$

The dispersion relation $\omega_P(k_P)$ is schematically depicted in Fig. 12.13. The result of mixing optical lattice vibrations and electromagnetic waves is that two branches appear. For small k_P the polarization waves of the upper branch exhibit phononlike behavior with the vibrational frequency ω_{LO}, and for large k_P they exhibit photonlike behavior with the velocity $c\varepsilon_{\text{el}}^{-1/2}$. In the lower branch small k_P are associated with a photonlike behavior with the velocity $c\varepsilon_{\text{st}}^{-1/2}$, and large k_P with phononlike behavior with the vibrational frequency ω_{TO}. It must be added here that by means of the abovementioned relations the vibrational coordinate $\mathfrak{q}_P$, too, can be quantized, that is, transformed into the operator $\hat{\mathfrak{q}}_P$. By virtue of (12.98), $\hat{\mathfrak{q}}_P(k_p)$ is proportional to the field-strength operator given in (12.103).

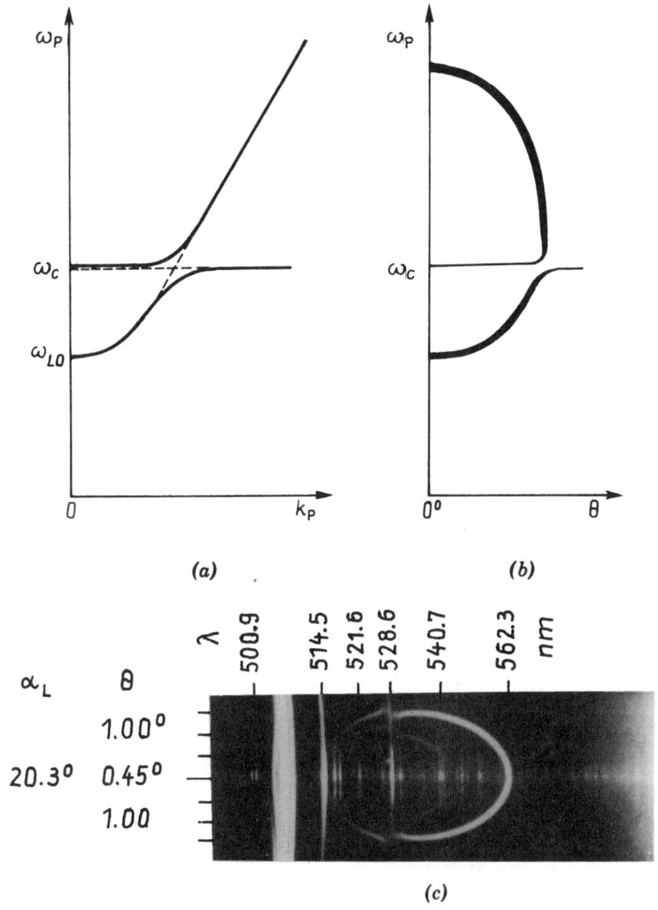

Fig. 12.14. Splitting of a polariton branch (after Ref. 12.23): (*a*) Splitting of the upper polariton branch caused by a combination vibration (ω_C). (*b*) The polariton frequency ω_P plotted versus the scattering angle θ at a fixed angle α_L. (*c*) Fermi-resonance splitting observed for $LiIO_3$: spontaneous Raman scattering by *o*-polaritons (exterior ellipse) and *e*-polaritons (interior ellipse).

Equation (12.104) and Figure 12.13 give an insight into the basic features of dispersion relations of phonon polaritons. In real systems more complicated structures can appear, mostly caused by the anisotropy of the medium and/or by the influence of further vibrations. Let us sketch the following example. If a dipole-active combination vibration (frequency ω_c) or an overtone exists within the frequency range of the upper polariton branch, this branch splits into two subbranches (see Fig. 12.14*a*). Due to the conservation of energy and of momentum this leads to a dependence of the polariton frequency ω_P on the scattering angle θ, as illustrated in Fig. 12.14*b*. For sufficiently high anisotropy, which is influenced by the angle α_L between the optical axis and the propagation direction of the pump laser, the abovementioned effect can be experimentally verified with the help of spontaneous Raman scattering by ordinary and by extraordinary polaritons of the upper dispersion branch (Ref. 23). Figure 12.14*c* shows the instructive experimental result for $LiIO_3$.

12.2.2. Interaction of the External Radiation Field with Phonons and Polaritons

The medium under study is assumed to be both infrared-active and Raman-active. This requirement can be met by crystals with point groups without inversion symmetry, for instance, by $LiNbO_3$. Let us use a spatially one-component representation for describing the fundamental nonlinear effects.

We start by discussing the interaction process on the basis of the simple classical formalism. According to (12.85) the essential properties of a medium that is unaffected by the external radiation field are described by the Hamiltonian density $\mathfrak{H}^F$, which consists of terms quadratic in $\mathfrak{q}$ and/or $\mathfrak{E}$. For determining interaction processes that give rise to stimulated light scattering (as described in the preceding section), in forming the total Hamiltonian density a third-order term must be added to $\mathfrak{H}^F$, which is of the form

$$\mathfrak{H}^C = -\frac{\alpha'}{2}\mathfrak{q}E^2, \tag{12.105}$$

where the vector symbol is again omitted [cf. discussion preceding (12.97)]. This term is of the same structure as (12.44), and it is responsible for the Raman activity of the medium; α' represents the derivative of the polarization with respect to the coordinate $\mathfrak{q}$. From the total Hamiltonian density $\mathfrak{H}^F + \mathfrak{H}^C$ relations for $\mathfrak{q}$ and P can be derived in a way analogous to that used in connection with (12.85). The resulting equation of motion for $\mathfrak{q}$ is extended by a damping term proportional to $d\mathfrak{q}/dt$. It plays a role analogous to that of the corresponding term in the Equation (12.7) for molecular vibrations. Thus for a given spatial coordinate z the following relations result:

$$\frac{d^2}{dt^2}\mathfrak{q} + \Gamma\frac{d}{dt}\mathfrak{q} + \omega_{\mathrm{TO}}^2\mathfrak{q} = \sigma E + \frac{\alpha'}{2}E^2, \tag{12.106}$$

$$P = \varepsilon_0\chi_{\mathrm{el}}E + \sigma\mathfrak{q} + \alpha'\mathfrak{q}E. \tag{12.107}$$

From a comparison of these relations with (12.87) and (12.88) it becomes apparent that now quadratic terms containing products of $\mathfrak{q}$ and E appear. The solution $\mathfrak{q}(t; z)$ with given $E(t; z)$ is determined from (12.106) and is inserted into (12.107). Thus one obtains

$$P(t, z) = \varepsilon_0\chi_{\mathrm{el}}E(t, z) + \sigma^2\int_0^\infty d\tau\, G(\omega_{\mathrm{TO}}, \Gamma; \tau)E(t-\tau, z) + P^{(\mathrm{NL})}(t, z), \tag{12.108}$$

where the nonlinear polarization part is given by

$$P^{(\mathrm{NL})}(t, z) = \frac{\sigma\alpha'}{2}\left\{\int_0^\infty d\tau\, G(\omega_{\mathrm{TO}}, \Gamma; \tau) E^2(t-\tau, z) + 2E(t, z)\int_0^\infty d\tau\, G(\omega_{\mathrm{TO}}, \Gamma; \tau) E(t-\tau, z)\right\} + \frac{|\alpha'|^2}{2} E(t, z)\int_0^\infty d\tau\, G(\omega_{\mathrm{TO}}, \Gamma; \tau) E^2(t-\tau, z). \quad (12.108\mathrm{a})$$

The function $G(\omega_{\mathrm{TO}}, \Gamma; \tau)$ is the time response of a damped harmonic oscillator with frequency ω_{TO} and damping constant Γ. As a result of the interaction term (12.105) in the polarization a nonlinear second-order term in the field strength appears, which is proportional to $\sigma\alpha'$. In addition, there appears a nonlinear third-order term, which is proportional to $|\alpha'|^2$. On substituting the polarization $P(t, z)$ from (12.108) into the wave equation (12.96), the latter can be written in the form

$$L[E] = -\mu_0 \frac{\partial^2}{\partial t^2} P^{(\mathrm{NL})}, \quad (12.109)$$

where $L[E]$ is assumed to contain all terms linear in the field strength. As it could be expected, one obtains with the ansatz

$$E = \tfrac{1}{2}\bar{E}_P^{(-)} e^{i(k_p z - \omega_p t)} + \mathrm{c.c.} \quad (12.110)$$

the dispersion relation (12.101) for polaritons from the requirement that $L[E]$ be zero for $\Gamma \to 0$.

Let us now consider the interaction between a laser wave (with frequency ω_L, wave number $k_{L\bullet}$, wave amplitude $\bar{E}_L^{(-)}$), a scattered wave (ω_S, $k_{S\bullet}$, $\bar{E}_S^{(-)}$), and a polarization wave (ω_P, $k_{P\bullet}$, $\bar{E}_P^{(-)}$). To this end we superpose the field strengths arising from these three waves and insert the sum into (12.109); then this equation splits into three equations for the three waves mentioned above. Due to the existence of the nonlinear polarization part $P^{(\mathrm{NL})}$, an interaction between the three waves arises which may be accompanied by coherent amplification or attenuation. The value of the amplification factors depends critically on the ratio of the frequencies in the electric field strength to ω_{TO}. The conditions to be satisfied in order to obtain stationary simultaneous solutions of the three wave equations following from (12.109) are given by

$$\omega_L = \omega_S + \omega_P \quad (12.111\mathrm{a})$$

$$k_{L\bullet} = k_{S\bullet} + k_{P\bullet}, \quad (12.111\mathrm{b})$$

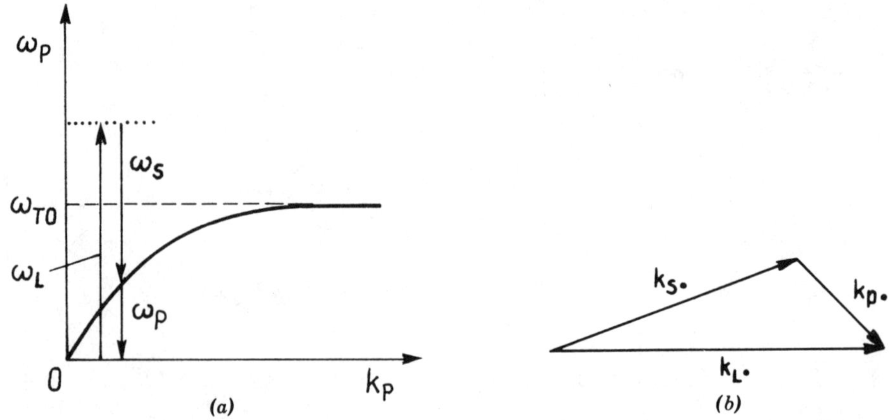

Fig. 12.15. Relation between frequencies (a) and wave numbers (b) for scattering by polaritons (lower dispersion branch).

which physically means the conservation of energy and momentum. Figure 12.15 illustrates how the relations between frequencies and wave numbers are satisfied for the lower branch of the dispersion relation $\omega_P(k_P)$.

The following consideration deals with the influence of *damping terms* and *relaxation times*, and thus we pass on from the classical interpretation used so far to the semiclassical one. In general the insertion of a damping term (proportional to $\partial E/\partial t$) into the wave equation of the field strength leads to more complicated conditions. Nevertheless in this case it can be concluded that, if the dispersion relation and the relations (12.111) are satisfied, the amplification coefficient of the wave scattered into the direction $k_{S\bullet}$ exhibits a peak.

In particular we shall discuss coherent polariton-type molecular vibrations. The interaction operator for one molecule both infrared and Raman-active is set equal to

$$\hat{H}'^C = -\hat{d}(\mathfrak{q})E - \tfrac{1}{2}\hat{\alpha}(\mathfrak{q})E^2 \tag{12.112}$$

for one vibration with the vibration coordinate $\mathfrak{q}$; $\hat{d}$ and $\hat{\alpha}$ are the operators of the dipole moment and the polarizability, respectively, both of which depend on the vibrational coordinate $\mathfrak{q}$. When combining all the molecules in the unit volume, one has to use the total interaction operator $\hat{H}^C$ instead of $\hat{H}'^C$. Then $\hat{H}^C$ yields the expectation value

$$P = \left\langle -\frac{\partial}{\partial E}\hat{H}^C \right\rangle = \varepsilon_0 \chi_{\text{el}} E + \langle \hat{\sigma} \rangle \mathfrak{q} + \langle \hat{\alpha}' \rangle \mathfrak{q} E + O(\mathfrak{q}^2) \tag{12.113}$$

for the polarization with $O(\mathfrak{q}^2)$ an order symbol. So the equation is the analogue of (12.107); $\hat{\sigma}$ and $\hat{\alpha}'$ are the coefficients of $\mathfrak{q}$ in an expansion for $\gamma\hat{d}$ and $\gamma\hat{\alpha}$ in terms of $\mathfrak{q}$, where γ is the number density of the molecules. (We assume that no permanent dipole moment exists).

Using (12.112), the equation of motion for the molecular density operator can be established. It contains the matrix elements of $\hat{d}$ and $\hat{\alpha}$, the field strength E, the transition frequency ω_{10}, and the associated transverse relaxation time τ_{10}. The expectation values $\langle\hat{\sigma}\rangle$ and $\langle\hat{\alpha}'\rangle$ can be determined with the help of the density operator. Thus, from (12.112) an equation of motion for the vibrational coordinate $\mathfrak{q}$ can be derived. Altogether there result two equations for $\mathfrak{q}$ and P, which are of the same structure as (12.106) and (12.107), and by means of which the formal treatment of the problem can be continued in the same way as described below these two classical equations. All the parameters, however, are now explained in terms of atomic quantities: ω_{TO} is replaced by ω_{10}; σ^2 is proportional to the square $|d_{10}|^2$ of the dipole moment. Introducing a damping term with the damping constant Γ_P into the wave equation for E_P, we obtain the dispersion relation

$$k_P^2 = \left\{\zeta_1 n_P^2 + \frac{1}{\omega_{10}^2 - \omega_P^2}\left(\zeta_2|d_{10}|^2\omega_{10}\gamma + \zeta_3\frac{n_P\Gamma_P}{\tau_{10}}\right)\right\}\omega_P^2 \quad (12.113\text{a})$$

for coherent polariton-type molecular vibrations, where n_P is the refractive index $\sqrt{\varepsilon_{\text{el}}}$ at the frequency ω_P and the quantities $\zeta_1, \zeta_2, \zeta_3$ contain universal constants. It is obvious that (12.113a) is of the same structure as the dispersion relation in the undamped case following from (12.101); but now the damping term Γ_P and the relaxation time τ_{10} enter the term with the factor $\omega_P^2/(\omega_{10}^2 - \omega_P^2)$.

The foregoing classical and semiclassical treatment provides completely prescribed temporal functions for the electromagnetic oscillations and lattice vibrations. But in reality, there are stochastic processes, which are of great importance under the experimental conditions imposed insomuch as the generation of coherent Stokes and polariton waves frequently takes place "from fluctuations." This requires a fully quantum-theoretical description.

We again start from a three-wave interaction and assume the validity of the relations (12.111). Then the Hamiltonian of the uncoupled subsystems according to

$$\hat{H}^F = \hbar\omega_L\left(\hat{a}_L^\dagger\hat{a}_L + \frac{\hat{I}}{2}\right) + \hbar\omega_S\left(\hat{a}_S^\dagger\hat{a}_S + \frac{\hat{I}}{2}\right) + \hbar\omega_P\left(\hat{a}_P^\dagger\hat{a}_P + \frac{\hat{I}}{2}\right) \quad (12.114)$$

is composed of contributions of the laser, Stokes, and polariton waves. To obtain the total Hamiltonian, the interaction operator

$$\hat{H}^C = \mathfrak{C}\hat{a}_L\hat{a}_S^\dagger\hat{a}_P^\dagger + \mathfrak{C}^*\hat{a}_L^\dagger\hat{a}_S\hat{a}_P \quad (12.115)$$

must be added to $\hat{H}^F$. A comparison with the interaction operator (12.51) tailored to Raman scattering by molecular vibrations displays the same structure when the operators of the molecular vibration are replaced by the respective polariton operators.

For determining the quantized interaction operator (12.115) the classical description can be used, and for polaritons one may proceed as follows. In a three-wave ansatz the interaction system of the Hamiltonian density is calculated to be

$$\mathfrak{H}^C = -\mathfrak{K}_1 \bar{E}_L^{(-)} \bar{E}_S^{(+)} \bar{\mathfrak{q}}_P^{(+)} - \mathfrak{K}_2 \bar{E}_L^{(-)} \bar{E}_S^{(+)} \bar{E}_P^{(+)} + \text{c.c.} \qquad (12.116)$$

The first term follows directly from the nonlinear interaction term given in (12.105), if the three-wave ansatz is inserted and contributions rapidly varying with time are neglected. The second term is an extension of (12.105). It is responsible for the nonlinear contribution to the polarization amplitudes that are formed by partial differentiation of the Hamiltonian density with respect to the field-strength amplitudes. Thus we have

$$\bar{P}_P^{(\text{NL})(-)} = -\frac{\partial \mathfrak{H}^C}{\partial \bar{E}_P^{(+)}} = \mathfrak{K}_2 \bar{E}_L^{(-)} \bar{E}_S^{(+)}. \qquad (12.117)$$

So far, for simplicity, we dropped the explicit vector notation. In reality, if we restrict to optically isotropic media (cubic crystals), neither the tensor character of $\mathfrak{K}_1$ and $\mathfrak{K}_2$ nor the vector character of the field strength effectively plays a role. Passing over from the classical wave amplitudes to the corresponding operators, we have $\bar{E}_L^{(-)} \propto \hat{a}_L$, $\bar{E}_S^{(-)} \propto \hat{a}_S$, $\bar{\mathfrak{q}}_P^{(-)} \propto \hat{a}_P$, $\bar{E}_P^{(-)} \propto \hat{a}_P$; the proportionality factors for the electromagnetic waves can be taken from (2.51), and for the polariton waves from (12.98) and (12.103). Hence we obtain for the density of the Hamiltonian operator

$$\hat{\mathfrak{H}}'^C = \mathfrak{C}' \hat{a}_L \hat{a}_S^\dagger \hat{a}_P^\dagger + \text{h.c.}; \qquad (12.118)$$

$\mathfrak{C}'$ contains $\mathscr{k}_1, \mathscr{k}_2, \omega_L, \omega_S, \omega_P, k_L, k_S, k_P$, characteristics of the function $\omega_P(k_P)$, and the volume V of the periodicity cell; it is connected with the expansion of the electromagnetic fields and the vibrational coordinates in traveling plane waves.

In our treatment the wave amplitudes in the volume V were assumed to be constant. For real material (nonzero damping of the polariton wave) and real processes (amplification of the Stokes wave) exact constancy of the wave amplitudes cannot be assumed. Then the linear dimensions of the periodicity cell have to be chosen less than the reciprocal coefficients of absorption and amplification. Provided the condition (12.111b) is rigorously satisfied, in the interaction part the transition from the Hamiltonian density (12.118) to the Hamiltonian operator (12.115) is simply achieved by multiplying $\hat{\mathfrak{H}}'^C$ by V.

Since $\mathfrak{C}'$, because of the V dependence of the wave amplitudes, is proportional to $V^{-3/2}$, $\mathfrak{C}$ is proportional to $V^{-1/2}$. If $\Delta k_\bullet = k_{L\bullet} - k_{S\bullet} - k_{P\bullet}$ is unequal to zero, $\hat{\mathfrak{H}}^C$ cannot be obtained simply by multiplication of $\hat{\mathfrak{H}}'^C$ by V as before, but it results from the integral

$$\int_V d^3r_\bullet \, e^{\pm i \Delta k_\bullet r_\bullet} \hat{\mathfrak{H}}'^C.$$

Thus in general, the momentum mismatch $\Delta k_\bullet$ enters into the interaction operator.

12.2.3. Specific Processes and Applications

On the basis of the foregoing results we shall discuss the determination of the Stokes amplification factor in the three-wave interaction and the important problem of the generation of Stokes waves from fluctuations. (see Section 12.2.3.1). In Section 12.2.3.2 the particular coherence properties connected with Raman scattering will be dealt with. The scattering of photons by polaritons and phonons provides important insights into the properties of matter, which may be used for the invention of tunable coherent radiation sources (see Section 12.2.3.3).

12.2.3.1. Amplification and Generation of Stokes, Phonon, and Polariton Waves. First we determine the gain coefficient g_S of the Stokes wave in the three-wave interaction between laser, stokes, and polariton waves as a function of the parameters of the atomic system. Under the same conditions as were used in the semiclassical consideration for calculating the dispersion relation (12.113a) for the polariton wave, which is connected with molecular vibrations, one obtains

$$g_S = \zeta \left\{ \frac{\gamma |\alpha_{10}|^2 \tau_{10} |\bar{E}_L^{(-)}|^2}{n_S \lambda_S} \right\} \left\{ \frac{\omega_{10}/\omega_P}{1 + \Gamma_P n_P \tau_{10} \left(\omega_{10}^2 - \omega_P^2 \right) / 2\omega_P^2 c F} \right\}. \quad (12.119)$$

In this way it is possible to connect the amplification factor of the three-wave interaction with the polarizability element α_{10}, the transition frequency ω_{10}, the relaxation time τ_{10}, and the damping rate Γ_P. The quantity $F = F(\omega_{10}, \omega_P)$ is the second summand in the braces of (12.113a); ζ contains universal constants. The first factor in braces in (12.119) corresponds to the common Stokes amplification factor [cf. (12.60)] of the (ordinary) Raman effect of molecular vibrations. The second factor characterizes the particular conditions of the scattering by polaritons: for small ω_P^2 the second term in the denominator behaves as ω_P^{-2}; for $\omega_P^2 \approx \omega_{10}^2$ it behaves as $(\omega_{10} - \omega_P)^2$.

Second, we treat the important process of the *generation of Stokes quanta, polaritons, and phonons "starting from fluctuations."* This requires that the fully quantum-theoretical description from Section 12.2.2 must be used. Suppose that the initial state $|\psi(t_0)\rangle$ is determined by fixed particle numbers n_L, n_S, n_P in the laser, the Stokes, and the polariton mode. Thus we have

$$|\psi(t_0)\rangle = |n_L\rangle |n_S\rangle |n_P\rangle. \tag{12.120}$$

We see from (12.115) that the interaction operator $\hat{H}^C$ for this initial state $|\psi(t_0)\rangle$ leads to a nonvanishing transition probability only if the final state is $|n_L + 1\rangle |n_S - 1\rangle |n_P - 1\rangle$ or $|n_L - 1\rangle |n_S + 1\rangle |n_P + 1\rangle$. According to (A.123), for the first-mentioned final state the rate for the annihilation of one Stokes quantum and one polariton reads

$$\frac{dP_-}{dt} = \frac{2\pi}{\hbar^2} |\mathfrak{C}|^2 (n_L + 1) n_S n_P \delta(\omega_L - \omega_S - \omega_P), \tag{12.121a}$$

and for the second-mentioned final state the rate for the generation of a Stokes quantum and a polariton reads

$$\frac{dP_+}{dt} = \frac{2\pi}{\hbar^2} |\mathfrak{C}|^2 n_L (n_S + 1)(n_P + 1) \delta(\omega_L - \omega_S - \omega_P). \tag{12.121b}$$

Thus the net rate of the generation of Stokes quanta and polaritons is

$$\frac{dP}{dt} = \frac{dP_+}{dt} - \frac{dP_-}{dt}$$

$$= \frac{2\pi}{\hbar^2} |\mathfrak{C}|^2 [n_L(n_S + n_P + 1) - n_S n_P] \delta(\omega_L - \omega_S - \omega_P). \tag{12.122}$$

When considering damping, the delta function is to be replaced according to

$$\delta(\omega_L - \omega_S - \omega_P) \to \frac{\tilde{\Gamma}_P / 2\pi}{(\omega_L - \omega_S - \omega_P)^2 + (\tilde{\Gamma}_P/2)^2} \tag{12.123}$$

where $\tilde{\Gamma}_P$ is the width of the corresponding line shape function.

From (12.122) it is obvious that the net rate dP/dt for $n_L > n_S, n_P$ always attains positive values. Even if in the initial state there exist neither Stokes quanta nor polaritons, but only laser quanta, ($n_L > 0$), Stokes quanta and polaritons are generated because of the spontaneous decay of laser quanta. This process is the basis of an important experimental variant of polariton scattering, namely, when a medium is irradiated by an intense laser wave which evokes in it a spontaneous scattering process, from which a coherent Stokes wave and a polariton wave can originate. However, we achieve a better correspondence with the real conditions at the beginning of the process if we do not work with states of fixed particle number, as we did in (12.120), but identify the laser wave with a coherent state [cf. (2.120b)] and the Stokes wave and the polaritons with chaotic states [cf. (2.125b)]. The density operator at time t_0 is then given by

$$\hat{\rho}(t_0) = \hat{\rho}_L \hat{\rho}_S \hat{\rho}_P, \tag{12.124}$$

where

$$\hat{\rho}_L = |\alpha_L\rangle\langle\alpha_L|, \qquad \hat{\rho}_{S,P} = \sum_{n_{S,P}} \frac{(\bar{n}_{S,P})^{n_{S,P}}}{(1+\bar{n}_{S,P})^{n_{S,P}+1}} |n_{S,P}\rangle\langle n_{S,P}|. \tag{12.124a}$$

Here $|\alpha_L\rangle$ is the coherent state of the laser mode, with $|\alpha_L|^2 = \bar{n}_L$ the mean photon number, whereas $\bar{n}_{S,P}$ is the mean number of Stokes photons or polaritons, respectively. The perturbation calculation yields the generation rate

$$\frac{\Delta\langle \hat{a}_S^\dagger \hat{a}_S\rangle}{\Delta t} = \frac{\langle \hat{a}_S^\dagger \hat{a}_S\rangle_{t_0+\Delta t} - \langle \hat{a}_S^\dagger \hat{a}_S\rangle_{t_0}}{\Delta t}$$

$$= \frac{\Omega}{V}\{\bar{n}_L(\bar{n}_S + \bar{n}_P + 1) - \bar{n}_S \bar{n}_P\}, \tag{12.125}$$

where

$$\Omega = \frac{V 2\pi}{\hbar^2} |\mathfrak{C}|^2 \delta(\omega_L - \omega_S - \omega_P) \text{ or } \frac{V 2\pi}{\hbar^2} |\mathfrak{C}|^2 g(\omega_L - \omega_S - \omega_P). \tag{12.125a}$$

Equation (12.125) reveals that the Stokes photon generation rate for chaotically distributed Stokes quanta and polaritons depends on the mean particle numbers in the same way as on the particle numbers for states of fixed particle number. Thus the conclusions drawn from (12.122) can be transferred in a simple way to the more general case. The rates of polariton generation and laser-quantum annihilation can be determined in analogy to the generation rate of Stokes quanta.

Up to now we have studied the generation and annihilation of photons and polaritons in a (small) volume V. Now we intend to discuss the spatiotemporal behavior in an extended region, the particles being assumed to move in the z direction. We pass from particle numbers to spatial number densities which are functions of the time t and the propagation coordinate z.

The particle numbers n and mean particle numbers $\bar{n}$ are divided by the basic volume V, which leads to the number density γ. The balance for the temporal variation in a volume element yields for each sort L, S, or P of quanta the expression

$$\left(\frac{\partial \gamma}{\partial t}\right)_{L,S,P} = \left(-\bar{v}\frac{\partial \gamma}{\partial z} + w - \mathfrak{V}\right)_{L,S,P}. \tag{12.126}$$

The first term on the right-hand side represents the temporal variation of the number density in a volume element as a result of the passage of the particles through the boundary surfaces perpendicular to the direction of propagation, $\bar{v}$ being the group velocity. The second term stands for the net generation rate

$$w_{L,S,P} = \frac{1}{V}\frac{\Delta\langle \hat{a}^{\dagger}_{L,S,P}\hat{a}_{L,S,P}\rangle}{\Delta t}$$

related to the volume V. The third term denotes the loss rate due to the interaction with a dissipative system. For light quanta we may set $\mathfrak{B} = \bar{v}\ell_a\gamma$, with ℓ_a the absorption coefficient. For polaritons, since they are excitation states of the medium, $\mathfrak{B} = \beta(\gamma_P - \gamma_P^e)$, where γ_P^e is assumed to be the equilibrium value. The quantities $\bar{v}\ell_a$ and β have the character of reciprocal lifetimes. Since the generation rate $w_{L,S,P}$ in general contains the quantities $\gamma_L, \gamma_S, \gamma_P$ in a coupled form, the simultaneous treatment of the three sorts of quanta leads, according to (12.126), to a set of coupled partial differential equations.

We emphasize an essential difference between the present treatment and the treatment using classical equations: here the net generation rates $w_{L,S,P}$ in themselves contain spontaneous contributions. It may be remarked that also the Raman scattering by long-wave optical phonons can be treated according to the above description of Raman scattering by polaritons. This is indicated by the analogy of the quantization rules given in Section 12.2.1 for long-wave optical phonons and polaritons as well as by the similarity in the structure of the interaction operators [cf. (12.51) and (12.115)].

Turning to *real experimental conditions*, we start with the case in which the two incident electromagnetic waves are assumed to emerge from two intense lasers, and in which the polariton density is much less than the photon density. According to (12.125) we have

$$w_S = \Omega\gamma_L\gamma_S \tag{12.127}$$

for the net generation rate of Stokes quanta. Thus we have as the differential equation for γ_S

$$\frac{\partial\gamma_S}{\partial t} = -\bar{v}_S\frac{\partial\gamma_S}{\partial z} + \Omega\gamma_L\gamma_S - \bar{v}_S(\ell_a)_S\gamma_S. \tag{12.128}$$

For stationary conditions one obtains

$$\frac{d\gamma_S}{dz} = \left\{\frac{\Omega}{\bar{v}_S}\gamma_L - (\ell_a)_S\right\}\gamma_S. \tag{12.129}$$

The expression in braces represents the effective Stokes amplification coefficient. It attains a positive value only if the laser photon density is larger than $\bar{v}_S(k_a)_S/\Omega$. Given a large absorption coefficient $(k_a)_S$ and a small interaction constant $\mathfrak{C}$, a high laser density is needed for amplification.

Now we deal with Raman scattering by long-wave optical phonons. For reasons discussed above, we can use the set of equations (12.126) and index the excitation state by ph (phonon) instead of P (polariton). Because of the annihilation of laser photons accompanied by the generation of Stokes quanta and polaritons, we have to use $w_S = w_{\text{ph}} = -w_{\text{L}}$; thus there results the system

$$\frac{\partial \gamma_L}{\partial t} = -\bar{v}_L \frac{\partial \gamma_L}{\partial z} - \Omega\left[\gamma_L\left(\gamma_S + \gamma_{\text{ph}} + V^{-1}\right) - \gamma_S \gamma_{\text{ph}}\right] - \bar{v}_L(k_a)_L \gamma_L,$$

$$\frac{\partial \gamma_S}{\partial t} = -\bar{v}_S \frac{\partial \gamma_S}{\partial z} + \Omega\left[\gamma_L\left(\gamma_S + \gamma_{\text{ph}} + V^{-1}\right) - \gamma_S \gamma_{\text{ph}}\right] - \bar{v}_S(k_a)_S \gamma_S,$$

$$\frac{\partial \gamma_{\text{ph}}}{\partial t} = \Omega\left[\gamma_L\left(\gamma_S + \gamma_{\text{ph}} + V^{-1}\right) - \gamma_S \gamma_{\text{ph}}\right] - \beta_{\text{ph}}\left(\gamma_{\text{ph}} - \gamma_{\text{ph}}^e\right). \qquad (12.130)$$

The term with the spatial derivative of γ_{ph} could be neglected because of the relative smallness of $\bar{v}_{\text{ph}}$. Here γ_{ph}^e is the equilibrium value of the phonon density. This set of equations will be evaluated for stationary conditions following the pattern given in Ref. 12.24. The expression $(\gamma_S + \gamma_{\text{ph}} + V^{-1})$ on the right-hand sides can be written in the form $[\gamma_S + (\gamma_{\text{ph}} - \gamma_{\text{ph}}^e) + (\gamma_{\text{ph}}^e + V^{-1})]$. The term containing $\Omega\gamma_L(\gamma_{\text{ph}}^e + V^{-1})$ represents an amplification process starting from fluctuations. The term containing $\Omega\gamma_L(\gamma_{\text{ph}} - \gamma_{\text{ph}}^e)$ expresses a parametric instability. Under stationary conditions, as an important result of the set (12.130), the space dependence of the density of the Stokes photons can be derived; we have

$$\gamma_S(z) = \tilde{\gamma}(e^{g_S' z} - 1) + \gamma_S(0) e^{g_S z} \qquad (12.131)$$

if $\gamma_S(z)$ is well below the saturation value. Here

$$g_S' = \frac{\Omega S_L(0)}{\hbar \omega_L \bar{v}_S \bar{v}_L} \frac{1}{1 - S_L(0)/S_0} - (k_a)_S \qquad (12.131a)$$

is the effective power gain coefficient for the Stokes wave. The quantity

$$S_0^{-1} = \frac{\Omega}{\hbar \omega_L \bar{v}_L \beta_{\text{ph}}} \qquad (12.131b)$$

corresponds to an instability of the Stokes amplification which may appear if the laser power $S_L(0)$ attains appropriate values at the entrance of the medium.

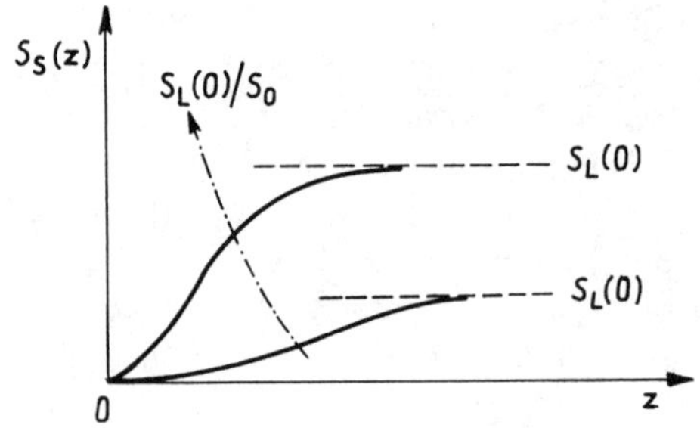

Fig. 12.16. Schematic representation of the Stokes intensity $S_S(z)$ for different values of the parameter $S_L(0)/S_0 < 1$.

The quantity $\tilde{\gamma}$ is determined by

$$\tilde{\gamma} = \left(\gamma_{\text{ph}}^{e} + V^{-1}\right)\left(1 + \frac{(k_a)_S}{g_S'}\right). \tag{12.131c}$$

The first term on the right-hand side of (12.131) represents the amplified spontaneous contribution; it indicates the extent to which a Stokes wave is being built up "from fluctuations" even if the medium is not irradiated by a Stokes wave. For z values small enough that $g_S' z \ll 1$, the spontaneous effect predominates. The second term on the right-hand side of (12.131) describes that contribution to the z-dependent Stokes quantum density which results when a Stokes wave is incident on the medium. Figure 12.16 illustrates schematically the Stokes intensity $S_S(z)$ as a function of z for various values of the parameter $S_L(0)/S_0$ less than unity. The curves are of the same character; however the larger the value of $S_L(0)/S_0$, the lower the z value at which saturation $S_S \to S_L(0)$ is reached. Saturation occurs at a propagation length of $z_{sa} \approx \bar{v}_S/\beta_{\text{ph}}$, which in practice attains values from 10^{-4} to 10^{-2} m.

The set of equations (12.130) yields the representation of three experimentally observable stages of Raman scattering. For low laser intensities $[S_L(0)/S_0 \lesssim 0.1]$ spontaneous Raman scattering predominates, the Stokes intensity at the exit from the sample being proportional to $S_L(0)$. For medium laser intensities $[0.1 \lesssim S_L(0)/S_0 \lesssim 0.9]$ an exponential dependence of the Stokes intensity on $S_L(0)$ results. If the laser intensity $S_L(0)$ increases further, a rapid increase of the Stokes intensity up to a value close to the saturation value follows. We see that S_0^{-1} is an important parameter; although it is, according to (12.131b), determined by internal parameters of the system, its value can more advantageously be found by fitting the theoretical results to the empirical findings.

In an interpretation of the resultant intensity function it is of course necessary to insure that the assumptions underlying the calculation are obeyed. In particular the influence of other nonlinear interaction processes (e.g., self-focusing) as well as of nonstationary phenomena was left unconsidered in the foregoing discussion.

Let us sketch finally some basic features of Raman scattering by optical phonons and polaritons under nonstationary conditions (see, e.g., Ref. 38, Vol. 2, Chapter 3.222). These phenomena can be described by semiclassical means if terms representing fluctuation forces are added to the equations of motion for the temporally prescribed functions [the situation is analogous to that in the case of the basic laser equations (7.58a, b, c); these fluctuation forces can in principle be justified only by a fully quantum-theoretical treatment].

Basic concepts for the *nonstationary* Raman scattering by optical phonons and polaritons are presented in Refs. 12.25, 12.26, 12.27. The scattering process exhibits typically nonstationary features if the duration T_L of the laser pulse is of the order of the phase relaxation time and if the scattering fields propagate nonsynchronously. The basic equations for nonstationary Raman scattering have to be established under the condition that the crystal vibrations are driven by a coherent radiation field and by fluctuation forces arising from a dissipative system. The correlation time of the corresponding vibrational fluctuations can be assumed to be small compared with the pulse duration T_L. Thus the fluctuation correlators may be regarded as delta functions. In establishing the equations of motion the vibrations are assumed to be Raman-active and infrared-active, so that the interaction operators given in Section 12.2.2 can be used. In deducing the specific nonstationary features, the transformation of the equations of motion into equations for the so-called nonstationary Fourier amplitudes by use of a four-dimensional Fourier transform (three space coordinates and one time coordinate) is helpful. Thus it becomes possible to determine quantities directly accessible to measurement, such as the transient radiant flux density, the instantaneous power spectrum, and the time-dependent energy spectrum. The nonstationary Stokes gain, which can be obtained from the considerations presented above, depends on the degree of synchronization of the propagation of laser and Stokes pulses as well as on the dispersion properties of phonons and polaritons. For normal dispersion the Stokes pulse outruns the laser pulse, whereas the polariton pulse is delayed; as in the stationary case, the gain coefficient is proportional to the interaction length. (In other cases a saturated nonstationary gain occurs that does not increase with increasing interaction length.)

12.2.3.2. Coherence Properties. In 1967 an instructive example of the interrelation between coherence properties and stimulated Raman effect was given by Freedhoff (Ref. 12.28). She showed that the positive generation rate for stimulated anti-Stokes scattering by molecules distributed stochastically in gases and liquids can only be explained by assuming the existence of sufficiently coherent fields. A positive anti-Stokes gain occurs only in those directions in which momentum conservation is satisfied [cf. (12.68) and the subsequent discussion] and if a certain inequality holds for the phases of the pump, the Stokes, and the anti-Stokes wave. This requirement can be met only if waves with small phase fluctuations are involved.

When discussing the Raman effect of the polarizable molecules, as the typical interaction operator a three-boson operator was derived in (12.51). It contains—in the case of the generation of Stokes quanta—the annihilation operator for laser photons, the creation operator for Stokes quanta, and the creation operator for molecular excitations. The same structure became apparent in the case of scattering by elementary excitations, such as polaritons and phonons, by the interaction Hamiltonian in (12.115), where, of course, the creation operator for the molecular vibrational quantum was replaced by the creation operator for polaritons or phonons. We generally write for the decisive interaction operator for Raman scattering

$$\hat{H}_S^C = -\hbar\chi'_S\hat{a}_L\hat{a}_S^\dagger\hat{a}_e^\dagger + \text{h.c.}, \tag{12.132}$$

whose L, S, and e are the indices for the pump laser, the Stokes, and the elementary excitation mode; χ'_S is the interaction parameter for the Stokes quantum generation. A comparison between $\hat{H}_S^C$ and the interaction operator (11.77) for parametric amplification reveals the same structure if the signal and idler modes are identified with the Stokes and excitation modes. This is true in particular because the elementary excitations under study are described by bosonlike operators. The structural equivalence suggests that the conclusions about the coherence properties in parametric amplification and stimulated Raman scattering will be similar; a few examples will demonstrate this.

The first conclusion concerns the coherence properties of the elementary excitation mode (with the energy $\hbar\omega_e = \hbar\omega_L - \hbar\omega_S$). We assume that two intense laser waves, the pump wave and the Stokes wave, are fed into the crystal. This assumption allows us to replace the operators $\hat{a}_L(t)$ and $\hat{a}_S(t)$ in the Heisenberg picture by complex c-numbers according to

$$\hat{a}_{L,S}(t) \to \alpha_{L,S}\exp[-i\omega_{L,S}t], \tag{12.133}$$

where $\alpha_{L,S}$ represents the complex amplitudes. At $t = 0$ the elementary mode is assumed to be in the vacuum state. Under these conditions, in the elementary excitation mode the coherent state $|\alpha_e(t)\rangle$ with the complex amplitude

$$\alpha_e(t) = i\chi'_S\alpha_L\alpha_S^*t \tag{12.134}$$

results from the equations of motion for $t > 0$. This means that from the very beginning of the two-beam interaction process a state with full coherence occurs. One could infer from this result that, even for small t, in a good approximation the classical description, say in the form of amplitude equations, holds instead of the more complicated quantum-theoretical formalism. The analysis of the situation on the basis of the solutions of the eigenvalue equation of the electric field operator, which was discussed in Section 2.3.1.3,

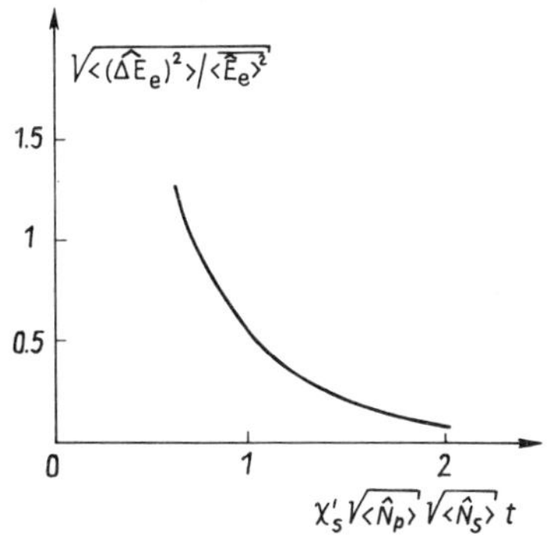

Fig. 12.17. Expectation value of reduced r.m.s. field fluctuations versus normalized time.

allows an unambiguous comparison between classical quantities and relations for measurable values derived from quantum-theoretical quantities (Ref. 12.29). For the scattering by polaritons, physically relevant knowledge can be gained from considering the fluctuations of the field strength. In Fig. 12.17 the expectation value of the r.m.s. field fluctuations—related to the spatially averaged value $\overline{\langle E_e \rangle^2}$—is plotted versus time. At the beginning of the interaction process large field fluctuations due to unavoidable quantum fluctuations occur. With increasing time the field fluctuations decrease and tend to zero; thus for larger t the electric field associated with the elementary excitation may be regarded as a prescribed function of time, and thus a classical description becomes possible. Figure 12.17 reveals that the lower bound of the time interval, in which a classical description may be allowed, decreases with increasing interaction constant χ'_S and with increasing mean photon numbers in the pump and the Stokes mode.

As in the case of frequency conversion (cf. Section 11.4.2), we shall consider certain correlation functions to gain an insight into the coherence behavior of the waves taking part in the Raman scattering process. On the basis of the interaction Hamiltonian (12.133) the equations of motion are established, and their solutions expanded in powers of the interaction time t, where it is assumed that the interaction is switched on at $t = 0$. Suppose the excitation mode to be initially in a chaotic state with the mean number $\langle \hat{N}_e(0) \rangle$ of excitations, and the pump and Stokes mode to be in a coherent states. Then for the Stokes mode the correlation function $\Delta^2\Gamma_S(t)$ in the short-time approximation (cf. Section 11.4) is given by

$$\Delta^2\Gamma_S(t) = \left(\frac{\hbar\omega_S}{2\varepsilon_0 V} \right)^2 2|\chi'_S|^2|\alpha_L|^2|\alpha_S|^2\left(1 + \langle \hat{N}_e(0) \rangle t^2 + \cdots \right). \quad (12.135)$$

The form and meaning of $\Delta^2\Gamma_S$ are shown in (11.83). The right-hand side of (12.135) reveals the tendency to bunching with increasing t. The mutual correlation behavior between different modes can be derived from the function

$$\Delta\Gamma_L(t)\,\Delta\Gamma_S(t) \equiv \frac{\hbar^2\omega_L\omega_S}{4\varepsilon_0^2V^2}\Big[\big\langle \hat{a}_L^\dagger(t)\hat{a}_S^\dagger(t)\hat{a}_L(t)\hat{a}_S(t)\big\rangle - \big\langle \hat{a}_L^\dagger(t)\hat{a}_L(t)\big\rangle\big\langle \hat{a}_S^\dagger(t)\hat{a}_S(t)\big\rangle\Big]. \quad (12.136a)$$

Under the abovementioned initial conditions one obtains

$$\Delta\Gamma_L(t)\,\Delta\Gamma_S(t) = -\frac{\hbar^2\omega_L\omega_S}{4\varepsilon_0^2V^2}|\chi_S'|^2|\alpha_L|^2|\alpha_S|^2\big(1 + 2\langle \hat{N}_e(0)\rangle t^2 + \cdots\big). \quad (12.136b)$$

The negative value on the right-hand side exhibits an anticorrelation effect between the pump and the Stokes mode.

Let us remark that the evolution of the nonstationary spontaneous and stimulated Raman scattering is noticeably affected by those fluctuation correlators which arise from the relaxation processes in the medium. These correlators depend on properties and the state of the medium (relaxation times, temperature) and on external influences (e.g. the field strengths of incident radiation). A complete treatment of the coherence behavior requires one to take into account these influences (Refs. 12.25, 12.30).

12.2.3.3. Investigation and Exploitation of Material Properties. The results from phonon and polariton scattering yield important insights into the material properties of crystals and liquids (molecules). On the one hand a relationship is established between the quantities accessible to measurement and the atomic quantities, which for example is evidenced by (12.119) for the Stokes gain coefficient. On the other hand we are now in a position to determine important macrophysical optical quantities, such as characteristic parameters in nonlinear susceptibilities and in dispersion and relaxation relations.

The discussion of the relation (12.131) reveals the importance of the parameter S_0^{-1} for describing the nonlinear behavior of the Stokes signal intensity. This behavior can be described quantitatively by fitting S_0^{-1} to empirical findings, as obtained for example for nitrogen (Ref. 12.24). In certain cases, from polariton scattering, optical data in those wavelength regions can be obtained that can otherwise only be covered by extrapolation; so IR absorption coefficients and refractive indices in the region of strong polariton dispersion have been determined. Of special importance are the lifetime measurements of excitations of lattice vibrations. By changing the direction of incidence and polarization with respect to the orientation of the crystal and by measuring the angular distribution of the emitted radiation, the excitation

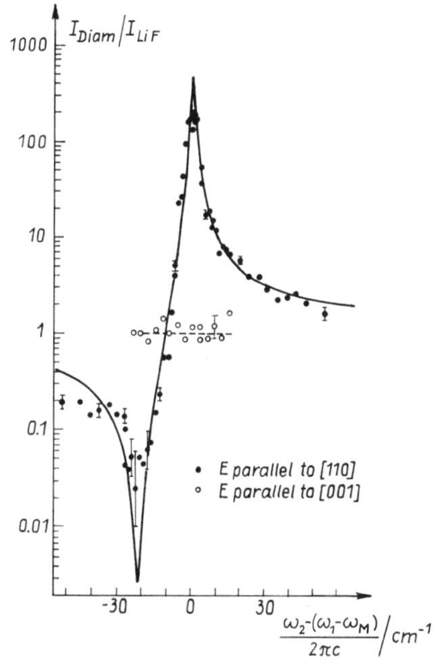

Fig. 12.18. Normalized anti-Stokes intensity for scattering by phonon vibrations of diamond (after Ref. 12.16).

states in the medium (polaritons, optical phonons) for the particular above-mentioned crystal symmetry can be determined.

For determining the material properties of crystals the methods of active Raman scattering described in Section 12.1.4.3 can be successfully used. Two examples will show how this is done.

In Ref. 12.16 the application of the CARS method to diamond was described. The frequency difference of the two lasers with the frequencies ω_1 and ω_2 was tuned over the frequency range of the optical phonon of diamond. As a result, the coherent anti-Stokes wave with frequency $\omega_A = 2\omega_1 - \omega_2$ was observed which arises from the scattering of the wave with frequency ω_1 by the coherently driven optical-phonon vibrations. In Fig. 12.18 the relative signal intensity of the anti-Stokes wave is given as a function of $\omega_2 - (\omega_1 - \omega_M)$. In the resonance region the nonlinear susceptibility is equal to the sum of the nonresonant (real) susceptibility and the resonant (complex) Raman susceptibility, which gives rise to the maximum and minimum shown in the figure. We see that according to the particular polarization state, the resonant or the nonresonant contribution in the signal intensity is favored.

In Refs. 12.31 and 12.32 the Raman-induced Kerr effect was first considered and experimentally tested in crystals by taking into account the phonon–photon interaction, that is, RIKE by polaritons. The circularly polarized laser pump wave yields results similar to those in liquids as discussed in Section 12.1.4.3. RIKE with an elliptically polarized laser pump wave in optically active and anisotropic crystals shows that also in more complicated cases the

RIKE method is capable of yielding information about the nonresonant term in the third-order susceptibility. Through experiments on the 817-cm^{-1} A (LO) mode of the uniaxial crystal $LiIO_3$ an upper bound of the amount of the relative nonresonant term was determined.

Of great practical importance is the possibility of using polariton scattering to construct *tunable radiation sources* in the infrared and far infrared regions. To explain these devices we rely on the conservation laws of energy and momentum as shown in Fig. 12.15. The use of a resonator for the scattered radiation allows us to choose the direction of the scattered wave relative to the laser wave. When the dispersion relation of the photons and polaritons is satisfied, at a given angle θ between $k_{L\bullet}$ and $k_{S\bullet}$ only one Stokes wave with fixed frequency and thus only one particular polariton wave is capable of oscillating. The angle θ can be adjusted by the experimenter; by doing so he can sweep over the dispersion curve for polaritons (cf. Fig. 12.15a) and thereby select definite frequencies ω_S and ω_P according to $\omega_L = \omega_S + \omega_P$. Such tuning is of particular advantage in the photonlike part of the lower branch; in materials especially suitable for such purposes a variation of θ by a few degrees is accompanied by a variation of the polariton frequency by a factor of about 5. In the experimental studies on $LiNbO_3$ in Ref. 12.33, the stimulated process was initiated by a ruby laser (wavelength $\lambda = 0.694$ μm, power $P_L \approx 1$ MW, pulse duration ≈ 20 ns). The generated Stokes wave and the electromagnetic part of the generated polariton wave were coupled out. With the Stokes wave radiation powers of $\approx 0.5P_L$ have been achieved, with the radiation in the far infrared region resulting from the polariton wave, powers of the order of $10^{-5}P_L$. Here the conversion rate is much below the limit determined by the Manley–Rowe relation, since there is strong IR absorption connected with an intense mechanical-wave part. Variation of the angle in the region of $0.5° \lesssim \theta \lesssim 4°$ leads to tuning over a wavelength region 250 μm $> \lambda_P >$ 50 μm. Thus coherent waves at 200 μm can be coupled out with radiation powers of more than 50 W.

12.3. STIMULATED BRILLOUIN SCATTERING

In the preceding sections we studied stimulated Raman scattering of electromagnetic radiation interacting with vibrations of molecules and with crystal vibrations on the optical branch. There exists an analogous phenomenon in which radiation is scattered by vibrations of the *acoustical* branch of ordered or disordered condensed matter. This type of scattering is called *Brillouin scattering*.

When quasimonochromatic intense radiation sources became available after the invention of lasers, numerous investigations on thermal Brillouin scattering were performed; the scattering of electromagnetic radiation by thermally excited pressure waves was observed. In 1964, by means of Q-switched lasers, Chiao, Townes, and Stoicheff (Ref. 12.34) discovered stimulated Brillouin scattering, which is connected with the generation of coherent hypersonic waves. For the description of this process we may proceed along the same lines as those employed for the stimulated Raman effect given in Sections 12.1 and 12.2. Since the spatial-transformation behavior has already been treated there, we confine ourselves here to a (classical) one-dimensional model. The basic relationships are dealt with in Section 12.3.1; there we demonstrate that—in analogy to the spontaneous and the stimulated Raman effect—the thermal (i.e., the nonstimulated) effect and the stimulated effect can be treated by one and the same model. Section 12.3.2 is devoted to basic experimental results and applications.

12.3.1. Fundamentals of Thermal and Stimulated Brillouin Scattering

First let us deal with the case where the pressure waves taking part in the light scattering are not induced by electromagnetic radiation (i.e., the nonstimulated effect). If alterations of the density or other distortions appear in a condensed medium, the permittivity ε and thus the electric polarization are changed. (We remark that alterations of ε due to entropy and orientation fluctuations are ignored in the following.) We assume the space–time dependence of the electromagnetic and pressure waves to be given as a function of z and t. This means that these waves propagate in the direction of the z axis; in particular, the distortion $\partial u/\partial z$ characterizing the elastic state of matter is a function of z and t only. The pressure σ (or the tension) is connected with the distortion by

$$-\sigma = \Lambda \frac{\partial u}{\partial z}, \tag{12.137}$$

where the parameter Λ is the elastic modulus. The relation

$$\varepsilon(\sigma) = \varepsilon_{(0)} + \varepsilon_{(1)}\sigma, \qquad \text{where} \quad \varepsilon_{(0)} = \varepsilon(\sigma = 0), \quad \varepsilon_{(1)} = \left(\frac{d\varepsilon}{d\sigma}\right)_{\sigma=0} \tag{12.138}$$

gives the alteration of the permittivity due to the pressure. Terms higher than first order in σ are not needed and are therefore ignored. This relation enables us to give the polarization as a function of the pressure; one obtains

$$P = \varepsilon_0(\varepsilon - 1)E = \varepsilon_0(\varepsilon_{(0)} - 1)E + \varepsilon_0\varepsilon_{(1)}\sigma E. \tag{12.139}$$

The term containing the factor σE represents the coupling between the electric field strength and the pressure. Let us estimate typical values for the coupling parameter $\varepsilon_{(1)}$. We have

$$\frac{d\varepsilon}{d\rho} \sim \frac{\varepsilon_{(0)} - 1}{\rho}, \qquad \frac{d\rho}{d\sigma} \sim \frac{\rho}{\Lambda}, \tag{12.139a}$$

where ρ is the mass density of the medium. Consequently in the frequency region of visible light we have $\varepsilon_{(1)} \sim \Lambda^{-1}$. According to the microphysical structure of the medium, typical values of Λ^{-1} lie between 10^{-11} and 10^{-9} m^2/N.

Let us assume that there is one pressure wave (frequency ω_v, vibrational amplitude $\hat{\sigma}(\omega_v) = \bar{\sigma}(\omega_v)e^{+ik(\omega_v)z}$) and one laser wave (frequency ω_L, field-strength amplitude $\hat{E}(\omega_L) = \bar{E}(\omega_L)e^{+ik(\omega_L)z}$). Because of the term $\varepsilon_0\varepsilon_{(1)}\sigma E$ in (12.139), these waves give rise to two polarization waves with frequencies

$$\omega_{sc} = \omega_L + b_1\omega_v, \tag{12.140}$$

which are shifted with respect to the laser frequency ω_L. These two polarization waves yield frequency-shifted scattered radiation, where the Stokes wave is characterized by $b_1 = -1$, the anti-Stokes wave by $b_1 = +1$. The amplitudes of these polarization waves are

$$\hat{P}(\omega_L + \omega_v) = \frac{\varepsilon_0\varepsilon_{(1)}}{2}\bar{\sigma}(\omega_v)\bar{E}(\omega_L)e^{+i[k(\omega_L)+k(\omega_v)]z}, \tag{12.141a}$$

$$\hat{P}(\omega_L - \omega_v) = \frac{\varepsilon_0\varepsilon_{(1)}}{2}\bar{\sigma}^*(\omega_v)\bar{E}(\omega_L)e^{+i[k(\omega_L)-k(\omega_v)]z}. \tag{12.141b}$$

Analogous reasoning to that for stimulated Raman scattering leads to the relation

$$k(\omega_{sc}) = k(\omega_L) + b_1k(\omega_v) \tag{12.142}$$

for the wave number of the scattered light. The ratio $\omega_v/k(\omega_v)$ is the phase velocity v of the pressure wave. The frequency ω_v is very small compared to the frequencies ω_L and ω_{sc} of the electromagnetic waves; therefore we may assume that the laser and the scattered wave have the same value of the phase velocity, denoted by w.

Now for given ω_L and $k(\omega_L) = w^{-1}\omega_L$ (which is assumed to be positive, so that the laser wave propagates in the z direction) we can calculate the values of ω_{sc}, $k(\omega_{sc})$, ω_v, and $k(\omega_v)$. We have to distinguish between two cases according to the sign of $k(\omega_{sc})$. In the case $k(\omega_{sc}) > 0$ (observation of the light scattered in the direction of positive z) one obtains $\omega_v = 0$; this scattered light is not frequency-shifted. In the case $k_{sc} < 0$ (backward scattering) one obtains

$$\omega_v = 2\frac{v}{w}\omega_L, \tag{12.143}$$

where the relation $v/w \ll 1$ is taken into account. Here a frequency shift of the scattered wave with respect to the laser wave appears, which is proportional to the laser frequency and to the ratio between the phase velocities of the pressure wave and the electromagnetic waves. Since $v/w \sim 10^{-5}$ and

$\omega_L \sim 10^{16}$ s^{-1}, the frequency ω_v of the pressure wave turns out to be of the order of 10^{11} s^{-1}. In this case the wave numbers

$$k(\omega_{sc}) = -\frac{\omega_L}{w}, \qquad k(\omega_v) = -b_1\frac{2\omega_L}{w} \tag{12.144}$$

result, where again $v/w \ll 1$ is used. Obviously the absolute value of the wave number of the pressure wave is twice as large as the wave numbers of the light waves $k(\omega_L)$, $k(\omega_{sc})$. If—say by thermal excitation—pressure waves with various frequencies and wave numbers are excited, then, according to (12.143) and (12.144), only one wave with a certain frequency ω_v and wave number $k(\omega_v)$ contributes to the light scattering.

If arbitrary propagation directions of the waves are taken into consideration instead of the one-dimensional treatment (12.142) the vector relation

$$k_\bullet(\omega_{sc}) = k_\bullet(\omega_L) + b_1 k_\bullet(\omega_v) \tag{12.145}$$

must be applied. Thus one obtains the expression

$$\omega_v = 2\frac{v}{w}\omega_L \sin\tfrac{1}{2}\theta_{sc} \tag{12.146}$$

for the frequency of the pressure wave, where θ_{sc} is the angle between $k_\bullet(\omega_L)$ and $k_\bullet(\omega_{sc})$. This relation reveals that the frequency shift of (12.143), which has been calculated in the one-dimensional treatment of backward scattering radiation, is the maximum value that can be reached.

Now let us pass to *stimulated Brillouin scattering*, in which the pressure waves taking part in the scattering process are induced by the light waves themselves. The corresponding mechanism can be discussed analogously to stimulated Raman scattering.

The consideration is based on that polarization term $P_{\sigma E}$ in (12.139)—the second term on the right-hand side—which reflects the coupling between pressure and electromagnetic field. From this term the density of the interaction energy $\mathscr{H}^C$ can be derived:

$$\mathscr{H}^C = -\int_0^E dE' \, P_{\sigma E}(\sigma; E'). \tag{12.147}$$

For a given field function $E(t)$ the polarization $P_{\sigma E}$ attains its value, according (12.139), within a very small time, during which there is no noticeable alteration of the pressure σ. Thus the pressure σ may be considered to be constant in evaluating the integral. The reasoning is the same as in the case of molecular vibrations [cf. (12.5)]: the motion of the nuclei or the atomic cores due to acoustic vibrations proceeds more slowly than the buildup of the electronic state in condensed matter for a given E field. Using (12.139) one obtains

$$\mathscr{H}^C = \tfrac{1}{2}\varepsilon_0\varepsilon_{(1)}\Lambda\frac{\partial u}{\partial z}E^2. \tag{12.148}$$

The theory of elasticity yields the following expression for the Lagrange density of a free elastic system:

$$\mathscr{L}^F = \frac{\rho}{2}\left(\frac{\partial u}{\partial t}\right)^2 - \frac{\rho}{2}v^2\left(\frac{\partial u}{\partial z}\right)^2, \tag{12.149}$$

where the first term on the right-hand side expresses the density of the kinetic energy, and the second term the density of the potential energy. The phase velocity of the pressure and the distortion waves is $v = \sqrt{\Lambda/\rho}$. Using the general field formalism [cf. (A.85)], from the total Lagrange density

$$\mathscr{L} = \mathscr{L}^F - \mathscr{H}^C \tag{12.149a}$$

the equation of motion for the field can be obtained. By virtue of the transformation (12.137) one finally arrives at the equation of motion

$$\frac{\partial^2\sigma}{\partial t^2} - v^2\frac{\partial^2\sigma}{\partial z^2} + \Gamma_v\frac{\partial\sigma}{\partial t} = -\frac{\varepsilon_0\varepsilon_{(1)}}{2\rho}\Lambda^2\frac{\partial^2 E^2}{\partial z^2} \tag{12.150}$$

for the pressure σ. Note that the third term on the left-hand side was added to the equation of motion resulting from (12.149a) in order to take losses into account. The wave equation for the field E is given by

$$\frac{\partial^2 E}{\partial t^2} - w^2\frac{\partial^2 E}{\partial z^2} = -\frac{w^2}{c^2}\varepsilon_{(1)}\frac{\partial^2}{\partial t^2}(\sigma E). \tag{12.151}$$

Here the phase velocity w is assumed to be a real quantity; the absorption losses of the electromagnetic waves are neglected.

The pair of equations (12.150) and (12.151) are coupled by the nonlinear terms; this reflects the fact that there is an interdependence between pressure and electromagnetic waves. Let us now study the result of this coupling in the case of monochromatic waves. The assumption of one single electromagnetic wave, say a laser wave [amplitude $\hat{E}(\omega_L)$, frequency ω_L] does not suffice to predict the phenomena expected. In this case it becomes apparent in (12.150) that the driving term on the right-hand side, because of E squared, would lead to a σ wave with the frequency $2\omega_L$; this frequency, however, would drastically exceed the frequency range of acoustic waves (the d.c. pressure occurring simultaneously is of no interest). Obviously, in addition to the laser wave with frequency ω_L, one electromagnetic wave with another frequency must take part. This wave will be called the Stokes wave [amplitude $\hat{E}(\omega_S)$, frequency

ω_S]. The difference frequency $\omega_L - \omega_S$ is assumed to lie in the frequency range of the pressure waves. The pressure will be represented by one single wave [amplitude $\sigma(\omega_v)$, frequency ω_v].

To satisfy the differential equations (12.150) and (12.151) and to obtain maximum σ excitation, the rapidly varying temporal factors as well as the rapidly varying spatial factors on both sides must agree. These requirements lead to the relations

$$\omega_S = \omega_L - \omega_v, \tag{12.152a}$$

$$k(\omega_S) = k(\omega_L) - k(\omega_v), \tag{12.152b}$$

which reveal that in the case of the stimulated Brillouin scattering, the same conditions hold for frequencies and wave numbers as in the case of thermal Brillouin scattering [cf. (12.140) and (12.142)]. In particular it can be concluded from (12.142) that the maximum frequency shift ω_v in the case of the stimulated effect exactly agrees with the value derived for the thermal effect [cf. (12.143)]. Using the above-introduced three-wave ansatz under stationary conditions, the differential equations (12.150) and (12.151) yield the relations

$$\frac{d\bar{\sigma}(\omega_v)}{dz} + \frac{\Gamma_v}{2v}\bar{\sigma}(\omega_v) = +\frac{i\varepsilon_0\varepsilon_{(1)}\Lambda^2 k(\omega_v)}{8\rho v^2}\bar{E}(\omega_L)\bar{E}^*(\omega_S), \tag{12.153a}$$

$$\frac{d\bar{E}(\omega_S)}{dz} = -\frac{i\varepsilon_{(1)}\omega_S w}{4c^2}\bar{E}(\omega_L)\bar{\sigma}^*(\omega_v), \tag{12.153b}$$

$$\frac{d\bar{E}(\omega_L)}{dz} = +\frac{i\varepsilon_{(1)}\omega_L w}{4c^2}\bar{E}(\omega_S)\bar{\sigma}(\omega_v) \tag{12.153c}$$

for the wave amplitudes $\bar{\sigma}(\omega_v)$, $\bar{E}(\omega_S)$, and $\bar{E}(\omega_L)$, where a slow variation in z is supposed, so that all terms containing the second derivative with respect to z may be neglected.

Let us remark that the coupling between electromagnetic and acoustic waves —treated so far on the basis of classical waves—can also be described quantum-theoretically. This can be done in analogy to Section 12.2.2, where the interaction between photons and optical phonons was dealt with. In the case of acoustic waves, the interaction between photons and acoustic phonons must be considered. The relations (12.152a, b) reflect the conservation of energy and momentum of these particles.

12.3.2. Applications

Let us discuss some basic aspects of application. Equation (12.153a) shows that intense laser radiation leads to the generation of pressure waves with high intensity. Utilizing stimulated Brillouin scattering, the propagation of intense

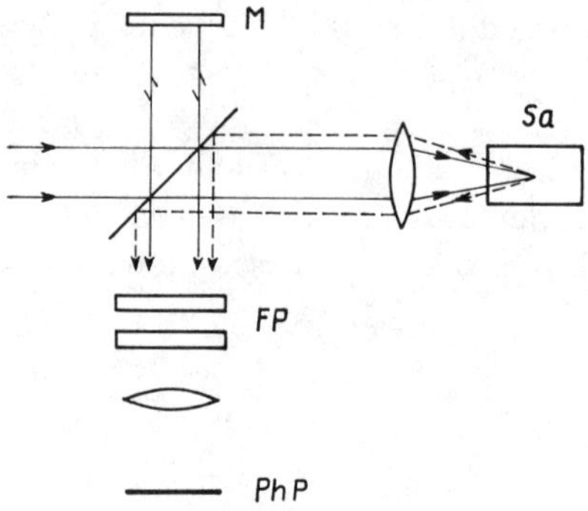

Fig. 12.19. Scheme for measurement arrangement for stimulated Brillouin scattering: Sa, sample; M, mirror; FP, Fabry–Perot interferometer; PhP, Photographic plate.

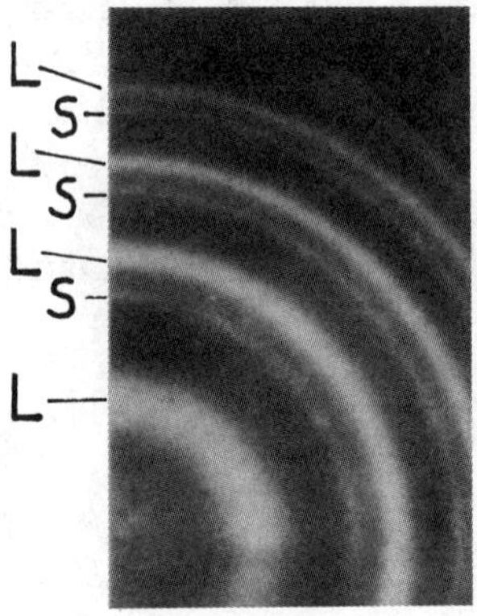

Fig. 12.20. Stimulated Brillouin scattering by quartz (after Ref. 12.34). Fringes assigned to laser (L) and Stokes (S) radiation.

hypersonic waves in crystals and in amorphous material can be induced and observed. The scheme of the experimental device used in Ref. 12.34 is given in Fig. 12.19. The incident laser radiation generates in the sample the scattered radiation, which can be studied as a function of the direction of propagation. By using a beam splitter and a Fabry–Perot interferometer it becomes possible to compare directly the laser radiation with the radiation backward-scattered by the sample. The experimental output of stimulated Brillouin scattering by quartz is shown in Fig. 12.20. The fringes assigned to Stokes radiation are denoted by S, while the fringes assigned to laser radiation are denoted by L. The irradiance of the laser radiation was of the order of 10^{15} W/m^2. In these experiments very intense pressure waves were generated and sometimes damaged the medium.

The connection between the velocity v of the sound waves and the frequency shift can be demonstrated very instructively for mixtures of liquids: on changing the concentration of the various components continuously, the velocity v and thus the frequency shift vary.

Both in crystals and in liquids the frequency shift calculated from (12.152) showed good agreement with the experimental findings.

The system (12.153) of differential equations yields quasistationary solutions; they represent a good approximation to experimental conditions in case the duration of the light pulses is much greater or greater than the buildup time of pressure waves in the medium. The buildup time is given by the reciprocal of the damping constant Γ_v, which is twice the product of the sound velocity and the sound absorption coefficient. For liquids at room temperature the buildup time is of the order of 10^{-9} s. If the damping length of an acoustic wave can be assumed to be small with respect to the length characterizing the change of light amplitudes due to attenuation or amplification processes, one obtains the solution

$$\bar{\sigma}(z) = -\frac{i\varepsilon_0\varepsilon_{(1)}\Lambda^2 k(\omega_v)}{4\rho v\Gamma_v}\bar{E}(\omega_L, z)\bar{E}^*(\omega_S, z) \tag{12.154}$$

for the pressure-wave amplitude. This means that the pressure amplitude at z depends only on the field-strength amplitudes at this position. According to (12.139) this also holds for the polarization. In this case the Brillouin susceptibility $\chi^{(3)}_{\mathrm{Br}}$ characterizing the connection between polarization and field strength is—in analogy to the Raman susceptibility from (12.22)—given by

$$\chi^{(3)}_{\mathrm{Br}}(\omega_S; \omega_S, \omega_L, -\omega_L) = -\frac{i\varepsilon_0\varepsilon^2_{(1)}\Lambda^2\omega_v}{8v^2\rho\Gamma_v}. \tag{12.155}$$

Typical numerical values of $\omega_v = 6 \times 10^{10}\ \mathrm{s}^{-1}$, $v = 10^3$ m/s, $\rho = 10^3$ kg/m^3, $\Gamma_v = 10^9\ \mathrm{s}^{-1}$, $\varepsilon_{(1)} \sim \Lambda^{-1}$ lead to a Brillouin susceptibility of the order of 10^{-19} m^2/V^2.

Let us now discuss solutions under the conditions explained in establishing (12.154). Substituting the pressure amplitude (12.154) into (12.153b) and into (12.153c), one obtains two differential equations for $\bar{E}(\omega_S, z)$ and $\bar{E}(\omega_L, z)$. These equations can be rewritten as differential equations for the time-averaged values $\overline{S(\omega_S, z)}^t$ and $\overline{S(\omega_L, z)}^t$ of the Poynting vectors, which are known to be proportional to $|\bar{E}(\omega_S, z)|^2$ and $|\bar{E}(\omega_L, z)|^2$. At one end of the cuvette (at $z = 0$) the laser wave is fed into the sample; at the other end (at $z = l$) a (relatively weak) counterpropagating wave with the frequency $\omega_S = \omega_L - \omega_v$ is fed in.

First we consider small-signal amplification, where the relation $\overline{S(\omega_L, z)}^t = \overline{S(\omega_L, 0)}^t$ is satisfied. There we have the simple solution

$$\overline{S(\omega_S, z)}^t = \overline{S(\omega_S, l)}^t e^{(l-z)g_{\mathrm{Br}}} \tag{12.156}$$

with g_{Br} the Brillouin amplification coefficient

$$g_{\mathrm{Br}} = \frac{\varepsilon^2_{(1)}k(\omega_v)\Lambda^2 w^2\omega_S}{4c^2\Gamma_v\rho v}\overline{S(\omega_L, 0)}^t. \tag{12.157}$$

The power of the scattered wave increases exponentially with increasing $l - z$. With the help of gain measurements the quantity $g_{\mathrm{Br}}/\overline{S(\omega_L,0)}^t$, a characteristic of the material under study, has been determined, and for small-signal amplification the observed values are in good agreement with the calculated values.

Now we will turn to the more general case beyond small-signal amplification; here the decrease of the laser power due to the scattering process must be taken into account. The system (12.153b, c) of differential equations leads, by virtue of (12.154), in the case of large $g_{\mathrm{Br}}l$ values to the limiting solution

$$\frac{\overline{S(\omega_L,l)}^t}{\omega_L} = 0, \qquad \frac{\overline{S(\omega_S,0)}^t}{\omega_S} - \frac{\overline{S(\omega_S,l)}^t}{\omega_S} = \frac{\overline{S(\omega_L,0)}^t}{\omega_L}. \quad (12.158)$$

This case reflects the situation where all the incident laser photons, whose number is proportional to $\overline{S(\omega_L,0)}^t/\omega_L$, are converted into backward-scattered photons of the Brillouin wave.

The solution (12.156) is based on the assumption of an incident weak backward-propagating wave of frequency ω_S, which on the one hand generates a pressure wave according to (12.154) and on the other hand is amplified itself. However, it is also possible to explain the buildup of a stimulated Brillouin Stokes wave by amplification of thermally scattering radiation with the help of arguments like those used for the explanation of the buildup of a stimulated Stokes wave from spontaneous Raman scattering at the end of Section 12.1.3.1. In this case a stimulated Brillouin wave can be accomplished with only one incident laser wave. If $g_{\mathrm{Br}}l$ attains a sufficiently large value, then because of the amplification of the Stokes wave at the expense of the laser wave, the laser power at the end of the cuvette (at $z = l$) nearly vanishes. This means that the medium in the cuvette acts on the incident laser wave as a mirror, where, however, the reflected radiation displays a (relatively small) frequency shift; the effective reflectivity of this mirror increases from 0 to 1 with increasing laser pump power.

This reflection behavior can be utilized for Q-switching, as first reported in Ref. 12.35. Moreover, it should be mentioned that this inelastic photon scattering via Brillouin effect is found to be one of the most promising processes for nonlinear optical phase conjugation, which will be discussed in Subsection 14.2.1.

From the above numerical estimation of the conditions for quasistationary behavior it can be concluded that they are violated if the intensity changes considerably in intervals below 10^{-8} s. Then, in view of (1.215), the explanation of experiments requires one to introduce further terms into the differential equations (12.153) (in particular, terms containing the second time derivative must be added). Solutions of such systems are, for instance, dealt with in Ref. 12.36.

12.4. SPIN-FLIP PROCESSES AND STIMULATED RAMAN SCATTERING

From observations of radiation-excited spin-flip processes involving quasifree charge carriers in semiconductors one can gain specific information on the material. Fundamentals of these phenomena will be dealt with in Section 12.4.1. Applications will be treated in Section 12.4.2.

12.4.1. Fundamentals of Spin-Flip Processes

The energy eigenstates of charge carriers in semiconductors are degenerated. In (quasi-) stationary magnetic fields these degenerate levels are split up into the Landau levels, whose energy differences correspond to the cyclotron frequency, and into sublevels that arise from different orientations of the electron spin. By interaction with electromagnetic radiation, transitions between levels of different spin orientation can occur; they are called *spin-flip processes*.

Our considerations refer to spin flip of electrons in the conduction band (index C) near the band edge. It is assumed that an electron of the lowest Landau level of the conduction band undergoes, by virtual transition via the valence band (index V), a spin flip according to the scheme

$$|n = 0, \uparrow\rangle_C \to (V\text{-band}) \to |n = 0, \downarrow\rangle_C. \tag{12.159}$$

The arrows in the state vectors designate the direction of the spin; n is the quantum number of the Landau level. A more precise theory must take into account, among other facts, the occupation of the bands as a function of the electron concentration and of the magnetic field strength as well as the occurrence of two-electron processes. The effective Hamiltonian

$$\hat{H}^{\text{eff}} = \frac{1}{2\tilde{m}}(\hat{p}_\bullet + \hat{I}eA_\bullet)^2 - \hat{\mu}_\bullet H_\bullet + R(\hat{p}_\bullet, H_\bullet) \tag{12.160}$$

makes it possible to treat the problem (in good approximation) as a one-electron problem (Ref. 12.37). Here $\hat{p}_\bullet$ and $\hat{\mu}_\bullet$ are the operators of the momentum and the magnetic dipole moment of the electron; $\tilde{m}$ is the effective mass. The quantities $A_\bullet$ and $H_\bullet$ denote the (classical) vector potential of a radiation field and the magnetic field. If no radiation is incident, the following energy eigenvalues result:

$$\mathscr{E}_{k,n,s} = \mathscr{E}_G + \frac{\hbar^2 k^2}{2\tilde{m}} + \frac{\hbar e \mu_0}{2\tilde{m}}(2n+1)|H_\bullet| + g\mu_B s|H_\bullet| + \tilde{R}(k, |H_\bullet|). \tag{12.161}$$

Here e is the elementary charge, μ_0 is the magnetic permeability of free space,

μ_B is the Bohr magneton, and $\mathscr{E}_G$ is the energy gap. The quantum numbers $n = 0, 1, 2, \ldots$ denote the Landau levels, and the quantum numbers $s = \pm\frac{1}{2}$ designate the spin levels. The terms $R(\hat{p}, |H_\bullet|)$ and $\tilde{R}(k; |H_\bullet|)$ summarize the contributions to the energy that are nonparabolic with respect to $\hat{p}$ and k, respectively, as well as the contributions that are nonlinear with respect to the magnetic field strengh. The effective mass $\tilde{m}$ and the g factor can be determined from the characteristic parameters of the conduction band and the valence band. Under irradiation of the sample by pump radiation, which is assumed to be a laser wave of frequency ω_L, scattered radiation of frequency

$$\omega_S = \omega_L - \frac{|g||\mu_B||H_\bullet|}{\hbar} \tag{12.162}$$

can be observed that results from the abovementioned spin-flip process. The differential cross section for this spontaneous Stokes scattering is given by

$$\left(\frac{d\sigma}{d\Omega}\right)_{\uparrow\downarrow} \approx \left(\frac{e^2}{4\pi\varepsilon_0 m_e c^2}\right)^2 \left(\frac{g}{g_{\text{fr}}}\right)^2 \left(\frac{\hbar\omega_L}{\mathscr{E}_G}\right)^2. \tag{12.163}$$

Here m_e and g_{fr} are the mass and the g factor of the free electron. Typical values measured with InSb are

$$\mathscr{E}_G = 0.23eV, \qquad \tilde{m} \approx \tfrac{1}{70} m_e,$$

$$g(B=0) = -48, \qquad g(B = 10\,\text{V s/m}^2) = -35. \tag{12.164}$$

The first factor on the right-hand side of (12.163) is the cross section for Thomson scattering by free electrons. The second and the third factor are determined by specific properties of the semiconductor and the laser frequency. If the radiation of a CO_2 laser (10.6 μm) is used for excitation, the values of (12.164) lead to a cross section for Stokes scattering of the order of 10^{-27} m^2/sr. This value is higher than that of typical cross sections for transitions between Landau levels, and it is substantially higher than typical cross sections for spontaneous Raman scattering by molecules, which are of the order of 10^{-32} m^2/sr.

12.4.2. Applications

As in molecular Raman scattering [cf. (12.61)], stimulated Raman oscillations on the basis of spin-flip processes will be possible only if the gain coefficient is large enough to compensate for the optical losses. In the case of spin-flip Stokes scattering, the frequency of which is given in (12.162), there results the following estimate for the gain coefficient:

$$g_S \sim \left(\frac{d\sigma}{d\Omega}\right)_{\uparrow\downarrow} \frac{n_e}{\omega_S^3} \frac{S_L}{\Delta\omega} G(\mathscr{E}_F, |H_\bullet|, \mathscr{T}). \tag{12.165}$$

Here n_e is the charge carrier density, $\Delta\omega$ is the line width for spontaneous scattering, and SL is the laser power. The function G takes account of band filling and thus depends on the Fermi energy $\mathscr{E}_F$, the magnetic field strength, and the temperature $\mathscr{T}$. At $B = 5\ \mathrm{V\,s/m^2}$ the factor g_S/S_L is found to be of the order of $10^{-7}\ \mathrm{m^{-1}/W\,m^{-2}}$. This value is large enough to compensate for the intrinsic optical losses of the material and to generate stimulated radiation at pump intensities of some $10^9\ \mathrm{W\,m^{-2}}$.

Let us now discuss some typical experimental results (see Refs. 12.38, 12.39, 12.40). By the use of a Q-switched CO_2 laser (power 1.5 kW, pulse duration 100 ns, repetition frequency 120 Hz), pulse powers near 100 W have been obtained. Under these conditions the wavelength of scattered radiation was tuned between 11 and 13 μm by varying $|B_\bullet|$ between 1.5 and 10 $\mathrm{V\,s/m^2}$. In addition to the comparatively high power conversion rate, which is of the order of a few per cent, the high accuracy of the frequency tuning is remarkable. For instance, at $\lambda = 12\ \mu$m a wavelength accuracy better than 0.4 nm was observed. The relative line width of the scattered radiation was smaller than 3×10^{-5}. (Using more favorable experimental conditions with regard to the pump laser and the magnetic field, relative line widths smaller than 10^{-10} have been measured with the help of heterodyne detection.) In particular, because of these properties the tunable spin-flip laser proves to be a radiation source of great interest for very high-resolution spectral investigations of gases in the infrared region.

The occupation of the spin bands influences the stimulated effect most favorably if the Fermi level is situated above the lower subband ($n = 0$, $s = +\frac{1}{2}$) and below the upper subband ($n = 0$, $s = -\frac{1}{2}$), so that the latter is nearly unoccupied. For a given value of n_e the stimulated process starts only at magnetic fields that exceed a certain minimum value $H_{\min}(n_e)$. This results mainly from the fact that the upper subband is nearly unoccupied at high field strength (provided the temperature is sufficiently low), so that a comparatively large number of electrons can flip from the highly occupied lower subband into the upper one by spin-flip processes. The smaller the charge carrier density is, the smaller is the field strength at which the level occupation is favorable for the occurrence of spin-flip transition. Of course the electron density n_e must not become so small that the gain coefficient drops to a point [see (12.165)] at which the losses can no longer be compensated for.

An increase of the pump frequency ω_L yields an increase of the gain, as is obvious from the last factor in (12.163) and from the influence of this factor on the gain coefficient g_S of (12.165). A decrease of n_e has not only the abovementioned negative influence on g_S, but in addition some positive influences, which may give rise to compensation or even overcompensation for losses; these positive influences result from the increasing mobility and from the fact that the band edge becomes more pronounced with decreasing n_e. By the last-mentioned effect the momentum-matching condition

$$2k_{L\bullet} = k_{S\bullet} + k_{A\bullet} \tag{12.166}$$

for stimulated anti-Stokes radiation (ω_A, $k_{A\bullet}$) can be satisfied with sufficient accuracy. If the experimental conditions are altogether favourable for the spin-flip process, then for n-InSb (dimensions $2 \times 2 \times 4$ mm^3) with small charge carrier densities ($n_e = 1 \times 10^{15}$ cm^{-3}) and at small fields ($B = 0.05$ V s m^{-2}) excited by a cw CO laser ($\lambda = 5.3\ \mu$m, $P = 3$ W), the first Stokes line (with a power of about 0.2 W) can be generated, as well as the second Stokes line and the first anti-Stokes line (both with a power of 0.015 W).

REFERENCES

12.1. E. J. Woodbury and W. K. Ng, *Proc. IRE* **50**, 2347 (1962).

12.2. P. A. M. Dirac, *Proc. Roy. Soc. London Ser. A* **114**, 710 (1927).

12.3. M. Schubert and B. Wilhelmi, *Introduction to Nonlinear Optics*, Vol. 2, Teubner, Leipzig, 1978, Appendix 2 (German).

12.4. R. Chiao and B. P. Stoicheff, *Phys. Rev. Lett.* **12**, 290 (1964).

12.5. H. J. Weigmann, M. Pfeiffer, A. Lau, and K. Lenz, *Opt. Comm.* **12**, 231 (1974).

12.6. R. A. McLaren and B. P. Stoicheff, *Appl. Phys. Lett.* **16**, 140 (1970).

12.7. W. Werncke, J. Klein, A. Lau, K. Lenz, and G. Hunsalz, *Opt. Comm.* **11**, 159 (1974).

12.8. R. W. Terhune and P. D. Maker; *Phys. Rev. A* **137**, 801 (1965).

12.9. V. S. Letokhov, *Laser Spectroscopy*, Akademie, Berlin, 1977 (German).

12.10. S. A. J. Druet and J.-P. E. Taran, *Progr. Quant. Electron.* **7**, 1 (1981).

12.11. P. D. Maker and R. W. Terhune, *Phys. Rev.* **137**(3A), A801 (1965).

12.12. M. D. Levenson, *Phys. Today* **30**(5), 44 (1977).

12.13. A. Lau, H.-J. Weigmann, W. Werncke, K. Lenz, and M. Pfeiffer, *Time-Resolved Vibrational Spectroscopy*, Academic, 1983, p. 353.

12.14. M. Pfeiffer, A. Lau, W. Werncke, and L. Holz, *Opt. Comm.* **41**, 363 (1982).

12.15. F. R. Aussenegg, J. Brandmüller, M. E. Lippitsch, and W. Nitsch, *Opt. Comm.* **37**, 59 (1981).

12.16. M. D. Levenson, C. Flytzanic, and N. Bloembergen, *Phys. Rev. B* **6**, 3962 (1972).

12.17. D. Heiman, R. W. Hellwarth, M. D. Levenson, and Graham Martin, *Phys. Rev. Lett.* **36**, 189 (1976).

12.18. M. D. Levenson and J. J. Song, *J. Opt. Soc. Am.* **66**, 641 (1976).

12.19. J. Bergmann and M. Schubert, *Exp. Techn. Phys.* **27**, 37 (1979).

12.20. A. Haug, *Theory of Solid-State Physics*, Deuticke, Wien, 1964, Vol. 1, §46 (German).

12.21. Ch. Kittel, *Quantum Theory of Solid-State Physics*, Oldenbourg. Wien, 1970, Chapter 3 (German).

12.22. R. Claus, L. Merten, and J. Brandmüller, *Light Scattering by Phonon-Polaritons*, Springer Tracts in Modern Physics, Vol. 75, 1975.

12.23. K. D. Kneipp, R. Kühmstedt, and H. E. Ponath, *Exp. Techn. Phys.* **XXI**, 403 (1973).
12.24. M. Sparks, *Phys. Rev. Lett.* **32**, 450 (1974).
12.25. H. E. Ponath and M. Schubert, *Ann. d. Phys.* **33**, 413 (1976).
12.26. J. Herrmann and H.-E. Ponath, *Ann. d. Phys.* **33**, 427 (1976).
12.27. J. Herrmann, H. E. Ponath, and M. Schubert, in *Proceedings of ICO-10*, Prague, 1975, p. 249.
12.28. H. S. Freedhoff, *J. Chem. Phys.* **47**, 2554 (1967).
12.29. M. Schubert and W. Vogel, *Wiss. Zeitschr. Univ. Jena Math.-Nat.* **27**, 179 (1978).
12.30. H. E. Ponath and M. Schubert, *Ann. d. Phys.* **34**, 456 (1977).
12.31. J. Bergmann, K. Kneipp, and H. E. Ponath, *Phys. Stat. Sol. B* **80**, K55 (1977).
12.32. J. Bergmann, K. Kneipp, and H. E. Ponath, *Kvant. Elektron.* **4**, 3570 (1977).
12.33. S. S. Sussmann, B. C. Johnson, J. M. Yarborough, H. E. Puthoff, and R. H. Pantell; in *Proceedings of the Symposium on Submillimeter Waves*, Brooklyn, 1970, p. 211.
12.34. R. Y. Chiao, C. H. Townes, and B. P. Stoicheff, *Phys. Rev. Lett.* **12**, 592 (1964).
12.35. D. Pohl, *Phys. Lett.* **24A**, 239 (1967).
12.36. E. E. Hagenlocker, R. W. Minck, and W. G. Rado, *Phys. Rev.* **154**, 226 (1967).
12.37. Ch. Kittel, *Quantum Theory of Solids*, Oldenbourg, Wien, 1970, Kap. 14 (German).
12.38. C. K. N. Patel and E. D. Shaw, *Phys. Rev. B* **3**, 1279 (1971).
12.39. C. K. N. Patel, *Appl. Phys. Lett.* **19**, 400 (1971).
12.40. B. S. Wherrett and W. J. Firth, *IEEE J. Quant. Electron.* **QE-8**, 865 (1972).

13

Optical Bistability

A physical system is called *bistable* if in response to one value of a stationary input signal its output signal can take on one of two possible values, the value actually taken on depending on the prehistory of the system and its fluctuations. The phenomenon of bistability, which can only occur in nonlinear systems with feedback, has already been known for a long time in mechanics and electromagnetics, where it has been extensively studied. We employ a simple model for explaining the basic principle of bistability, which, moreover, can be used for introducing some concepts required for the considerations we want to present below (see Fig. 13.1*a*). In this model the input signal S_{in} acts on the system and produces the signal S, a fraction r of which can be fed back into the system, where it causes a change of an internal parameter $\mathscr{P}$, whereas the contribution $S_{out} = (1 - r)S$ appears at the output.

Let us first consider the setup with the feedback loop disconnected. In this case we assume the internal signal transmission $\tilde{\mathscr{T}}_D = S/S_{in}$ of the device to be constant, that is, to be independent of the signal power and to depend only on an internal parameter $\mathscr{P}$ whose value can be controlled externally. The overall transmission is defined by $\mathscr{T}_D = S_{out}/S_{in}$. Let the dependence $\tilde{\mathscr{T}}_D = f(\mathscr{P})$ be nonlinear and exhibit the characteristic shown in Fig. 13.1*b*; it can be seen that we restrict our consideration to the simple case in which, with the parameter $\mathscr{P}$ kept constant, the output signal depends linearly on the input signal. Now we switch on the feedback and assume—this is again the simplest case—that according to

$$\mathscr{P} = \mathscr{P}_0 + \mathscr{P}_1 \cdot rS, \tag{13.1}$$

the internal parameter $\mathscr{P}$ changes linearly with the feedback signal rS. Thus the signal transmission of the feedback system satisfies the equation

$$\tilde{\mathscr{T}}_D = f\left[\mathscr{P}_0 + \mathscr{P}_1 \cdot rS_{in}\tilde{\mathscr{T}}_D\right], \tag{13.2}$$

which is in general a transcendental one. Graphically, its solutions can be readily obtained as the intersections of the nonlinear characteristic $\tilde{\mathscr{T}}_D(\mathscr{P})$ and the straight line $\mathscr{P} = \mathscr{P}_0 + (\mathscr{P}_1 rS_{in})\tilde{\mathscr{T}}_D$ whose slope, if the system parameters $\mathscr{P}_1$ and r are kept constant, depends on the input signal only.

From Fig. 13.1*c* possible solutions can be obtained for several values of the input signal. Obviously three solutions for the signal transmission result in a certain range of the input signal, whereas in the remaining ranges there exist

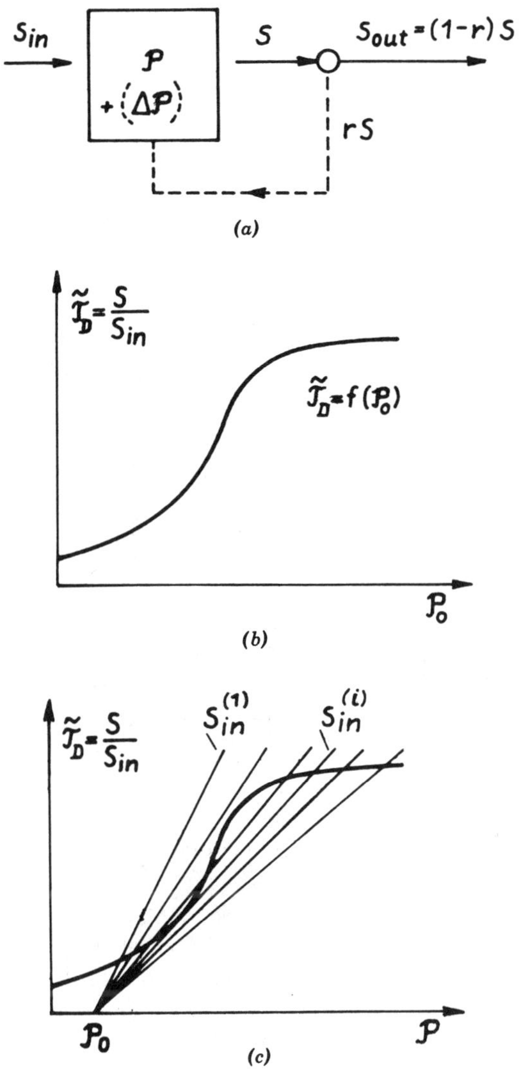

Fig. 13.1. Bistability. (a) Scheme of the device: S_{in}, input signal; $S_{\text{out}} = (1 - r)S$, output signal; rS, feedback signal; $\mathscr{P} = \mathscr{P}_0 + \Delta\mathscr{P}$, internal parameter of the system, which is changed by the feedback signal. (b) Signal transmission for zero feedback. (c) Signal transmission with feedback: For given input signal the resulting transmission is given by the intersection of the corresponding straight line with the nonlinear function $\tilde{\mathscr{T}}_D = f(\mathscr{P})$.

unambiguous solutions. Figure 13.2 shows the plot of the resulting solutions for the output signal

$$S_{\text{out}} = \mathscr{T}_D S_{\text{in}} = (1 - r)\tilde{\mathscr{T}}_D S_{\text{in}} \tag{13.3}$$

versus the input signal S_{in}. This steady-state curve is S-shaped. In the dashed part $dS_{\text{out}}/dS_{\text{in}} < 0$ and therefore the system does not pass through this part

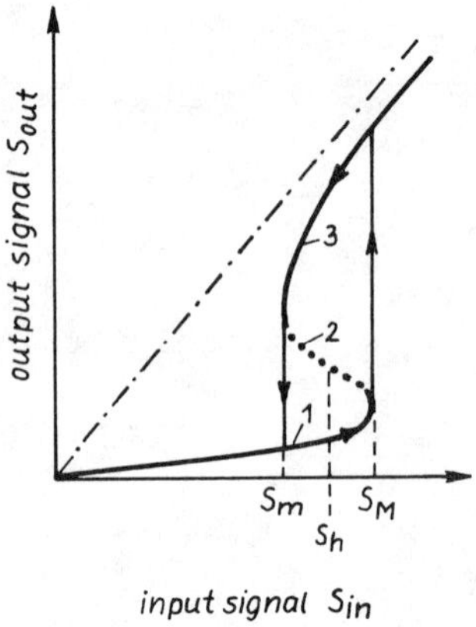

Fig. 13.2. Output signal of a bistable device as a function of the input signal S_{in}: 1, 3, stable states of the device; 2, unstable state; S_M, S_m, upper and lower discontinuity signals; S_h, holding signal.

of the curve, according to general stability considerations. If the increase of S_{in} starts from small values, at $S_0 = S_M$ the system jumps from the lower stable state (1) to the upper stable state (3). If then the input signal is decreased, the return does not happen until the value $S_{in} = S_m < S_M$ is attained. This means that the feedback gives rise to a bistable element that in a certain region of the input signal has two transmission states. This is what we call a *hysteresis loop*.

From the theoretical-physics standpoint bistability belongs to the class of cooperative phenomena in open systems in stationary equilibrium, where the transitions can be considered to be phase transitions. [From what we have discussed so far we may find a formal analogy with other phase transitions. The above relation between input and output signal recalls, in its form, the state function $p(V)$ for the pressure and volume of van der Waals gases in certain temperature ranges.] The connection between bistability and other phenomena of structure formation and organization accounts for the great theoretical interest in this phenomenon (see e.g., Refs. 13.1–13.3). Moreover, bistable elements have important applications in signal processing and storage. For such applications the operating point of the device described above can be stationarily controlled by a so called holding signal S_h where $S_m < S_h < S_M$, and by using appropriate pulses switching between the two transmission states becomes possible.

After the discovery of nonlinear optical phenomena, the question arose as to the conditions under which bistabilities may occur in optical systems and whether they can be applied to information processing and storage of information. Such logical elements based on optical effects will be of particular relevance in the future, since in communication lines increasing use is being made of optical channels because of their high transmission capacity.

Seidel (Ref. 13.4), who had performed analogous investigations in the microwave region, was the first to propose in 1969 the construction of bistable optical devices. In 1976 Gibbs, McCall, and Venkatesan (Ref. 13.5) succeeded in constructing a bistable optical device based on a Fabry–Perot interferometer containing sodium vapor as a optical material. In such experiments the

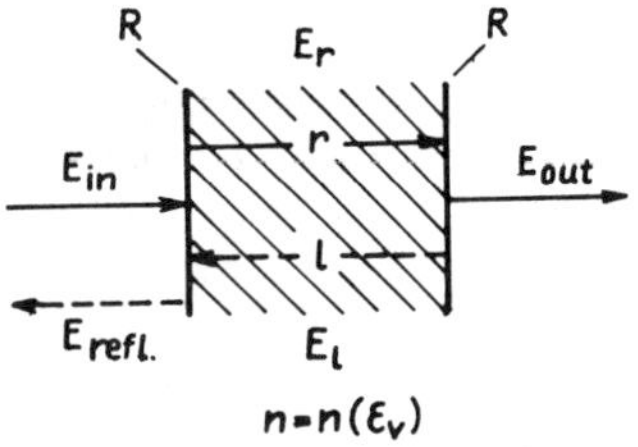

Fig. 13.3. Fabry–Perot resonator containing a nonlinear dispersive medium: $E_{r,l}$, field strength of the waves propagating in the right and left hand direction within the resonator, respectively; E_{in}, E_{out}, field strength of the input and output wave, respectively; $n(\mathcal{E}_V)$, refractive index as a function of the energy density $\mathcal{E}_V$ of the field within the resonator.

feedback of the signal is provided by the interferometer, and the refractive index of the sample, which varies with the light intensity in the interferometer as a result of saturation phenomena, serves as an internal parameter upon which the transmission of the interferometer depends (see Fig. 13.3).

In the general case the refractive index is complex near resonance, and depending on the actual conditions, the intensity dependence either of the real or of the imaginary part may play the decisive role. Hence we speak of *dispersive* or *absorptive optical bistability*. Until now in most of the experiments the dispersive nature has been predominant, and therefore we restrict our preliminary discussion to it.

The type of experiment mentioned above refers to bistable systems that rest on optical action principles throughout. This is what we mean by *intrinsic optical bistability*. If however, by way of a detector, the optical feedback signal is transformed into a signal of different type, for example into an electrical one, that is used, say by means of the Pockels effect, to control the sample transmission, then we speak of *hybrid optical bistability*. In 1973 Kastalskij (Ref. 13.6) suggested such a hybrid system in which an electrooptical crystal experiences changes in refractive index proportional to the intensity of the output. Smith and Turner (Ref. 13.7) constructed such a device in 1977.

In the following we concentrate entirely on intrinsic optical bistability. (For a comparison between intrinsic and hybrid optical bistabilities, see, e.g., Refs. 13.8 and 13.9 and the references given there.) We start with a simple description of intrinsic dispersive optical bistability in resonators. Then Section 13.2 will be concerned with appropriate nonlinear optical materials and their parameters, in particular the intensity-dependent refractive index and the characteristic relaxation times that influence the transient behavior of the devices. In Section 13.3 we study transient phenomena in optical bistability, and in Section 13.4 we present some important experimental results.

13.1. INTRINSIC DISPERSIVE OPTICAL BISTABILITY IN RESONATORS

From the experimental point of view, a Fabry–Perot interferometer (see Fig. 13.3) containing a nonlinear optical material is the simplest bistable optical system. Let us mention however that an exact theoretical treatment can be

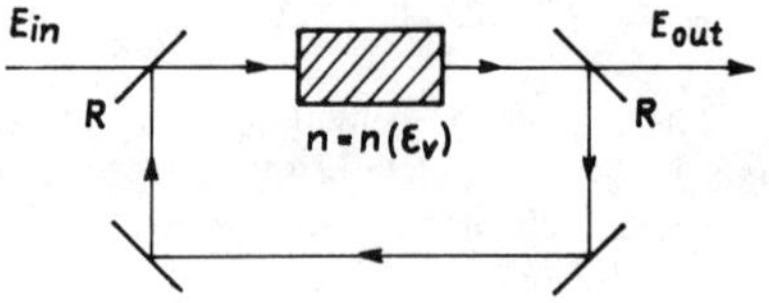

Fig. 13.4. Ring resonator containing a nonlinear optical sample. The upper mirrors have the (energy) reflectivity R and the transmission $1 - R$; the lower mirrors have reflectivity 1.

carried out in a much simpler way when using a *ring resonator* (see, e.g., Refs. 13.3, 13.10, 13.11), since in this case, in contrast to the Fabry–Perot interferometer, the nonlinear optical material is only traversed in one direction (see Fig. 13.4 and the treatment of ring lasers in Section 7.1). As a result, assuming weak output coupling, a spatial field distribution that varies only slowly exists in the nonlinear material, whereas in the Fabry–Perot resonator the standing-wave field produces a spatially modulated refractive index that causes Bragg reflections. In our simple model we ignore this spatial inhomogeneity also in the Fabry–Perot resonator of arbitrary finesse; that means we merely take into account spatial mean values of the squared field strength and the refractive index and neglect reflections by the refractive-index grating. Therefore our results, when applied to the Fabry–Perot resonator, are of qualitative character only.

In the following we describe optical bistability in the Fabry–Perot resonator in analogy to the general model presented in the introduction, but with the feedback now being inherent in the system. The transmission $\mathscr{T}_D \equiv \mathscr{T}_{\mathrm{FP}}$ of the Fabry–Perot resonator was already dealt with in Section 1.1.3 [cf. (1.35) and Fig. 1.7]. In Fig. 13.5 we present the transmission as a nonlinear function of the refractive index n, here assumed to be real; n serves as an internal parameter of the system. At very low input intensities let the material in the Fabry–Perot resonator have the refractive index $n^{(0)}$. The choice of the environmental conditions, such as the temperature, allows the external adjustment of the value of $n^{(0)}$. Therefore in this figure we start from two different small-signal values for $n^{(0)}$. Let the refractive index n depend on the mean square of the field strength in the resonator, which is proportional to the mean energy density $\mathscr{E}_V$. In the case of small outcoupling losses, $\mathscr{E}_V$ is related to the outcoupled photon flux density according to

$$I_{\mathrm{out}} = \mathscr{T}_{\mathrm{FP}} I_{\mathrm{in}} \approx \frac{1}{2} \frac{c}{n^{(0)}} (1 - R) \mathscr{E}_V \frac{1}{\hbar\omega}. \tag{13.4}$$

Now we again start from a mathematically simple, namely linear relation between the refractive index and the mean square of the field strength; we obtain

$$n = n^{(0)} + \left[\frac{2}{1 - R} n^{(2)} I_{\mathrm{in}} \right] \mathscr{T}_{\mathrm{FP}}, \tag{13.5}$$

where $n^{(2)}$ is a nonlinear optical parameter of the refractive index, introduced

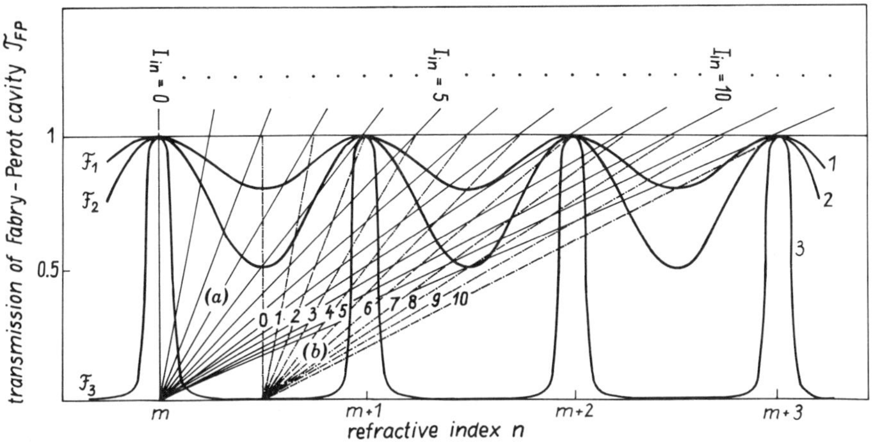

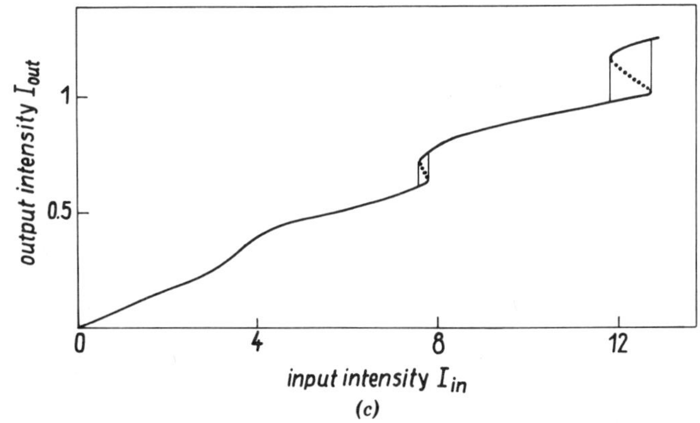

Fig. 13.5. Graphical solutions for the transmission and output intensity of Fabry–Perot interferometers containing Kerr-type nonlinear materials. (a, b) Transmission of Fabry–Perot cavities versus refractive index (full lines) according to (1.35) for three values of the cavity finesse (curves 1, 2, 3), and refractive index versus transmission (dashed lines) according to (13.5) for ten values of the input intensity, which is given in arbitrary units: (a) maximum small-signal transmission; (b) minimum small-signal transmission. Possible states of the nonlinear device are given as the intersections between the full and the dashed lines. (c–f) Output intensity I_{out} versus input intensity I_{in} for (c) maximum small-signal transmission and finesse $\mathcal{F}_1$; (d) maximum small-signal transmission and finesse $\mathcal{F}_2$; (e) minimum small-signal transmission and finesse $\mathcal{F}_1$; and (f) minimum small-signal transmission and finesse $\mathcal{F}_2$.

for a plane wave of photon flux I by the relation $n = n^{(0)} + n^{(2)}I$ (cf. Section 6.3.1). We are interested in particular in the linear relation between the refractive index and the transmission of the Fabry–Perot resonator. For two values of $n^{(0)}$ and several values of I_{in} the straight lines of (13.5) are plotted in Fig. 13.5. The number of possible values of the transmission $\mathcal{T}_D$ and hence of the output intensity I_{out} for a fixed value of the input intensity I_{in} depends on

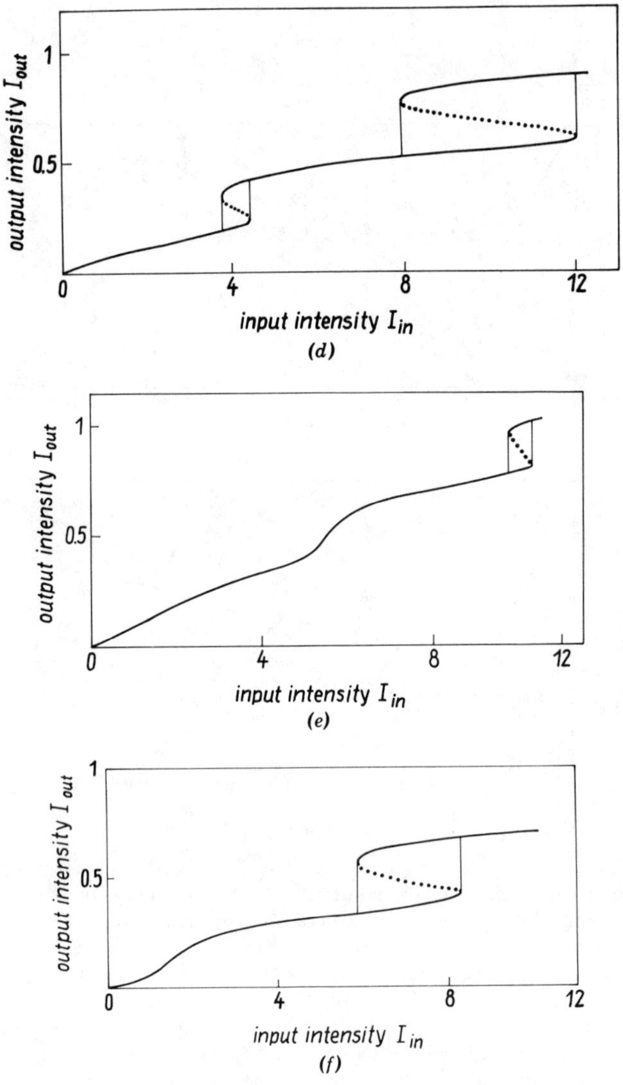

Fig. 13.5. (*Continued*).

the form of the transcendental relation $\mathcal{T}_{FP}(n)$ and on the small-signal value of the refractive index.

Let us consider a system in which we have maximum transmission for small signals (Fig. 13.5*a*). In the case of low input intensities I_{in} there is only one intersection of the curve $\mathcal{T}_D(n)$ with the straight line (13.5) corresponding to I_{in}, namely on the descending branch of the mth mode. With increasing input intensity the first intersection moves to the right. The intensity value beyond which several intersections appear is determined by the shape of the function

$\mathscr{T}_D(n)$, which depends on the finesse of the resonator. On increasing the input intensity of the resonator with the smaller finesse (curve 1 of Fig. 13.5*a*), several intersections appear for the first time in the order $m + 2$. At higher finesse (curve 2), however, several intersections appear at relatively small signals.

A quantitative change in the picture of intersections occurs if we do not start from a maximum of small-signal transmission but from a minimum (see Fig. 13.5*b*). This change can be obtained for instance by variation of temperature. From Fig. 13.5 it is also apparent that at very high intensities an increasing number of intersections may occur. This means that not only bistability but also multistability can be achieved. Figures 13.5*c–f* give plots of the solutions for the output intensity $I_{out} = \mathscr{T}_{FP} I_{in}$ which were obtained (graphically) from Fig. 13.5*a*, *b*. From Fig. 13.5*c*, which refers to the low-finesse resonator ($\mathscr{F}_1$ in Fig. 13.5*a*), we see that at the beginning the output intensity undergoes a stepwise change. By providing an appropriate holding intensity such an element can be operated in the steep part of the characteristic as an *optical amplifier*; the amplification may take very large values. Operation in the nonlinear part of the characteristic enables us to achieve, according to the specific functional relationship between I_{out} and I_{in}, *discrimination*, *limitation*, or *stabilization of signals* as well as shaping of light pulses.

Beyond the order $m + 2$ the solutions become multivalued. In the order $m + 2$ the situation is very similar to that outlined in Fig. 13.1. We get three intersections of the line (13.5) with the nonlinear function $\mathscr{T}_{FP} = \mathscr{T}_{FP}(n)$. Of these three solutions only those two are stable that possess maximum or minimum output intensity. With higher finesse such multivalued solutions already result at lower input intensities, namely, with displacements of only about one order from the small-signal point.

After this introductory discussion of the graphical solution, we investigate more closely the region in which bistabilities occur at the smallest possible intensity. For this purpose we expand the reciprocal transmission function $\mathscr{T}_{FP}^{-1}$ from (1.35) in the vicinity of the transmission maximum in terms of $\delta = (2\pi/\lambda_{vac})[2\, Ln \cos\theta - (2\, Ln \cos\theta)_{max}]$ (cf. Section 1.1.3.2) and obtain, up to the second order in δ,

$$I_{in} = \left(1 + \frac{R}{(1 - R)^2}\delta^2\right) I_{out} \tag{13.6}$$

as a relation between the input intensity I_{in} and the output intensity I_{out}, where $R = \sqrt{R_1 R_2}$ with R_1 and R_2 the reflectivities of the interferometer plates. Corresponding to the relation (13.5) between n and I_{out} given above, we here assume a linear relation between δ and I_{out}:

$$\delta - \delta(0) = \beta I_{out}, \tag{13.7}$$

where

$$\beta = \frac{2\pi}{\lambda_{vak}} 2L(\cos\theta)\left(\frac{2}{1-R}\right)n^{(2)} \tag{13.8a}$$

and

$$\delta(0) = \frac{2\pi}{\lambda_{vak}}\{2L(\cos\theta)n^{(0)} - [2L(\cos\theta)n]_{\max}\} \tag{13.8b}$$

We insert (13.7) into (13.6) and obtain

$$I_{\text{in}} = I_{\text{out}} + \frac{R}{(1-R)^2}\left[\beta^2 I_{\text{out}}^3 + 2\beta\delta(0)I_{\text{out}}^2 + \delta^2(0)I_{\text{out}}\right]. \tag{13.9}$$

To discuss this nonlinear relation we calculate the extrema of the function $I_{\text{in}} = f(I_{\text{out}})$, which lie at

$$I_{\text{in}}^{\text{ext}} = -\frac{2}{3}\frac{\delta(0)}{\beta} \pm \sqrt{\left[\frac{R}{(1-R)^2}\delta^2(0) - 3\right]\Big/\left[9\frac{R}{(1-R)^2}\beta^2\right]}. \tag{13.10}$$

Obviously extrema appear only for

$$\delta^2(0) > \frac{3(1-R)^2}{R}, \tag{13.11}$$

where with $\beta > 0$ only the value $\delta(0) < 0$ is possible. Under this condition we obtain from (13.11)

$$-\delta(0) > \frac{\sqrt{3}\,\pi}{\mathcal{F}}, \tag{13.12}$$

where $\mathcal{F}$ is the finesse of the resonator. The appearance of two extreme values implies that the nonlinear characteristic $I_{\text{out}} = I_{\text{out}}[I_{\text{in}}]$ is an S-shaped one and that in a certain range of values of the input intensity there exist three solutions for the output intensity. Here again the middle branch of the solution can be excluded from considerations of physical stability, since for it $dI_{\text{out}}/dI_{\text{in}} < 0$.

Now it is easy to determine the input intensity required to force the system to move from the small-signal point with $\delta = \delta(0)$ to the center of the bistability region, which may be defined by the position of the turning point, $d^2I_{\text{in}}/dI_{\text{out}}^2 = 0$. By differentiating (13.9) and making use of (13.12) we obtain

$$I_h \approx \frac{2}{3}\frac{-\delta(0)}{\beta} \gtrsim \frac{2\pi}{\sqrt{3}\,\mathcal{F}\beta} \tag{13.13}$$

for this holding input intensity I_h. [Furthermore we see from (13.10) that the two extrema are symmetrically situated with respect to I_h.]

Note that feedback in a nonlinear optical system can also be due to the dependence of the shift of the quasiresonant spectral line with respect to the laser frequency on a certain internal parameter, say temperature.

13.2. NONLINEAR OPTICAL MEDIA FOR BISTABLE DEVICES

In Section 13.1 it was shown that in an optical system with a nonlinear relationship between transmission and refractive index bistability may occur if, via nonlinear optical effects, the refractive index depends on the field strength of the light. It was sufficient to assume a linear dependence of the refractive index on the square of the field strength and hence on the intensity. Such an intensity dependence of the refractive index may be caused by various nonlinear optical one- or multiphoton processes, such as for example were described in the preceding chapters. Let us summarize here the essentials of some nonlinear optical effects with respect to their usefulness in generating optical bistability. At this point it will again be mentioned that it is not only possible to generate *dispersive* bistabilities but also *absorptive* ones, and because of this the intensity dependence of the real part of the complex refractive index as well as that of its imaginary part is of interest.

In Section 4.4 estimations on the intensity dependence of the electronic part of the refractive index in the nearly absorption-free region were carried out, and they provided values of $\tilde{n}^{(2)} \sim 10^{-18}\,\mathrm{cm^2/V^2}$ and $n^{(2)} \sim 10^{-34}\,\mathrm{cm^2\,s}$. The advantages we are offered by utilizing this electronic effect are the short response time, which in typical liquids and solids amounts to $\sim 10^{-14}$ s, and the low optical losses. The main disadvantage is that in accordance with the values of $n^{(2)}$ the required holding and signal intensities are extremely high. Thus, for example, for $\mathscr{F} \sim 100$ and $2Ln^{(0)}\cos\theta/\lambda_{\mathrm{vac}} \sim 10^3$ holding intensity of $I_h \gtrsim 10^{27}\,\mathrm{cm^{-2}\,s^{-1}}$ is needed.

Somewhat larger values of $\tilde{n}^{(2)}$ ($\sim 10^{-14}\,\mathrm{cm^2/V^2}$) can be obtained with the *optical Kerr effect*, which arises from the orientation of molecules with anisotropic polarizability ellipsoids in the light field [see, e.g., Refs. 38 (Vol 1, Section 2.3), 13.12]. In this case, for molecules with a large energy gap between the ground level and the first excited electronic level, the liquid samples are almost loss-free in the visible and near ultraviolet spectral regions. The response time of the Kerr effect, however, is longer than it is in the nonresonant electronic effect, since it is determined by the reorientation of molecules against the frictional resistance of their environment, that means, by the Debye relaxation. For small molecules the reorientation time is in the order of a few picoseconds; for example, it is about 2 ps for CS_2. For most applications such response times in the picosecond region hardly imply any limitations.

Absorptive and dispersive saturation in one-photon processes near resonance are of great importance, since for achieving the required nonlinearity comparatively small fields are sufficient. To begin, estimations of this nonlinearity in an ensemble of two-level systems will be carried out. For the complex susceptibility of the two-level systems with homogeneous broadening, in the quasi-stationary case, according to (4.53), we have

$$\chi^{(1)}(\omega) \equiv \frac{\bar{P}(\omega)}{\varepsilon_0 \bar{E}(\omega)} = -\frac{\tau_{10}|d_{10}|^2/(\hbar\varepsilon_0)}{\Delta\cdot\tau_{10} + i}\gamma_I, \tag{13.14a}$$

where τ_{10}, d_{10}, Δ, and γ_I are the transverse relaxation time, the transition moment, the resonance detuning $\omega_{10} - \omega$, and the density of the occupation-number inversion, respectively. Nonlinearities occur in cases in which the intensity dependence of the occupation-number inversion has to be taken into account. For the latter we have

$$\gamma_I = \frac{\gamma_I^e}{1 + (I/I_S)g_L(\omega - \omega_{10})\pi/\tau_{10}} \tag{13.14b}$$

where

$$I_S = \frac{J_S}{\hbar\omega_{10}} = \frac{n^{(0)}\varepsilon_0 c\hbar^2}{2\tau_{10}T_{10}|d_{10}|^2\hbar\omega_{10}} \tag{13.14c}$$

is the saturation photon flux density of the transition, $\gamma_I^e \equiv -\gamma$ is the equilibrium value of the occupation-number inversion density in the field-free sample, and $g_L(\Delta)$ is the Lorentzian line-shape function. For $I \ll I_S$ in (13.14b) we can confine ourselves to the first order when expanding in powers of I/I_S. Thus from (13.14a) a complex refractive index results whose real and imaginary parts depend linearly on the intensity. Hence, according to the particular value of the resonance detuning Δ, either absorptive or dispersive bistabilities can be produced, or a mixture of both.

Let us here discuss the case of dispersive bistabilities, and therefore assume $\Delta\cdot\tau_{10} \gg 1$ in our estimations. Under these conditions, if the contributions of the quasiresonant transition to the refractive index is much less than unity (as for example is typical for gases), we obtain from (13.14)

$$n^{(2)} \approx -\frac{\tau_{10}^2 T_{10}|d_{10}|^4(\hbar\omega_{10})}{\hbar^3\varepsilon_0^2 c(n^{(0)})^2}\cdot\frac{\Delta\cdot\tau_{10}}{\left[(\Delta\cdot\tau_{10})^2 + 1\right]^2}\gamma \approx -\frac{T_{10}|d_{10}|^4\hbar\omega_{10}\gamma}{\hbar^3\varepsilon_0^2 c(n^{(0)})^2\tau_{10}\Delta^3}.$$

With parameter values that are characteristic of strong (due to the pressure of a buffer gas homogeneously broadened) electronic transitions in gases ($T_{10} \sim 10^{-8}$ s, $\tau_{10} \sim 10^{-9}$ s, $d_{10} \sim 10^{-29}$ A s m, $\gamma \sim 10^{24}$ m^{-3}, $\Delta\cdot\tau_{10} \sim 10$), we obtain $n^{(2)}/\hbar\omega \sim 10^{-2}$ cm^2/W as an estimate. (Note that in low-pressure gases the inhomogeneous broadening is predominant, which requires the calculations to be modified; cf. Chapter 8.)

From this it is obvious that even at low intensities pronounced nonlinearities can be achieved, which give rise to optical bistability. But the time required for the stationary regime in the material to build up is determined by the relaxation times T_{10} and τ_{10}. One might think of increasing $n^{(2)}$ by using arbitrarily slowly relaxing transitions. This, however, is not possible if fast devices are required.

Attention must also be paid to the fact that in applying devices based on quasiresonant one-photon processes a certain absorption loss has to be tolerated in order to achieve saturation of the dispersion. This however means that in all cases the absorption losses have to be taken into account in the equation of motion for the field in the resonator. Calculations for two-level systems show that both absorptive and dispersive bistability can only occur for

$$C \equiv \frac{k_a L}{2(1 - R)} > 4, \tag{13.15}$$

where k_a is the small-signal absorption coefficient (cf. Section 13.3 and Ref. 13.3). This means that in bistable systems based on saturable two level systems, the two limiting cases $k_a \to 0$ and $1 - R \to 0$ cannot be studied independently.

The transition in two-level systems, which we have dealt with so far, is an important special case of saturable one-photon transitions. Of course the saturation of one-photon transitions is of great importance also in more complex atomic systems (cf. the comments on the saturation behavior in Section 8.1). As a result of the saturation of a single transition, the contribution to the real and the imaginary part of the refractive index arising from other transitions, (e.g., from those between the upper state of the saturated transition and still higher states) can be changed, too.

A simple example may again serve for illustrating the behavior of the atomic system. On exciting semiconductor electrons from the valence band to the conduction band, quasifree electrons and holes are generated, which can yield a substantial contribution to the susceptibility that is proportional to the interband saturation (for more details on changes of susceptibilities—in particular with consideration of exciton transitions—in typical semiconductors as preconditions for bistable devices, see, e.g., Refs. 13.8, 13.9, 13.13–13.15).

It may even happen that, outside the saturation region of the quasiresonant transition, the one-photon absorption will give rise to a nonlinear dependence of the complex refractive index on the intensity. This may be due to dependence of material parameters on the dissipated energy. The fast relaxation of the excited atomic level causes a temperature increase of the sample, which, in the stationary regime, is proportional to the light intensity and which, in addition, is dependent on the thermal conductivity of the sample and its environment. Thus as a consequence of the temperature increase an intensity dependence of the refractive index and the absorption coefficient results. Note that here the transient behavior is no longer primarily determined by the relaxation times of the atomic systems, but by the transport phenomena, which in general lead to relatively slow reactions of such devices.

13.3. TRANSIENT RESPONSE OF BISTABLE DEVICES

So far we have only discussed the stationary response of bistable devices, but it is their transient behavior that attracts particular interest, for two reasons. First, besides the sensitivity, the transient response is decisive for the possibility of applying an actual system to information processing and storage, since for such an application only devices with ultrafast response can be used. Second, in the transient mode there occur extraordinary physical phenomena.

The boundary between the stationary and the nonstationary region is mainly determined by the relaxation times of the nonlinear optical sample and by the transient time of the "empty resonator." The relaxation times of the materials involved can vary over very wide ranges; for examples see Section 13.2. In particular we can observe processes that proceed—as is desirable—on the picosecond and subpicosecond time scales. The response time of the "empty resonator" for a Fabry–Perot arrangement, in which the losses are solely determined by the outcoupling at the two mirrors with reflectivity R, is given for small losses according to (1.45) by

$$T_{\mathscr{E}} \approx \frac{L}{c(1-R)} .$$

To obtain values of $T_{\mathscr{E}}$ in the order of 1 ps, the resonator length must be of the order of 10 μm if we assume $1 - R \sim 0.1$. For such a short length, according to (13.13) in combination with (13.8b), very large nonlinear optical susceptibilities or very high intensities are required in order to reach the bistable region. As materials with short relaxation times and large nonlinear optical susceptibilities certain semiconductors may be chosen, as for example GaAs ($\tilde{\tilde{n}}^{(2)} \approx 0.4$ cm^2/kW, where $\tilde{\tilde{n}}^{(2)} = n^{(2)}/\hbar\omega$) and InSb ($\tilde{\tilde{n}}^{(2)} \approx 1.0$ cm^2/kW). The effective relaxation times of semiconductors and, in particular, of semiconducting thin films depend heavily on the techniques applied in producing and doping them (see Refs. 13.16, 13.17).

For the following study let us employ a simple model in which the transient behavior, both of the resonator and of the atomic systems, is of importance. We assume an ensemble of atomic two-level systems that do not interact with one another to be in a resonator with low outcoupling losses, and only the resonator mode μ, which is in exact resonance with the electric field ($\omega_\mu = \omega$), to be excited. We furthermore assume that this mode is in exact resonance with the atomic transition ($\omega_\mu = \omega_{10}$); this means we restrict our consideration to the case of absorptive bistability.

As in Section 1.4, we introduce the mode function $E_{\mu\bullet}(r_\bullet)$, which describes the spatial structure of the field in the resonator, and obtain for the one-mode field

$$E_\bullet(r_\bullet, t) = -\frac{1}{\sqrt{\varepsilon_0}} p_\mu(t) E_{\mu\bullet}(r_\bullet), \tag{13.16a}$$

where

$$p_\mu(t) = -\sqrt{\varepsilon_0}\, \tfrac{1}{2}\tilde{E}_\mu(t) e^{-i\omega_\mu t} + \text{c.c.} \tag{13.16b}$$

describes the mode excitation as a function of time. From (1.227b) we obtain

$$\frac{d}{dt}\tilde{E}_\mu = -\frac{1}{T_A}\tilde{E}_\mu + \frac{i\omega_\mu}{2\varepsilon_0} K_P \tilde{P}_\mu, \tag{13.17a}$$

where the $\tilde{P}_\mu$ are time-dependent polarization amplitudes defined by the mode expansion

$$P_\bullet(r_\bullet, t) = P_\bullet(r_\bullet) \sum_\mu \left[\tfrac{1}{2}\tilde{P}_\mu e^{-i\omega_\mu t} + \text{c.c.}\right]; \tag{13.17b}$$

$$K_p = \int dV P_\bullet(r_\bullet) E_{\mu\bullet}(r_\bullet) \tag{13.17c}$$

is a coupling factor (for the special case that the entire polarization $P_\bullet(r_\bullet, t)$ has the same spatial structure as the electric field of the μth mode, the relation $K_p = 1$ follows from the normalization of the mode functions).

The first term on the right-hand side of (13.17a) describes the change in mode excitation because of losses arising from outcoupling, scattering, or absorption. We assume that the outcoupling at the two mirrors with the reflectivity $R_1 = R_2 = R$ predominates and set, according to (1.45),

$$T_A = 2T_{\mathscr{E}} \approx T_{\text{rt}} \frac{1}{1 - R}, \tag{13.18}$$

where T_{rt} is the round-trip time of the resonator, that is, the time which is required for a complete circuit of light in the resonator. The second term on the right-hand side of (13.17a) describes the change of the field due to the action of the polarization excited in the resonator.

One thing that has not been taken into account so far is the change in mode excitation owing to the coupling-in of an incident field with the wave amplitude $\overline{E}_{\text{in}}$. Coupling-in through one of the mirrors yields an additional contribution to the change in mode excitation, which we can write in the form $(\sqrt{1 - R}\, K_{\text{in}}/T_{\text{rt}})\overline{E}_{\text{in}}$, where K_{in} is a numerical factor of the order of unity, which for the Fabry–Perot resonator (Fig. 13.3) is approximately one-half and for the ring resonator (Fig. 13.4) is one. With $K_P = 1$, and if coupling in and outcoupling are weak, we obtain

$$\frac{d}{dt}\tilde{E}_\mu = -\frac{1-R}{T_{\text{rt}}}\left[\tilde{E}_\mu - \frac{K_{\text{in}}}{\sqrt{1-R}}\bar{E}_{\text{in}}\right] + \frac{i\omega_\mu}{2\varepsilon_0}\tilde{P}_\mu \tag{13.19a}$$

This equation has to be solved, together with the equations of motion for the atomic systems, for the spatially averaged density of the occupation-number inversion $\tilde{\gamma}_I(t)$ and the polarization amplitude $\tilde{P}_\mu(t)$ of the material in the resonator. Under the assumptions mentioned we obtain from (4.101) and (4.102)

$$\frac{d}{dt}\tilde{P}_\mu = -\frac{1}{\tau_{10}}\tilde{P}_\mu - \frac{i}{\hbar}|d_{10}|^2\tilde{\gamma}_I\tilde{E}_\mu \tag{13.19b}$$

and

$$\frac{d}{dt}\tilde{\gamma}_I = -\frac{1}{T_{10}}(\tilde{\gamma}_I - \tilde{\gamma}_I^e) - \frac{i}{2\hbar}\left(\tilde{P}_\mu\tilde{E}_\mu^* - \tilde{P}_\mu^*\tilde{E}_\mu\right), \tag{13.19c}$$

where d_{10} is the transition moment of the quasiresonant transition, and τ_{10} and T_{10} are the transverse and the longitudinal relaxation time, respectively. [If necessary, d_{10} can be replaced by an effective value that includes correction factors required for taking into consideration orientation averaging (cf. Section 4.5) and the deviations between the local and macroscopic fields (cf. Section 1.2). Unless P_μ and E_μ have the same spatial structure, in (13.19) the coupling factor K_p and additional numerical factors appear that result from the integrals over products of the respective spatial functions.]

We are only interested in the contribution of the resonant transition to the polarization. [If one intends to determine the influence of all the nonresonant transitions on the optical parameters and thus on the mode structure in an adequate way, one has to start from (1.228) instead of (1.227b) and from a relation with the structure of (1.229).] With real amplitude $\tilde{E}_\mu$ the contribution of the resonant transition to the polarization amplitude is purely imaginary, which provides, with $\tilde{P}_\mu := i\tilde{P}$ and $\tilde{E}_\mu := \tilde{E}$, the following equations to start from:

$$\frac{d}{dt}\tilde{E} = -\frac{1-R}{T_{\text{rt}}}\left[\tilde{E} - \frac{K_{\text{in}}}{\sqrt{1-R}}\bar{E}_{\text{in}}\right] - \frac{\omega_{10}}{2\varepsilon_0}\tilde{P}, \tag{13.20a}$$

$$\frac{d}{dt}\tilde{P} = -\frac{1}{\tau_{10}}\tilde{P} - \frac{1}{\hbar}|d_{10}|^2\tilde{\gamma}_I\tilde{E}, \tag{13.20b}$$

$$\frac{d}{dt}\tilde{\gamma}_I = -\frac{1}{T_{10}}(\tilde{\gamma}_I - \tilde{\gamma}_I^e) + \frac{1}{\hbar}\tilde{P}\tilde{E}. \tag{13.20c}$$

These equations correspond to those obtained in the mean-field approximation in Ref. 13.3.

At this point we again want to add some remarks on the stationary regime, as so far we have restricted our consideration to the case of dispersive bistability. Following Ref. 13.3, we seek the stationary solutions of (13.20b) and (13.20c) for the polarization and the inversion (cf. Section 8.1). We substitute these solutions into (13.20a), whereby we obtain the nonlinear state equation. In this state equation we pass from the internal field with the strength $\tilde{E}$ according to

$$\overline{E}_{\text{out}} = K_{\text{out}}\sqrt{1 - R}\,\tilde{E}$$

to the output field, where K_{out} is a numerical factor that, as described above, takes into account the actual coupling conditions of the resonator. Hence the nonlinear state equation takes the form

$$y = x + \frac{2C}{1 + x^2}x, \tag{13.21a}$$

where

$$y = \frac{K_{\text{in}}/\sqrt{2}}{\sqrt{1 - R}} \frac{\sqrt{T_{10}\tau_{10}}\,|d_{10}|}{\hbar}\overline{E}_{\text{in}}, \tag{13.21b}$$

$$x = \frac{1}{\sqrt{2}\,K_{\text{out}}\sqrt{1 - R}} \cdot \frac{\sqrt{T_{10}\tau_{10}}\,|d_{10}|}{\hbar}\overline{E}_{\text{out}}, \tag{13.21c}$$

and

$$C = \frac{\omega_{10}\tau_{10}}{4\varepsilon_0\hbar n^{(0)}}|d_{10}|^2\left(\frac{T_{\text{rt}}}{1 - R}\right)\gamma = \frac{1}{4}\frac{cT_{\text{rt}}}{(1 - R)}k_a, \tag{13.21d}$$

and where $k_a = \sigma_{01}^{(1)}\gamma$ is the maximum small-signal absorption coefficient of the material. For a Fabry–Perot-type resonator we obtain $T_{\text{rt}} \approx 2L/c$, and thus (13.21d) takes the form $C = Lk_a/2(1 - R)$.

We see that with saturable two-level absorbers in the mean-field approximation the nonlinearity in the state equation $\overline{E}_{\text{in}} = f(\overline{E}_{\text{out}})$ is determined by a single parameter, namely by C, which was mentioned at the end of Section 13.2. A discussion of the function $y(x)$ of (13.21a) shows that regions with $dy/dx < 0$ and thus bistability can only occur if $C > 4$, whereby (13.15) is verified for the case of absorptive bistability. For $C > 4$ we get a functional relationship between x, where $x \propto \overline{E}_{\text{out}}$, and y, where $y \propto \overline{E}_{\text{in}}$, as shown in Fig. 13.2. For $C \leq 4$ unambiguous functions $x = x(y)$ result.

Starting from the stationary solutions, we study the transient behavior. We assume a small variation $\delta\overline{E}_{\text{in}}$ of the input field strength $\overline{E}_{\text{in}}$ with $\delta\overline{E}_{\text{in}} \ll \overline{E}_{\text{in}}$

and analyze the variations $\delta\tilde{P}$, $\delta\tilde{\gamma}_I$, and $\delta\overline{E}_{\text{out}}$ as functions of the time. From (13.20) we obtain a system of linear ordinary differential equations, which enable us to calculate the time-dependent variations. It reads

$$\frac{d}{dt}\,\delta\tilde{E} = -\frac{1-R}{T_{\text{rt}}}\left[\delta\tilde{E} - \frac{K_{\text{in}}}{\sqrt{1-R}}\,\delta\overline{E}_{\text{in}}\right] - \frac{\omega_{10}}{2\varepsilon_0}\,\delta\tilde{P}, \qquad (13.22\text{a})$$

$$\frac{d}{dt}\,\delta\tilde{P} = -\frac{1}{\tau_{10}}\,\delta\tilde{P} - \frac{1}{\hbar}|d_{10}|^2(\tilde{\gamma}_I\,\delta\tilde{E} + \tilde{E}\,\delta\tilde{\gamma}_I), \qquad (13.22\text{b})$$

$$\frac{d}{dt}\,\delta\tilde{\gamma}_I = -\frac{1}{T_{10}}\,\delta\tilde{\gamma}_I + \frac{1}{\hbar}(\tilde{P}\,\delta\tilde{E} + \tilde{E}\,\delta\tilde{P}). \qquad (13.22\text{c})$$

Now we seek solutions of the form

$$\delta\tilde{E}(t) = \sum_{j=1}^{3} (\delta\tilde{E})_j e^{-\lambda_j t},$$

$$\delta\tilde{P}(t) = \sum_{J=1}^{3} (\delta\tilde{P})_j e^{-\lambda_j t},$$

$$\delta\tilde{\gamma}_I(t) = \sum_{j=1}^{3} (\delta\tilde{\gamma}_I)_j e^{-\lambda_j t},$$

where the $(\delta\tilde{E})_j$, $(\delta\tilde{P})_j$, $(\delta\tilde{\gamma}_I)_j$, and λ_j are time-independent parameters. The reciprocals of the real parts of λ_j are effective relaxation times of the system for passage to the new stationary state. We are particularly interested in the longest of these relaxation times, since it mainly determines the response time of the system; the parameter λ_j belonging to it is denoted by Λ. The evaluation is simplest for a high-quality resonator, for which we assume

$$\frac{T_{\text{rt}}}{1-R} \gg \tau_{10}, T_{10}$$

In this case Λ is approximately real, which implies a monotonic time evolution, and we obtain

$$\Lambda = \frac{1-R}{T_{\text{rt}}}\frac{dy}{dx} \qquad (13.23)$$

(cf. Ref. 13.3).

This presentation shows that, in our simple, linearized theory, Λ vanishes at the two discontinuity points (x_m, y_m) and (x_M, y_M), since there (for $C > 4$)

the extrema of the function $y(x)$ appear, which correspond to infinitely large values of the effective response time of the system. If, starting from small input signals, we are approaching the discontinuity point (y_M, x_M) on the lower branch, then Λ goes to zero, as does the function $\sqrt{y_M^2 - y^2}$. Analogously, if, starting from large input signals, we are approaching (y_m, x_m) on the upper branch, then Λ goes to zero, as does the function $\sqrt{y^2 - y_m^2}$. This means that, as happens in other phase transitions, a critical slow down occurs with optically bistable systems. To avoid extremely slow system response, the input field strengths of the switching pulses must be sufficiently far from the discontinuity points. Exact studies reveal that in general a deviation of the input fields of about 10% is sufficient.

In the opposite case of a resonator with low quality, where we assume

$$\frac{T_{\text{rt}}}{1-R} \ll \tau_{10}, T_{10},$$

the parameter Λ is complex; this means the system approaches the stationary state with damped oscillations, and at high input field strength the oscillation frequency of the transient process is approximately given by the Rabi frequency.

A more exact analysis of the transient regime must, of course, not be restricted to the consideration of a single mode only, since in reality the neighboring off-resonance modes can be excited as well and mode coupling may occur. In such an analysis, among other phenomena, self-pulsing has been found. It results when the system, starting from an unstable point, evolves into a periodic regime, the limit cycle, with a periodic sequence of output pulses resulting from the continuous input signal (see Refs. 13.2, 13.9, 13.18). This means that with given y, either the system can pass from the unstable point to the stable state on the lower branch, or the self-pulsing regime may appear.

In systems with multistability another type of nonstationarity may occur: *optical chaos*, where a chaotic sequence of pulses arises from the continuous input signal. Chaotic behavior has been predicted and observed in several types of optical systems, including optically bistable devices. The first theoretical investigations of optical chaos were presented by Ikeda et al. (13.19, 13.20), and the first observations of it were reported in Refs. 13.21, 13.22.

Having arrived at this point, we want to draw attention to an inherent shortcoming of our theoretical treatment. In our considerations we have disregarded the fact that owing to the dependence of the refractive index on the field strength, not only the optical length but also the feedback of the resonator varies. This effect becomes most obvious when investigating a Fabry–Perot resonator that consists of an uncoated thin platelet of the nonlinear material. Here the reflectivity of the plate depends on the field strength via the refractive index. A consistent treatment of this problem is given in Ref. 13.23. However, in most cases important in practice the approximation used by us is justified, because $n^{(2)}I_{\text{in}} \ll n^{(0)}$.

Finally, in concluding our simple theoretical considerations it should be emphasized that we have employed a classical or semiclassical approach throughout and assumed nonfluctuating signals and nonfluctuating internal parameters. This is indeed justified provided that we are studying strong, coherent fields at the input and that we are not interested in the details of signal fluctuations. It should be noted, however, that in certain cases—for instance, in the critical slowdown, and in the transitions to self-pulsing regimes and to chaotic behavior—quite small fluctuations of the input signal and of the nonlinear medium may give rise to macroscopic effects. It is especially in such quantum-optical systems where questions arise concernig the existence and characterizations of quantum chaos (cf.A.2.5.2). For detailed quantum-statistical treatments of optical bistability see, e.g., Refs. 13.3, 13.24–13.30.

13.4. EXPERIMENTAL STUDIES

As already mentioned, intrinsic optical bistability was observed to occur for the first time with a Fabry–Perot resonator containing sodium vapor as a nonlinear material (Ref. 13.5; see Fig. 13.6 and Fig. 13.7). If buffer gases are used for suppressing the inhomogeneous broadening in favor of homogeneous broadening and if the device is irradiated at frequencies close to the D_1 transition, the saturable absorber can be described in good approximation with the aid of identical two-level systems. In Ref. 13.5 dispersive bistability was observed with such devices at rather low input powers of about 10 mW. In Ref. 13.31 an absorptive bistability was detected with the holding power about one order of magnitude higher. In both cases the switching time was some 10^{-7} s. Altogether, we may say that for small outcoupling losses ($1 - R \sim 0.01$) good agreement between observations and the results of the mean-field model, as it is presented in Section 13.3, can be achieved if the actual transverse structure of the wave fields is taken into account (see Ref. 13.32).

The first bistable solid-state system to be studied was a Fabry–Perot resonator with ruby, where also fairly low holding powers ($\sim$ 10 mW), but (as

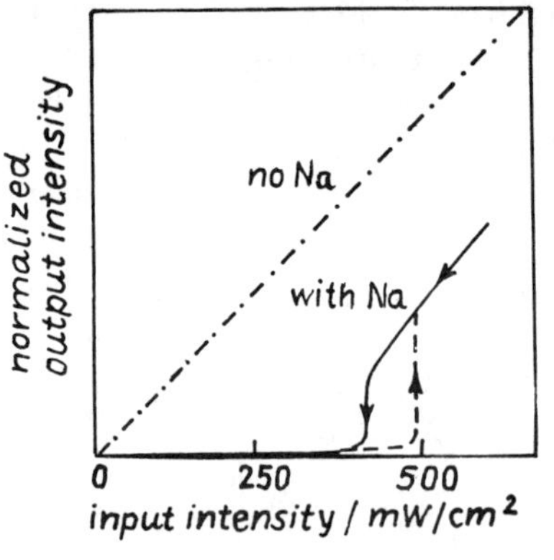

Fig. 13.6. Output signal versus input signal for dispersive optical bistability in sodium vapor (after Ref. 13.8).

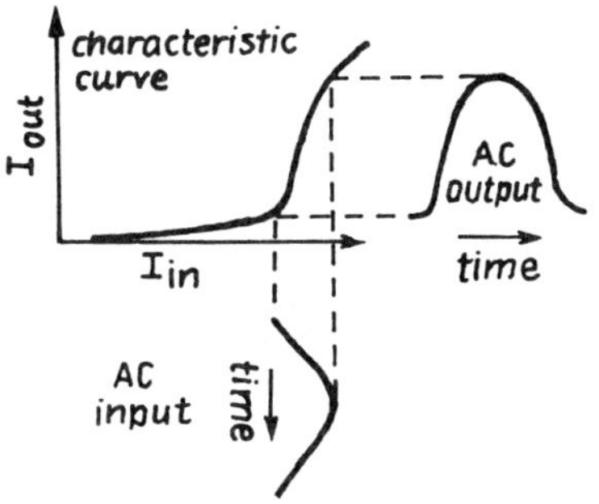

Fig. 13.7. Differential gain in sodium vapor (after Ref. 13.8). A gain at 1kHz of about 2 has been observed.

a result of the long relaxation time of the R transition) only very slow switching (in about 10 ms) could be achieved (Ref. 13.8). Moreover, this arrangement served as an optical discriminator in which input signals exceeding certain values give rise to high transmissions, and thus to strong output signals. As a consequence, high-intensity pulses could be reliably discriminated from smaller input signals.

For potential future application experimental studies on semiconductors, for example on GaAs (Refs. 13.33–13.35), InSb (Refs. 13.13, 13.14, 13.36), and Te (Ref. 13.15), are of special importance. In semiconductors, because of their high nonlinear susceptibilities, the use of very short resonators is possible at comparatively low holding intensities. An important step forward in these studies was made in Refs. 13.33, 13.34, where optical bistability was achieved in a 4.1-μm layer of GaAs between 0.2-μm $Al_{0.42}Ga_{0.58}As$ windows (see Fig. 13.8*a*, *b*, *c*). In this layer the nonlinear optical contribution to the refractive index mainly originates from the free-exciton resonance in GaAs, which, particularly at low temperatures ($\lesssim$ 100 K) results in a strong, sharp spectral line. Here off-resonance bistability was achieved in the dispersive region. The light incident on the sample generates free charge carriers that, as a result of intraband relaxation, rapidly thermalize in the valence band as well as in the conduction band. This carrier generation and thermalization is accompanied by changes in absorption, and hence by changes in dispersion. The holding intensity amounts to about 1 mW/μm^2 at $\lambda = 589$ nm, which, with a 15-μm spot diameter, requires a holding power of about 0.2 W. By improving focusing it seems possible to reduce the holding power down to a few milliwatts (Refs. 13.8, 13.9). The arrangement employed allowed one to switch the system from the "off state" into the "on state" by using pulses with a duration of $\tau_L \approx 200$ ps and with an energy per unit area of 0.1×10^{-12} J/μm^2. As the switch-on time is mainly determined by the pulse duration, for shorter light pulses switch-on times of about 1 ps may be possible. For short resonator lengths the switchoff time is mainly determined by the relaxation processes in the semiconductor, and in the given example it is about 40 ns. But in suitably manufactured and appropriately doped materials relaxation times and thus switchoff times on the picosecond scale may be reached (cf. Section 13.2).

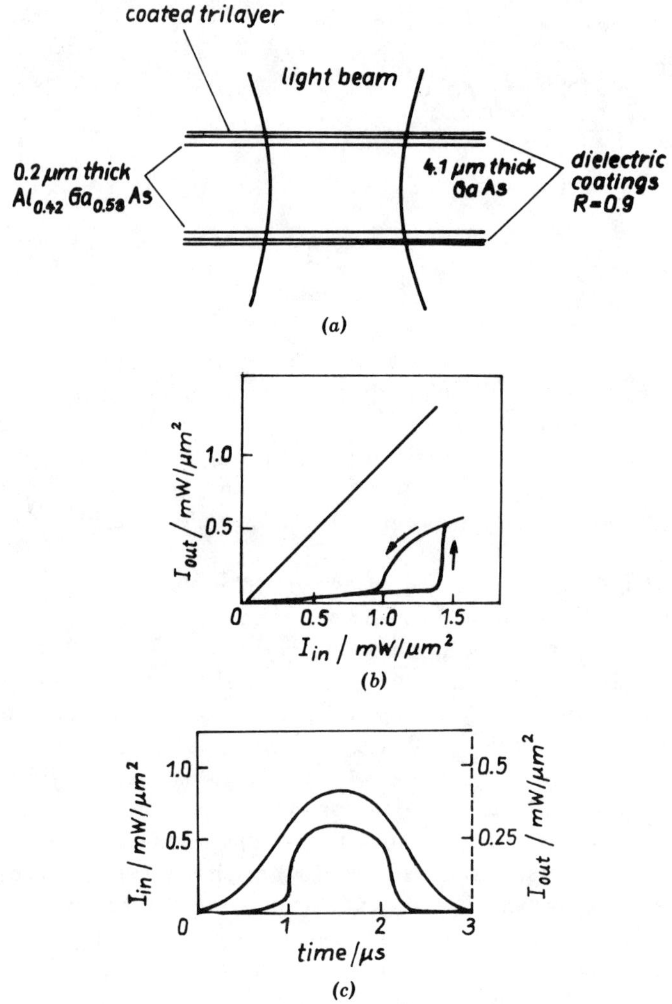

Fig. 13.8. Optical bistability in GaAs (after Ref. 13.8): (*a*) GaAs–GaAlAs coated trilayer, (*b*) output signal versus input signal, (*c*) pulse shaping in an optical bistable device.

In Section 13.2 we showed that nonlinear dispersive behavior can be achieved not only in one-photon absorption processes but also by means of other nonlinear optical effects. As one example let us mention the possibility of achieving dispersive bistability by means of two-photon transitions (Ref. 13.37). Other examples, in our opinion far more important, concern the dispersive bistability in liquids exhibiting the optical Kerr effect (Refs. 13.38, 13.39). In these liquids, because of the fast orientational relaxation of small molecules, short switching times in the picosecond region can be achieved, but the required intensities are rather high. In Ref. 13.38 optical bistability was achieved by means of the Kerr liquids CS_2 and nitrobenzene, using the light of a Q-switched ruby laser. The transient processes resulting from cavity ringing were studied in detail, and good agreement could be obtained between theory and experiment even with respect to the dynamical processes.

With the aim of achieving bistability, not only are Kerr liquids being investigated in resonators, but there are also experiments and theoretical treatments of Kerr liquids as nonlinear reflectors. The possibility of obtaining optical bistability with reflection at the interface between a nonlinear and a linear optical material was first discussed in Ref. 13.40. We assume that light goes from a linear optical medium with refractive index n_1 into a medium with small-signal refractive index $n_2^{(0)}$ smaller than n_1 ($n_2^{(0)} < n_1$) and with positive nonlinearity ($n_2^{(2)} > 0$). In this case, at low intensities total reflection occurs for incidence angles above the critical angle. This means that in the second medium there appears only a wave that is heavily damped in the direction normal to the surface. If the input intensity is increased, the refractive index in the second medium also increases as a consequence of the growing intensity of that wave, which couples light into the second medium. Hence, at a certain critical value of the input intensity a sudden change from total reflection to partial transmission is to be expected. Also in such systems, as in resonators, the combination of nonlinearity and feedback may give rise to bistable behavior. But subsequent theoretical investigations have revealed that one has to be cautious in interpreting the results from Ref. 13.40, since the plane waves assumed there are not stable as a result of self-focusing (see Refs. 13.41–13.48). In Ref. 13.49 experiments are reported that studied these phenomena with a CS_2–glass interface. The critical intensity for the transition from total reflection to partial transmission could be determined, and the nonlinear characteristic was recorded. With pulsed excitation it was possible to observe a hysteresis loop in the reflectivity of the entire device.

REFERENCES

13.1. H. Haken, *Synergetics, an Introduction*, 2nd ed., Springer, Berlin (West), Heidelberg, New York, 1978.

13.2. G. Nicolis and I. Prigogine, *Self-Organization in Non-equilibrium Systems*, Wiley, New York, 1977.

13.3. R. Bonifacio and L. A. Lugiato, "Theory of Optical Bistability," in R. Bonifacio (Ed.), *Dissipative Systems in Quantum Optics*, Springer, Berlin (West), Heidelberg, New York, 1982, p. 61.

13.4. H. Seidel, U.S. Patent 3,610,731, 1969.

13.5. H. M. Gibbs, S. L. McCall, and T. N. C. Venkatesan, *Phys. Rev. Lett.* **36**, 1135 (1976).

13.6. A. A. Kastalskij, *Sov. Phys. Semicond.* **7**, 635 (1973).

13.7. P. W. Smith and E. H. Turner, *Appl. Phys. Lett.* **30**, 280 (1977).

13.8. H. M. Gibbs, S. L. McCall, and T. N. C. Venkatesan, *Opt. News* 5, 6 (1979).

13.9. S. L. McCall and H. M. Gibbs, "Optical Bistability," in R. Bonifacio (Ed.), *Dissipative Systems in Quantum Optics*, Springer, Berlin (West), Heidelberg, New York, 1982, p. 93.

13.10. R. Bonifacio and L. A. Lugatio, *Opt. Comm.* **19**, 172 (1976).

13.11. R. Roy and M. S. Zubairy, *Phys. Rev. A* **21**, 274 (1980).

13.12. S. Kielich, *Molecular Nonlinear Optics*, Nauka, Moscow, 1981 (Russian).

13.13. D. A. B. Miller, S. D. Smith, and A. Johnston, *Appl. Phys. Lett.* **35**, 658 (1979).

13.14. D. A. B. Miller, S. D. Smith, and B. S. Wherrett, *Opt. Comm.* **35**, 221 (1980).

13.15. G. Staupendahl and K. Schindler, in *Proceedings of the Second International Symposium Ultrafast Phenomena in Spectroscopy* (*UPS80*), Reinhardsbrunn, 1980; *Opt. Quant. Electron.* **14**, 157 (1982).

13.16. D. H. Auston, P. R. Smith, A. M. Johnson, W. M. Augustiniak, J. C. Bean, and D. B. Fraser, in R. M. Hochstrasser, W. Kaiser, and C. V. Shank (Eds.), *Picosecond Phenomena*, Springer, Berlin (West), Heidelberg, New York, 1980, p. 71.

13.17. D. H. Auston, P. Lavallard, N. Sol, and D. Kaplan, *Appl. Phys. Lett.* **36**, 66 (1980); **37**, 371 (1980); **38**, 47 (1981).

13.18. R. Bonifacio, M. Gronchi, and L. A. Lugiato, *Opt. Comm.* **30**, 129 (1979).

13.19. K. Ikeda, *Opt. Comm.* **30**, 257 (1979).

13.20. K. Ikeda, H. Daido, and O. Akimoto, *Phys. Rev. Lett.* **45**, 709 (1980).

13.21. H. M. Gibbs, F. A. Hopf, D. L. Kaplan, and R. L. Shoemaker, *Phys. Rev. Lett.* **46**, 474 (1981).

13.22. H. M. Gibbs, F. A. Hopf, D. L. Kaplan, M. W. Derstine, and R. Shoemaker, *Soc. Photo-Opt. Instrum. Eng.* **317**, 297 (1981).

13.23. J. Goll and H. Haken, *Phys. Stat. Sol. b* **101**, 489 (1980).

13.24. R. Bonifacio, M. Gronchi, and L. A. Lugiato, *Phys. Rev. A* **18**, 2266 (1978).

13.25. F. T. Arecchi and A. Politi, *Opt. Comm.* **29**, 361 (1979).

13.26. R. F. Gragg, W. C. Schieve, and A. D. Bulsara, *Phys. Rev. A* **19**, 2052 (1979).

13.27. A. Zardeck, *Phys. Rev. A* **22**, 1664 (1980).

13.28. L. A. Lugatio, *Nuovo Cimento* **50B**, 89 (1979).

13.29. P. D. Drummond, K. J. McNeil, and D. F. Walls, *Phys. Rev. A* **22**, 1672 (1980).

13.30. F. A. Hopf and P. Meystre, *Opt. Comm.* **29**, 235 (1979).

13.31. W. J. Sandle and A. Gallagher, *Phys. Rev. A* **24**, 2017 (1981).

13.32. K. G. Weyer, M. Wiedenmann, M. Rateike, W. R. MacGillivray, P. Meystre, and H. Walther, *Opt. Comm.* **37**, 426 (1981).

13.33. H. M. Gibbs, S. L. McCall, T. N. C. Venkatesan, A. C. Gossard, A. Passner, and W. Wiegmann, in *Digest of 1979 IEEE/OSA Conference on Laser Engineering and Applications*, IEEE, New York, 1979: in H. Walther and K. W. Rothe (Eds.), *Laser Spectroscopy IV*, Springer, Heidelberg, Berlin (West), New York, 1979.

13.34. H. M. Gibbs, T. N. C. Venkatesan, S. L. McCall, A. Passner, A. C. Gossard, and W. Wiegmann, *Appl. Phys. lett.* **34**, 511 (1979).

13.35. C. V. Shank, R. L. Fork, R. F. Leheny, and J. Shah, *Phys. Rev. Lett.* **42**, 112 (1979).

13.36. S. D. Smith, D. A. B. Miller, B. S. Wherrett, in *Proceedings of UPS 80*, Reinhardsbrunn, 1980, p. 425.

13.37. E. Giacobino, M. Devand, F. Biraben, and G. Grynberg, *Phys. Rev. Lett.* **45**, 434 (1980).

13.38. S. L. McCall, and H. M. Gibbs, *J. Opt. Soc. Am.* **68**, 1378 (1978).

13.39. T. Bischofsberger, and Y. R. Shen, *Appl. Phys. Lett.* **32**, 156 (1978); *Phys. Rev. A* **19**, 1169 (1979).

13.40. A. E. Kaplan, *Zh. Eksper. Teor. Fiz. Pisma* **24**, 114 (1976); *Sov. Phys. JETP* **45**, 896 (1977).

13.41. B. B. Boiko, I. Z. Dzhilavdar, and N. S. Petrov, *J. Appl. Spectrosc.* **23**, 1511 (1975).

13.42. A. E. Kaplan, *Kvant. Elektron.* **8**, 95 (1978); *Radio Phys. Quant. Electron.* **22**, 229 (1979).

13.43. A. E. Kaplan, *IEEE J. Quant. Electron.* **QE-17**, 336 (1981).

13.44. A. A. Kolokolov and A. I. Sukov, *Radio Phys. Quant. Electron.* **21**, 1013 (1978).

13.45. N. N. Rosanov, *Opt. i Spektr.* **47**, 337 (1979).

13.46. D. Marcuse, P. W. Smith, and W. J. Tomlinson, *J. Opt. Soc. Amer.* **70**, 594 (1970).

13.47. D. Marcuse, *Appl. Opt.* **19**, 3130 (1980).

13.48. W. J. Tomlinson, *Opt. Lett.* **5**, 323 (1980).

13.49. P. W. Smith, J. P. Herman, W. J. Tomlinson, and P. J. Moloney, *Appl. Phys. Lett.* **35**, 846 (1979); *IEEE J. Quant. Electron.* **QE-17**, 340 (1981).

14

Nonlinear Optical Phase Conjugation

Nonlinear optical phase conjugation is a subfield of the more general optical phase conjugation, which historically got its thrust from the task of restoring the wavefronts of distorted optical beams to their original unaberrated state. At the end of the sixties and the beginning of the seventies it was verified theoretically and experimentally (for instance in Refs. 14.1, 14.2, 14.3) that a real-time holographically generated pseudoscopic image is capable of compensating for phase aberrations. However, the first works reporting on the application of nonlinear optical interaction in the field of optical phase conjugation were published later on in 1972 by Zel'dovich, Nosach, and coworkers (Refs. 14.4, 14.5). They recognized and utilized the fact that in the case of stimulated Brillouin scattering the amplitude of the backscattered electromagnetic wave can be regarded as the complex conjugate of the amplitude of the laser pump, so that the scattered wave may serve as "conjugate wave." Later on it was recognized that there are, in principle, many nonlinear optical effects that are appropriate for the generation of waves possessing this complex-conjugate property; of particular importance is degenerate four-wave mixing.

Nonlinear optical phase conjugation has numerous potential applications, the most promising ones being adaptive optics and the possibility of building an optical resonator with advantageous properties. Some review papers have been given on nonlinear optical phase conjugation (see, e.g., Ref. 14.6).

We shall deal with the basic properties of phase-conjugate fields in Section 14.1. Various mechanisms of nonlinear optical phase conjugation will be discussed in Section 14.2, and applications are dealt with in Section 14.3.

14.1. PROPERTIES OF PHASE-CONJUGATED FIELDS

Let us define the ideal phase-conjugate wave $E_{\mathrm{pc}\bullet}(r_\bullet, t)$ of an original wave $E_{o\bullet}(r_\bullet, t)$. The original wave is assumed to be a monochromatic wave (frequency ω_o) propagating in the $+z$ direction; it is described by

$$E_{o\bullet}(r_\bullet, t) = \tfrac{1}{2}\bar{E}_{o\bullet}(r_\bullet)e^{-i(\omega_o t - k_o z)} + \text{c.c.}, \qquad (14.1)$$

where $\bar{E}_{o\bullet}(r_\bullet)$ is the slowly varying complex wave amplitude and k_o the wave number. The phase-conjugate wave or "conjugate replica" $E_{\mathrm{pc}\bullet}$ is represented by the usual wave formula

$$E_{\mathrm{pc}\bullet}(r_\bullet, t) = \tfrac{1}{2}\bar{E}_{\mathrm{pc}\bullet}(r_\bullet)e^{-i(\omega_{\mathrm{pc}}t - k_{\mathrm{pc}}z)} + \text{c.c.}, \tag{14.2}$$

where the following relations hold with respect to the terms of the original wave:

$$\omega_{\mathrm{pc}} = \omega_o, \; k_{\mathrm{pc}} = -k_o, \qquad \bar{E}_{\mathrm{pc}\bullet}(r_\bullet) = \bar{E}^*_{o\bullet}(r_\bullet). \tag{14.3}$$

In reference to the original wave, the phase-conjugate wave $E_{\mathrm{pc}\bullet}$ is defined as a counterpropagating wave with a complex-conjugate amplitude. A physical system that produces a phase-conjugate wave from an original wave is termed a *phase-conjugate mirror* or a *phase conjugator.*

Inserting the relations (14.3) into the right-hand side of (14.2), one obtains

$$\begin{aligned} E_{\mathrm{pc}\bullet}(r, t) &= \tfrac{1}{2}\bar{E}^*_{o\bullet}(r_\bullet)e^{-i(\omega_o t + k_o z)} + \text{c.c.} \\ &= \tfrac{1}{2}\bar{E}_{o\bullet}(r_\bullet)e^{-i(-\omega_o t - k_o z)} + \text{c.c.} = E_{o\bullet}(r_\bullet, -t). \end{aligned} \tag{14.4}$$

The phase-conjugate wave moves in the $-z$ direction with phase reversed relative to the original wave. This is equivalent to leaving the spatial part of the original wave unchanged and reversing the sign of t. Since $\partial^2 E_{\mathrm{pc}\bullet}/\partial t^2$ is equal to $\partial^2 E_{o\bullet}/\partial t^2$, the basic equation for electromagnetic waves,

$$\nabla \times (\nabla \times E_\bullet) = -\frac{1}{v^2}\frac{\partial^2 E_\bullet}{\partial t^2}, \tag{14.5}$$

is satisfied by the phase-conjugate wave $E_{\mathrm{pc}\bullet}$ as well as by the original wave $E_{o\bullet}$. In view of the relation (14.4), the phase-conjugate wave may be called the *time reverse* of the original wave.

One may illustrate the operation of a phase-conjugate mirror by the following schematic representations. Figure 14.1 shows both an ordinary mirror and a phase-conjugate mirror illuminated by a point source. While the ordinary mirror reflects a diverging beam by redirecting the propagation direction, the phase-conjugate mirror throws back the light to the point source, where the equiphase surfaces of the phase-conjugate mirror produces a time-reversed image of the original wave. Fig. 14.2 shows the effect of a phase aberrator on the equiphase surfaces. The phase aberrator (for instance a piece of glass) causes a bulge in the equiphase surface of the original plane wave. In the case of an ordinary mirror the bulge depth is doubled after the wave has traversed the aberrator in reverse. The phase-conjugate mirror causes quite another behavior; the equiphase surfaces of the phase-conjugate wave correspond to the equiphase surfaces of the original wave; the distorting element causes a compensation.

However, we must stress that the behavior illustrated in Figs. 14.1 and 14.2 only characterizes the mode of operation of a phase-conjugate mirror in the limiting case of infinite transverse extent. The fidelity of compensation by the

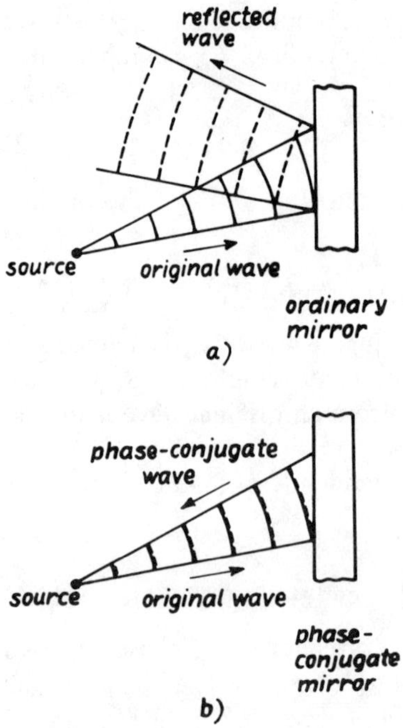

Fig. 14.1. (*a*) Ordinary and (*b*) phase-conjugate mirror illuminated by a point source. Solid curves: equiphase surfaces of the original wave. Dashed curves: equiphase surfaces of the reflected and phase-conjugate waves.

phase-conjugate wave is limited by the effective aperture of the conjugator, since the laws of diffraction also hold for phase-conjugate waves. Under real conditions of a finite transverse extent l of the conjugator, diffractive effects cause deviations from the depicted (ideal) equiphase surfaces of the phase-conjugate waves. Thus, by diffractive effects, a minimum transverse scale $d_{\min}$ of resolvable distortion elements that can be compensated exists. The length $d_{\min}$ is of the order of $\lambda a/l$, where λ is the wavelength, and a is the distance of the aberrating element from the phase-conjugate mirror.

It should be mentioned that the above compensation procedure based on double-pass geometry can be also used in the case of polarization-dependent aberrations and if the medium possesses losses (without transverse variation). Even nonlinear phase aberrations can be corrected in this way under suitable conditions.

There are also conjugators that are capable of producing a *forward-going* replica of the original wave:

$$E_{f\,\mathrm{pc}\bullet}(r_\bullet, t) = \tfrac{1}{2}\bar{E}^*_{o\bullet}(r_\bullet)\,e^{-i(\omega_o t - k_o z)} + \text{c.c.}. \tag{14.6}$$

As in the case of the backward-going phase-conjugate wave (14.4), the equiphase surfaces are reversed, whereas in contrast to (14.4) the propagation direction of the original wave is retained.

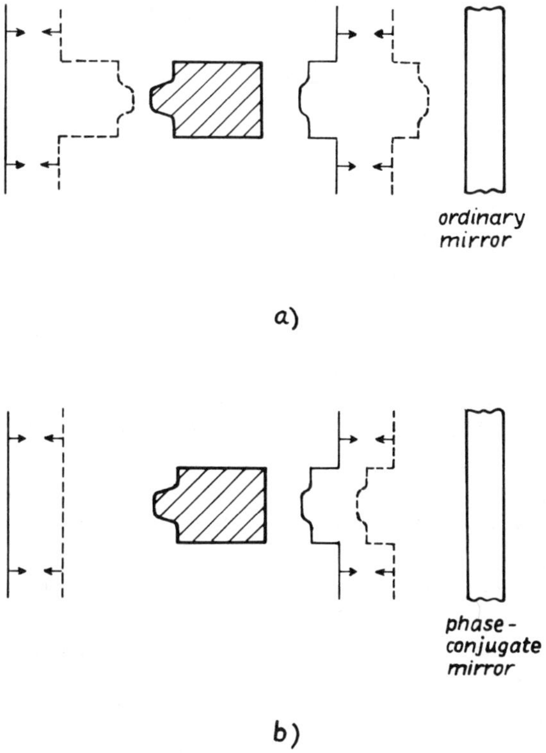

Fig. 14.2. The effect of passage through a phase aberrator (cross-hatched), before and after reflection from (*a*) an ordinary mirror and (*b*) a phase-conjugate mirror, of a wavefront (equiphase surface) of the original wave.

14.2. NONLINEAR OPTICAL MECHANISMS FOR PHASE CONJUGATION

One may distinguish between two types of nonlinear optical mechanisms for phase conjugation. On the one hand there are nonlinear interactions where the states of the medium prior to and after the interaction differ (e.g., stimulated Brillouin scattering). On the other hand there are nonlinear interactions where the initial and final states of the medium are the same (e.g., four-wave mixing). The former mechanisms will be treated in Section 14.2.1, the latter in Section 14.2.2.

14.2.1. Mechanisms Connected with a Change of the State of the Medium

Mechanisms that may be utilized for nonlinear optical phase conjugation are stimulated Brillouin scattering (Section 12.3), stimulated Raman scattering

(Sections 12.1 and 12.2), and Rayleigh wing scattering. For practical reasons, the Brillouin scattering is the most promising mechanism.

The results given in Section 12.3.2 on stimulated Brillouin scattering can be interpreted in the following way: If the medium is irradiated by an intense laser wave propagating in the $+z$ direction, a backward-going scattered electromagnetic wave and a forward-going acoustic wave are generated. The scattered optical wave is frequency-downshifted with respect to the incident wave. The quantitative results are given in (12.141a, b); the wave number of the Stokes wave is downshifted on the order of 1 cm^{-1}.

From the relations for the wave amplitudes of stimulated Brillouin scattering given in Subsection 12.3.1, one can conclude that the just-discussed collinear process involving the counterpropagating original and scattered waves has the largest spatial gain of all the possible scattering processes, and thus its buildup is favored. The practical importance of this fact was emphasized as early as 1972 by Zel'dovich and coworkers (14.4). They proceeded as follows. The original (laser) wave described by the field amplitude $\hat{E}(\omega_L, z) = \bar{E}(\omega_L, z)\exp[+ik(\omega_S)z]$ is assumed to satisfy the amplitude equation. The difference between the wave number of the original wave $[k(\omega_L)]$ and that of the backward-scattered Stokes wave $[k(\omega_S)]$ is ignored. An expansion of the sought Stokes amplitude $\bar{E}(\omega_S, z)\exp[+ik(\omega_S)z]$ in orthogonal solution functions of the amplitude equations reveals that the term that corresponds to the solution of the laser wave is precisely the one that increases fastest with increasing interaction length. Thus, in the case of a sufficiently large interaction length there results

$$\bar{E}(\omega_S, z) = A(z)\bar{E}^*(\omega_L, z). \tag{14.7}$$

Obviously under these conditions the backward-scattered wave has the properties of the phase-conjugate wave, aside from the (relatively small) frequency shift.

This process can be made to proceed with high efficiency, near unity. This is especially true under circumstances, where the conjugator operates in a multimode optical-wave-guide configuration. We give the following examples of the realization of phase conjugation via stimulated Brillouin scattering: Irradiation of SF_6 by pulsed radiation of a C_3F_7I laser ($\lambda = 1.315$ μm) led to reflectivity values between 30 and 80% (Refs. 14.6, 14.8, 14.9). Irradiation by pulsed radiation of a Nd : YAG laser ($\lambda = 1.06$ μm) and of a ruby laser ($\lambda = 0.694$ μm) on CS_2 and CH_4 led to reflectivity values between 10 and 90% (Refs. 14.10, 14.4).

Let us now discuss some disadvantages and goals concerning the utilization of stimulated Brillouin scattering. First the deficiencies: From Section 12.3 it becomes apparent that stimulated Brillouin scattering—like stimulated Raman scattering—is a process characterized by a threshold; this means that a threshold input intensity is needed for the process to proceed. Furthermore, the inherent generation of the acoustic wave leads, because of the conservation of energy, to a nonlinear reflectivity of the phase-conjugate mirror with a value less than unity.

Brillouin scattering does have the following advantage: Because of its passive nature, no activation of the medium by radiation is required. The arrangement is relatively simple, since only one optical beam, the original wave, is needed. This fact appears promising for high-power pulsed laser systems (e.g. for laser fusion experiments). The frequency disparity of the original and the phase-conjugate waves does, in general, not play an important role because of its relative smallness.

For stimulated Raman scattering the frequency disparity between original and phase-conjugated wave is considerably greater ($\sim 1000\ \text{cm}^{-1}$) than for stimulated Brillouin scattering. Phase conjugation on the basis of stimulated Raman scattering and Rayleigh wing scattering are discussed, for instance, in Refs. 14.11 and 14.12, respectively.

14.2.2. Mechanisms Without Change of the State of the Medium

In the processes to be treated now, the nonlinear medium, in which the interaction of the various waves occurs, is left in the same state as it was prior to the process.

The most important one of these processes is *degenerate four-wave mixing*; utilizing this mechanism, either a forward-going or a backward-going phase-conjugate wave can be generated (according to the geometry used). There are no particular requirements on the symmetry properties of the medium. For the generation of the forward-going wave, certain phase-matching conditions must be fulfilled, whereby some application problems may arise. In the case of the backward-going wave to be discussed now, no phase-matching conditions appear; this gives rise to a greater latitude in applications.

Four-wave mixing is a nonlinear process, in which three waves (two pump waves and the original wave) are mixed to generate the phase-conjugate wave. If all the waves have the same frequency ω, the process is called *degenerate* four-wave mixing. The pump waves are assumed to be counterpropagating plane waves whose intensities are high with respect to that of the original wave. The original wave (wave amplitude $\overline{E}_o$) enters the medium at an arbitrary angle with respect to the pump waves. These three waves couple in the medium and lead to the nonlinear polarization

$$P_{\text{pc}}(r_\bullet, t) = \tfrac{1}{2}\varepsilon_o \chi^{(3)} \overline{E}_1(r_\bullet)\overline{E}_2(r_\bullet)\overline{E}_o^*(r_\bullet)e^{-i(\omega t - k_{\text{pc}\bullet} r_\bullet)} + \text{c.c.} \quad (14.8)$$

of the sought fourth wave, to be considered as the phase-conjugate wave. $\overline{E}_1$ and $\overline{E}_2$ are the wave amplitudes of the pump waves; vector sum $(k_{1\bullet} + k_{2\bullet})$ of the wave vectors of the pump waves vanishes. Therefore we have for the wave number of the fourth wave $k_{\text{pc}\bullet} = k_{1\bullet} + k_{2\bullet} - k_{o\bullet} = -k_{o\bullet}$. Thus, the field strength $E_{\text{pc}}(r_\bullet, t)$ corresponding to the polarization $P_{\text{pc}}(r_\bullet, t)$ represents a backward-going phase-conjugate replica of the original wave. Since this process is independent of the input angle of the original wave, its operation comes closest to that of an ideal phase-conjugate mirror, and therefore there it has the widest range of applicability. Under off-resonance conditions, the third-order susceptibility $\chi^{(3)}$ may be regarded as a constant nondispersive parameter, while in the near-resonance case $\chi^{(3)}$ may have a strong frequency dependence in a range of the order of the natural line width (cf. Section 3.3.2.1).

The influence of the pump wave amplitudes on the right-hand side of (14.8) reveals that the amplitude of the phase-conjugate wave may exceed the amplitude of the original wave, so that an *amplifying* phase-conjugate mirror can be realized. From the analyses given in Refs. 14.13 and 14.14 the following conclusions can be drawn. The original wave is taken to be traveling in the z direction and assumed to enter the phase-conjugate mirror with the interaction length L at $z = 0$. Neglecting the depletion of the pump waves as well as linear losses, the amplitude equations of the original and the phase-conjugate wave become

$$\frac{d\overline{E}_{\text{pc}}}{dz} = -i\varkappa^* \overline{E}_o^*, \qquad \frac{d\overline{E}_o^*}{dz} = -i\varkappa \overline{E}_{\text{pc}} \tag{14.9}$$

with the interaction parameter $\varkappa^*$ being proportional to the product $\omega\chi^{(3)}\overline{E}_1\overline{E}_2$. These equations lead (with a reflection factor A) to the relation

$$\overline{E}_{\text{pc}}(z=0) = A\overline{E}_o^*(z=0). \tag{14.9a}$$

This means that a phase-conjugate wavefront of each plane-wave component of the original wave exists. Furthermore, one obtains the power reflectivity

$$R = \tan^2(|\varkappa|L) \tag{14.10}$$

of the phase-conjugate wave with respect to the original wave. This means that an effective *amplification* of the phase-conjugate wave takes place if $|\varkappa|L$ exceeds $\pi/4$. As $|\varkappa|L$ tends to $\pi/2$, the reflectivity tends to an infinite value.

Let us now discuss some aspects of four-wave mixing that require a quantum-theoretical analysis. To this end we have to establish the interaction operator characterizing the process, where we assume off-resonance conditions. We may proceed analogously to the description of the generation of the sum frequency. In the classical case the polarization amplitude at the sum frequency ω_{sum} was given by

$$\hat{P}(\omega_{\text{sum}}) \propto \chi^{(2)}\hat{E}(\omega_1)\hat{E}(\omega_2)$$

where ω_1, ω_2 are the frequencies of the pump waves [cf. (4.28)]. The corresponding interaction part of the Hamiltonian function was given by

$$H^C \propto \left[\chi^{(2)}\hat{E}(\omega_1)\hat{E}(\omega_2)\hat{E}^*(\omega_{\text{sum}}) + \text{c.c.}\right]$$

[cf. (11.27)], because $H^C \propto [\hat{P}(\omega_{\text{sum}})\hat{E}(\omega_{\text{sum}}) + \text{c.c.}]$. We may proceed analogously in describing the four-wave mixing; from the nonlinear polarization given in (14.8) we obtain the interaction part

$$H^C \propto \left[\hat{E}_1\hat{E}_2\hat{E}_o^*\hat{E}_{\text{pc}}^* + \text{c.c.}\right]. \tag{14.11}$$

Passing over to quantized quantities, one obtains an interaction Hamiltonian $\hat{H}^C$ containing terms with four-boson operator products. Let us discuss that part $\hat{H}^C_{\text{FWM}}$ of the interaction operator responsible for the generation of a phase-conjugate wave via four-wave mixing (FWM). We have

$$\hat{H}^C_{\text{FWM}} \propto \hat{a}_1\hat{a}_2\hat{a}_o^\dagger\hat{a}_{\text{pc}}^\dagger. \tag{14.12}$$

We now want to discuss the transition rate $\Delta p_{i\to f}/\Delta t$ of the process with the help of Fermi's "golden rule" (A.123). We have

$$\frac{\Delta p_{i\to f}}{\Delta t} \propto \left|\langle n_1', n_2', n_o', n_{\text{pc}}'|\hat{H}^C_{\text{FWM}}|n_1, n_2, n_o, n_{\text{pc}}\rangle\right|^2, \tag{14.13}$$

where the unprimed photon occupation numbers $n_1, n_2, n_o, n_{\text{pc}}$ designate the initial state and the respective primed numbers designate the final state in the various modes. As can be seen from (14.12) the transition rate is nonzero only if

$$n_1' = n_1 - 1, \quad n_2' = n_2 - 1, \quad n_o' = n_o + 1, \quad n_{\text{pc}}' = n_{\text{pc}} + 1. \tag{14.14}$$

That means the four-wave mixing process must be understood as a process where two pump photons, one of mode 1 and one of mode 2, are annihilated, whereas one photon of the phase-conjugate wave and one photon of the original wave are created. Under the conditions we have the energy changes $-\hbar\omega_1, -\hbar\omega_2, +\hbar\omega_o, +\hbar\omega_{\text{pc}}$ in the four modes; because $\omega_1 = \omega_2 = \omega_o = \omega_{\text{pc}}$, conservation of energy holds. The momentum changes in the four modes are $-\hbar k_{1\bullet}, -\hbar k_{2\bullet}(= +\hbar k_{1\bullet}), +\hbar k_{o\bullet}, +\hbar k_{\text{pc}\bullet}(= -\hbar k_{o\bullet})$; thus we have momentum conservation too. The relations (14.13) and (14.14) readily lead to the rate

$$\frac{\Delta p_{i\to f}}{\Delta t} \propto n_1 n_2\left(n_o n_{\text{pc}} + n_o + n_{\text{pc}} + 1\right). \tag{14.15}$$

This analysis reveals that the depletion of the pump waves yields an amplification of the original wave and generation (or amplification) of the phase-conjugate wave. Even the possibility of optical oscillation arises if all required conditions, in particular the compensation or overcompensation for losses, are satisfied; the transition rate is nonzero even in the case that n_o and n_{pc} are zero. Furthermore, the above discussion reveals that, because of the conservation of momentum, no force (radiation pressure) is transferred by the interacting waves to an ideal phase-conjugate mirror. This property is in striking contrast to the mode of operation of an ordinary mirror.

The near-resonant case of four-wave mixing has also been analyzed, and aspects of spatial modulation of the polarization and the formation of volume holograms taken into account (see, e.g., Ref. 14.15); note that hints on these processes were already given in Section 9.3. Furthermore, nearly degenerate four-wave mixing has been studied (Ref. 14.16) under the condition $\omega_1 = \omega_2 = \omega$, $\omega_o = \omega + \Delta\omega$, $\omega_{pc} = \omega - \Delta\omega$, with $\Delta\omega/\omega$ much less than unity. Due to these frequency deviations, a nonzero phase mismatch now occurs.

There exists many experimental verifications of phase conjugation via degenerate four-wave mixing. We give some examples of various wavelength values and reflectivities of the phase-conjugate mirrors:

Pump Laser (Pulsed)		Nonlinear Medium		Reflectivity	
Matl.	λ (μm)	Matl.	Length (cm)	(%)	Ref.
CO_2	10	SF_6	2	37	14.17
Nd : YAG	1.06	Si	0.1	180	14.18
Ruby	0.694	CS_2	40	> 200	14.19

A typical geometry for the generation of the phase-conjugate wave divides the output radiation of the pump laser into three beams, two of which form the counterpropagating planar pump waves, while the third acts as the original wave.

At the end of the preceding subsection some shortcomings and remedies of phase conjugation via stimulated Brillouin scattering were sketched. These deficiencies, mainly frequency disparity and threshold nature either do not appear in degenerate four-wave mixing or do not play an important role. As an example four-wave mixing has been demonstrated for very weak original waves down to intensities of the order of 10^{-6} W/cm^2 (Ref. 14.20). However, the removal of these deficiencies is coupled with the necessity of activating the medium by pump waves, which involves larger experimental effort.

Recently, self-pumped and self-starting phase conjugators have been demonstrated (Refs. 14.21, 14.22), in which both pumping fields are generated from the original wave in the nonlinear medium (barium titanate) without external mirrors. It is worth noting that this device behaves as simply as one based on stimulated Brillouin scattering with respect to the generation of the phase-conjugate wave. This is schematically illustrated in Fig. 14.3.

Another mechanism for phase conjugation without changing the state of the nonlinear medium is three-wave mixing, where the phase-conjugate wave is formed from two collinear copropagating waves: a pump wave (index 1) and the original wave (index o); the procedure follows the pattern of difference frequency generation described in (11.14b) or of degenerate parametric amplification (cf. Section 11.4). There results for the wave with the difference

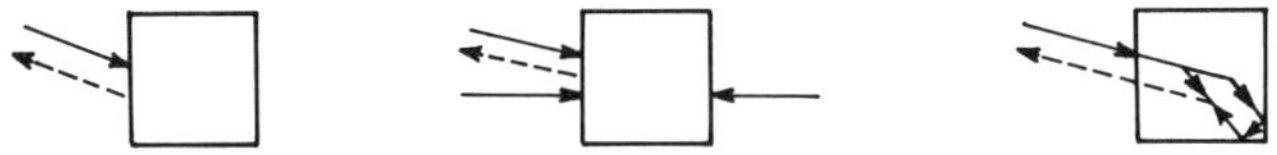

Fig. 14.3. Phase-conjugation scheme via (*a*) stimulated Brillouin scattering, (*b*) traditional degenerate four-wave mixing, and (*c*) self-pumped degenerate four-wave mixing with internal reflection (after Ref. 14.22).

frequency ω_d

$$E_d \propto \left[\chi^{(2)}\bar{E}_1(r_\bullet)\bar{E}_o^*(r_\bullet)e^{i(\omega_d t - k_d z)} + \text{c.c.}\right], \tag{14.16}$$

where ω_d is equal to $\omega_1 - \omega_o$ and $k_d(\omega_d)$ is equal to $k(\omega_1) - k(\omega_o)$. In case $\omega_o = \omega_1/2$ we have $\omega_d = \omega_1/2$ and the difference-frequency wave E_d radiates at the frequency of the original wave. The wave number of this wave becomes $k_d = k(2\omega_o) - k(\omega_o)$. If $k(2\omega_o) - k(\omega_o) = k(\omega_o)$, then the wave vector of the difference-frequency wave E_d agrees with that of the original wave in absolute value and direction. This means that in this case the difference-frequency wave (14.16) corresponds to the forward-going phase-conjugate wave $E_{f\,\text{pc}}$ of the original wave. If the momentum matching condition is not fulfilled exactly, the influence of the momentum mismatch at a given interaction length L must be studied. In the case of arbitrary propagation directions of the waves we have to examine whether

$$\left|L_\bullet\left[k_{f\,\text{pc}\bullet}(\omega_o) - k_{1\bullet}(2\omega_o) + k_{o\bullet}(\omega_o)\right]\right| \ll 1; \tag{14.17}$$

here $L_\bullet$ is the (directed) interaction length. As a rule this phase constraint causes a limitation upon the permissible angles and thus upon the spatial information.

Experiments on successful phase conjugation via three-wave mixing with a lithium formate conjugator are reported in Ref. 14.23, where the original and the pump waves were respectively the fundamental from a Nd:YAG laser (1.06 μm) and its second harmonic (0.53 μm). The forward-going original and phase-conjugate waves were distinguished by linear polarization discrimination. Also the phase-matching limitation was demonstrated.

To conclude the present subsection we list a few mechanisms also being capable of producing phase conjugation without change of the state of the medium: degenerate six-wave mixing, transient optical excitation processes such as photon echoes (see Section 8.4.3), local modifications of the susceptibilities caused by interacting fields, spatial modulation of the current density in a plasma, and local absorption of radiation.

14.3. APPLICATIONS

Let us sketch some applications of nonlinear optical phase conjugation that have been proposed or that have been already demonstrated experimentally.

Contributions to nonlinear laser spectroscopy can be obtained by investigating the behavior of phase conjugators. In (14.10) the nonlinear reflectivity of a phase conjugator operating on the basis of four-wave mixing is given. Obviously the investigation of the mode of operation of the conjugator gives an insight into optical third-order processes of the medium and its atomic systems. Also the dependence on external influences such as pressure, temperature, and external fields can be studied.

Adaptive optics must be regarded as one of the most promising applications of nonlinear optical phase conjugation (Ref. 14.24), since a conjugator is capable of restoring aberrated waves to their original state, even in the case of a severe aberration. This application is based on the complex conjugation of the spatial part of the original wave by using an arrangement with a double pass (cf. Fig. 14.2*b*) through the system (medium) that causes the phase distortion by one or more effects such as spatially varying reflective index, poor optical quality, or atmospheric turbulences. The far-field photographs in Fig. 14.4 (from Ref. 14.25) of the original (unperturbed) laser beam (A), of the beam (B) after passing through a random aberrator (an etched glass plate), and of the phase-conjugate beam (C) after passing back through the aberrator impressively reveal the possibility of achieving almost perfect phase correction. While the beam in B obviously deviates severely from the beam in A, the beam in C agrees with that in A within the measurement error. These results were obtained by using a Q-switched ruby laser and a semiconductor-doped glass as nonlinear medium.

The advantage of such a procedure for the generation of intense, aberration-free beams is illustrated in Fig. 14.5. The target is illuminated with

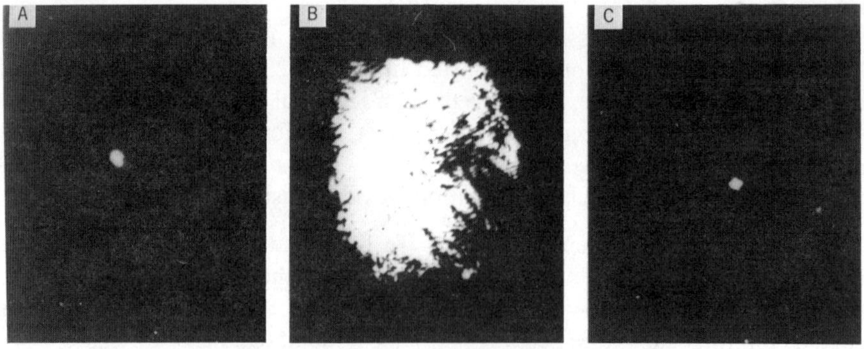

Fig. 14.4. Far-field photographs of the original laser beam (A), of the beam after passing through the aberrator (B), and of the phase-conjugated beam (C) (from Ref. 14.25).

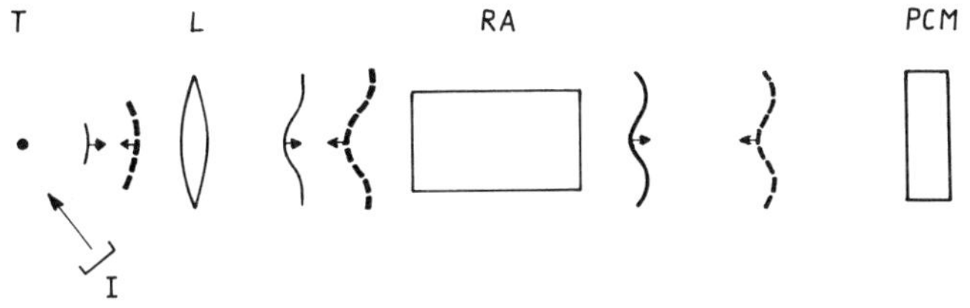

Fig. 14.5. Phase conjugation applied to the illumination of a target: T, target; L, lens; RA, radiation amplifier; PCM, phase-conjugate mirror; I, illuminator.

(relatively weak) radiation of a laser source. Some of the light reflected from the target (the original wave) passes through the lens, which represents the focusing element of the optical system. Let us assume a real optical system consisting of passive elements and an active light amplifier; then the incident wave accumulates phase distortions, so that its equiphase surfaces may severely deviate from those of a wave propagating through perfect optical elements. After phase conjugation the distorted wave propagates back through the optical system. Due to the phase-conjugation process, an amplified (intense) aberration-free beam is focused on the target [of course the beam is diffraction-limited (cf. Section 14.1) because of finite transverse extent].

This procedure can be successfully applied even in the case of a moving target (within a certain range of velocity) and in the case where the original wave has a noticeable amplitude variation over the equiphase surface; the optimum focusing is automatically maintained by phase conjugation. This must be regarded as an important advantage of nonlinear optical phase conjugation over the conventional approach of wavefront correction, which is not capable of compensating for amplitude variations. The above illumination procedure may be of particular importance in laser fusion systems, since in this case the target is irradiated from several directions and there is a very complex problem of beam focusing.

Another example is the image transmission through multimode optical fibers. Over a long propagation distance the transmitted image information can be degraded by phase modal dispersion. Under certain circumstances the image information can be restored after phase conjugation and propagation back through the first waveguide or a second similar fiber. An instructive example is given in Ref. 14.26, where degenerate four-wave mixing was used for restoration. The demonstration was accomplished by utilizing a multimode step-index fiber (diameter 85 μm, length 1.75 m) and a krypton laser ($\lambda = 0.647$ μm). The beam carrying the information is imaged at the input end of the transmission fiber. The fiber output is directed into the nonlinear medium (barium titanate), where the phase-conjugate signal is generated, which then retraverses the same fiber. Figure 14.6 gives oscilloscope traces of far-field intensity distributions: (a) the input beam with the profile of an apodized Gaussian beam, (b) the output from the fiber, revealing the severe aberrations

14.17. R. C. Lind, D. G. Steel, M. B. Klein, R. L. Abrams, C. R. Giuliano, and R. K. Jain, *Appl. Phys. Lett.* **34**, 457 (1979).

14.18. R. K. Jain, M. B. Klein, and R. C. Lind, *Opt. Lett.* **4**, 328 (1979).

14.19. D. M. Pepper, D. Fekete, and A. Yariv, *Appl. Phys. Lett.* **33**, 41 (1978).

14.20. J. Feinberg and R. W. Hellwarth, *Opt. Lett.* **5**, 519 (1980); **6**, 257 (1981).

14.21. J. O. White, M. Cronin-Golomb, B. Fischer, and A. Yariv, *Appl. Phys. Lett.* **40**, 450 (1982).

14.22. K. R. MacDonald and J. Feinberg, *J. Opt. Soc. Am.* **73**, 548 (1983).

14.23. P. V. Avizonis, F. A. Hopf, W. D. Bambeger, S. F. Jacobs, A. Tomita, and K. H. Womack, *Appl. Phys. Lett.* **31**, 435 (1977).

14.24. C. R. Giuliano, *Phys. Today*, Apr. 1981, p. 27.

14.25. R. C. Lind and C. R. Giuliano, presented at Conference on Laser Engineering and Applications, Washington, June 1979; in Ref. 14.24.

14.26. G. J. Dunning and R. C. Lind, *Opt. Lett.* **7**, 558 (1982).

14.27. H. P. Yuen and J. H. Shapiro, *Opt. Lett.* **4**, 334 (1979).

14.28. R. E. Slusher, L. W. Hollberg, B. Yurke, J. C. Mertz, and J. F. Valley, *Phys. Rev. Lett.* **55**, 2409 (1985).

Appendix A

Compilation of Quantum-Theoretical Definitions and Relations

For the explanation and description of nonlinear optical phenomena and multiphoton processes, it is necessary to make use of quantum theory to a certain extent. In the so-called semiclassical theory the atomic systems are described quantum-mechanically, while the interacting radiation field is described classically. However, an adequate description of the electromagnetic radiation field requires the full application of quantum theory; therefore, also aspects of quantum field theory must be learned.

In this appendix we compile basic quantum-theoretical definitions, relations, and applications, which are often needed in this book. This compilation will enable us to make direct reference to quantum-theoretical means and results without interrupting the discussion of nonlinear optical problems. We shall interpret the physical meaning of the compiled quantities and relations, but to give mathematical proofs is beyond the scope of this Appendix. More detailed information about this subject can be obtained from several textbooks (e.g., Refs. A.1–A.4).

A.1. DIRAC FORMULATION OF QUANTUM THEORY

In the Dirac formulation the basic relations for the quantization, the eigenvalues and eigenstates, the expectation values, and the temporal development of states and dynamical variables are given in a compact, "representation-free" form, which nevertheless yields the possibility of using a convenient representation for the treatment of a specific problem.

A.1.1. States, Dynamical Variables, Observables

In this subsection we shall separately describe the physical meaning of the basic quantum-theoretical quantities and their mathematical properties.

A.1.1.1. The Physical Meaning of the Basic Quantities. At a certain time t a given physical system can be characterized by a *state*, which the system has reached as a result of an actually performed experiment or as a result of an imaginary experiment. Such a physical state is mathematically represented by a Hilbert-space vector $|\psi\rangle$ with a positive norm $\langle\psi|\psi\rangle > 0$. Here $|\psi\rangle$ is the so-called ket vector, and $\langle\psi|$ is the corresponding bra vector in the dual Hilbert space (to be described later on). It is postulated that with respect to any obtainable result of measurement, a physical system can be completely described by the vector $|\psi\rangle$. The operator relations of the dynamical variables and observables belonging to the physical system characterize the properties of the Hilbert space, in particular the basis set by which the Hilbert space is spanned (see Section A.1.3). The state vectors $|\psi\rangle$ obey all general axioms and rules for the proper elements of a Hilbert space, such as linearity, complexity, hermiticity of the metric, separability, and completeness. Assume that $\{|\beta_j\rangle\}$ is a complete and orthonormalized set of basis vectors (spanning the Hilbert space) with the properties

$$\sum_j |\beta_j\rangle\langle\beta_j| = \hat{I}, \qquad \langle\beta_j|\beta_{j'}\rangle = \delta_{jj'}. \tag{A.1}$$

Then an arbitrary vector $|\psi\rangle$ can be expanded in terms of the $|\beta_j\rangle$ by

$$|\psi\rangle = \sum_j \langle\beta_j|\psi\rangle|\beta_j\rangle, \tag{A.2}$$

where $\langle\beta_j|\psi\rangle$ is a complex function of the discrete index j.

Important physical problems require the introduction of generalized elements of the Hilbert space into the theory. Such elements (*Dirac vectors*) can be interpreted as the result of a limiting process applied to proper Hilbert-space elements, where the discrete index j in $|\beta_j\rangle$ goes over into a continuous variable and the countability of the basis vectors $|\beta_j\rangle$ is abandoned. The completeness and orthonormality are now given by

$$\int dj\, |\beta_j\rangle\langle\beta_j| = \hat{I}, \qquad \langle\beta_j|\beta_{j'}\rangle = \delta(j-j'), \tag{A.3}$$

where the Kronecker symbol $\delta_{jj'}$ is replaced by the delta function $\delta(j-j')$. The expansion in terms of the basis vectors is given by

$$|\psi\rangle = \int dj\, \langle\beta_j|\psi\rangle|\beta_j\rangle, \tag{A.4}$$

where $\langle\beta_j|\psi\rangle$ is a complex function of the real continuous variable j.

The state vector $|\psi\rangle$ has no direct physical significance; but with the help of $|\psi\rangle$, the expressions for measurable quantities can be formulated (see Section A.1.2).

Dynamical variables are represented by linear operators $\hat{A}$ that act on Hilbert-space vectors. In the strict sense, different symbols for a dynamical variable and its quantum-theoretical representative should be used, but in practice, it is sufficient to designate a quantum-theoretical dynamical variable by its Hilbert-space operator (which will always be denoted by a caret). In correspondence with the respective classical equations, quantum-theoretical equations of motion can be formulated for dynamical variables (see Section A.1.4). With the help of an orthonormalized, complete basis system $\{|\beta_j\rangle\}$, the operator $\hat{A}$ can be represented by

$$\hat{A} = \sum_{j,j'} \langle \beta_j|\hat{A}|\beta_{j'}\rangle |\beta_j\rangle\langle\beta_{j'}|, \tag{A.5}$$

where the matrix elements $\langle\beta_j|\hat{A}|\beta_{j'}\rangle$ can be regarded as the components of $\hat{A}$. A linear operator is completely defined by its action on the vectors of a complete set of orthonormal vectors.

Observables $\hat{L}$ are special dynamical variables that are measurable in the following sense. Measurement means interaction of a measuring instrument (which is characteristic of the observable $\hat{L}$) with the physical system, whereby a measuring value results. An operator $\hat{L}$ represents an observable if and only if the following equations are fulfilled:

$$\hat{L} = \hat{L}^\dagger, \qquad \text{(hermiticity)}, \tag{A.6a}$$

$$\langle l|l'\rangle = \begin{cases} \delta_{l,l'}, \\ \delta(l - l'), \end{cases} \qquad \text{(orthonormality)}, \tag{A.6b}$$

$$\hat{I} = \begin{cases} \sum_l |l\rangle\langle l|, \\ \int dl\, |l\rangle\langle l|, \end{cases} \qquad \text{(completeness)}, \tag{A.6c}$$

where the eigenvalue equation

$$\hat{L}|l\rangle = l|l\rangle \tag{A.6d}$$

holds for the eigenstates $|l\rangle$ and the real eigenvalues l. The hermitian operator $\hat{L}$ [whose mathematical properties will be defined by (A.34)] can be represented by its eigenstates and eigenvalues by means of the so-called spectral representation:

$$\hat{L} = \sum_l l|l\rangle\langle l| \text{ and/or } \int dl\, l|l\rangle\langle l|, \qquad \text{respectively.} \tag{A.6e}$$

In the case of degeneracy an additional summation over all degenerate states belonging to an eigenvalue l must be performed; this leads to

$$\hat{I} = \sum_{l} \sum_{\lambda=1}^{\Lambda_l} |l_\lambda\rangle\langle l_\lambda|, \tag{A.6f}$$

where Λ_l is the order of degeneracy of the eigenvalue l.

In contrast to the representatives of classical physical quantities, the Hilbert-space operators representing the quantum-theoretical dynamical variables obey a noncommutative algebra. While the former are called c-numbers (commutable numbers), the latter are called q-numbers. There are so-called *quantization relations* between the Hilbert-space operators, which are formulated in terms of commutators

$$[\hat{A}, \hat{B}] \equiv \hat{A}\hat{B} - \hat{B}\hat{A} \tag{A.7a}$$

or anticommutators

$$[\hat{A}, \hat{B}]_+ \equiv \hat{A}\hat{B} + \hat{B}\hat{A}. \tag{A.7b}$$

Two examples may be given where fundamental properties are expressed by such relations. Firstly, let $\hat{x}_j$ and $\hat{p}_j$ be the Cartesian vector components of the position operators and the corresponding momentum operators of a system of pointlike particles. Then the commutation relations

$$[\hat{x}_j, \hat{p}_{j'}] = i\hbar\delta_{jj'}\hat{I}, \qquad [\hat{x}_j, \hat{x}_{j'}] = \hat{0}, \qquad [p_j, p_{j'}] = \hat{0} \tag{A.8}$$

hold; the physical content of these relations is associated with the basic assumption of homogeneity in the position space. Secondly, let $\hat{c}_B^\dagger$ and $\hat{c}_F^\dagger$ be the creation operators, and $\hat{c}_B$ and $\hat{c}_F$ the annihilation operators of bosons (subscript B) and fermions (subscript F). Then the commutation relations

$$[\hat{c}_B, \hat{c}_B^\dagger] = \hat{I}, \qquad [\hat{c}_B, \hat{c}_B] = \hat{0} \qquad [c_B^\dagger, c_B^\dagger] = \hat{0} \tag{A.9a}$$

hold for bosons and the anticommutation relations

$$[\hat{c}_F, \hat{c}_F^\dagger]_+ = \hat{I}, \qquad [\hat{c}_F, \hat{c}_F]_+ = \hat{0}, \qquad [\hat{c}_F^\dagger, \hat{c}_F^\dagger]_+ = \hat{0} \tag{A.9b}$$

hold for fermions.

A.1.1.2. Mathematical Properties of State Vectors and Linear Operators. Having described the physical meaning of states, dynamical variables, and observables in the Dirac formulation of quantum theory, we shall now give important rules for their mathematical representatives.

The structure of the vector space $\mathscr{H}$ is given by the axiomatic relations between its points—the proper Hilbert-space vectors inclusive of their generalized elements mentioned in Section A.1.1.1. In the following this vector space will be briefly called a "Hilbert space."

Linearity and complexity of the Hilbert space mean that the addition of two vectors and the multiplication of a vector by a complex number lead to an element of the Hilbert space. Here the following relations, expressing associativity and commutativity, hold:

$$|\psi\rangle + |\varphi\rangle = |\varphi\rangle + |\psi\rangle, \tag{A.10}$$

$$(|\psi\rangle + |\varphi\rangle) + |\gamma\rangle = |\psi\rangle + (|\varphi\rangle + |\gamma\rangle), \tag{A.11}$$

$$c(|\psi\rangle + |\varphi\rangle) = c|\psi\rangle + c|\varphi\rangle, \tag{A.12}$$

$$c(d|\psi\rangle) = (cd)|\psi\rangle. \tag{A.13}$$

The zero vector $|0_v\rangle$, which must not be mistaken for the c-number zero, is defined by

$$|\psi\rangle + |0_v\rangle = |\psi\rangle. \tag{A.14}$$

Furthermore,

$$\hat{0}|\psi\rangle = |0_v\rangle. \tag{A.15}$$

The hermiticity of the metric is expressed by the property that to every two vectors $|\psi\rangle$ and $|\varphi\rangle$ a scalar product $\langle\psi|\varphi\rangle$—a complex number—is assigned. This way of forming a product obeys the following relations:

$$\langle\psi|(|\varphi\rangle + |\gamma\rangle) = \langle\psi|\varphi\rangle + \langle\psi|\gamma\rangle, \tag{A.16}$$

$$\langle\psi|\varphi\rangle = \langle\varphi|\psi\rangle^*. \tag{A.17}$$

The norm $\| |\psi\rangle\|^2 \equiv \langle\psi|\psi\rangle$ of a Hilbert-space vector is greater than or equal to zero, with equality only for $|\psi\rangle = |0_v\rangle$. The symbol $\langle\psi|$ in the scalar product $\langle\psi|\varphi\rangle$ may be given an independent meaning, namely as a vector (bra vector) of the dual Hilbert space.

Furthermore, a basis-vector set with the properties (A.1) or (A.3) is stated to exist, which can serve for the representation of each vector $|\psi\rangle$ of the vector space $\mathscr{H}$ according to (A.2) or (A.4). If a vector depends on a continuous parameter b, and hence $|\psi\rangle = |\psi(b)\rangle$ holds, differentiation with respect to b can be carried out according to

$$\frac{d}{db}|\psi(b)\rangle = \lim_{\varepsilon\to 0}\frac{|\psi(b+\varepsilon)\rangle - |\psi(b)\rangle}{\varepsilon}. \tag{A.18}$$

In Section A.1.1.1 we have defined the vector space $\mathscr{H}$. Now we can define the operators of this space. Assume that to each ket vector $|\psi\rangle$ of the vector space a certain ket vector $|\varphi\rangle$ is assigned. This can be understood as follows: $|\varphi\rangle$ results from the action of a certain operator $\hat{F}$ on $|\psi\rangle$, which is written as

$$|\varphi\rangle = \hat{F}|\psi\rangle. \tag{A.19}$$

Two operators are equal ($\hat{F}_1 = \hat{F}_2$) if for arbitrary $|\psi\rangle$

$$\hat{F}_1|\psi\rangle = \hat{F}_2|\psi\rangle$$

holds, where the same domain of definition is assumed.

The formation of sums and products of operators is given by

$$(\hat{F} + \hat{G})|\psi\rangle = \hat{F}|\psi\rangle + \hat{G}|\psi\rangle, \qquad \hat{F}(\hat{G}|\psi\rangle) = (\hat{F}\hat{G})|\psi\rangle. \tag{A.20}$$

In the Dirac formulation of quantum theory, it is possible to restrict ourselves to linear operators. In addition to the relations above, we have for linear operators

$$\hat{F}(|\psi_1\rangle + |\psi_2\rangle) = \hat{F}|\psi_1\rangle + \hat{F}|\psi_2\rangle. \tag{A.21}$$

$$\hat{F}(c|\psi\rangle) = c(\hat{F}|\psi\rangle). \tag{A.22}$$

$$(\hat{F} + \hat{G})\hat{H} = \hat{F}\hat{H} + \hat{G}\hat{H} \quad \text{and} \quad \hat{F}(\hat{G} + \hat{H}) = \hat{F}\hat{G} + \hat{F}\hat{H}. \tag{A.23}$$

Linear operators commute with complex numbers.

In analogy to the action of an operator on a ket vector [cf. (A.19)], the action of an operator on a bra vector can be defined by

$$\langle\sigma| = \langle\varphi|\hat{F}. \tag{A.24}$$

Here $\hat{F}$ implies an assignment between the bra vectors $\langle\varphi|$ and $\langle\sigma|$. There holds the identity

$$\langle\psi|(\hat{F}|\varphi\rangle) = (\langle\psi|\hat{F})|\varphi\rangle, \tag{A.25}$$

so that the parentheses can be omitted and both sides can be written in the form $\langle\psi|\hat{F}|\varphi\rangle$.

The convergence behavior of a sequence of operators is expressed by

$$\{\hat{F}_n\} \underset{n\to\infty}{\to} \hat{F} \quad \text{if} \quad \{\hat{F}_n|\psi\rangle\} \underset{n\to\infty}{\to} \hat{F}|\psi\rangle \quad \text{for all } |\psi\rangle. \tag{A.26}$$

Important limiting values of operators concern derivatives. If an operator depends on a real parameter b, it can be differentiated with respect to b according to

$$\frac{d}{db}\hat{F}(b) = \lim_{\varepsilon\to 0} \frac{\hat{F}(b+\varepsilon) - \hat{F}(b)}{\varepsilon}. \tag{A.27}$$

The derivative of the operator function $\hat{G} = G(\hat{F})$ with respect to $\hat{F}$ can be introduced as follows:

$$\frac{d}{d\hat{F}}G(\hat{F}) = \lim_{\varepsilon \to 0} \frac{G(\hat{F} + \varepsilon\hat{I}) - G(\hat{F})}{\varepsilon} \tag{A.28}$$

In an analogous way partial derivatives can also be obtained if $\hat{G}$ depends on more than one operator. Then we have

$$\frac{\partial}{\partial\hat{F}}G(\hat{F}, \hat{K}) = \lim_{\varepsilon \to 0} \frac{G(\hat{F} + \varepsilon\hat{I}, \hat{K}) - G(\hat{F}, \hat{K})}{\varepsilon}. \tag{A.28a}$$

These partial derivatives can be used to formulate commutation relations of variables that are dependent on other variables. An important example is the following one. Let $G = G(\hat{x}_1, \ldots, \hat{x}_m, \hat{p}_1, \ldots, \hat{p}_n)$ be a variable depending on the vector components of the position and momentum operators of the particles of a system [cf. explanation of (A.8)]; then one has

$$[\hat{x}_j, \hat{G}] = i\hbar\frac{\partial\hat{G}}{\partial\hat{p}_j}, \qquad [\hat{p}_j, \hat{G}] = -i\hbar\frac{\partial\hat{G}}{\partial\hat{x}_j} \tag{A.29}$$

if the partial derivatives exist.

Let us now consider some special operators which play an important role. The zero operator $\hat{0}$ and the identity operator $\hat{I}$ satisfy the relations

$$\hat{0}|\psi\rangle = |0_v\rangle, \tag{A.30a}$$

$$\hat{I}|\psi\rangle = |\psi\rangle \tag{A.30b}$$

for an arbitrary vector $|\psi\rangle$; $\hat{I}$ commutes with any other operator.

The possibility of forming the power $(\hat{F})^n$, with n a nonnegative integer, is provided by (A.20) as

$$(\hat{F})^n = \hat{F}(\hat{F})^{n-1}, \tag{A.31}$$

where $(\hat{F})^0 = \hat{I}$. Since it is possible to add operators, to multiply them by complex numbers, and to raise them to powers, one is allowed to introduce nth-degree polynomials in $\hat{F}$. Thus we have

$$P_n(\hat{F}) = c_0\hat{I} + c_1\hat{F} + c_2(\hat{F})^2 + \cdots + c_n(\hat{F})^n. \tag{A.32}$$

Given convergence, operator functions may be defined for $n \to \infty$ in terms of power series; for example $\exp\hat{F} = \sum_{n=0}^{\infty}(n!)^{-1}(\hat{F})^n$.

Reciprocal operators $\hat{F}^{-1}$ are defined by

$$\hat{F}^{-1}\hat{F}|\psi\rangle = |\psi\rangle \tag{A.33}$$

This leads to the operator equation $\hat{X}\hat{F} = \hat{I}$ with the solution $\hat{X} = \hat{F}^{-1}$.

An operator $F^\dagger$ is called the *hermitian adjoint* operator of $\hat{F}$ if

$$|\psi\rangle = \langle\varphi|\hat{F}^\dagger \tag{A.34}$$

holds whenever $|\psi\rangle = \hat{F}|\varphi\rangle$. Important rules of forming the Hermitian adjoint are

$$(\hat{F}\hat{G})^\dagger = \hat{G}^\dagger\hat{F}^\dagger, \qquad (c\hat{F})^\dagger = c^*\hat{F}^\dagger, \quad \text{and} \quad (|\psi\rangle\langle\varphi|)^\dagger = |\varphi\rangle\langle\psi|. \tag{A.35}$$

Those operators for which $\hat{F} = \hat{F}^\dagger$ are called *hermitian*; in this case

$$\langle\varphi|\hat{F}|\psi\rangle = \langle\psi|\hat{F}|\varphi\rangle^*, \tag{A.36}$$

and $\langle\varphi|\hat{F}|\varphi\rangle$ is a real number.

An operator $\hat{U}$ is called *unitary* if

$$\hat{U}\hat{U}^\dagger = \hat{U}^\dagger\hat{U} = \hat{I} \quad \text{and so} \quad \hat{U}^{-1} = \hat{U}^\dagger. \tag{A.37}$$

If any vectors $|\psi\rangle$ of $\mathscr{H}$ and operators $\hat{G}$ on $\mathscr{H}$ are subjected to a unitary transformation according to $|\psi'\rangle = \hat{U}|\psi\rangle$ and $\hat{G}' = \hat{U}\hat{G}\hat{U}^{-1}$, all scalar products remain invariant. Thus we have

$$\langle\psi'|\varphi'\rangle = \langle\psi|\varphi\rangle, \qquad \langle\psi'|\hat{G}'|\varphi'\rangle = \langle\psi|\hat{G}|\varphi\rangle. \tag{A.38}$$

By summing all diagonal elements $\langle\beta_j|\hat{G}|\beta_j\rangle$ of an operator (formed with the basis $\{|\beta_j\rangle\}$), the trace $\mathrm{Tr}\{\hat{G}\}$ is obtained. Its numerical value is independent of the basis chosen.

A special kind of operator is the so-called *dyadic product* of a ket and a bra vector, $|\psi\rangle\langle\varphi|$. In particular, one has

$$\mathrm{Tr}\{[|\psi\rangle\langle\psi|]\hat{G}\} = \langle\psi|\hat{G}|\psi\rangle. \tag{A.39}$$

As is apparent from the relations (A.6), the eigensolutions of hermitian operators, as representatives of observables, are of great importance. The manifold of the eigenstates can be regarded as a basis set of the Hilbert space. The real eigenvalues play—as we shall demonstrate in Section A.1.2—the role of measurable values. A polynomial $P_n(\hat{L})$ has the eigenvalue equation

$$P_n(\hat{L})|l\rangle = P_n(l)|l\rangle. \tag{A.40}$$

provided that for $\hat{L}$ the eigenvalue equation $\hat{L}|l\rangle = l|l\rangle$ holds.

To achieve solutions of practical problems it is often advantageous to transform the eigenvalue equation into an algebraic equation or into an integral equation (or differential equation) by use of a familiar complete basis system $\{|\beta_j\rangle\}$. We will sketch the first of the two procedures. The representation of the eigenvalue equation (A.6d) in terms of $\{|\beta_j\rangle\}$ leads to

$$\sum_{j'} \{\langle\beta_j|\hat{L}|\beta_{j'}\rangle - l\,\delta_{jj'}\}\langle\beta_{j'}|l\rangle = 0 \qquad \text{for all } j, \tag{A.41}$$

where the $\langle\beta_j|l\rangle$ are the components of the eigenket $|l\rangle$. From this system of linear equations the eigenvalues l and the components of $|l\rangle$ must be determined. Because of the requirement for $|l\rangle$ to be normalizable, only nontrivial solutions are allowed; thus

$$\det\{\langle\beta_j|\hat{L}|\beta_{j'}\rangle - l\,\delta_{jj'}\} = 0 \tag{A.42}$$

must be satisfied. This is the secular equation of the problem, from which the eigenvalues l can be determined.

If a physical system can be decomposed into independent physical subsystems, it is useful to consider the Hilbert space $\mathscr{H}$ of the total system as a product space consisting of subspaces. The essential features can be discussed by considering a product space composed of only two subspaces ($\mathscr{H}_1$, $\mathscr{H}_2$). Let $|\psi_1\rangle$ be a vector of $\mathscr{H}_1$, and $|\psi_2\rangle$ be a vector of $\mathscr{H}_2$; then we can form the vector $|\psi_1\psi_2\rangle$ of the total space. For this vector

$$|\psi_1\psi_2\rangle = |\psi_1\rangle|\psi_2\rangle = |\psi_2\rangle|\psi_1\rangle. \tag{A.43}$$

This is the so-called *direct product* (or tensor product) of the two single vectors, which is commutative. One has distributivity:

$$|\psi_1\psi_2\rangle = |\varphi_1\psi_2\rangle + |\gamma_1\psi_2\rangle \qquad \text{if} \quad |\psi_1\rangle = |\varphi_1\rangle + |\gamma_1\rangle. \tag{A.44}$$

The action of an operator $\hat{G}_1$ which only acts on vectors of the subspace $\mathscr{H}_1$ is given by

$$\hat{G}_1|\psi_1\psi_2\rangle = |\varphi_1\psi_2\rangle \tag{A.45}$$

if $\hat{G}_1|\psi_1\rangle = |\varphi_1\rangle$.

The properties of independent subspaces can be used for a simple treatment of the eigenvalue problem when dealing with more complex problems. Let

$$\hat{G} = \sum_j \hat{G}_j \tag{A.46}$$

be an operator assigned to the total space, whereas the operators $\hat{G}_j$ are assigned to single, disjoint subspaces. The $\hat{G}_j$ are assumed to commute with each other: $[\hat{G}_j, \hat{G}_{j'}] = \hat{0}$. Let

$$\hat{G}_j|g_j\rangle = g_j|g_j\rangle. \tag{A.47}$$

be the eigenvalue equations for the operators of the jth subspace. Then we have

$$\hat{G}|g\rangle = g|g\rangle. \tag{A.48}$$

for the operator $\hat{G}$ of the total system, where the eigenvalue g and the eigenket $|g\rangle$ are given by

$$g = \sum_j g_j, \qquad |g\rangle = |g_1\rangle|g_2\rangle \cdots |g_j\rangle \cdots. \tag{A.49}$$

A.1.2. Description of the Physical Measurement

In this subsection the quantities introduced above will be associated with experimentally accessible quantities.

If the observable represented by the operator $\hat{L}$ is measured in the sense of Section A.1.1, the only values that can be obtained are those of the eigenvalue spectrum of $\hat{L}$. Let $\mathscr{H}_{\mathscr{S}}$ be a subspace of the Hilbert space with the assemblage $\{l\}_{\mathscr{S}}$ of eigenvalues. Then the probability of measuring one eigenvalue out of $\{l\}_{\mathscr{S}}$ is given by

$$p_{\mathscr{S}} = \frac{\langle\psi|\hat{P}_{\mathscr{S}}|\psi\rangle}{\langle\psi|\psi\rangle}, \tag{A.50}$$

where $\hat{P}_{\mathscr{S}}$ is the projection operator assigned to the subspace $\mathscr{H}_{\mathscr{S}}$, and $|\psi\rangle$ is the state of the system just before the measurement. The state immediately after the measurement process is $|\psi\rangle_{\mathscr{S}} = \hat{P}_{\mathscr{S}}|\psi\rangle$.

The formula (A.50) covers important special cases. If $\hat{L}$ has a discrete spectrum, the projection operator associated with a certain eigenvalue l is $|l\rangle\langle l|$. Thus the probability of measuring the value l is given by

$$p_l = \frac{|\langle\psi|l\rangle|^2}{\langle\psi|\psi\rangle}. \tag{A.50a}$$

In the case of a purely continuous spectrum, the projection operator associated with the eigenvalue interval $(l, l + \Delta l)$ is

$$\int_l^{l+\Delta l} dl'\, |l'\rangle\langle l'|.$$

Therefore the probability of measuring a value within this interval is

$$p_{(l,\, l+\Delta l)} = \langle\psi|\psi\rangle^{-1} \int_l^{l+\Delta l} dl' |\langle\psi|l'\rangle|^2. \tag{A.50b}$$

The result (A.50) allows the calculation of the arithmetic average or *expectation value* of the observable $\hat{L}$:

$$\langle \hat{L} \rangle = \frac{\langle \psi | \hat{L} | \psi \rangle}{\langle \psi | \psi \rangle}. \tag{A.51}$$

Insertion of (A.6e) into the right-hand side gives the expression

$$\bar{l} = \sum_{l} l \frac{|\langle \psi | l \rangle|^2}{\langle \psi | \psi \rangle} \quad \text{or} \quad \int dl\, l \frac{|\langle \psi | l \rangle|^2}{\langle \psi | \psi \rangle}, \tag{A.52}$$

which directly reveals the properties of mean values. Note that the projection operator $\hat{P}_{\mathscr{S}}$ can be regarded as the observable assigned to the measurable probability $p_{\mathscr{S}}$ in (A.50). In general the measurement of $\hat{L}$ is associated with fluctuations of the measured values. A measure for this fundamental quantum phenomenon is the mean-square deviation, which is defined as the expectation value of the operator $(\widehat{\Delta L})^2 \equiv (\hat{L} - \langle \psi | \hat{L} | \psi \rangle)^2$:

$$\overline{(l - \bar{l})^2} = \langle \psi | \psi \rangle^{-1} \langle \psi | (\widehat{\Delta L})^2 | \psi \rangle. \tag{A.53}$$

In the case of two arbitrary observables $\hat{L}_1$ and $\hat{L}_2$ there holds the inequality

$$\frac{\langle \psi | (\widehat{\Delta L_1})^2 | \psi \rangle}{\langle \psi | \psi \rangle} \cdot \frac{\langle \psi | (\widehat{\Delta L_2})^2 | \psi \rangle}{\langle \psi | \psi \rangle} \geq \langle \psi | \psi \rangle^{-2} \langle \psi | \frac{[\hat{L}_1, \hat{L}_2]}{2i} | \psi \rangle^2 \tag{A.54}$$

for an arbitrary state $|\psi\rangle$. Since the two factors on the left-hand side of (A.54) are measurable quantities (mean-square deviations), it becomes obvious that the commutator

$$\hat{K} \equiv [\hat{L}_1, \hat{L}_2] \tag{A.55}$$

must have physical relevance. If $\hat{K}$ vanishes, the operators $\hat{L}_1$ and $\hat{L}_2$ are called *compatible*. In this case both the operators possess a complete, orthonormal set of common eigenstates. To interpret this we have to discuss two measuring processes. Suppose that after a measurement of the observable $\hat{L}_1$ the measured value l_1 and the state $|l_1\rangle$ are obtained, and after a subsequent measurement of the observable $\hat{L}_2$ the measured value l_2 and the state $|l_2\rangle$ are obtained. If one is dealing with compatible operators, a repeated measurement of the observable $\hat{L}_1$ performed immediately after the second measurement certainly yields the measured value l_1 again. In this sense two compatible operators do not interfere with each other in the measurement. In the case of noncompatible observables, where the right-hand side of (A.54) does not vanish, there is no state $|\psi\rangle$ that is simultaneously an eigenstate of both the observables $\hat{L}_1$ and $\hat{L}_2$. In this case a measurement of $\hat{L}_1$ performed after that of $\hat{L}_2$ does not lead back to $|l_1\rangle$, as happened with compatible operators.

On measuring the mean squared deviations of $\hat{L}_1$ and $\hat{L}_2$ in an ensemble of systems in the same state $|\psi\rangle$, one finds that their product is greater or equal to $\langle\psi|\psi\rangle^{-2}\langle\psi|([\hat{L}_1, \hat{L}_2]/2i)|\psi\rangle^2$. For this reason the inequality (A.54) is called the *uncertainty relation*. The uncertainty (mean squared deviation) in the $\hat{L}_1$ measurement is related to the uncertainty in the $\hat{L}_2$ measurement. Compatible observables can be measured simultaneously without uncertainty; incompatible observables cannot. As a special but important case of noncompatibility, we identify $\hat{L}_1$ with $\hat{x}$ and $\hat{L}_2$ with $\hat{p}$; substituting (A.8) into the right-hand side of (A.54), the Heisenberg position–momentum uncertainty relation

$$\left\langle \left(\widehat{\Delta x}\right)^2 \right\rangle \left\langle \left(\widehat{\Delta p}\right)^2 \right\rangle \geq \frac{\hbar^2}{4} \tag{A.56}$$

is obtained.

The relation (A.51) reveals that physically relevant expectation values do not vary

if the vector $|\psi\rangle$ is replaced by $c|\psi\rangle$, where c is a nonvanishing complex number,

if all the operators $\hat{A}$ and all the vectors $|\psi\rangle$ are replaced by

$$\hat{A}' = \hat{U}\hat{A}\hat{U}^{-1}, \tag{A.57a}$$

$$|\psi'\rangle = \hat{U}|\psi\rangle, \tag{A.57b}$$

where $\hat{U}$ is a unitary operator with $\hat{U}^{-1} = \hat{U}^\dagger$. The unitary transformation (A.57a) applied to (A.55) leads to the statement that the commutation relations also have an invariant form with respect to a unitary transformation $\hat{U}$.

A.1.3. Construction of the Vector Space of a Physical System

Let us assume that all the dynamical variables of a physical (quantum) system are known, as well as all the relations between them. With this knowledge the Hilbert space belonging to the physical system can be constructed—in particular, the vector basis that spans the space. Let us select from the entire assemblage of observables a complete set of observables, all pairs of which are compatible; we denote them by $\hat{L}_1, \ldots, \hat{L}_j, \ldots, \hat{L}_g$. Here the expression "complete set" means: if in addition to these observables an observable $\hat{L}$ exists commuting with all the $\hat{L}_1, \ldots, \hat{L}_g$, then it is not independent; that means a function $\hat{L} = L(\hat{L}_1, \ldots, \hat{L}_g)$ must exist. By virtue of the assumption

$$\left[\hat{L}_j, \hat{L}_{j'}\right] = \hat{0} \qquad \text{for} \quad j, j' = 1, 2, \ldots, g, \tag{A.58}$$

all the observables $\hat{L}_j$ can be simultaneously determined precisely (cf. Section A.1.2). Therefore each set of eigenvalues $\{l_1, \dots, l_g\}$ defines an eigenket $|l_1, \dots, l_j, \dots, l_g\rangle$ in the Hilbert space $\mathscr{H}$ of the physical system. Varying each of the values $\{l_1, \dots, l_g\}$ over the entire range of the eigenvalues of $\{\hat{L}_1, \dots, \hat{L}_g\}$, the assemblage $\{|l_1, \dots, l_j, \dots, l_g\rangle\}$ represents a complete orthogonal set of basis vectors in $\mathscr{H}$. The orthogonality and normalization of these vectors are given by

$$\langle l_1, \dots, l_j, \dots, l_g | l_1', \dots, l_j', \dots, l_g' \rangle = \delta(l_1, l_1') \cdots \delta(l_j, l_j') \cdots \delta(l_g, l_g'), \tag{A.59}$$

where $\delta(l_j, l_j')$ must be identified with the Kronecker symbol in the case of discrete variables and with the delta function in the case of continuous variables. The completeness is expressed by

$$\hat{I} = \sum_{\dots, l_j, \dots} |l_1, \dots, l_j, \dots, l_g\rangle\langle l_1, \dots, l_j, \dots, l_g|, \tag{A.60}$$

where the sum must be replaced by the integral $\int dl_1 \cdots dl_j \cdots dl_g$ in the case of continuous variables. The quantity g is called the number of quantum-theoretical degrees of freedom.

The simultaneous measurement of all the observables $\hat{L}_1, \dots, \hat{L}_j, \dots, \hat{L}_g$ yields the maximum information about the physical quantum system. An arbitrary state $|\psi\rangle$ can be expanded in terms of the $|l_1, \dots, l_j, \dots, l_g\rangle$ by

$$|\psi\rangle = \sum_{\dots, l_j, \dots} \langle l_1, \dots, l_j, \dots, l_g | \psi\rangle | l_1, \dots, l_j, \dots, l_g\rangle. \tag{A.61}$$

Generally, the effect of an operator $\hat{G}$ (representing an arbitrary dynamical variable) on an arbitrary state $|\psi\rangle$ in $\mathscr{H}$ must be defined. This requires, by (A.61), the definition of the vector $\hat{G}|\dots, l_j, \dots\rangle$ in the Hilbert space.

A.1.4. The Temporal Behavior of a Physical System

In the preceding text theoretical relations have been formulated for a fixed time. Now we consider how the state and the dynamical variables change in time. The essential quantity for describing the time behavior is the Hamiltonian $\hat{H}$ of the system. This quantity can be time-independent (if a conservative system is under consideration) or time-dependent (if the system is influenced by external forces). If in a time interval (t_0, t) no disturbance caused by a measurement occurs, then the state vector $|\psi(t)\rangle$ for the system obeys the equation of motion

$$i\hbar \frac{d}{dt} |\psi(t)\rangle = \hat{H}(t) |\psi(t)\rangle. \tag{A.62}$$

By means of a unitary evolution operator $\hat{U}$ we can describe the state $|\psi(t)\rangle$ at a time $t \geq t_0$ with the help of the initial state $|\psi(t_0)\rangle$ by

$$|\psi(t)\rangle = \hat{U}(t, t_0)|\psi(t_0)\rangle, \tag{A.63}$$

where the evolution operator satisfies the equation

$$i\hbar \frac{d}{dt}\hat{U}(t, t_0) = \hat{H}(t)\hat{U}(t, t_0), \tag{A.64}$$

with $\hat{U}(t_0, t_0) = \hat{I}$.

By means of successive approximation one obtains from (A.64) the solution

$$\hat{U}(t, t_0) = \hat{I} + \sum_{n=1} (i\hbar)^{-n} \int_{t_0}^{t} dt_1 \int_{t_0}^{t_1} dt_2 \cdots \int_{t_0}^{t_{n-1}} dt_n\, \hat{H}(t_1) \cdots \hat{H}(t_n) \tag{A.65}$$

in the form of a series. Since the Hamiltonian operators $\hat{H}(t_j)$ need not commute at different times, it is necessary to pay attention to the ordering of the operator product. If the Hamiltonian $\hat{H}$ is independent of time, the expression (A.65) reduces to

$$\begin{aligned} \hat{U}(t, t_0) &= \hat{I} + \sum_{n=1} (i\hbar)^{-n}(n!)^{-1}(t - t_0)^n \hat{H}^n \\ &\equiv \exp\left[-\frac{i}{\hbar}(t - t_0)\hat{H}\right]. \end{aligned} \tag{A.66}$$

Altogether, we can conclude that the state vector $|\psi(t)\rangle$, according to (A.63), evolves in an exactly predictable manner, provided no disturbance caused by a measurement process occurs.

Up to this point the description of the mathematical representatives of the states and dynamical variables has been given in the so-called *Schrödinger picture*. This type of description is characterized by the following features: The dynamical time dependence, involved in $\hat{U}(t, t_0)$, is carried by the state vector $|\psi(t)\rangle$, which follows from (A.63) and (A.64). However, the dynamical time dependence does not enter into the operators. In the operators only the so-called explicit time dependence may occur—if external forces act on the system. (It must be mentioned that the density operator to be discussed in Section A.1.5 differs from this behavior; it contains the dynamical time dependence.) If we want to stress that the quantities are in the Schrödinger picture, they will be indicated by the subscript S, which leads to the following state notation and operator notation:

$$|\psi\rangle_S = \hat{U}(t, t_0)|\psi(t_0)\rangle, \qquad \hat{A}_S. \tag{A.67}$$

The statement (A.57a, b) expresses that the physically relevant expectation values do not change if the operators and state vectors are subjected to a unitary transformation. This also holds for the unitary evolution operator of time development. If all state vectors $|\psi\rangle_S$ and operators $\hat{A}_S$ in the Schrödinger picture undergo the unitary transformation $\hat{U}^{-1}(t, t_0)$ defined by (A.64), we arrive at the operators and state vectors in the *Heisenberg picture* (subscript H):

$$|\psi\rangle_H = \hat{U}^{-1}(t, t_0)|\psi\rangle_S, \qquad \hat{A}_H = \hat{U}^{-1}(t, t_0)\hat{A}_S\hat{U}(t, t_0). \tag{A.68a}$$

The inverse transformation is given by

$$|\psi\rangle_S = \hat{U}(t, t_0)|\psi\rangle_H, \qquad \hat{A}_S = \hat{U}(t, t_0)\hat{A}_H\hat{U}^{-1}(t, t_0), \tag{A.68b}$$

where the Schrödinger and the Heisenberg picture are assumed to coincide at $t = t_0$; in particular, from a comparison of (A.68a) with (A.63) it follows that $|\psi\rangle_H = |\psi(t_0)\rangle_S$.

The commutation relation (A.55) has the same form in both pictures:

$$[\hat{A}_S, \hat{B}_S] = \hat{K}_S \leftrightarrow [\hat{A}_H, \hat{B}_H] = \hat{K}_H. \tag{A.69}$$

The transformation relations between the Schrödinger and the Heisenberg picture can be utilized to derive the equations of motion for the operators in the Heisenberg picture. One obtains

$$\frac{d}{dt}\hat{A}_H = \frac{1}{i\hbar}[\hat{A}_H, \hat{H}_H] + \hat{U}^{-1}\left(\frac{\partial}{\partial t}\hat{A}_S\right)\hat{U}. \tag{A.70}$$

The second term on the right-hand side vanishes if the operator $\hat{A}_S$ possesses no explicit time dependence. For a conservative system the Hamiltonian in the Heisenberg picture coincides with that in the Schrödinger picture, and it holds $\hat{H}_H = \hat{H}_S$.

Let us add some remarks on the picture notation in this book. For simplicity, time-independent operators without picture subscript are to be understood as being in the Schrödinger picture, whereas time-dependent operators without picture subscript are to be understood as being in the Heisenberg picture. If the character of the picture must be displayed explicitly, the subscripts S and H will be used.

A.1.5. The Density Operator

In Section A.1.1 it was postulated that the state vector $|\psi\rangle$ completely describes a physical system with respect to every obtainable result of measurement. In Section A.1.3 the procedure for gaining this maximum information connected with $|\psi\rangle$ was presented. However, one may actually lack knowledge of the physical system under consideration. This may result from "weak preparation" of the physical system under realistic experimental conditions; for instance, if such a great number of degrees of freedom exists that it is practically impossible to make as many measurements as required for a complete determination of the state $|\psi\rangle$ of the system. In the case of weak preparation we may use the concept of an ensemble of systems whose members are of the same type as the physical system under consideration, but differ with respect to their states $|\psi\rangle$. A probability distribution $p_{|\psi\rangle}$ over these states $|\psi\rangle$ expresses the incomplete knowledge of the considered physical system.

The quantities $p_{|\psi\rangle}$ have the properties of probabilities, namely

$$0 \leq p_{|\psi\rangle} \leq 1, \qquad \sum_{|\psi\rangle} p_{|\psi\rangle} = 1. \tag{A.71}$$

In the following we shall assume that the state vectors are normalized to unity. If an observable $\hat{L}$ is measured, the ensemble average of the expectation values $\langle\psi|\hat{L}|\psi\rangle$ is given by

$$\overline{\overline{\langle\hat{L}\rangle}} = \sum_{|\psi\rangle} p_{|\psi\rangle}\langle\psi|\hat{L}|\psi\rangle. \tag{A.72}$$

Substituting the expression for the expectation value as given in (A.52), one obtains

$$\overline{\overline{\langle\hat{L}\rangle}} = \sum_{|\psi\rangle} p_{|\psi\rangle}\sum_{l} l|\langle l|\psi\rangle|^2. \tag{A.72a}$$

This relation contains two averaging procedures: on the one hand, the familiar quantum-theoretical one described by the probabilities $|\langle l|\psi\rangle|^2$, and on the other hand, the procedure originating from incomplete knowledge of the physical system described by the probabilities $p_{|\psi\rangle}$. Making use of the relation (A.39) we may write the ensemble mean value in the form

$$\overline{\overline{\langle\hat{L}\rangle}} = \mathrm{Tr}\left\{\left[\sum_{|\psi\rangle} p_{|\psi\rangle}|\psi\rangle\langle\psi|\right]\hat{L}\right\}. \tag{A.72b}$$

Thus the ensemble mean value of the measured values depends on the observable $\hat{L}$ as well as on the operator in brackets, namely

$$\hat{\rho} = \sum_{|\psi\rangle} p_{|\psi\rangle}|\psi\rangle\langle\psi|, \tag{A.73}$$

which represents the physical system. This operator is called the *density operator*. It has the properties

$$\hat{\rho} = \hat{\rho}^{\dagger}, \tag{A.74a}$$

$$\mathrm{Tr}\{\hat{\rho}\} = 1. \tag{A.74b}$$

As it is customary, we denote the ensemble mean value $\overline{\overline{\langle \hat{L} \rangle}}$ by $\langle \hat{L} \rangle$ in the following; finally we obtain

$$\langle \hat{L} \rangle = \mathrm{Tr}\{\hat{\rho}\hat{L}\}. \tag{A.75}$$

If at least two nonvanishing values $p_{|\psi\rangle}$ exist, the system is said to be in a *mixed state* which is an incoherent mixture of pure states. If $p_{|\psi\rangle} = 1$ for a fixed state $|\psi\rangle$, the system is described by $\hat{\rho} = |\psi\rangle\langle\psi|$ and exhibits all properties of a pure state.

The projection operator $|l\rangle\langle l|$ can be regarded as a special observable $\hat{L}$; its mean value

$$\mathrm{Tr}\{\hat{\rho}|l\rangle\langle l|\} = \langle l|\hat{\rho}|l\rangle \tag{A.76}$$

represents the probability of measuring the value l.

The temporal behavior of the density operator is described by the differential equation

$$i\hbar\frac{d}{dt}\hat{\rho}_S(t) = \left[\hat{H}_S(t), \hat{\rho}_S(t)\right]. \tag{A.77}$$

This equation takes the place of the relation (A.62) for the pure state $|\psi(t)\rangle$. The operator $\hat{\rho}_S(t)$ in (A.77) has to be understood as an operator in the Schrödinger picture. Unlike the representatives of dynamical variables described in Section A.1.4, this operator carries the dynamical time dependence. By successive approximation, from (A.77) the solution

$$\hat{\rho}_S(t) = \hat{\rho}_S(t_0) + \sum_{n=1} (i\hbar)^{-n}\int_{t_0}^{t} dt_1 \cdots \int_{t_0}^{t_{n-1}} dt_n \times\left[\hat{H}_S(t_1), \left[\cdots, \left[\hat{H}_S(t_n), \hat{\rho}_S(t_0)\right]\cdots\right]\right] \tag{A.78}$$

can be obtained.

For application, the quantity $\hat{\rho}_S(t)$ is in general evaluated from $\hat{\rho}_S(t_0)$ with the help of (A.77), (A.78). The initial value $\hat{\rho}_S(t_0)$ is determined by the special experimental conditions. In the case of equilibrium, $\hat{\rho}_S(t_0)$ can be determined by the requirement for maximum entropy, where a certain knowledge of the system may be expressed by a number of expectation values.

In correspondence with the classical interpretation, the relation between the entropy σ per single system and the density operator $\hat{\rho}$ is given by

$$\sigma = -k_B\mathrm{Tr}\{\hat{\rho}\ln\hat{\rho}\}. \tag{A.79}$$

The m expectation values

$$\langle \hat{L}_1 \rangle = \mathrm{Tr}\{ \hat{\rho} \hat{L}_1 \}, \ldots \ldots, \langle \hat{L}_j \rangle = \mathrm{Tr}\{ \hat{\rho} \hat{L}_j \}, \ldots \ldots, \langle \hat{L}_m \rangle = \mathrm{Tr}\{ \hat{\rho} \hat{L}_m \} \tag{A.80}$$

may express the knowledge of the system gained from real (or imaginary) experiments. If σ takes its maximum value under the secondary conditions (A.80) and the general condition (A.74b), one obtains, by means of the variational calculus,

$$\hat{\rho} = \frac{\exp\left[-\lambda_1 \hat{L}_1 - \cdots - \lambda_j \hat{L}_j - \cdots - \lambda_m \hat{L}_m\right]}{\mathrm{Tr}\left\{\exp\left[-\lambda_1 \hat{L}_1 - \cdots - \lambda_j \hat{L}_j - \cdots - \lambda_m \hat{L}_m\right]\right\}}, \tag{A.81}$$

where the λ_j are the Lagrange multipliers, which have to be determined from the relations (A.80).

A.1.6. Aspects of Quantum Field Theory

To facilitate the quantum-theoretical treatment of fields, we start by discussing some aspects of the *classical* field theory.

First we deal with the simple case of a real, one-component field $\psi(r_\bullet, t)$ depending on the position vector $r_\bullet$ and on the time t. The field-theoretical features are to some extent analogous to those of point mechanics, and therefore we may refer to this well-known formalism. This can be done by regarding the continuous field system, in a first step, as a discrete system of point mechanics whose basic equations yield the basic equations for the field by a subsequent limiting process. First the entire space is subdivided into cells, the volume of the jth cell being ΔV_j around the position vector $r_{j\bullet}$. The coordinate $q_j(t)$ of a point-mechanical system, where j denotes the jth degree of freedom, is assigned to the mean value ψ_j of the field ψ in the cell j; an analogous connection is assumed for the time derivative of q_j and $\partial\psi/\partial t$. Thus we have

$$q_j(t) \to \psi_j(t) \equiv \frac{1}{\Delta V_j} \int_{\Delta V_j} d^3 r_\bullet \, \psi(r_\bullet, t),$$

$$\dot{q}_j(t) \to \dot{\psi}_j(t) \equiv \frac{1}{\Delta V_j} \int_{\Delta V_j} d^3 r_\bullet \, \frac{\partial}{\partial t} \psi(r_\bullet, t). \tag{A.82}$$

The Lagrangian function L belonging to the field can be represented with the help of average values $\tilde{L}_j$ assigned to the individual cells:

$$L = \sum_j \Delta V_j \tilde{L}_j. \tag{A.83}$$

$\tilde{L}_j$ depends on ψ_j, $\dot{\psi}_j$ and on the differences between ψ_j and the corresponding mean field values in adjacent cells. These differences (divided by the cell length) describe the coupling between the cells. In the limiting case of infinitesimally small ΔV_j, the field $\psi(r_\bullet, t)$ can be thought of as a point-mechanical system with a noncountable number of degrees of freedom. Then the Lagrangian function L of the field is given by

$$L = \int d^3 r_\bullet \, \tilde{L}, \tag{A.84}$$

where the Lagrangian density

$$\tilde{L} = \tilde{L}\left(\psi, \frac{\partial \psi}{\partial t}, \frac{\partial \psi}{\partial x_1}, \frac{\partial \psi}{\partial x_2}, \frac{\partial \psi}{\partial x_3}\right), \tag{A.84a}$$

with $x_1 = x$, $x_2 = y$, $x_3 = z$, depends on $\psi(r_\bullet, t)$ and on its temporal and spatial derivatives. Here external influences are not taken into account; therefore explicit time dependence in L can be omitted. The Lagrangian equation of point mechanics, $(d/dt)\,\partial L/\partial \dot{q}_j - \partial L/\partial q_j = 0$, leads to the equation of motion

$$\left\{\frac{\partial}{\partial t} \frac{\partial \tilde{L}}{\partial(\partial \psi/\partial t)}\right\} - \left\{\frac{\partial \tilde{L}}{\partial \psi} - \sum_{l=1} \frac{\partial}{\partial x_l} \frac{\partial \tilde{L}}{\partial(\partial \psi/\partial x_l)}\right\} = 0 \tag{A.85}$$

for the field, where the first and second terms in (A.85) correspond to $(d/dt)\,\partial L/\partial \dot{q}_j$ and $-\partial L/\partial q_j$, respectively. The canonical formalism of point mechanics uses, besides the coordinate q_j, the canonically conjugate momentum $p_j = \partial L/\partial \dot{q}_j$. Analogously, the canonically conjugate field $\Pi(r_\bullet, t)$ is formed by

$$\Pi(r_\bullet, t) = \frac{\partial \tilde{L}}{\partial(\partial \psi/\partial t)}. \tag{A.86}$$

Both the quantities $\psi(r_\bullet, t)$ and $\Pi(r_\bullet, t)$ determine the dynamical state of the field. The Hamiltonian density $\tilde{H}$ is given by

$$\tilde{H} = \Pi \frac{\partial \psi}{\partial t} - \tilde{L}, \tag{A.87}$$

the Hamiltonian function by

$$H = \int d^3 r_\bullet \, \tilde{H}. \tag{A.88}$$

Let us now sketch the more general field theory for a complex field function $\psi(r_\bullet, t)$, where $\psi \neq \psi^*$. A separation into the real and the imaginary part can be avoided if ψ and ψ^* are regarded as independent quantities. In this case the Lagrangian density contains ψ and ψ^* as well as the temporal and spatial derivatives of both these field quantities.

Our field-theoretical considerations will particularly be applied to the electromagnetic radiation field. Therefore, let us pass from the one-component field to multicomponent fields. This can be achieved by an extension of the results just discussed. Let the field with the components $\psi_1, \psi_2, \ldots$ be denoted by $\{\psi_A(r_\bullet, t)\}$, where ψ_A is the Ath field component. The Lagrangian density $\tilde{L}$ now depends on the field components $\psi_1, \psi_2, \ldots$ and on their temporal and spatial derivatives. The equation of motion for ψ_A agrees with (A.85), where everywhere in this equation ψ must be replaced by ψ_A. For each component ψ_A, the canonically conjugate field $\Pi_A(r, t)$ is defined by

$$\Pi_A(r_\bullet, t) = \frac{\partial \tilde{L}}{\partial(\partial\psi_A/\partial t)}. \tag{A.89}$$

The Hamiltonian density is given by

$$\tilde{H} = \sum_A \Pi_A \frac{\partial \psi_A}{\partial t} - \tilde{L}. \tag{A.90}$$

According to E. Noether's theorem, each continuous symmetry transformation is associated with a conservation law for a certain physical quantity. Because of the homogeneity of the four-dimensional Minkowski space, two conserved quantities are the total energy of the field

$$H = \int d^3r_\bullet \left[\left(\sum_A \Pi_A \frac{\partial \psi_A}{\partial t}\right) - \tilde{L}\right], \tag{A.91}$$

and the total linear momentum of the field, whose lth vector component is given by

$$G_l = (-1)\int d^3r_\bullet \sum_A \Pi_A \frac{\partial \psi_A}{\partial x_l}. \tag{A.92}$$

Because of the isotropy of the Minkowski space, another conserved quantity is the total angular momentum of the field with the vector components J_{12}, J_{23}, J_{31}, where the vector component $J_{ll'}$ is given by

$$J_{ll'} = (-1)\int d^3r_\bullet \sum_A \Pi_A \left(x_l \frac{\partial}{\partial x_{l'}} - x_{l'} \frac{\partial}{\partial x_l} \right)\psi_A$$

$$+(-1)\int d^3r_\bullet \sum_{A,\,B} \Pi_A(\delta_{lA}\delta_{l'B} - \delta_{lB}\delta_{l'A})\psi_B. \tag{A.93}$$

Obviously the first term on the right-hand side is dependent on the point of reference (origin) and must be regarded as the *orbital angular momentum*, whereas the second term (here formulated for a vector field) is independent of the point of origin and must be regarded as the *intrinsic part of the total angular momentum*.

Let us now proceed to the *quantization of the field*. To this end, the component $\psi_A(r_\bullet, t)$ of the classical field is replaced by the operator $\hat{\psi}_A(r_\bullet, t)$. Thus also the Lagrangian density becomes an operator $\hat{\tilde{L}}$, which depends on the components $\hat{\psi}_A(r_\bullet, t)$ and their temporal and spatial derivatives. The operator $\hat{\Pi}_A(r_\bullet, t)$ of the canonically conjugate field is formed by differentiating $\hat{\tilde{L}}$ with respect to the time derivative of $\hat{\psi}_A$ according to (A.89). In correspondence to the relations (A.90) and (A.91), the operator $\hat{\tilde{H}}$ of the Hamiltonian density and the Hamiltonian operator $\hat{H}$ are formed by means of $\hat{\psi}_A(r_\bullet, t)$ and $\hat{\Pi}_A(r_\bullet, t)$. The operators $\hat{\psi}_A(r_\bullet, t)$ and $\hat{\Pi}_A(r_\bullet, t)$ as well as those constructed from them have to be regarded as operators in the Heisenberg picture. External influences are not taken into account; therefore the given time dependence has to be regarded as merely a dynamical one; consequently, the partial derivative $\partial/\partial t$ here means the derivative with respect to the dynamical time dependence. To make particular reference to the Heisenberg picture by means of the subscript H is unnecessary in this case.

From the field operators $\hat{\psi}_A$ and $\hat{\Pi}_A$ general variables $\hat{F}$ can be constructed. This for example applies to the total energy, the linear momentum, and the total angular momentum of the field; here the quantities ψ_A and Π_A in the relations (A.91), (A.92), and (A.93) must be replaced by the operators $\hat{\psi}_A$ and $\hat{\Pi}_A$, whose ordering must be preserved.

The temporal evolution of a general variable $\hat{F}$, which is supposed not to depend on t explicitly, is given, according to (A.70), by

$$\frac{d}{dt}\hat{F}(t) = \frac{1}{i\hbar}[\hat{F}(t), \hat{H}(t)]. \tag{A.94}$$

As a special case this relation also provides the equation of motion for $\hat{\psi}_A(r_\bullet, t)$ and $\hat{\Pi}_A(r_\bullet, t)$, if $\hat{F}$ is identified with $\hat{\psi}_A$ or $\hat{\Pi}_A$.

The *quantization relations* for the field and its canonically conjugated field prove to be of crucial importance for the properties of the quantum system. They can be formulated with the help of the commutation relations

$$[\hat{\psi}_A(r'_\bullet, t), \hat{\Pi}_{A'}(r''_\bullet, t)] = i\hbar\,\delta_{AA'}\,\delta^{(3)}(r'_\bullet - r''_\bullet)\hat{I},$$
$$[\hat{\psi}_A(r'_\bullet t), \hat{\psi}_{A'}(r''_\bullet, t)] = \hat{0}, \qquad [\hat{\Pi}_A(r'_\bullet, t), \hat{\Pi}_{A'}(r''_\bullet, t)] = \hat{0}, \tag{A.95}$$

or with the help of the anticommutation relations

$$[\hat{\psi}_A(r'_\bullet, t), \hat{\Pi}_{A'}(r''_\bullet, t)]_+ = i\hbar\,\delta_{AA''}\,\delta^{(3)}(r'_\bullet - r''_\bullet)\hat{I},$$
$$[\hat{\psi}_A(r'_\bullet, t), \hat{\psi}_{A'}(r'_\bullet, t)]_+ = \hat{0}, \qquad [\hat{\Pi}_A(r'_\bullet, t), \hat{\Pi}_{A'}(r''_\bullet, t)]_+ = \hat{0}. \tag{A.96}$$

The commutator $[\cdots, \cdots]$ and the anticommutator $[\cdots, \cdots]_+$ are defined by (A.7a, b); $\delta^{(3)}(r'_\bullet - r''_\bullet)$ is the ordinary three-dimensional delta function $\delta(x' - x'')\,\delta(y' - y'')\,\delta(z' - z'')$. All quantities in the relations above apply at the same time.

The type of quantization relations chosen is connected with the corpuscular interpretation of the quantized field: commutation relations of the type (A.95) have to be used in the case of boson systems, anticommutation relations of the type (A.96) have to be used in the case of fermion systems. It should be noted that the first relation in (A.95) reveals an analogy with a relation of quantum (point) mechanics: The point-mechanical operators $\hat{q}_{j'}(t)$ and $\hat{p}_{j''}(t)$ obey the commutation relation

$$\left[\hat{q}_{j'}(t), \hat{p}_{j''}(t)\right] = i\hbar\, \delta_{j'j''} \hat{I}. \tag{A.97}$$

This equation can easily be related to the quantum-field-theoretical equation

$$\left[\hat{\psi}(r'_\bullet, t), \hat{\Pi}(r''_\bullet, t)\right] = i\hbar\, \delta^{(3)}(r'_\bullet - r''_\bullet)\hat{I} \tag{A.98}$$

for a real, one-component field, since $\hat{\psi}(r'_\bullet, t)$ corresponds to $\hat{q}_{j'}(t)$ and $\hat{\Pi}(r''_\bullet, t)$ corresponds to $\hat{p}_{j''}(t)$, as shown above; instead of the Kronecker symbol, the three-dimensional delta function must be used now.

Any field component $\psi(r_\bullet, t)$ of a classical field can be expanded in terms of a complete, orthonormal set of functions $\{v_\mu(r_\bullet, t)\}$, which in general facilitates the practical performance of calculations and their physical interpretability. Thus we have the expansion

$$\psi(r_\bullet, t) = \sum_\mu \left[a_\mu v_\mu(r_\bullet, t) + b^*_\mu v^*_\mu(r_\bullet, t)\right]. \tag{A.99}$$

The quantities a_μ and b_μ represent the expansion amplitudes for the field, which is in general complex ($b^*_\mu = a^*_\mu$ has to be written if the field is real). Under certain circumstances the discrete variable μ must be replaced by a continuous one and the sum by an integral over μ.

The decision which system of functions $\{v_\mu\}$ can advantageously be used depends on the symmetry and boundary conditions to be satisfied by the field. To illustrate this, let us give two examples. Firstly, in the case of spatial periodic conditions the use of a Fourier series over plane waves $v_\mu(r_\bullet, t) = \exp[i(k_{\mu\bullet} r_\bullet - \omega_\mu t)]$ is of advantage. Secondly, the assemblage of eigenfunctions of the one-particle Schrödinger equation represents a complete orthonormal system, the boundary conditions being given by the potential function contained in the Hamiltonian; one has

$$v_\mu(r_\bullet, t) = u_\mu(r_\bullet)\exp\left[-\frac{i}{\hbar}\mathscr{E}_\mu t\right], \tag{A.100}$$

where $u_\mu(r_\bullet)$ satisfies the time-independent Schrödinger equation in the position representation, that is,

$$H_{\text{Schröd}}\left(r_\bullet, \frac{\hbar}{i}\nabla_\bullet\right)u_\mu(r_\bullet) = \mathscr{E}_\mu u_\mu(r_\bullet). \tag{A.101}$$

Let us revert to the general case of the expansion (A.99). It is possible to transfer into the expansion amplitudes a_μ and b_μ those parts from $v_\mu(r_\bullet, t)$ that carry the time dependence. So the expansion

$$\psi(r_\bullet, t) = \sum_\mu\left[a_\mu(t)u_\mu(r_\bullet) + b_\mu^*(t)u_\mu^*(r_\bullet)\right] \tag{A.102}$$

is obtained, where the functions $u_\mu(r_\bullet)$ are time-independent functions.

In the quantization of the field $\psi(r_\bullet, t)$—that is, the transition from $\psi(r_\bullet, t)$ to $\hat{\psi}(r_\bullet, t)$—the operator character on the right-hand side of (A.102) is assumed by the quantities $a_\mu(t)$ and $b_\mu(t)$; thus one has

$$\hat{\psi}(r_\bullet, t) = \sum_\mu\left[\hat{a}_\mu(t)u_\mu(r_\bullet) + \hat{b}_\mu^\dagger(t)u_\mu^*(r_\bullet)\right]. \tag{A.103}$$

The operators $\hat{a}_\mu(t)$ and $\hat{b}_\mu(t)$ are, like the operator $\hat{\psi}(r, t)$, operators in the Heisenberg picture. An expansion analogous to that for $\hat{\psi}(r_\bullet, t)$ can also be performed for the canonically conjugate field $\hat{\Pi}(r_\bullet, t)$. The operators $\hat{a}_\mu(t)$ can be interpreted as annihilation operators of particles whose properties are determined by $u_\mu(r_\bullet)$. Analogous statements apply to the operators $\hat{b}_\mu(t)$, but they refer to a different type of particles (the antiparticles). From the commutation relations (A.95) or the anticommutation relations (A.96) for the fields $\hat{\psi}(r_\bullet, t)$ and $\hat{\Pi}(r_\bullet, t)$, conclusions can be drawn for the operators $\hat{a}_\mu(t)$, $\hat{a}_\mu^\dagger(t)$, $\hat{b}_\mu(t)$, $\hat{b}_\mu^\dagger(t)$. In the former case these operators satisfy the commutation relations (A.9a) for boson creation and annihilation operators; in the latter case they satisfy the anticommutation relations (A.9b) for fermion creation and annihilation operators.

Sometimes field quantization is called "second quantization." This term is based on the example of the quantization of the Schrödinger field mentioned above. The wave function $\psi(r_\bullet, t)$ is the solution of the Schrödinger equation

$$i\hbar\frac{\partial\psi(r_\bullet, t)}{\partial t} = H_{\text{Schröd}}\left(r_\bullet, \frac{\hbar}{i}\nabla_\bullet\right)\psi(r_\bullet, t), \tag{A.104}$$

whose establishment with the help of the commutation relations for $\hat{r}_\bullet$ and $\hat{p}_\bullet$ [in the position representation $r_\bullet$ and $(\hbar/i)\nabla_\bullet$] can be regarded as the first quantization; correspondingly the transition from $\psi(r_\bullet, t)$ to $\hat{\psi}(r_\bullet, t)$ may be called the second quantization.

A.2. TREATMENT OF BASIC PHYSICAL PROBLEMS

In this section we shall treat some general problems whose results are needed for the description of the radiation field and the interaction between radiation and matter.

A.2.1. The interaction picture

Let the Hamiltonian $\hat{H}$ of the total system under consideration be decomposed into an operator $\hat{H}^F$ representing the "free system" and an operator $\hat{H}^C$ representing the interaction (coupling), so that we have

$$\hat{H} = \hat{H}^F + \hat{H}^C. \tag{A.105}$$

The operator $\hat{H}^F$ is supposed to be time-independent, while the operator $\hat{H}^C$ may depend on time. The procedure we are going to discuss offers advantages for the treatment of important problems, namely those where the total system consists of subsystems with a certain coupling. In the case of two subsystems the Hamiltonian of the total system is

$$\hat{H} = \hat{H}^{F_1} + \hat{H}^{F_2} + \hat{H}^{(1-2)}, \tag{A.106}$$

where $\hat{H}^{F_1}$ and $\hat{H}^{F_2}$ are the energy operators of the subsystems if the coupling between them is neglected. $\hat{H}^{(1-2)}$ is the interaction operator; it must be identified with $\hat{H}^C$. The operator $\hat{H}^{F_1} + \hat{H}^{F_2}$ is identical with the energy operator $\hat{H}^F$ of the total free system. By means of the time-independent operator $\hat{H}^F$ from (A.105), the unitary operator

$$\hat{U}^F(t, t_0) = \hat{I}\exp\left[-\frac{i}{\hbar}(t - t_0)\hat{H}^F\right] \tag{A.107}$$

can be formed.

In Section A.1.2 it has been stated that physically relevant values and relations do not vary if all state vectors and operators are subjected to a unitary transformation. This equally holds for unitary time-development operators, as can be seen from the connection between the Schrödinger and Heisenberg pictures in (A.68). The effect of the unitary transformation $(\hat{U}^F)^{-1}$ defined by (A.107) on quantities in the Schrödinger picture leads to quantities in the interaction picture (with the subscript I):

$$|\psi\rangle_I = \left[\hat{U}^F(t, t_0)\right]^{-1}|\psi\rangle_S, \tag{A.108a}$$

$$\hat{A}_I = \left[\hat{U}^F(t, t_0)\right]^{-1}\hat{A}_S\hat{U}^F(t, t_0), \tag{A.108b}$$

where the interaction picture and the Schrödinger picture are assumed to coincide at $t = t_0$. These relations lead to the equation of motion

$$i\hbar \frac{d}{dt} |\psi(t)\rangle_I = \hat{H}_I^C(t) |\psi(t)\rangle_I \tag{A.109}$$

for the state vector in the interaction picture, if within the interval (t_0, t) no disturbance due to a measurement process occurs. Furthermore, the equation of motion

$$\frac{d}{dt}\hat{A}_I = \frac{1}{i\hbar}[\hat{A}_I, \hat{H}_I^F] + [\hat{U}^F]^{-1}\left(\frac{\partial}{\partial t}\hat{A}_S\right)\hat{U}^F \tag{A.110}$$

is obtained for the operator $\hat{A}_I$ in the interaction picture. The equation

$$i\hbar \frac{d}{dt}\hat{\rho}_I = \left[\hat{H}_I^C, \hat{\rho}_I\right] \tag{A.111}$$

governs the time dependence of the density operator $\hat{\rho}_I(t)$ of the system [cf. Section A.1.5].

The relation (A.108b) reveals that the matrix elements of $\hat{A}_I$ can be formed in a rather simple way by means of the energy eigenvectors of the free system $|\mathscr{E}_j^F\rangle$, which often can be regarded as known quantities. Taking into account (A.107), one obtains

$$\langle \mathscr{E}_j^F | \hat{A}_I(t) | \mathscr{E}_{j'}^F \rangle = \langle \mathscr{E}_j^F | \hat{A}_S | \mathscr{E}_{j'}^F \rangle \exp\left(\frac{i}{\hbar}(t - t_0)\left(\mathscr{E}_j^F - \mathscr{E}_{j'}^F\right)\right). \tag{A.112}$$

The equation (A.109) shows that the time dependence of the state vector in the interaction picture is not determined by the total Hamiltonian, but by the interaction Hamiltonian. The treatment of important problems often allows the assumption of a small interaction operator in the sense of the time-dependent perturbation theory, which will be sketched in Section A.2.2. Compared with the Schrödinger and Heisenberg pictures, the interaction picture is a so-called intermediate picture, where the dynamical time dependence is carried partly by the state vectors and partly by the operators of the dynamical variables.

In mathematical analogy to the approach taken in Section A.1.4, the above differential equations yield the following results:

$$|\psi(t)\rangle_I = \hat{U}^C(t, t_0) |\psi(t_0)\rangle_I, \tag{A.113}$$

where $\hat{U}^C$ obeys the differential equation

$$i\hbar \frac{d}{dt}\hat{U}^C = \hat{H}_I^C \hat{U}^C. \tag{A.114}$$

The solutions for $\hat{U}^C$, $|\psi\rangle_I$, and $\hat{\rho}_I$ are given by

$$\hat{U}^C(t, t_0) = \hat{I} + \sum_{n=1} (i\hbar)^{-n} \int_{t_0}^{t} dt_1 \cdots \int_{t_0}^{t_{n-1}} dt_n \, \hat{H}_I^C(t_1) \ldots \hat{H}_I^C(t_n), \quad \text{(A.114a)}$$

$$|\psi(t)\rangle_I = |\psi(t_0)\rangle + \sum_{n=1} (i\hbar)^{-n} \int_{t_0}^{t} dt_1 \cdots$$

$$\times \int_{t_0}^{t_{n-1}} dt_n \hat{H}_I^C(t_1) \ldots \hat{H}_I^C(t_n) |\psi(t_0)\rangle, \quad \text{(A.115)}$$

$$\hat{\rho}_I(t) = \hat{\rho}_I(t_0) + \sum_{n=1} (i\hbar)^{-n} \int_{t_0}^{t} dt_1 \cdots$$

$$\times \int_{t_0}^{t_{n-1}} dt_n \left[\hat{H}_I^C(t_1), \left[\cdots, \left[\hat{H}_I^C(t_n), \hat{\rho}_I(t_0) \right] \right] \cdots \right]. \quad \text{(A.116)}$$

A.2.2. Time-Dependent Perturbation Theory

In principle, the temporal behavior of the state vector $|\psi(t)\rangle_I$ is completely determined by the relation (A.115). For important realistic problems, however, the given infinite series cannot be added up to a compact expression. In this case one has to be satisfied with breaking off the summation after a finite number of terms. One speaks of time-dependent nth-order perturbation theory if the summation is carried out up to the term with the index n. For estimating the range of validity of this approximation calculus it is advantageous to know the relation between the terms with the indices $n + 1$ and n from (A.115). If the vectors and operators are represented by their matrix elements with the energy eigenvectors $|\mathscr{E}_j^F\rangle$ of the free system, a relation between c-numbers is obtained that can be estimated by using the means of normal analysis for series. Obviously, the ratio of the terms with the indices $n + 1$ and n has the order of magnitude

$$\tilde{s} = \left| \hbar^{-1}(t - t_0) \langle \mathscr{E}_j^F | \hat{H}^C | \mathscr{E}_{j'}^F \rangle \right|. \quad \text{(A.117)}$$

This means that in carrying out the time-dependent perturbation theory, only a few steps are needed if $\tilde{s} \ll 1$; hence the product of the interaction strength expressed by $\left| \langle \mathscr{E}_j^F | \hat{H}^C | \mathscr{E}_j^F \rangle \right|$ and the interaction duration $t - t_0$ must be much less than $\hbar$.

It should be noted that the terms of the sum on the right-hand side of (A.115) can be obtained by solving the following hierarchy of differential equations:

$$i\hbar\frac{d}{dt}|\psi^{(0)}(t)\rangle_I = 0$$

$$i\hbar\frac{d}{dt}|\psi^{(1)}(t)\rangle_I = \hat{H}_I^C(t)|\psi^{(0)}(t)\rangle_I,$$

$$\vdots$$

$$i\hbar\frac{d}{dt}|\psi^{(n)}(t)\rangle_I = \hat{H}_I^C(t)|\psi^{(n-1)}(t)\rangle_I. \tag{A.118}$$

The term $|\psi^{(n)}(t)\rangle_I$ in the sum (A.115) can be obtained by iteration beginning with the time-independent solution of the first differential equation in (A.118).

A.2.3. Transition Probabilities and Rates

Let the assumptions of the interaction picture apply to a certain system, and the operator $\hat{H}^C$ describing the interaction (coupling) be time-independent. Suppose that the system initially occupies one of the eigenstates $|\mathscr{E}_i^F\rangle$ of the free operator $\hat{H}^F$ at the time $t = 0$. Then the existence of an interaction operator causes the system to be in another state at a later time $t > 0$. This also causes the measurable quantities to vary with time. The probability of finding the system at time t in a final state $|\mathscr{E}_f^F\rangle$ (which is also assumed to be an eigenstate of $\hat{H}^F$) is

$$p_{i\to f}(t) = \left|\langle\mathscr{E}_f^F|\psi(t)\rangle_I\right|^2. \tag{A.119}$$

In the following we shall apply the time-dependent perturbation theory in first order. Thus we get approximately

$$p_{i\to f}(t) = \hbar^{-2}t^2\left|\langle\mathscr{E}_f^F|\hat{H}^C|\mathscr{E}_i^F\rangle\right|^2\operatorname{sinc}^2\left[\frac{\left(\mathscr{E}_f^F - \mathscr{E}_i^F\right)t}{2\hbar}\right]. \tag{A.120}$$

The transition probability is time-dependent and depends on the transition energy $\mathscr{E}_f^F - \mathscr{E}_i^F$ and the matrix element $\langle\mathscr{E}_f^F|\hat{H}^C|\mathscr{E}_i^F\rangle$. The maximum value of $p_{i\to f}$ is attained at $\mathscr{E}_f^F = \mathscr{E}_i^F$; nevertheless, in the case of quasiresonance (i.e., $\mathscr{E}_f^F \approx \mathscr{E}_i^F$), values of similar magnitude are reached. In the case of nonresonance (if $|\mathscr{E}_f^F - \mathscr{E}_i^F|t/2\hbar \gg \pi$) the probability $p_{i\to f}(t)$ takes on relatively small values.

Many applications require the consideration of the change of $p_{i \to f}$ over a finite time interval. Therefore, we consider the expression

$$\frac{\Delta p_{i \to f}}{\Delta t} = \frac{p_{i \to f}(t) - p_{i \to f}(0)}{t - 0} \tag{A.121}$$

Because $p_{i \to f}(0) = 0$, we arrive at

$$\frac{\Delta p_{i \to f}}{\Delta t} = \frac{2\pi}{\hbar^2} \left| \langle \mathscr{E}_f^F | \hat{H}^C | \mathscr{E}_i^F \rangle \right|^2 \frac{\left| \exp\left[i\hbar^{-1} \left(\mathscr{E}_f^F - \mathscr{E}_i^F \right) t \right] - 1 \right|^2}{2\pi t \left(\mathscr{E}_f^F - \mathscr{E}_i^F \right)^2 \hbar^{-2}}. \tag{A.122}$$

This expression will be discussed under the condition of sufficiently large t values; then

$$\frac{\Delta p_{i \to f}}{\Delta t} = \frac{2\pi}{\hbar^2} \left| \langle \mathscr{E}_f^F | \hat{H}^C | \mathscr{E}_i^F \rangle \right|^2 \delta\left[\hbar^{-1} \left(\mathscr{E}_f^F - \mathscr{E}_i^F \right) \right] \tag{A.123}$$

is obtained. This rate is time-independent.

A topic of much interest is the transition between a discrete initial state $|\mathscr{E}_i^F\rangle$ and a manifold of final states with nearly the same energy $\mathscr{E}_f^F$. Then the total transition probability

$$P(t) = \sum_{|\mathscr{E}_f^F\rangle} p_{i \to f}(t), \tag{A.124}$$

where $\mathscr{E}_f^F \approx \mathscr{E}_i^F$, must be considered. Making use of (A.120), the time-independent rate of the total transition probability is approximately given by

$$\frac{\Delta P}{\Delta t} = \frac{2\pi}{\hbar} \sigma\left(\mathscr{E}_f^F \right) \left| \langle \mathscr{E}_f^F | \hat{H}^C | \mathscr{E}_i^F \rangle \right|^2, \tag{A.125}$$

where $\sigma(\mathscr{E}_f^F)$ is the state density, that is the number of final states per unit energy. The relation (A.125) is called Fermi's "golden rule." It can be applied under the conditions given for (A.123) if, in addition, the functions $\sigma(\mathscr{E}_f^F)$ and $\langle \mathscr{E}_f^F | \hat{H}^C | \mathscr{E}_i^F \rangle$ are assumed to be sufficiently slowly varying functions in the $\mathscr{E}_f^F$ region. An analogous formula holds if the transition between a manifold of closely spaced initial states and a discrete final state is considered.

The range of validity of the formulas (A.123) and (A.125) is determined by the value of $\Delta t \equiv t - 0$. On the one hand Δt is not allowed to attain too large values because of the assumptions of the time-dependent perturbation theory of the first order. On the other hand the formulas (A.123) and (A.125) were derived under the assumption of Δt values exceeding the rise time of internal transient processes. In practice, the expressions (A.123) and (A.125) are often used as differential quotients in rate equations, where the rise time of the involved quantities (caused by external influences) is expected to exceed considerably the rise time of internal transient processes. This technique is called *coarse-graining in the time domain*.

A.2.4. Eigenvalue Problem of the Position Operator

The operators $\hat{x}$ and $\hat{p}$ representing the one-dimensional position of a particle and its momentum can be assumed to exhibit hermiticity and satisfy the commutation relations (A.8). Hence there hold the relations

$$\hat{x} = \hat{x}^{\dagger}, \qquad \hat{p} = \hat{p}^{\dagger}, \qquad [\hat{x}, \hat{p}] = i\hbar\hat{I},$$
$$[\hat{x}, \hat{x}] = \hat{0}, \qquad [\hat{p}, \hat{p}] = \hat{0}. \tag{A.126}$$

Since $\hat{x}$ represents an observable, we can assume the existence of at least one eigenvalue x with the eigenstate $|x\rangle$ satisfying the eigenvalue equation

$$\hat{x}|x\rangle = x|x\rangle. \tag{A.127}$$

Let us introduce the so-called translation operator $\hat{X}(s)$ by

$$\hat{X}(s) = \exp\left[-\frac{i}{\hbar}\hat{p}s\right], \tag{A.128}$$

where s is taken to be an arbitrary real number. Making use of (A.29), we arrive at

$$[\hat{x}, \hat{X}(s)] = s\hat{X}(s). \tag{A.129}$$

This expression leads to

$$\hat{x}(\hat{X}(s)|x\rangle) = (x + s)(\hat{X}(s)|x\rangle) \tag{A.130}$$

Therefore the vector $\hat{X}(s)|x\rangle$ must be an eigenvector of $\hat{x}$ with the eigenvalue $x + s$. Since s is an arbitrary real number, $\hat{x}$ has a continuous eigenvalue spectrum extending over the whole real x axis. An arbitrary eigenket $|x\rangle$ can be generated by the action of $\hat{X}(x)$ on $|x = 0\rangle$,

$$|x\rangle = \hat{X}(x)|x = 0\rangle. \tag{A.131}$$

From these results the components of the Hilbert-space vectors and Hilbert-space operators in terms of the eigenstates $|x\rangle$ can be derived. If $\hat{F} = F(\hat{x}, \hat{p})$ is an operator function of $\hat{x}$ and $\hat{p}$, then

$$\langle x'|F(\hat{x}, \hat{p})|x''\rangle = F\left(x', \frac{\hbar}{i}\frac{\partial}{\partial x'}\right)\langle x'|x''\rangle, \tag{A.132}$$

representation, which reads

$$\left(-\frac{\hbar^2}{2m}\frac{d^2}{dx^2}+\frac{m\omega^2}{2}x^2-\mathscr{E}_n\right)\langle x|\mathscr{E}_n\rangle = 0; \tag{A.145a}$$

here $\langle x|\mathscr{E}_n\rangle$ is the eigenket $|\mathscr{E}_n\rangle$ in the $\hat{x}$ representation.

In a similar way to the energy operator, other variables depending on $\hat{x}$ and $\hat{p}$ can be formed with the help of the operators $\hat{a}^\dagger$ and $\hat{a}$. For instance, the dipole-moment operator can be given in the form

$$\hat{d} = -e\sqrt{\frac{\hbar}{2m\omega}}\,(\hat{a}^\dagger + \hat{a}) \tag{A.146}$$

if the particle considered is an electron with charge $-e$.

Classically, the motion of a particle in a potential with harmonic x dependence, which is described by the Hamiltonian (A.135), is a harmonic oscillation with frequency ω. Quantum-theoretically, energy values with the equal differences $\hbar\omega$ result. This can be interpreted in the following way: When the quantum system occupies the state $|n\rangle$, then n vibrational quanta of the same kind exist; each of them has the vibrational energy $\hbar\omega$. The state $|n=0\rangle$ has no quantum and is called the vacuum state (it should be not mixed up with the zero vector). The states $\{|n\rangle\}$ form the occupation-number representation; $\hat{N}$ is the occupation-number operator. The relations given in (A.143) lead to the interpretation that by the action of $\hat{a}^\dagger$ or $\hat{a}$ one vibrational quantum is created or annihilated. Therefore $\hat{a}^\dagger$ is called the *creation operator* and $\hat{a}$ the *annihilation operator*. The zero-point energy $\hbar\omega/2$ exists independently of the number n.

The motion of the constituents (atoms, ions) of molecules and solids can be described by the Hamiltonian

$$\hat{H} = \sum_{j=1}\frac{\hat{p}_{j\bullet}^2}{2m_j} + U(\hat{r}_{1\bullet},\ldots,\hat{r}_{j\bullet},\ldots). \tag{A.147}$$

In some important cases the potential function U can be sufficiently well described by a harmonic approximation (an expansion up to the second order in the position coordinates). Then, by a transformation to the normal coordinates, the Hamiltonian operator can always be given the form

$$\hat{H} = \sum_{\alpha=1}^{3M}\hat{H}_\alpha, \qquad \text{where} \quad \hat{H}_\alpha = \hbar\omega_\alpha\left(\hat{a}_\alpha^\dagger\hat{a}_\alpha + \tfrac{1}{2}\hat{I}\right). \tag{A.148}$$

This expresses the superposition of noninteracting one-dimensional harmonic oscillators, the so-called vibrational modes. The frequencies ω_α are dependent on the M masses m_j and on the linear force constants contained in U. The operators $\hat{a}_\alpha^\dagger$ and $\hat{a}_\alpha$ again have the character of creation and annihilation operators. They satisfy the commutation relations

$$[\hat{a}_\alpha, \hat{a}_{\alpha'}^\dagger] = \delta_{\alpha\alpha'}\hat{I}, \qquad [\hat{a}_\alpha, \hat{a}_{\alpha'}] = \hat{0}, \qquad [\hat{a}_\alpha^\dagger, a_{\alpha'}^\dagger] = \hat{0}. \qquad \text{(A.149)}$$

From these relations it can be inferred that the occupation-number operator $\hat{N}_\alpha$ of the individual mode of vibration with index α has the eigenvalues $n_\alpha = 0, 1, 2, 3, \ldots$. This result, namely an arbitrary number of quanta in the same dynamical state with the energy $\hbar\omega_\alpha$, is a characteristic property of bosons.

A.2.5.2. Description on the Basis of Fermion Operators. There is another class of particles, the fermions, which in contrast to the bosons have the property that a given state can either be empty or be occupied by only one particle. Electrons, for instance, are fermions.

We first discuss a one-electron atom. Let the atom have the energy states $|\mathscr{E}_j\rangle$. Its Hamiltonian can then be represented by

$$\hat{H} = \sum_j \mathscr{E}_j \hat{N}_j, \qquad \text{(A.150)}$$

where $\mathscr{E}_j$ is the energy eigenvalue and $\hat{N}_j$ is the occupation-number operator of the jth state. The operator $\hat{N}_j$ can be constructed as

$$\hat{N}_j = \hat{b}_j^\dagger \hat{b}_j, \qquad \text{(A.151)}$$

where the operators $\hat{b}_j^\dagger$, $\hat{b}_j$ obey the anticommutation relations

$$[\hat{b}_j, \hat{b}_j^\dagger]_+ = \hat{I}, \qquad [\hat{b}_j, \hat{b}_j]_+ = \hat{0}, \qquad [\hat{b}_j^\dagger, \hat{b}_j^\dagger]_+ = \hat{0}. \qquad \text{(A.152)}$$

These relations directly imply that the operator $\hat{N}_j$ possesses only the eigenvalues

$$n_j = 0, 1 \qquad \text{(A.153)}$$

with the eigenstates $|0_j\rangle$ and $|1_j\rangle$. The action of $\hat{b}_j^\dagger$ and $\hat{b}_j$ on the kets $|0_j\rangle$, $|1_j\rangle$ yields

$$\hat{b}_j^\dagger|0_j\rangle = |1_j\rangle, \qquad \hat{b}_j^\dagger|1_j\rangle = |0_v\rangle,$$

$$\hat{b}_j|0_j\rangle = |0_v\rangle, \qquad \hat{b}_j|1_j\rangle = |0_j\rangle. \qquad \text{(A.154)}$$

The first and the fourth equation show that $\hat{b}_j^\dagger$ can be interpreted as a creation operator and $\hat{b}_j$ as an annihilation operator. The action of $\hat{b}_j^\dagger$ on $|1_j\rangle$ and that of $\hat{b}_j$ on $|0_j\rangle$ lead to the zero vector of the Hilbert space.

Let us now proceed to the system of states of the atom. The creation and annihilation operators which belong to different states with indices j and j' obey the anticommutation relations

$$\left[\hat{b}_j, \hat{b}_{j'}^\dagger\right]_+ = \delta_{jj'}\hat{I}, \ \left[\hat{b}_j, \hat{b}_{j'}\right]_+ = \hat{0}, \ \left[\hat{b}_j^\dagger, \hat{b}_{j'}^\dagger\right]_+ = \hat{0}. \tag{A.155}$$

A vacuum state $|\psi_0\rangle$ can be postulated that has the property

$$\hat{b}_j|\psi_0\rangle = |0_v\rangle \qquad \text{for an arbitrary index } j. \tag{A.156}$$

With its help an arbitrary atomic eigenstate can be expressed by

$$|1_j\rangle = \hat{b}_j^\dagger|\psi_0\rangle. \tag{A.157}$$

The notation $|1_j\rangle$ characterizes a state where the electron occupies the state $|\mathscr{E}_j\rangle$, while all the other states $|\mathscr{E}_{j'\neq j}\rangle$ are empty. The occupation-number operator $\hat{N}_j$ obeys the relation

$$\hat{N}_j|1_{j'}\rangle = \begin{cases} |0_v\rangle & \text{for} \quad j' \neq j, \\ |1_j\rangle & \text{for} \quad j' = j. \end{cases} \tag{A.158}$$

The summation of all the $\hat{N}_j$ gives the identity operator

$$\sum_j \hat{N}_j = \hat{I}. \tag{A.159}$$

Let us now consider operators of the form

$$\hat{B}_{kl} = \hat{b}_k^\dagger \hat{b}_l, \qquad \text{where} \quad k \neq l. \tag{A.160}$$

By virtue of (A.154) these operators obey the relation

$$\hat{B}_{kl}|1_j\rangle = \begin{cases} |0_v\rangle & \text{for} \quad l \neq j, \\ |1_k\rangle & \text{for} \quad l = j. \end{cases} \tag{A.161}$$

It may be said that the state $|1_l\rangle$ "flips" into the state $|1_k\rangle$ if the operators B_{kl} acts on it. Therefore the operators $\hat{B}_{kl}$ are called *flip operators*. Their matrix elements are given by

$$\langle 1_{j'}|\hat{B}_{kl}|1_j\rangle = \delta_{j'k}\delta_{lj}. \tag{A.162}$$

From this expression all properties of the operators $\hat{B}_{kl}$ can be derived; for instance, $\hat{B}_{kl} = \hat{B}_{lk}^\dagger$.

The statements above for the one-electron atom can be transferred to general atomic systems (multi-electronic systems). We start from a general atomic system with the orthonormalized energy eigenkets $|\mathscr{E}_j\rangle$, which need not necessarily be one-electron states. Corresponding to the energy state $|\mathscr{E}_j\rangle$ an occupation state $|1_j\rangle$ is introduced for each j, which means that precisely the state $|\mathscr{E}_j\rangle$ is occupied. The states $|1_j\rangle$ satisfy the orthonormality relation

$$\langle 1_j | 1_{j'} \rangle = \delta_{jj'}. \tag{A.163}$$

The occupation-number operators $\hat{B}_{kk}$ with the properties

$$\hat{B}_{kk}|1_j\rangle = \begin{Bmatrix} |0_v\rangle & \text{for} \quad k \neq j \\ |1_k\rangle & \text{for} \quad k = j \end{Bmatrix}, \qquad \sum_k \hat{B}_{kk} = \hat{I} \tag{A.164}$$

are introduced, by which the Hamiltonian can be represented in the form

$$\hat{H} = \sum_k \mathscr{E}_k \hat{B}_{kk}. \tag{A.165}$$

Moreover, the flip operators $\hat{B}_{kl}$ with $k \neq l$ are introduced, satisfying the equations

$$\hat{B}_{kl}|1_j\rangle = \begin{Bmatrix} |0_v\rangle & \text{for} \quad l \neq j \\ |1_k\rangle & \text{for} \quad l = j \end{Bmatrix}, \qquad \langle 1_{j'} | \hat{B}_{kl} | 1_j \rangle = \delta_{j'k}\delta_{lj}. \tag{A.166}$$

According to (A.70) the equation of motion for $\hat{B}_{kl}$ is given by

$$i\hbar \frac{d}{dt}(\hat{B}_{kl})_H = \left[(\hat{B}_{kl})_H, \hat{H}_H\right]. \tag{A.167}$$

By (A.165) the right-hand side equals $(\mathscr{E}_l - \mathscr{E}_k)(\hat{B}_{kl})_H$. Thus (A.167) leads to

$$(\hat{B}_{kl}(t))_H = (\hat{B}_{kl}(t_0))_H \exp\left[i\hbar^{-1}(\mathscr{E}_k - \mathscr{E}_l)(t - t_0)\right]. \tag{A.168}$$

Since an arbitrary operator $\hat{G}$ of the atomic system can be represented by means of the operators $\hat{B}_{kk}$ and $\hat{B}_{kl}$ according to

$$\hat{G} = \sum_{jj'} \langle \mathscr{E}_j | \hat{G} | \mathscr{E}_{j'} \rangle \hat{B}_{jj'}, \tag{A.169}$$

and since the properties of $\hat{B}_{kk}$ and $\hat{B}_{kl}$ are completely determined by (A.164), (A.166), (A.168), the origin of the $\hat{B}_{kk}$ and $\hat{B}_{kl}$ in creation and annihilation operators can mostly be ignored in the practical treatment of problems.

A special, but important case is the treatment of a *two-level system*. Its Hamiltonian is

$$\hat{H} = \mathscr{E}_1 \hat{B}_{11} + \mathscr{E}_2 \hat{B}_{22}, \tag{A.170}$$

where $\mathscr{E}_2 > \mathscr{E}_1$. The flip operator $\hat{B}_{12}$ will be denoted by $\hat{B}$. Thus we have

$$\hat{B}_{11} = \hat{B}\hat{B}^\dagger, \quad \hat{B}_{22} = \hat{B}^\dagger\hat{B}, \quad \hat{B}_{12} = \hat{B}, \quad \hat{B}_{21} = \hat{B}^\dagger. \tag{A.171}$$

The anticommutation relations

$$[\hat{B}, \hat{B}^\dagger]_+ = \hat{I}, \qquad [\hat{B}, \hat{B}]_+ = \hat{0}, \qquad [\hat{B}^\dagger, \hat{B}^\dagger]_+ = \hat{0} \tag{A.172}$$

hold. The Hamiltonian is

$$\hat{H} = (\mathscr{E}_2 - \mathscr{E}_1)\hat{B}^\dagger\hat{B} + \mathscr{E}_1\hat{I}. \tag{A.173}$$

By appropriate fixing of the zero-point energy the term $\mathscr{E}_1\hat{I}$ can be omitted and the Hamiltonian takes the shortened form

$$\hat{H} = (\mathscr{E}_2 - \mathscr{E}_1)\hat{B}^\dagger\hat{B}. \tag{A.173a}$$

The dipole operator that characterizes the transition between the two levels is

$$\hat{d} = d_{21}\hat{B}^\dagger + d_{12}\hat{B} \qquad \text{with} \quad d_{jj'} = \langle\mathscr{E}_j|\hat{d}|\mathscr{E}_{j'}\rangle. \tag{A.174}$$

Sometimes it is of advantage to describe the two-level system with the help of operators that obey similar quantization relations to those obeyed by the spin operators that are assigned to the vector components of the spin momentum of a particle (or to linear combinations of them). This description is called the *energy spin model*. Energy spin operators $\hat{s}^+, \hat{s}^-, \hat{s}_z$ are introduced, which are connected with the operators $\hat{B}$ and $\hat{B}^\dagger$ by

$$\hat{s}^+ \equiv \hat{B}_{21} = \hat{B}^\dagger, \qquad \hat{s}^- \equiv \hat{B}_{12} = \hat{B},$$

$$2\hat{s}_z = \hat{B}_{22} - \hat{B}_{11} = \hat{B}^\dagger\hat{B} - \hat{B}\hat{B}^\dagger. \tag{A.175}$$

The first two equations refer to quantities by which the dipole moment is to be described, whereas the last one represents an operator that concerns the occupation-number inversion of the two states; the energy operator is essentially determined by the product $\hat{s}^+\hat{s}^-$. The quantization relations for the $\hat{s}$ operators can be easily obtained from the quantization relations for $\hat{B}$ and $\hat{B}^\dagger$. For example,

$$[\hat{s}^+, \hat{s}^-] = 2\hat{s}_z, \qquad [\hat{s}^-, \hat{s}^-]_+ = [\hat{s}^+, \hat{s}^+]_+ = \hat{0},$$

$$[\hat{s}_z, \hat{s}_z]_+ = \frac{\hat{I}}{2}. \tag{A.176}$$

Let us close the treatment of two-level systems by some remarks on their connection with *quantum chaos*. The recently increased understanding of classical Hamiltonian systems, which are nonintegrable and thus exhibit chaotic dynamical behavior, leads to the question of a natural extension to the quantum properties of these systems. This problem is termed quantum chaos. The systems of a two-level atom (flip operators $\hat{B}^\dagger$, $\hat{B}$; transition frequency ω_{21}) coupled to a single quantized electromagnetic mode (modal operators $\hat{a}^\dagger$, $\hat{a}$; frequency ω) may be regarded as a basic model. This system has the Hamiltonian operator

$$\hat{H} = \hbar\omega\hat{a}^\dagger\hat{a} + \hbar\omega_{21}\hat{B}^\dagger\hat{B} + \kappa(\hat{s}^+ + \hat{s}^-)(\hat{a} + \hat{a}^\dagger), \tag{a.177}$$

where the third term represents the (dipole) interaction. The classical limit of this model is nonintegrable; in this limit chaotic dynamical behavior occurs. The quantum calculation of energy levels and occupation probabilities of the lower atomic level as a function of ω/ω_{21} shows typical results for systems, which are weakly and strongly chaotic in the classical limit (cf. Ref. A.5).

REFERENCES

A.1 P. A. M. Dirac, *The Principles of Quantum Mechanics*, 4th ed., Clarendon, Oxford, 1958.

A.2 A. Messiah, *Quantum Mechanics*, Vols. I, II, North-Holland, Amsterdam, 1961.

A.3 M. Schubert and G. Weber, *Quantum Theory—Fundamentals, Methods, Applications*, Vols. I, II, VEB Deutscher Verlag der Wissenschaften, Berlin, 1980 (German).

A.4 C. Itzykson and C. B. Zuber, *Quantum Field Theory*, McGraw-Hill, New York, 1980.

A.5 R. Graham and M. Höhnerbach, *Phys. Lett.* **101A**, 61 (1984).

General References

1. S. A. Akhmanov, Ju. E. Djakov, and A. S. Chirkin, *Introduction to Statistical Radiophysics and Optics*, Nauka, Moscow, 1981 (Russian).
2. S. A. Akhmanov and R. V. Khokhlov, *Problems of Nonlinear Optics*, Moscow, 1965 (Russian).
3. L. Allen and J. H. Eberly, *Optical Resonance and Two-Level Atoms*, Wiley, New York, London, Sydney, Toronto, 1975.
4. P. A. Apanasevich, *Theory of the Interaction between Light and Matter*, Izd. Nauka i Tekhnika, Minsk, 1977 (Russian).
5. (a) F. T. Arecchi and E. O. Schulz-Dubois (Eds.), *Laser Handbook*, Vols. 1, 2, North Holland, Amsterdam, 1972; (b) M. L. Stitch (Ed.), *Laser Handbook*, Vol. 3, North Holland, Amsterdam, 1979.
6. G. C. Baldwin, *An Introduction to Nonlinear Optics*, Plenum, New York, 1969.
7. N. Bloembergen, *Nonlinear Optics*, Benjamin, New York, Amsterdam, 1965.
8. M. Born and E. Wolf, *Principles of Optics*, Pergamon, Oxford, 1970.
9. P. N. Butcher, *Nonlinear Optical Phenomena*, Ohio State Univ., Columbus, 1965.
10. W. Demtröder, *Laser Spectroscopy*, Springer, Berlin (West), Heidelberg, New York, 1981.
11. R. W. Ditchburn, *Light*, 3rd ed., Academic, London, 1976.
12. J. H. Eberly, Ed., *Multiphoton Processes*, Wiley, New York, 1978.
13. M. S. Feld and V. S. Letokhov (Eds.), *Coherent Nonlinear Optics*, Springer, Berlin (West), Heidelberg, New York, 1980.
14. D. Cotter, D. C. Hanna, and M. A. Yuratich, *Nonlinear Optics of Free Atoms and Molecules*, Springer Series in Optical Sciences, Vol. 17, Springer, Berlin (West), Heidelberg, New York, 1979.
15. H. G. Heard (Ed.), *Laser Parameter Measurements Handbook*, Wiley, New York, 1968.
16. W. Heitler, *The Quantum Theory of Radiation*, Oxford University Press, New York, 1956.
17. S. Kielich, *Molecular Nonlinear Optics*, Nauka, Moscow, 1981 (Russian).
18. J. R. Klauder and E. C. G. Sudarshan, *Fundamentals of Quantum Optics*, Benjamin, New York, Amsterdam, 1968.
19. D. N. Klyshko, *Photons and Nonlinear Optics*, Nauka, Moscow, 1980 (Russian).
20. W. Kleen and R. Müller, *Laser*, Springer, Berlin (West), Heidelberg, New York, 1969.
21. V. S. Letokhov, *Laser Spectroscopy*, Akademie, Berlin, 1977 (German).
22. V. P. Chebotayev and V. S. Letokhov, *Nonlinear Laser Spectroscopy*, Springer Series in Optical Sciences, Vol. 4, Springer, Berlin (West), Heidelberg, New York, 1977.
23. A. K. Levine (Ed.), *Lasers. A Series of Advances*, Vols. 1–4, Dekker, New York, 1966–1971.
24. W. H. Louisell, *Quantum Statistical Properties of Radiation*, Wiley, New York, 1973.
25. W. H. Louisell, *Radiation and Noise in Quantum Electronics*, McGraw-Hill, New York, San Francisco, Toronto, London, 1964.

26. R. Loudon, *The Quantum Theory of Light*, Clarendon, Oxford, 1973.

27. M. Dunn and A. Maitland, *Laser Physics*, North-Holland, Amsterdam, London, 1969.

28. B. Mandel and E. Wolf, *Selected Papers on Coherence and Fluctuations of Light*, Vols. 1, 2, Dover, New York, 1970.

29. R. H. Pantell and H. E. Puthoff, *Fundamentals of Quantum Electronics*, Wiley, New York, London, Sydney, Toronto, 1969.

30. H. Paul, *Nonlinear Optics*, Vols. 1, 2, Akademie, Berlin, 1973 (German).

31. H. Paul, *Laser Theory*, Vols. 1, 2, Akademie, Berlin, 1969 (German).

32. (a) J. Peřina, *Coherence of Light*, Van Nostrand Reinhold, London, New York, Cincinnati, Toronto, Melbourne, 1971; (b) J. Peřina, *Quantum Statistics of Linear and Nonlinear Optics*, Reidel, Dordrecht, 1984.

33. R. J. Pressley, *Handbook of Lasers with Selected Data on Optical Technology*, Chemical Rubber, Cleveland, 1971.

34. H. Rabin and C. L. Tang (Eds.), *Quantum Electronics*, Vol. 1, Academic, New York, 1975.

35. F. H. Read, *Electromagnetic Radiation*, Wiley, New York, 1980.

36. B. Saleh, *Photoelectron Statistics*, Springer Series in Optical Sciences, Vol. 6, Springer, Berlin (West), Heidelberg, New York, 1978.

37. W. E. Lamb, M. Sargent, and M. O. Scully, *Laser Physics*, Addison-Wesley, London, 1974.

38. M. Schubert and B. Wilhelmi, *Introduction to Nonlinear Optics*, Vol. 1, Teubner, Leipzig, 1971; Vol. 2, Teubner, Leipzig, 1978 (German).

39. F. P. Schäfer, "Dye Lasers," in *Topics in Appl. Physics*, Vol. 1, 2nd ed., Springer, Berlin (West), Heidelberg, New York, 1978.

40. A. E. Siegman, *An Introduction to Lasers and Masers*, McGraw-Hill, New York, 1971.

41. A. Stimson, *Photometry and Radiometry for Engineers*, Wiley-Interscience, New York, 1974.

42. O. Svelto, *Principles of Lasers*, Hayden, London, New York, 1976.

43. C. De Witt (Ed.), *Quantum Optics and Electronics*, Lectures delivered at Les Houches during the 1964 session of the summer school of theoretical physics, University of Grenoble, Gordon & Breach, New York, 1965.

44. A. Yariv, *Quantum Electronics*, Wiley, New York, 1975.

45. M. Young, *Optics and Lasers: An Engineering Approach*, Springer Series in Optical Sciences, Vol. 5, Springer, Berlin (West), Heidelberg, New York, 1977.

46. J. E. Midwinter and F. Zernike, *Applied Nonlinear Optics*, Wiley, New York, Sydney, Toronto, London, 1973.

47. V. A. Zubov, *Measurement of the Parameters of Laser Radiation*, Nauka, Moscow, 1973 (Russian).

Index